# MOLECULAR-GENETIC AND STATISTICAL TECHNIQUES FOR BEHAVIORAL AND NEURAL RESEARCH

# MOLECULAR-GENETIC AND STATISTICAL TECHNIQUES FOR BEHAVIORAL AND NEURAL RESEARCH

Edited by

ROBERT T. GERLAI

ELSEVIER

ACADEMIC PRESS

An imprint of Elsevier

Academic Press is an imprint of Elsevier
125 London Wall, London EC2Y 5AS, United Kingdom
525 B Street, Suite 1800, San Diego, CA 92101-4495, United States
50 Hampshire Street, 5th Floor, Cambridge, MA 02139, United States
The Boulevard, Langford Lane, Kidlington, Oxford OX5 1GB, United Kingdom



**Library of Congress Cataloging-in-Publication Data**
A catalog record for this book is available from the Library of Congress

**British Library Cataloguing-in-Publication Data**
A catalogue record for this book is available from the British Library

ISBN: 978-0-12-804078-2



*Publisher:* Nikki Levy
*Acquisition Editor:* Natalie Farra
*Editorial Project Manager:* Kristi Anderson
*Production Project Manager:* Poulouse Joseph
*Designer:* Mark Rogers

Typeset by TNQ Books and Journals

*In memory of Dr. John C. Roder (July 17, 1950–January 6, 2018),*
*a pioneer of modern neurobehavioral genetics,*
*my mentor and friend.*

# CONTENTS

## Section II: Searching for New Genes: Natural Genetic Variation

# CONTRIBUTORS

**Hanifa J. Abu Toamih Atamni**
Tel Aviv University, Tel Aviv, Israel

**Antoine Adamantidis**
Department for BioMedical Research, University of Bern, Bern, Switzerland

**Eva Adamová**
The University of Tennessee Health Science Center, Memphis, TN, United States

**Alejandra I. Aguirre**
IBCN, University of Buenos Aires and CONICET, Ciudad de Buenos Aires, Argentina

**Vladimir N. Babenko**
Institute of Cytology and Genetics, Siberian Branch of Russian Academy of Sciences, Novosibirsk, Russia

**Maria V. Baez**
IBCN, University of Buenos Aires and CONICET, Ciudad de Buenos Aires, Argentina

**Jon Beckwith**
Harvard Medical School, Boston, MA, United States

**Douglas M. Bowden**
National Primate Research Center at the University of Washington, Seattle, WA, United States

**Camron D. Bryant**
Boston, MA, United States

**Jarryd M. Campbell**
Mayo Clinic, Rochester, MN, United States

**Karl J. Clark**
Mayo Clinic, Rochester, MN, United States

**John C. Crabbe**
Oregon Health & Science University, and VA Portland Health Care System, Portland, OR, United States

**M. Imad Damaj**
Richmond, VA, United States

**Priscila Darakjian**
Oregon Health & Science University, Portland, OR, United States

**Fernando P.M. De Villena**
University of North Carolina at Chapel Hill, Chapel Hill, NC, United States

**Mara Dierssen**
Centre for Genomic Regulation (CRG), Barcelona, Spain; Universitat Pompeu Fabra (UPF), Barcelona, Spain; CIBER de Enfermedades Raras (CIBERER), Barcelona, Spain

**Evan Dong**
National Primate Research Center at the University of Washington, Seattle, WA, United States

**Mark F. Dubach**
National Primate Research Center at the University of Washington, Seattle, WA, United States

**Louis Y. El Khoury**
Mayo Clinic, Rochester, MN, United States

**Alberto L. Epstein**
UMR INSERM U1179 – Université de Versailles Saint Quentin en Yvelines, UFR des Sciences de la Santé "Simone Veil", Montigny le Bretonneux, France

**Martin T. Ferris**
University of North Carolina at Chapel Hill, Chapel Hill, NC, United States

**Anna G. Galyamina**
Institute of Cytology and Genetics, Siberian Branch of Russian Academy of Sciences, Novosibirsk, Russia

**Robert T. Gerlai**
University of Toronto Mississauga, Mississauga, ON, Canada

**Terri L. Gilbert**
GilHou Science Communication, Seattle, WA, United States

**Su Guo**
University of California San Francisco, San Francisco, CA, United States

**David Hanwell**
University of Toronto, Toronto, ON, Canada

**Christina A. Harrington**
Oregon Health & Science University, Portland, OR, United States

**Robert Hitzemann**
Oregon Health & Science University, Portland, OR, United States; Veterans Affairs Medical Center, Portland, OR, United States

**Robert J. Huber**
Trent University, Peterborough, ON, Canada

**Ovidiu D. Iancu**
Oregon Health & Science University, Portland, OR, United States

**Fuad A. Iraqi**
Tel Aviv University, Tel Aviv, Israel

**Diana A. Jerusalinsky**
IBCN, University of Buenos Aires and CONICET, Ciudad de Buenos Aires, Argentina

**Zhengping Jia**
The Hospital for Sick Children, Toronto, ON, Canada; University of Toronto, Toronto, ON, Canada

**Byron C. Jones**
The University of Tennessee Health Science Center, Memphis, TN, United States

**Sulev Kõks**
University of Tartu, Tartu, Estonia; Estonian University of Life Sciences, Tartu, Estonia; Prion OÜ, Tartu, Estonia

**Gea Kõks**
University of Tartu, Tartu, Estonia; Prion OÜ, Tartu, Estonia

**Irina L. Kovalenko**
Institute of Cytology and Genetics, Siberian Branch of Russian Academy of Sciences, Novosibirsk, Russia

**Natalia N. Kudryavtseva**
Institute of Cytology and Genetics, Siberian Branch of Russian Academy of Sciences, Novosibirsk, Russia

**Vivek Kumar**
Bar Harbor, ME, United States

**Celeste Leung**
The Hospital for Sick Children, Toronto, ON, Canada; University of Toronto, Toronto, ON, Canada

**Peter V. Lovell**
Oregon Health and Sciences University, Portland, OR, United States

**Anthony R. McIntosh**
Rotman Research Institute, Baycrest Centre, Toronto, ON, Canada

**Shannon McWeeney**
Oregon Health & Science University, Portland, OR, United States; Oregon National Primate Research Center, Portland, OR, United States

**Adam Melgoza**
University of California San Francisco, San Francisco, CA, United States

**Claudio V. Mello**
Oregon Health and Sciences University, Portland, OR, United States

**Megan K. Mulligan**
The University of Tennessee Health Science Center, Memphis, TN, United States

**Michael A. Myre**
University of Massachusetts Lowell, Lowell, MA, United States

**Klotilda Narkaj**
University of Toronto, Toronto, ON, Canada

**Lydia Ng**
Allen Institute for Brain Science, Seattle, WA, United States

**Danton H. O'Day**
University of Toronto, Toronto, ON, Canada; University of Toronto Mississauga, Mississauga, ON, Canada

**Denesa Oberbeck**
Oregon Health & Science University, Portland, OR, United States

**Ilse S. Pienaar**
Imperial College London, London, United Kingdom; University of Sussex, Falmer, United Kingdom

**Robin Pierce**
Tilburg University, Tilburg, The Netherlands

**Firyal Ramzan**
University of Toronto Mississauga, Mississauga, ON, Canada

**Petra Ritter**
Charité — Universitätsmedizin Berlin & Berlin Institute of Health & Bernstein Center for Computational Neuroscience, Berlin, Germany

**Belén Sancristóbal**
Centre for Genomic Regulation (CRG), Barcelona, Spain

**David Schlessinger**
National Institute on Aging, National Institutes of Health, Baltimore, MD, United States

**Cornelia Schöne**
Department for BioMedical Research, University of Bern, Bern, Switzerland

**Robert Searles**
Oregon Health & Science University, Portland, OR, United States

**Puneet Sharma**
Imperial College London, London, United Kingdom

**Alexei A. Sharov**
National Institute on Aging, National Institutes of Health, Baltimore, MD, United States

**Dmitry A. Smagin**
Institute of Cytology and Genetics, Siberian Branch of Russian Academy of Sciences, Novosibirsk, Russia

**Ana Solodkin**
Anatomy & Neurobiology and Neurology, UC Irvine Health, Irvine, CA, USA

**Gilda Stefanelli**
University of Toronto Mississauga, Mississauga, ON, Canada

**Leon Stefanovski**
Charité — Universitätsmedizin Berlin & Berlin Institute of Health & Bernstein Center for Computational Neuroscience, Berlin, Germany

**Cindy Tao**
University of Toronto Mississauga, Mississauga, ON, Canada

**Ibrahim Tastekin**
Centre for Genomic Regulation (CRG), Barcelona, Spain

**Joe Z. Tsien**
Augusta University, Augusta, GA, United States; Banna Biomedical Research Institute, Yunnan Academy of Science and Technology, Yunnan, China

**Nikki Walter**
Oregon Health & Science University, Portland, OR, United States; Oregon National Primate Research Center, Portland, OR, United States

**Brandon J. Walters**
The Hospital for Sick Children, Toronto, ON, Canada

**J.T. Westwood**
University of Toronto, Mississauga, ON, Canada

**Robert W. Williams**
The University of Tennessee Health Science Center, Memphis, TN, United States

**Morgan Wirthlin**
Oregon Health and Sciences University, Portland, OR, United States; Carnegie Mellon University, Pittsburgh, PA, United States

**Christina Zheng**
Oregon Health & Science University, Portland, OR, United States

**Joelle Zimmermann**
Rotman Research Institute, Baycrest Centre, Toronto, ON, Canada

**Iva B. Zovkic**
University of Toronto Mississauga, Mississauga, ON, Canada

# PREFACE

The use of recombinant DNA technologies has revolutionized neuroscience research over the past 5 decades, created the field of biotechnology, and had a dramatic impact on translational research and medicine. Products resulting from the use of these technologies are found in laboratories, pharmacies, hospitals, and supermarkets all around the world. Discoveries made possible by recombinant DNA technologies propelled basic research, engaged the media, and fascinated lay people and scientists alike.

The first edition of *Molecular-Genetic and Statistical Techniques for Behavioral and Neural Research* provides an informative account of some of the most exciting molecular and recombinant DNA techniques of the 21st century as they are employed for the analysis of brain function and behavior. The information presented in this book is critical for clinicians, scientists, researchers, graduate students, and advanced undergraduate students, as well as professors interested in teaching related disciplines. The book provides a clear presentation of the frontiers of basic research, as well as translationally relevant techniques that are used by neurobehavioral geneticists.

The first edition of *Molecular-Genetic and Statistical Techniques for Behavioral and Neural Research* includes chapters on neuroinformatics, genetic and neurobehavioral databases and data mining, examples on the analysis of natural genetic variation, approaches to studying gene expression for understanding brain function and behavior, principles and applications of forward (mutagenesis) and reverse genetics (gene targeting), as well as discussions about the ethical use of animals in neurobehavioral genetics research. It is written by leading international experts of the field, and it is edited by a behavior geneticist who has been working in the biopharmaceutical and biotechnology research industry as well as in academic research for over 3 decades.

# ACKNOWLEDGMENTS

This book represents the dedicated and diligent work of the authors who wrote its chapters, but it would also not have been possible without the help from the numerous anonymous referees who provided their expert advice. I would also like to recognize the help and encouragement of the staff of the Elsevier editorial office, particularly Ms. Kristi Anderson and Dr. Natalie Farra. I thank Dr. Wim E. Crusio, my long-time colleague and friend, for initiating this book project and for contributing to and reviewing its first plans. Last, I must recognize the, most often, positive role my family played: my wife Julia and daughter Flora who put up with me working many late nights, and my son Mark, whose inquisitive questions always reminded me that the next generation of scientists are already knocking on the door. This book is for all of them and for everyone else who would like to learn about the genetics of behavior.

# Why Should We Use Genetics in the Analysis of Brain Function and Behavior? Practical and Theoretical Considerations

**Robert T. Gerlai**
University of Toronto Mississauga, Mississauga, ON, Canada

Charles Darwin in his book *On the Origin of Species* (1859)[1] already realized that behavioral characteristics are not uniquely different from other phenotypical features in that they also must be subject to the forces of natural selection and evolve. Although the scientific field we call genetics today did not exist back then, and although the role of genes and structure of DNA were discovered only about a 100years later, Darwin's theory of evolution implied that behavioral characteristics must have been under the influence of genes. This inherent assumption turned out to be correct and led to the development of a new field of science, behavior genetics. The birth of behavior genetics is usually dated to 1960, the year when the first book specifically focusing on questions addressed by this discipline was published.[2] In this book, the reader could find methods and applications of how one could study the effects of genes on a variety of behavioral traits, from perceptual processes to personality and temperament. The methods centered around what we now call quantitative genetics, a branch of classical genetics that is extended from Mendelian genetics to characteristics that cannot be classified into discrete categories but instead show continuous variation. Indeed, quantitative genetics was the main method of the behavioral geneticist for several years to come. It allowed scientists to conclude about genetic effects that underlie observed phenotypical differences. Quantitative genetics investigated the genetic architecture,[3] i.e., it allowed one to estimate the source of genetic variance, e.g., whether it was due to additive genetic effects or to dominance-recessive allelic interaction-related genetic effects (intralocus interaction) or to epistatic effects (interlocus interaction). It allowed one to estimate the number of segregation units (likely genes with major effects), and it even allowed one to dissect phenotypical correlation and to estimate purely genetic and purely environmental components of correlation.[4] What it did not allow one to do was to actually identify the gene, and to understand and/or to manipulate its biochemical or biologic function. That is, the gene itself remained an esoteric concept for behavioral geneticists.

Molecular biology changed all this with the discovery of restriction endonucleases (e.g., Ref. 5), a discovery that revolutionized genetics, and biology in general. Restriction endonucleases allowed scientists to create recombinant DNA,[6,7] and in doing so, to discover and manipulate genes and to probe their function. The past 5 decades have seen an amazing evolution of this technology. Genome editing in the 21st century is not a practice restricted to only a select few elite laboratories anymore. Service provider companies can custom design DNA, or genetically engineer laboratory organisms, for practically any research purpose according to the specifications of the biologist user. The biotechnology and biopharmaceutical research industry routinely utilize such technologies, and academic researchers all over the world employ and continuously develop increasingly powerful and novel methods, all falling under the category of the recombinant DNA approach. Behavior genetics has also benefited from this revolution.

Recombinant DNA technologies now allow one to identify genes that show natural genetic variance in brain function, or to employ random mutagenesis and identify novel mutants with altered behavioral phenotypes and the genes that carry the mutation, an approach called forward genetics. Recombinant DNA methods now allow us to specifically target genes whose nucleotide sequence has been known but whose function in the brain is yet to be discovered, an approach called reverse genetics. They make it possible for us to check how our manipulation of behavior affected gene function, e.g., leading to transient or lasting changes in gene expression. But the story does not end there. Twenty-first century molecular biology now allows scientists to genetically manipulate the functioning of specific neuronal cell types, or circuits, giving unprecedented level of spatial resolution for the scientist. The genetic manipulation can also be conducted with extremely high temporal resolution: activation or deactivation of neurons can now be achieved at subsecond-level speed. Last, unlike in the past, genetic manipulation is now not restricted to a select few laboratory species, such as rodents, zebrafish, or the fruit fly, but it may be performed in a large variety of other animals too.

Capturing this revolution is not an easy task. A novice, and perhaps even the expert, in this field may be overwhelmed by the sheer volume of publications, the large number of methods, and the even larger number of their applications. The current book cannot present this vast literature in a comprehensive way. But what it does attempt to do is to showcase some of the most powerful and most exciting developments of the fast-evolving field of behavior genetics, or as we now call it, neurobehavioral genetics. In doing so, the authors of our chapters, who are leading experts within their fields, will guide the readers, helping them make the best choices about how they could use a variety of molecular and classical genetic methods to study the effects of genes on brain function and behavior.

From pharmacology to brain imaging, there are numerous elegant nongenetic methods in the tool set of the neuroscientist. Why should we bother with genetic methods or care about genetic effects on behavior and brain function? There are multiple reasons. One, because genes represent the most basic functional units of biologic organization, and thus manipulating them may allow us to understand how things are built and how things work in the organism. Two, manipulation of genes provides one of the most precise ways with which we can probe the functioning of the organism. We understand their structure, we have an amazing array of methods to modify them, and unlike environmental factors whose number is practically infinite, we have to deal with only 20–30 thousand of them. Three, comparison of nucleotide sequences across different species is perhaps the most precise way to discover similarities and differences among species. In other words, genetics allows us to translate findings from one species to another, including findings obtained with laboratory species to humans, which may lead to appropriate modeling and mechanistic understanding of human disorders. Last, the fourth reason is that without genetic effects, behavioral characteristics would not have evolved. Genes are the target of evolution, and behavior is the output of the brain, an organ that has been shaped by natural selection. Although a particular behavior and a gene that influences it are perhaps at the two most extreme levels of the multilayered biologic organization, evolutionary biology has taught us that they are inherently tied together. Thus, the preceding phenogenetic and phylogenetic arguments will likely keep the seemingly distant fields of psychology, neuroscience, and genetics married for a long time. The subsequent chapters of this book attest that this marriage has been a happy one, producing increasingly sophisticated methods and exciting discoveries.

## REFERENCES

1. Darwin CR. *On the Origin of Species by Means of Natural Selection, or the Preservation of Favoured Races in the Struggle for Life*. 1st ed. London: John Murray; 1859:502.
2. Fuller JL, Thompson WR. *Behavior Genetics*. New York: John Wiley & Sons, Inc.; 1960:396.
3. Crusio WE, van Abeelen JH. The genetic architecture of behavioural responses to novelty in mice. *Heredity*. 1986;56:55–63.
4. Gerlai R, Crusio WE. Organization of motor and posture patterns in paradise fish (*Macropodus opercularis*): environmental and genetic components of phenotypical correlation structures. *Behav Genet*. 1995;25: 385–396.
5. Roulland-Dussoix D, Boyer HW. The *Escherichia coli* B restriction endonuclease. *Biochim Biophys Acta*. 1969;195:219–229.
6. Cohen SN, Chang AC, Boyer HW, Helling RB. Construction of biologically functional bacterial plasmids in vitro. *Proc Natl Acad Sci USA*. 1973;70:3240–3244.
7. Jackson DA, Symons RH, Berg P. Biochemical method for inserting new genetic information into DNA of Simian Virus 40: circular SV40 DNA molecules containing lambda phage genes and the galactose operon of *Escherichia coli*. *Proc Natl Acad Sci USA*. 1972;69:2904–2909.

## FURTHER READING

1. Grant SG, O'Dell TJ, Karl KA, Stein PL, Soriano P, Kandel ER. Impaired long-term potentiation, spatial learning, and hippocampal development in fyn mutant mice. *Science*. 1992;258:1903–1910.
2. Silva AJ, Stevens CF, Tonegawa S, Wang Y. Deficient hippocampal long-term potentiation in alpha-calcium-calmodulin kinase II mutant mice. *Science*. 1992;257:201–206.

SECTION I

# Neuroinformatics, Computational Models and Data Analysis

CHAPTER 1

# Neurological Biomarkers and Neuroinformatics: The Role of The Virtual Brain

**Ana Solodkin[1], Joelle Zimmermann[2], Anthony R. McIntosh[2], Leon Stefanovski[3], Petra Ritter[3]**

[1]Anatomy & Neurobiology and Neurology, UC Irvine Health, Irvine, CA, USA; [2]Rotman Research Institute, Baycrest Centre, Toronto, ON, Canada; [3]Charité – Universitätsmedizin Berlin & Berlin Institute of Health & Bernstein Center for Computational Neuroscience, Berlin, Germany

## ADDING INFORMATION TO BRAIN IMAGING DATA VIA BRAIN NETWORK MODELING

The neuroscience community is immersed in the collection of large datasets to provide greater sensitivity for understanding brain function and dysfunction. Such initiatives span normal function (Human Connectome Project), life-span including normal and abnormal function (UK Biobank),[1,2] development (NIH Pediatric Database), brain disorders such as Alzheimer disease (ADNI),[3] and psychiatry (RDoC: Research Domain Criteria Project). While these initiatives provide the necessary empirical foundation, what are lacking are the quantitative tools to relate these multiple datasets to "reconstruct" the brain and provide the link between these data and those from a single person. The technologic advances in brain imaging yield unprecedented information on structure and function, yet these data are severely underutilized for clinical decision-making.

While in many cases group analyses can show statistically significant results between young and old or between healthy and diseased populations, strong underlying intersubject and intrasubject variability[4] preclude accurate individual predictions. That is, MR imaging–derived markers are of small predictive value at an individual level.[2] While multiple investigations of potential imaging-based markers in preclinical studies are ongoing, clinical trials are hampered by their limited scope, as new innovative tools are needed to allow an exact assessment of the degree of disease pathology, including dynamic processes. In this chapter, we propose the advancement of current state-of-the-art approaches by using The Virtual Brain (TVB) simulation modeling platform, a novel method that explicitly infers network mechanisms at multiple spatial and temporal scales.

TVB is an *open-source* whole-brain simulation platform that integrates empirical data of different modalities and spatiotemporal scales to construct comprehensive and biologically plausible models of brain network dynamics (http://www.thevirtualbrain.org/). Simplifying, TVB simulates mathematically driven brain activity on the three-

*Molecular-Genetic and Statistical Techniques for Behavioral and Neural Research*
ISBN 978-0-12-804078-2, https://doi.org/10.1016/B978-0-12-804078-2.00001-5

dimensional map of an individual brain and its connections to fit empirically measured brain signals (from raw signals to electroencephalography [EEG], magnetoencephalography [MEG], and functional magnetic resonance imaging [fMRI]). By the systematic exploration of the relation between structural anatomic tracts and biophysical parameters producing the dynamics (spatiotemporal characterizations), TVB allows the construction of the individual's brain model within a unifying framework. One of the defining features of TVB is that it builds a unique Virtual Brain for each subject based on his/her own neuroimaging data within a plausible anatomic framework.[5] This is of relevance since during aging and in various brain disorders, most studies have reported structural and functional changes focusing at the group level[6–8] with highly variable intersubject relationships between structural abnormalities (lesions, etc.) and functional deficits. At the same time, similar deficits can be associated with different locations of damage where lesions involving large areas of the brain may result in only subtle deficits, whereas less extensive lesions may affect a number of cognitive domains. The basis for such interindividual variability is poorly understood because we do not understand the underlying functional brain dynamics associated with structural changes. TVB as a multiscale approach fills the gap by linking from cellular events to whole brain dynamics reflected in the modeling parameters, providing an opportunity to uncover basic mechanisms associated with brain function and dysfunction.

Our preliminary work using this multiscale modeling approach has demonstrated the explanatory power of connectome-based, large-scale brain models[9,10] and has facilitated the elucidation of the interrelationship between the structural and functional connectomes highlighting intersubject differences.[9–11] Furthermore (Fig. 1.1), we have also shown the algorithms to achieve optimal individual simulations (fit) of functional connectivity matrices based on their respective individual anatomic connectivity matrices.[9] Finally, we have tested the dependency of functional connectomes and brain dynamics derived from damage to the structural connectome in stroke[10–12] and epilepsy.[13,14] Because of the ability for TVB to generate personalized "Virtual Brains" via the simulation of functional signals, the model parameters associated with each healthy subject or patient can then be correlated to individual clinical phenotypes (e.g., motor function to cognition).

TVB houses a library of local models, which catalog the biophysical parameters that produce different empirical brain states or network patterns.[9,15] Table 1.1 lists the repertoire of neural mass models available in TVB and their state variables along biophysically relevant parameters. Because of the insight that TVB provides on these latent biophysical variables is invisible to noninvasive brain imaging techniques, it acts de facto as a computational "microscope." TVB modeling is able to simulate a plethora of brain states and neural behaviors from raw data (e.g., neuronal firing) to brain signals detected as EEG or fMRI signals. Furthermore, these regimes are detected in condition-specific contexts (condition-specific regimes), enabling the model system to associate specific parameter

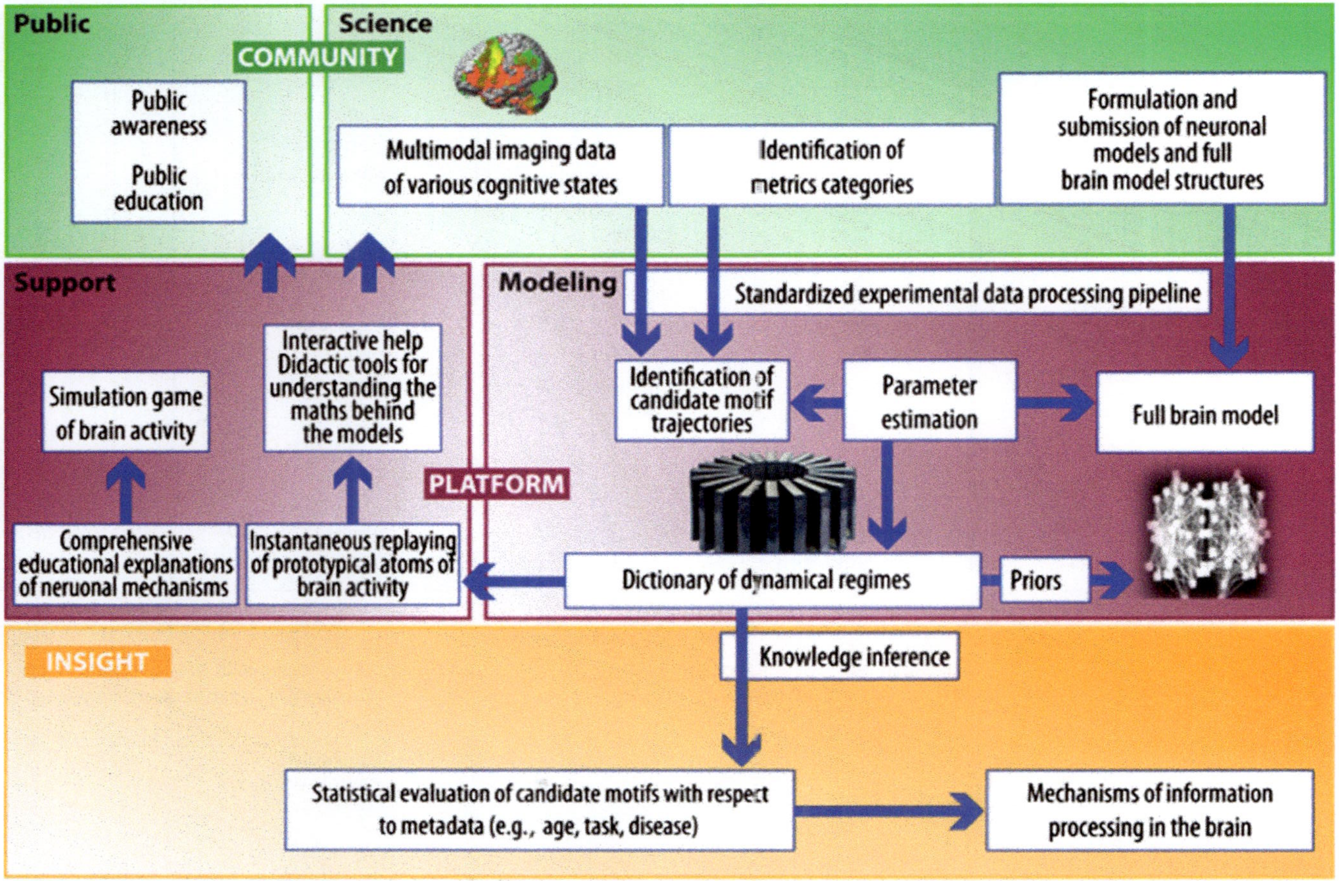


**Figure 1.1** ***The Virtual Brain discovery technology.*** The repertoire of tool chains, neuronal models, motif dictionaries, and educational support is continuously developed and advanced. This scheme shows the interrelationship of the components involved in the technologic development of The Virtual Brain platform.

**Table 1.1** Overview of exemplary neural mass models included in the TVB repertoire. d, dimensionality, NM, Neural mass. Parameters with a physiological corresponding surrogate are listed along with the parameter descriptions and formulas.[23] See also pertinent references[82–87].

| Neural Mass Model | d | NM | State variable | Physiological surrogate parameters | Characteristics | Equations |
|---|---|---|---|---|---|---|
| Generic 2d Oscillator (FitzHugh-Nagumo) | 2 | 1 | $V$ — function of membrane potential<br>$W$ — recovery variable | $I$ — system excitability | Multistable in excitable and oscillatory behavior | $\dot{V} = d\tau\left[-fV^3 + eV^2 + gV + \alpha W + \gamma I\right]$<br>$\dot{W} = \frac{d}{\tau}\left[cV^2 + bV - \beta W + a\right]$ |
| Wilson and Cowan | 2 | 2 | $I$ — inhibitory activity<br>$E$ — excitatory activity | $C_x$ — coupling coefficients<br>$\tau_i$, $\tau_e$ — membrane time constants<br>$r_i$, $r_e$ — refractory time<br>$P$, $Q$ — external stimuli | Simplest model with 2 NM, multistable, contains excitatory loops (Wilson and Cowan, 1972) | $f(E) = c_{ee}E - c_{ei}I + P - \theta_e$<br>$f(I) = c_{ie}E - c_{ii}I + Q - \theta_i$<br>$\dot{E} = \frac{1}{\tau_e}(-E + (k_e - r_eE)S_e(f(E))$<br>$\dot{I} = \frac{1}{\tau_i}(-I + (k_i - r_iI)S_i(f(I))$ |
| Wong, Wang, and Deco (Reduced Wong and Wang) | 1 | 1 | $S$ — average synaptic gating variable | $\tau_s$ — NMDA decay<br>$J_N$, $w$ — excitatory recurrent connection<br>$I_0$ — effective external input | Alpha activity, monostable for low activity (Wong and Wang, 2006) | $x_k = wJ_NS_k + I_0 + J_N\Gamma(S_k, S_ju_{kj})$<br>$H(x_k) = \frac{ax_k - b}{1 - \exp(-d(ax_k - b))}$<br>$\dot{S}_k = -\frac{S_k}{\tau_s} + (1 - S_k)H(x_k)\gamma$ |

| | | | | | | |
|---|---|---|---|---|---|---|
| Jansen and Rit (Zetterberg, Jansen, and Rit) | 12 | 3 | $S$ – average population firing rate | $\gamma_x$ – connectivity constants<br>$H_e$, $H_i$ – maximum amplitudes<br>$\kappa_e$, $\kappa_i$ – postsynaptic response constant<br>$e_0$ – half of maximum mean firing rate<br>$\rho_1$ – mean firing threshold potential | Cytoarchitectonic based, alpha and beta activity, evoked potentials, epileptic waves (Zetterberg, Kristiansson, et al., 1978) | $\dot{\nu}_{1_k} = y_{1_k}$<br>$\dot{y}_{1_k} = H_e K_e(\gamma_1 S[\nu_{6_k}]) + \gamma_{1T}\left(U + \varepsilon_1 + \Gamma\left(\nu_{6_j}\right) + W_\varsigma S\left[\nu_{6_j}\right]\right) - 2_{k_e} y_{1_k} + k_e^2 \nu_{1_k}$<br>$\dot{\nu}_{2_k} = y_{2_k}$<br>$\dot{y}_{2_k} = H_e K_e(\gamma_2 S[\nu_{1_k}]) + \gamma_{2T}\left(P + \varepsilon_2 + \Gamma\left(\nu_{6_j}\right) + W_\varsigma S\left[\nu_{6_j}\right]\right) - 2_{k_e} y_{2_k} + k_e^2 \nu_{2_k}$<br>$\dot{\nu}_{3_k} = y_{3_k}$<br>$\dot{y}_{3_k} = H_i K_i(\gamma_4 S[\nu_{7_k}]) - 2_{k_i} y_{3_k} + k_i^2 \nu_{3_k}$<br>$\dot{\nu}_{4_k} = y_{4_k}$<br>$\dot{y}_{4_k} = H_e K_e(\gamma_3 S[\nu_{6_k}]) + \gamma_{3T}\left(Q + \varepsilon_3 + \Gamma\left(\nu_{6_j}\right) + W_\varsigma S\left[\nu_{6_j}\right]\right) - 2_{k_e} y_{4_k} + k_e^2 \nu_{4_k}$<br>$\dot{\nu}_{5_k} = y_{5_k}$<br>$\dot{y}_{5_k} = H_i K_i(\gamma_5 S[\nu_{7_k}]) - 2_{k_i} y_{5_k} + k_i^2 \nu_{5_k}$<br>$\dot{\nu}_{6_k} = y_{2_k} - y_{3_k}$<br>$\dot{\nu}_{7_k} = y_{4_k} - y_{5_k}$ |
| Stefanescu and Jirsa 3D | 18 | 2 | $\xi$ – membrane potential<br>$\eta$ – fast ion channel flow<br>$\tau$ – slow ion channel flow<br>$\alpha$, $\beta$, $\gamma$ – idem for the inhibitory population | $a$, $b$, $c$, $d$ – fast ion channel constants<br>$r$ – slow ion channel constant<br>$s$ – bursting strength<br>$IE$, $II$ – excitatory and inhibitory input current<br>$K_{11}$, $K_{12}$, $K_{21}$ – coupling parameters<br>$X_1$, $X_2$ – mean field potentials | Adaptable because of 3 modes (Stefanescu and Jirsa[24]; Stefanescu and Jirsa, 2011) | $\dot{\xi}_i = \eta - a\xi_i^3 + b\xi_i^2 - \tau_i + K_{11}\left(\sum_{k=1}^{3} A_{ik}\xi_k - \xi_i\right) - K_{12}\left(\sum_{k=1}^{3} B_{ik}\alpha_k - \xi_i\right) + IE_i$<br>$\dot{\eta}_i = c_i - d\xi_i^2 - \eta_i$<br>$\dot{\tau}_i = rs\xi_i - r\tau_i - m_i$<br>$\dot{\alpha}_i = \beta_i - e_i\alpha_i^3 + f_i\alpha_i^2 - \gamma_i + K_{21}\left(\sum_{k=1}^{3} C_{ik}\xi_k - \alpha_i\right) + II_i$<br>$\dot{\beta}_i = h_i - p_i\alpha_i^2 - \beta_i$<br>$\dot{\gamma}_i = rs\alpha_i - r\gamma_i - n_i$ |

*Continued*

**Table 1.1** Overview of exemplary neural mass models included in the TVB repertoire. d, dimensionality, NM, Neural mass. Parameters with a physiological corresponding surrogate are listed along with the parameter descriptions and formulas.[23] See also pertinent references[82–87].—cont'd

| Neural Mass Model | d | NM | State variable | Physiological surrogate parameters | Characteristics | Equations |
|---|---|---|---|---|---|---|
| Larter and Breakspear (3D chaotic oscillator) | 3 | 2 | $V$ – membrane potential<br>$Q_V$ – excitatory firing rate | $g_x$ – ion channel conductances<br>$V_x$ – Nernst potentials<br>$Q_{Vmax, Zmax}$ – maximal firing rates<br>$r_{NMDA}$ – NMDA to AMPA ratio<br>$d_x$ – ion channel threshold variances<br>$T_x$ – ion channel thresholds<br>$Z_T$ – mean inhibitory threshold potential<br>$\delta_x$ – population threshold variances<br>$I$ – subcortical input strength<br>$a_x$ – coupling factors | Highly structured, chaotic system, nonoscillatory, differentiated ion channel parameters (Breakspear and Stam, 2005) | $\dot{V}_k = -\Big(g_{Ca} + (1-C)r_{NMDA}a_{ee}Q_V + Cr_{NMDA}a_{ee}\langle Q_V\rangle^k m_{Ca}(V - V_{Ca})$ $- g_K W(V - V_K) - g_L(V - V_L)$ $-\big(g_{Na}m_{Na} + (1-C)a_{ee}Q_V + Ca_{ee}\langle Q_V\rangle^k\big)(V - V_{Na})$ $-a_{ie}ZQ_Z + a_{ne}I$<br>$\dot{W}_k = \phi\frac{m_K - W}{\tau_K}$<br>$\dot{Z}_k = b[a_{ni}I + a_{ei}VQ_V]$<br>$Q_V = Q_{V_{max}}\left(1 + \tanh\left(\frac{V_k - V_T}{\delta_V}\right)\right)$<br>$Q_Z = Q_{Z_{max}}\left(1 + \tanh\left(\frac{Z_k - Z_T}{\delta_Z}\right)\right)$ |

profiles to condition-specific physiologic spatiotemporal behaviors. The overreaching goal is to provide proof of concept that TVB can in fact identify aging and disease biomarkers with predictive potential for disease diagnosis and progression based on underlying biophysical processes.

Finally, part of the approach of TVB is that it involves a broad community leading to an organized workflow of systematic knowledge inference going from the public (educational) domain[16] to basic science research and clinical settings (Fig. 1.2).

## BRIDGING SCALES WITHIN THE VIRTUAL BRAIN

Brain function derives from the complex interactions of the underlying microscopic structure, yet the description of a systems behavior, especially brain function from the bottom-up alone, is very limited in part due to redundancy of the basic elements (neuronal groups). In other words, it is not possible to deduct large-scale behavior based solely on aggregating "independent" small elements. The same can be said of the link between cellular and molecular levels. Classic neuroscience studies, focused at the level of single neurons and their physical interaction, do not use a bottom-up approach from the molecular level. That is, even when specific microstructure of axons, dendrites, glial components, and vesicles, for example, can be captured with very high resolution,[18] this microcomplexity does not enable capturing information at the superordinate level, as the interaction of neurons and neuronal ensembles develop local functional patterns described by functional clusters and other nonrandom features.[19] The new field of network science makes use of mathematical methods to describe these functional patterns.[20] In practical terms, functional modeling of individual human brains is achieved by defining and explicitly linking different network scales with some dimension reduction (to avoid excessive computational demands). In this context, choosing the right coarseness within each scale is an important step in developing multiscale models.[21] For example, brain regions referred to as "nodes" in the modeling context are associated with the determination of their *local dynamics* via neural mass models, and their size has to be carefully determined. A sensitive description by Breakspear and Jirsa define a neural mass as a group of neurons with equal or similar in- and outgoing connections and a common function (similar to the definition determining the threshold of micro- and macrocolumns, i.e., populations up to $10^4$ or $10^6$ neurons).[22] In our case, the TVB platform provides a wide range of multiscaling levels based on a variety of basic biophysical considerations implemented by access to a variety of local models (Table 1.1). A given model can in principle reproduce the neural population activity at the magnetic resonance imaging (MRI) resolution level with about 50,000 nodes, each maintaining approximately 5 million neurons. In addition, larger, functionally or anatomically driven parcellations are useful if the focus is on large-scale functional network features.[9,21] TVB also introduces an intermediate mesoscopic scale, which pools local connectivity features,

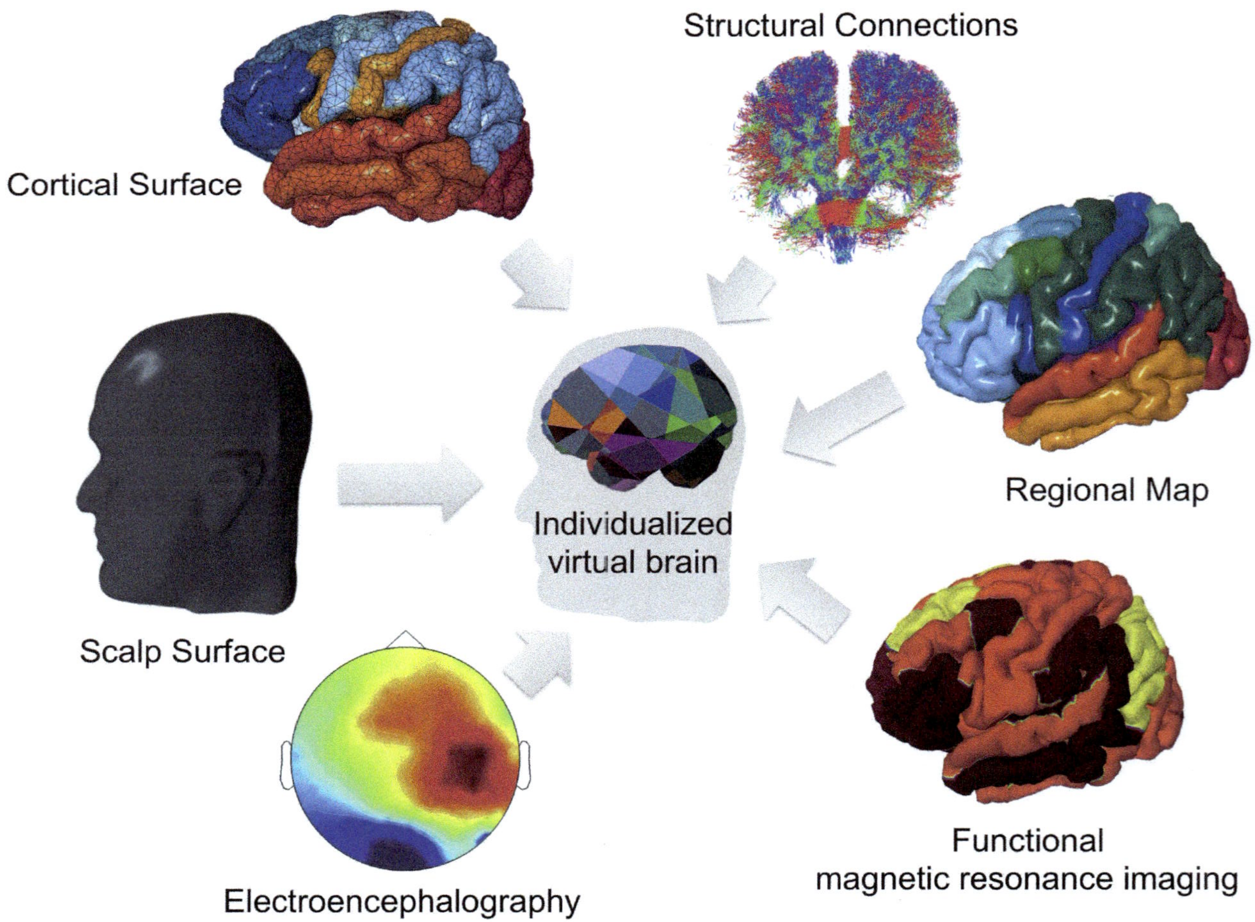


**Figure 1.2** ***Knowledge generation with personalized Virtual Brains.*** Multimodal imaging data are used to constrain biophysically grounded full brain models. Brain regions (regional map, parcellation) are represented by neural field models. Individualized brain network models can reveal the origin of interindividual differences in brain function and therefore are potential powerful tools for precision medicine. *(With permission from Stefanovski L, Ghani A, McIntosh AR, Ritter P. Linking connectomics and dynamics in the human brain. e-Neuroforum. 2016;7(3):64–70.)*

practically building a bridge between the macro- and the microscale. The mesoscale has no specific biological surrogate so far and is an operational construction to model microscopic population features in a large-scale network. Because of the fluid transitions between scales, it is necessary to define levels clearly, according to the respective case. But in general (Fig. 1.3), the following can be stated:

1. The microscale covers the irreducible elements of the network from the connectional point of view, i.e., single neurons and their membrane potentials. Functionally, it models individual action potentials and the microscopic circuits between single neurons. In the neural mass models of TVB, this scale is simplified as a neural mass; the circuits are represented, for instance, by the intrinsic interaction between the population of excitatory and inhibitory neurons.[22,23]
2. The mesoscale is a surrogate intermediate scale that describes relevant population features, both in terms of function and structure. For example, it mathematically defines the average interaction between neighboring nodes by physical and functional laws[22] or prominent modes of activity at the population level in a neural mass (e.g., in Stefanescu and Jirsa).[24]
3. The macroscale relates to the entire brain. It refers to emerging spatiotemporal neuronal activity modes resulting from local and global interactions of neuronal populations across the brain. With respect to structural features the macroscale refers to the structural connectome derived from diffusion tensor imaging (DTI) tractography and represents major white matter fiber tracts.

There are no clear boundaries between the scales because of the fluid transitions from one scale to the next, suggesting that they are linked. Hence, TVB represents the microscopic level with the biophysically based equations of the neural mass models and its cellular and subcellular surrogate parameters where neural masses themselves represent the mesoscale (i.e., local field potentials and population firing rates of a network of several thousand if not million neurons). Likewise, distant large-scale connections (intracortical or projection) can be considered macroscopic, but regional tracts (e.g., between cortical layers or columns within a specific region) are mesoscopic. That is, scales are context dependent (Fig. 1.3), and the close integration of these scales is one of the most enduring features of TVB, as it fosters the determination of brain dynamics via faithful simulations of brain signals.

Thus, TVB captures brain dynamics at the coarse scale of interacting neuronal populations (see evolution equation, Fig. 1.4), where local models provide a reduced or simplified description. These local models (Table 1.1) highlight different aspects of neuronal dynamics. Some are derived from models based on microscopically coupled neurons. For example, the Stefanescu-Jirsa 2D and 3D models are derived from networks of single-neuron FitzHugh-Nagumo and Hindmarsh-Rose[79] models, which in turn are derived from the famous single-neuron Hodgkin-Huxley model. In other words, TVB can be considered a "top-down" approach, where simulations are done at a coarser level

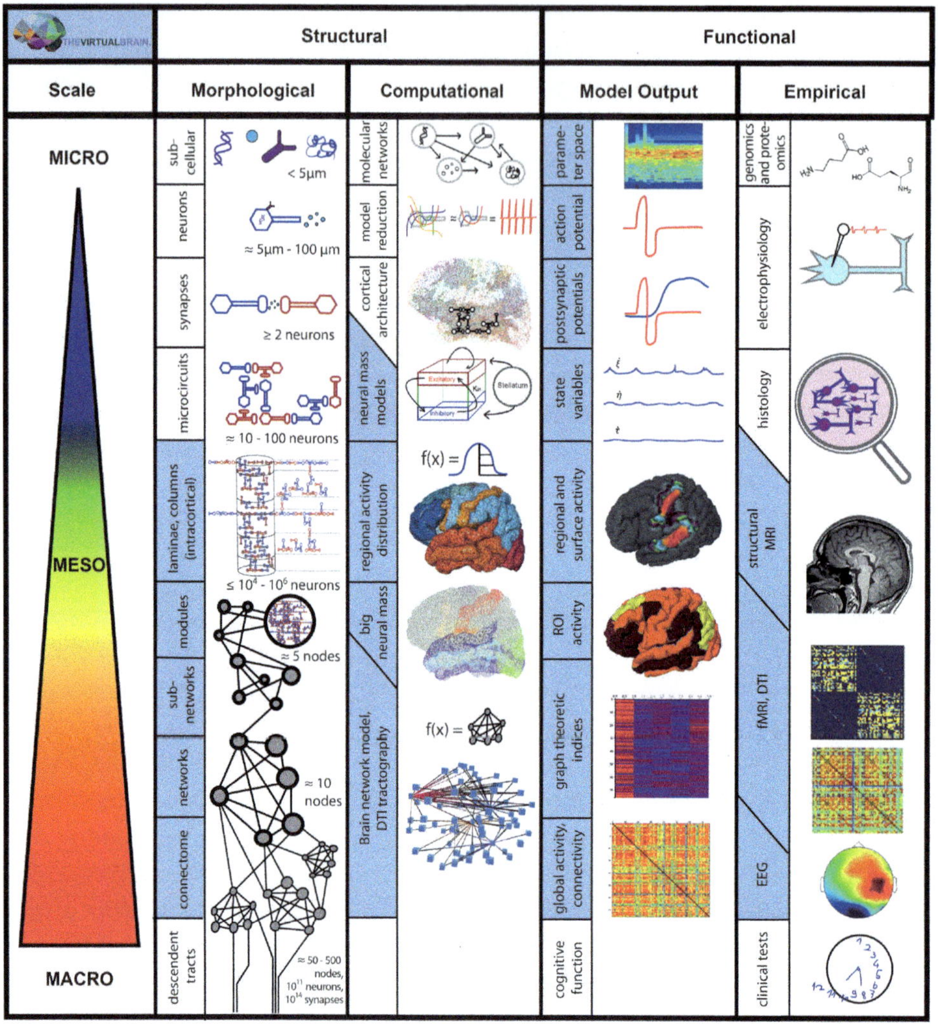


**Figure 1.3** ***Multiple scale dimensions in The Virtual Brain (TVB).*** Graphic representation of scales (structural and functional) associated with brain dynamics. The aspects colored with *blue underground* mark the scales formally represented in TVB. The *triangle* to the left is a reminder of the lack of clear border between scales. *Morphologic level*: Lower border of resolution is represented by columns as functional units. Specific cortical-brainstem tracts (corticospinal or medial lemniscus) are currently not established for TVB. *Computational basement*: As an artificial mathematic level, it is widely represented in TVB and (indirectly) models the whole brain. Structures below the neural masses have no direct representation in TVB. However, microscale features are essential in the development of TVB (e.g., in the mathematic reduction of microscopic models). *Model output*: In dependence of the used neural mass model, activity can be represented to the level of neuron populations in the form of subcellular features (as fitted parameters) and from the neuronal scale as spikes, action potentials, oscillations, etc. At the large-scale level, TVB has extensive options (monitors) representing activity (e.g., BOLD activity, EEG, MEG, and derived computations as the distribution of frequency powers). *Empirical measurements*: Because of its base the lower threshold of TVB is the resolution of MRI, as also for the network components.

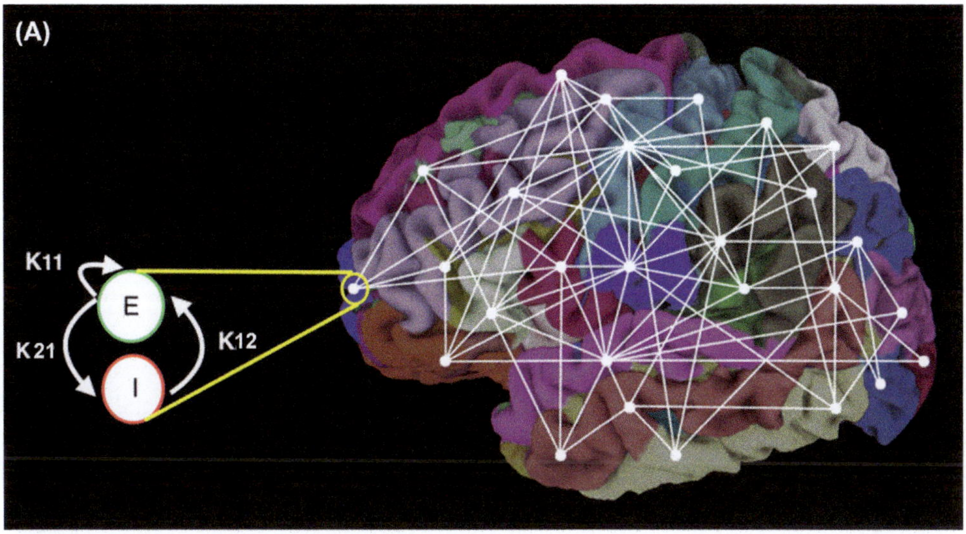


**(B)**

$$\frac{dx_i(t)}{dt} = N(x_i(t)) + G\sum_j SC_{ij}x_j(t - d_{ij}^{glob}) + L\sum_j LC_{ij}x_j(t - d_{ij}^{loc}) + I_i(t) + v_i(t)$$

Differential operator that maps the system of equations into its derivatives
State operator for each neural mass i (i.e., neural mass model).
Vector of state variables for each region i, e.g., synaptic activity, firing rates, mean field
Global coupling scaling factor
Structural connectivity matrix (global coupling)
Time delays
Local connectivity
Injected input
Noise

**Figure 1.4** ***Integration of scales within The Virtual Brain (TVB) via evolution differential equations.*** (A) Graphic representation of global and local parameters based on the Stefanescu-Jirsa 3D (SJ3D). Each brain region (node seen as *white circles*) is represented by the SJ3D neural mass model that consists of inhibitory and excitatory population mean field dynamics that can infer mesoscopic population behaviors such as excitation-inhibition balance and firing rates, as well as intrinsic coupling strength between inhibitory and excitatory mean fields indicated here by parameters $K_{11,\ 12,\ 21}$. Global parameters are derived from the associations (*white lines*) between the nodes. (B) Generic evolution equation of The Virtual Brain. TVB computes the evolution of a dynamic system—the brain—via differential equations. On the left side, it has a differential operator, and on the right, a vector of state variables provides a phase flow from where the system trajectory is integrated. One of the vectors (turquoise) represents the state variables for the local model (e.g., synaptic activity, firing rate, mean field potentials). Additionally, parameters derived from anatomic tracts map the current state of the system to the next (representing the interaction between nodes). Hence, these parameter symbolize global variables (purple and blue). An important global parameter (G, pink), a numeric scale reflecting the balance between local and global dynamics, varies according to the strength of anatomic connections. Additional elements in the equation (light and darker green) represent input simulation and noise to help to reproduce naturalistic biologic responses.

at a whole brain level. In contrast, other modeling approaches, such as the Human Brain Project (https://www.humanbrainproject.eu/), are "bottom-up," where simulations are done at subcellular and cellular levels to later infer activity at coarser scales. Independent of the approach, the ultimate goal based on noninvasive imaging is to ultimately understand the links between genes, proteins, and molecules to individual neuron physiology to neuronal population behaviors and eventually large-scale dynamics.

Databases such as the UK Biobank or the Human Connectome Project—in addition, to brain imaging—provide detailed data on genetic traits, proteomics, and metabolomics along with information on lifestyle and environment. Hence, in the not so far future, it seems we will be able to link TVB to cellular and subcellular scales on the one hand and—similarly important—environmental and social factors on the other.

Interestingly, at each level of description—from molecules to brain and to society—we are dealing with complex systems where rules and insights of nonlinear systems dynamics could apply. As each of these scales may have their peculiarities and specific rules, a new frontier is emerging to identify similarities and discrepancies among them, integrating them in complex manifolds.

## PRAGMATICS OF MODELING IN THE VIRTUAL BRAIN

The sole empirical input to the TVB platform is the structural connectivity matrix derived from the individual subject's tractography. Based on this input, TVB ***simulates*** brain signals (e.g., field potentials) by integrating global dynamics with local (mesoscopic) models that determine the neuronal dynamics *within* brain regions. Following, EEG or fMRI blood oxygen level–dependent (BOLD) signals are derived from the generated field potentials. For a comprehensive list of the state variables and free parameters for the local models currently included in the TVB platform, see Table 1.1.

Empirical data required for modeling with TVB (Fig. 1.5) are as follows:

1. High-resolution T1-weighted anatomic images covering the whole brain. This allows for the anatomic location of brain regions. Regional brain parcellations are derived from this MR acquisition.[9,21,25]
2. Diffusion-weighted imaging (e.g., DTI) allows the determination of individual structural brain connectomes via tractography algorithms. The two metrics used for each tract are weight of connection and length of connection.
3. Functional data, typically, falls into one of these three possibilities: resting state fMRI (rsfMRI)[9,26,27] or intra- or extracranial electroencephalographic recordings.[9,28] Additionally, it is possible to use multiple modalities (e.g., simultaneous EEG-fMRI).[29]

Pragmatically, individual empirical structural and functional data can be directly extracted and formatted using an automated imaging data processing pipeline developed for TVB.[25] This TVB preprocessing pipeline[25] is already accelerated with parallel computing, and it is publicly accessible (http://www.nsgportal.org). The resulting

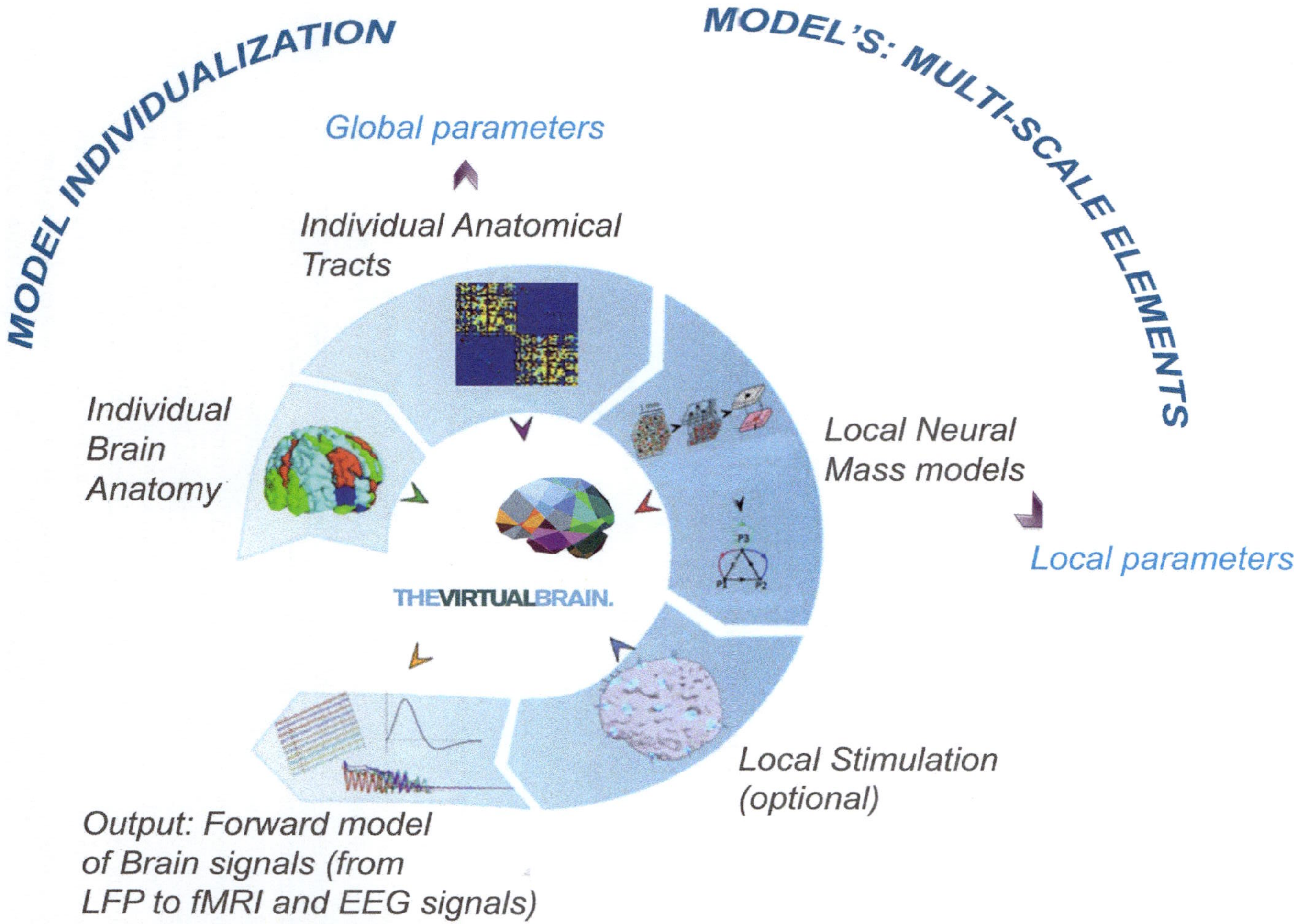


**Figure 1.5** ***The Virtual Brain (TVB) modeling workflow.*** Graphic representation depicting the elements necessary for performing TVB modeling (represented by the central figure of brain nodes). (1) Empirical inputs include the individual cortical geometry and the respective structural connectome generated from T1-weighted magnetic resonance imaging acquisitions (allowing for brain parcellation and DTI tractography). (2) Selection of local (neural mass) model that will allow for the exploration and fitting of biophysical parameters. (3) The platform has an additional option of adding stimulation to any brain nodes. (4) Once parameter values are fitted, the simulated output is obtained. This can be a raw brain signal (local field potentials or neuronal firing rates) or EEG/MEG and BOLD signals. (5) Finally, the efficacy of the simulation is calculated by correlating it to the empirical signals (not depicted in the figure). *(Modified from Kringelbach ML, McIntosh AR, Ritter P, Jirsa VK, Deco G. The rediscovery of slowness: exploring the timing of cognition. Trends Cogn Sci. 2015.)*

structural and functional matrices from the pipeline can then be directly uploaded to the TVB platform.

In summary, large-scale brain network models in general, and TVB in particular, have few critical components: anatomic connectivity, neural (local) mass models, and brain lesions (patient populations). First, TVB uses the structural high-resolution T1-weighted MRI data to create a custom brain surface, and it uses diffusion-weighted MRI data to infer the anatomic connections among brain areas. Second, neural mass models describing the network nodes (brain regions) are used to determine the dynamic nature of the interactions in a biologically realistic manner. Third, an evolution equation integrating the connectivity between regions and the activity within regions calculates global brain activity over time, generating simulated brain signals for each region. Finally, acquisition of functional data (fMRI) permits testing the accuracy of the simulated signals generated by the model (Fig. 1.5).

The TVB modeling workflow includes the following sequential steps performed for each individual subject (example based on the stroke work):

1. *Importing* of the two subject-specific structural connectivity matrices (weights and lengths) into the TVB platform.
2. *Selection of the local neural mass model*: The platform includes a variety of 12 neural population models. These can go from phenomenological models (e.g., Generic 2D, Epileptor) to those simulating local field potentials or synaptic activity (e.g., Wong-Wang, Stefanescu-Jirsa) and firing rates (e.g., Wilson-Cowan, Jansen-Rit). For example, in the stroke studies presented here, we use the Stefanescu-Jirsa 3D model,[24] as it permits characterization of the direct association between neural mass responses (field potentials) and rsfMRI.[30,31]
3. *Parameter Space Estimation*: Most of the modeling focuses on determining the optimal global and local parameters for each subject. This includes two steps. (a) *Parameter space exploration*: We use heat maps of global variance (mean variance of time series across all brain regions) to constrain the range of values for each model parameter. The desired range has high global variance flanked by bifurcation points[22] (Fig. 1.6). (b) *Parameter fitting*: The derived optimal values result in the best fit between the empirical and simulated signals based on the following metrics: Global parameters: conduction velocity and global coupling; Local parameters: K12 (inhibitory on excitatory coupling), K21(excitatory on inhibitory coupling), and K11 (excitatory on excitatory coupling). The Stefanescu-Jirsa 3D model has several additional parameters (i.e., affecting ion channels, bursting strength, input current, and excitability[10]; see Table 1.1) that are kept at default values unless the simulated signals are not optimal.
4. *Stochastic network simulation*: Based on the optimal values obtained, we generate field potentials. White noise with Gaussian amplitude (mean = 0, standard deviation = 1) is added to each node, and numeric integration of the system is performed using a stochastic Heun's method,[32] with an integration step size of 0.0122 ms.

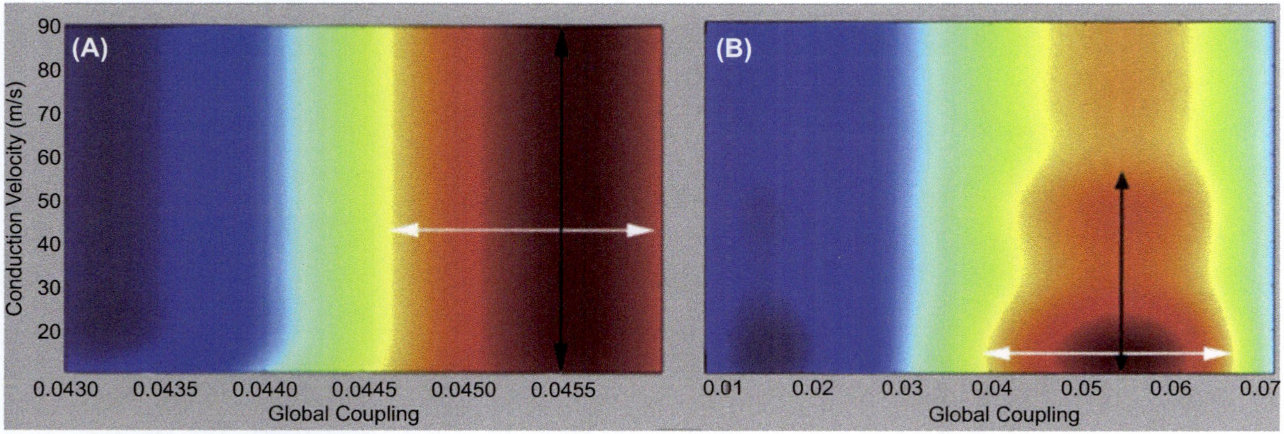


**Figure 1.6** ***Parameter space mapping.*** These heat maps represent variance from a healthy subject (A) and a stroke patient (B). The plots show the global variance of the model's dynamics as a function of changing parameter values for conduction velocity (*y* axis) and global coupling (*x* axis). A noteworthy feature is that topography on the stroke case clearly differs from the parameter map associated with healthy controls. These differences are most noticeably in the areas of higher variance (*dark red*) flanked by bifurcation points (*white arrows*). These values are essential for parameter exploration and the final parameter fitting.

5. *Determining BOLD signals*: These forward models can then be derived from the field potentials using a hemodynamic response function implemented with a gamma kernel[23] with the same duration and sampling rate (TR) as the empirical rsfMRI acquisition.
6. *Assessing the accuracy of the simulated time series*: We compare the empirical and simulated BOLD time series in terms of amplitude, frequency, and phase. Amplitude is calculated from the highest and lowest peaks present in the time series across all regions. Frequency is computed via Fast Fourier transform using Matlab's "fft" function with a sampling frequency of 0.5 Hz to determine the range, profile, and peak frequencies.[9] We assess phase by comparing the functional connectivity matrices of the simulated and empirical time series via Pearson's correlation.
7. *Excluding overfitting and ensuring model identifiability*: The number of free parameters in the model does raise concerns of overfitting and identifiability (i.e., whether there is convergence to a single optimal solution). To this end, we examine convergence to a global extreme, and then, with a fivefold cross-validation, we exclude overfitting. For each subset (fold), the functional (predicted) data are randomly divided into an 80% sample as training segment and a 20% sample as testing segment.

Note that some of these computations—both in the preprocessing pipeline and in the TVB modeling per se—are very demanding in terms of processor time and memory demands, requiring novel computational infrastructures[33] and high-performance computing.[34] Fortunately, more efficient approaches to perform some of the computational tasks are in development.[35]

## MODELING WITH THE VIRTUAL BRAIN IN THE CLINICAL CONTEXT

### Stroke

Focal damage to the brain results in disruption of a distributed network of connections (white matter tracts) initiated in the damaged region(s), making network analyses particularly useful in studying the widespread effect of brain injury[36–38] (for a review, see Ref. 39). Our approach focuses on using methods from network neuroscience to contribute to a personalized data-driven approach to neurological care (i.e., precision medicine[40]). We hope that by building computational models integrating data from clinics and from basic research, we can develop a mechanistic understanding of the disease process,[40,41] and in the context of the recent push to collect large datasets, to develop an approach to making useful diagnostic and therapeutic inferences from such complex data.

As previously mentioned, multiscale brain network models are gaining increasing attention[42–44] because they produce biologically relevant interpretations and operate via *simulation* of brain activity as a mean to understand its properties. That is, the formal link between different observational levels is established by numeric *simulations* resulting from the integration of *local* and *global* dynamics,[45] where the biophysical parameters at the local level describe the properties of small populations of neurons, and at global level, they represent biological mechanisms governing interactions between brain regions.

### *Relationship Between Recovery After Stroke and Modeled Biophysical Parameters*

In our previous work,[10] we sought to determine changes in brain dynamics after stroke by modeling local and global parameters. In this study, motor performance was assessed before, after, and long after (6—12 months) a therapeutic intervention for hand motor skill, yet the modeling was solely done based on the imaging before therapy. Based on individual structural connectomes, we showed that patients with stroke demonstrated a consistent reduction in conduction velocity, increased local dynamics, and local excitability (via a decrease in inhibitory coupling). Importantly, we found a negative relationship between local excitation and motor recovery and a positive correlation between local dynamics and motor recovery (Fig. 1.7). Thus, our TVB modeling revealed a disrupted poststroke system favoring excitation-over-inhibition and local-over-global dynamics, consistent with existing literature on stroke mechanisms.[46—48] This work is an important step demonstrating the potential of TVB to determine individualized biophysical changes associated with stroke recovery. In a second study,[11] we aimed to relate these results to existing literature and subjected each structural connectome to graph theoretic analysis to determine the relationship between modeled global coupling and the organization of anatomic tracts.[49,50] For this, we calculated degree centrality, betweenness centrality, and global efficiency.[51] Linear regression analysis demonstrated that our TVB parameter global coupling was negatively correlated with global efficiency ($P = .038$), but not with degree centrality or betweenness centrality. In other words, the larger influence of local dynamics seen through changes in global coupling parameter is closely associated with a decreased efficiency of the system in stroke.

### *The Virtual Therapy*

In an additional study,[12] we piloted the question of "Virtual Therapy" (VT), elicited within TVB by changing specific simulation parameters toward control values. In this study, we hypothesized that the normalization of parameter values to those found in healthy controls would result in the reestablishment of healthier brain dynamics even in the presence of a damaged brain. After performing VT (i.e., normalizing the parameter to control levels), 17 of 20 stroke subjects showed an increased correlation of resting state functional connectivity with the resting state functional connectivity of an average of healthy control participants (Fig. 1.8). The percent gain ranged from 5.0% to 85.7%, and the combination of parameters that produced the largest improvement in brain dynamics included global coupling and local excitatory on inhibitory coupling. In other words, the change in these parameters produced normalization of the balance between global and local dynamics and excessive excitation.

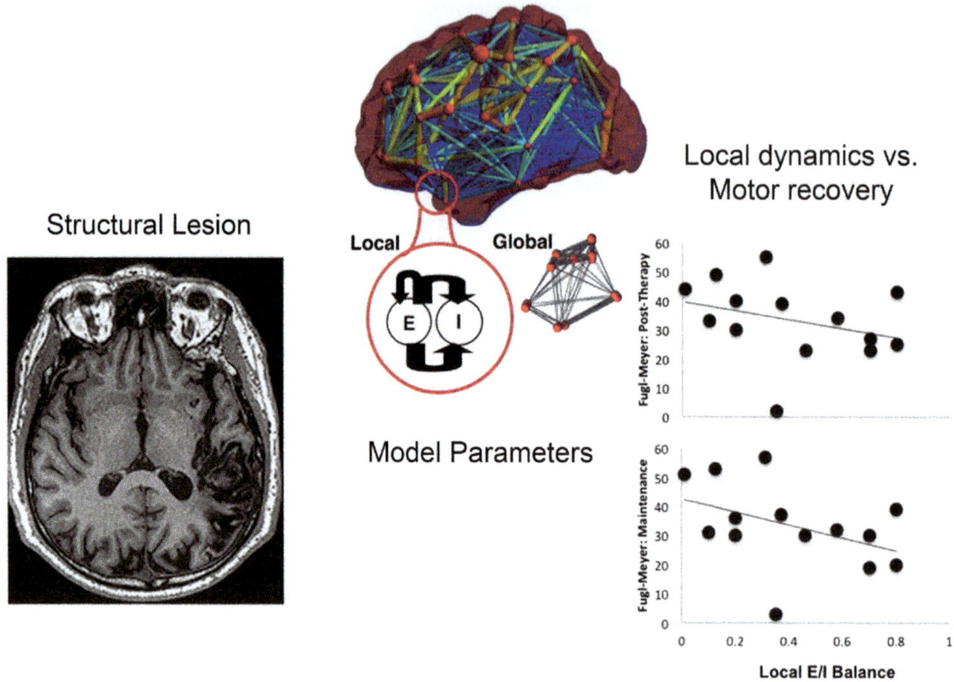


**Figure 1.7** ***The Virtual Stroke Brain.*** The figure highlights salient aspects of TVB modeling to generate the Virtual Stroke Patient Brain. First, the modeling involves the determination of the effects of structural damage to brain dynamics (represented by the axial T1-weighted image on the left). Second, the assessment involves local and whole brain dynamics (center figure); finally, the scatter plots to the right represent the close relationship between local dynamics (balance of excitatory-inhibitory coupling) with motor function after therapy (upper graph) and 1 year after therapy (lower graph). The higher the dynamic change was, the worse was the motor recovery.

## Neurodegeneration: Alzheimer Disease

Alzheimer disease is a complex degenerative process of the brain, which begins in ventromedial cortical limbic regions before it spreads to the rest of the brain.[52,53] The long, chronic degeneration with a loss of a critical number of neurons leads to Alzheimer dementia (AD), which is characterized by an ongoing loss of memory and higher cognitive functions. Remarkably, one of the hallmark neuropathologic markers of AD (neurofibrillary tangles) disrupts axonal integrity, suggesting AD is a classic disconnection syndrome. Indeed, AD is marked by cognitive dysfunction emerging from altered brain network activity.[54–56] Interestingly, alterations across similar brain networks are also implicated in cognitive impairments seen in Parkinson disease (PD),[57] where cognitive dysfunction can appear in the final stage of the idiopathic Parkinson syndrome.[58] Determining differences and similarities between AD and PD poses a significant challenge because of overlapping affected brain networks captured by changes in large-scale brain

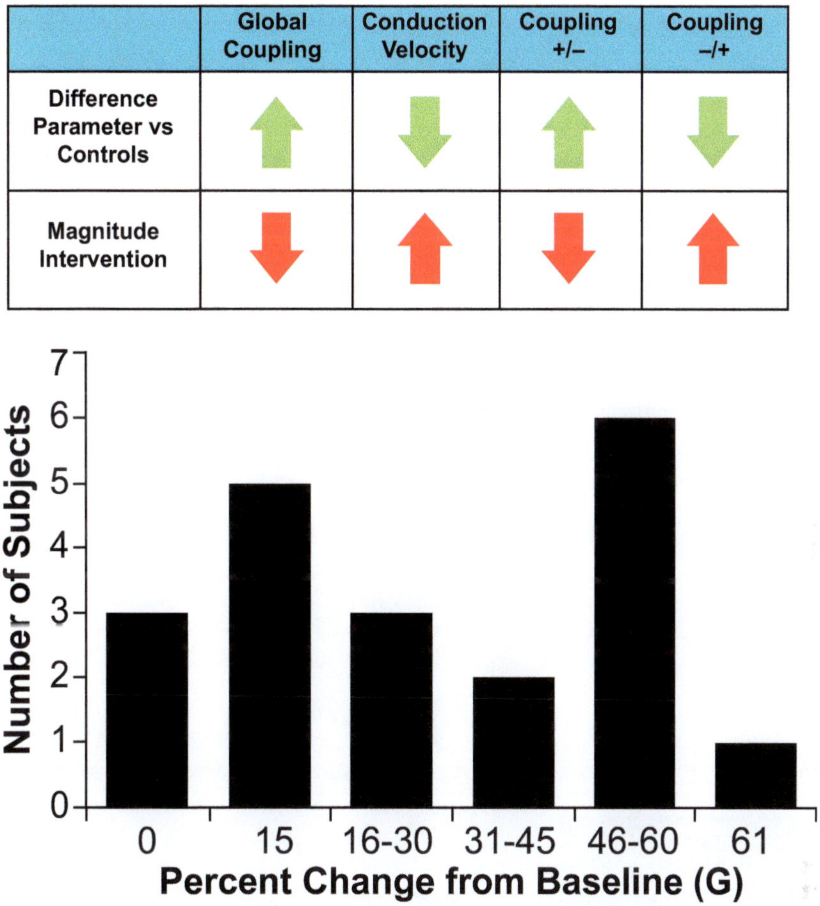


**Figure 1.8** ***The Virtual Therapy.*** (A) Example of the Virtual Therapy (VT) implementation for one stroke case. Parameters were normalized from the baseline value to the healthy control average. This process was repeated for each patient. (B) Frequency histogram showing the number of stroke cases stratified by percent of change in the correlation of functional connectomes from baseline after the VT in relation to healthy controls. In all but three cases, the functional connectivity matrices after the therapy were closer to those in healthy controls. Changing the parameter values in the stroke subjects to mimic those in controls resulted in healthier brain dynamics even when the damage did not change.

network studies.[59] To develop new therapeutic approaches to dementia, an initial question to resolve is to determine individuals vulnerable to develop dementia in preclinical stages before degeneration is extensive.

As TVB is a connectivity-based simulator targeting neural populations[9] underscored by structural and functional connectivity changes (Matthews et al., 2013; Sun et al., 2014)[80,81], along with local changes in brain dynamics, we recently initialized a study to determine the specific global and local biophysical parameters associated with the generation of the AD clinical phenotype. That is, to understand the functional pathology in AD, we are currently investigating biophysical TVB parameters associated with cognitive profiles from healthy controls, subjects with mild cognitive impairment (MCI), and Alzheimer disease.

For this, structural connectomes derived from 124 subjects (16 AD, 35 MCI, and 73 healthy controls) from the Sydney Memory and Aging Study were included in the modeling. Based on individual anatomic connections, personalized Virtual Brain models were constructed using the Reduced Wong-Wang neuronal mass model included in the TVB repertoire. The model successfully simulated rsfMRI time series for each subject. To determine the relationship between optimized model parameters and clinical phenotype, neuropsychologic cognitive measures were integrated in a single score and correlated with parameter measures also integrated in a single score (Fig. 1.9).

Our preliminary results not only hint to consistent changes in biophysical parameters, but they also show a consistent positive relationship with the clinical phenotype. Based

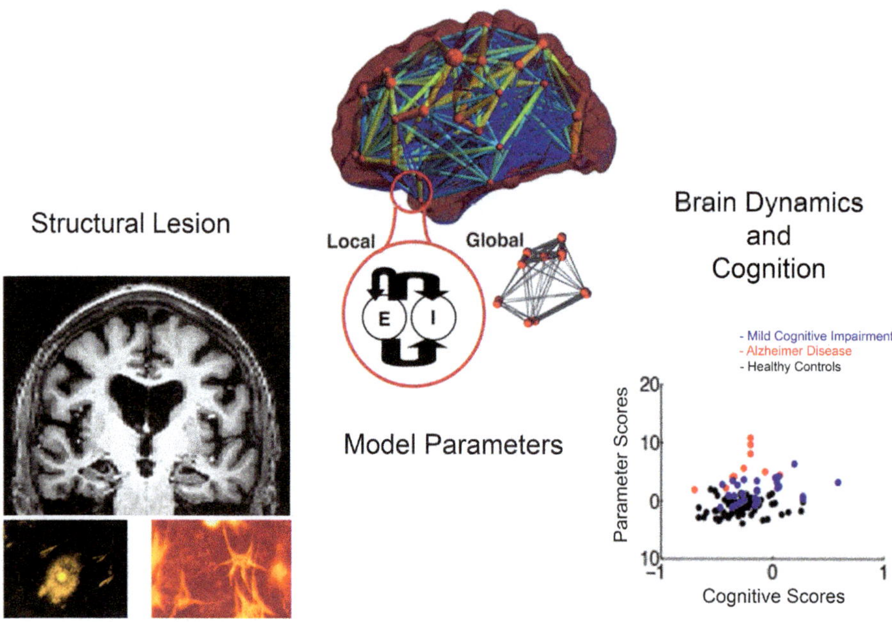


**Figure 1.9** ***The Virtual Alzheimer Brain.*** The figure highlights salient aspects of The Virtual Brain modeling to generate the Virtual Alzheimer Patient Brain. First, the modeling involves the determination of the effects of structural degeneration to brain dynamics (represented by the coronal T1-weighted image on the left along two microphotographs of neuritic plaques and neurofibrillary tangles). Second, the assessment involves local and whole brain dynamics (center figure); finally, the scatter plot to the right represents the relationship between brain dynamics (local and global combined) with combined cognitive scores. Note the relationship between these two metrics for healthy controls (*black dots*), Alzheimer cases (*red dots*), and cases with amnesic mild cognitive impairment (*blue dots*).

on this last feature, we are currently expanding the analysis to determine the statistical membership of MCI subjects to either the AD or the healthy control groups, as this is a crucial step to determine converters to AD and, hence, pave the way in the development of novel therapeutic interventions.

## SUMMARY: THE ROLE OF NEUROINFORMATICS IN BIOMARKER DISCOVERY

The goal of personalized medicine is that patients receive the right treatment, at the right dose, and the right time. To achieve this, pertinent individual information associated with disease is needed, which explains the popularity of searching for proper "biomarkers."

The use of biomarkers in clinical practice and clinical research ranges from the assessment of disease severity and diagnosis to surrogate endpoints in clinical trials. In consequence, the term has been associated with several related definitions. For instance, in the most general sense, a biomarker has been described as a "surrogate clinical measure that ideally reflects underlying biological processes on the cellular/molecular level that can be used as an indicator of disease state"[60]; or as a "broad subcategory of signs which can be measured accurately and reproducibly,"[61] and a "characteristic that is objectively measured and evaluated as an indicator of normal biological processes, pathogenic processes, or pharmacological responses to a therapeutic intervention" (NIH Biomarker definitions Working group).[62] Based on these definitions, the essential features of a biomarker can be inferred (for an excellent review, see Ref. 61). First, it should show consistency at an *individual* patient level, determining diagnosis, disease progression, or responses to treatment.[63,64] A widely used example in medicine is the diagnosis and treatment of diabetes based on the concentration of hemoglobin A1c, which reflects degree of glucose metabolism consistently increased in diabetes.[65,66] Second, related to the first point, is that it should provide clinically relevant information, by accurately predicting clinical outcome. Third, a biomarker should be part of the disease's pathologic pathway. Because of their utility in medical practice, researchers in the neurologic field are actively seeking biomarkers of brain injury as a means to personalize treatment and track individual recovery. However, it is essential to distinguish between true biomarkers to variables reflecting simple correlations with behavioral states, even if they are sensitive, specific, minimally invasive, and relatively easy and inexpensive to measure.[63,67] The many studies involving a variety of metrics as putative neurologic biomarkers typically fall into one of three categories: (1) indirectly correlated with disease; (2) directly correlated with disease at a systems level; or (3) directly related to molecular and cellular disease processes. From these, only the second and third categories fulfill the definition of a true biomarker.

In this context, several clinical imaging modalities, such as computed tomography, magnetic resonance spectroscopy, MRI, ultrasonography, gamma scintigraphy, and positron emission tomography (PET), are prominent in these efforts. For example, various neuroimaging measures have identified correlations with different aspects of diseases for neurologic disorders such as Alzheimer disease,[68] epilepsy,[69] PD,[70] and amyotrophic lateral sclerosis.[71] However, none of these candidate biomarkers can be used clinically because they show a high rate of type 1 and type 2 diagnostic errors; that is, although many identified measures may show correlations with disease within a large population, they lack the precision, specificity, and reliability necessary as evaluation at the *individual* patient level. Let us consider, as an example, hippocampal atrophy that has been considered a prominent candidate for a surrogate measure of neurodegeneration in Alzheimer disease. Interestingly, the prediction rate is as low as 65%,[72–74] due probably to the absence of a direct relation to underlying pathologic processes. Because these correlative measures are not individualized and have no direct relation with the pathophysiology of diseases, they cannot be considered biomarkers.

Conversely and in addition to system-level methods, other venues have also been explored for viable biomarkers of neurologic diseases, including genetics,[63] immunologic factors,[75] molecular imaging,[63] protein, antibodies,[64] ribonucleic acid (RNA), or metabolite levels detected in blood or cerebrospinal fluid samples. These methods, more closely related to the disease process, lack in many instances a relation to a clinical phenotype and hence do not "capture the efficacy of treatments."[76] In other words a good candidate biomarker not only requires a direct link with basic biological processes associated with the pathologic mechanisms of the disease but also requires a clear association with clinical phenotypes.

Using TVB to model brain dynamics after stroke meets a number of basic requirements for biomarker research. First, the approach is physiologic rather than correlative; that is, the basic elements of the model are biologic signals rather than statistical regularities (e.g., by contrast with functional imaging). Second, the method is multiscale rather than focused on a single level of description. Thus, unlike measures at the level of the individual neuron or local neuronal field (e.g., cortical electrodes, microdialysis) or at the level of anatomic region (e.g., MRI, PET), TVB models span levels to create a unifying description possibly associated with pathologic processes. Third, TVB modeling has individual predictive value since it is based on highly individualized structural connectivity across the brain. Typically, brain imaging studies in recovery show population-based correlations, but due to enormous variability across individuals, they are not applicable for precision (personalized) medicine. The critical and unique features of our approach are that our models are individualized, multiscale, and biologically grounded.

Even though specific criteria to assess imaging-based biomarkers are lacking, the path to determine if the parameter profiles derived from the modeling can be used as such is just beginning, as strict evaluation is needed, including selection before discovery phase, characterization/validation strategies, and robust analyses techniques used in clinical trials.[77] Biomarkers with ideal specificity/sensitivity are hard to find; therefore the combination of different model parameters may be useful. Importantly, the process has to include a large number of patients because low sample sizes make correct biomarker identification difficult. Finally, translational basic research testing the validity of disease mechanisms uncovered by our modeling needs to be tested.[78]

In summary, we are confident that our patient parameter profiles and data-driven models will provide knowledge and methodology for the systematic determination of treatment strategies (e.g., therapeutic targets) and patient management procedures, including stratification of patients prior to treatment. The development and use of virtual patient models to improve patient care represents an important new approach to precision medicine in neurology.

## GLOSSARY OF TERMS

| Term | Definition |
|---|---|
| Betweenness centrality | Graph theoretic value on how central a node is in the whole network, i.e., number of shortest paths going through this node |
| Biomarkers | Empirical measure with a direct link with basic biologic processes associated with the pathologic mechanisms of a disease and a clear association with clinical phenotypes |
| Bottom-up modeling | Modeling initialized in the smallest elements and then moving up to larger elements (e.g., from cellular to whole brain) |
| Brain dynamics | Rules that govern the processes in the brain |
| Conduction velocity | Speed of transmission of neural signals in white matter. In TVB, it is computed as time delays in long-range white matter tracts |
| Connectomics | Science that deals with the connectome, e.g., as genomics deals with the genome. A connectome is the entirety of connections in a brain |
| Coupling function | Mathematic algorithm determining the influence of a brain region over another or the influence of neuronal ensembles over others |
| Degree centrality | A graph theoretic value for centrality reflecting how many neighbors or strong connections a node (brain region) has |
| Deterministic simulation | Method to run a simulation without random influences where each step is determined by the state before |
| Excitatory-inhibitory coupling | Representative for neural transmission from excitatory neurons to inhibitory neurons. It represents also the count and activity of excitatory neurotransmitters and its receptors |
| Functional connectivity dynamics | A special aspect of the functional connectome that focuses on temporal changes. Mathematically, it represents the first derivative of functional connectivity |
| Functional connectome | Entirety of functional brain connections assessed via covariance matrices |

*Continued*

| Term | Definition |
|---|---|
| Global coupling | Numeric function to rescale incoming activity between brain regions before entering the local dynamic equations of the model |
| Global dynamics | Dynamics that affect the whole brain. Global dynamics influence, e.g., the transmission in large-scale connections or superordinate brain states of activity. In TVB, it is represented by global coupling and conduction velocity |
| Global efficiency | A graph theoretic value on the effective communication within a network. Average of inverse path lengths |
| Graph analysis | Mathematic analysis of complex networks based on graph theory |
| Inhibitory-excitatory coupling | Influence of local neural transmission from inhibitory neurons to excitatory neurons. Number and activity of inhibitory neurotransmitters and its receptors |
| Local dynamics | Dynamics at the local regional level. In TVB, it refers to model parameters that affect the activity in a neural mass model |
| Macroscopic level, macroscale | Brain scale of the entire brain and its parts: regions, tracts, and their connectivity |
| Mesocopic level, mesoscale | Brain scale between the macro- and the microscale representing neuronal populations and their functional characteristics |
| Microscopic level, microscale | Brain scale of irreducible elements and their characteristic behavior |
| Model parameters | Free/unknown variables in a function or equation, which makes it to a family of functions. With each free parameter, the solution space of an equation gets an additional dimension |
| Multiscale modeling | A modeling approach that combines aspects of different brain levels, from microscopic features and single neurons to global dynamics |
| Network modeling | Approach to design an artificial structure (model) that captures aspects of reality, in this case, brain dynamics |
| Neural mass | A population of neurons with similar or equal in- and outgoing connections and a common function |
| Neural mass model | Model of a neural mass including several neuronal subpopulations. Smallest element in a mesoscopic model, focusing on those populations |
| Neuroinformatics | Science that examines neural structure and function via information-technology methods |
| Overfitting | Mathematic phenomenon that results when a model has too many degrees of freedom. Modeling artifacts |
| Parameter space estimation | Calculation of probable parameter values based on specific conditions |
| Parameter space fitting | Mathematical and informatics approach on complex systems: computer resources simulate possible combinations (parameter space exploration) to find the best fitting |

| Term | Definition |
|---|---|
| Parcellation | Anatomic schemes to divide the brain into meaningful subregions (nodes) |
| Path length | Graph theoretic value for a node associated with the smallest number of paths needed to reach other nodes |
| Pipeline | A standardized and automatized workflow, consisting of informatics scripts or programs |
| Precision medicine | Medical model that proposes the customization of healthcare, with medical decisions, practices, and/or products tailored to individual patients |
| Simulation | Mathematic and informatics approach to generate data, driven by a model, exhibiting features from biologic systems |
| Spatiotemporal scales | Metrics defined by space (size and localization) and by time (static or dynamic) |
| Stochastic simulation | Simulation that traces dynamics changing stochastically (noise) with certain probabilities |
| Structural connectome | Entirety of existing anatomic tracts in brain. Currently can be computed via DTI tractography |
| The Virtual Brain (TVB) | Multiscale brain simulation platform: www.thevirtualbrain.org |
| The Virtual Therapy | Assessment of the effects of a potential therapy within TVB |
| Top-down modeling | Opposite to bottom-up modeling |
| Tractography | A visualization of white matter tracts of the brain |
| Whole brain level | Approach covering all brain regions and all scales |
| Workflow | Standardized sequence of processing steps. If automatized, it is a pipeline. If mathematically formalized, it is an algorithm |

## REFERENCES


1. Sudlow C, Gallacher J, Allen N, et al. UK biobank: an open access resource for identifying the causes of a wide range of complex diseases of middle and old age. *PLoS Med.* 2015;12(3):e1001779.
2. Miller KL, Alfaro-Almagro F, Bangerter NK, et al. Multimodal population brain imaging in the UK Biobank prospective epidemiological study. *Nat Neurosci.* 2016;19(11):1523–1536.
3. Hendrix JA, Finger B, Weiner MW, et al. The worldwide Alzheimer's disease neuroimaging initiative: an update. *Alzheimers Dement.* 2015;11(7):850–859.
4. Mueller S, Wang D, Fox MD, et al. Individual variability in functional connectivity architecture of the human brain. *Neuron.* 2013;77(3):586–595.
5. Bezgin G, Solodkin A, Bakker R, Ritter P, McIntosh AR. Mapping complementary features of cross-species structural connectivity to construct realistic "virtual brains". *Hum Brain Mapp.* 2017;38(4): 2080–2093.
6. Price CJ, Friston KJ. Functional imaging studies of neuropsychological patients: applications and limitations. *Neurocase.* 2002;8(5):345–354.
7. Corbetta M, Ramsey L, Callejas A, et al. Common behavioral clusters and subcortical anatomy in stroke. *Neuron.* 2015;85(5):927–941.
8. Zimmermann J, Ritter P, Shen K, Rothmeier S, Schirner M, McIntosh AR. Structural architecture supports functional organization in the human aging brain at a regionwise and network level. *Hum Brain Mapp.* 2016;37(7):2645–2661.

9. Ritter P, Schirner M, McIntosh AR, Jirsa VK. The virtual brain integrates computational modeling and multimodal neuroimaging. *Brain Connect.* 2013;3(2):121–145.
10. Falcon MI, Riley JD, Jirsa V, McIntosh AR, Chen EE, Solodkin A. Functional mechanisms of recovery after chronic stroke: modeling with the virtual brain. *eNeuro.* 2016b;3(2).
11. Falcon MI, Riley JD, Jirsa V, et al. The virtual brain: modeling biological correlates of recovery after chronic stroke. *Front Neurol.* 2015;6:228.
12. Falcon MI, Jirsa V, Solodkin A. A new neuroinformatics approach to personalized medicine in neurology: the virtual brain. *Curr Opin Neurol.* 2016a;29(4):429–436.
13. Jirsa VK, Stacey WC, Quilichini PP, Ivanov AI, Bernard C. On the nature of seizure dynamics. *Brain.* 2014;137(Pt 8):2210–2230.
14. Jirsa VK, Proix T, Perdikis D, et al. The virtual epileptic patient: individualized whole-brain models of epilepsy spread. *Neuroimage.* 2017;145(Pt B):377–388.
15. Sanz Leon P, Knock SA, Woodman MM, et al. The virtual brain: a simulator of primate brain network dynamics. *Front Neuroinform.* 2013;7:10.
16. Matzke H, Schirner M, Vollbrecht D, et al. TVB-EduPack-an interactive learning and scripting platform for the virtual brain. *Front Neuroinform.* 2015;9:27.
17. Stefanovski L, Ghani A, McIntosh AR, Ritter P. Linking connectomics and dynamics in the human brain. *e-Neuroforum.* 2016;7(3):64–70.
18. Kasthuri N, Hayworth KJ, Berger DR, et al. Saturated reconstruction of a volume of neocortex. *Cell.* 2015;162(3):648–661.
19. Song S, Sjostrom PJ, Reigl M, Nelson S, Chklovskii DB. Highly nonrandom features of synaptic connectivity in local cortical circuits. *PLoS Biol.* 2005;3(3):e68.
20. Fornito A, Bullmore ET. Connectomics: a new paradigm for understanding brain disease. *Eur Neuropsychopharmacol.* 2015;25(5):733–748.
21. Proix T, Spiegler A, Schirner M, Rothmeier S, Ritter P, Jirsa VK. How do parcellation size and short-range connectivity affect dynamics in large-scale brain network models? *Neuroimage.* 2016;142:135–149.
22. Breakspear M, Jirsa VK. Neuronal dynamics and brain connectivity. In: *Handbook of Brain Connectivity.* Springer; 2007:3–64.
23. Sanz-Leon P, Knock SA, Spiegler A, Jirsa VK. Mathematical framework for large-scale brain network modeling in the virtual brain. *Neuroimage.* 2015;111:385–430.
24. Stefanescu RA, Jirsa VK. A low dimensional description of globally coupled heterogeneous neural networks of excitatory and inhibitory neurons. *PLoS Comput Biol.* 2008;4(11):e1000219.
25. Schirner M, Rothmeier S, Jirsa VK, McIntosh AR, Ritter P. An automated pipeline for constructing personalized virtual brains from multimodal neuroimaging data. *Neuroimage.* 2015;117:343–357.
26. Kringelbach ML, McIntosh AR, Ritter P, Jirsa VK, Deco G. The rediscovery of slowness: exploring the timing of cognition. *Trends Cogn Sci.* 2015;19(10):616–628.
27. Deco G, Kringelbach ML, Jirsa V, Ritter P. The dynamics of resting fluctuations in the brain: metastability and its dynamical cortical core. *bioRxiv.* 2016. https://doi.org/10.1101/065284.
28. Jirsa VK, Proix T, Perdikis D, et al. The virtual epileptic patient: individualized whole-brain models of epilepsy spread. *Neuroimage.* 2016;145:377–388.
29. Schirner M, McIntosh AR, Jirsa V, Deco G, Ritter P. Bridging multiple scales in the human brain using computational modelling. *bioRxiv.* 2016. https://doi.org/10.1101/085548.
30. Magri C, Schridde U, Murayama Y, Panzeri S, Logothetis NK. The amplitude and timing of the BOLD signal reflects the relationship between local field potential power at different frequencies. *J Neurosci.* 2012;32(4):1395–1407.
31. Pan WJ, Thompson GJ, Magnuson ME, Jaeger D, Keilholz S. Infraslow LFP correlates to resting-state fMRI BOLD signals. *Neuroimage.* 2013;74:288–297.
32. Mannella R. Quasisymplectic integrators for stochastic differential equations. *Phys Rev E Stat Nonlin Soft Matter Phys.* 2004;69(4 Pt 1):041107.
33. Gee T, Kenny S, Price CJ, et al. Data warehousing methods and processing infrastructure for brain recovery research. *Arch Ital Biol.* 2010;148(3):207–217.

34. Small SL, Wilde M, Kenny S, Andric M, Hasson U. Database-managed grid-enabled analysis of neuroimaging data: the CNARI framework. *Int J Psychophysiol.* 2009;73(1):62–72.
35. Saggio ML, Ritter P, Jirsa VK. Analytical operations relate structural and functional connectivity in the brain. *PLoS One.* 2016;11(8):e0157292.
36. Jirsa VK, Sporns O, Breakspear M, Deco G, McIntosh AR. Towards the virtual brain: network modeling of the intact and the damaged brain. *Arch Ital Biol.* 2010;148(3):189–205.
37. McIntosh AR, Ghilardi F, Sporns O. Overview: integrating computational, cognitive and clinical expertise to understand brain network recovery. *Arch Ital Biol.* 2010;148(3):185–187.
38. Cabral J, Kringelbach ML, Deco G. Exploring the network dynamics underlying brain activity during rest. *Prog Neurobiol.* 2014;114:102–131.
39. Grefkes C, Fink GR. Connectivity-based approaches in stroke and recovery of function. *Lancet Neurol.* 2014;13(2):206–216.
40. Duffy DJ. Problems, challenges and promises: perspectives on precision medicine. *Brief Bioinform.* 2016; 17(3):494–504.
41. Wolkenhauer O, Auffray C, Brass O, et al. Enabling multiscale modeling in systems medicine. *Genome Med.* 2014;6(3):21.
42. Eliasmith C, Trujillo O. The use and abuse of large-scale brain models. *Curr Opin Neurobiol.* 2014;25: 1–6.
43. Grillner S. Megascience efforts and the brain. *Neuron.* 2014;82(6):1209–1211.
44. Sompolinsky H. Computational neuroscience: beyond the local circuit. *Curr Opin Neurobiol.* 2014;25: xiii–xviii.
45. Deco G, Jirsa VK, Robinson PA, Breakspear M, Friston K. The dynamic brain: from spiking neurons to neural masses and cortical fields. *PLoS Comput Biol.* 2008;4(8):e1000092.
46. Butefisch CM, Netz J, Wessling M, Seitz RJ, Homberg V. Remote changes in cortical excitability after stroke. *Brain.* 2003;126(Pt 2):470–481.
47. Kim JH, Lee HW, Cohen LG, Park KD, Choi KG. Motor cortical excitability in patients with post-stroke epilepsy. *Epilepsia.* 2008;49(1):117–124.
48. Clarkson AN, Carmichael ST. Cortical excitability and post-stroke recovery. *Biochem Soc Trans.* 2009; 37(Pt 6):1412–1414.
49. Bullmore E, Sporns O. Complex brain networks: graph theoretical analysis of structural and functional systems. *Nat Rev Neurosci.* 2009;10(3):186–198.
50. Sporns O. From simple graphs to the connectome: networks in neuroimaging. *Neuroimage.* 2012;62(2): 881–886.
51. van den Heuvel MP, Sporns O. Network hubs in the human brain. *Trends Cogn Sci.* 2013;17(12): 683–696.
52. Alzheimer A. Über eigenartige Krankheitsfälle des späteren Alters. *Zeitschrift für die gesamte Neurologie und Psychiatrie.* 1911;4(1):356–385.
53. Braak H, Braak E. Neuropathological staging of Alzheimer-related changes. *Acta Neuropathol.* 1991; 82(4):239–259.
54. Hyman BT, Van Hoesen GW, Damasio AR, Barnes CL. Alzheimer's disease: cell-specific pathology isolates the hippocampal formation. *Science.* 1984;225(4667):1168–1170.
55. Van Hoesen GW, Solodkin A. Cellular and system neuroanatomical changes in Alzheimer's disease. *Ann N Y Acad Sci.* 1994;747:12–35.
56. Greicius M. Resting-state functional connectivity in neuropsychiatric disorders. *Curr Opin Neurol.* 2008;21(4):424–430.
57. Rektorova I. Resting-state networks in Alzheimer's disease and Parkinson's disease. *Neurodegener Dis.* 2014;13(2–3):186–188.
58. Gomperts SN. Lewy body dementias: dementia with Lewy bodies and Parkinson disease dementia. *Continuum (Minneap Minn).* 2016;22(2 Dementia):435–463.
59. Fornito A, Zalesky A, Breakspear M. The connectomics of brain disorders. *Nat Rev Neurosci.* 2015; 16(3):159–172.
60. Burke E, Cramer SC. Biomarkers and predictors of restorative therapy effects after stroke. *Curr Neurol Neurosci Rep.* 2013;13(2):329.
61. Strimbu K, Tavel JA. What are biomarkers? *Curr Opin HIV AIDS.* 2010;5(6):463–466.
62. FDA-NIH. *BEST (Biomarkers, Endpoints, and Other Tools) Resource.* 2016.

63. Agrawal M, Biswas A. Molecular diagnostics of neurodegenerative disorders. *Front Mol Biosci.* 2015;2:54.
64. Berger T, Reindl M. Antibody biomarkers in CNS demyelinating diseases – a long and winding road. *Eur J Neurol.* 2015;22(8):1162–1168.
65. Larsen ML, Horder M, Mogensen EF. Effect of long-term monitoring of glycosylated hemoglobin levels in insulin-dependent diabetes mellitus. *N Engl J Med.* 1990;323(15):1021–1025.
66. Caveney EJ, Cohen OJ. Diabetes and biomarkers. *J Diabetes Sci Technol.* 2011;5(1):192–197.
67. Harris VK, Sadiq SA. Biomarkers of therapeutic response in multiple sclerosis: current status. *Mol Diagn Ther.* 2014;18(6):605–617.
68. Mintun MA. Utilizing advanced imaging and surrogate markers across the spectrum of Alzheimer's disease. *CNS Spectr.* 2005;10(11 suppl 18):13–16.
69. Walker LE, Mirza N, Yip VL, Marson AG, Pirmohamed M. Personalized medicine approaches in epilepsy. *J Intern Med.* 2015;277(2):218–234.
70. Miller DB, O'Callaghan JP. Biomarkers of Parkinson's disease: present and future. *Metabolism.* 2015; 64(3 suppl 1):S40–S46.
71. Verstraete E, Foerster BR. Neuroimaging as a new diagnostic modality in amyotrophic lateral sclerosis. *Neurotherapeutics.* 2015;12(2):403–416.
72. Frisoni GB, Weiner MW. Alzheimer's disease neuroimaging initiative special issue. *Neurobiol Aging.* 2010;31(8):1259–1262.
73. Wolz R, Julkunen V, Koikkalainen J, et al. Multi-method analysis of MRI images in early diagnostics of Alzheimer's disease. *PLoS One.* 2011;6(10):e25446.
74. Coupe P, Fonov VS, Bernard C, et al. Detection of Alzheimer's disease signature in MR images seven years before conversion to dementia: toward an early individual prognosis. *Hum Brain Mapp.* 2015; 36(12):4758–4770.
75. Karim AR, Jacob S. Immunological markers in neurological disorders. *Ann Clin Biochem.* 2012; 49(Pt 1):29–43.
76. Fleming TR, Powers JH. Biomarkers and surrogate endpoints in clinical trials. *Stat Med.* 2012;31(25): 2973–2984.
77. Drucker E, Krapfenbauer K. Pitfalls and limitations in translation from biomarker discovery to clinical utility in predictive and personalised medicine. *EPMA J.* 2013;4(1):7.
78. Zolg W. The proteomic search for diagnostic biomarkers: lost in translation? *Mol Cell Proteomics.* 2006; 5(10):1720–1726.
79. Hindmarsh JL, Rose RM. A model of neuronal bursting using three coupled first order differential equations. *Proc R Soc B Biol Sci.* 1984;221(1222):87–102.
80. Matthews PM, Filippini N, Douaud G. Brain structural and functional connectivity and the progression of neuropathology in Alzheimer's disease. *J Alzheimers Dis.* 2013;33(Suppl 1):S163–172. https://doi.org/10.3233/JAD-2012-129012.
81. Sun Y, Yin Q, Fang R, Yan X, Wang Y, Bezerianos A, Sun J, et al. Disrupted functional brain connectivity and its association to structural connectivity in amnestic mild cognitive impairment and Alzheimer's disease. *PLoS One.* 2014;9(5):e96505. https://doi.org/10.1371/journal.pone.0096505.
82. FitzHugh R. Impulses and physiological states in theoretical models of nerve membrane. *Biophys J.* 1961;1:445–466.
83. Nagumo J, Arimoto S, Yoshizawa S. An active pulse transmission line simulating nerve axon. *Proc IRE.* 1962;50:2061–2070.
84. Zetterberg LH, Kristiansson L, Mossberg K. Performance of a model for a local neuron population. *Biol Cybern.* 1978;31:15–26.
85. Jansen BH, Rit VG. Electroencephalogram and visual evoked potential generation in a mathematical model of coupled cortical columns. *Biol Cybern.* 1995;73:357–366.
86. Breakspear M, Stam CJ. Dynamics of a neural system with a multiscale architecture. *Philos Trans R Soc L B Biol Sci.* 2005;360:1051–1074.
87. Wong K-F. A recurrent network mechanism of time integration in perceptual decisions. *J Neurosci.* 2006;26:1314–1328.
88. Stefanescu RA, Jirsa VK. Reduced representations of heterogeneous mixed neural networks with synaptic coupling. *Phys Rev E - Stat Nonlinear Soft Matter Phys.* 2011;83.

CHAPTER 2

# Informatics for Interoperability of Molecular-Genetic and Neurobehavioral Databases

**Douglas M. Bowden, Mark F. Dubach, Evan Dong**
National Primate Research Center at the University of Washington, Seattle, WA, United States

## INTRODUCTION

### Neuroinformatics and Bioinformatics Are Different

Geneticists entering the field of neurobehavioral research confront a lack of neuroscientific databases with processing power to match that of the online gene databases that have been standard tools of genomics and proteomics research for decades. Any geneticist who has sequenced part of an unidentified gene can go on the internet, screen the sequence against any number of gene libraries worldwide, and obtain answers to fundamental questions, such as, "What genes are known to contain this sequence? In what species are the genes found? In what organs and cell types are they expressed? With what biochemical, physiological, or behavioral functions are they associated? In what diseases do they play a role?"

Neuroscientists have no system of databases with comparable search, collative, and analytic capacities. Why not? Because digital processing requires a unique identifier for every data element and every combination of data elements treated as independent objects in computations. Geneticists are blessed in that the fundamental elements of their discipline are five well-defined nucleotides of DNA and RNA structure and ~20 amino acids of protein structure. Furthermore, those fundamental elements are organized in a one-dimensional linear code that is readily analyzed by the kinds of cryptographic techniques that were developed half a century ago to break codes of secret wartime communications.

One can think of a DNA or RNA sequence as a "word" consisting of a particular ordering of four nucleotide "letters" and the primary structure of a protein as a particular sequence of some 20 amino acids. With such a system, the transcription of the DNA code to RNA and the translation to protein structure are straightforward. When the World Wide Web came online two decades ago, all aspiring bioinformaticists spoke the same language. They had a standard nomenclature for nucleotides and amino acids, which they programmed into their database collating and analysis routines. The ontologic relations of words (letter sequences as names) to concepts (longer letter sequences as

*Molecular-Genetic and Statistical Techniques for Behavioral and Neural Research*
ISBN 978-0-12-804078-2, https://doi.org/10.1016/B978-0-12-804078-2.00002-7

definitions) to entities (DNA, RNA, and protein molecules) were unambiguous, and the linear organization of the genome and proteome corresponded to the linear organization of human language, so the one-to-one translation from DNA reality to symbolic representation of reality was straightforward. Ontology, the specifier of relations of entities to concepts and concepts to names, posed no challenge to interoperability.

The entities and concepts of neuroanatomy are qualitatively and quantitatively quite different. They are qualitatively different in that the entities are three-dimensional cells and collections of cells organized in larger three-dimensional patterns. The most informative concepts are visual representations of those three-dimensional objects. The challenge to the *neuroscientist* is to identify the patterns, determine which are meaningful, i.e., correlated with specific functions, and to translate the three-dimensional code of structural reality into the one-dimensional, linear code of human language. The job of the *neuroinformaticist* is to assist investigators in the communication of neuroanatomic data. It is to codify the relations of diverse names and structural concepts in an ontology that is comprehensive, that captures the diversity of structures and models of brain organization, and that represents them by a standard nomenclature. Neuroscientists can communicate among themselves with imperfect terminology because when they encounter ambiguity they can request an explanation of the concepts behind the words of others. Computers cannot. The text name or numeric code for each concept must be unique.

The elements of neuroanatomic discourse also differ quantitatively from those of genomics and proteomics. If one considers the functionally distinct structural elements of the central nervous system, such as cortical areas and subcortical nuclei, to be the "letters" of the neural code and their arrangement into superstructures and pathways as "words," the code consists of some 500−1000 structure-letters (compared to the 5−20 of genomics and proteomics) organized into a system of some 25,000−100,000 connection-words organized in three-dimensional space.[1] Furthermore, in the language code of neuroanatomy, few structures have a single legitimate name, in the sense of "a term commonly used by domain experts." Because several hundred of the universally recognized structures of the nervous system were defined and named independently by multiple anatomists in different countries over the course of centuries, each structure now has an average of six legitimate English and Latin names.[2]

The qualitative and quantitative differences in the elemental entities of neuroanatomy compared to genomics and proteomics have great implications for the computerization of databases where data are classified, or indexed, on the basis of the neural structure from which it is recorded. During the past decade, informaticists have devoted considerable effort to developing a standard neuroanatomic nomenclature. The dominant strategy has been to engage the scientific community in distributed cooperative ventures, such as Open Biological Ontologies to formulate and promulgate best practices for ontology creation. They have developed languages for formal representations of knowledge, such as the Web Ontology Language (OWL). And they have developed computer applications, such as Protégé, to standardize and facilitate the development of such ontologies.[3]

Neuroinformaticists who have attempted to develop a comprehensive ontology of neuroanatomy in accord with such efforts have not, however, enjoyed great success.[4] Such approaches worked well for categorical models of brain organization, e.g., for topographic classification based on dissection.[5,6] They have not been found practical for models that classify structures by connectivity, by internal histologic structure using stains for various ligands, and by physiologic or behavioral functions.[7] The reasons, which range from technical to philosophical, we address in the Discussion section. Here, we describe a different approach to comprehensive codification of an ontology and standard nomenclature for the domain of neuroanatomy.

## History of the NeuroNames Standard Nomenclature and Neuroanatomic Ontology

The simplicity of the NeuroNames' standard nomenclature masks the enormity of a 25-year effort in ontology development. The goal was to codify the language by which neuroscientists currently classify neuroanatomically annotated data for automated digital processing. We started with the brain of the macaque monkey,[2] the focus of our own research interest. Later, we expanded the project to include the brains of the four species most widely studied by neuroscientists: the human, macaque, rat, and mouse. Also, in the beginning, we referred to the standard nomenclature as a "default vocabulary," which we needed for unambiguous definitions in BrainInfo, our portal to neuroanatomy on the Web (http://braininfo.org).[8–10] Later, as we came to appreciate the growing need of the multidisciplinary digitally oriented neuroscience community for a comprehensive nomenclatural standard, and as we learned from prolonged experience the time-consuming nature of producing such a standard, we resolved to offer the NeuroNames default vocabulary as a standard nomenclature for download from the Web.[8,11] We felt that no neuroscientific colleague should have to retrace our tedious path through dozens of neuroanatomic atlases and textbooks and hundreds of journal articles to capture the current conceptual organization and language of neuroanatomy.

A comprehensive ontology must include not only the ~1000 primary structures of classical neuroanatomy (i.e., structures defined by gross dissection and by the gray–white histologic dichotomy), but also the ~2000-and-growing number of structures and combinations of structures found in neuroanatomic models based on other techniques. The goal was not to create a "better" ontology and vocabulary for neuroanatomists but simply to codify the language of neuroanatomy as it exists and evolves. It was to provide a concise, readily accessible resource for scientists from disciplines other than neuroanatomy who need to obtain information about specific structures without having to take a course in neuroanatomy or consult multiple books and journals every time they encounter an unfamiliar neuroanatomic term: a one-stop terminological resource for specialists in other disciplines, such as genetics, molecular biology, pharmacology, psychology, and psychiatry.

The NeuroNames ontology is designed to accommodate all of the English and Latin *names* (words and multiword terms) used by neuroscientists to reference all neuroanatomic *concepts* (word definitions and images) in the neuroscience literature of the last half century. Every name is classified as either a *standard name* or a *synonym* for the concept it represents. Some synonyms (multiple names for the same structure) are also *homonyms* (same name for different structures). Later the names for ~1000 structures encountered in neuroanatomic textbooks of six other languages were incorporated: French, German, Italian, Indonesian, Russian, and Spanish. Since 1992, the entire NeuroNames name set has been a regularly updated Source Vocabulary for the National Library of Medicine's Unified Medical Language System.[12] PubMed uses the synonym lists of NeuroNames to retrieve abstracts of articles in which authors have used different names for a structure than the search term submitted by the user.

The earliest version of NeuroNames was programmed in Hypercard[13] and was distributed on computer diskettes in 1990.[14] It was updated and expanded for distribution on compact discs and was published on the Web in 2002.[11] Those versions pertained only to the macaque brain as represented in the classical structural model, i.e., the hierarchic, closed partitive model of structures identified either by dissection or Nissl stain and organized in accord with the largely hierarchic model of the Encephalon division of *Nomina Anatomica.*[15] In that version, we established the basic rules for selecting standard terms for structures. They were the following: give preference to English and anglicized Latin terms; give preference to terms with high use frequency as judged by the number of PubMed abstracts in which they appear; avoid eponyms, e.g., select "insula," not "island of Reil"; avoid awkward Latin spellings, e.g., select "locus ceruleus," not "locus coeruleus"; consider short as better than long, e.g., select "amygdala," not "amygdaloid nuclear complex"; and avoid homonyms, i.e., use no standard name to represent more than one structure.[16]

In 1995, we had extended the classical macaque hierarchy to include classical structures of the human brain,[2] and in 1996, we added the rat brain equivalents of classical brain structures found in humans and macaques.[17] Shortly thereafter, we added structures in the rat and mouse brain that are not found in humans and macaques. In recent years, we have expanded the ontology to accommodate a number of hierarchic models, in addition to the classical model where structures are grouped by proximity. Most pertinent to the interests of geneticists and molecular biologists, we incorporated Swanson's functional model of the rat central nervous system,[18] where structures are grouped by a combination of developmental, connectional, and functional features as well as classical structural features. Swanson's functional model applies equally to the rat and mouse brain and, except for cerebral cortical areas, is largely applicable to the human and macaque as well.

We have thus designed NeuroNames to codify relations among all of the neuroanatomic *names* and *concepts* of *structures* and *models* that one encounters in the neuroscientific literature. To date, 2017, we have coded the relations of some 3000 structures and models in the central nervous systems of the four most studied species and the relations of some 18,000 names in eight languages to the structures and to each other.[19] The NeuroNames ontology of neuroanatomic structures provides a standard nomenclature to support interoperability among databases representing the methodologies of different neuroscience disciplines and different species. Specifically, it can be used to mediate communication among neuroscientific databases representing such topic domains as neurogenetics, neurophysiology, and behavioral and cognitive neurosciences.

## METHODS

Once one has a standard nomenclature, developing an interoperable system of databases is relatively straightforward. The key component, the hub of the system, is a *mediator*, a digital mechanism for translating between the disparate nomenclatures of individual databases.[16] The nomenclature of a database is represented by *metadata*, the set of terms by which the neuroanatomic locations of data in a particular database are identified. The function of the mediator is to translate, at the most granular level possible, the metadata of one database into that of another. The mediator also has a record of the kinds of data stored in each of the databases, i.e., the *attributes* of neural structure and function studied by creators of each database.

One can think of the mediator as containing the following for each database: (1) a look-up table of equivalent structure names, which contains a complete list of the *local names* of brain structures in the database's metadata and the NeuroNames *standard name* for each structure, and (2) a list of attributes of neuroanatomic structures for which the database has data, e.g., gene expression, neurotransmitter concentration, histochemical stains, magnetic resonance imaging (MRI) activation, neuron firing frequencies, etc.

The first step in development of a mediator is to create a look-up table of NeuroNames equivalents and attributes studied for each database. Since NeuroNames includes all of the English and Latin names of some 3000 structures, the database manager can expect the vast majority of names submitted to be processed automatically. A small number, however, may require adjudication by personal contact between the manager and the curator of the mediator's nomenclature. If the database's name for a structure is a previously unrecognized synonym for an existing concept in the NeuroNames ontology, the curator adds the new name to NeuroNames as a synonym and lists the standard name as its equivalent.

If the requested name is a homonym, that is, a synonym for multiple structures in the NeuroNames ontology, the mediator presents to the database manager the standard names and definitions of all structures to which the name might apply. The manager indicates which definition corresponds to the database's use of the term, and the corresponding standard name is entered into the table as the NeuroNames equivalent.

A very common example of homonymy is the term "hippocampus," which is used by three different segments of the neuroscience community for related, but quite different structures. Investigators working with the relatively low-resolution recordings obtained by noninvasive imaging, e.g., MRI, functional magnetic resonance imaging (fMRI), or positron emission tomography (PET), are likely to mean the *hippocampal formation*. If their data consist of high-resolution observations from microscopic studies of stained sections for, say, gene expression or neurotransmitters, they more likely mean the *CA fields*, one of several components of the hippocampal formation. And if they are comparative neuroanatomists, particularly if educated in the classical neuroanatomy of European tradition, their concept is of the *archicortex*, which includes the hippocampal formation and its forward extension through the fasciola cinerea, the supracallosal gyrus on the upper surface of the corpus callosum, and the paraterminal gyrus rostral to the hypothalamus.[20]

If a name in the database metadata represents a concept not included in the NeuroNames ontology, and if the manager can cite a peer-reviewed publication defining the concept, the curator will add the name and the concept definition to the mediator's version of the NeuroNames ontology and list the local name as the standard name. If the concept is idiosyncratic to the requestor's database, the curator will return the standard name of the next larger structure in the NeuroNames ontology. For example, if the unrecognized concept represents a previously undefined part of the medial mammillary nucleus, the curator will list "medial mammillary nucleus" as the NeuroNames equivalent. That will allow local data mapped to the smaller structure to be collated with the closest lower resolution data for analysis.

When the manager of one database wishes to collect and collate data from multiple others, he or she contacts the mediator and requests a list and description of other databases that have information on attributes of interest. They select databases that appear most likely to contain data relevant to their own. The mediator returns, for each target database, a table of the requestor's terms and the corresponding names in the target database. The collating database imports data sets from the other databases and uses the look-up tables to translate the local structure names of the other sites into their own.

The NeuroNames standard nomenclature and synonyms in XML format can be downloaded from BrainInfo at http://braininfo.org/Nnont.aspx. Definitions are accessible by manual search at http://braininfo.org/SearchByName.aspx.

## DISCUSSION

A decade ago, informaticists of the Neuroscience Information Framework (NIF), an organization supported by the Blueprint Consortium of NIH institutes and agencies that support neuroscientific research, set out to create a "web-based structure for identifying, locating, relating, accessing, integrating, and analyzing information from the neuroscience research enterprise."[21] Essential to that effort was a vocabulary for describing resources and their content and a concept-based query system that could be used to search for neuroscience information across diverse resources. For that purpose, they used widely accepted tools and best practices of informatics, which were under development for the Semantic Web.[22] That approach uses tools of descriptive logic to define a domain of knowledge by creating an ontology of concepts and compiling a nomenclature of standard names contributed by domain experts.

After several years of dedicated attempts, the NIF team concluded that the descriptive logic approach was less than adequate for their needs.[4] "Early in our efforts to build ontologies, we tied ourselves in knots trying to capture all of the class-level rules that define something like [for example] a Purkinje neuron." They identified two sources of difficulty. First, "…ontology languages such as OWL are not particularly good at dealing with numerical values, such as ranges or probabilities." An ontology of neuroanatomic structure must accommodate mathematically defined geometric patterns of such attributes as gene expression, neurotransmitter concentration, and fMRI activation, for example. Such patterns are defined digitally by arrays of $x$, $y$, and $z$ coordinates of voxels or pixels, each of which is tagged for the expression of various attributes.

The second source of difficulty they identified was that "much of what we observe is colored by the experimental procedures used to prepare and image biological specimens." Our interpretation of that issue is that an ontology useful for classifying data neuroanatomically must be tied to *operational definitions*, that is, the definitions of structures as the distribution patterns of attributes, such as gene expression, etc., must include the basic experimental procedures used to detect them. Furthermore, to be useful for automatic interoperability and machine processing, a digital ontology must codify not only the relations of structures to names in the standard nomenclature of the mediator but also must codify the relations between all synonyms used in different databases and names in the standard nomenclature. Since changes in terminology can occur over time, the ontology must accommodate *evolution of preferred names* of structural concepts. The NeuroNames ontology was designed to address each of those issues: quantification, operational definitions, and terminological evolution.

## Quantification of Visual Pattern: One Image Is Worth a Thousand Words

The structural entities of neuroanatomy are inherently three-dimensional. The most precise neuroanatomic *concepts* are of visual patterns representing structures and meaningful combinations of structures in the brain. The neuroscientist's challenge is to determine which data patterns are correlated with each other and with specific neural and behavioral functions. The logical principle that makes neuroanatomy the common language of the multiple neuroscientific disciplines is the Aristotelian intuition that two bodies cannot occupy the same space at the same time. The important corollary for neuroscientists is that multiple physical and functional attributes found in the same part of the brain may usefully be considered to represent a common entity. The greater the number of overlapping attributes, such as patterns of cyto-myeloarchitecture, gene expression, unit firing, and fMRI activation under specific stimulus conditions, the greater the utility of considering them to represent unique functional entities.

Descriptive logic works best with one or a few attributes that are either present or absent. Few biologic attributes, however, are all-or-none in character. The data provided by modern instrumentation are quantitative measurements of intensities, e.g., of gene expression, which can vary over wide ranges. Considerable useful information resides in the values of measurements, so the ontology must be able to accommodate variation in the values. And the overlap of spatial patterns is never exact, so one needs statistical criteria to judge the significance of overlapping patterns. To test for anatomic overlap between data patterns obtained by different methods, one maps the disparate patterns to a standard, scaled, reference atlas of the brain. One needs both the mathematic tools to extract maximal information from such data and an ontology capable of accommodating quantitative categorization of such information.

Information theory allows one to calculate the relative value of image versus text representations of brain structure. The potential information conveyed by an MR image of a brain is proportional ($=\log_2$) to the number of voxels in the MRI times the number of dimensions (attributes) on which the brain is evaluated times the average number of possible values (data) per dimension. For a mouse brain represented in an MRI of $512 \times 1024 \times 512$ voxels[29] coded for 256 gray-levels on a single dimension representing density of hydrogen ions, the maximal potential information would be $10^{36}$ bits. If each voxel were coded on two dimensions, for hydrogen ion density, which reflects differences between gray (cellular) matter and white (myelinated axonal) matter, and expression of a particular gene, the maximal information content would be $10^{37}$ bits.

The potential information conveyed by words about the same brain is proportional to the number of named primary structures (concepts) times two (presence or absence). For a model of brain organization that includes ~1000 structures coded for two states the maximal information content would be $10^{11}$ bits. If each structure were coded for a second dimension, the maximal content would be $10^{12}$ bits. Thus, a digital processing system that can work with quantitative representation of images is considerably more powerful than a system that works only with words.

Early on, we defined the features necessary for a reference brain atlas to serve as a tool of quantitative neuroanatomy[7] (Table 2.1). We identified features that distinguish such an atlas from the dozens of brain atlases of different species published between 1950 and 1990. The basic format of such conventional atlases was a set of some ~50 gray-level images of Nissl-stained coronal cross-sections or cross-sectional line diagrams representing the most prominent landmarks and partial boundaries between structures. The most

**Table 2.1** Comparison of conventional and informatics-oriented brain atlases

| Conventional brain atlas | Informatics-oriented brain atlas |
|---|---|
| Printed format | Digital format |
| One hemisphere | Both hemispheres; symmetric |
| Coronal sections | Multiple planes of section |
| Structures designated by labels at centers | Structures defined by boundaries |
| Many areas undefined | Comprehensively segmented |
| Revised editions at 10–20 year intervals | Continuously updated |
| Approximate mapping of data to structures | Unambiguous mapping of data to structures |
| Few tools for manual mapping of data | Many tools for automated mapping of data |
| Stereotaxis referenced to a few cranial or commissural landmarks | Spatial indexation (mapping) referenced to hundreds of local anatomical landmarks |
| No mechanism for display of mapped data | Overlays for comparison of multimodal data |
| Query by atlas-specific structure names | Query by any name or by spatial coordinates |
| High resolution in two dimensions | High resolution in three dimensions |

salient structures were identified by arrows pointing from names in the margin or by acronymic labels applied to their centers. In 1996, we published the first informatics-oriented "template atlas" for quantitative neuroanatomy in the long-tailed macaque (*Macaca fascicularis*). It appeared in several formats: as a journal article,[23] as a spiral bound, flat-lying book for lab use,[24] and as a book in hardcover with a compact disc containing printable coronal templates for mapping data manually.[25] The printable templates are also published on the BrainInfo website as a set of downloadable pages suitable for mapping data for presentation or publication.[26]

Today, neuroscientists have access to several informatics-oriented brain atlases of the four most studied mammalian species: human, macaque, rat, and mouse. Most pertinent for the neurogenetics community is the comprehensively segmented mouse brain atlas published on the Web by the Allen Institute for Brain Science (AIBS). In 2008 the AIBS launched its reference atlas of the mouse brain, which is based on closely spaced Nissl-stained coronal and sagittal sections. In 2011, that atlas was updated to enable interactive online exploration of gene expression data and to provide three-dimensional annotation for informatics analysis and viewing in the Brain Explorer 3-D viewer.[27] Since then the institute has released a number of tools allowing users to exploit the quantitative capabilities of the atlas in analyzing the neuroanatomic distributions and interrelations of some 20,000 genes expressed in the mouse brain.[28]

In 2011, we created an MRI-based atlas of the Waxholm canonical mouse brain suitable for semiautomated mapping of data obtained by any methodology.[29] More recently, we have incorporated the Waxholm mouse brain atlas and a similarly prepared atlas of the macaque brain into the NeuroMaps Mapper, a semiautomated Java application, which enables one to map data to a standard high-resolution MRI of either the mouse brain or the macaque brain.[30] The Mapper is a stand-alone application that allows one to map data to cross-sections in any plane or to a three-dimensional surface image of the reference MRI in any orientation. It allows one to edit the size, pixel resolution, colors, and labels in the image to publisher's specifications and to download the resultant figure in a format suitable for presentation or publication.[31] The macaque atlas is comprehensively segmented and annotated in accord with the NeuroNames standard nomenclature, so the anatomic structures to which data are mapped are immediately evident. A second Web-accessible, MRI-based macaque brain atlas segmented according to the NeuroNames nomenclature is the Stanford Research Institute's INIA19 Template and NeuroMaps Atlas.[32]

Informatics-oriented brain atlases of the mouse and the macaque are now in use by individual labs and institutes, but they have yet to see great application as mediators of interoperability among multiple databases. The most sophisticated tools for interoperability of multiple databases have emerged from consortium arrangements between laboratories that use noninvasive imaging methods, such as MRI, fMRI, and PET imaging in studies of the human brain. There investigators import massive three-dimensional datasets created by diverse imaging methods in which every voxel is coded for intensity of activation by gray level. They warp an MRI image of each subject to a standard MRI image (the reference atlas) by matching landmarks, such as the boundary between the brain and its cerebrospinal fluid envelope, and/or by iterative matching of voxels by gray level. Then, they warp an image in which voxels are coded for a data variable to the same form. When the experimental image is warped to the standard image, the data are "dragged" to overlie the corresponding voxels in the atlas. Voxels in the atlas image are coded digitally for the cortical or nuclear structure to which they belong; those in the experimental image are coded by gray level reflecting the intensity of an attribute measured by the imaging device, such as blood oxygenation reflecting neural activation in an fMRI or receptor binding of an injected neurochemical in a PET scan. Warping the experimental image to the atlas image allows every voxel to be tagged with both the structure name and the experimental attribute. The standardized images from any number of different subjects acquired by any number of different imaging systems in any number of different databases can be combined into a single three-dimensional database for analysis of neuroanatomic correlations between the distribution patterns of multiple experimental variables.

Using such techniques, cognitive neuroscientists are able to combine scans from multiple databases to study, for example, the visceral, cognitive, and emotional correlates of activation of a particular cortical area, such as the insula,[33] or the combination of cortical areas activated by tasks related to a particular cognitive concept, such as "theory of mind."[34] In mouse studies, researchers at the AIBS have accomplished similar feats of quantitative "interoperability" within a single institution by mapping a database of hundreds of thousands of slides, which are stained by in situ hybridization for expression of different genes, to their standard Mouse Brain Atlas.[26,27] The AIBS process for mapping between histologic images is similar to that for mapping between MRI and PET images, except that, working with the virtually identical brains of an inbred mouse strain, the investigators do not have to warp experimental images to the standard; adequate spatial registration is achieved simply by overlaying surface boundaries. Similar MRI-based digital atlases have become available for two different widely studied rat strains: the Wistar strain[35] and the Sprague–Dawley strain.[36] The latter has been augmented to enable automated brain-wide mapping of axonal connections demonstrated by neurochemical labeling in microscopic sections.[37]

Groups of scientists who are able to map their data to a common atlas, or to provide their database of raw data to others who map it to a common atlas, are best advised to do so. That is not always possible, however. A person or institution wishing to combine information from published studies in which the raw image data are no longer accessible, or the data are from species so different in brain structure that the validity of image mapping from one to the other cannot be evaluated, between human and mouse for example, must rely on text mapping, i.e., correlating the anatomic locations of data from different sources on the basis of the names of structures. The cerebral cortex of primates and rodents are so different in the number and configuration of component structures that the feasibility and utility of forcing a map of one image to the other are questionable. Text mapping is considerably less powerful than image mapping because it works with semiprocessed data from which variance due to differences between subjects and between methods of raw data acquisition cannot be quantified.

The kind of mediator described in this chapter is designed to maximize the information extractable from databases merged by either image mapping or text mapping. It enables one to preserve information that would otherwise be lost due to differences in terminology for the same structure, to segregate information that would otherwise introduce distortion due to use of the same name for different structures, and to render explicit the identity of structures that neuroanatomists regard as equivalent between species. The past two decades have seen great progress in efforts to standardize neuroanatomic terminology. Paxinos, Mai, and colleagues have published brain atlases and neuroanatomic textbooks in which a consistent effort has been made for standardization of nomenclature for equivalent structures within and across human,[38] macaque,[39] rat,[40] and mouse[41] species. Swanson has published a definitive account of the historical origins and meanings of terminology for the most widely recognized structures across species up to mid-19th century, when the explosion of neuroanatomic concepts based on staining, connectivity, and functional criteria began.[5] Integration of the products of those efforts into the NeuroNames ontology is a continuing process.

## Operational Definitions: Scientists Are Boundary Hunters

Operational definition is the sine qua non of the scientific approach to knowledge. Every report of a scientific experiment, regardless of discipline, contains a methods section, which states explicitly the parameters of the research effort: the objects of study, the variables measured, and the instruments used to measure them. Those objective factors define real-world contextual boundaries within which other scientists can expect to observe the same phenomena. They allow peers to test and judge independently the validity of the measures and the range of conditions within which the author's observations can be predicted to hold true. For neuroanatomy, the most important parameters in operational definitions are the species and developmental stage of the subjects and the methodology, e.g., dissection, cutting and staining, or noninvasive imaging. One's conceptual model of brain structure and function includes the ranges, in terms of species along a branch of the phylogenetic tree, research techniques, and the boundaries of patterns in a brain atlas where evidence of structural and functional entities are observed.

In the NeuroNames digital ontology, operational definitions are the text representations of concepts of structures, and character strings are text representations of names of concepts. Stated formally: A *digital ontology* of data coded for neuroanatomic location is "*human-readable text and human-recognizable visual images specifying relationships between the names of a nomenclature and the operationally defined structural concepts they represent.*"[2] It is based on the definition of a digital ontology for descriptive logic published under the auspices of the Stanford Knowledge Systems Laboratory, in 1993,[42] with the addition of *visual images* as a source of conceptual knowledge, and with the added specification of "*operationally defined*" to render it applicable to quantitative scientific knowledge as well as descriptive logic knowledge.

One reason the descriptive logic approach to ontology development failed to meet the needs of the NIF was that a definition suitable for descriptive logic processing only describes the "essence" of an observed phenomenon, what the phenomenon "is" in terms of yes-no attributes, features which the entity either has or does not have. An operational definition of a structure describes not only the essence but the parameter conditions under which attributes can be observed and the boundaries of the structure as identified by different methods.

A structure concept defined by a single attribute is called "monothetic." Defined by more than a single attribute, a concept is called "polythetic." Descriptive logic works best with monothetic concepts, because the more attributes associated with a class of entities, the more difficult it becomes to define it for descriptive logic processing. At the same time, the more attributes with overlapping boundaries, the more confidence the neuroscientist has in the utility of considering them to represent a common entity. An ontology designed for quantitative neuroanatomy allows one to apply statistical techniques to evaluate the spatial variation in boundaries of patterns of multiple attributes as dimensions.

Another weakness of the descriptive logic approach has to do with its handling of differing *model concepts* of how the smallest, primary, structures of the brain combine into functionally significant groups. Neuroanatomists are largely agreed as to the identities of the ~1000 primary structures of the mammalian brain. Their preferred names for structures and combinations of structures, however, vary greatly. It has been said that "an academic would rather use his colleague's toothbrush than use his terminology."[43] Part of the reason is a scientist's natural attachment to the vocabulary of the instructors and textbooks from which they learned neuroanatomy and to which they contribute in the course of their own careers. A more substantive reason is that neuroanatomists have developed different conceptual *models* of how the nervous system is organized, and they use different terms for the same structure in different models to convey the conceptual context of their communications about the structure.

Neuroscientists know the structures, models, and nomenclature appropriate to their area of expertise. Brain surgeons use the classical neuroanatomic model in which structures are organized hierarchically by proximity; they decide where to cut on the basis of nearby landmarks. Developmental neuroscientists organize structures by the parts of the embryonic nervous system from which the structures grow; they explain certain patterns of structural abnormality in the adult brain based on the embryonic neuromeres from which aberrant structures derived. Neurohistologists group them by common staining characteristics, neuropharmacologists by patterns of neurotransmitter distribution, and geneticists by the common expression of genes or gene products. Different models of how the primary structures of the brain are organized reflect different operational definitions and are useful for different purposes. As combinations of concepts, models are better judged in terms of *utility* for a given purpose than as *true* or *untrue* representations of entities. In any case, a comprehensive ontology must accommodate multiple models of nervous system organization.

Descriptive logic informaticists handle the multiple models problem by creating separate ontologies for different models.[4] That solution is satisfactory for neuroscientists who, as humans, easily resolve ambiguity by context. It is not sufficient for a computerized mediator of interoperability among databases that use the same name, "hippocampus," to reference three different models. To resolve ambiguities in a digital processing system, accurate translation between databases requires that, (1) all concepts of structures and models are represented in a single ontology; and (2) each concept is represented by a unique definition and unique standard name that are the same in all models where it occurs. Fig. 2.1 shows the hierarchical scheme by which the NeuroNames ontology meets both criteria. A thorough explication of the principles underlying the design is found in *NeuroNames: an ontology for the BrainInfo portal to neuroscience on the web*.[44] Here, we simply describe the links by which a mediator navigates to determine the conceptual relations between structures that appear in multiple models based on any number of different methodologies.

The NeuroNames ontology is a merger of categorical and partitive hierarchies. For readers familiar with descriptive logic terminology, a categorical hierarchy is equivalent to an "is-a" hierarchy; the superimposed partitive hierarchy is equivalent to a "has-a [part]" hierarchy. The duality reflects an important feature of the NeuroNames ontology,

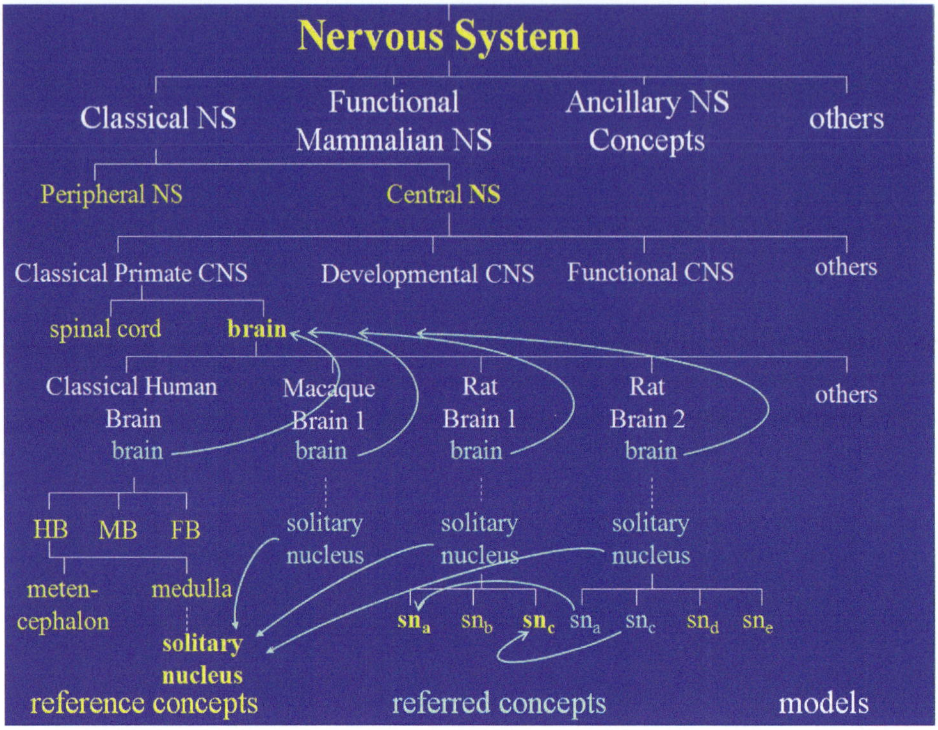


**Figure 2.1** ***NeuroNames ontology of the nervous system.*** A hypothetical set of brain models (one human, one macaque, and two rat models) are integrated with a subset of concepts from the NeuroNames ontology to illustrate relations among reference concepts (yellow), referred concepts (light blue), and models (white). *CNS*, central nervous system; *FB*, forebrain; *HB*, hindbrain; *MB*, midbrain; *NS*, nervous system; $sn_{a\text{-}e}$, subnuclei of solitary nucleus in two hypothetical atlases of the rat brain. *Dashed lines* indicate intervening levels of hierarchy. *Arrows* indicate identity of referred concepts in models with their reference concepts. All instances of "brain" in light blue share the standard name and definition of the reference concept "brain" in yellow. The coding of subnuclei $sn_{a\text{-}c}$ in Rat Brain 1 as reference concepts and of $sn_a$ and $sn_c$ as referred concepts in Rat Brain 2 indicates that the concepts were first found, coded, and defined in the ontology from the Rat Brain 1 atlas. *(With permission from Bowden DM, Song E, Kosheleva J, Dubach MF. NeuroNames: an ontology for the BrainInfo portal to neuroscience on the web.* Neuroinformatics. *2012;10:97–114.)*

namely, that the name of the model itself, a categorical concept, is different from the name of the top concept in the partitive hierarchy that defines it. The lower middle row of Fig. 2.1 shows the tops of the hierarchies of four partitive models of the entity "brain": the Classical Human Brain model of the NeuroNames ontology and hypothetical models of a macaque brain and two rat brains. All share the generic concept definition of "brain" as the "intracranial part of the central nervous system demonstrated by dissection." A definition should appear at only one location in the ontology relational database schema.[44] In this example, the definition is attached to "brain" in the row above the "brain" models of different species. Each model represents a somewhat different concept of the brain as a composite structure in the craniums of different species and strains. Thus each model has a different name: Classical Human Brain, Macaque Brain 1, Rat Brain 1, etc. The common identity of the models is represented by their belonging to the same category, "brain," and by the fact that the top concept in the model hierarchies is the same "brain" for all. Their differences are represented in the unique names of the models and the composition and organization of parts below the top concepts (not shown).

## How the NeuroNames Standard Nomenclature Tracks the Evolution of Preferred Terms

The task of neuroscientists is innovative conceptualization and testing of structure-function relations in the brain. Their job is to grapple with nature and to develop concepts and models of entities that they detect by carefully designed experiments, the "operations" of operational definitions. Until a preponderance of evidence obtained by different methods indicates the utility of considering data acquired from the same general region of the brain as representing a single entity, a digital ontology must treat them as distinct entities with distinct names and definitions that reflect the differences of conditions under which they are identified.

In the early days of NeuroNames development the cortical area now bearing the standard name "area 47/12" was represented by several different concepts with the species and/or authors included in the standard name to distinguish them from one another. Only in the late 1990s, 80 years after the first reports by Brodmann[45] and Mauss,[46] were we able to say with confidence that area 47 of Brodmann,[45] defined by *cytoarchitectonics*, using Nissl stain in the human; area 12 of Brodmann,[45] in roughly the same area demonstrated by the same method in the guenon monkey; area 12 of Mauss,[46] defined by *myeloarchitecture* in macaque and guenon monkeys; area 12 of Carmichael and Price,[47] defined by *nine different stains* in three species of macaque; area 47/12 of Öngür et al.[48] defined by *five stains* in the human; and area 47/12 activated, as defined by various *noninvasive imaging techniques*, when humans engage in various cognitive tasks,[49] all represent the same structure. All of the studies detected an area at about the same location in the lateral orbital gyrus extending to a horizontal strip of the inferior frontal gyrus. As diverse supportive data accumulated, it became possible to drop the names of the species and the investigators who defined it, and simply refer to it as "area 47/12."

Until scientists converge on the opinion that an area of about the same shape, location, and internal structure is detected by multiple methods, however, a digital ontology must keep species, methodology, and/or name of investigator in the definition. A digital ontology must accommodate not only current knowledge but accumulating data postulated to represent better ways of conceptualizing the structural and functional organization of the brain.[44]

In NeuroNames, every concept is represented by a *permanent numeric ID* for the word definition thereof. The permanent ID for a concept allows curators of different databases to keep the names preferred by their research group without creating ambiguity of definition for the mediator. Furthermore, it allows the curator of NeuroNames to codify the evolution of neuroanatomic language as neuroscientists abandon a name, which has become a homonym for multiple slightly different conceptual models of the same entity, in favor of a name unique to the predominantly accepted model. A good example of the latter phenomenon is evolution in usage of the term "rhinencephalon" for the brain component of the olfactory system during the last 60 years (Fig. 2.2).

In the 1950s and 1960s the term "rhinencephalon" was used by many different authors to represent several slightly different models of the olfactory system. Beginning in the 1960s, a series of thorough studies of the system, which the investigators uniformly referred to as "olfactory cortex," gained acceptance as the definitive model.[50] Use of the term "rhinencephalon" gradually declined over the next two decades. By the 2000s the term "rhinencephalon" was all but extinct; even the two less well-defined alternative terms, "olfactory lobe" and "olfactory brain," appeared more often.

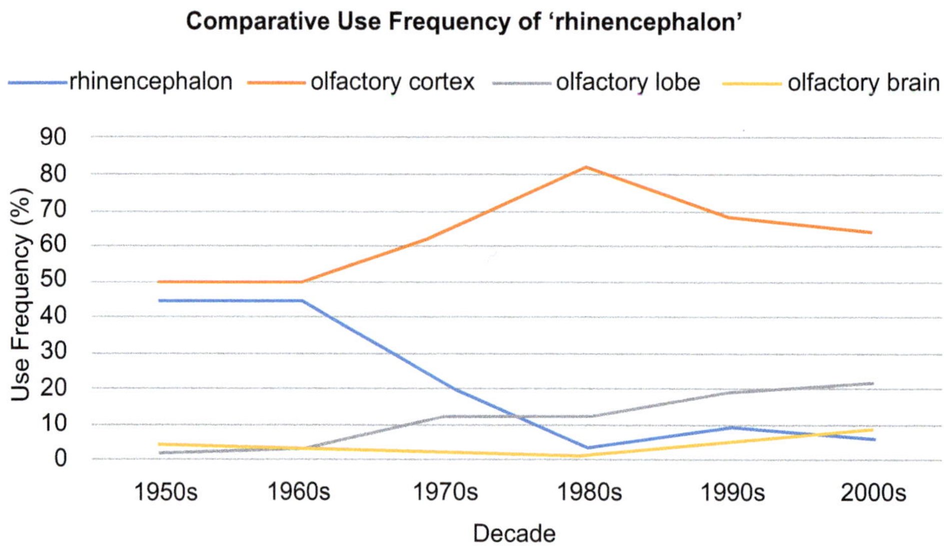


**Figure 2.2** ***Use frequencies of names for olfactory cortex.*** The use frequencies of a name represent the proportional numbers of PubMed citations in which it appeared in each decade between 1950 and 2010.[49] During the two decades between the 1960s and 1980s, the use frequency of the operationally defined term "olfactory cortex" displaced the variably defined term "rhinencephalon," the use frequency of which decreased to the range of two other imprecisely defined terms, "olfactory lobe" and "olfactory brain."

The definition and name "rhinencephalon" were not changed in the ontology,[51] but the name was reclassified. It became not only the standard name for the ambiguous concept, but also a synonym for "olfactory cortex," and the concept "olfactory cortex" was substituted for "rhinencephalon" in models that earlier included the ambiguous concept.

Having described NeuroNames and its uses, we should make clear what it is not and what it cannot do. It is not a tool of neuroscientific discovery. It is a tool for efficient discovery of reliable existing scientific information. It is an informatics tool more similar to the tools of library science than of neuroscience. One cannot extract from the NeuroNames ontology any information about the brain that has not already been discovered by a scientist and recorded in the literature.

The hierarchical lists of NeuroNames *standard terms and synonyms* are mainly useful for unambiguous indexing of data and knowledge. A catalog of the operational definitions of structural concepts in the NeuroNames *ontology* is more equivalent to a rapid-access neuroanatomic encyclopedia than to a textbook, which covers in depth the portion of the knowledge space that current authors consider most important. One consults a neuroanatomic encyclopedia to obtain a concise description of the current state of knowledge, not to find as yet undiscovered knowledge about the organization of the brain. Presumably that is the reason that the U.S. National Library of Medicine was the earliest major scientific institution to build NeuroNames into the citation and indexing system for PubMed.

NeuroNames will always be a work in progress. Nevertheless, it is currently, and promises to be for the foreseeable future, the most comprehensive, internally consistent, scientifically valid starting point for any system that requires indexing neuroanatomic data and knowledge. We trust that it will prove to be a valuable tool for establishing interoperability among molecular-genetic and neuroscientific databases. The standard terms and synonyms are freely available for download at http://braininfo.rprc.washington.edu/NeuroNames.xml. The definitional ontology is freely accessible for viewing at http://braininfo.org.

In summary, we have presented a method for removing a key obstacle to interoperability between molecular-genetic and neurobehavioral databases, namely, inconsistency of neuroanatomic terminology. Each resource database combines data by mapping data from a number of individuals to a species-specific brain reference atlas and coding neuroanatomic locations according to the nomenclature of the reference atlas. When one resource undertakes to collate data from multiple other databases, it translates the neuroanatomic nomenclatures of the other databases through a mediator into its own nomenclature. That enables it to use its own data processing tools to analyze the combined data sets. The translation of terminology is performed using the standard nomenclature of the NeuroNames neuroanatomic ontology.

## ACKNOWLEDGMENTS

This work is supported by the National Institutes of Health (NIH) Office of Research Infrastructure Programs to the University of Washington (UW) under award number P51OD010425. NeuroNames is maintained as a part of the BrainInfo website by the Neuroscience Information Framework (NIF) at the University of California at San Diego (UCSD), which is supported by the Blueprint Consortium of the NIH (NIDA) under grant 3U24DA039832-01S1. It is cosponsored by the UW, NIF, and the International Neuroinformatics Coordinating Facility (INCF), Karolinska Institute, Stockholm.

## REFERENCES

1. Swanson LW, Lichtman JW. From Cajal to connectome and beyond. *Annu Rev Neurosci*. 2016;39: 197–216.
2. Bowden DM, Martin RF. NeuroNames brain hierarchy. *Neuroimage*. 1995;2:63–83.
3. Garcia A, O'Neill K, Garcia LJ, et al. Developing ontologies in decentralized settings. *Nat Preced*; 2009. http://precedings.nature.com/documents/3231/version/1.
4. Larson SD, Martone ME. Ontologies for neuroscience: what are they and what are they good for? *Front Neurosci*. 2009;3(1):60–67.
5. Swanson LW. *Neuroanatomical Terminology: A Lexicon of Classical Origins and Historical Foundations*. Oxford: Oxford University Press; 2015.
6. Swanson LW, Bota M. Foundational model of structural connectivity in the nervous system with a schema for wiring diagrams, connectome, and basic plan architecture. *Proc Natl Acad Sci USA*. 2010; 107(48):20610–20617.
7. Bowden DM, Dubach MF. Neuroanatomical nomenclature and ontology [chapter 2]. In: Koslow S, Subramaniam S, eds. *Databasing the Brain: Data to Knowledge*. Hoboken, NJ: John Wiley & Sons; 2005: 27–45.
8. Bowden DM, Dubach MF. BrainInfo [chapter 18]. In: Kötter R, ed. *A Practical Guide to Neuroscience Databases and Associated Tools*. Norwell, NY: Kluwer Academic Publishers; 2002:259–273.
9. Bowden DM, Dubach M. *BrainInfo: A Portal to Neuroanatomy on the Web*. University of Washington; 2005. http://braininfo.rprc.washington.edu/pdf/For%20Steinbusch%20on%20Web%20final.pdf.
10. Bowden DM. BrainInfo. In: *Encyclopedia of Computational Neuroscience*. Berlin: Springer; 2013.
11. Bowden DM, Dubach MF. NeuroNames 2002. *Neuroinformatics*. 2002;1:43–59.
12. Tuttle M, Sheretz D, Olson N, et al. Using Meta-1 – the first version of the UMLS metathesaurus. *Proc 14th Symp Comp Appl Med Care*. 1990;14:131–135.
13. Wikipedia. *HyperCard*. Wikipedia – The Free Encyclopedia. https://en.wikipedia.org/wiki/HyperCard.
14. Martin RF, Dubach J, Bowden DM. NeuroNames: human/macaque neuroanatomical nomenclature. In: *Symposium on Computer Applications in Medical Care: Standards in Medical Information*. Los Alamitos, CA: IEEE Computer Society Press; 1990:1018–1019.
15. International Anatomical Nomenclature Committee. *Nomina Anatomica*. Baltimore: Williams & Wilkins; 1983.
16. Bowden DM, Dubach M, Park J. Creating neuroscience ontologies [chapter 4]. In: Crasto CJ, ed. *Methods in Molecular Biology: Neuroinformatics*. Totawa, NJ: Humana Press; 2007:67–87.
17. Bowden DM, Martin RF. A digital Rosetta stone for primate brain terminology. In: Bloom FE, Björklund A, Hökfelt T, eds. *Handbook of Chemical Neuroanatomy: The Primate Nervous System, Vol 13, Part 1*. Amsterdam: Elsevier Science; 1997:1–35.
18. Swanson LW. *Brain Maps: Structure of the Rat Brain*. 3rd ed. Amsterdam: Elsevier Academic Press; 2004.
19. BrainInfo. *NeuroNames Nomenclature of Mammalian Neuroanatomy*. University of Washington, Seattle, and University of California at San Diego. http://braininfo.rprc.washington.edu/Nnont.aspx.
20. BrainInfo. *Hippocampus*. University of Washington, Seattle, and University of California at San Diego. http://braininfo.org/centraldirectory.aspx?ID=3157.

21. NIF. *The History of NIF*. Neuroscience Information Framework. https://neuinfo.org/about/organization.
22. Wikipedia. *Semantic Web*. Wikipedia — The Free Encyclopedia. https://en.wikipedia.org/wiki/Semantic_Web.
23. Martin RF, Bowden DM. A stereotaxic template atlas of the macaque brain for digital imaging and quantitative neuroanatomy. *Neuroimage*. 1996;4:119–150.
24. Martin RF, Bowden DM. *Template Atlas of the Primate Brain*. Primate Information Center, University of Washington; 1996.
25. Martin RF, Bowden DM, Wu J, Dubach MF, Robertson JE. *Primate Brain Maps: Structure of the Macaque Brain*. Amsterdam: Elsevier; 2000.
26. BrainInfo. *Macaque Atlas Templates for Mapping*. University of Washington, Seattle, and University of California at San Diego. http://braininfo.rprc.washington.edu/PrimateBrainMaps/atlas/Mapcorindex.html.
27. AIBS. *Allen Mouse Brain Atlas — Interactive Viewer*. Allen Institute for Brain Science. http://atlas.brain-map.org/atlas?atlas=1#atlas=1&plate=100960224&structure=549&x=5279.738246372768&y=3744.014325823103&zoom=-3&resolution=11.97&z=5.
28. AIBS. *Allen Brain Atlas — Data Portal*. Allen Institute for Brain Science. http://mouse.brain-map.org/search/index.
29. Bowden DM, Johnson GA, Zaborsky L, et al. A symmetrical Waxholm canonical mouse brain for NeuroMaps. *J Neurosci Methods*. 2011;195:170–175.
30. BrainInfo. *NeuroMaps Mapper*. University of Washington, Seattle, and University of California at San Diego. http://braininfo.org/MapViewData.aspx.
31. BrainInfo. *NeuroMaps Mapper*. University of Washington, Seattle, and University of California at San Diego. http://braininfo.rprc.washington.edu/MapViewData.aspx.
32. Rohlfing T, Kroenke CD, Sullivan EV, et al. The INIA19 template and NeuroMaps atlas for primate brain image parcellation and spatial normalization. *Front Neuroinform*. 2012;6(27):1–15.
33. Kurth F, Zilles K, Fox PT, Laird AR, Eickhoff SB. A link between the systems: functional differentiation and integration within the human insula revealed by meta-analysis. *Brain Struct Funct*. 2010; 214(5):519–534.
34. Schurz M, Radua J, Aichhorn M, Richlan F, Perner J. Fractionating theory of mind: a meta-analysis of functional brain imaging studies. *Neurosci Biobehav Rev*. 2014;42:9–34.
35. Johnson GA, Calabrese E, Badea A, Paxinos G, Watson C. A multidimensional magnetic resonance histology atlas of the Wistar rat brain. *Neuroimage*. 2012;62:1848–1856.
36. Papp EA, Leergaard TB, Calabrese E, Johnson GA, Bjaalie JG. Waxholm space atlas of the Sprague Dawley rat brain. *Neuroimage*. 2014;97C:374–386.
37. Papp EA, Leergaard TB, Csucs G, Bjaalie JG. Brain-wide mapping of axonal connections: workflow for automated detection and spatial analysis of labeling in microscopic sections. *Front Neuroinform*. 2016; 10. https://doi.org/10.3389/fninf.2016.00011.
38. Mai JK, Paxinos G, eds. *The Human Nervous System*. Amsterdam: Elsevier; 2012.
39. Paxinos G, Huang X-F, Petrides M, Toga AW. *The Rhesus Monkey Brain in Stereotaxic Coordinates*. 2nd ed. Amsterdam: Elsevier Academic Press; 2009.
40. Paxinos G, Watson C. *The Rat Brain in Stereotaxic Coordinates*. 6th ed. Amsterdam: Elsevier Academic Press; 2009.
41. Franklin KBJ, Paxinos G. *The Mouse Brain in Stereotaxic Coordinates*. 3rd ed. Amsterdam: Elsevier Academic Press; 2008.
42. Gruber T. *What Is an Ontology?* Stanford Knowledge Systems Laboratory. http://www-ksl.stanford.edu/kst/what-is-an-ontology.html.
43. Gray JA, Felden J, Rawlins DRH, Smith AD. The neuropsychology of schizophrenia. *Behav Brain Sci*. 1991;14:1–84.
44. Bowden DM, Song E, Kosheleva J, Dubach MF. NeuroNames: an ontology for the BrainInfo portal to neuroscience on the web. *Neuroinformatics*. 2012;10:97–114.
45. Brodmann K. *Brodmann's 'Localisation in the Cerebral Cortex'*. Garey LJ, trans. and ed. London: Imperial College Press; 1999.

46. Mauss T. Die faserarchitektonische Gliederung der Grosshirnrinde bei den niederen Affen. *J Psychol Neurol.* 1908;13:263–325.
47. Carmichael ST, Price JL. Architectonic subdivision of the orbital and medial prefrontal cortex in the macaque monkey. *J Comp Neurol.* 1994;346(3):366–402.
48. Öngür D, Ferry AT, Price JL. Architectonic subdivision of the human orbital and medial prefrontal cortex. *J Comp Neurol.* 2003;460(3):425–449.
49. PubMed. Home – PubMed – NCBI. https://www.ncbi.nlm.nih.gov/pubmed.
50. Price JL. Olfactory system [chapter 29]. In: Paxinos G, ed. *The Human Nervous System.* San Diego: Academic Press; 1990:979–998.
51. BrainInfo. *Rhinencephalon*. University of Washington, Seattle, and University of California at San Diego. http://braininfo.rprc.washington.edu/centraldirectory.aspx?ID=2078.

CHAPTER 3

# The Allen Brain Atlas: Toward Understanding Brain Behavior and Function Through Data Acquisition, Visualization, Analysis, and Integration

**Terri L. Gilbert[1], Lydia Ng[2]**
[1]GilHou Science Communication, Seattle, WA, United States; [2]Allen Institute for Brain Science, Seattle, WA, United States

## MAPPING GENE EXPRESSION IN THE BRAIN

With the sequencing of the human genome,[1] it became possible to systematically study the genetic underpinning of the components, systems, and processes involved in thought, memory, behavior, and even consciousness. The Allen Institute for Brain Science began as a project to map gene expression in the brain, with the goal of systematically building a more complex understanding of the components, circuits, function, and ultimately the information processing occurring in both the healthy and diseased brain.

This chapter will describe the collection of the data to understand the structure and function of the brain, the tools developed to explore and visualize these data, and how the integration of these data and tools has led to the efforts currently underway at the Allen Institute for Brain Science to understand the higher order functions of the brain.

### Allen Mouse Brain Atlas Data Generation

The original online product of the Allen Institute for Brain Science was a genome-wide, high-resolution, colorimetric, in situ hybridization (ISH) map of gene expression in the adult, male C57BL/6J mouse. This atlas contains data for approximately 20,000 distinct mouse genes. All the known genes were assayed in the sagittal plane, while only ~4000 genes were probed in the coronal plane. The processes for riboprobe generation, ISH, image acquisition, and data processing were automated in a high-throughput pipeline.[2]

While the automated high-throughput pipeline process allows for the generation of an immense amount of data heretofore unseen on such a time scale, it is the informatics[3] overlaid on the data that makes this atlas such a powerful tool (Fig. 3.1). Mapping the gene expression data into a three-dimensional (3D) reference framework allowed for the breakthroughs in the field of genomics to be spatially associated with neuroanatomy.[4] Not only did this allow scientists to be able to leverage the vast expertise of neuroanatomists who had previously done much of the hard work mapping the brain, it allowed for

*Molecular-Genetic and Statistical Techniques for Behavioral and Neural Research*
ISBN 978-0-12-804078-2, https://doi.org/10.1016/B978-0-12-804078-2.00003-9

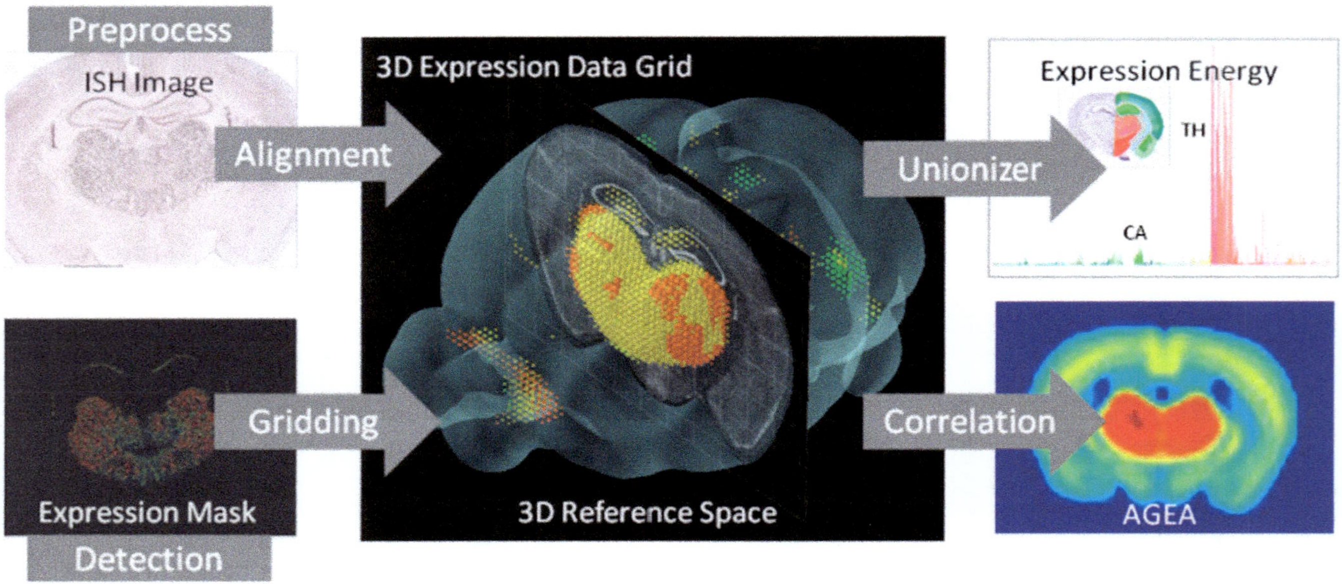


**Figure 3.1** ***Informatics data processing for the adult mouse brain atlas.*** The Alignment module registers each in situ hybridization (ISH) image to the common coordinates of a three-dimensional (3D) reference model. The Expression Gridding module produces an expression summary in 3D for downstream analysis. The Structure Unionizer module generates structure-based statistics by combining or "unionizing" grid voxels with the same 3D structural label from the hierarchic reference atlas. Further downstream, the grid data are used to compute gene-to-gene correlations and voxel-to-voxel correlations to support "Find Correlates" (similarity search) and Anatomic Gene Expression Atlas (AGEA) functions. The ISH image shows the gene *Plekhg1*.

direct querying of the gene expression data itself to begin to unravel fundamental questions about brain structure and function.

The Allen Mouse Brain Atlas is the standard for discovering and exploring gene expression in the mouse brain with more than 2000 scientific publications citing the original article[2] and more than 1000 scientific publications citing "mouse.brain-map.org." With more than 15,000 unique visits to the mouse gene expression dataset per month more than 10 years after the first data release, this open source tool can be considered a foundational resource for neuroscientists.[5]

While this work captured the known genes at the time, one limitation of the data was it assayed gene expression at a single time point (postnatal day 56) in a genetically identical strain of male mice that were housed similarly, to avoid environmental disruptions that might alter gene expression. It is well-known that gene expression patterns and intensity can alter over development and with experience, and two studies were conducted to demonstrate this phenomenon. Even with a limited number of genes, altering the sleeping conditions of mice[6] or assaying the same genes in different mouse strains[7] results in significant changes in gene expression.

## The Allen Reference Atlas

The original reference space for the adult mouse brain, the Allen Reference Atlas, was first released in 2005 to allow users of the adult mouse gene expression database to compare gene expression patterns against neuroanatomic structures. This reference space was updated several times over the years with increasingly detailed neuroanatomic delineations (Fig. 3.2) and will be described in more detail later in this chapter. The original reference space was created from approximately 500 sections of a single mouse. Each 25-μm section was Nissl stained to visualize the cell bodies of all the neurons in the section. Structural delineations were determined by in-house anatomists using the visual cytoarchitecture and drawn on a single hemisphere in 132 sections at 100-μm intervals in the coronal plane, and on 21 sections at intervals of 200 μm in the sagittal plane. A print version of this static reference atlas was published in 2008[8] and was the primary reference atlas for the gene expression data until 2011 (version 2) when the atlas was upgraded to allow digital exploration of the structures, the Nissl stains, and the gene expression and projection data.

The nomenclature for the Allen Reference Atlas was a modified version of the Swanson rat brain atlas,[9] incorporating ontologies from Hof et al.[10] and from two informatics-based neuroanatomy resources, the Brain Architecture Management System[11] and the NeuroNames portal.[12]

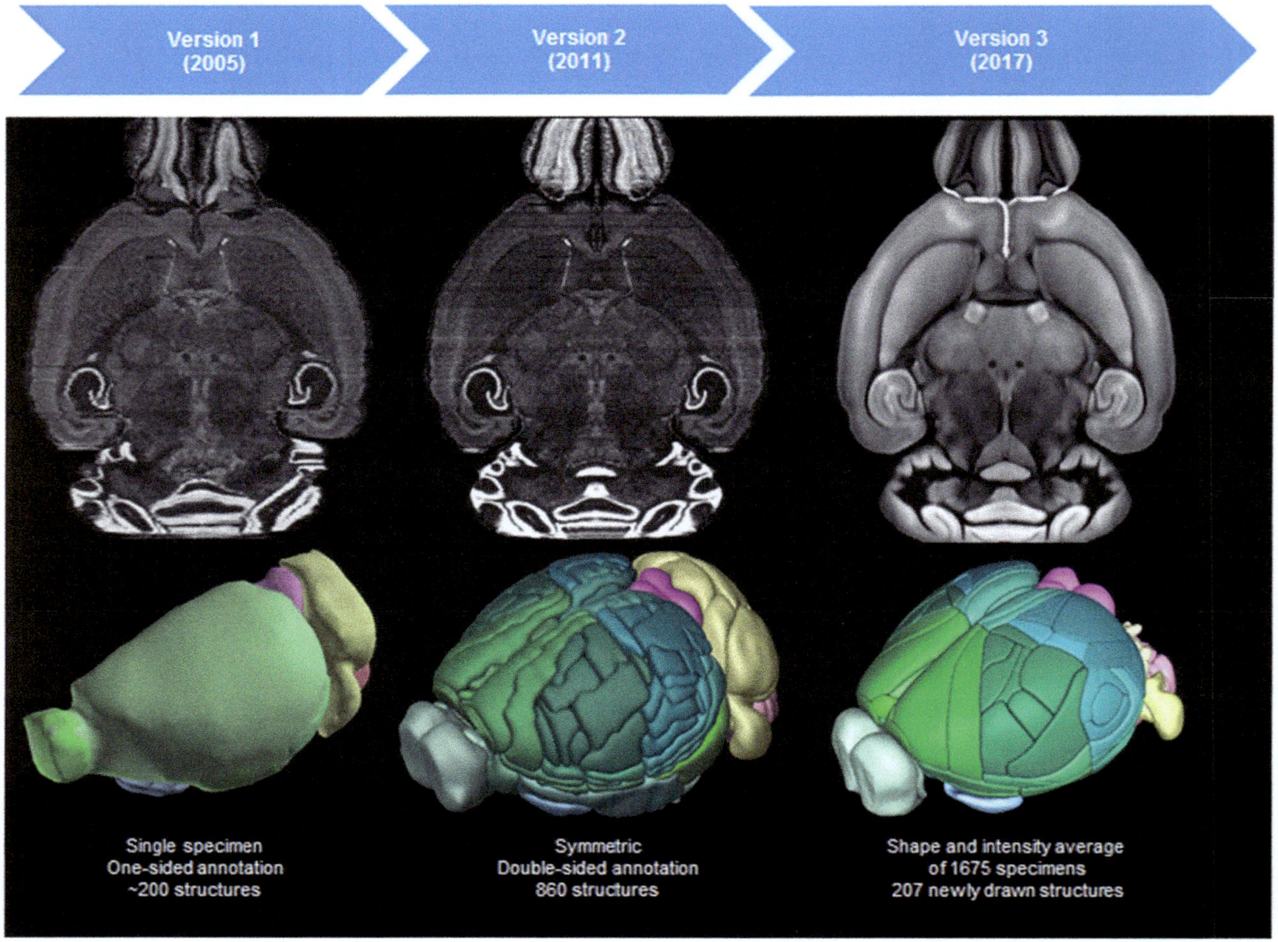


**Figure 3.2** ***Evolution of the mouse common coordinate framework (CCF).*** The first version (2005) supported the Allen Mouse Brain Atlas and was based upon the Allen Reference Atlas specimen. A second version (2011) supported the Allen Mouse Brain Connectivity Atlas where a double-sided and more deeply annotated framework was needed. Version 3 (2015) of the CCF is based on a population average of 1675 specimens and as of June 2017 has 207 newly drawn 3D structures, including nine higher visual areas.

## Allen Adult Mouse Brain Informatics

Integrating the gene expression data and a digital neuroanatomic atlas provided a framework from which the data could be queried to probe the relationships between gene expression, structure—and presumably the function—in the brain.

Due to the background staining inherent in ISH assays, an adaptive thresholding algorithm[3] sensitive to process variation and artifacts was implemented to allow the user to determine the probability of positive expression for a gene of interest (Expression Mask in Fig. 3.1). Each image was divided into a 200 μm × 200 μm grid. For each division, pixel-based statistics of the sum of the number of expressing pixels and sum of expressing pixel intensity were collected. From these statistics measures were obtained for the following:

- expression density = sum of expressing pixels/sum of all pixels in division
- expression intensity = sum of expressing pixel intensity/sum of expressing pixels
- expression energy = sum of expressing pixel intensity/sum of all pixels in division

The expression energy—which is visualized as a color matrix where higher expression is indicated in the yellow and red hues and lower expression in the cooler blue to green colors—takes into account the intensity of the signal as well as the density of cells.[13]

The Expression Gridding module (Gridding in Fig. 3.1) operates on a per-image-series basis and was designed as a low-resolution 3D summary of the gene expression, as well as to enable the comparison of data between different experiments. The expression data grids are used for downstream search and analysis.

From Fig. 3.1, the Alignment module computes the transforms that rotate each two-dimensional (2D) image to form a consistent 3D volume per experiment. Each per-image 2D expression grid is smoothed and rotated to form a 3D grid. Finally, $z$-direction smoothing is applied to the 3D grid, which is then transformed into the standard reference space.

Expression statistics can be computed for each structure in the reference atlas by combining or "unionizing" grid voxels with the same 3D structural label (Unionize in Fig. 3.1). While the reference atlas is annotated at the lowest level of the ontology tree, statistics were computed at upper level structures by combining measurements of the hierarchic children to obtain statistics for the "parent" structure. The end result is per structure expression density, intensity, and energy values for each image series, which are stored in the database and used in the web application to display expression summary bar graphs.

One of the practical applications of informatically processing this data is to enable the user to visualize gene expression and neuroanatomy in an interactive 2D and 3D framework. The image viewer allows navigation and viewing of all the images in an experiment, viewing multiple experiments at a time, and even comparing the ISH images to the drawn reference atlases. This flexibility of viewing each experiment also comes with some advanced viewing and navigation features, such as synching all the images on the page to a similar location (Fig. 3.3).

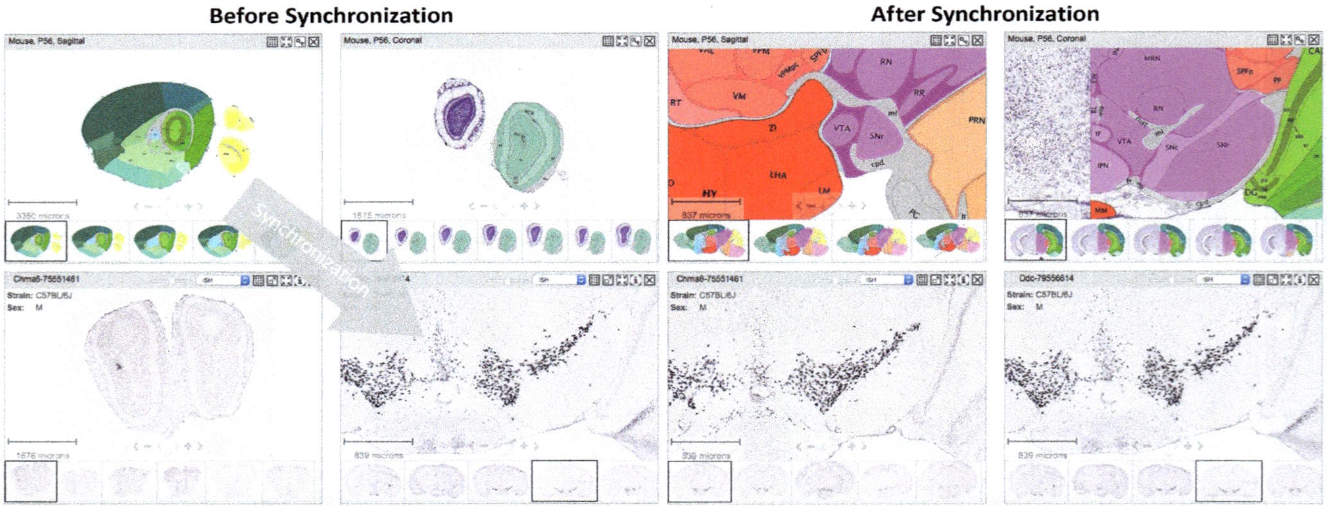


**Figure 3.3** ***Point-based image synchronization.*** Multiple image series in the image viewer can be synchronized to the same approximate location both in sagittal and coronal planes. Before and after synchronization screenshots showing genes *Dpp6* and *Myo16* and the relevant coronal and sagittal plates of the Allen Reference Atlas. Gene *Myo16* shows enriched expression in the medial habenula (MH).

A novel expression grid search service was implemented to allow users to instantly search over the ~25,000 experiments to find genes with specific expression patterns:

- The "Differential Search" function allows users to find genes that have higher expression in one structure (or set of structures) compared to another structure (or set of structures).
- The "Correlative Search" function enables the user to find genes that have a similar spatial expression profile to a seed gene when compared over a user-specified domain.

## Anatomic Gene Expression Atlas

The Anatomic Gene Expression Atlas (AGEA) integrates the gene expression profiles of the 4376 genes assayed in the coronal plane with the spatial voxels of the 3D common reference space.[14] AGEA provides the user with three research features:

1. Clusters mode: a hierarchic clustering technique that bins similar putative anatomic structures based on transcriptome profiles;
2. Correlation mode: a correlation map based on average gene expression profiles; and
3. Gene Finder mode: a differential search tool that allows users to search for enhanced gene expression at the voxel level rather than at the structure level.

The top panel in the AGEA visualization (Fig. 3.4 and mouse.brain-map.org/agea) illustrates the spatial extent of the mouse brain generated from the Nissl-stained sections of the original Allen Reference Atlas colored according to the annotated structures. The red crosshairs pinpoint a 200 μm × 200 μm × 200 μm voxel in the brain.

In "Correlation" mode, the bottom panel is an interactive visualization of correlation maps in the coronal, sagittal, and horizontal planes rendered as false color images. Initially, the crosshairs in the bottom panel mirror the location from the top panel. Clicking on any other location in the brain from the bottom panel returns the Pearson's correlation coefficient between the top and bottom structures. Correlation is computed over a "gene vector" whose elements represent the expression energy for a gene at the voxel of interest. Correlation maps were generated for each possible seed voxel (~51,000).

In "Clusters" mode, the bottom panel displays a data-driven hierarchic spatial organization of the brain computed from the correlation maps. A simple, binary-tree clustering approach was used to capture the various scales of spatial co-expression. To initialize, all voxels were assigned to the root node of the tree. Each descending node is bifurcated into two nodes to achieve the maximal dissimilarity between the two groups of voxels based on correlation values. This is repeated until each node contains a single voxel.

In "Gene Finder" mode, users can generate a list of genes that contribute to enhanced gene expression in the region pinpointed by the crosshairs. This form of spatial search utilizes the Differential Search service with the target regions delineated by the yellow and red regions of the correlation map contrasted with the blue regions. Changing the threshold using the slider bar increases or decreases the target region.

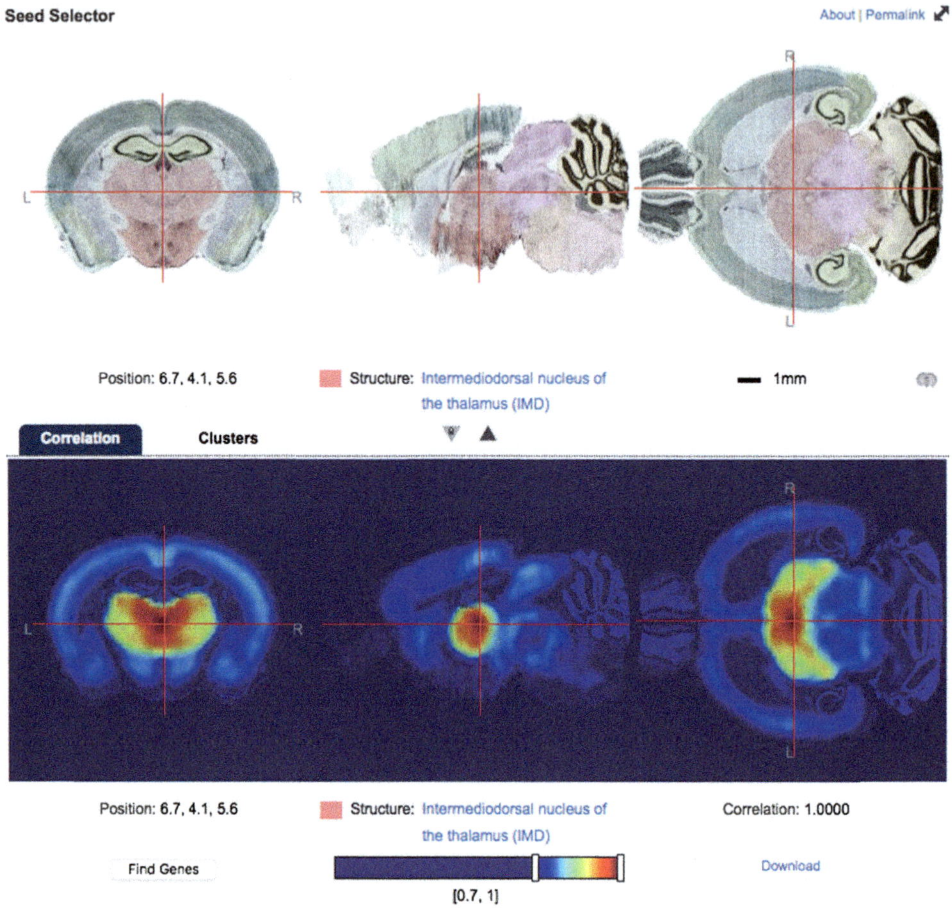


**Figure 3.4** ***Anatomic Gene Expression Atlas (AGEA).*** Images in the upper panel show three orthogonal views of the Nissl-based Allen Reference Atlas with crosshairs used to select the "seed" voxel. The corresponding AGEA correlation map shown in the lower panel illustrates the average gene expression correlation between the seed voxel and other voxels in the 3D brain. Users can threshold the correlation mapping to change the dynamic range by the slider control shown.

## Gene Expression Across Development

The utility of the Allen Brain Atlas resources to understand gene expression patterns in the central nervous system was enhanced with the addition of each new resource that incorporated gene expression across developmental stages.

### *Allen Mouse Spinal Cord Atlas*

The Allen Mouse Spinal Cord Atlas[15] was created out of a demand from the spinal cord community, as there was no digital atlas equivalent. This atlas is a genome-wide ISH atlas of the mouse spinal cord at two developmental stages: juvenile (postnatal day 4) and adult (postnatal day 56), and each experiment contains up to 16 images.

Unlike the Allen Mouse Brain Atlas, which includes a 3D brain space, only rudimentary informatics were developed for this atlas. The cord in the adult atlas and the spinal column in the juvenile atlas were sampled in such a way that a coarse spinal location was made possible alongside the relative gene expression at both developmental stages. Reference atlases were created for this dataset and include histologic sections (Nissl and acetyl cholinesterase, AChE) and an annotated atlas drawn by Drs. Charles Watson and Gulgun Kayalioglu.

The lack of a 3D space representing the mouse spinal cord made performing quantitative determinations of gene expression impractical; therefore each gene in this dataset was manually curated to quantify locations with enriched gene expression. This resource is accessible online at spinal.brain-map.org.

## Allen Developing Mouse Brain Atlas

The Allen Developing Mouse Brain Atlas was built on the foundation of the adult mouse atlas, and it was an attempt to gather gene expression patterns of a panel of important genes to better understand the changes across development. Gene expression was assayed at seven developmental stages in C57BL/6J mice: four embryonic stages (embryonic days 11.5, 13.5, 15.5, and 18.5) and three postnatal stages (postnatal days 4, 14, and 28).[16] This atlas contains assays, primarily in the sagittal plane, from ~2100 genes. These genes were determined by a scientific advisory committee and were considered, at the time, to be the most important and relevant developmental and disease genes to survey.

### *Developing Mouse Reference Atlases*

To provide the same unique juxtaposition of gene expression and anatomy that the Allen Mouse Brain Atlas provided, sagittal reference atlases at each developmental stage were created by Dr. Luis Puelles.[17] These reference atlases were drawn to varying depths of complexity depending upon the developmental stage.

The informatics tools built for this atlas include 3D spaces for each developmental stage and algorithms that allow for synching the images not only to similar locations in space but to those spaces across development (Fig. 3.5).

Due to the very small size of the prenatal samples coupled with the precise locations of some transcriptionally important genes, the search algorithms were insufficient to find very localized, even intense staining. To allow for better searching of this dataset, the gene expression patterns for every gene were manually annotated for gene expression intensity, cell density, and pattern of expression[16] in each of the embryonic stages. Unique search tools were created: an advanced search tool utilizing the manual annotation (Annotation Search), as well as a Temporal Search, Anatomic Search, and a Correlative Search. Each of these services are described in the documentation at developingmouse.brain-map.org.

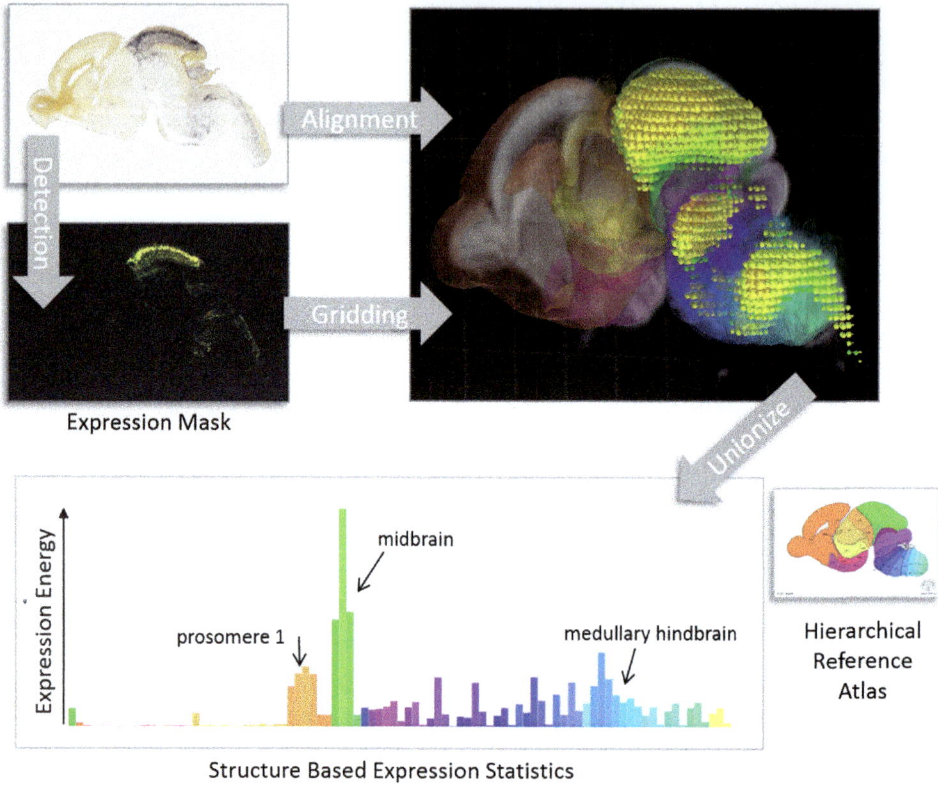


**Figure 3.5** ***The developing mouse informatics data processing pipeline.*** The Alignment module registers each in situ hybridization (ISH) image to the common coordinates of a 3D reference model. The Expression Gridding module produces an expression summary in 3D for downstream analysis. The Structure Unionizer module generates structure-based statistics by combining or "unionizing" grid voxels with the same 3D structural label from the hierarchic reference atlas. Further downstream the grid data is used to compute gene-to-gene correlations and voxel-to-voxel correlations to support developmental Anatomic Gene Expression Atlas functions. The ISH image shows gene *Tcfap2b* at age E18.

One common issue in neuroanatomy is the lack of shared nomenclatures. The Allen Institute mouse atlases are not immune to this ongoing discourse and utilize two distinct ontologies. Since all the annotated references atlases are registered into the postnatal day 56 mouse reference framework, the two ontologies can be directly compared, allowing for translation between the two nomenclatures (Fig. 3.6).

All of the image data is available for direct download from the website (brain-map.org) image viewers; and entire image series, ontologies, unionized grid files, as well as search and other services are also available from the Application Programming Interface (API, http://www.brain-map.org/api/index.html), a RESTful Model Access interface.

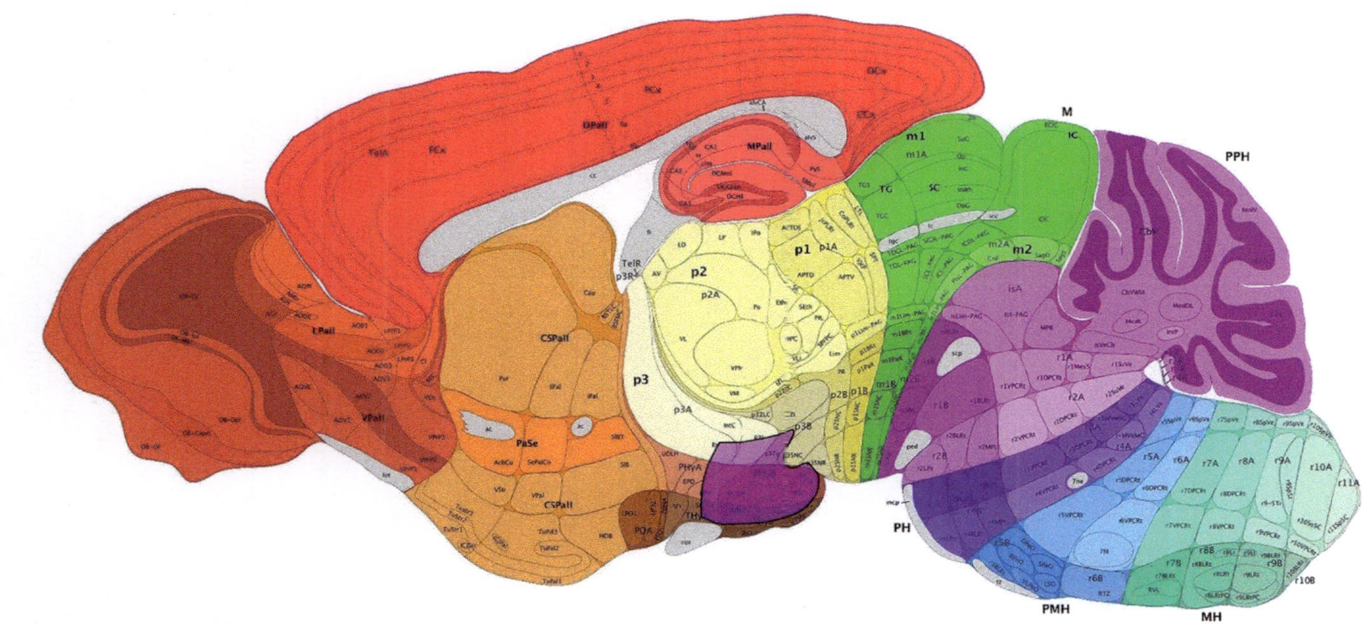


**Figure 3.6** ***Translating between the two mouse ontologies.*** The P56 stage for bcth the developing mouse (shown) and adult mouse brain reference atlases can be used to translate between the two Allen ontologies. The Lateral Hypothalamic Area (outlined in purple) from the adult mouse ontology is overlaid on the developing mouse brain atlas.

## Comparative Gene Expression

While mapping gene expression in the mouse brain was underway, ISH studies were also being conducted in the human brain[18] and included a study to probe cortical expression of the top 1000 queried genes in the Allen Mouse Brain Atlas, and a study probing 55 neurotransmitter and marker genes in the subcortex. These gene expression surveys demonstrated that while some genes show species-specific gene expression patterns, the majority of genes showed a similar expression pattern in the mouse and human brain.[19] Projects to comprehensively map gene expression in the human brain were conducted concurrently with a program that utilized several intersectional strategies to generate transgenic mouse lines expressing high levels of genetic constructs to explore these genetic similarities in mouse and human.

## Allen Human Brain Atlas Data Collection

Creation of the Allen Mouse Brain Atlas required assaying the brains of approximately 3000 genetically identical mice. Prior to the commercialization of microarray chip analysis, performing this kind of a systematic study to map gene expression in the human brain was not feasible. With the advent of microarray analysis, the Allen Human Brain Atlas project was designed to create a comprehensive, quantitative map of gene expression profiled across the entire adult brain, with the emphasis on anatomically complete coverage in a small number of high-quality, clinically unremarkable brains.

A tissue imaging, processing, and data collection pipeline was generated to extract RNA from anatomically annotated tissue samples for microarray analysis.[20] Postmortem brains were acquired under stringent conditions to collect neurotypical or "normal control" brains, imaged by magnetic resonance (MR), slabbed, frozen, and shipped to the institute within 30 h postmortem. Each tissue slab underwent ISH and histologic imaging and was sampled either by manual macrodissection or laser microdissection and the RNA extracted and used to generate labelled cRNA probes for hybridization to custom 64K Agilent microarrays.

In such a large-scale microarray project, thousands of samples were processed during an approximately 3-year time span, likely introducing systematic biases into the dataset (drifts due to age of reagents or array chips, reagent lot changes, vendor acquisitions, RNA integrity among samples, RNA quality due to different sample capturing methods, etc.). To minimize the effects of these nonbiologic biases while keeping biologic variance intact, the data were normalized so that within- and across-brain comparisons primarily revealed differences and similarities that were biologically relevant.[21]

By the completion of this project, there were data collected from approximately 500 tissue samples per hemisphere, extracted from eight hemispheres of six donor brains, all available for mining from the portal at human.brain-map.org.

## BrainSpan Atlas of the Developing Human Brain

In 2009, a consortium consisting of the Allen Institute for Brain Science, Yale University, the University of California, Los Angeles, and Harvard Medical School was awarded funding from President Obama's American Recovery and Reinvestment Act to map gene expression in the human brain across development.

This study included transcriptome profiling by RNA sequencing and exon microarray analysis, histology and ISH, neuroimaging, and epigenetic assays on 16 brain regions across 11 human developmental stages,[22] all of which is available to the public from brainspan.org. The adult reference atlas created for this resource to understand the brain context in which gene expression and structures were located was published in print format and is the most detailed human digital atlas ever published.[23] The backbone of the Allen Human Brain Atlas ontology and nomenclature is derived from NeuroNames[24] (http://braininfo.rprc.washington.edu/). Some names have been altered to be consistent with the use of English rather than Latin names. In some cases, where a structure is more commonly referred to by the Latin term, that name has been used (e.g., gyrus rectus, telencephalon). The nomenclature within this ontology incorporates the work of the most widely recognized experts in specific subregions of the brain. For example, the ontologies and nomenclatures for the cerebellum and thalamus are derived from MR imaging Atlas of the Human Cerebellum[25] and The Thalamus,[26] respectively. More information on the methods behind creating these atlases can be found in the documentation available from human.brain-map.org and brainspan.org.

## Quantitative Gene Expression Informatics

While the qualitative image data used to map gene expression in the mouse brain provides a visually satisfying context for gene expression, the methodology used to collect the quantitative expression values in humans required visualization of a different sort. The massive donor/structure versus expression value matrix is best illustrated using a heatmap visualization. Interactive heatmaps with the gene/probe on the vertical axis and the donor and structure as two color-coded column headers display the data. For the Allen Human Brain Atlas, gene expression data are represented in z-score, a tool for visualization that colors the average expression intensity for the probe over all the brains and samples black (in the red-green color map) and highlights structures with expression higher than average in red and lower than average in green. The RNA sequencing data are displayed in the normalized expression intensity values. More information on interacting with the data through the heatmaps is available from Help at human.brain-map.org or brainspan.org.

Download of the data matrices for any of the quantitative datasets are available from the download pages in each of the respective products or directly from the heatmaps. The API (http://www.brain-map.org/api/index.html) also provides access to these data, ontologies, imaging files, and services.

## Genetic Geography of the Brain

Collecting a comprehensive survey of brain data—as in the genome-wide gene expression databases at the Allen Institute for Brain Science—allows for integrating the data in such a way that associations can be revealed that may have been missed when viewed from a narrower perspective. An example visualization of data integration is available from the online portal (casestudies.brain-map.org/ggb) and is a scientific vignette with an interactive figure that explores differential gene expression across the structures of the six human brains from the Allen Human Brain Atlas. One noteworthy feature of displaying data in this fashion is a clear demonstration of similar gene expression patterns across the neocortex, the amygdala, and the cerebellum (old cortex).[27] This patterning suggests that the basic architecture across the cortex is similar; therefore, exploring the functional operation of the cortex (or unit of cortex) may be the most valuable avenue to understand the complexities of cognition and behavior.

# GENETIC TOOLS: TRANSGENIC MOUSE LINES

The Allen Institute human cortex study in which ~1000 genes were surveyed at cellular resolution demonstrated approximately 21% species-related differential expression, patterning that may contribute to the uniqueness of human brain function.[19] That almost 80% of the gene expression patterns were similar in both mouse and human suggest there may be conserved cortical functions in the two species. The genetic tools that afford the ability to manipulate gene expression and function in the mouse provide an ideally suited methodology to investigate cell type–specific circuits and cellular function in the brain.

The Allen Institute has created a collection of mice that express fluorescent probes and functional tools at high levels in selected cell populations and that are designed to study neural circuits and brain function.[28] To create transgenic lines capable of expressing markers and genetic tools in a range of specific cell types throughout the brain, various driver lines were combined with driver-dependent reporter lines. The transgenic mouse lines generated by the Allen Institute are deposited to common repositories, such as The Jackson Laboratory, for distribution to the scientific community.

## Cre-Dependent Reporter Lines

The Allen Institute for Brain Science created a pipeline for reporter line generation that produced Rosa26-CAG-based Cre-dependent mouse lines that express various proteins at high levels and in a cell type–specific fashion in a wide range of cell types within the brain. To achieve finer control and therefore further restrict transgene expression to an even smaller group of cells, Rosa26-targeted reporter lines in which transgene expression is regulated by the activity of two independent recombinases were created. Example mouse lines using this strategy are the Ai65 line—in which expression of tdTomato requires the activity of both Cre and Flp recombinases—and Ai66, where both Cre and Dre recombinases are needed.[29]

Although many of the Allen Institute Rosa26-based reporter lines express functional levels of genetically encoded fluorescent probes and tools, an alternative expression strategy was developed that would offer the potential for both increased specificity and enhanced levels of transgene expression. Much of the transgenic mouse–line generation currently underway at the Allen Institute for Brain Science produces TIGRE reporter lines that exhibit higher expression than comparable Rosa26 lines, greater specificity, and low background (Cre-independent expression) signal.

Expression and/or functional characterization of the Cre-dependent reporter lines can be found at Ai2-Ai14,[30] Ai27, Ai32, Ai35, Ai39,[31] Ai38[32] and Ai62, Ai65, Ai66, Ai78, Ai82, Ai87, Ai95, and Ai96.[33]

## Driver Lines

In the interest of creating transgenic lines capable of expressing markers and genetic tools in a range of specific cell types throughout the brain, Cre, CreERT2, dCre, tTA2, and FlpO driver lines using either BAC or knock-in approaches were generated.[34]

An automated pipeline was generated to systematically characterize the expression patterns of these driver and reporter mouse lines and is available online at connectivity.brain-map.org/transgenic. Currently, characterization data are available for approximately 25 Cre-dependent reporter lines generated at the Allen Institute and more than 200 Cre/CreERT2 or other driver lines, generated in-house or by other investigators using various expression strategies.

The focus of the work being conducted at the Allen Institute for Brain Science has shifted from developing large-scale mapping surveys to creating resources to understand how the components of the brain, from cell types to circuits, give rise to its function. Projects currently underway include understanding connectivity in the brain at the micro- and mesoscale, creating a data-driven multimodal cellular taxonomy, and efforts to understand the information processing in the functioning, behaving brain. The next section describes the data and tools that were available as of June 2017 from the data portal at brain.map.org.

## Allen Mouse Brain Connectivity Atlas

The Allen Mouse Brain Connectivity Atlas is a 3D, high-resolution map of long-range neuronal projections in the mouse brain. Built upon an array of transgenic mice genetically engineered to target specific cell classes, the Allen Mouse Brain Connectivity Atlas comprises a unique compendium of axonal projections from selected neuronal populations mapped using a standardized platform to generate a comprehensive database of neural projections. In addition, the projection data are connected to the Allen Institute's gene expression and reference atlases and thus provide opportunities to help understand how genes contribute to connectivity formation and function.

Using a viral delivery method to induce the expression of fluorescent protein, the anterograde tracer was injected into almost 500 sites, resulting in baseline projection mapping from each structure in the brain. Subsequent experiments utilized an infection strategy that required the presence of Cre-recombinase for fluorescent protein expression. This strategy targeted specific populations of neurons to better categorize cell type–specific projections.

Each of the specimens in the Allen Mouse Brain Connectivity Atlas was imaged using a customized, serial two-photon (STP) tomography system[35] that couples high-speed, two-photon microscopy with automated vibratome sectioning. STP tomography yields a series of inherently prealigned images amenable for precise 3D spatial mapping.

There are currently over 2500 projection experiments available for mining either from the website (connectivity.brain-map.org) or via the Allen Software Development Kit (Allen SDK): http://alleninstitute.github.io/AllenSDK/. The Allen SDK is a Python-based resource that was designed to provide easier access to the data and informatics available from the Allen Institute for Brain Science.

## Allen Mouse Brain Connectivity Informatics

While registration of gene expression data into the first version of the reference space (see Fig. 3.2) was adequate, the second iteration of the Allen Mouse Common Coordinate Framework (CCF) was created in 2011 to fulfil the scientific goals of the Allen Mouse Brain Connectivity Atlas where a double-sided and more deeply annotated digital framework was needed.

Similar to the Allen Mouse Brain Atlas, the informatics data processing pipeline for the Allen Mouse Brain Connectivity Atlas consists of the following components: a preprocessing module, a 3D reference model, an alignment module, a detection module, a gridding module, and a structure unionizer module (Fig. 3.7). The output of the pipeline is quantified signal values at a grid voxel level and a structure level according to the Allen Reference Atlas ontology. The grid level data are used downstream to provide on-the-fly data and correlative search services and to support visualization of projection pathway relationships.[36]

## Large-Scale Systems Neuroscience

The Allen Institute for Brain Science is currently involved in a 10-year, milestone-driven plan to catalog the components of the mouse cortex and to develop experimental and modeling platforms for studying the information processing in the brains of awake, behaving mice and to construct abstract and biophysically realistic models of cortical networks.[37] To facilitate this effort, a next generation CCF was needed to support integration of data generated as part of this plan.

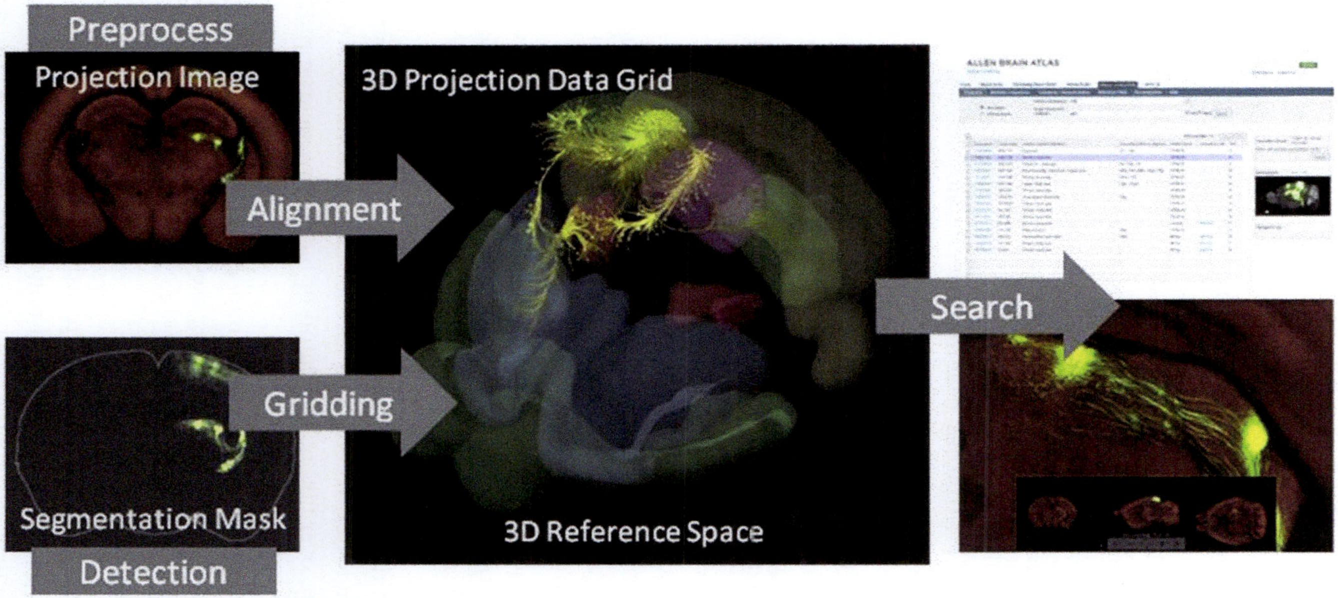


**Figure 3.7** ***The informatics data processing pipeline for the Allen Mouse Brain Connectivity Atlas.*** The Alignment module registers each projection image to the common coordinates of a 3D reference model. The Detection module separates signal from background. The Gridding module produces a 3D summary of the detected signal for each data set. Further downstream, the gridded data is used for data and correlation searches. The projection image above shows terminal and passing fiber signal in the thalamus arising from viral infection in the primary visual area.

### *Common Coordinate Framework*

The anatomic template of the CCF v3 is a shape and background signal intensity average of 1675 specimens from the Allen Mouse Brain Connectivity Atlas.[38] The population average was created through an iterative process, involving 12-parameter affine registration of specimens, averaging many brains over multiple cycles.[36] As of October 2017, the CCF consists of 662 annotated structure volumes, including gray matter, white matter, and ventricles. Overall, 242 cortical and 330 subcortical gray matter, 82 fiber tract, and 8 ventricle and associated structure volumes were delineated natively in three dimensions.

## Building a Multimodal Cellular Taxonomy

The plan to catalog the cellular components and circuits in the visual cortex requires collecting a large amount of high-quality data from diverse modalities and analyzing these data to determine uniquely distinguishing features. Ideally the data from each modality would show a similar pattern of discrimination that would make determining distinguishing features of a cell type uniform and standardized. Building on the gene expression data from the Allen Mouse Brain Atlas, the transgenic mouse lines created from those studies, and the advances in single-cell assays (including RNA sequencing and cell sorting), a pilot study was conducted to determine if cell types could be distinguished in the primary visual cortex from single-cell gene expression profiles.[39] Not only did this study distinguish 49 distinct cell types based on gene expression, but electrophysiological and morphologic indicators seemed to corroborate the clusters determined by the expression data.

Teams at the Allen Institute are currently in the data collection phase of this project[37] producing data from a standardized pipeline to characterize electrical properties and morphology of individual neurons from brain slices, coupled with the generation of point neuron and biophysical models of these cells. While a unique and laudable practice at the Allen Institute is to release data on the web portal once it passes quality control, that means only rudimentary data analysis has been performed, and a cell taxonomy is slated for a future date. From the portal (celltypes.brain-map.org), you can view or download single-cell electrophysiological recordings (~1000 cells), morphology from a subset of these cells (~300 cells), over 600 point neuron or biophysical neuron models, and the raw RNA sequencing data for ~16,000 cells.

The Allen SDK (http://alleninstitute.github.io/AllenSDK/cell_types.html) also provides easy access to downloading and analyzing these data.

## Allen Brain Observatory

Another project underway at the Allen Institute for Brain Science is designed to understand the information processing that is occurring in the visual cortex of the awake, behaving mouse. The intent of the project is not only to understand the neural coding but also to bring together scientists from around the globe to participate in an endeavor that would create standardized formats and allow data sharing on a grand scale.[40] The first product released from this project is aimed at understanding visual coding and contains the first standardized in vivo survey of physiological activity in the mouse visual cortex, featuring representations of visually evoked calcium responses from GCaMP6-expressing neurons in selected cortical layers, visual areas, and Cre lines.

### *Visual Coding*

The design of the Visual Coding pipeline is to build a data matrix of quantitated cell responses, based on brain location (primary and higher visual areas), cell class (cortical layers), and visual stimuli (synthetic and natural). The raw data captured in this project are movies recording the change in fluorescence of individual neurons in response to the visual stimuli. To provide a searchable database, analysis has been conducted from each recording session to isolate the signal from individual cells, calculate the change in fluorescence over time during stimulus presentation, and then create visualizations that describe the cellular responses. To date, there are 199 experiments available to view from observatory.brain-map.org/visualcoding with a total of almost 40,000 cell responses. Navigating this dataset is best accomplished by first exploring the overview page including the Visual Stimuli, which introduce the cellular visualizations created to describe the responses and the documentation to understand the study design and protocols. Exploring the population dynamics is achieved by searching the Experiments tab, while looking for cells with particular tuning properties can be accomplished from the search tool in the Cells tab.

The ideal way to interact with this dataset is from the Allen SDK (http://alleninstitute.github.io/AllenSDK/brain_observatory.html) in which Python notebooks have been created to work with the data.

## COMMUNITY USE OF THE ALLEN INSTITUTE FOR BRAIN SCIENCE RESOURCES

Perhaps the greatest significance of the Allen Institute for Brain Science data and tools is that they are a freely available resource the community uses in the pursuit of an understanding of the brain. Studies incorporating the Allen Human Brain Atlas data have discovered biologic processes underpinning the developmental changes that may be involved in brain disorders in adolescents,[41,42] brain regions involved in migraine pathophysiology,[43] and genes involved in speech and language processing.[44] The accessibility of the data through the application programming interface also supports the creation of online tools designed to foster discovery. An automated tool using text-mining, meta-analysis, and machine-learning techniques was created to process functional imaging studies to make connections between cognitive states and brain regions (neurosynth.org).[45] This tool incorporated the Allen brain data and allows for gene expression to be associated with cognitive states and brain structures (see neurosynth.org/genes/pdyn as an example). Brainscope (http://www.brainscope.nl/brainscope) is a network analysis visualization tool that allows for exploration of the Allen human gene expression data that elegantly shows relationships between gene coexpression modules and biologic function.[46]

## CONCLUSION

The reductionist approach taken over the last decade and a half to understand the complexities of the brain has led to the creation of cumulative and elegant biologic strategies to probe the higher order functions of the brain. Coupled with the collection of vast quantities of data, the informatics built around navigating, visualizing, and probing correlations have provided new pathways to tackle the questions raised by studying the most complex thing in the known universe.

## ACKNOWLEDGMENTS

A most profound acknowledgement to Paul G. Allen for his vision and his generosity in funding such a worthy endeavor as the Allen Institute. The Human Brain Reference Atlas was supported by Award Numbers RC2MH089921 (PIs: Ed Lein & Michael Hawrylycz, Allen Institute for Brain Science), RC2MH090047 (PI: James A. Knowles, University of Southern California), and RC2MH089929 (PI: Nenad Sestan, Yale University) from the National Institute of Mental Health.

## REFERENCES

1. Lander ES, Linton LM, Birren B, et al. Initial sequencing and analysis of the human genome. *Nature.* 2001;409(6822):860–921.
2. Lein ES, Hawrylycz MJ, Ao N, et al. Genome-wide atlas of gene expression in the adult mouse brain. *Nature.* 2007;445(7124):168–176.
3. Ng L, Pathak S, Kuan C, et al. Neuroinformatics for genome-wide 3D gene expression mapping in the mouse brain. *IEEE ACM Trans Comput Biol Bioinform.* 2007;4(3):382–393.
4. Yushkevich PA, Avants BB, Ng L, et al. 3D mouse brain reconstruction from histology using a coarse-to-fine approach. In: *Biomedical Image Registration.* Springer; 2006:230–237.
5. Koch C, Jones A. Big science, team science, and open science for neuroscience. *Neuron.* 2016;92(3):612–616.
6. Thompson CL, Wisor JP, Lee CK, et al. Molecular and anatomical signatures of sleep deprivation in the mouse brain. *Front Neurosci.* 2010;4:165.
7. Morris JA, Royall JJ, Bertagnolli D, et al. Divergent and nonuniform gene expression patterns in mouse brain. *Proc Natl Acad Sci USA.* 2010;107(44):19049–19054.
8. Dong HW. *The Allen Reference Atlas: A Digital Color Brain Atlas of the C57Bl/6J Male Mouse.* John Wiley & Sons Inc; 2008.
9. Swanson LW. *Brain Maps III: Structure of the Rat Brain: An Atlas with Printed and Electronic Templates for Data, Models, and Schematics.* Gulf Professional Publishing; 2004.
10. Hof PR. *Comparative Cytoarchitectonic Atlas of the C57BL/6 and 129/Sv Mouse Brains.* Elsevier; 2000.
11. Bota M, Dong H-W, Swanson LW. Brain architecture management system. *Neuroinformatics.* 2005;3(1):15–47.
12. Bowden DM, Song E, Kosheleva J, Dubach MF. NeuroNames: an ontology for the BrainInfo portal to neuroscience on the web. *Neuroinformatics.* 2012;10(1):97–114.

13. © 2011 Allen Institute for Brain Science. *Allen Mouse Brain Atlas*. Allen Institute for Brain Science; 2011. Available from: mouse.brain-map.org. http://help.brain-map.org/display/mousebrain/Documentation.
14. Ng L, Bernard A, Lau C, et al. An anatomic gene expression atlas of the adult mouse brain. *Nat Neurosci*. 2009;12(3):356–362.
15. © 2008 Allen Institute for Brain Science. *Allen Spinal Cord Atlas*; 2008. Available from: mousespinal.brain-map.org. http://mousespinal.brain-map.org/docs.html.
16. Thompson CL, Ng L, Menon V, et al. A high-resolution spatiotemporal atlas of gene expression of the developing mouse brain. *Neuron*. 2014;83(2):309–323.
17. Puelles L, Harrison M, Paxinos G, Watson C. A developmental ontology for the mammalian brain based on the prosomeric model. *Trends Neurosci*. 2013;36(10):570–578.
18. © 2010 Allen Institute for Brain Science. *Allen Human Brain Atlas*; 2010. Available from: human.brain-map.org/ish/Search.
19. Zeng H, Shen EH, Hohmann JG, et al. Large-scale cellular-resolution gene profiling in human neocortex reveals species-specific molecular signatures. *Cell*. 2012;149(2):483–496.
20. Hawrylycz MJ, Lein ES, Guillozet-Bongaarts AL, et al. An anatomically comprehensive atlas of the adult human brain transcriptome. *Nature*. 2012;489(7416):391–399.
21. © 2013 Allen Institute for Brain Science. *Allen Human Brain Atlas-Microarray Normalization Whitepaper*; 2013. Available from: http://human.brain-map.org/static/docs.
22. Miller JA, Ding SL, Sunkin SM, et al. Transcriptional landscape of the prenatal human brain. *Nature*. 2014;508(7495):199–206.
23. Ding SL, Royall JJ, Sunkin SM, et al. Comprehensive cellular-resolution atlas of the adult human brain. *J Comp Neurol*. 2016;524(16):3127–3481.
24. Bowden DM, Dubach MF. NeuroNames 2002. *Neuroinformatics*. 2003;1(1):43–59.
25. Schmahmann JD, Doyon J, Petrides M, Evans AC, Toga AW. *MRI Atlas of the Human Cerebellum*. Academic Press; 2000.
26. Jones EG. *The Thalamus*. Springer Science & Business Media; 2012.
27. Hawrylycz M, Miller JA, Menon V, et al. Canonical genetic signatures of the adult human brain. *Nat Neurosci*. 2015;18(12):1832–1844.
28. © 2010 Allen Institute for Brain Science. *Transgenic Characterization*; 2010. Available from: connectivity.brain-map.org/transgenic.
29. Anastassiadis K, Fu J, Patsch C, et al. Dre recombinase, like Cre, is a highly efficient site-specific recombinase in *E. coli*, mammalian cells and mice. *Dis Model Mech*. 2009;2(9–10):508–515.
30. Madisen L, Zwingman TA, Sunkin SM, et al. A robust and high-throughput Cre reporting and characterization system for the whole mouse brain. *Nat Neurosci*. 2010;13(1):133–140.
31. Madisen L, Mao T, Koch H, et al. A toolbox of Cre-dependent optogenetic transgenic mice for light-induced activation and silencing. *Nat Neurosci*. 2012;15(5):793–802.
32. Zariwala HA, Borghuis BG, Hoogland TM, et al. A Cre-dependent GCaMP3 reporter mouse for neuronal imaging in vivo. *J Neurosci*. 2012;32(9):3131–3141.
33. Madisen L, Garner AR, Shimaoka D, et al. Transgenic mice for intersectional targeting of neural sensors and effectors with high specificity and performance. *Neuron*. 2015;85(5):942–958.
34. Harris JA, Hirokawa KE, Sorensen SA, et al. Anatomical characterization of Cre driver mice for neural circuit mapping and manipulation. *Front Neural Circ*. 2014;8:76.
35. Ragan T, Kamelin P, Sutin J. Two-photon autofluorescence spectroscopy of mouse brains using serial two-photon tomography. *Biophys J*. 2013;104(2):342a.
36. Kuan L, Li Y, Lau C, et al. Neuroinformatics of the Allen Mouse Brain Connectivity Atlas. *Methods*. 2015;73:4–17.
37. Hawrylycz M, Anastassiou C, Arkhipov A, et al. Inferring cortical function in the mouse visual system through large-scale systems neuroscience. *Proc Natl Acad Sci USA*. 2016;113(27):7337–7344.
38. Oh SW, Harris JA, Ng L, et al. A mesoscale connectome of the mouse brain. *Nature*. 2014;508(7495):207–214.
39. Tasic B, Menon V, Nguyen TN, et al. Adult mouse cortical cell taxonomy revealed by single cell transcriptomics. *Nat Neurosci*. 2016;19(2):335–346.
40. Koch C, Reid RC. Neuroscience: observatories of the mind. *Nature*. 2012;483(7390):397–398.

41. Vértes PE, Rittman T, Whitaker KJ, et al. Gene transcription profiles associated with inter-modular hubs and connection distance in human functional magnetic resonance imaging networks. *Philos Trans R Soc Lond B Biol Sci*. 2016;371(1705).
42. Whitaker KJ, Vértes PE, Romero-Garcia R, et al. Adolescence is associated with genomically patterned consolidation of the hubs of the human brain connectome. *Proc Natl Acad Sci USA*. 2016;113(32): 9105–9110.
43. Eising E, Huisman SMH, Mahfouz A, et al. Gene co-expression analysis identifies brain regions and cell types involved in migraine pathophysiology: a GWAS-based study using the Allen Human Brain Atlas. *Hum Genet*. 2016;135(4):425–439.
44. Bohland J, Myers EM, Kim E. An informatics approach to integrating genetic and neurological data in speech and language neuroscience. *Neuroinformatics*. 2014;12(1):39–62.
45. Yarkoni T, Poldrack RA, Nichols TE, Van Essen DC, Wager TD. Large-scale automated synthesis of human functional neuroimaging data. *Nat Methods*. 2011;8:665.
46. Huisman SMH, van Lew B, Mahfouz A, et al. BrainScope: interactive visual exploration of the spatial and temporal human brain transcriptome. *Nucleic Acids Res*. 2017;45(10):e83.

CHAPTER 4

# ExAtlas: Online Tool to Integrate Gene Expression and Gene Set Enrichment Analyses

**Alexei A. Sharov, David Schlessinger**
National Institute on Aging, National Institutes of Health, Baltimore, MD, United States

## INTRODUCTION

Studying biological mechanisms of neuronal function is among the major challenges of contemporary neuroscience.[1] We still do not know specific functions of many of the >200 types of neurons, and their relations to behavior (e.g., memory and associative learning) remain phenomenological. The problem of identifying neuronal functions is important for detecting and treating pathologies associated with neurodegenerative diseases, personality disorders, and/or aging.[2] In this respect, global gene expression profiling appears to be one of the most informative exploratory approaches in the study of neural cells and tissues.[3–5] This method quantifies the activity of all genes in the genome, and hence, provides an instantaneous snapshot of major molecular processes occurring in the cells. Furthermore, neural tissues are highly heterogeneous, including various types of neurons and glial cells. Thus, single-cell expression profiling is increasingly coming to the fore in neuroscience to assess molecular signatures of various cell types.[6–8]

Statistical analysis of gene expression profiles in cells and tissues is supported by versatile software applications.[9–12] These tools include public software Significance Analysis of Microarrays,[13] NIA Array Analysis,[14] Onto-Express,[15,16] TM4 Microarray Software Suite,[17] GenMAPP,[18,19] DAVID,[20] GSEA,[21] and commercial programs (Genomatix Suite and Pathway Studio). Meta-analysis programs include GeneMeta,[22] metaArray,[23] and MAMA.[24] However, these software tools do not support several important functions in gene expression analysis. First, software for the analysis of gene set enrichment or overlap (e.g., DAVID or GSEA) can handle only one gene expression profile at a time. However, a comprehensive analysis requires a matrix of enrichment scores where multiple expression profiles are compared with multiple Gene Ontology (GO) annotations or other gene sets. Second, no programs have been available to analyze the degree of correlation between all gene expression profiles in different data sets. Such analyses can help to find similarities and differences between tissue-specific gene expression in mouse and human, between stages of development, or between the effects of manipulation of transcription factors (TFs) and tissue-specific expression

*Molecular-Genetic and Statistical Techniques for Behavioral and Neural Research*
ISBN 978-0-12-804078-2, https://doi.org/10.1016/B978-0-12-804078-2.00004-0

profiles.[25] Finally, available software cannot extract data from public databases such as GEO[26] or ArrayExpress.[27] Although "Microarray retriever" allows downloading data from GEO and ArrayExpress,[28] it does not have support for statistical analysis.

As a tool that overcomes these limitations, we report an online public software ExAtlas (https://lgsun.irp.nia.nih.gov/exatlas). It integrates various programs for statistical assessment and visualization of gene expression data.[29] The main feature of the software is its capacity to compare different data sets in a global manner (i.e., all components in data set A are compared to all components in data set B). The methods include evaluation of multiprofile gene set enrichment and gene set overlap; new method for global correlations between gene expression data sets and identification of coregulated genes using Expected Proportion of False Positives (EPFP),[30] and meta-analysis of paired gene expression profiles. The suite includes searches for data series in the GEO/NCBI database based on submitted keywords and automated extraction of gene expression profiles that are formatted for subsequent statistical analysis. Annotations of microarray platforms are updated to match the latest gene symbols and names in the NCBI database. ExAtlas includes standard statistical analysis including analysis of variance (ANOVA), pairwise comparison of expression profiles, false discovery rate (FDR), and principal component analysis (PCA), and it generates graphic outputs: heat maps (heatmaps), scatterplots, bar charts, and interactive three-dimensional plots. Several public data sets (e.g., GNF, BrainSpan, GO, KEGG, GAD phenotypes) are preloaded, updated regularly, and on hand for functional annotation. We illustrate the utility of ExAtlas to facilitate gene expression analysis for neural research and provide a tutorial with step-by-step instructions for processing sample data (see Exercises (Tutorial) section).

## EXPERIMENTAL APPROACHES TO GENE EXPRESSION PROFILING

A brief outline of experimental methods for global gene expression profiling and primary data processing can indicate some of the potential problems with data quality. The human body includes several hundred cell types, each characterized by the active transcription of an array of the full complement of 21,000 protein-coding and >18,000 noncoding genes.[31] The cell-specific gene expression profile can be represented as a vector of the abundance of mRNA molecules of all genes in the cells. The counts of mRNA molecules are highly dynamic, varying with cell- and condition-specific rates of transcription and degradation. The gene expression profiles are relatively easy to quantify for cultured cells (e.g., fibroblasts, stem cells, carcinoma culture), but cell types in the body are always mixed, and it may be difficult to separate them. Thus, most gene expression data are available for whole organs or portions of them. Laser capture microdissection[32] is widely used to isolate specific cell types, and dissociation of cells from a tissue followed by cell sorting for specific gene markers or single-cell sequencing are increasingly used to isolate cells of a specific type.[33–35] These methods have potential problems including cell stress during isolation, high error in measuring gene expression based on small amounts of RNA in samples, or the need for classification of transcriptomes of individual cells. Consequently, to achieve optimum results, gene expression profiles of each cell type have to be evaluated many times followed by meta-analysis. Methods of RNA isolation and preprocessing strongly affect transcriptome profiles. Thus, identical techniques should be applied to all samples within a project. Preprocessing often includes separation of specific fractions of RNA, such as poly-A-containing transcripts, with magnetic beads carrying complementary poly-T primers. Poly-A enrichment generates relatively clean samples of mostly spliced RNA. Another preprocessing method targets the removal of bulk ribosomal RNA and repetitive sequences; such "total RNA" samples contain unspliced RNA and noncoding RNAs that can increase the yield of information but also may add noise to gene expression measurement. Preprocessing may include reverse transcription to make cDNA copies of RNA molecules, PCR amplification, addition of linker sequences, labeling with fluorescent tags, etc.

Two main methods are used for global gene expression profiling: microarrays and RNA-seq. The microarray grids bear 40,000–200,000 spots, each carrying synthesized DNA oligos 25–60 bp long that match specific genes in the genome. Oligo sequences are generally selected from unique regions of each gene/potential transcript. Then cDNA is labeled fluorescently (with green Cy3 or red Cy5) and hybridized to the microarray. The intensity of fluorescence at each spot in the array is then measured by laser scanners at high resolution (e.g., 2 μ), and the average signal intensity in each spot is scored as the relative expression of the corresponding gene.

RNA-seq is usually based on shredding transcripts into short fragments (e.g., 200 bp long) and then sequencing individual fragments using Next Generation Sequencing methods. A flow cell can be loaded with up to 100 million DNA molecules (depending on the brand of the sequencer), which can be sequenced simultaneously in one run. DNA fragments are sequenced by adding single fluorescent-marked nucleotides to the growing complementary DNA strand, and the flow cell is laser scanned after each nucleotide is added. After a number of cycles (typically 50), all images are analyzed by computer to reconstruct 50–150 nucleotide sequences for all fragments in the flow cells. In one option, the fragments are sequenced from both ends (paired-end sequencing), which can help to validate locations and detect alternative splicing. The sequences are then mapped back to the genome, and the relative expression of each gene is evaluated by the number of matching fragments per 1 M fragments mapped to the genome and 1 Kb of transcript length (FPKM). RNA-seq data are usually processed by CuffLinks[36] or Bioconductor.[37]

## PREPROCESSING OF GENE EXPRESSION DATA

Experiments on global gene expression usually include multiple samples, for example, several biological and technical replications for treatment and control. Some data sets include >1000 samples (e.g., for various tumors, tissues, or cell types). Thus, the first task is to combine data from many files corresponding to individual samples. Gene expression data obtained with microarrays are relatively easy to combine because the order of oligos in all samples is the same. This can be done manually in Excel if the number of arrays is small. However, for combining large sets of arrays, it is necessary to use computer programs for data extraction (e.g., arrayjoin in NIA Array Analysis).[14] Gene expression profiles generated with RNA-seq have no fixed order of transcripts (or genes), so they can be combined only using appropriate software programs (e.g., Bioconductor).[38]

Data combined from multiple samples must be normalized to make gene expression values comparable between samples (columns). Microarray data obtained with two-dye competitive hybridization can be normalized by signal intensity of a common reference sample.[14] Other normalization methods for microarrays include median subtraction, $z$-value transformation of log-transformed data,[39] and the quantile method[40]; the latter method is available as an option in ExAtlas. RNA-seq data is usually normalized by transformation to FPKM values, as these are assumed to be proportional to the numbers of mRNA molecules per cell.[36] However, this method of normalization creates difficulties for certain algorithms of statistical analysis. For example, ExAtlas uses error models for correcting error variance in ANOVA assuming that measurement error is inversely related to the level of gene expression (see Data Management in ExAtlas section later). This assumption does not hold for FPKM values, because short genes may have high FPKM values even if the number of matching sequence reads is small; this tends to cause high measurement errors. Thus, it is better to normalize fragment counts only by the total number of fragments mapped to the genome.

Gene expression values are always nonnegative, and their variance tends to increase with increasing average. Thus, log-transformation is commonly used before statistical analysis to equalize variances in genes with low and high expression.[14,39] However, there is a caveat because logarithm is undefined for negative values (e.g., if a microarray signal is below the background) or zero value (e.g., in RNA-seq data). Thus, such values need to be replaced with positive, low, random pseudo-values. Log-transformed data can be further processed by z-transformation[39]; however, information about the fold change in gene expression in this case is thereby lost. Thus, in ExAtlas, log-transformation of data is used as such for data preprocessing.

## OVERVIEW OF THE STATISTICAL ANALYSIS OF GENE EXPRESSION PROFILES IN EXATLAS

ExAtlas is designed primarily to facilitate global statistical comparison of various projects on gene expression profiling and gene set annotation. We designate as "global" the analysis of all combinations between data components in two projects and the presentation of the results in the form of a table/matrix. The analysis builds on standard

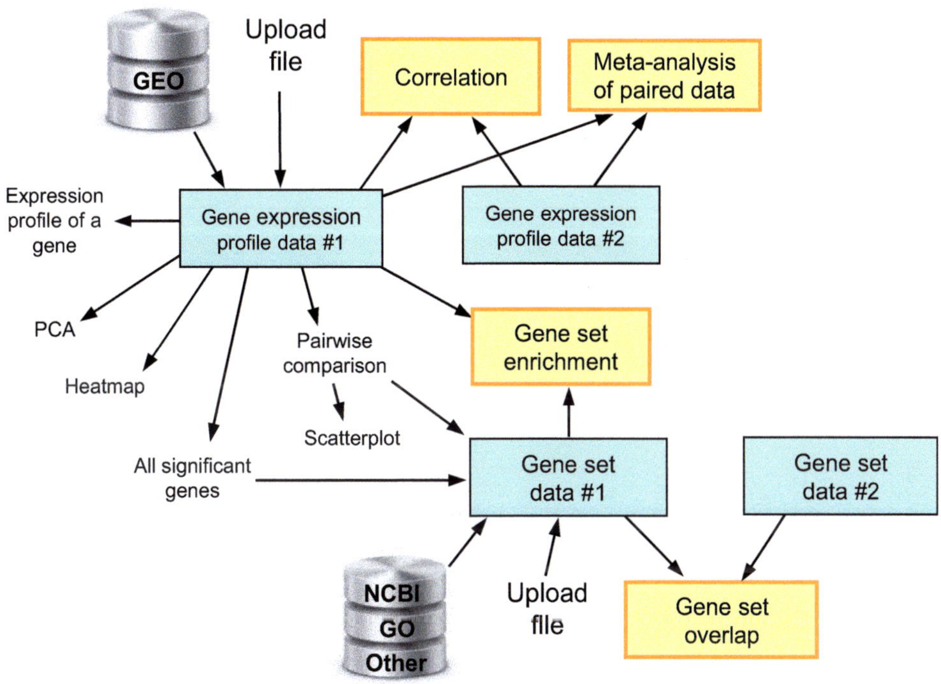


**Figure 4.1** The workflow in ExAtlas. Two types of data files (*blue boxes*)—gene expression profiles and gene sets—are uploaded or retrieved from online databases. Tools for comparison of two or more data sets (meta-analysis) are shown by *yellow boxes*.

statistical methods applied to each data set, such as ANOVA, pairwise comparison between tissues or cell types, PCA, plotting of heatmaps, scatterplots, and expression profiles for individual genes. These software components were adopted with minor modifications from NIA Array Analysis.[14] In addition, new components include tools for cross-project comparison, meta-analysis, as well as searching and uploading gene expression profiles from the GEO database (Fig. 4.1). Gene expression data can also be uploaded manually; in this case, the data must be formatted in a tab-delimited text file and normalized before uploading to ExAtlas.[1] Preloaded public data sets (Table 4.1) improve the functionality of the software because they save time on data search, retrieval, and preprocessing. ExAtlas has an online help feature that provides with step-by-step instructions and annotated screen captures.

Global correlation analysis compares gene expression profiles in two different data sets and generates a table (matrix) of correlation coefficients and their significance (Fig. 4.2).

[1] A few preprocessing tools are available in ExAtlas: quantile normalization and replacement of low expression values with random pseudo-values.

**Table 4.1** Public data sets preloaded in ExAtlas

| Data set name | Data source | Species | Data type | Method | References | PMID |
|---|---|---|---|---|---|---|
| BrainSpan Atlas | BrainSpan (brainspan.org) | Human | Gene expression | RNA-seq | 41 | 24695229 |
| BrainSpan Atlas adult | BrainSpan (brainspan.org) | Human | Gene expression | RNA-seq | 41 | 24695229 |
| GNFv2 human tissues | BioGPS (biogps.org) GEO:GSE1133-GPL96 | Human | Gene expression | Microarrays | 42 | 18442421 |
| Human body index | Neurocrine Biosciences, Inc. GEO:GSE7307 | Human | Gene expression | Microarrays | 43 | 21794127 |
| Human cells RNAseq ENCODE | ENCODE Caltech GEO:GSE23316 | Human | Gene expression | RNA-seq | 44 | 21037257 |
| Human tissues GSE3526 | Neurocrine Biosciences, Inc. GEO:GSE3526 | Human | Gene expression | Microarrays | 45 | 16572319 |
| RefEx CAGE FANTOM human tissues | Reference Expression dataset (RefEx http://refex.dbcls.jp) | Human | Gene expression | CAGE | 46 | 28850115 |
| IGC human tissues | International Genomics Consortium (intgen.org) GEO:GSE2109 | Human | Gene expression | Microarrays | | |
| Illumina body map v2 | EMBL European Bioinformatics Institute GEO: GSE30611 | Human | Gene expression | RNA-seq | | |

| | | | | | | |
|---|---|---|---|---|---|---|
| Knockdown (KD) 57 TFs GM19238 | Washington University GEO: GSE50588 | Human | Gene expression | Microarrays | 47 | 24603674 |
| KD 400 genes human | University of Southern Denmark GEO: GSE27871 | Human | Gene expression | Microarrays | 48 | 22121215 |
| GNFv2 mouse tissues | BioGPS (biogps.org) GSE1133-GPL1073 PMID:18442421 | Mouse | Gene expression | Microarrays | 42 | 18442421 |
| GNFv3 mouse tissues | BioGPS (biogps.org) GSE10246 PMID: 18442421 | Mouse | Gene expression | Microarrays | 42 | 18442421 |
| RefEx CAGE FANTOM mouse tissues | Reference Expression dataset (RefEx http://refex.dbcls.jp) | Mouse | Gene expression | CAGE | 46 | 28850115 |
| NIA induction 137 TFs mouse | National Institute on Aging GEO: GSE16375 | Mouse | Gene expression | Microarrays | 25,49 | 22355682, 19796622 |
| NIA induction 186 TFs mouse | National Institute on Aging GEO: GSE16375, GSE72350 | Mouse | Gene expression | Microarrays | 25,49,50 | 22355682, 27150017, 19796622 |
| NIA KD 100 TFs mouse | National Institute on Aging GEO: GSE26520 | Mouse | Gene expression | Microarrays | 51 | 23462645 |
| NIA preimplantation embryo | National Institute on Aging GEO: GSE936 | Mouse | Gene expression | Microarrays | 52 | 14723852 |
| Rat bodymap v2 | Fudan Univresity GEO:53960 | Rat | Gene expression | RNA-seq | 53 | 24510058 |

*Continued*

**Table 4.1** Public data sets preloaded in ExAtlas—cont'd

| Data set name | Data source | Species | Data type | Method | References | PMID |
|---|---|---|---|---|---|---|
| FlyAtlas *Drosophila* | University of Glasgow GSE:GSE7763 | Drosophila fly | Gene expression | Microarrays | 54 | 17534367 |
| Macaque brain GSE31613 | Allen Institute for Brain Science GEO: GSE31613 | Rhesus monkey | Gene expression | Microarrays | 55 | 22445337 |
| Dog tissues GSE20113-matrix | National Cancer Institute GEO: GSE20113 | Dog | Gene expression | Microarrays | 56 | 21655323 |
| Pig tissue GSE0898 | Universitat Autonoma Barcelona GEO: GSE10898 | Pig | Gene expression | Microarrays | 57 | 18416811 |
| Gene Ontology (GO) | GO Consortium (geneontology.org) | Many species | Gene set | Curated annotation | 58 | 25428369 |
| GAD MGI phenotypes | Genetic Association Database PMID: 20092628 | Human, Mouse | Gene set | Curated annotation | 59 | 20092628 |
| MSIGDB genesets | Broad Institute (broadinstitute.org/gsea) | Many species | Gene set | Curated annotation | 21 | 16199517 |
| KEGG pathways | Extracted from MSIGDB (see earlier) | Many species | Gene set | Curated annotation | same | same |
| BIOCARTA pathway | Extracted from MSIGDB (see earlier) | Many species | Gene set | Curated annotation | same | same |
| REACTOME pathways | Extracted from MSIGDB (see earlier) | Many species | Gene set | Curated annotation | same | same |

| | | | | | | |
|---|---|---|---|---|---|---|
| GNF2-coregulated | Extracted from MSIGDB (see earlier) | Many species | Gene set | Curated annotation | same | same |
| PID pathways | Extracted from MSIGDB (see earlier) | Many species | Gene set | Curated annotation | same | same |
| MIR targets | Extracted from MSIGDB (see earlier) | Many species | Gene set | Curated annotation | same | same |
| Protein interaction | BioGRID, physical interaction (thebiogrid.org) | Many species | Gene set | Curated annotation | 60 | 25428363 |
| TF Targets human ENCODE Q10 | encodedcc.sdsc.edu/ftp | Human | Gene set | ChIP-seq | 61 | 25164757 |
| TF Targets mouse | National Institute on Aging (compiled from literature) | Mouse | Gene set | ChIP-seq | 30 | 24918633 |
| StarBase v2 protein-RNA | Sun Yat-sen University (starbase.sysu.edu.cn) | Human, Mouse | Gene set | CLIP-seq | 62 | 21037263 |

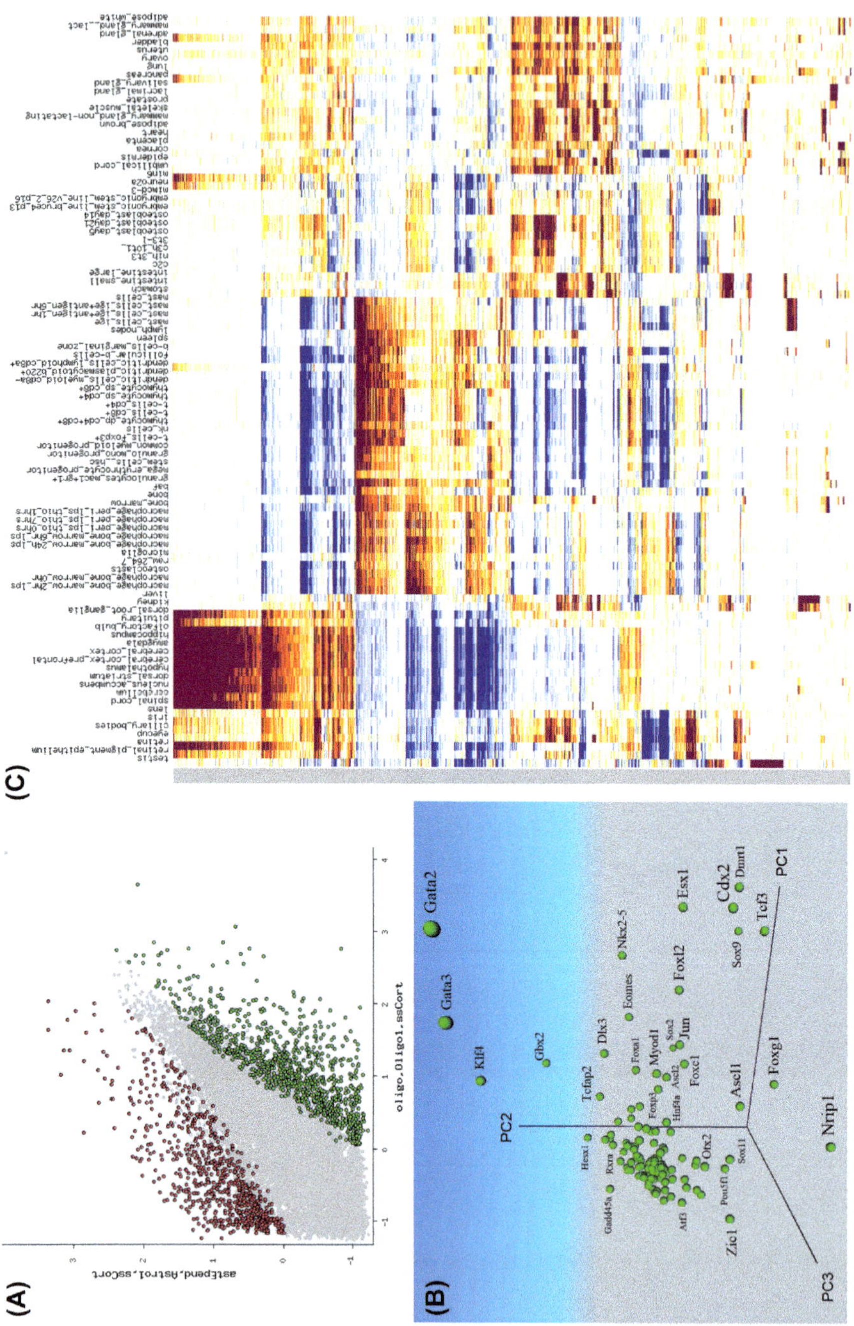


**Figure 4.2** Analysis of a single data set with gene expression profiles in ExAtlas. (A) Scatterplot of log-transformed gene expression in glial cell types: astrocytes and oligodendrocytes based on single-cell RNA-seq.[35] (B) Principal component analysis of gene expression change after induction of 137 transcription factors in mouse ESCs (VRML interactive format).[25] (C) Heatmap of tissue-specific gene expression in the GNF database: columns represent mouse tissues and organs, and rows correspond to individual genes; *red colors* of different intensity show increased gene expression, whereas *blue colors* of different intensity show decreased gene expression.

Also, histograms can be generated for individual rows or columns. For each pair of samples with correlated expression profiles, a set of coregulated genes is identified. Meta-analysis of paired gene expression profiles is another approach for integrating data from multiple sources. The goal is to find common differentially expressed genes in multiple pairs of expression profiles from different data sets or from the same data set. The advantage of this method is that it can detect significant change in gene expression even if no significance was found in comparisons of individual pairs of expression profiles.[73]

In addition to gene expression profiles, ExAtlas uses another important type of data file, sets of genes (or gene sets) united by some common features. For example, GO gene sets are characterized by common molecular function or localization in cells.[63] Several databases are available that characterize genes involved in signaling pathways: KEGG,[64] BIOCARTA (NCI/NIH), and REACTOME,[65] whereas the GAD database summarizes genes associated with mammalian phenotypes.[59] Gene sets are used for functional annotation of differentially expressed genes. For example, if a set of upregulated genes in a certain experiment or cell type largely overlaps (i.e., is significantly more correlated than expected by random) with a gene set of known genes associated with Transforming growth factor (TGF) beta signaling, then a hypothesis can be proposed for further testing that TGF beta was active in the experiment (or cell type). Instead of using a fixed cutoff for identifying differentially expressed genes (i.e., fixed *p-value*, FDR, and/or fold change), it is possible to use set enrichment analysis (GSEA), which takes advantage of the rank order of genes sorted by their expression change.[21] ExAtlas implements both approaches: analysis of gene set overlap and gene set enrichment (see Gene Set Overlap and Gene Set Enrichment section), which can be applied to all samples and all gene sets. A matrix of association between samples and gene sets is generated and displayed as a color-coded table. Also, histograms can be plotted for individual rows or columns. Additional graphic output is available to compare individual samples and gene sets: Venn diagram (for gene set overlap) and rank order plot (for gene set enrichment). Many public gene set databases are preloaded and ready for use in ExAtlas (Table 4.1). Users can upload their own gene set data or save differentially expressed genes as a new gene set file. For example, all gene expression profiles in a data set can be compared with a selected baseline profile (e.g., median gene expression in all tissues or cell types, or expression in control cell type), saving the differentially expressed genes as a new gene set file for further analysis.

## DATA MANAGEMENT IN EXATLAS

New data can be added to ExAtlas either via import from the GEO database or by manual upload of tab-delimited text files. Specific keywords (e.g., "kidney" or "heart"), authors, or accession numbers are used for searches, and information is returned on cognate series and samples. Samples from one or multiple GEO data series can be selected and saved in a user profile. Sample names can be edited for clarity and to specify replications. In particular, ExAtlas requires that replications have precisely identical names. Data files and annotations of microarray platforms are then retrieved from GEO automatically, and gene expression information for selected samples is extracted and compiled into a new file. If all selected samples utilize the same array platform, then gene expression data are stored for each probe of the array. Otherwise, data obtained with different array platforms are combined using official gene symbols. In this case, only one probe is used for each gene, which is selected based on the highest expression value. Gene symbols are verified using the RefSeq (NCBI) database. Extracted data are normalized by the quantile method.[40]

RNA-seq data cannot be retrieved from the GEO because there is no standard format for storing processed RNA-seq results. Processing of the raw RNA-seq data requires extensive computing and is not implemented in ExAtlas. Gene expression data (including RNA-seq) can be uploaded manually as a tab-delimited table, where column headers are sample names, and row headers (in the first column) are probe names, transcript names (e.g., GenBank or Ensembl accession numbers), or gene symbols.

Users can upload, download, delete, or edit all data files in their account: gene expression profiles, gene sets, microarray platform annotations, lists of GEO samples, and output files that carry the results of analyses. Data file name, description, and the names of sample or gene sets can be changed if necessary. Gene expression data and gene sets can be copied to a new file or appended to an existing file. Gene-specific batch-normalization can be used to combine two or more data sets with gene expression profiles. For example, if two data sets include the same tissue or cell type (e.g., liver or fibroblasts), then median expression levels for this common tissue or cell type are set equal in two data sets for each gene. The quality of gene expression data is assessed by several criteria, including expression of housekeeping genes and reproducibility of gene expression values in each sample. Samples with low quality can then be marked and removed from the data set. A list of genes can be uploaded for functional annotation and can then be saved as a gene set if needed. Any tab-delimited table can be uploaded as "output file" for generating a heatmap.

## ANALYSIS OF A SINGLE DATA SET WITH GENE EXPRESSION PROFILES

ANOVA is used in ExAtlas to evaluate the variability of log-transformed gene expression values between replications. First, ANOVA is applied to each individual probe (or gene), and then the error variance (also known as mean square error) is adjusted to the results obtained for other genes.[14] Error variance may appear very low by pure chance for some genes because the number of replications is often small and the number of genes/probes is very large. This can result in a large number of false positives, so error models are used to get a better estimate for the true error variance than the error variance estimated from data (we call this "empirical error variance"). Adjustment is done by estimating the average error variance, which is the value expected in genes expressed at a certain level. All genes are sorted based on the average log-expression in all samples, and then the average error variance is estimated in a sliding window of 500 genes. The top 1% of genes with highest empirical error variances are not counted in this calculation. Poorly expressed genes generally have higher error variances than highly expressed genes. However, their variation could be underestimated due to data processing (e.g., using cutoff thresholds, rounding, or pseudo-values). Thus in ExAtlas, the average error variance is forced to increase monotonically with decreasing expression of genes below the median. According to the default error model, adjusted error variance equals the maximum of the empirical error variance and the average error variance. Other error models are available when users select the option "Run ANOVA again." They include using empirical error variance (i.e., no adjustment), average error variance, Bayesian correction of error variance,[36] and the maximum between empirical error variance and its Bayesian correction.

ExAtlas can run ANOVA even for data sets with no replications (this option is not available in the earlier NIA Array Analysis tools). In this case, the error variance is estimated based on the half-normal probability plot method.[66] We assume that at least half of gene log-expression values represent random deviations from the median. Then the standard deviation, $\sigma$, of random effects can be approximated by the median absolute deviation (median of $|x_{i,j} - m_i|$, where $x_{i,j}$ is log-expression of gene $i$ in sample $j$, and $m_i$ is median log-expression of gene $i$), divided by 0.675, which is the median of an inverse cumulative standard half-normal distribution. The error variance in ANOVA is then set to $\sigma^2$. This method is applied to each of the 500 genes in a sliding window that is shifted across the set of all genes sorted by their average log-expression.

Adjusted error variance, $\sigma^2$, estimated by ANOVA can then be used for pairwise comparison of expression profiles in two tissues or cell types. To evaluate whether the difference between average expression values in two tissues is statistically significant, the $z$-value is estimated:

$$z = \frac{M_1 - M_2}{\sqrt{\sigma^2(1/n_1 + 1/n_2)}}, \tag{4.1}$$

where $M_i$ and $n_i$ are the mean value and sample size for cell type $i$. Then the $p$-value is

$$p = 2\left(1 - F^{-1}(z)\right), \tag{4.2}$$

where $F$ is the cumulative standard normal distribution. The use of $z$-values instead of Student $t$-value is justified by the error model adjustment, which is based on the data from many genes, so the number of degrees of freedom is large. Although $p$-values are suitable to evaluate the expression of individual genes, they cannot be used to identify all genes with changed expression because the task requires multiple hypothesis testing. For example, if a microarray has 60,000 probes, then c.3000 probes satisfy the condition $P < .05$ even if the samples compared are biologically identical (i.e., taken from the same cultured cells or tissues). To avoid such large numbers of false positives, it is necessary to account for multiple hypothesis testing, using FDRs instead of $p$-values. FDR is estimated as shown:

$$\mathrm{FDR}_i = \frac{p_i \cdot N}{r_i}, \tag{4.3}$$

where $p_i$ is a $p$-value for probe/gene $i$, $N$ is the total number of probes (or genes) in the microarray, and $r_i$ is the rank order of gene $i$ (gene with the highest $z$-value has $r_i = 1$).[67] In addition, $\mathrm{FDR}_i$ values are forced to increase monotonically with rank $r_i$. To compare gene expression profiles in two different tissues or cell types, a scatterplot of log-transformed expression values for all genes is generated in ExAtlas, showing sets of significantly upregulated and downregulated genes (Fig. 4.2A). The difference in the expression of a gene is considered significant if both conditions are true: FDR is smaller than the FDR threshold (0.05 as default) and the fold change is greater than the fold change threshold (2 as default).

Besides simple pairwise comparison of gene expression profiles, ExAtlas supports bulk comparison of gene expression profiles of all tissue (cells types) with a common baseline, which can be the median expression profile or any selected tissue or cell type (e.g., control). Criteria for differential expression include FDR, fold change, and specificity. The latter optional criterion is measured by a $z$-value that compares expression in a given tissue with the average expression in other tissues that are not correlated with the tested tissue. Tissues are considered correlated if the log-transformed multidimensional distance of their log-transformed expression profiles is $<1/3$ of the maximum distance between tissues measured by PCA. Specificity of a gene is significant if $z > 2$. For finding genes with high levels of specificity, users can select thresholds for $z$-values $>2$.

Heatmap and PCA are used to explore gene expression profiles of all tissues or cell types in a data set (Fig. 4.2B and C). A heatmap is a table (matrix) where the individual values contained are coded by color and intensity. Both methods are applied to genes with significant change of expression as defined by the choice of thresholds for FDR and fold change, measured as the maximum difference in expression values between tissues. For faster processing, users can select more stringent conditions (i.e., a lower FDR and/or higher fold change) to reduce the number of genes analyzed. By default, both the heatmap and PCA are applied to averages of log-transformed gene expression in each tissue, but there is an option to use individual replications for analysis. PCA determines major correlations for all gene expression profiles in a data set. Users can check the box "PC gene clustering" to identify genes that are correlated positively and negatively with each principal component. In the heatmap, both columns and rows are sorted by hierarchical clustering. However, if the number of genes is >2000, then rows (genes) are clustered by PCA: first by positive and negative values of each PC that exceed 1/4 of the maximum value, then by combinations of values in each pair of PCs (both >1/4 of the maximum value and the ratio between PCs < threefold). Residual genes that matched neither to individual PCs nor to their pairs are placed at the end of the heatmap and sorted by PC1.

## CORRELATION BETWEEN GENE EXPRESSION PROFILES

Correlation between gene expression profiles may be indicative of functional similarity between cells or tissues, so it can be used as a tool for functional annotation of transcriptome changes. For example, the function of various TFs can be inferred from the expression change of downstream genes after induction of these TFs in cultured cells. Previously, 137 transgenic lines of embryonic stem cells (ESCs) with inducible transgenic individual TFs integrated into the genome were developed.[25] Then the change of gene expression 2 days after the induction of each TF was compared using correlation analysis with tissue-specific gene expression profiles in the GNF database.[42] Later, we repeated this analysis in a slightly modified way with ExAtlas (Fig. 4.3). In this example, positive correlation indicates that the induction of a specific TF shifts the expression of genes toward levels that are typical for a specific tissue. Neural-specific genes were upregulated after the induction of *Ascl1*, *Sox9*, and *Foxg1*; endoderm-specific genes were upregulated after the induction of *Hnf4a*, *Gata2*, and *Gata3*; and muscle-specific genes were upregulated after the induction of *Myod1* or *Mef2c*.[25] The advantage of this correlation analysis as compared to alternative methods (e.g., gene set enrichment) is that it accounts for both upregulated and downregulated genes and does not require any arbitrarily chosen criteria for selection of tissue-specific gene sets.

Genes that do not show significant change of expression may add noise to correlation analyses. Thus, it is reasonable to analyze only those genes with significant change of expression in both data sets. ExAtlas estimates correlation as follows: first, the best probe is selected for each gene based on the highest value of $F$-statistic (ANOVA). Second, genes with significant change of expression are identified in each data set based on selected thresholds of FDR and fold change, and then genes that are significant in both data sets are used for correlation analysis (Fig. 4.3A). Third, gene expression change is estimated as a difference of log-transformed expression value in each tissue and median expression profile or other user-defined baseline. Finally, Pearson (or Spearman) correlation between gene expression changes in two data sets is estimated within the set of common significant genes (Fig. 4.3B). This procedure is repeated

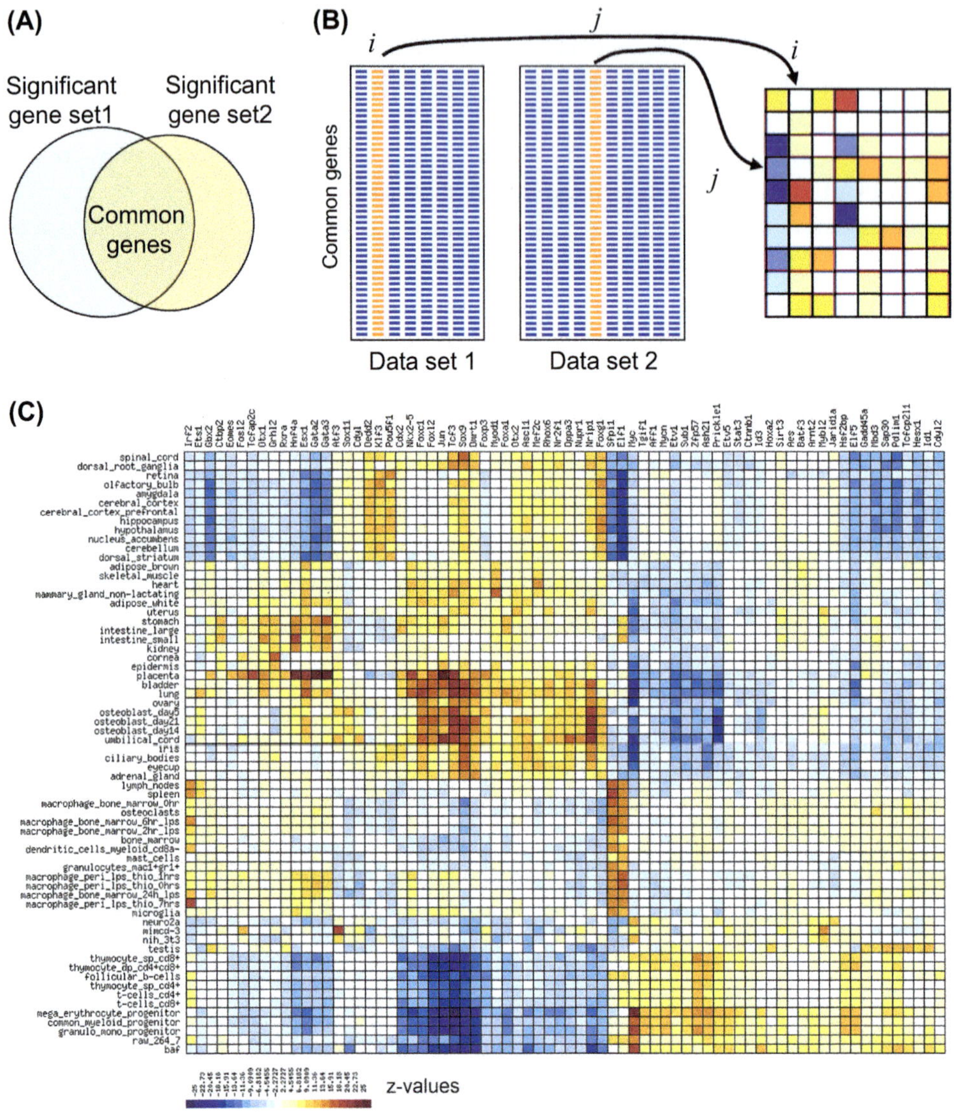


**Figure 4.3** Analysis of correlation between gene expression profiles in two data sets: gene expression change after induction of 137 individual transcription factors (TFs) in mouse ESCs[25] and tissue-specific gene expression in the GNF database.[42] (A) Correlation analysis is based on genes that show significant changes of expression in both data sets compared to the median. (B) Correlation is estimated for each column *i* (induced TF) in the first data with each column *j* (tissue) in the second data set, and the result is recorded in the cell (*i,j*) of the output table. (C) Color-coded table of *z*-values that shows statistically significant correlations (if $|z| > 2$); columns, induced TFs; rows, tissues or organs.

for each combination of tissues (or cell types) in both data sets until the table (matrix) of correlation values is filled. The significance of correlation is assessed by $z$-values: $z = r/\text{sqrt}[(1 - r^2)/(n - 2)]$, where $r$ is the correlation and $n$ is the number of genes used for estimation. By default, a color-coded table of $z$-values is displayed for the analysis of correlation patterns (Fig. 4.3C). The tables of correlation values and the number of coregulated genes (see later) can also be plotted.

If the correlation between gene expression profiles is positive and significant ($P \leq .05$), then we expect that the number of genes upregulated in both profiles or downregulated in both profiles (called "coregulated" genes) is larger than expected in pairs of noncorrelated gene expression profiles. To identify coregulated genes and evaluate their statistical significance, users check a box "Identify coregulated genes" before doing correlation analysis. Statistical significance is evaluated by EPFP,[30] which is similar to FDR but not estimated from $p$-values. The algorithm for finding coregulated genes is based on the analysis of data points in the positive quadrant (i.e., $x > 0$ and $y > 0$) for upregulated genes, and negative quadrant (i.e., $x < 0$ and $y < 0$) for downregulated genes, where $x$ and $y$ are expression values of a gene (i.e., log ratio vs. median or other baseline) in samples compared from data sets #1 and #2, respectively. Because the processing of upregulated and downregulated genes is identical except for the direction of expression change, we describe the algorithm only for the upregulated genes. First, gene expression change values are replaced by their rank order in the positive quadrant. If the null hypothesis holds (i.e., no correlation), then genes are expected to have a uniform random distribution (Fig. 4.4). However, if gene expression profiles are positively correlated, then more genes/points can be found near the upper right corner of the plot (Fig. 4.4B). To estimate EPFP($rx$,$ry$) for a gene with rank orders $rx$ and $ry$ in the first and second expression profile, respectively, the density of dots/genes is estimated in a dark-shaded rectangle in Fig. 4.4B, and compared to the density of dots in two adjacent light-shaded rectangles. The EPFP equals the density of dots in the light-shaded area divided by the density of dots in the dark-shaded area. Because we have two light-shaded rectangles, EPFP is estimated twice, and we then select the largest value (to be conservative in assessments). Because EPFP may not monotonically decrease with the increasing rank of genes, it is forced to decrease monotonically. In particular, if EPFP($rx_1$,$ry_1$) > EPFP($rx$,$ry$), $rx_1 > rx$, and $ry_1 > ry$, then EPFP($rx_1$,$ry_1$) is set equal to EPFP($rx$,$ry$). All genes with EPFP below a selected threshold are considered coregulated (Fig. 4.4). To identify oppositely coregulated genes (i.e., upregulated genes in file #1 that are downregulated in file #2), the direction of change in file 2 is set to "Reversed." The gene expression change in file 2 is thus inverted (i.e., multiplied by $-1$), and oppositely coregulated genes can be identified in the same way as shown in Fig. 4.4.

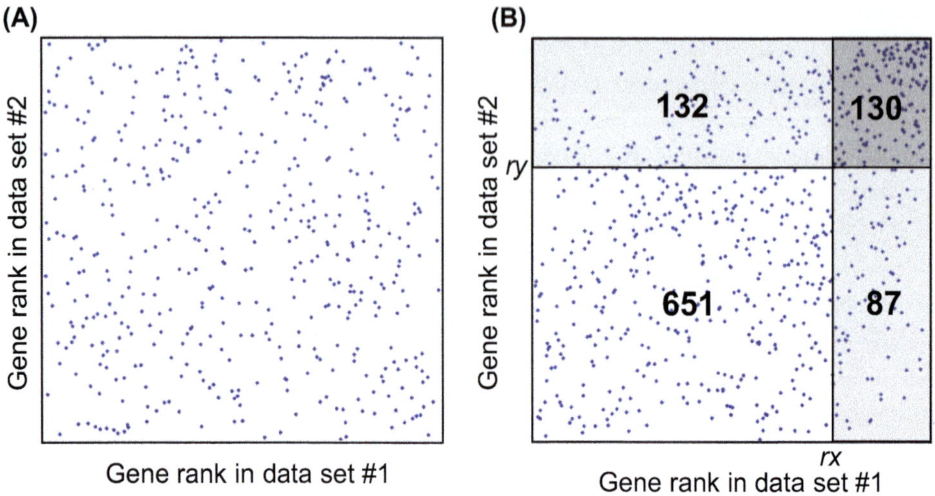


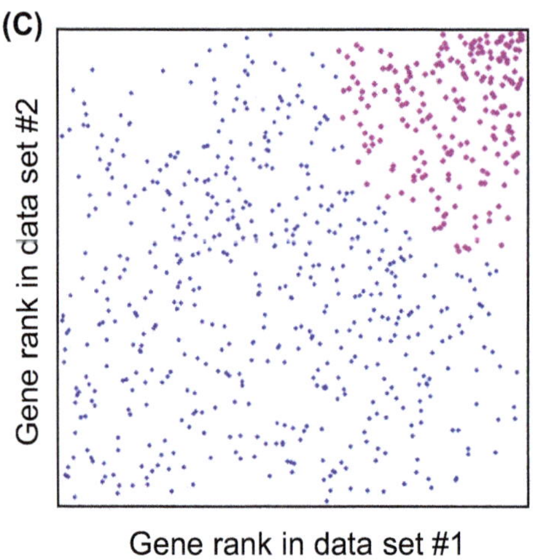


**Figure 4.4** Estimating the expected proportion of false positives (EPFP) for positively coregulated genes: (A) scatterplot of gene expression ranks in the positive quadrant if there is no correlation. (B) The same plot if gene expression profiles are correlated; numbers indicate gene counts in each area. The density of dots/genes in the *dark-shaded rectangle* is compared to the density of dots in two *light-shaded rectangles* used as baseline, generating two estimates of EPFP for a gene at the lower left corner of the *dark-shaded rectangle* (with expression rank orders $rx = 651 + 132 = 783$ and $ry = 651 + 87 = 738$): $\text{EPFP}_1 = 0.281$ and $\text{EPFP}_2 = 0.237$. The greater value is then selected: EPFP = 0.281. (C) All coregulated genes with EPFP $\leq 0.3$ are highlighted in *magenta*.

## GENE SET OVERLAP AND GENE SET ENRICHMENT

Gene set overlap and gene set enrichment analyses are used to evaluate if specific gene sets (such as GO or KEGG) are overrepresented among upregulated and/or downregulated genes. For example, if a set of upregulated genes, $S_1$, is identified using threshold values of FDR and/or fold change, then we can test if this set is enriched in genes that belong to a set of annotated genes, $S_2$. The proportion of genes from set $S_2$ in the genome genes equals $p = n_2/N$, where $n_2$ and $N$ are the numbers of genes in $S_2$ and in the whole genome, respectively. The proportion of genes from set $S_2$ within set $S_1$ is equal $q = m/n_1$, where $m$ is the number of common genes in $S_1$ and $S_2$, and $n_1$ is the number of genes in $S_1$. Genes that are common in sets $S_1$ and $S_2$ belong to the overlap of these two sets (i.e., $S_1 \cap S_2$); thus, the goal of the analysis is to check if the overlap can (or cannot) be explained by a random draw of genes. In particular, we can say that genes from set $S_2$ are enriched in the set $S_1$ if the overlap of genes in sets $S_1$ and $S_2$ is greater than can be explained by random choice. Assuming that the number of common genes has a hypergeometric probability distribution, the $z$-value for enrichment is

$$z = \frac{q - p}{\sqrt{p \cdot (1 - p) \cdot (N - n_1)/n_1/(N - 1)}}. \tag{4.4}$$

The $p$-value and FDR are then estimated from $z$-values using Eqs. (4.2) and (4.3), and a Venn diagram is plotted to visually inspect the degree of gene set overlap.

Gene set enrichment analysis is similar to the method of gene set overlap, but it has more statistical power because it does not require defining parameters for delineating sets of differentially expressed genes. In particular, gene set enrichment analysis can find significant associations with functional gene sets even if there are no significantly upregulated genes based on standard criteria (e.g., FDR $\leq 0.05$ and fold change $\geq 2$). Gene set enrichment analysis uses the order of genes sorted by the change of their expression to identify the best threshold to delineate the set of differentially expressed genes. Among various approaches to gene set enrichment analysis, we selected Parametric Analysis of Gene Enrichment (PAGE)[68] because of its simplicity and reliability.[59] The algorithm of PAGE determines whether the mean log-expression change, $x_{set}$, in genes that belong to a functionally annotated gene set $S$ is significantly greater than expected from the mean and standard deviation of log-expression change in all genes ($x_{all}$ and $SD_{all}$, respectively). The z-value for testing the null hypothesis is

$$z = \frac{(x_{set} - x_{all}) \cdot \sqrt{n_{set}}}{SD_{all}}, \tag{4.5}$$

where $n_{set}$ is the number of genes in set $S$. In ExAtlas, this method is modified by applying Eq. (4.5) to the subset of $n$ top upregulated genes and another subset of $n$ top downregulated genes rather than to all genes. The number of genes $n$ can be selected by the user (as a default, $n$ is equal to 1/4 of all genes). Based on this modification, some gene sets (e.g., genes with H3K27me3 histone methylation in their promoters) appear significantly enriched among both upregulated and downregulated genes.[69] We used this approach to generate a matrix of enriched GO terms among upregulated genes after induction of individual TFs in mouse ESCs.[25]

Rank order plot (or rank-plot) is used to visualize the results of gene set enrichment analysis (Fig. 4.5A and B). Genes are sorted according to their expression change from

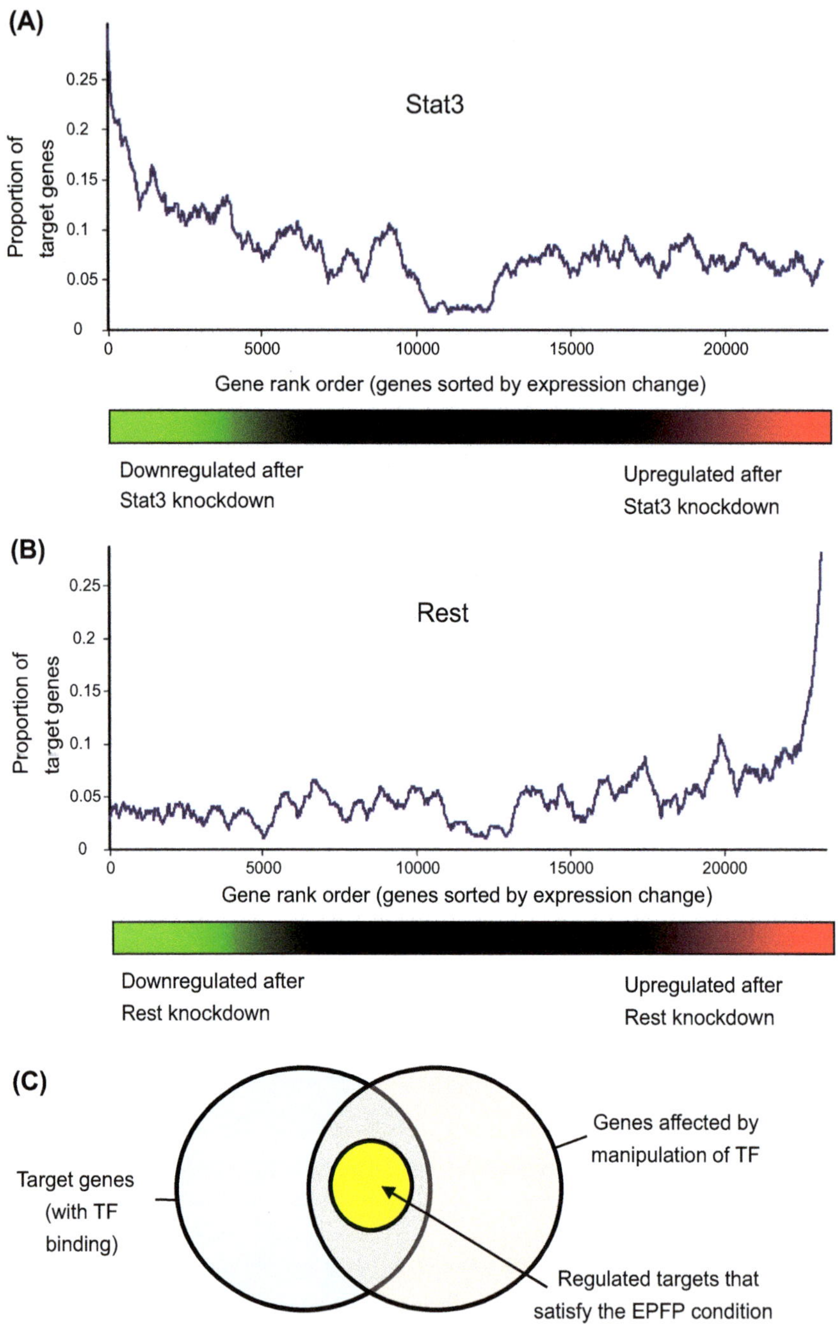


**Figure 4.5** Gene set enrichment analysis of regulated targets of transcription factors (TFs). Rank plots of the proportion of target genes (with TF binding) among genes that were downregulated or upregulated after the knockdown of TFs *Stat3* (A) and *Rest* (B). All genes were sorted by the change of their expression, and the proportion of target genes was estimated in a sliding window of 300 genes. The increased proportion of targets is observed in downregulated genes in (A) and in upregulated genes in (B), because *Stat3* is an activator and *Rest* is a repressor. (C) Regulated targets of TFs represent a subset of genes within the overlapping sector of TF targets and genes affected by manipulation of a TF.

downregulated at the left to upregulated at the right, and then the proportion of genes from the set of annotated genes, *S*, is estimated in a sliding window of 300 genes. Fig. 4.5A and B shows two examples of enrichment of TF targets among upregulated and downregulated genes in mouse ESCs after the knockdown (KD) of these TFs with shRNA.[51] Sets of target genes of individual TFs include all genes with TF binding sites in their promoters and enhancers, based on published ChIP-seq experiments. The rank-plot for *Stat3*, which is a major TF controlling pluripotency in mouse ESCs, shows enrichment of targets among genes that were downregulated after the KD of *Stat3* (on the left in Fig. 4.5A), which is consistent with the activating role of this TF. In contrast, the rank-plot for a repressor of neural differentiation, *Rest*, shows enrichment of targets among upregulated genes (on the right in Fig. 4.5B).

If gene set enrichment is statistically significant among upregulated genes ($P \leq .05$), then we know that some genes in the annotated gene set are more abundant among upregulated genes than expected by random. However, PAGE does not provide information on the EPFP in these genes. Most studies of downstream effects of TFs assume that the set of regulated target genes is the intersection of two subsets: a subset of all target genes and another of genes that changed their expression following manipulation of a TF (Fig. 4.5C). However, the intersection may include a large number of false positive genes that changed their expression not because of TF binding to promoters or enhancers but from other causes (e.g., second-order effects, compensation, etc.). Thus, a smaller subset of genes within the intersection needs to be delineated to bring EPFP below an acceptable threshold (Fig. 4.5C). To identify enriched genes and evaluate their statistical significance by EPFP, users check a box "Identify associated genes" before doing gene set enrichment analysis. EPFP is calculated sequentially for different thresholds of gene expression change as a ratio of the proportion of targets among "control" genes (i.e., not affected by TF manipulation) to the proportion of targets among genes that responded to TF manipulation. The smallest fold change threshold that yields EPFP below the accepted level (e.g., 0.3), is then used to delineate the set of regulated targets of TFs.

Methods of gene set analyses described earlier assume that all genes in a set are equally weighted. However, experimental and mining methods used for generating gene sets often provide additional quantitative information for individual genes. Genes in a set may differ by $z$-score, $p$-value, expression change, number of sequence reads in ChIP-seq peaks, or in HITS-CLIP assay, and other criteria that may appear useful for the analysis of gene set enrichment. Thus, an option to handle gene sets with additional quantitative attributes was added to ExAtlas. If the box "Use attributes" is checked, then genes with low scores are transiently excluded from the analysis for varying score thresholds until the highest statistical significance ($z$-value) is achieved. This approach is applied both to gene set enrichment (PAGE) and to gene set overlap (hypergeometric test).

The advantage of ExAtlas in the analysis of gene set enrichment and overlap compared to other software packages is that all expression profiles in a data set are

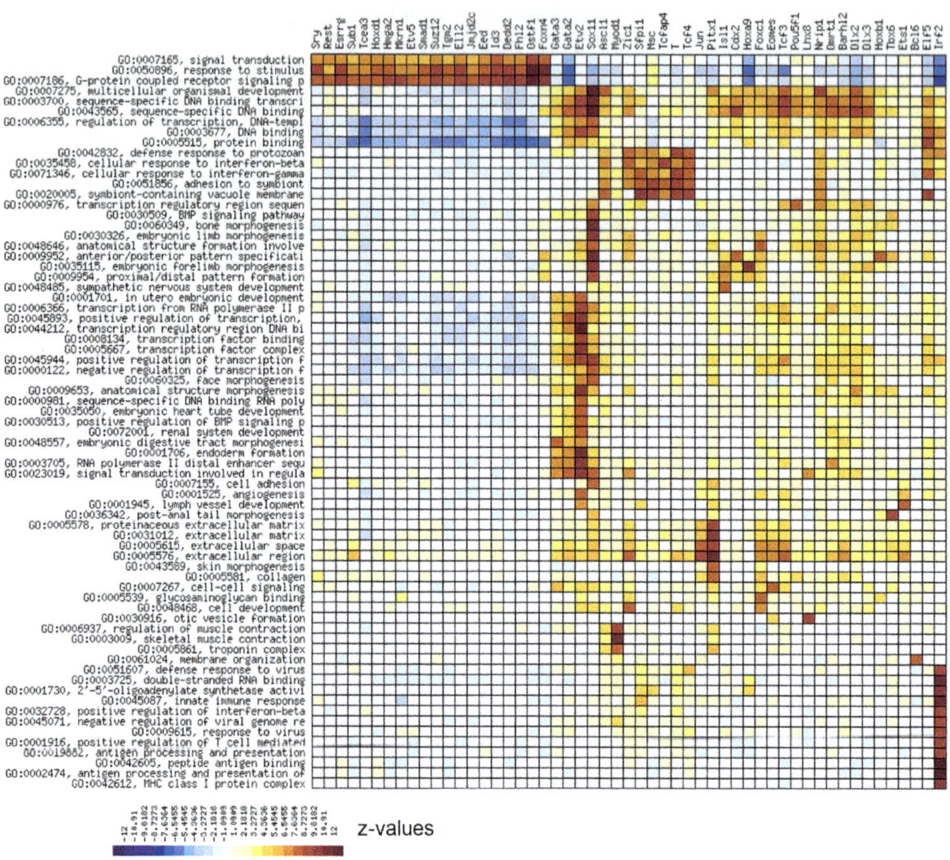


**Figure 4.6** Gene set enrichment scores (PAGE) for Gene Ontology (GO) annotated sets of genes among upregulated genes in mouse ESCs after induction of 137 individual transcription factors (TFs).[25] Columns, induced TFs; rows, GO annotations.

analyzed in one step, and the results are displayed as a table with columns corresponding to gene expression profiles in various tissues or cell types and rows corresponding to annotated gene sets. These tables are plotted with color-coded values to simplify visual inspection of associations between gene expression profiles and annotated gene sets (Fig. 4.6).

## META-ANALYSIS OF PAIRED GENE EXPRESSION PROFILES

The goal of meta-analysis is to use information from multiple independent studies to increase the power of statistical inference and reduce the number of false positives. ExAtlas applies four frequently used methods of meta-analysis: Fisher's, $z$-score, fixed effect, and random effect to evaluate the difference in gene expression between two tissues or cell types. The first three methods assume that all studies implement exactly the same methodology (e.g., same cell lines, equipment, and reagents). In practice, however, methods often do not match between experiments, so the random effect seems to be most relevant method. Fisher's method combines log-transformed $p$-values from $m$ studies and generates a chi-square statistic with $2m$ degrees of freedom.[70] The $z$-score method combines $z$-scores (i.e., the ratio of mean effect to its $SD$) of different studies with weights equal to square root of sample size.[71] The fixed effect method estimates a weighted sum of effects (i.e., log ratio of gene expression change), where weights are equal to $1/\sigma^2$ with $\sigma^2$ as the error variance. The random effect method takes into account the variance of heterogeneity between studies, $T^2$, which is added to the variance of individual effects.[72]

Meta-analysis in ExAtlas starts by selecting gene expression data to be combined. Pairs of tissues (or cell types) are compiled from different data files or the same file. It is even possible to compile gene expression data from different species, and in that case, gene orthologs are compared. A list of all expression profile pairs used in meta-analysis can be saved for future retrieval and modification. The analysis is done using all four methods, and a table with the number of significantly overexpressed and underexpressed genes for each method is displayed. Differentially expressed genes can be saved as gene sets for functional annotation.

## EXERCISES (TUTORIAL)

### Starting an ExAtlas Session and Accessing Gene Expression Data

Open ExAtlas website https://lgsun.grc.nia.nih.gov/exatlas in your browser. For online help, follow the hyperlink "A guide to ExAtlas." Click button "START USING EXATLAS" to log in as a guest. Alternatively, you can register to create a personal account where you can keep your data and results of analysis. In the next screen, use the top pull-down menu (i.e., File type = "Expression profiles") to select the data set "public-GNFv3_mouse_tissues," which is a public data set on tissue-specific gene expression profiles in mice obtained by Novartis Foundation.[42] If you cannot find this data set, check species name. If it is not mouse, select mouse as a species. After selecting the data set, click corresponding button "Open" and the data set will appear in a new tab. It is recommended to set browser options to allow pop-up messages and open new pages as additional tabs rather than as new windows. In Section 3 of the web page named "Find genes," enter gene symbol "Rest," check the box "Sort expression values," and click the "Search" button to display a histogram of expression values of gene *Rest* in different mouse tissues. *Rest* is a TF that represses neural-related genes. Thus, it has very low expression in all neural tissues (on the right side of the histogram). The highest expression is seen in ESCs on the left side. Expression values are log-transformed ($\log_{10}$) signal intensities in microarrays. When done, you can close the tab with the histogram.

## Comparing Gene Expression Values in Two Tissues

Return to the browser tab where "public- GNFv3_mouse_tissues" data are open. In Section 2 of the web page named "Pairwise comparison," select data "adipose brown" (i.e., brown fat) and compare it with "adipose white" (i.e., white fat). Then click the button "Scatter-plot" to open the graph in the new tab. Each dot represents one gene; red dots are genes overexpressed in the brown fat and green dots are genes underexpressed in brown fat as compared to white fat. Click on individual points to display a gene and explore its expression profile. At the top of the screen, you can find hyperlinks to the lists of overexpressed genes and underexpressed genes. Click on the hyperlink to open a list of overexpressed genes. The table shows gene symbols, gene names, and expression statistics. Probe names are hyperlinks to expression profiles of corresponding genes. At the top of the page, find a hyperlink "Table as tab-delimited text," which can be used to download the list of genes as plain text. Later, you can open this file in Excel for further analysis. Text output is provided in a similar way for other types of analysis in ExAtlas.

A list of significant genes can be saved as a new gene set for future comparison with other data sets: at the bottom of the web page, fill in gene set name and description, then select a file where to save or chose "New file" and click on the button "Save genes." Return to the top of the web page and find a tool for functional annotation of the list of genes, which starts with "Use gene set" pull-down menu. The default gene set is GO annotations file named "public-GO_geneset_10090." The number at the end of file name is a species ID for mouse in the NCBI database. To find GO annotations enriched in a given list of genes, click on the button "Overlap analysis." On the next screen, click on the button "Get profile." A histogram will appear showing top significant GO annotations, among which "mitochondrion" has the highest $z$-score. This is expected because brown fat cells are enriched with mitochondria.

## Plotting Gene Expression Heatmap

A heatmap displays gene expression profiles in all tissues (or cell types) in a data set, and in this respect, it is a more powerful graphic tool than a scatterplot, which compares only two tissues. Before using the button "Make heatmap," review the parameters of analysis. Heatmap includes only genes that show statistically significant change in expression (FDR $\leq 0.05$ and expression change $\geq$ twofold, as default). The number of significantly changed genes (18,700 in our case) can be found below the description of the data set. The larger the number of significant genes, the more time is required to plot the heatmap. Thus, consider reducing the number of genes by selecting a higher threshold of fold change $= 4$. Then, click the button "Make heatmap" and wait for the appearance of the heatmap table.

## Principal Component Analysis

Before using the button "PCA," review the parameters of analysis. Similar to heatmap, PCA includes only genes that show statistically significant change in expression. To see genes associated with each principal component, check the box "PC gene clusters." To analyze individual replications, check the box "Show replications." After clicking the button "PCA," wait for the analysis to finish, and then explore the graph. To see PCA in three dimensions, install VRML viewer FreeWRL (http://freewrl.sourceforge.net) and configure the browser to associate ".wrl" files with the VRML browser. Click on graphs of gene clusters associated with principal components to see lists of genes. Then use GO annotations to explore biological meaning of individual principal components.

## Uploading Data From GEO Database

From the main menu, click "Find samples in GEO," and on the next screen, type in comma-separated search terms "neuron,astrocyte,glia." Also enter the accession number of array platform GPL1261 because it is better to analyze data obtained with the same platform. Click on the button "Search," and then explore multiple pages of results. Use buttons to go to the next page or previous page. Check samples from data series GSE9566, GSE42607, and GSE17784, and then save these sample information in a file "samples_neuron_astrocyte_glia." Change sample names so that they match exactly in replications (case-sensitive). For example, rename "Callosal_Neurons_E18_rep1_430 2.0 array" as "Callosal_Neurons_E18." If some sample names are not informative, click on the sample name to open the sample description in the GEO database; then modify the sample name to make it clear. After editing sample names, click the button "Refresh list" to make sure that all changes are saved. Type in the name for gene expression data set to be generated (e.g., "matrix_neuron_astrocyte_glia") and click "Generate matrix." A waiting screen will appear, where you can check the status of your task. Click on the "Log file" hyperlink to see the progression of data extraction and normalization. Refresh the screen after some time until you see line "Task finished" at the end of the log file. Then, close the log file tab and click "Check task" in the waiting screen. Examine the heatmap and PCA of the loaded data set. Return to the main menu and click the "Refresh" button; then make sure that the new file with gene expression data appears in the pull-down menu.

## Upload a Custom File With Gene Expression

Open GEO database (https://www.ncbi.nlm.nih.gov/geo) and search for GSE17784 series. Open GSE17784 series page click on "Series matrix file(s)." Download file GSE17784-GPL1261_series_matrix.txt.gz and unzip it using gzip utility (gzip can be downloaded from ftp.gnu.org/gnu/gzip). Open the file in Excel, copy line 32 (!Sample title) to line 71 (ID_REF), delete lines 1–70. Edit column headers: remove "_rep1_430 2.0 array" (hint: use "Replace" tool inExcel) and do the same for "rep2_430 2.0 array" and "rep3_430 2.0 array." Remove the last line "!Series_matrix_table_end," and save the file as tab-delimited text "neurons_GSE17784.txt." In ExAtlas main menu, make sure that the species name is mouse, and then click "Upload data file." Click "Browse," select file "neurons_GSE17784.txt" and then select platform GPL1261 in the pull-down menu below. Because the data are log2-transformed, select transformation "log2." Then click "Upload file." After the data are uploaded, file name "neurons_GSE17784.txt" will appear in the pull-down menu for expression profile data. Open this file and make a heatmap to see gene expression patterns. Check the box "Show replications" and do PCA to examine how well do replications match each other. Click "Data quality" and examine results. Sample "Corticospinal_Motor_Neurons_P14," replication 2, shows poor correlation with expression of housekeeping genes (top histogram). The same sample has high global standard deviation: S.D.(glob) in the table. To remove this low-quality sample, check the corresponding box and click "Remove checked probes." Reopen file "neurons_GSE17784.txt" and click button "Best probe" at the bottom to remove redundant probes. The number of probes is then reduced from 45,036 to 21,487. Besides gene expression profiles, you can upload other types of data: gene sets, lists of GEO samples, lists of gene symbols, output tables, and array platform annotations. For these types of data you do not need to select array platform or transformation.

## Finding Genes Significantly Upregulated and Downregulated

In the main menu, change the species to Human (*Homo sapiens*). Open gene expression dataset "public-BrainSpan_Atlas_adult" and click button "Significant genes…". If you want to find genes that are not only significantly different from the median expression (or other tissue/cells) but also specific among other samples, then change parameter "Specific gene filter" from 0 to 3. Edit the name of output gene set file as "BrainSpan_Atlas_adult-signif2_spec3" and click "Save significant genes." The results page displays a histogram showing the number of specific upregulated (orange) and downregulated (blue) genes. The largest sets of specific genes are found for cerebellum followed by medulla, striatum (STR), and hippocampus. To see lists of genes, select a gene set (e.g., "CBC_F_8-15yrs_up") and click the button "Display genes."

## Correlation Between Sets of Gene Expression Profiles

This task is designed to evaluate correlations between any gene expression data sets, including data sets that belong to different species. As an example, we consider the data on gene expression change in mouse ESCs 2 days after the induction of individual

TFs. In ExAtlas, change the species to Mouse and open the data set "public-NIA_induction_186TFs_mouse." It is interesting to see if gene expression changes are directed toward expression profiles in specific tissue or cell types. Thus, analyze correlations with the tissue-specific gene expression profiles in GNF database: click the button "Correlations..." and select File 2 "public-GNFv3_mouse_tissues." Check the box "Identify coregulated genes," and change "Fold change threshold" for File 1 from 2 to 1.5. Expression change 2 days after induction of TFs is not very strong, so we would like to account for genes with expression change below twofold. Click "Estimate correlation matrix" and monitor the progression of calculations in the log file. When finished, click on the button "Plot output table." The color-coded table shows z-values for correlation between each combination of transcription profiles in two data sets. To make colors brighter, change the "Maximum value" to 20 (at the bottom of the screen) and click "Re-plot the matrix." Close the table and explore correlation for a specific TF *Ascl1*. In Section 2 of the web page "Profiles of rows, columns, and cells," select column "Ascl1," check the box "Add graphics," and then click "Get profile." The histogram will appear that shows tissues with highest correlation to the effects of *Ascl1*. Many of them are neural-related tissues (dorsal root ganglion, hypothalamus, spinal cord). In the table below the histogram, click "hypothalamus" and examine the list of coregulated genes that are upregulated both in hypothalamus and after induction of *Ascl1*. The scatterplots at the bottom of the page show coregulated genes as magenta dots.

## Gene Set Enrichment Analysis

Open the data set "public-NIA_knockdown_100TFs_mouse" that represents gene expression change in mouse ESCs 3 days after repression of individual TFs by shRNA. Gene set enrichment analysis can be used to determine if upregulated and downregulated genes in this data set are enriched in targets of TFs (targets are genes with TF binding to promoter or enhancer). Click the button "Gene set analysis..." and select gene set file "public-TF_targets_mouse_ESC" as File 2 for analysis. Check the box "Identify associated genes" and click the button "Gene set analysis." In the new screen, plot the output table for upregulated genes ("Select data type to plot" is set to "z-value_upregulated"). The table shows that targets of *Rest* are enriched in genes upregulated after the KD of *Rest*, which is consistent with known repressive function of this TF. Close the table, set "Select data type to plot" to "z-value_downregulated" and click "Plot output table." This table indicates that targets of pluripotency-related factors *Pou5f1* (*Oct4*), *Sox2*, and *Stat3* are enriched in genes that were downregulated after the KD of these factors. To see targets of *Stat3* that are positively regulated by *Stat3*, close the table, select data type "z-value_downregulated" in Section 2, select column "Stat3," check the box "Add graphics," and click the button "Get profile." Below the histogram, find

"Stat3_targets_ESC_15k" at the top of the table. Click on this hyperlink to see the list of genes. At the bottom of the page below the table of genes, find the rank-plot for *Stat3*, which clearly shows the enrichment of *Stat3* targets among downregulated genes (at the left).

## Significance of Gene Set Overlap

In this example, we use GO gene sets to annotate genes that were upregulated or downregulated in certain regions of the brain (see section 10.7). From the main menu in ExAtlas, change species to Human, and open previously saved gene set "BrainSpan_Atlas_adult-signif2_spec3." If you do not see it in the pull-down list, click the "Refresh" button. Use Section 2 "Analyze overlap with another gene set file," and click the "Overlap analysis" button. Monitor task progression in the log file. Then click button "Check your task," and in the next screen, click the button "Plot output table." GO categories associated with visual learning, locomotory behavior, learning, and memory are enriched in genes upregulated in female STR. Gene sets "Neural crest cell migration" and "synaptic transmission, cholinergic" are enriched in medulla-specific genes. Interestingly, some samples are enriched in genes associated with "response to amphetamine" and "response to ethanol," which indicates that samples were taken from drug- and alcohol-affected individuals. Close the table, then select "Down-regulation" in Section 1 and plot the table for downregulated genes. Interestingly, the previously found alcohol-related sample shows downregulation of genes associated with synaptic function and learning.

## Meta-analysis of Pairwise Comparison

Open gene expression data set "public-GNFv2_human_tissues"; in Section 2 (Pairwise comparison), select data "Spinal_cord" and compare it with "Whole_brain." The goal is to find genes that are differentially expressed in spinal cord versus brain in multiple data sets. Thus, we click the button "Meta-analysis" and try finding comparable samples in other data sets. In Section 4 of the web page, entitled "Add a pair of expression profiles," select file "public-Human_tissues_GSE3526," then select data "spinal_cord" and compare it to "frontal lobe". Click the button "Add data." The next data set is "public-IGC_human_tissues," where we take "Spinal cord" compared to "Brain." Finally, we can add a pair of tissues from mice: first we change species to mouse (Section 4 of the web page), select file "public-GNFv3_mouse_tissues," data "spinal_cord" to be compared with "cerebral_cortex." Now, in Section 2 of the web page, click the button "Start analysis." The results appear in a new tab as a table that shows the number of genes that are consistently upregulated and downregulated in the spinal cord as compared to the brain in all pairs of data based on four commonly used methods of meta-analysis. Click on the number of upregulated genes detected by random effect method (it usually gives the most conservative results). The table shows statistical results for each gene in the table and log ratio of expression in each data pair. Note that the third data pair (IGC_human_tissues) has many negative log ratio values (outliers), which possibly indicates the lower quality of samples. GO annotations enriched in spinal cord genes include many immune-related gene sets because the brain is an immune-privileged organ. The same analysis of genes upregulated in the brain shows the enrichment of GO gene sets associated with synapses, GABA signaling, learning, and memory.

## CONCLUSIONS

ExAtlas software offers a suite of statistical tools for gene expression analysis and visualization. The main feature of ExAtlas compared to other software is the capacity to compare multicomponent data sets in one run and generate results for all combinations as a table/matrix. Data from the GEO/NCBI database can be searched, downloaded, and used immediately for statistical analysis. Another advantage is that gene expression analysis is seamlessly integrated with gene set enrichment tests. Thus, there is no need to switch between different software tools and reformat data for compatibility. ExAtlas does not require any programming or database management skills, and it can therefore be used by biologists with little or no expertise in bioinformatics.

## ACKNOWLEDGMENTS

We thank Dr. M.S.H. Ko (Keio University, School of Medicine, Tokyo) for initiating projects on manipulation of transcription factors in mouse ESCs, which facilitated the development of ExAtlas. We thank Dawood Dudekula, Charles Weber, and Valere Binet for web server configuration and maintenance. Also, we thank users of ExAtlas who provided important feedback on the functionality of various components. This research was supported entirely by the Intramural Research Program of the National Institutes of Health, National Institute on Aging.

## REFERENCES

1. Slotnick SD. *Cognitive Neuroscience of Memory*. Cambridge, New York: Cambridge University Press; 2017.
2. Hof PR, Mobbs CV, eds. *Handbook of the Neuroscience of Aging*. Amsterdam, Boston: Elsevier; 2009.
3. Crow M, Paul A, Ballouz S, Huang ZJ, Gillis J. Exploiting single-cell expression to characterize co-expression replicability. *Genome Biol.* 2016;17:101.
4. Ganglberger F, Kaczanowska J, Penninger JM, Hess A, Buhler K, Haubensak W. Predicting functional neuroanatomical maps from fusing brain networks with genetic information. *Neuroimage*. 2017. https://doi.org/10.1016/j.neuroimage.2017.08.070.
5. Maussion G, Diallo AB, Gigek CO, et al. Investigation of genes important in neurodevelopment disorders in adult human brain. *Hum Genet.* 2015;134(10):1037–1053.
6. French L, Ma T, Oh H, Tseng GC, Sibille E. Age-related gene expression in the frontal cortex suggests synaptic function changes in specific inhibitory neuron subtypes. *Front Aging Neurosci.* 2017;9:162.
7. Kee N, Volakakis N, Kirkeby A, et al. Single-cell analysis reveals a close relationship between differentiating dopamine and subthalamic nucleus neuronal lineages. *Cell Stem Cell.* 2017;20(1):29–40.

8. Shah S, Lubeck E, Zhou W, Cai L. In situ transcription profiling of single cells reveals spatial organization of cells in the mouse hippocampus. *Neuron*. 2016;92(2):342–357.
9. Grant GR, Manduchi E, Stoeckert Jr CJ. Analysis and management of microarray gene expression data. *Curr Protoc Mol Biol*. 2007. https://doi.org/10.1002/0471142727.mb1906s77 [Chapter 19: Unit 19.6].
10. Grewal A, Lambert P, Stockton J. Analysis of expression data: an overview. *Curr Protoc Bioinform*. 2007. https://doi.org/10.1002/0471250953.bi0701s17 [Chapter 7: Unit 7.1].
11. Mehta JP, Rani S. Software and tools for microarray data analysis. *Methods Mol Biol*. 2011;784:41–53.
12. Tseng GC, Ghosh D, Feingold E. Comprehensive literature review and statistical considerations for microarray meta-analysis. *Nucleic Acids Res*. 2012;40(9):3785–3799.
13. Tusher VG, Tibshirani R, Chu G. Significance analysis of microarrays applied to the ionizing radiation response. *Proc Natl Acad Sci USA*. 2001;98(9):5116–5121.
14. Sharov AA, Dudekula DB, Ko MS. A web-based tool for principal component and significance analysis of microarray data. *Bioinformatics*. 2005;21(10):2548–2549.
15. Khatri P, Draghici S. Ontological analysis of gene expression data: current tools, limitations, and open problems. *Bioinformatics*. 2005;21(18):3587–3595.
16. Khatri P, Voichita C, Kattan K, et al. Onto-tools: new additions and improvements in 2006. *Nucleic Acids Res*. 2007;35(Web Server issue):W206–W211.
17. Saeed AI, Sharov V, White J, et al. TM4: a free, open-source system for microarray data management and analysis. *Biotechniques*. 2003;34(2):374–378.
18. Dahlquist KD, Salomonis N, Vranizan K, Lawlor SC, Conklin BR. GenMAPP, a new tool for viewing and analyzing microarray data on biological pathways. *Nat Genet*. 2002;31(1):19–20.
19. Doniger SW, Salomonis N, Dahlquist KD, Vranizan K, Lawlor SC, Conklin BR. MAPPFinder: using Gene Ontology and GenMAPP to create a global gene-expression profile from microarray data. *Genome Biol*. 2003;4(1):R7.
20. Huang da W, Sherman BT, Lempicki RA. Bioinformatics enrichment tools: paths toward the comprehensive functional analysis of large gene lists. *Nucleic Acids Res*. 2009;37(1):1–13.
21. Subramanian A, Tamayo P, Mootha VK, et al. Gene set enrichment analysis: a knowledge-based approach for interpreting genome-wide expression profiles. *Proc Natl Acad Sci USA*. 2005;102(43): 15545–15550.
22. Gentleman R, Ruschhaupt M, Huber W, Lusa L. *Meta-analysis for Microarray Experiments*; 2014:1–14. http://www.bioconductor.org/packages/release/bioc/vignettes/GeneMeta/inst/doc/GeneMeta.pdf.
23. Choi H, Shen R, Chinnaiyan AM, Ghosh D. A latent variable approach for meta-analysis of gene expression data from multiple microarray experiments. *BMC Bioinform*. 2007;8:364.
24. Zhang Z, Fenstermacher D. An introduction to MAMA (meta-analysis of microarray data) system. *Conf Proc IEEE Eng Med Biol Soc*. 2005;7:7730–7733.
25. Correa-Cerro LS, Piao Y, Sharov AA, et al. Generation of mouse ES cell lines engineered for the forced induction of transcription factors. *Sci Rep*. 2011;1:176.
26. Barrett T, Troup DB, Wilhite SE, et al. NCBI GEO: archive for functional genomics data sets—10 years on. *Nucleic Acids Res*. 2011;39(Database issue):D1005–D1010.
27. Kolesnikov N, Hastings E, Keays M, et al. ArrayExpress update-simplifying data submissions. *Nucleic Acids Res*. 2014.
28. Ivliev AE, t Hoen PA, Villerius MP, den Dunnen JT, Brandt BW. Microarray retriever: a web-based tool for searching and large scale retrieval of public microarray data. *Nucleic Acids Res*. 2008;36(Web Server issue):W327–W331.
29. Sharov AA, Schlessinger D, Ko MS. ExAtlas: an interactive online tool for meta-analysis of gene expression data. *J Bioinform Comput Biol*. 2015;13(6):1550019.
30. Sharov AA, Nishiyama A, Qian Y, et al. Chromatin properties of regulatory DNA probed by manipulation of transcription factors. *J Comput Biol*. 2014;21(8):569–577.
31. Yip KY, Cheng C, Bhardwaj N, et al. Classification of human genomic regions based on experimentally determined binding sites of more than 100 transcription-related factors. *Genome Biol*. 2012;13(9): R48.
32. Kim T, Lim CS, Kaang BK. Cell type-specific gene expression profiling in brain tissue: comparison between TRAP, LCM and RNA-seq. *BMB Rep*. 2015;48(7):388–394.

33. Reddy AS, O'Brien D, Pisat N, et al. A comprehensive analysis of cell type-specific nuclear RNA from neurons and glia of the brain. *Biol Psychiatry*. 2017;81(3):252–264.
34. Takeuchi M, Yamaguchi S, Sakakibara Y, et al. Gene expression profiling of granule cells and Purkinje cells in the zebrafish cerebellum. *J Comp Neurol*. 2016.
35. Zeisel A, Munoz-Manchado AB, Codeluppi S, et al. Brain structure. Cell types in the mouse cortex and hippocampus revealed by single-cell RNA-seq. *Science*. 2015;347(6226):1138–1142.
36. Trapnell C, Roberts A, Goff L, et al. Differential gene and transcript expression analysis of RNA-seq experiments with TopHat and Cufflinks. *Nat Protoc*. 2012;7(3):562–578.
37. Huber W, Carey VJ, Gentleman R, et al. Orchestrating high-throughput genomic analysis with Bioconductor. *Nat Methods*. 2015;12(2):115–121.
38. Gao D, Kim J, Kim H, et al. A survey of statistical software for analysing RNA-seq data. *Hum Genom*. 2010;5(1):56–60.
39. Cheadle C, Vawter MP, Freed WJ, Becker KG. Analysis of microarray data using Z score transformation. *J Mol Diagn*. 2003;5(2):73–81.
40. Bolstad BM, Irizarry RA, Astrand M, Speed TP. A comparison of normalization methods for high density oligonucleotide array data based on variance and bias. *Bioinformatics*. 2003;19(2):185–193.
41. Miller JA, Ding SL, Sunkin SM, et al. Transcriptional landscape of the prenatal human brain. *Nature*. 2014;508(7495):199–206.
42. Lattin JE, Schroder K, Su AI, et al. Expression analysis of G protein-coupled receptors in mouse macrophages. *Immunome Res*. 2008;4:5.
43. Wirth H, Loffler M, von Bergen M, Binder H. Expression cartography of human tissues using self organizing maps. *BMC Bioinform*. 2011;12:306.
44. Raney BJ, Cline MS, Rosenbloom KR, et al. ENCODE whole-genome data in the UCSC genome browser (2011 update). *Nucleic Acids Res*. 2011;39(Database issue):D871–D875.
45. Roth RB, Hevezi P, Lee J, et al. Gene expression analyses reveal molecular relationships among 20 regions of the human CNS. *Neurogenetics*. 2006;7(2):67–80.
46. Ono H, Ogasawara O, Okubo K, Bono H. RefEx, a reference gene expression dataset as a web tool for the functional analysis of genes. *Sci Data*. 2017;4:170105.
47. Cusanovich DA, Pavlovic B, Pritchard JK, Gilad Y. The functional consequences of variation in transcription factor binding. *PLoS Genet*. 2014;10(3):e1004226.
48. Hurley D, Araki H, Tamada Y, et al. Gene network inference and visualization tools for biologists: application to new human transcriptome datasets. *Nucleic Acids Res*. 2012;40(6):2377–2398.
49. Nishiyama A, Xin L, Sharov AA, et al. Uncovering early response of gene regulatory networks in ES cells by systematic induction of transcription factors. *Cell Stem Cell*. 2009;5(4):420–433.
50. Yamamizu K, Sharov AA, Piao Y, et al. Generation and gene expression profiling of 48 transcription-factor-inducible mouse embryonic stem cell lines. *Sci Rep*. 2016;6:25667.
51. Nishiyama A, Sharov AA, Piao Y, et al. Systematic repression of transcription factors reveals limited patterns of gene expression changes in ES cells. *Sci Rep*. 2013;3:1390.
52. Hamatani T, Carter MG, Sharov AA, Ko MS. Dynamics of global gene expression changes during mouse preimplantation development. *Dev Cell*. 2004;6(1):117–131.
53. Zhou Y, Zhu S, Cai C, et al. High-throughput screening of a CRISPR/Cas9 library for functional genomics in human cells. *Nature*. 2014;509(7501):487–491.
54. Chintapalli VR, Wang J, Dow JA. Using FlyAtlas to identify better *Drosophila melanogaster* models of human disease. *Nat Genet*. 2007;39(6):715–720.
55. Bernard A, Lubbers LS, Tanis KQ, et al. Transcriptional architecture of the primate neocortex. *Neuron*. 2012;73(6):1083–1099.
56. Briggs J, Paoloni M, Chen QR, Wen X, Khan J, Khanna C. A compendium of canine normal tissue gene expression. *PLoS One*. 2011;6(5):e17107.
57. Ferraz AL, Ojeda A, Lopez-Bejar M, et al. Transcriptome architecture across tissues in the pig. *BMC Genom*. 2008;9:173.
58. Gene Ontology Consortium. Gene Ontology Consortium: going forward. *Nucleic Acids Res*. 2015; 43(Database issue):D1049–D1056.

59. Zhang Y, De S, Garner JR, Smith K, Wang SA, Becker KG. Systematic analysis, comparison, and integration of disease based human genetic association data and mouse genetic phenotypic information. *BMC Med Genom*. 2010;3:1.
60. Chatr-Aryamontri A, Breitkreutz BJ, Oughtred R, et al. The BioGRID interaction database: 2015 update. *Nucleic Acids Res*. 2015;43(Database issue):D470–D478.
61. Boyle AP, Araya CL, Brdlik C, et al. Comparative analysis of regulatory information and circuits across distant species. *Nature*. 2014;512(7515):453–456.
62. Yang JH, Li JH, Shao P, Zhou H, Chen YQ, Qu LH. starBase: a database for exploring microRNA-mRNA interaction maps from Argonaute CLIP-Seq and Degradome-Seq data. *Nucleic Acids Res*. 2011; 39(Database issue):D202–D209.
63. Dimmer EC, Huntley RP, Alam-Faruque Y, et al. The UniProt-GO annotation database in 2011. *Nucleic Acids Res*. 2011;40(Database issue):D565–D570.
64. Kanehisa M, Goto S, Sato Y, Kawashima M, Furumichi M, Tanabe M. Data, information, knowledge and principle: back to metabolism in KEGG. *Nucleic Acids Res*. 2013;42(Database issue):D199–D205.
65. Matthews L, Gopinath G, Gillespie M, et al. Reactome knowledgebase of human biological pathways and processes. *Nucleic Acids Res*. 2009;37(Database issue):D619–D622.
66. Dyson G, Wu J. Use of the half-normal probability plot to identify significant effects for microarray data. In: *Joint Statistical Meeting (JSM) 2002*. 2002. New York, NY.
67. Benjamini Y, Hochberg Y. Controlling the false discovery rate – a practical and powerful approach to multiple testing. *J Roy Stat Soc B*. 1995;57:289–300.
68. Kim SY, Volsky DJ. PAGE: parametric analysis of gene set enrichment. *BMC Bioinform*. 2005;6:144.
69. Sharov AA, Nishiyama A, Piao Y, et al. Responsiveness of genes to manipulation of transcription factors in ES cells is associated with histone modifications and tissue specificity. *BMC Genom*. 2011;12:102.
70. Fisher RA. *Statistical Methods for Research Workers*. Edinburgh, UK: Oliver and Boyd; 1925.
71. Mosteller F, Bush RR. Selected quantitative techniques. In: Lindzey G, ed. *Handbook of Social Psychology*. vol. 1. Cambridge, MA: Addison Wesley; 1954:289–334.
72. DerSimonian R, Laird N. Meta-analysis in clinical trials. *Control Clin Trials*. 1986;7(3):177–188.
73. Waldron L, Riester M. Meta-analysis in gene expression studies. *Methods Mol Biol*. 2016;1418: 161 176.

# CHAPTER 5

# Computational Models: How Do They Help to Understand Neurologic Diseases?

**Belén Sancristóbal[1], Ibrahim Tastekin[1], Mara Dierssen[1,2,3]**
[1]Centre for Genomic Regulation (CRG), Barcelona, Spain; [2]Universitat Pompeu Fabra (UPF), Barcelona, Spain; [3]CIBER de Enfermedades Raras (CIBERER), Barcelona, Spain

## INTRODUCTION

Neurologic disorders and their sequelae are currently estimated to affect more than a billion people worldwide, with a high social and economic burden. However, despite the vast amount of data accumulated since the last century, we have not been able to unravel the causal relationship between the pathology of neurologic diseases and their molecular, cellular, and circuit-level underpinnings. This is in part due to the large number of variables and factors affecting the structure and function of the nervous system that makes it almost impossible for them to be considered in a common framework. To understand how brain function is affected and to find effective treatments, we need to integrate findings at different scales of activity, ranging from molecular cascades to network function. Then, it would hopefully be possible to link the genetic etiology to the neuronal alterations that comprise the properties of neural networks.

## THE CHALLENGE OF MODELING BRAIN COMPLEXITY

The assertion "neuroscience is data-rich yet theory-poor"[2] is grounded in the elaborated and complex machinery of the brain, whose function cannot be derived from simple linear summation of its parts. This coins the concept of emergence in which the behavior, function, and/or other properties of the system are more than the sum of the system's parts at any particular level or across multiscale levels of organization. In fact, such system properties can emerge from complex patterns of underlying subsystems. A myriad of variables (ion channels, neuromodulators, time-evolving natural stimuli, synaptic efficacies, or delays) contribute to each single neural process (firing activity, synaptic plasticity, synchronization) in a way that is often unknown.

The brain's computational power is thus organized at different spatial scales from single neurons to local neural networks and macroscopic networks formed by interregional connections. A high variety of cellular and neuronal types, coupled through specific

*Molecular-Genetic and Statistical Techniques for Behavioral and Neural Research*
ISBN 978-0-12-804078-2, https://doi.org/10.1016/B978-0-12-804078-2.00005-2

feedforward and feedback connections, disclose the complex structure of the brain. The presence of these dense interconnections among diverse neuronal subtypes is a defining property of the network, so both physiological features of neurons and their synapses shape its activity.

In addition, all these cellular elements show coexisting dynamics that have different time scales: the fastest processes correspond to the activation of the membrane ionic channels responsible for the generation of action potentials, slower time scales on the order of several milliseconds (or even seconds during deep sleep) modulate synchronization of thousands of neurons producing oscillations, and the slowest processes standing for long-term plasticity lie on the range of minutes or hours.[3] Last but not least, the brain's electrical activity exhibits a wide variety of rhythmic and transient dynamics, whose specific physiological functions are sometimes obvious, as our daily awake-sleep transition,[4] but usually remain elusive. Temporal scaling is evident across the inherent rhythms of brain activity and relates to different cognitive capacities, but also the pattern of neuronal connections in the brain changes with learning and memory through the process of synaptic plasticity on both short (seconds to minutes) and long (hours to days to months) timescales.

Interactions among different scales[5] change the responses, effective coupling, and synchronization of individual neurons and neuronal networks.[6] Eventually, the functional outcome of the whole brain's activity (i.e., memory formation, thoughts, and emotions) is far more ambiguous to quantify.

Such variety and amount of data is conceptually hard to be conceived, and the mechanistic links between neural processes taking place at different scales are difficult to uncover. Therefore, it is unclear what can be inferred from one level of integration and be extrapolated to the next, so most of our "causal assumptions" along the bottom-up (from microscopic to macroscopic) or the top-down (from macroscopic to microscopic) axes are simply correlational. For instance, alterations in a patient's behavior or neurologic function are correlated with changes in brain electrical activity and/or structure. In turn, disturbances in cellular/molecular function are assumed to lie under such abnormalities, considering the structural properties of a neuron as a manifestation of systems-level behavior of a fundamentally unknown network of genes and proteins, which ultimately constrains neuronal function.

Thus, even though we assume that gathering more information would increase our explanatory power, biophysical, molecular, or electrophysiological findings from animal models used to fulfill the gaps in our knowledge of the pathophysiology of human neurologic disorders still do not provide for causal inferences. In many cases, they only increase the available information, but they does not per se improve the level of interpretation of the gathered data.

## COMPUTATIONAL MODELS AND THEIR CORRESPONDENCE TO COMPLEX BIOLOGIC REALITY

Computational models have become an important tool in the study of the nervous system since they are particularly powerful tools to take up these challenges.[7] It is important to realize the different types of models that can provide different tools and approaches for our questions: descriptive models summarize large amounts of experimental data, thereby intended to describe phenomena but not to explain them. Mechanistic models, instead, address the question of how the systems operate on the basis of known anatomy and physiology. Interpretive models use computational and mathematic principles to explore the significance of nervous system function, addressing the question of why nervous systems operate as they do. The ease of control of the variables defined in a computational model of the brain (or a part of it) makes it suitable for simplification of a high-dimensional system. On the other hand, a custom-made model forces the explicit formulation of mathematic relations between time-varying quantities (gene expression, transmembrane potentials, firing rates, etc.) and parameter dependencies (dendritic structure, connectivity, transmission velocities, etc.) to explain the observations (patterns of activation of distant brain areas, tremors, motor disturbances, oscillations, etc.). Therefore, models can serve as a test for experimentally based hypotheses and predict potential mechanisms that might underlie the pathologies of human neural diseases. Models can also suggest which operations at the neural network level are disturbed, exceeding the single-neuron causes.[8] However, models also have limitations. The large number of variables and factors affecting the structure and function of the nervous system makes it almost impossible to consider all in a single model, and most computational models only consider a few specific biologic variables, which is sometimes viewed as a limitation for explaining biologic reality. In fact, it is sometimes difficult to identify the appropriate level of modeling for a particular problem. To allow the integrative power of multilevel modeling a more detailed model is not necessarily superior, but it must be detailed enough at the lower level while simple enough to yield clear results at the higher level.

The first contribution of computational neuroscience is owed to the work of Alan Hodgkin and Andrew Huxley. Although far from neurologic disorders, their work is a clear example of how modeling assumptions can guide neuroscience research. Hodgkin and Huxley built a model for the generation of neuronal action potential assuming, inter alia, deterministic ion channel kinetics.[9] The proposed set of equations for the dynamics of membrane ion channels were first derived from voltage clamp experiments on the squid giant axon. They predicted specific structural and functional properties for these channels, such as their permeability, selectivity, voltage dependency, and gating,[10] and their quantitative model based on the nonlinear dynamics of ionic conductance was able to explain transmembrane current flows. Importantly, their work promoted an extensive research that provided experimental confirmation of their work years later.[11]

Nowadays, their original mathematic formalism is still being applied in all conductance-based models of single neurons.

In the following sections, we will discuss seminal contributions of computational neuroscience to the field of neurologic diseases. We will first introduce the type of data and analyses used to characterize brain activity. Then, we will focus on the achievements of particular models of Alzheimer (AD) and Parkinson disease (PD). Finally, we will introduce Down syndrome (DS) as a novel field of study for computational neuroscience that has so far remained unexplored. To conclude, we will provide a critical perspective of the benefits of applying neural models to the study of neurologic diseases.

## WHAT ARE OUR READOUTS OF NEUROLOGIC DISORDERS?

Neurologic diseases are assumed to be the consequence of molecular and cellular disturbances. The cholinergic hypothesis of AD,[12] which associates the cognitive decline with cholinergic cell loss and synaptic dysfunction in the basal forebrain,[13] or the dopaminergic (DA) hypothesis of PD are classical examples. In intellectual disability, synaptic pathology and dendritic spine dysgenesia are pathognomonic anatomopathologic signs. However, the connection between these molecular and cellular phenotypes with the altered cognition and disruptive behavior is not straightforward. For instance, dendritic spine alterations are common to different disorders (such as AD, epilepsy, or developmental disorders like fragile X or DS), but the corresponding neuronal phenotypes vary largely and affect different brain regions, being even layer-specific in some cases. Although the genetic basis of the wide spectrum of neurologic disorders may converge on a set of core molecular pathways,[14] their final impact on the neuronal phenotypes may depend on multiple cell autonomous and cell nonautonomous factors, including environmental influences or individual differences in susceptibility to neurologic insult.[15]

Besides, there are limitations also in the data gathering. For example the histopathology data are mainly collected from postmortem human tissues and may be influenced by postmortem interval, symptomatic heterogeneity, and the patient's medical history. Moreover, as neuropathologic studies are performed in advanced stages of the disease, they reflect later stages of the process and do not distinguish between cause and effect.

On the other hand, functionality aspects of the affected brain can be tracked in vivo by the use of noninvasive techniques such as scalp electroencephalography (EEG), which captures the electrical activity emerging from cortical neural assemblies. Other noninvasive techniques, such as magnetoencephalography (MEG), which shows a better spatial resolution than EEG signals, or functional magnetic resonance imaging (fMRI) are also clinically relevant. EEG and MEG have a higher temporal resolution than fMRI and are therefore widely used to analyze the presence of temporal patterns, although the localization from the scalp recordings of the electrical sources, i.e., how the current flow is anatomically distributed and oriented, is not unequivocally determined (inverse problem).[16] Only certain neural diseases, such as drug-resistant cases of epilepsy, allow the use of invasive electrodes that increase spatial resolution.[17]

EEG/MEG signals show a wide frequency bandwidth capturing fast and slow components of neural network activity. In the central nervous system, neural activity is rhythmic due to the capacity of neural tissue to generate oscillations. Local oscillatory activity generally arises from recurrent inhibitory feedback connections that result in the synchronization of firing patterns among thousands of neurons,[18] enabled by neural connectivity through chemical synapses (and the often-forgotten gap junctions[19]). Modulation of such synchronization occurs with a periodicity that depends on the time scale of the underlying mechanisms (after hyperpolarization, excitatory-inhibitory balance, etc.) and the type of neurons involved.[20] These mesoscopic oscillations contribute to the overall EEG signal,[21] and they become prominent under different tasks or cognitive states, giving rise to a rich association between rhythms and functions in specific brain regions.[22] For instance, local oscillations in the gamma range (30−100 Hz) are supposed to mediate perceptual binding[23] and communication between brain regions,[24] while theta oscillations (4−12 Hz) in specific brain regions have been associated to episodic memory.[25] Broader oscillations are a typical fingerprint of brain activity of active state[26] and deep sleep.[27] Modifications of the temporal structure of EEG signals are also signatures of abnormal brain activity[28−31] in neurologic disorders. As compared to healthy controls, particular frequency ranges may be altered, and they have been proposed as readouts or biomarkers that could be used for diagnoses or prognostic purposes. Examples are enhanced, slow frequencies in the delta and theta (below 8 Hz) in the resting state of PD patients, as detected in MEG recordings,[32,33] and EEG changes reported in AD,[33−35] left temporal lobe epilepsy,[36] or Huntington disease (HD) patients.[37]

These human studies allow establishing phenomenological correlations between rhythms and cognitive function. However, they do not provide insight into the microscopic brain, thus impeding one's ability to build a bottom-up description of the emergence of brain rhythms to understand mechanistic relationships between electrophysiology and cognition. Animal models have proven to be useful to circumvent the limitations of human studies to gather structural and molecular data. Experimental manipulations of animal models, such as optogenetics or in vivo imaging techniques, have helped characterize the variety of dynamic processes taking place in the brain. For instance, HD mouse models have shown an increased firing rate of medium spiny neurons of the striatum and a decreased correlation between bursts of cell pairs in freely behaving mice.[38,39] However, we still do not have any explanatory link between these results and the increase in the delta/theta power in HD patients mentioned before.[37]

In summary, pathophysiological effects—from genes to behavior—underlying human neurologic disorders have their macroscopic expression in imaging and scalp recordings, while the intracranial and microscopic dynamics remain largely in the realm of experimental animals. The interpretation of these data requires a multilevel modeling approach in which different models corresponding to different levels of organization, from subcellular to whole brain or organism, are constructed for the same phenomenon. These multilevel models could help linking observations at different levels of organization by experiments, which is one of the main challenges of modern neuroscience.

## NEUROLOGIC DISORDERS FROM LOCAL TO LARGE-SCALE NETWORKS

The recent discoveries in brain network organization and its changes in neurologic and psychiatric diseases, especially in dementia, epilepsy, and schizophrenia, suggest that brain disease involves both "local" and "global" pathology. Thus, integrating data from several sources is especially important to unravel the processes contributing to emergent properties of the network and allow for the comparison between numeric simulations and experimental results.[40,41] Here, we introduce the concept of functional network (FN) as an emergent property of interconnected neural circuits, and we present examples of FN in relation to neurologic disorders. Increasingly realistic models of large-scale structural and functional brain networks create opportunities to develop exact mathematic models of normal and disturbed brain network. By the end of this section, we hope that its potential use as a modeling outcome will be clear.

The rich repertoire of brain functions involves the coordination of the activity of distributed neural regions, which influence each other at overlapping or sequential times, allowing the emergence of interareal dynamic patterns.[42] Depending on the stimuli, task, or state, these FNs are defined according to the set of brain regions (or neural networks) that are coactivated. This large-scale mapping of the brain functional connectivity (FN, Fig. 5.1 right panel) differs from the underlying structural brain network (SN, Fig. 5.1 left panel) in that the links between areas (the nodes) are not necessarily sustained by physical (anatomic) connections.

The organization of the brain in feedback and feedforward connections between segregated cortical areas as well as with subcortical structures provides the basis for the large-scale SN description of the brain.[43] This SN information can be obtained using magnetic resonance imaging (MRI) and diffusion tensor imaging (DTI). The information that can be extracted from MRI refers to size and shape of the different brain regions and the amount of white versus gray matter of those structures. DTI provides detailed images of white matter interarea connections (Fig. 5.1 left panel) by tracing the diffusion of water molecules in biologic tissue along the neural fibers.[44]

On the other hand, to infer brain functional connectivity from EEG signals, temporal correlations[31] and phase coherence[32] quantification measures are used, since local electrical activity captured by the EEG cannot discern between activation patterns in afferent areas (which presynaptic areas induce activity) nor its efferent influences or temporal sequences (simultaneous or time-delayed activity). Correlations indicate the extent to

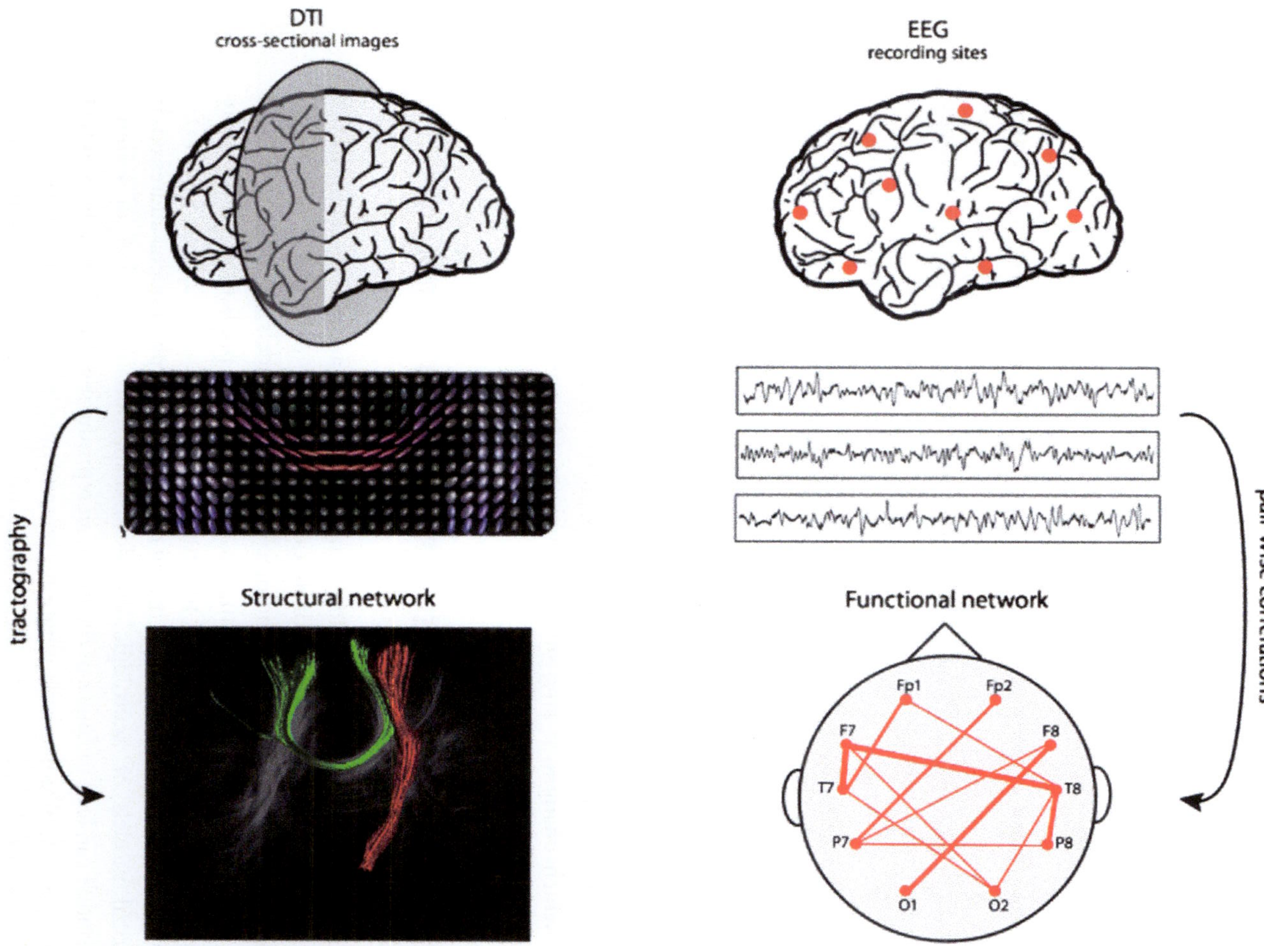


**Figure 5.1** ***Structural (SN) and functional (FN) brain networks.*** Left panel: The SN is obtained by diffusion tensor imaging (DTI) technique. The myelinated axonal tracts are reconstructed (tractography, middle and bottom) from cross-sectional scans (top) revealing the orientation of water molecules under a magnetic field. Left bottom: Primary motor cortex within the precentral gyrus, in *green*, and fibers of the upper motor neurons projecting out of the precentral gyrus ending in the brainstem, in *red*. Right panel: The FN is built from the local electrical activity (middle) captured at the electroencephalography (EEG) recording sites (top). Pairwise correlations quantify the strength of the functional connections (bottom). *(Left panel: Image used with permission of Mint Labs S.L. http://www.mint-labs.com (bottom) and Vesna Prchkovska (middle) PhD thesis: High Angular Resolution Diffusion Imaging: Processing and Visualization.)*

which pairs of time traces exhibit similar temporal patterns, and phase coherence determines the consistency of the relative phase difference across time or trials. Their presence is a sign of the effective relationship between areas that defines the strength of connections in the FN[45] (Fig. 5.1, right panel).

To a large extent, the SN, which can be defected in neurologic disorders, constrains the FN, although it does not unequivocally determine it, because functional links between neural populations can emerge with no physical connection. Conversely, coupled neural networks can have uncorrelated activity if their presynaptic sources differ. Therefore, we do not know yet how structure defines function, and making sense of the local electrical brain activity in terms of a functional large-scale network is extremely challenging. Computational models of networks composed of connected neural units—either individual neurons or assemblies—formally couple the structure and the activity, and the FN becomes the output of a simplified, yet certainly perturbable, SN. As an example of the relevance of FN in neurologic diseases, multiple brain dysfunctions have been detected during resting state both by correlations of EEG or fMRI signals in AD,[46] PD,[47] bipolar disorder, and schizophrenia.[48] These results are expected to indicate the neuronal network that is affected in each disorder. For instance, in AD, functional connectivity between the posterior cingulate cortex and the hippocampus is impaired,[49] while PD exhibits decreased correlations between the striatum and the thalamus, midbrain, pons, and cerebellum[47] and between the medial prefrontal cortex and the ventromedial caudate nucleus of the basal ganglia.[50] Both disorders have altered temporal coherence and correlation between distant neural assemblies,[51–53] but we do not know which changes, if any, in the underlying SN are responsible for the FN disturbances.

## WHICH COMPUTATIONAL MODELS ARE USEFUL IN NEUROLOGY AND WHY?

Computational models can be tremendously useful to understand the mechanisms that explain brain rhythms and behavioral disruptions from neural dynamics and structural changes. In the absence of a complete biologic description, models allow one to perturb a system in a selective and reversible manner to make and test predictions of the mechanisms underlying function and leading to neurologic diseases.

However, the selection of a suitable model depends on the phenomena addressed. In computational models of neurologic disorders, inappropriate selection of minimal neuronal models can lead to failure of the model in reproducing pathologic effects, which in turn can lead to misleading interpretations. For example, population network models (or neural mass models), i.e., a network in which nodes are modeled as effective neural assemblies and no single cell dynamics are included, are relevant for modeling based on

EEG data. In such cases, they uniquely enable the study of how particular anatomic connections across areas contribute to the functional connectivity and the effects of their disruption (reduction of connectivity, altered dendritic spine numbers, etc.). The functional connectivity can be superposed to the structural data available,[54] and a network model of mutually interacting areas can estimate the relationships between the structure and the arising dynamic FN.

On a more local scale, each brain area also shows specific intraareal connectivity patterns between the neurons they comprise. For instance, the six-layered neocortex[55,56] largely differs from the hippocampus or cerebellar structure, and its damage leads to different impairments.[57–59] Hence, the details of local synaptic connectivity need to be explored, and to this end, single-neuron-based network models, i.e., a network in which nodes are modeled as individual neurons, can shed light on local processing. However, modeling studies have an obvious limitation, as data from individual cells and also different cell types of human brain are not available, and models of the nervous system usually consider a single type of cell or interactions of two or three relevant types of cells, scarcely considering, for example, the active role of astroglia in various neuronal functions and synaptic plasticity. Also, large network models of different brain areas downscale the size of the network due to computational limitations, which may affect the reliability of the results, as the size of the network (number of computational units) has a major effect on network functionality.

Finally, at the single cell level, detailed conductance-based models (see Ref. [60], for instance) are computationally time consuming but are also best suited for more detailed simulations of thousands of neurons, where analogs of intracellular and local field potentials recordings are obtained. On the other hand, firing rate models[61] or neural mass models[40] can easily handle larger networks but at the expense of modeling averaged neuronal populations rather than single cell activity. Since in all cases network models need to implement structure, they offer a straightforward setup to understand the organizing rules of the brain.[62] In those areas where experimental data is scarce,[63] computational models can also elucidate the role of poorly investigated parameters, for example, how the relative dynamics between coupled brain areas is constrained by feedback and feedforward connections, as well as by the axonal delays and the degree of convergence and divergence of the axonal pathways.[64,65]

Neural models aim to make clinical predictions in human neural diseases,[66] and they can also guide experimental research to the relevant processes. Usually, physiological data—driven models use animal data, and they can also help to validate extrapolation of nonhuman studies results. To this aim, models have to reach beyond the confusing diversity of reality, but this requires choosing what to include and what to exclude given the model's assumptions and limitations.

## MODELING ALZHEIMER DISEASE

AD is a neurodegenerative disorder leading to short- and long-range wiring loss. Neurofibrillary tangles precede cognitive impairment and cause the death of pyramidal cortical neurons by losing the neural connectivity of different regions.[67] AD models have been developed at different levels, from the molecular level, to examine for the role of certain proteins in the development of the disease, linking amyloid beta buildup and tau tangle development, to the modeling of disease symptoms, using, for example, artificial neural networks that have the capability of simulating brain behavior under physiological and pathologic scenarios. However, one of the main questions in the field is why the same anatomic degradation simply leads to normal aging in some individuals, while in others, the neurodegeneration causes dementia. Network models applied to AD can reveal, in a simplified manner, how the observed FN alterations built on EEG, MEG, or fMRI correlations could emerge from the AD structural disconnection scenario.[68]

A good example is the model proposed by Pons et al.[69] to study the influence of different structural connectivity pathways on aberrantly aging-related functional states. EEG studies in AD patients show an attenuation of average power within the alpha band (7.5—13 Hz) and an increase of power in the theta band (4—7 Hz). Significant evidence has suggested that thalamocortical circuitry underpin the generation and modulation of alpha and theta rhythms. The authors aimed to analyze the contribution of the SN to the alpha rhythm alterations, characteristic of patients with mild cognitive impairment, relative to healthy elderly subjects. To this end, a network comprising different observed structural connections between anatomic cortical areas served as a model for the study of the emergent FNs. The network was composed of voxels (the nodes; see Box 5.1), meant to contain several cortical columns (see Fig. 5.2). Cortical regions from both hemispheres (insula, precuneus, temporal gyrus, etc.) were composed of several voxels, and their coupling was set according to real anatomic data. Each voxel was modeled by a neural mass model that consisted of three neural populations: pyramidal neurons (black triangles) and excitatory (black hexagons) and inhibitory (red circles) interneurons. A simplified thalamus provided extra input to those interconnected nodes, entrained to oscillate in the alpha band (7.5—12.5 Hz), the main frequencies of the resting state.

An analog of the EEG recording channels was built from a weighted sum of the net cortical input into the pyramidal population of the voxels, taking into account their spatial distance to the scalp electrode. Functional connectivity between pairs of electrodes was assessed by the phase lag index (PLI) introduced in Ref. [70], which is defined by the phase differences in the alpha frequency. High PLI scores imply a narrow distribution of the relative phases, i.e., a high reliability over time of the reciprocal dynamics of two signals. The short- and long-range connectivity strength between voxels was systematically varied and related to the resulting functional connectivity.

## Box 5.1 Neuronal networks can be modeled as graphs: general terms

Network science is a mixture of graph theory, dynamical systems theory, and statistics. The central idea in graph theory is to represent a complex set of relationships (in other words, a network) with a set of nodes and their connections. The basic concepts are these:

**Nodes**: functional units (minicolumns, neurons, etc.)
**Edges**: dependencies
**Node states**: measurable (observable) properties of the functional units that can be discrete or continuous, deterministic or stochastic
**Gates**: nodes with an associated function, its input and the resulting output
**Topology**: wiring, can be fixed or time-dependent
**Dynamics**: static, dynamic
**Synchrony**: synchronous, asynchronous
**Flow**: Quantity that is conveyed (flows) through the edges in a dynamic network

See Ref. [1] for review.

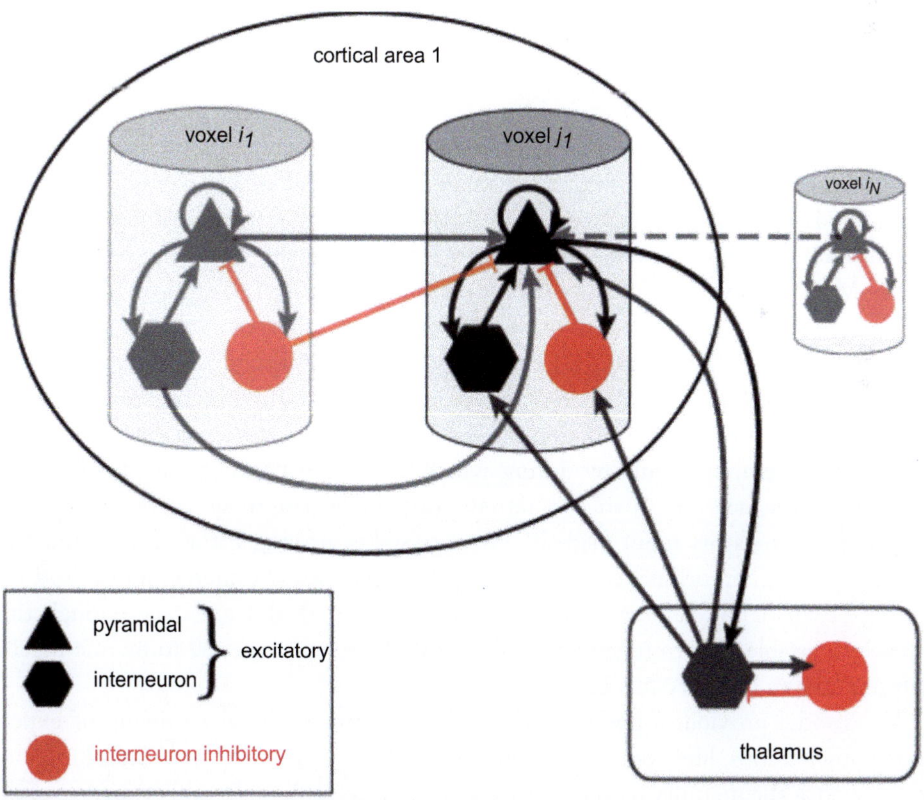


**Figure 5.2** ***Thalamocortical connectivity diagram.*** The geometric figures represent populations of excitatory (*black*) and inhibitory (*red*) neurons described altogether by a neural mass model. The coupling between voxels depends on whether they involve short-range connections (same cortical area, *solid lines*) or long-range connections (different cortical areas, *dashed lines*). Two interacting populations model the thalamus. *(Adapted from Pons AJ, et al. Relating structural and functional anomalous connectivity in the aging brain via neural mass modeling. Neuroimage. 2010;52(3):848–861.)*

Importantly, the model suggests that a decrease of the cortical inhibitory postsynaptic potential amplitude and an increase of the thalamocortical connectivity are responsible, respectively, for the slowing down of the alpha rhythm and increasing phase synchronization between cortical areas. Both results are typical of aging and cognitive impairment progress,[71] which often precedes AD.[72] Moreover, these results were compatible with reduced long-range cortico-cortical coupling (dashed line in Fig. 5.2), which is shown to degrade between association areas with normal aging.[73,74]

At a molecular level, another pathophysiological mechanism of the disease is the decrease of acetylcholine (ACh) levels due to cholinergic cell loss,[12] which has long been associated with AD possibly explaining recent memory loss, a salient characteristic of AD. This drop of ACh initially affects the entorhinal cortex and hippocampus[75] through mechanisms that are not yet fully understood.[76] Computational studies based on single-neuron networks instead of neural population models can integrate these fundamental neural mechanisms, targeting local neural computation as the functional integrative unit, which is particularly relevant in memory formation.[77]

Menschik and Finkel[78] proposed a network model for the CA3 hippocampal region (Fig. 5.3) to investigate the role of neuromodulation in memory degeneration. The network model was based on the 66-compartment hippocampal pyramidal cell model of Traub and colleagues and their 51-compartment interneuron interconnected with realistic AMPA-, NMDA-, and GABA(A)-mediated synapses, which is capable of synchronization in the gamma frequency range. Calcium channels were implemented with a conductance dependency on ACh dosage. As a result, pyramidal cells (black triangle) switch from intrinsically bursting to regular spiking with increasing ACh concentration. Memories were defined by the spatial firing pattern of the simulated pyramidal cells, which were driven by a theta (6–10 Hz) wave inhibitory input and oscillated within the gamma band (40–100 Hz) due to the inhibitory recurrent connections (pink circuit in Fig. 5.3) and the cholinergic external drive (gray projections in Fig. 5.3). Therefore, the gamma rhythm was nested within the theta cycle and corresponds to the temporal window during which the action potentials are fired. Within this interval, the network is able to retrieve one stable pattern of activity in response of a transient excitatory input made of a train of spikes arriving at the onset of the theta cycle (input form entorhinal cortex in Fig. 5.3). Decreased cholinergic inputs lead to a slowing down of the gamma frequencies, so the network did not have enough time to reach the stable pattern (or memory attractor). This was suggested to be one mechanism underlying memory loss in AD.

This model provides a good example of how computational modeling of experimental observations, like the reduced synthesis of ACh in AD patients,[12] links functional effects, such as the memory recalling process, to the electrophysiological properties of single cells and network behavior.

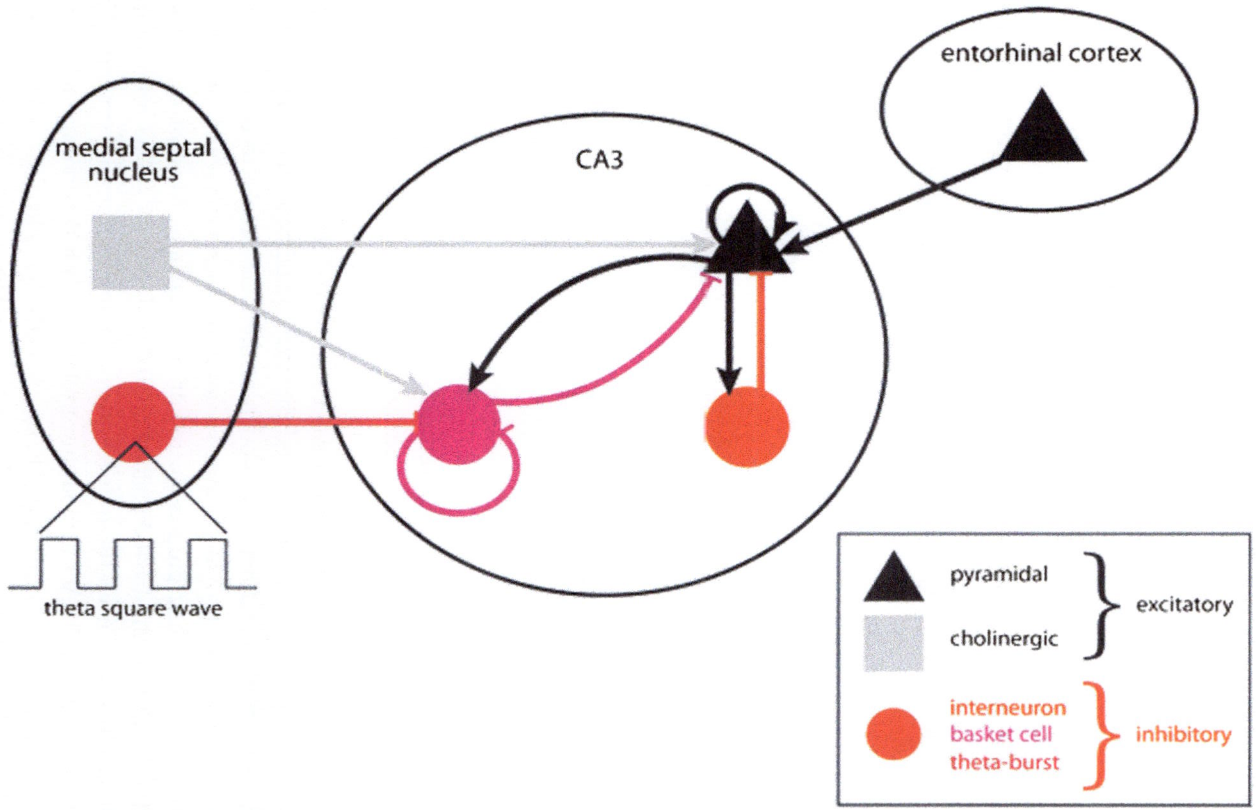


**Figure 5.3** ***CA3 hippocampal connectivity diagram.*** The geometric figures represent populations of excitatory (*black*) and inhibitory (*red*) neurons described by coupled individual neuron's model. *(Adapted from Menschik ED, Finkel LH. Cholinergic neuromodulation and Alzheimer's disease: from single cells to network simulations. Prog Brain Res. 1999;121:19–45.)*

Other groups (see the work of Hasselmo and coworkers[79]) have modeled multiple effects of acetylcholine: from local synaptic suppression (not affecting afferent inputs), neural depolarization, and suppression of adaptive currents. In their model, cholinergic suppression of the excitatory neurons modulates the ability of the network to store new information relative to the recall of old patterns by modifying the strength of recurrent transmission in CA3.

More recently, de Haan et al.[80] used models to investigate why some brain regions are more vulnerable to the neurodegenerative phenotypes. This is based on the published observation that deposition of the amyloid-β protein mainly occurs in brain regions that are highly connected to other regions. The authors tested the hypothesis that these "hub" regions are more vulnerable due to a higher neuronal activity level, using a computational model brain. Cortical brain regions were modeled as neural masses, each describing the average activity (spike density and spectral power) of a large number of interconnected excitatory and inhibitory neurons. The large-scale network consisted of 78 neural masses, connected according to a human DTI-based cortical topology. The authors simulated progressive damage to brain regions, reducing their activity, and they were able to reproduce hub vulnerability, loss of synchronization between areas, and AD-like changes in FN organization. An extensive review of computational models of AD is presented in Ref. [81].

## MODELS OF PARKINSON DISEASE

Neural network models have been also applied to the study of the mechanisms leading to tremor in PD. In PD, the motor manifestations are primarily linked to the selective, progressive loss of DA neurons in the substantia nigra pars compacta (SNpc).[82–84] However, it is still undetermined how this can eventually alter motor function, i.e., how the damaged brain circuit involved in movement causes tremor or bradykinesia.[85]

The contributions of biologically constrained computational models have been proposed to help our understanding of motor and cognitive deficits in PD. The loss of DA neurons innervating the striatum in PD, and the well-established role of DA in motor function, have enabled neural network models of the basal ganglia to derive concrete and testable predictions and provide a unified account of diverse and seemingly unrelated phenomena in PD, including progressive motor degeneration. Most of the proposed network models involve the basal ganglia circuitry (Fig. 5.4).[86,87] The primary input to the basal ganglia comes from the motor cortex and impinges on the striatum. These cells, which are mainly GABAergic neurons, project to both the internal and external part of the globus pallidus (GPint and GPext, respectively). The former is the output gate from the basal ganglia to the ventral lateral and anterior nucleus of the thalamus, which send feedback connections to the cortex.[88] This pathway is called the direct

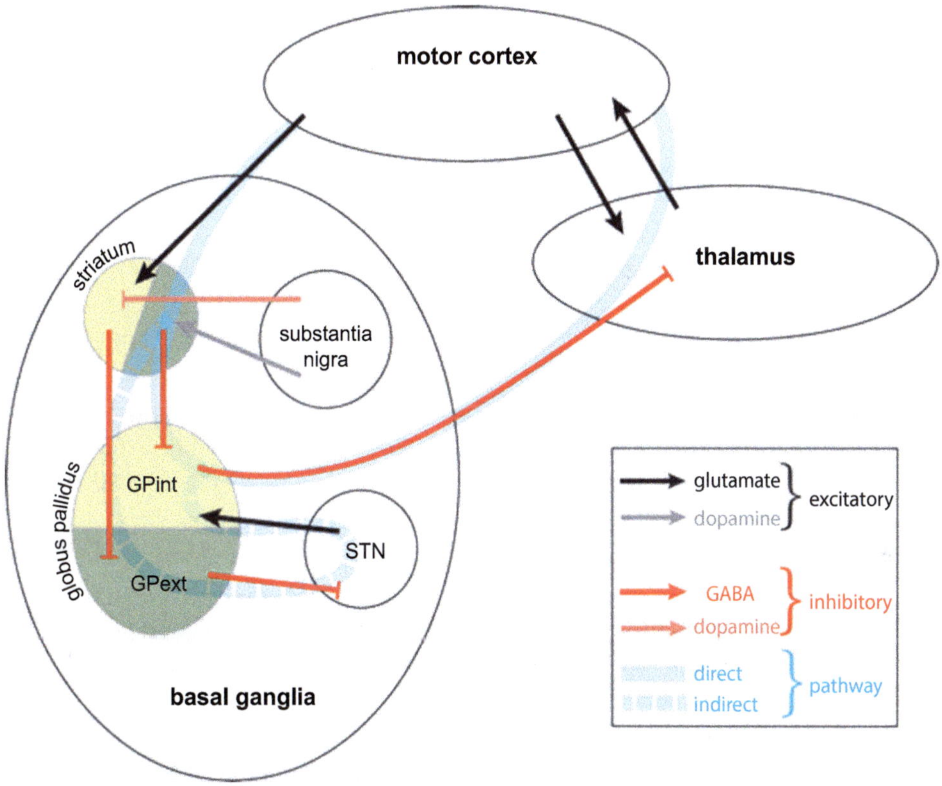


**Figure 5.4** ***Cortico-basal ganglia-thalamic-cortical loop diagram.*** Main structures and connection links considered in models of movement disturbances typical of Parkinson disease.

pathway (thick blue line in Fig. 5.4). The second, lies at the beginning of an inhibitory connectivity pathway in which GPext cells target the subthalamic nucleus (STN), whose axons synapse on the GPint neurons. This pathway is called indirect pathway (thick dashed blue line in Fig. 5.4).

The cortical input to the striatum is excitatory and strengthens inhibition onto the GPint, giving rise to a lower effective inhibition over the thalamus. Hence, activation of the direct pathway leads to an increased activity of the motor cortex through disinhibition of the thalamus, which switches on the muscles through the corticospinal connections. Similarly, inhibition on the GPext cells disinhibits the STN, giving rise to excitation of the GPint. Therefore, activation of the indirect pathway merges with the direct pathway inhibiting the thalamus and motor cortex.

Another important pathway to be considered is the nigrostriatal axons that target the striatum via dopamine release (gray and pink arrows in Fig. 5.4). Striatum cells that are presynaptic to the GPint are endowed with dopamine D1 receptors (excitatory response), while those that are presynaptic to GPext have dopamine D2 receptors (inhibitory response).[87] Hence, the ultimate effect of dopamine release is the increase of motor cortex activity trough the boost of the direct pathway and the dampening of the indirect pathway.

Although the different brain structures involved in control movement and the excitatory/inhibitory connectivity between them are well documented, there are several parameters that are largely undetermined. For instance, the strength of interareal synaptic coupling and their axonal delays are not known. Hence, results derived from neural network simulations need to be robust to relative parameter changes. Moreover, as shown next, network models elucidate how DA cell loss causing thalamic inhibition leads to specific movement patterns such as tremor or rigidity.

A first attempt to model tremor as the output of a neural circuit was presented in Ref. [89], where the authors assumed a simplified network of six units coupled through recurrent inhibitory interconnections. Excitation occurs as the net result of inhibiting an inhibitory node (disinhibition). The inhibitory interconnected loops contained in the whole network reflected the coexistence of the direct and indirect pathways described before. The network showed distinct dynamic states, from irregular chaotic membrane potential dynamics, associated with normal physiological tremor, to periodic oscillations or to fixed-point dynamics, related to PD, as the synaptic strength of some of the units was weakened. Hence, the decrease of certain synaptic efficacies represented, in a phenomenological manner, the loss of DA neurons in the substantia nigra of PD subjects. These results suggest that altered involuntary movement is an intrinsic dynamic regime arising in a minimal inhibitory feedback circuit version of Fig. 5.4. The chaotic regime, which corresponded to healthy behavior, bifurcates into pathologic behavior with the loss of particular structural connections (i.e., through simplification of the network).

A similar conclusion was obtained in the study of voluntary movements.[90] Here an abstract model of linear input summation and sigmoidal response function was used to describe the dynamics of each unit. The network consisted of a four-layer structure, including a closed-loop feeding the output back to the input layer and mimicking the cortico-basal ganglia-thalamic-cortical circuit. The synaptic weights were trained to minimize the error between the actual and desired output, which consisted of the typical triphasic muscle activation of movement around a single joint revealed by electromyography recordings.[91] The network shifted from a periodic attractor (regular oscillatory dynamics) to a fixed-point, stable solution (stationary behavior) as the excitability of the preoutput layer was reduced. The loss of DA neurons in PD was simulated by a decrease of the excitability threshold of the output presynaptic units, representing the overinhibited thalamus of PD patients. This parametric change led to abnormal slow oscillations prior to the nonmovement state and the loss of regular periodic movement.

These simplified modeling approaches are particularly relevant since they reveal that the rhythmic tremor is an emerging state of the network that can be tuned by synaptic efficacies. The fact that lesions or stimulation of distinct areas of the central nervous system can suppress tremor[92] supports this result, in contrast with a unique central pattern generator hypothesis. Moreover, the bifurcation scheme demonstrates that the onset of tremor behavior in PD arises abruptly at critical values of dopamine,[93] while it remains silent within certain ranges. This sudden change in the network dynamics has been widely explored to model the effects of deep brain stimulation, used as a therapy to suppress pathologic tremor, i.e., a destabilization of the oscillatory activity.[94,95]

Other studies[96] have assumed interconnected single neurons, representing the synchronized activity of cell assemblies, modeled by a detailed conductance-based model whose parameters were specific to GPext and STN. A single unit representing the thalamo-cortical-striatum tract simplified the input to both structures. Synaptic conductance to GPext and STN were modulated by dopamine, and the maximal conductance of the $Ca^{2+}$-activated voltage-independent after hyperpolarization $K^{+}$-current (AHP) was set either to the control or parkinsonian condition. Although multiple transitions from oscillatory to nonoscillatory behavior occurred along with dopamine increases, low dopamine levels triggered tremulous-like activity (bursting of STN and GPext neurons) by simultaneously strengthening the mutually interacting STN and GPext nodes and the feedback projections to both of them. Decreasing the AHP conductance brought the circuit back to the absence of tremor.

In a recent study, the STN-GPext circuit was also simulated using a conductance-based model with a different choice of ionic currents and considering synaptic plasticity.[97] Here, reciprocal connections between the two nuclei (inhibitory from GPext and excitatory from STN) were also dependent on dopamine, as well as the intrinsic excitability of GPext neurons. The power spectral density of the spike trains presented different frequency profiles as the strength of these conductances was changed. The parkinsonian state was defined according to electrophysiology results obtained in monkeys showing a decreased firing of GPext neurons and 8–15 Hz burst oscillations of STN neurons.[98] The model was able to reproduce this pattern of activity by reducing the excitability of GPext neurons, which in turn reduced short-term depression (see Fig. 2 in Ref. [97]) and led to a strong postinhibitory rebound firing that reoccurred rhythmically within the beta range. The decrease in the intrinsic activity of GPext was presented as a consequence of dopamine depletion[99] and, hence, a neuronal correlate of PD. Interestingly, the abnormal synchronized oscillations disappeared in the presence of an external out-of-phase oscillatory input to STN. The authors suggest that such current could have a cortical origin, as observed during voluntary movements,[100] whose relative delay to the STN oscillations could be monitored through the inhibitory projections of GPint to the thalamus.

## MODELS OF DOWN SYNDROME

Unlike other neural diseases, DS is prenatally initiated and produced by a genetic alteration, which in humans corresponds to an extra copy of at least part of chromosome 21 (HSA21). Besides the DS distinctive intellectual disability,[101,102] structural abnormalities have been reported in the brain, at the macroscopic (reduced volume and surface of specific brain areas such as the cortex or the hippocampus along with reduced cellularity, or hypoplasia of the cerebellum)[103] and the microscopic level, including altered connectivity with dendritic spine architecture alterations in newborns' motor[104] and visual[105] cortices, and dendritic atrophy compromising dendritic complexity in DS infants.[106] In DS adults, branching and dendritic length is significantly reduced.[107] This altered cell morphology perturbs neural network formation, probably undermining neural computation in a way that is still unknown despite its visible functional consequences. It is noteworthy that DS patients are prone to neurodegeneration and development of AD, suggesting the vulnerability of distorted brain structures.

Due to the complexity of gene interactions and the outnumbered molecular pathways activated by the extra copy of HSA21, there has been an increasing interest on the overexpression of those genes that were particularly able to recapitulate the DS phenotype.[108] Within a critical region proposed to be central to the disease,[109] the *Dyrk1A* gene encoding for a serine-threonine kinase, thought to be involved in neural growth and development[110] and shown to be 1.5-fold overexpressed in DS brains,[111] is a major candidate.[112]

Research on DS patients, similar to PD and AD, is also abundant and allows for model validation. Large-scale recordings of brain activity, such as EEG or fMRI, tightly correlate with cognitive performance and can be used to characterize the FN of DS brains. Alterations of dynamic patterns between specific brain regions can target the networks that are underperforming, and population network models can also help here, as previously reviewed, to link the FN to the SN. DS patients were shown to exhibit stronger synchronization between distributed networks (visual, limbic, sensorimotor, etc.) while watching a video, favoring short-range connected pairs. Interestingly, the results from this stimulus-driven activity were more heterogeneous across DS than across controls and the degree of internetwork correlations inversely varied with the subject's intelligence quotient.[113] In addition, during eyes-closed resting state, EEG cortical alpha waves were smaller in amplitude in DS than in control subjects.[114]

Despite the amount of experimental results available for data-driven models, DS has not drawn the attention of computational neuroscience. According to the existing data on DS patients (and DS mouse models) covering network architecture (dendrite morphology) and macroscopic functional changes (brain activity alterations), we claim that both single-neuron-based and population networks can boost our understanding of DS. Our group recently used a single-neuron-based model describing the cortical slow oscillations that spontaneously emerge during deep sleep.[60] We used a transgenic mouse model overexpressing *Dyrk1A (TgDyrk1A)* that has confirmed its relevance by reproducing typical structural and cognitive features of mental retardation in humans. Namely, *TgDyrk1A* mice recapitulate DS neuronal and cognitive phenotypes. Specifically, they exhibit defective cortical pyramidal cell morphology[115] and reduced axon growth in the hippocampus[116] along with impaired cognition.[117–119] Moreover, increased dosage of *Dyrk1A* recapitulates several phenotypes of the widely used Ts65Dn mouse model,[120,121] which contains a partial trisomy containing most of the orthologous genes of HSA21.[122] Therefore, *TgDyrk1A* mice offer a battery of data at multiple levels—molecular, cellular, and behavioral—that can be integrated by means of computational models aiming to understand DS neural networks.

Specifically, we showed excitation inhibition imbalance in these mice that might explain the decreased firing rates and gamma synchronization that we detected, which suggested an overinhibited prefrontal cortex.[123] However, immunohistochemical analysis only revealed a decreased inhibition of parvalbumin positive interneurons. Thus, we developed a computational model of a local cortical circuit with pyramidal cells and fast-spiking interneurons, incorporating NMDA, AMPA, and GABA synaptic kinetics, from a conductance-based model of single excitatory and inhibitory neurons, using the same parameters for the dynamics of the ionic channels and the membrane potential, as well as the connectivity pattern, as in Compte et al.[60] All neurons received an extra train of excitatory presynaptic potentials representing the synaptic current coming from areas other than the explicitly modeled neuronal network.

Using this model, we demonstrated that by changing the connectivity of the inhibitory population as observed in *TgDyrk1A* mice, we could recapitulate the observed decrease in inhibition on interneurons, and the neuronal activity deficits, including decreased gamma power and firing rate (Fig. 5.5). Hence, our computational approach suggests a causal relationship between two experimental observations; namely a reduction in the inhibitory-inhibitory connections (unlike any other connectivity disruption) accounts for the aberrant spontaneous gamma oscillations of the DS mouse model.[123]

In DS, since the genetic causes of the disease are well known, computational models can comprise another layer of biologic description, i.e., the genetic regulatory network, that have remained so far detached from the modeling of neural circuits.[124] Hence, models of DS might help to link systems biology and neuroscience.

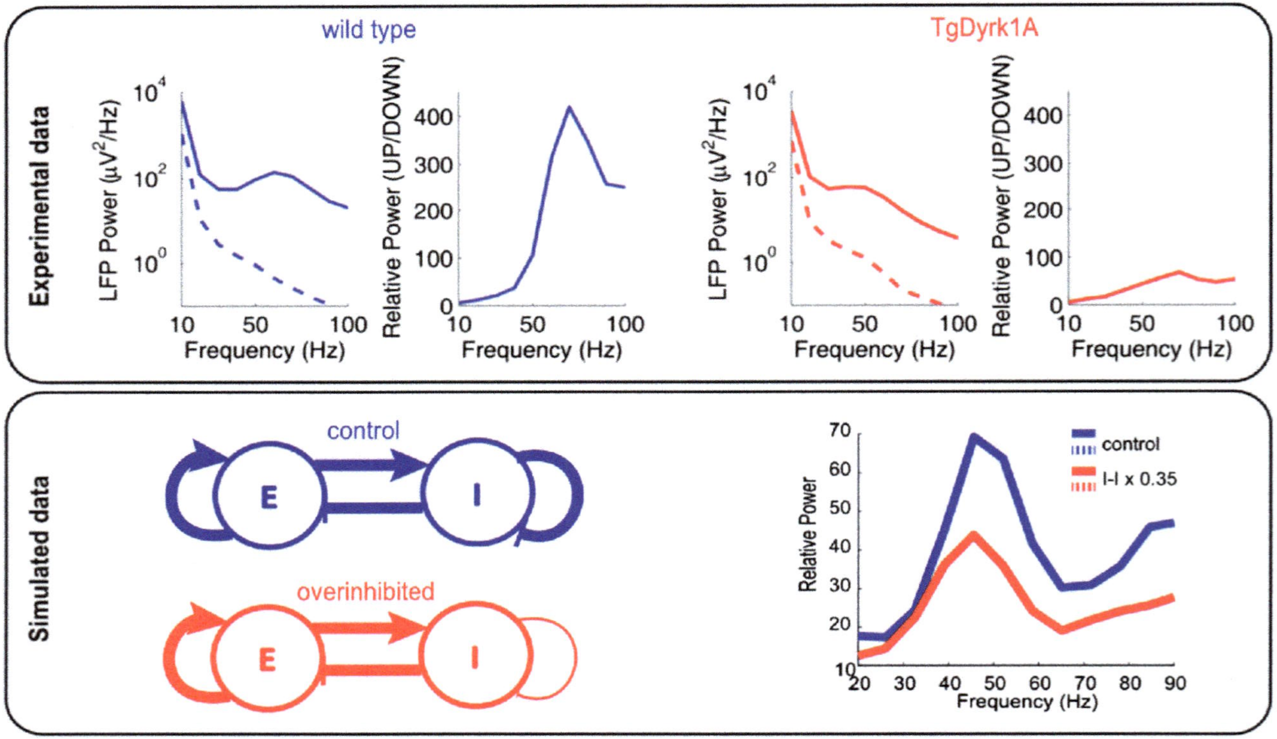


**Figure 5.5** ***Down syndrome neural network model.*** (Top) Data from a representative example of a wild type (*blue*) and a *TgDyrk1A* mouse (*red*) showing the power spectral density in the two characteristic states of the slow oscillation dynamics—the UP (*solid line*) and DOWN state (*dashed line*)—and relative power of the two. (Bottom) Population-based diagram of the neural network model. The overinhibited network (*red*) corresponds to a 35% reduction change of the inhibitory-inhibitory coupling with respect to the control network (*blue*). A reduction of the number of inhibitory synapses on inhibitory neurons leads to a decrease in power in the gamma band of the slow oscillations. *(Adapted from Ruiz-Mejias M, et al. Overexpression of* Dyrk1A*, a Down syndrome candidate, decreases excitability and impairs gamma oscillations in the prefrontal cortex. J Neurosci. 2016;36(13):3648–3659.)*

## CONCLUSIONS

Probably, one of the most ambitious goals of neuroscience is to trace all the steps from gene expression down to brain state in health and disease. In this instance, computational neuroscience requires a battery of supporting data obtained in the same experimental context and covering different levels of detail (neural types, neuromodulators, stimulus specific responses, etc.).[7] Models of neural diseases aim to establish a link between the biologic and pathologic abnormalities of the damaged brain. Regardless of the particularities of the model, all are network representations of computational units, i.e., excitable systems integrating several current sources. A neural network can be considered a function space that autonomously translates the domain of inputs into outputs. Due to the broad range of interacting variables (ions, neurotransmitters, proteins, etc.) and of brain capabilities (learning, memory, consciousness, etc.), it is extremely challenging to capture the connections between the micro- and macroscopic processes. Computational neuroscience can be experimentally tested, rather than mathematically proved. Thus, eventually, predictions of neural models need to be measurable. Moreover, the assumptions leading to a particular model should also be confirmed. However, it is commonly considered that correct predictions give support to the initial assumption. Modeling intrinsically implies simplifying a system to apprehend the basic mechanisms that do not rely on all levels of description. Very much like a metaphor that synthesizes an elusive concept, neural models propose a framework from where new knowledge can be developed and tested.

An important achievement in the field are the Hodgkin-Huxley equations for single-neuron membrane potential dynamics.[9] The original model was limited to the action potential generation in the squid giant axon, and its parameters were fitted to reproduce this particular case. Even though all variables have a physiological meaning, several assumptions and simplifications were done. For instance, the model consists of a point neuron, whose ionic channels are a phenomenological description of the averaged activity of thousands of them. Although still far from a wide-ranging model of the brain, it is a clear example of a computational model fed with experimental data. Moreover, extended to other neural types and to networks, it further helps to elaborate a hypothesis that will be validated (or rejected) by replicating (or not) observations.

The understanding of neural diseases is key to the improvement of human welfare and provides insight on brain function. A system's response to a given perturbation can reveal its dynamics and equilibrium states. As reviewed here, alterations in PD or AD have been successfully modeled and have targeted critical parameters for the onset of the network's abnormal activity. Moreover, the use of neural network models has extended the view that brain function is an emergent property of neuronal assemblies, ranging from local cortical columns to interareal recurrent connections. Hence, network models allow us to study in a noninvasive and harmless manner how different spatiotemporal scales are linked.[125] We have argued, as well, that DS offers a perfect framework to study the interplay between the genetic and the neural circuit.

Finally, we understand computational neuroscience is a valuable tool for framing data-driven hypothesis about the mechanisms regulating brain activity. In particular, neural diseases enable us to study the system under certain changes altering both local activity and the large-scale neural circuitry. Although there is still a big gap between cognition and the anatomic and biochemical knowledge of the brain, neural models help us to build a coherent interpretation of how single elements of computation (no matter whether it is a neuron or a whole neural population) can give rise to complex behavior such as orientation selectivity in the primary visual cortex.[126] It is true that successful model predictions are not sufficient to prove that the brain actually operates in *that* particular manner.[127] However, here, we have intended to review seminal papers in support of the idea that predictions made by models and submitted to experimental validation can open a path to build a theory of brain function.

## ACKNOWLEDGMENTS

The authors would like to thank A. Pons and J. Garcia-Ojalvo for fruitful discussion on their paper.[69] I would like to acknowledge the Jerome Lejeune Foundation and the Spanish Ministry of Economy and Competitiveness (SAF2016-79956-R)

## REFERENCES

1. Stam CJ. Modern network science of neurological disorders. *Nat Rev Neurosci*. 2014;15(10):683–695.
2. Churchland PS, Sejnowski T. The computational brain. In: *Computational Neuroscience*. Cambridge: MIT Press; 1992.
3. Han VZ, Grant K, Bell CC. Reversible associative depression and nonassociative potentiation at a parallel fiber synapse. *Neuron*. 2000;27(3):611–622.
4. Olbrich E, Claussen JC, Achermann P. The multiple time scales of sleep dynamics as a challenge for modelling the sleeping brain. *Philos Trans A Math Phys Eng Sci*. 2011;369(1952):3884–3901.
5. Barardi A, et al. Probing scale interaction in brain dynamics through synchronization. *Phil Trans R Soc B*. 2014;369(1653). https://doi.org/10.1098/rstb.2013.0533.
6. Sancristobal B, et al. Emergent bimodal firing patterns implement different encoding strategies during gamma-band oscillations. *Front Comput Neurosci*. 2013;7:18.
7. Skinner FK, Ferguson KA. Modeling oscillatory dynamics in brain microcircuits as a way to help uncover neurological disease mechanisms: a proposal. *Chaos*. 2013;23(4):046108.
8. Contrears-Vidal JL. Computer modeling in basal ganglia disorders. In: Litvan I, ed. *Atypical Parkinsonian Disorders: Clinical and Research Aspects*. Humana Press; 2005:95–109.
9. Hodgkin AL, Huxley AF. A quantitative description of membrane current and its application to conduction and excitation in nerve. *J Physiol*. 1952;117(4):500–544.
10. Catterall WA, et al. The Hodgkin-Huxley heritage: from channels to circuits. *J Neurosci*. 2012;32(41): 14064–14073.
11. Doyle DA, et al. The structure of the potassium channel: molecular basis of K+ conduction and selectivity. *Science*. 1998;280(5360):69–77.
12. Francis PT, et al. The cholinergic hypothesis of Alzheimer's disease: a review of progress. *J Neurol Neurosurg Psychiatry*. 1999;66(2):137–147.

13. McGeer PL, et al. Aging, Alzheimer's disease, and the cholinergic system of the basal forebrain. *Neurology*. 1984;34(6):741–745.
14. Murdoch JD, State MW. Recent developments in the genetics of autism spectrum disorders. *Curr Opin Genet Dev*. 2013;23(3):310–315.
15. Barulli D, Stern Y. Efficiency, capacity, compensation, maintenance, plasticity: emerging concepts in cognitive reserve. *Trends Cogn Sci*. 2013;17(10):502–509.
16. Baillet S. Forward and inverse problems of MEG/EEG. In: Jaeger D, Jung R, eds. *Encyclopedia of Computational Neuroscience*. New York: Springer; 2014:1–8.
17. Engel AK, et al. Invasive recordings from the human brain: clinical insights and beyond. *Nat Rev Neurosci*. 2005;6(1):35–47.
18. Timofeev I, et al. Neuronal synchronization and thalamocortical rhythms in sleep, wake and epilepsy. In: Noebels JL, et al., eds. *Jasper's Basic Mechanisms of the Epilepsies*. 2012. Bethesda, MD.
19. Tamas G, et al. Proximally targeted GABAergic synapses and gap junctions synchronize cortical interneurons. *Nat Neurosci*. 2000;3(4):366–371.
20. Grillner S, et al. Microcircuits in action–from CPGs to neocortex. *Trends Neurosci*. 2005;28(10): 525–533.
21. Buzsaki G, Draguhn A. Neuronal oscillations in cortical networks. *Science*. 2004;304(5679): 1926–1929.
22. Buzsaki G. *Rhythms of the Brain*. 2006.
23. Fries P. Neuronal gamma-band synchronization as a fundamental process in cortical computation. *Annu Rev Neurosci*. 2009;32:209–224.
24. Fries P. A mechanism for cognitive dynamics: neuronal communication through neuronal coherence. *Trends Cogn Sci*. 2005;9(10):474–480.
25. Klimesch W, Schimke H, Schwaiger J. Episodic and semantic memory: an analysis in the EEG theta and alpha band. *Electroencephalogr Clin Neurophysiol*. 1994;91(6):428–441.
26. Varela F, et al. The brainweb: phase synchronization and large-scale integration. *Nat Rev Neurosci*. 2001;2(4):229–239.
27. Amzica F, Steriade M. The K-complex: its slow (<1-Hz) rhythmicity and relation to delta waves. *Neurology*. 1997;49(4):952–959.
28. Kaper TJ, Kramer MA, Rotstein HG. Introduction to focus issue: rhythms and dynamic transitions in neurological disease: modeling, computation, and experiment. *Chaos*. 2013;23(4):046001.
29. Uhlhaas PJ, Singer W. Neural synchrony in brain disorders: relevance for cognitive dysfunctions and pathophysiology. *Neuron*. 2006;52(1):155–168.
30. Tsolaki A, et al. Electroencephalogram and Alzheimer's disease: clinical and research approaches. *Int J Alzheimers Dis*. 2014;2014:349249.
31. Al-Qazzaz NK, et al. Role of EEG as biomarker in the early detection and classification of dementia. *Sci World J*. 2014;2014:906038.
32. Stoffers D, et al. Slowing of oscillatory brain activity is a stable characteristic of Parkinson's disease without dementia. *Brain*. 2007;130(Pt 7):1847–1860.
33. Benz N, et al. Slowing of EEG background activity in Parkinson's and Alzheimer's disease with early cognitive dysfunction. *Front Aging Neurosci*. 2014;6:314.
34. Poza J, et al. Extraction of spectral based measures from MEG background oscillations in Alzheimer's disease. *Med Eng Phys*. 2007;29(10):1073–1083.
35. Gallego-Jutgla E, et al. A theta-band EEG based index for early diagnosis of Alzheimer's disease. *J Alzheimers Dis*. 2015;43(4):1175–1184.
36. Quraan MA, et al. Altered resting state brain dynamics in temporal lobe epilepsy can be observed in spectral power, functional connectivity and graph theory metrics. *PLoS One*. 2013;8(7):e68609.
37. Painold A, et al. Comparative EEG mapping studies in Huntington's disease patients and controls. *J Neural Transm*. 2010;117(11):1307–1318.
38. Miller BR, et al. Dysregulated information processing by medium spiny neurons in striatum of freely behaving mouse models of Huntington's disease. *J Neurophysiol*. 2008;100(4):2205–2216.
39. Estrada-Sanchez AM, et al. Dysregulated striatal neuronal processing and impaired motor behavior in mice lacking huntingtin interacting protein 14 (HIP14). *PLoS One*. 2013;8(12):e84537.

40. David O, Friston KJ. A neural mass model for MEG/EEG: coupling and neuronal dynamics. *Neuroimage*. 2003;20(3):1743−1755.
41. Babajani A, Nekooei MH, Soltanian-Zadeh H. Integrated MEG and fMRI model: synthesis and analysis. *Brain Topogr*. 2005;18(2):101−113.
42. McIntosh AR. Mapping cognition to the brain through neural interactions. *Memory*. 1999;7(5−6): 523−548.
43. Bressler SL, Kelso JA. Cortical coordination dynamics and cognition. *Trends Cogn Sci*. 2001;5(1): 26−36.
44. Jellison BJ, et al. Diffusion tensor imaging of cerebral white matter: a pictorial review of physics, fiber tract anatomy, and tumor imaging patterns. *AJNR Am J Neuroradiol*. 2004;25(3):356−369.
45. Jalili M, Knyazeva MG. Constructing brain functional networks from EEG: partial and unpartial correlations. *J Integr Neurosci*. 2011;10(2):213−232.
46. Stam CJ, et al. Small-world networks and functional connectivity in Alzheimer's disease. *Cereb Cortex*. 2007;17(1):92−99.
47. Hacker CD, et al. Resting state functional connectivity of the striatum in Parkinson's disease. *Brain*. 2012;135(Pt 12):3699−3711.
48. Mamah D, Barch DM, Repovs G. Resting state functional connectivity of five neural networks in bipolar disorder and schizophrenia. *J Affect Disord*. 2013;150(2):601−609.
49. Mevel K, et al. The default mode network in healthy aging and Alzheimer's disease. *Int J Alzheimers Dis*. 2011;2011:535816.
50. van Eimeren T, et al. Dysfunction of the default mode network in Parkinson disease: a functional magnetic resonance imaging study. *Arch Neurol*. 2009;66(7):877−883.
51. Selkoe DJ. Alzheimer's disease is a synaptic failure. *Science*. 2002;298(5594):789−791.
52. Coleman P, Federoff H, Kurlan R. A focus on the synapse for neuroprotection in Alzheimer disease and other dementias. *Neurology*. 2004;63(7):1155−1162.
53. Taverna S, Ilijic E, Surmeier DJ. Recurrent collateral connections of striatal medium spiny neurons are disrupted in models of Parkinson's disease. *J Neurosci*. 2008;28(21):5504−5512.
54. Hermundstad AM, et al. Structural foundations of resting-state and task-based functional connectivity in the human brain. *Proc Natl Acad Sci USA*. 2013;110(15):6169−6174.
55. Rempel-Clower NL, Barbas H. The laminar pattern of connections between prefrontal and anterior temporal cortices in the Rhesus monkey is related to cortical structure and function. *Cereb Cortex*. 2000;10(9):851−865.
56. Constantinople CM, Bruno RM. Deep cortical layers are activated directly by thalamus. *Science*. 2013; 340(6140):1591−1594.
57. Fauser S, et al. Disorganization of neocortical lamination in focal cortical dysplasia is brain-region dependent: evidence from layer-specific marker expression. *Acta Neuropathol Commun*. 2013;1(1):47.
58. Rossini L, et al. Altered layer-specific gene expression in cortical samples from patients with temporal lobe epilepsy. *Epilepsia*. 2011;52(10):1928−1937.
59. Hedreen JC, et al. Neuronal loss in layers V and VI of cerebral cortex in Huntington's disease. *Neurosci Lett*. 1991;133(2):257−261.
60. Compte A, et al. Cellular and network mechanisms of slow oscillatory activity (<1 Hz) and wave propagations in a cortical network model. *J Neurophysiol*. 2003;89(5):2707−2725.
61. Schaffer ES, Ostojic S, Abbott LF. A complex-valued firing-rate model that approximates the dynamics of spiking networks. *PLoS Comput Biol*. 2013;9(10):e1003301.
62. Park HJ, Friston K. Structural and functional brain networks: from connections to cognition. *Science*. 2013;342(6158):1238411.
63. De Schutter E. Why are computational neuroscience and systems biology so separate? *PLoS Comput Biol*. 2008;4(5):e1000078.
64. Sancristobal B, Vicente R, Garcia-Ojalvo J. Role of frequency mismatch in neuronal communication through coherence. *J Comput Neurosci*. 2014;37(2):193−208.
65. Barardi A, Sancristobal B, Garcia-Ojalvo J. Phase-coherence transitions and communication in the gamma range between delay-coupled neuronal populations. *PLoS Comput Biol*. 2014;10(7):e1003723.

66. Seeley WW, et al. Neurodegenerative diseases target large-scale human brain networks. *Neuron.* 2009; 62(1):42–52.
67. Delbeuck X, Van der Linden M, Collette F. Alzheimer's disease as a disconnection syndrome? *Neuropsychol Rev.* 2003;13(2):79–92.
68. Filippi M, Agosta F. Structural and functional network connectivity breakdown in Alzheimer's disease studied with magnetic resonance imaging techniques. *J Alzheimers Dis.* 2011;24(3):455–474.
69. Pons AJ, et al. Relating structural and functional anomalous connectivity in the aging brain via neural mass modeling. *Neuroimage.* 2010;52(3):848–861.
70. Stam CJ, Nolte G, Daffertshofer A. Phase lag index: assessment of functional connectivity from multi channel EEG and MEG with diminished bias from common sources. *Hum Brain Mapp.* 2007;28(11): 1178–1193.
71. Cantero JL, et al. Increased synchronization and decreased neural complexity underlie thalamocortical oscillatory dynamics in mild cognitive impairment. *Neuroimage.* 2009;46(4):938–948.
72. Petersen RC, et al. Mild cognitive impairment: clinical characterization and outcome. *Arch Neurol.* 1999;56(3):303–308.
73. Makris N, et al. Frontal connections and cognitive changes in normal aging rhesus monkeys: a DTI study. *Neurobiol Aging.* 2007;28(10):1556–1567.
74. Dickstein DL, et al. Changes in the structural complexity of the aged brain. *Aging Cell.* 2007;6(3): 275–284.
75. Pappas BA, et al. Choline acetyltransferase activity and cognitive domain scores of Alzheimer's patients. *Neurobiol Aging.* 2000;21(1):11–17.
76. Mesulam M. The cholinergic lesion of Alzheimer's disease: pivotal factor or side show? *Learn Mem.* 2004;11(1):43–49.
77. Mayford M, Siegelbaum SA, Kandel ER. Synapses and memory storage. *Cold Spring Harb Perspect Biol.* 2012;4(6).
78. Menschik ED, Finkel LH. Cholinergic neuromodulation and Alzheimer's disease: from single cells to network simulations. *Prog Brain Res.* 1999;121:19–45.
79. Hasselmo ME, Schnell E, Barkai E. Dynamics of learning and recall at excitatory recurrent synapses and cholinergic modulation in rat hippocampal region CA3. *J Neurosci.* 1995;15(7 Pt 2):5249–5262.
80. de Haan W, et al. Activity dependent degeneration explains hub vulnerability in Alzheimer's disease. *PLoS Comput Biol.* 2012;8(8):e1002582.
81. Duch W. Computational models of dementia and neurological problems. *Methods Mol Biol.* 2007;401: 305–336.
82. Hornykiewicz O. The discovery of dopamine deficiency in the parkinsonian brain. *J Neural Transm Suppl.* 2006;(70):9–15.
83. Hornykiewicz O. A brief history of levodopa. *J Neurol.* 2010;257(suppl 2):S249–S252.
84. Brichta L, Greengard P. Molecular determinants of selective dopaminergic vulnerability in Parkinson's disease: an update. *Front Neuroanat.* 2014;8:152.
85. Jankovic J. Parkinson's disease: clinical features and diagnosis. *J Neurol Neurosurg Psychiatry.* 2008;79(4): 368–376.
86. Mink JW. The basal ganglia and involuntary movements: impaired inhibition of competing motor patterns. *Arch Neurol.* 2003;60(10):1365–1368.
87. Wiecki TV, Frank MJ. Neurocomputational models of motor and cognitive deficits in Parkinson's disease. *Prog Brain Res.* 2010;183:275–297.
88. Golan DE, et al. Dopamine and control movement: Parkinson's disease. In: Rhyner S, ed. *Principles of Pharmacology: The Pathophysiologic Basis of Drug Therapy.* Philadelphia: Lippincott Williams & Wilkins; 2012.
89. Edwards R, Beuter A, Glass L. Parkinsonian tremor and simplification in network dynamics. *Bull Math Biol.* 1999;61(1):157–177.
90. Borrett DS, Yeap TH, Kwan HC. Neural networks and Parkinson's disease. *Can J Neurol Sci.* 1993; 20(2):107–113.
91. Berardelli A, et al. Single-joint rapid arm movements in normal subjects and in patients with motor disorders. *Brain.* 1996;119(Pt 2):661–674.

92. Deuschl G, et al. The pathophysiology of parkinsonian tremor: a review. *J Neurol.* 2000;247(suppl 5): V33–V48.
93. Crystal H, Finkel LH. Computational approaches to neurological disease. In: *Neural Modeling of Brain and Cognitive Disorders.* 1996:251–271.
94. Titcombe MS, et al. Dynamics of Parkinsonian tremor during deep brain stimulation. *Chaos.* 2001; 11(4):766–773.
95. Schiff SJ. Towards model-based control of Parkinson's disease. *Philos Trans A Math Phys Eng Sci.* 2010; 368(1918):2269–2308.
96. Dovzhenok A, Rubchinsky LL. On the origin of tremor in Parkinson's disease. *PLoS One.* 2012;7(7): e41598.
97. Shouno O, et al. Computational model of recurrent subthalamo-pallidal circuit for generation of parkinsonian oscillations. *Front Neuroanat.* 2017;11:21.
98. Tachibana Y, et al. Subthalamo-pallidal interactions underlying parkinsonian neuronal oscillations in the primate basal ganglia. *Eur J Neurosci.* 2011;34(9):1470–1484.
99. Chan CS, et al. HCN channelopathy in external globus pallidus neurons in models of Parkinson's disease. *Nat Neurosci.* 2011;14(1):85–92.
100. Turner RS, DeLong MR. Corticostriatal activity in primary motor cortex of the macaque. *J Neurosci.* 2000;20(18):7096–7108.
101. Rowe J, Lavender A, Turk V. Cognitive executive function in Down's syndrome. *Br J Clin Psychol.* 2006;45(Pt 1):5–17.
102. Ball SL, et al. Theoretical exploration of the neural bases of behavioural disinhibition, apathy and executive dysfunction in preclinical Alzheimer's disease in people with Down's syndrome: potential involvement of multiple frontal-subcortical neuronal circuits. *J Intellect Disabil Res.* 2010;54(4): 320–336.
103. Dierssen M. Down syndrome: the brain in trisomic mode. *Nat Rev Neurosci.* 2012;13(12):844–858.
104. Marin-Padilla M. Structural abnormalities of the cerebral cortex in human chromosomal aberrations: a Golgi study. *Brain Res.* 1972;44(2):625–629.
105. Takashima S, et al. Abnormal neuronal development in the visual cortex of the human fetus and infant with down's syndrome. A quantitative and qualitative Golgi study. *Brain Res.* 1981;225(1):1–21.
106. Becker LE, Armstrong DL, Chan F. Dendritic atrophy in children with Down's syndrome. *Ann Neurol.* 1986;20(4):520–526.
107. Takashima S, et al. Dendrites, dementia and the Down syndrome. *Brain Dev.* 1989;11(2):131–133.
108. Delabar JM, et al. Molecular mapping of twenty-four features of Down syndrome on chromosome 21. *Eur J Hum Genet.* 1993;1(2):114–124.
109. Korenberg JR, et al. Molecular definition of a region of chromosome 21 that causes features of the Down syndrome phenotype. *Am J Hum Genet.* 1990;47(2):236–246.
110. Park J, Song WJ, Chung KC. Function and regulation of Dyrk1A: towards understanding Down syndrome. *Cell Mol Life Sci.* 2009;66(20):3235–3240.
111. Dowjat WK, et al. Trisomy-driven overexpression of DYRK1A kinase in the brain of subjects with Down syndrome. *Neurosci Lett.* 2007;413(1):77–81.
112. Guimera J, et al. A human homologue of Drosophila minibrain (MNB) is expressed in the neuronal regions affected in Down syndrome and maps to the critical region. *Hum Mol Genet.* 1996;5(9): 1305–1310.
113. Anderson JS, et al. Abnormal brain synchrony in Down syndrome. *Neuroimage Clin.* 2013;2:703–715.
114. Babiloni C, et al. Cortical sources of EEG rhythms are abnormal in down syndrome. *Clin Neurophysiol.* 2010;121(8):1205–1212.
115. Martinez de Lagran M, et al. Dyrk1A influences neuronal morphogenesis through regulation of cytoskeletal dynamics in mammalian cortical neurons. *Cereb Cortex.* 2012;22(12):2867–2877.
116. Slepak TI, et al. Dyrk kinases regulate phosphorylation of doublecortin, cytoskeletal organization, and neuronal morphology. *Cytoskeleton (Hoboken).* 2012;69(7):514–527.
117. Altafaj X, et al. Neurodevelopmental delay, motor abnormalities and cognitive deficits in transgenic mice overexpressing Dyrk1A (minibrain), a murine model of Down's syndrome. *Hum Mol Genet.* 2001;10(18):1915–1923.

118. Martinez de Lagran M, et al. Motor phenotypic alterations in TgDyrk1a transgenic mice implicate DYRK1A in Down syndrome motor dysfunction. *Neurobiol Dis*. 2004;15(1):132–142.
119. Ahn KJ, et al. DYRK1A BAC transgenic mice show altered synaptic plasticity with learning and memory defects. *Neurobiol Dis*. 2006;22(3):463–472.
120. Souchet B, et al. Excitation/inhibition balance and learning are modified by Dyrk1a gene dosage. *Neurobiol Dis*. 2014;69:65–75.
121. De la Torre R, et al. Epigallocatechin-3-gallate, a DYRK1A inhibitor, rescues cognitive deficits in Down syndrome mouse models and in humans. *Mol Nutr Food Res*. 2014;58(2):278–288.
122. Davisson MT, et al. Segmental trisomy as a mouse model for Down syndrome. *Prog Clin Biol Res*. 1993;384:117–133.
123. Ruiz-Mejias M, et al. Overexpression of Dyrk1A, a Down syndrome candidate, decreases excitability and impairs gamma oscillations in the prefrontal cortex. *J Neurosci*. 2016;36(13):3648–3659.
124. Arron JR, et al. NFAT dysregulation by increased dosage of DSCR1 and DYRK1A on chromosome 21. *Nature*. 2006;441(7093):595–600.
125. Huys QJ, Moutoussis M, Williams J. Are computational models of any use to psychiatry? *Neural Netw*. 2011;24(6):544–551.
126. Sompolinsky H, Shapley R. New perspectives on the mechanisms for orientation selectivity. *Curr Opin Neurobiol*. 1997;7(4):514–522.
127. Sejnowski TJ, Koch C, Churchland PS. Computational neuroscience. *Science*. 1988;241(4871): 1299–1306.

CHAPTER 6

# The Use of Recombinant Inbred Strains in Systems Genetics and Functional Analyses in Behavioral Pharmacology

**Eva Adamová, Robert W. Williams, Byron C. Jones**
The University of Tennessee Health Science Center, Memphis, TN, United States

## INTRODUCTION

### Recombinant Inbred Strains of Rodents: What Are They?

Recombinant inbred (RI) strains are powerful tools for investigating the genetic basis of complex traits, such as behavior, physiological, neurochemical, and pharmacological phenotypes[1,2]. They have three major advantages relative to many other resources used to study the genetics of brain, behavior, and neuropharmacological responses. First, it is possible to use families of RI strains to map causal variants or gene loci that modulate behavior and responses to drug, almost always without the need to genotype any animals. Mapping used to be a struggle, but new resources now often make it practical to define variants into intervals as small as a few megabases. Second, it is possible to resample specific genotypes of animals many times and using different doses, routes, or time points. Not only can one reduce measurement error, but one can systematically study gene-by-treatment interactions effects. This advantage is critical in progressing to predictive models of drug action as a function of genotype and environmental factors. Third, it is possible to assemble large, coherent data sets on responses of precisely the same genotypes of animals to many different drugs and treatments and to define patterns of covariation between neuropharmacological responses, baseline behavioral difference, and any of thousands of other phenotypes already measured in an RI family. The assembly of data can last for decades, and in the case of the BXDs, we have pharmacogenetic data that extend back to 1973.[3] While the specific cases that were used in these many different studies will differ across many parameters, their genomes are comparatively stable. As we show here, prior data can be usefully compared and even combined with new data to define or refine hypotheses of drug action and effects.

How does one make an RI strain set? In the simplest situation, two fully inbred strains (e.g., C57BL/6J and DBA/2J, abbreviated B6 and D2 or just B and D) are crossed to make an $F_1$ generation (B6D2F1, where the strain of the female is listed first). These $F_1$s are in turn mated to produce $F_2$ progeny. Brother-sister matings of this type are

*Molecular-Genetic and Statistical Techniques for Behavioral and Neural Research*
ISBN 978-0-12-804078-2, https://doi.org/10.1016/B978-0-12-804078-2.00006-4

continued for 18 or more generations and for many separate B-by-D or BXD lines. In each generation, the parental B and D chromosomes recombine almost randomly in each line. At the same time, the lines become progressively inbred, fixing the pattern of recombinations. The result after many generations of inbreeding is a set of strains, each of which has fixed a unique set of B to D or D to B recombination events across the chromosomes. Each strain will harbor between 40 and 80 of these recombination events, about 2–4 per chromosome (19 autosomes in a mouse). The precise location of recombination events in each strain has to be determined by genotyping or sequencing, but this only needs to be done one time. In the case of the BXDs, we have defined about 5000 recombination events across 100 strains with precision of better than 1 Mb. These strain patterns of genotypes at markers can be compared to strain differences in phenotypes to map causal gene variants or QTLs (quantitative trait locus [or loci], a chromosome region containing one or more sequence variants that modulates the distribution of a variable trait measured in a sample of genetically diverse individuals from an interbreeding population).

## What Are the Uses for Recombinant Inbred Strains?

The first and perhaps the most obvious use for RI strains is identifying genetic markers that point to chromosomal regions and candidate genes that underlie variation in a phenotype. The process is straightforward; a trait such as locomotor activity measured across a set of RI strains can be compared with the pattern of inheritance of markers throughout the genome and across the same set of strains. These markers are now almost all single nucleotide polymorphisms (SNPs) that can be scored as −1 for B and as +1 for D. If the statistical association or linkage between variance in the trait and variants of the marker is strong, then this indicates with some defined probability that genes with allelic differences near the marker contribute to variance in that trait. This is called QTL analysis or QTL mapping.

A second application of RI strains is to study patterns of genetic correlations among phenotypes. As multiple animals are tested for a particular phenotype under a given experimental treatment, strain means become the index (expected value) for that phenotype using a specific protocol; random errors that can diminish reliability of the correlations are thus minimized. An investigator can evaluate the strength of association between an index phenotype and many other phenotypes, even those measured in different laboratories and using completely different methods. For example, variation in drug response can be compared to variation in levels of metabolites, proteins, or mRNA in the liver or brain. Correlations among strain means across studies will tend to be depressed due to unshared sources of technical and sampling error. In contrast, correlations among strain means within single studies will tend to be inflated due to shared errors and batch effects. But the hope and expectation is that a significant source of correlation will be generated by sets of gene variants that jointly modulate sets of traits either in the positive or negative direction. The genetic correlation may be caused by the pleiotropic action of gene variants on many traits at different scales of organization. Genetic correlational analysis can be extended to gene expression under experimental and control conditions for a particular organ, tissue, or cell type of interest. Correlational analysis of expression, per se, can be useful for revealing networks of molecules (genes, mRNAs, proteins, metabolites, even metagenomes) whose expression covary with phenotypes. Standard cautions apply: covariance among traits can be due to shared molecular mechanisms and processes (true gene pleiotropy) or covariance can be due to statistical error or to linkage of neighboring, but mechanistically independent gene variants. This latter effect is particularly serious and is described in more length by Gerlai[4] and in Wang[5] in the context of RI strains.

## GeneNetwork

GeneNetwork (GN, www.genenetwork.org) is an open web resource that enables scientists to search and extract published data for many groups of RI strains and other cohorts,[6] but it also includes analytic tools to map QTLs and study covariance among traits. GN currently contains more than 5000 phenotypes for the BXD family, as well as expression data from over 60 tissues and cells types, including many brain regions. Researchers can investigate complex networks of gene variants and their possible connections to behavior, pharmacological responses, baseline expression differences, and disease susceptibility.[5,7,8]

## Pharmacogenetic Analysis of the Behavioral Effects of Cocaine

Our goal in this section is to provide an example and guide to using GN with neuropharmacological data generated from RI strains and to illustrate the vitality and effective reuse of older data set (also see Mulligan et al.[6]). We revisit data from one of the first genetic study of the effects of cocaine on locomotor activity in the BXDs.[2] Effects of cocaine were measured at four doses as the difference between locomotor activity after a cocaine injection (the distance traveled in an open field) and the corresponding distance after a saline control injection (Fig. 6.1). Other phenotypes presented as difference between cocaine and saline include nose pokes (measure of exploration), rearings, and time in the center of the open field (a putative measure of anxiety-like behavior). Data were acquired for 23 BXD strains in a so-called open field (a box of about 25 cm on edge), and for a minimum of four animals per sex and dose (the full text of the paper with experimental detail is available here: http://preview.tinyurl.com/y9otc2ro).

Older data sets of this type are typically hard to obtain and error check; this comprises our ability to reanalyze data and reproduce results. What is interesting in

| | Dataset | Trait ID | Symbol | Description | Location | Mean | N Cases |
|---|---|---|---|---|---|---|---|
| 1 | BXDPublish | 10334 | COCofCent05F | Central nervous system, pharmacology, behavior: Cocaine response (5 mg/kg), anxiety assay, center time in an open field test, difference from saline control in females [sec] | -- | -28.132 | 23 |
| 2 | BXDPublish | 10333 | COCofCent05M | Central nervous system, pharmacology, behavior: Cocaine response (5 mg/kg, i.p.), anxiety assay, center time in an open field test, difference from saline control in males [sec] | -- | -35.999 | 23 |
| 3 | BXDPublish | 10326 | COCofCent15F | Central nervous system, pharmacology, behavior: Cocaine response (15 mg/kg i.p.), open field behavior, difference in time spent in the field center relative saline injection for females [sec] | -- | -18.446 | 23 |
| 4 | BXDPublish | 10335 | COCofCent15M | Central nervous system, pharmacology, behavior: Cocaine response (15 mg/kg i.p.), open field behavior, time in the field center (difference from saline control) in males [sec] | -- | -23.357 | 23 |
| 5 | BXDPublish | 10328 | COCofCent30F | Central nervous system, pharmacology, behavior: Cocaine response (30 mg/kg i.p.), open field behavior, center time difference from saline control in females [sec] | -- | -20.162 | 23 |
| 6 | BXDPublish | 10327 | COCofCent30M | Central nervous system, pharmacology, behavior: Cocaine response (30 mg/kg), open field behavior, center time (difference from saline) in males [sec] | -- | -17.270 | 23 |
| 7 | BXDPublish | 10330 | COCofCent45F | Central nervous system, pharmacology, behavior: Cocaine response (45 mg/k i.p.), open field center time relative to saline control in females [sec] | -- | -3.096 | 23 |
| 8 | BXDPublish | 10329 | COCofCent45M | Central nervous system, pharmacology, behavior: Cocaine response (45 mg/k i.p.), open field center time relative to saline control in males [sec] | -- | -12.251 | 23 |

**Figure 6.1** Effects of cocaine (5, 15, 30, and 45 mg/kg i.p.) measured as the difference between locomotor activity after a cocaine injection and the corresponding distance after a saline control injection.

this case is that the older data are easily accessible in GN. If we search in GN the BXD Published Phenotype database using the search string "Jones, Tarantino, Rodriguez" in the **Combined** search field, a total of 87 data sets are returned (GN Record ID 10249 to 10335). Like wine, these old data can become better as they age. We now have far better companion data set for reanalysis of these cocaine data. Genotypes used for mapping have improved significantly over the past 20 years. We now have full genome assemblies and know every gene on every chromosome. We also now have many new high-quality neuropharmacological data sets.[9] Finally, we also have gene expression data sets for amygdala, striatum, nucleus accumbens, hippocampus, hypothalamus, ventral tegmental area, prefrontal cortex, and many other brain regions and organs that are useful when evaluating candidate genes that may modulate drug responses.

Let's examine one index trait in particular: the time difference that male mice spend in the center of the field after a low-dose injection of cocaine relative to a saline control injection (BXD trait 10333). In general, mice display increased anxiety-like behaviors after such a low-dose injection, and we see this by a pronounced 36% (male mice) average decline in center time, but there are notable exceptions. BXD30 actually appears to increase center time after injection, and three other strains appear to be unaffected. This study examined change in open field behavior at four doses, so we can also efficiently evaluate change in the time spent in the center of the field as a function of dose. Higher doses are actually associated with relatively more time in the center compared to lower doses. This may be explained by the higher activity levels in mice given larger injections that counterbalances or even overcomes their anxiety. We point this out to emphasize that our perhaps not unreasonable supposition that the reduction in time spent in the center of the open field is due to an unobservable increase in anxiety is really just a convenient heuristic used to parse mouse behavior in response to cocaine. We will see, however, that QTL mapping can sometimes let us link relatively abstract or highly operational phenotypes back to specific loci, genes, and even mechanisms. If we are lucky, we may be able to define a set of genes and processes that are bound together and help us understand and interpret an amorphous term such as "anxiety."

The same trend is seen in females (trait 10334) with a mean reduction of about 28% in time spent toward the center of the open field. However, the correlation between sexes across 23 strains is close to zero ($r = 0.18$, NS) (Fig. 6.2). This indicates that the sexes should be treated as distinct in terms of genetic variation in response to cocaine. Averaging data by strain across sex would just confound a problem that is already hard to analyze. Given this finding, we should not expect to see the same or even similar QTLs between the sexes, as the shared genetic variance is <2%. This can be confirmed by mapping the male and female traits. The change in open field center time in males is associated with a strong QTL on Chr 1 at about 70 Mb (Fig. 6.3A). In contrast, female data (Fig. 6.3B) are not associated with any significant, or even suggestive loci.

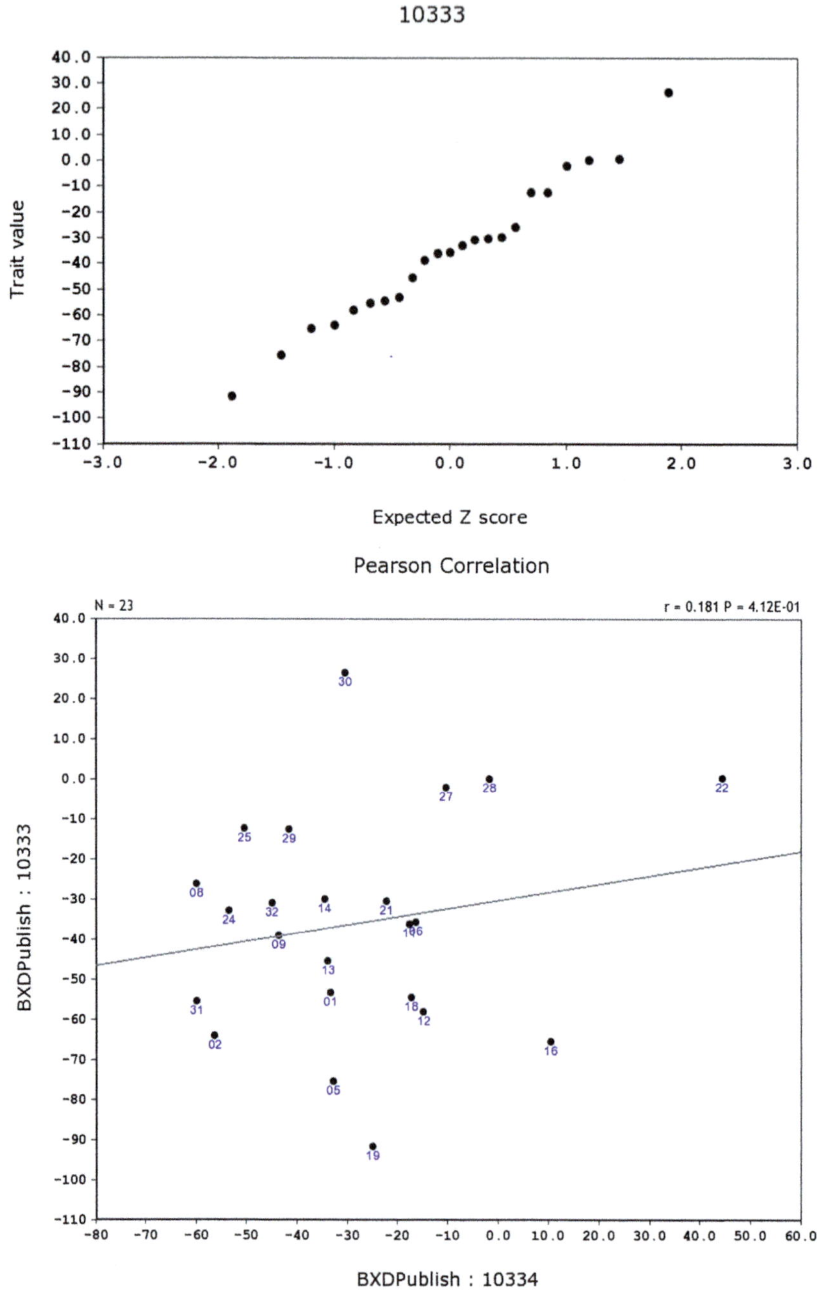


**Figure 6.2** Z score and Pearson correlation (between sexes) across 23 strains.

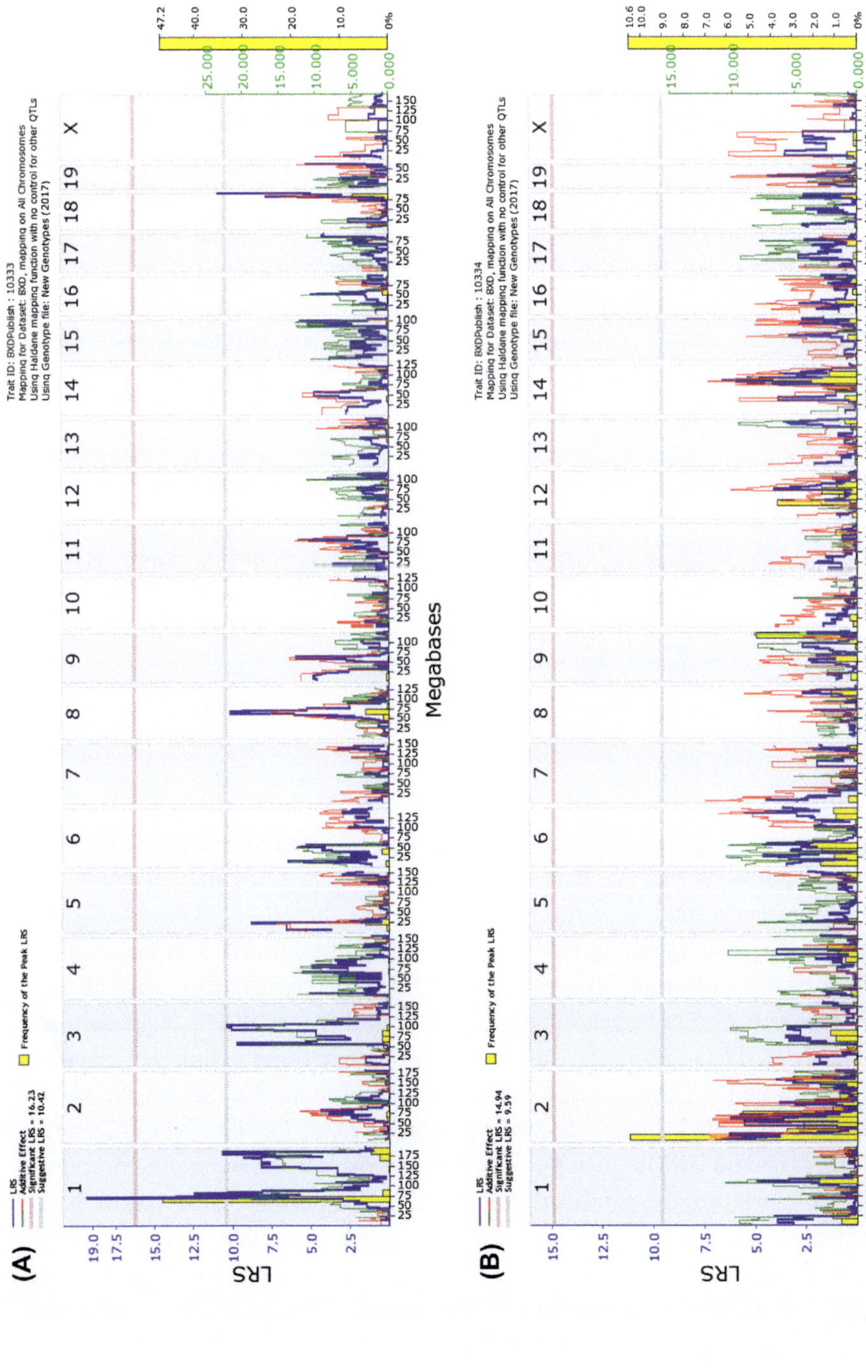


**Figure 6.3** (A) Genetic mapping of the effect of cocaine at 5 mg/kg in connection to the anxiety-like behaviors in males. Significant quantitative trait locus (QTL) observed on Chr 1. (B) Genetic mapping of the effect of cocaine at 5 mg/kg in connection to the anxiety-like behaviors in females. Significant QTL was not observed.

Given the small number of strains that went into the mapping ($n = 23$), we would like to evaluate the sensitivity of this QTL on Chr 1. How stable is the position of the locus; does its position change if we trim or winsorize values of some of the highest and lowest strains? In this particular case, winsorizing the top and bottom strains improves linkage to Chr 1. The likelihood ratio statistic score (used as a measurement of the association or linkage between differences in traits and differences in particular genotype markers) increases from 19.4 to 22.0. To winsorize, we clip the highest value from +26 to +1 and the lowest value from −92 to −76. Alternatively, we could just discard the two extreme strains. This has almost no effect on score. The yellow bars in the maps provide a summary of the results for 2000 bootstrap resamples of the original data. A bootstrap sample is a randomly drawn sample/resample that is taken from the original data set and that has the same number of samples as the original data set. GN applies bootstrap analysis to verify position and evaluate confidence intervals for QTLs. These bootstraps strongly support the robustness of this QTL—over 70% of the resamples map to this general part of Chr 1—about five times more than the next best regions, distal Chr 18. Note that for the females (Fig. 6.3B) the bootstrap bars are spread out over much of the genome, indicating that no one region is highly associated with the phenotype. The conclusion is that this QTL passes our initial critique and does deserve serious attention in terms of candidate genes that may influence responses to cocaine on Chr 1. In the next section, we will see how to use GN and complementary data sets to evaluate candidate genes in and around this QTL.

## Candidate Gene Analysis

There are many levels of candidate gene analysis. This is the first and perhaps single most critical step in progressing from locus to mechanism and to interventions or treatment. Having convinced ourselves that we have a significant QTL, we first need to estimate its size or statistical support interval of the chromosome that should be reviewed. A rule of thumb is to find the log of the odds (LOD) score at the QTL peak and then look under then peak down by 1.5−2.0 units on either side. LOD ratio provides a measure of the association between variation in a phenotype and genetic differences (alleles) at a particular chromosomal locus. LOD also provides a measurement of the strength of linkage between two markers and can be used to evaluate whether two or more markers are on the same chromosome. In Fig. 6.4, the LOD score peaks at just under 5.0 at 70 Mb. The LOD score is well below 3.0 at 68 and 73 Mb, giving us a 5-Mb support interval using this criterion. The bootstrap resamples are also informative in making this decision. The yellow bars depict the frequency with the peak score in a specific region. Note that more than 20% of the resamples actually fall between 65 and 68 Mb. And it is obvious why: BXD5, BXD19, and BXD22 all have recombinations in this interval. Their inclusion or exclusion in the bootstrap resamples account for proximal jumps in the location of the QTL peak.

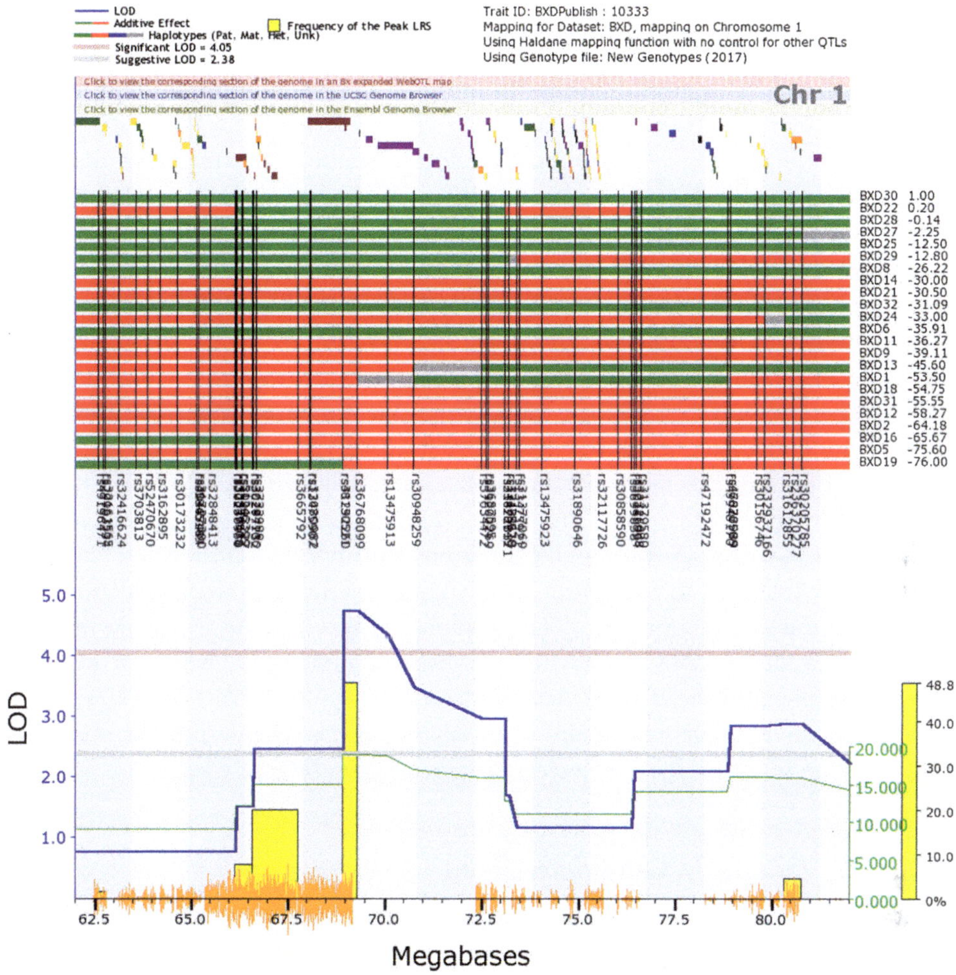


**Figure 6.4** Mapping on Chr 1, log of the odds (LOD) score at the quantitative trait locus (QTL) peak. LOD score peaks at just under 5.0 at 70 Mb. The labels below the red and green lines denote single nucleotide polymorphisms (SNP). Overlapping text indicates more than one SNP within a small interval.

Because we do not want to exclude viable candidate genes at this stage, we should expand our support interval to 66 Mb.

The second step in the analysis is to evaluate the density of sequence variants that distinguish *B* and *D* parental genomes. The orange hash along the *x* axis highlights parts of Chr 1 with many or few genetic differences. The support interval from 66 to 73 Mb can be divided into one large block (66–69.2 Mb) with high numbers of variants per megabase and a region with low numbers (69.2–72.2 Mb).

We are now well positioned to list, review, and filter candidate genes based on LOD score at their location, density of sequence differences, expression level in the brain, and any biologic priors. An example of a relevant prior might be a link in human studies of addiction or psychiatric disease. There are approximately 26 known coding genes in this interval: extending from *Mtap2* to *Tnp1*. To generate a list of these genes, perform the search "Mb = (Chr1 65.9 73.1)" in the BXD Brain RNA-seq database (see Mulligan et al.[6] for details on how to do this search). This region of the mouse genome corresponds to a segment of human Chr 2 from 209 to 217 Mb (the GRCh38 assembly of the human genome). Several genes in the interval have high central nervous system (CNS) expression, in mouse and human, including *Mtap2* (13 log2 units), *Unc80* (12 units), *Lancl1*, *Igfbp2*, *Rpl37a*, *Erbb4* (~10 units), and *Acadl* and *Fn1* (down to 8.5 units). At this stage, it is practical to review the literature for each gene and possible links to addiction. One of the fastest ways to do this is to review GeneWiki entries in GN (a Google Search with this structure "GeneWiki Errb4" will work, although a direct search in GN is also effective).

In our opinion, the most tantalizing candidate gene in the QTL is *Erbb4*, the neuregulin receptor. The reasons are simple: (1) it is a very large gene, (2) located precisely at the QTL peak (3) with high expression in many key brain regions, and a gene that has been linked (4) to psychiatric disease (schizophrenia) and to (5) GABAergic and glutamatergic neurotransmission.[10] The mapping results indicate that male mice that inherit the *D* allele in and around *Erbb4* evince less anxiety-like behavior than those that inherit the *B* allele. The predicated difference is substantial: at 5 mg/kg those, strains with the D allele evince almost no change in percentage time in the center, whereas those with the *B* allele decrease center time by 30% or more.

## Sequence Variants in Candidate Genes

What evidence do we have that there are interesting and possibly functional sequence variants in *Erbb4* that could control responses to cocaine? Are there missense mutations in the gene or genetic differences in expression among BXDs in different parts of the brain that could provide a distal cause and help us link to mechanism? *Errb4* is an enormous gene, just over 1.0 Mb in length, and it contains over 10,000 SNPs that differentiate C57BL/6J from DBA/2J. Fortunately, GN contains an SNP Browser tool that can efficiently filter these variants. Fig. 6.5 illustrates how to set up a query at www.genenetwork.org/webqtl/main.py?FormID=snpBrowser.

Enter the gene symbol, and trim the strain list ("Limit to") to the two parents of the BXDs. Also limit the request to SNPs with a missense of nonsynonymous effect on protein. The "Non-Redundant Only" check-box can be used to further simplify the list. In this case, the answer is there is only one known missense SNP in *Erbb4* that segregates among the BXDs. This SNP was initially named wt77-1-68118186 or MRS0130375, but is now formally called rs51992387. The variant is located at 68.071362 Mb in exon 25 and causes a highly conserved glutamate residue

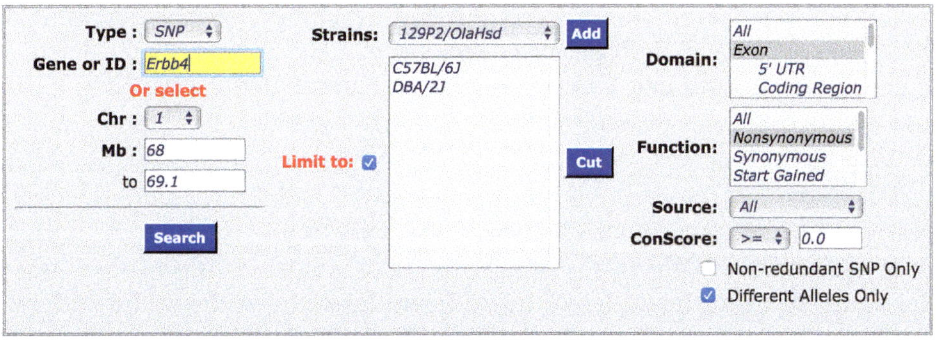


**Figure 6.5** Single nucleotide polymorphism (SNP) browser tool (*Errb4* gene) in GeneNetwork.

to be replaced with glycine between the *B* and *D* genotypes at codon 1014 (E1014G). This is a nonconservative substitution of a negatively charged glutamate for a small aliphatic glycine.

Differences in level of expression are a second potential source of variation in cocaine responses. Sequence differences across the entire gene and extending out to enhancers up to 0.5 Mb away may have effects on *Erbb4* expression. In general, the mutant *D* allele has lower expression, as much as fourfold lower in liver and hepatocytes (see *Erbb4* probe set 10355278). But there are exceptions, and expression of exon 12 is significantly higher in retinas of BXDs that inherit the *D* allele (*Erbb4* probe set 17223967). These differences in *Erbb4* mRNA levels can be mapped and in many cases are associated with large expression QTL (eQTLs) located precisely at the location of the *Erbb4* gene. These types of self-referential QTLs are called cis eQTLs. In the case of the *Erbb4* candidate gene, we therefore have one very interesting sequence difference in which the *D* allele is clearly a mutant and in which the D allele usually has lower expression.

In summarizing this simple workflow, we have revisited and reanalyzed older data on behavioral responses to cocaine using new genetic and genomic resources. The same could be done for hundreds of other behavioral data sets in GN. We have made a series of plausible assumptions that led to provisional findings, all of which could and should be retested. This is practical since these strains of mice, and over 70 new BXD strains, are readily available from the Jackson Laboratory. First, using new genotypes, we detected a QTL on Chr 1 in males (not females) that modulates an anxiety-associated phenotype at a cocaine dose of 5 mg/kg. Second, using vastly improved bioinformatic data and many new publications, we extracted a highly neurobiologically plausible candidate gene in the QTL interval.[11,12] Third, we examined sequence differences and expression differences in this top-ranked candidate. All of this is an appropriate subject for a discussion section and a review, but we certainly have not proved that a variant in *Erbb4* controls anxiety-like behavior in male mice. What we have done is generate a well-formulated hypothesis that extends from behavioral variation down to specific sequence variants. With a bit more work and reading,[13–15] it should be possible to propose specific mechanistic models; for example, the differential modulation of GABAergic neurotransmission by polymorphic ERBB4 protein in interneurons in the amygdala of the BXDs leads to predictable differences in anxiety[16] (see Bi et al.[16] for indirect support).

Before we leave this topic, we challenge the reader with one homework question: What phenotypes covary with variation in the expression of *Erbb4* in the amygdala and hippocampus of BXD mice? The relevant data for 50 to 100 strains are all in GN. Successfully answering this question[6] will, of course, lead to a set of secondary questions. The single most important secondary question is what is the false discovery rate when dealing with hundreds and thousands of correlations? This is still a hard question to answer, but the new version of GN2 will have tools that help address this issue.

In the next section, we perform one more minor miracle of data resurrection from the same study,[2] but in this case focusing on a very low-level phenotype (or endophenotype): the binding affinity of the crucial dopamine receptor 2 (DRD2) in striatum of the BXDs. The trait number is 10265 in the *BXD Published Phenotypes*, data set. Before reading further, you may want to see what sense you can make of these data and then compare to the subsequent text.

## DOPAMINE D2 RECEPTOR DENSITIES IN THE DORSAL STRIATUM

Dopamine (DA) is a major neurotransmitter in the mesolimbic and nigrostriatal systems. Cocaine increases dopaminergic neurotransmission of the midbrain dopamine neurons by inhibiting reuptake of DA by SLC6A3, the dopamine transporter.[17] There are two major dopamine receptors: DRD1 and DRD2.[18] In the mid-1990s, one of the most accurate ways to estimate expression of DRD2 protein was to measure binding of the radioactive ligand [$^{125}$I] epidepride to membrane fractions of dissected brain regions, such as the dorsal striatum. Epidepride also binds to DRD3, but this does not greatly complicate the picture since expression of D2R is at least 10 times higher in most brain regions, including striatum.

There are two independent studies of DRD2 protein expression in dorsal striatum: first, the 1999 paper we have been building on in this review and second, the study by Hitzemann.[19] The first question we can ask is how replicable are estimates of DRD2 levels generated in different laboratories using different methods (binding affinity in the case of trait 10264 vs. receptor autoradiography in the case of trait 10220) and using mice raised in somewhat different conditions? The correlation coefficient between males in the two studies is 0.66 with 14 strains in common ($P < .01$). This gives us some basis to trust the data sets and to move ahead with a genetic and even functional analysis of DRD2 expression in striatum. To avoid the trap of mouse chauvinism, we present trait

10265—the corresponding trait in females ($r = 0.67$ to males within the 1999 study, $P < .01$)—formally listed in GN with this description: CNS, pharmacology, protein expression: dopamine receptor 2 and 3 (DRD2/DRD3) maximum binding affinity (Bmax) to membrane fragments in dorsal striatum (caudate putamen) of females (125I-epidepride ligand) [fmol/mg wet weight].

The trait has an interesting distribution across 16 strains. On Fig. 6.6, the expected z score for a normally distributed trait with a mean beta max of 236 ± 104 fmol/mg (SD) is shown on the $x$ axis, and the actual value on the $y$ axis. A notable feature of this plot is the jump in values almost at the middle of the distribution (close to a z of 0), from 175 in BXD24 to 245 in BXD13. This type of discontinuity is what we expect if there is a single major effect locus modulating DRD2 protein levels. Even though we have data for a modest number of strains, if a trait is highly heritable, i.e., has a simple genetic architecture (as hinted at by this distribution), then we may be in luck.

In this case, mapping the strain distribution pattern of DRD2 expression, we detect a nominally significant QTL on distal Chr 15 (Fig. 6.7). Why qualify with the word "nominally"? Because when a trait has been measured in a small number of cases, the risk of false discovery can be high. We definitely need to run through the same routine that we did with the cocaine behavioral data to establish the sensitivity or robustness of the Chr 15 locus. The bootstrap analysis is highly encouraging: about 80% of resampled subsets (with replacement) map between 80 and 90 Mb. To the extent that we trust the original data, we can also trust the mapping to distal Chr 15. With this large QTL interval

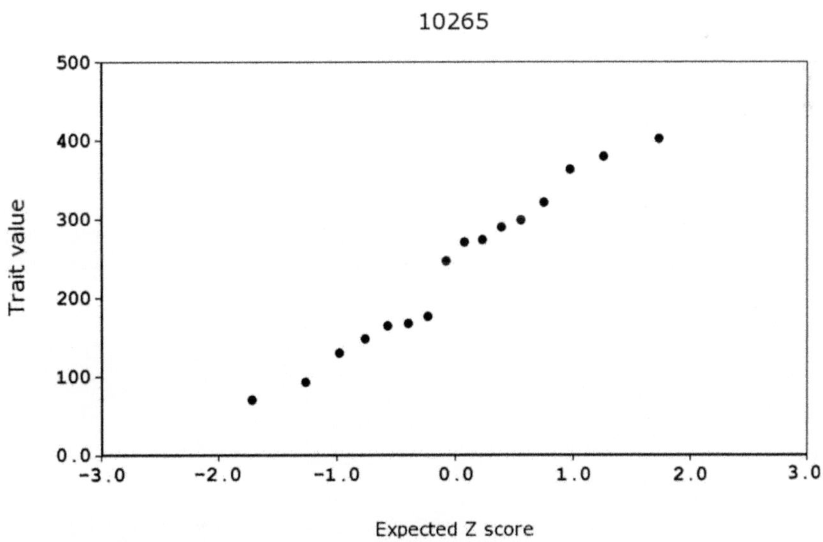


**Figure 6.6** Trait 10265 (in females), expected z score, DRD2/DRD3 maximum binding affinity to membrane fragments in dorsal striatum (caudate putamen).

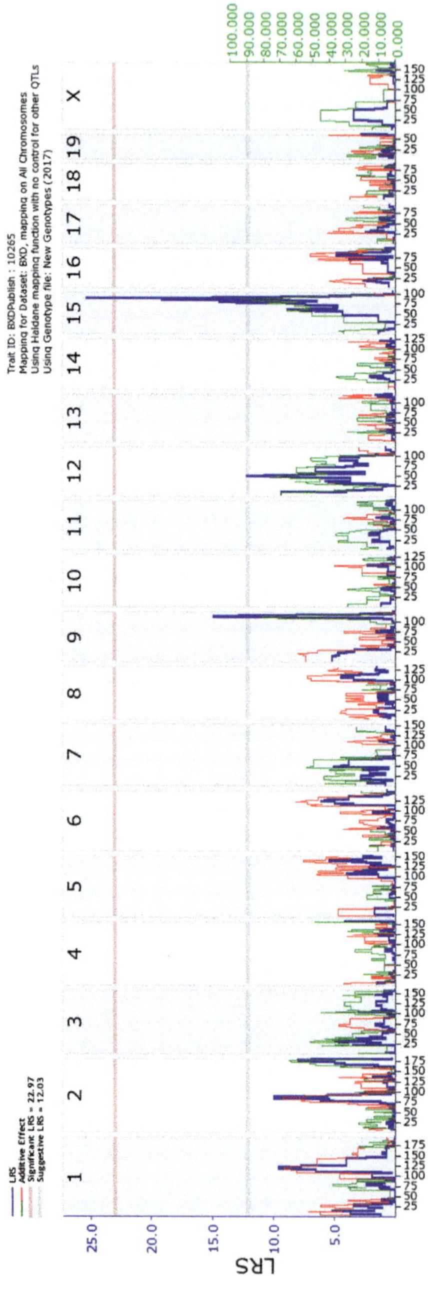


**Figure 6.7** Genetic mapping of trait 10265. The significant quantitative trait locus (QTL) is observed on Chr 15.

based on a very small sample, we immediately would like some additional support or validation. This locus should still be considered shaky. What can we do to corroborate a link to distal Chr 15 for DRD2 receptor and associated traits?

One way to corroborate this locus is to determine what other phenotypes (hopefully related) map to the same region of Chr 15. This general strategy is referred to as phenome-wide association or reverse complex trait analysis.[5,20] Instead of starting with a trait and proceeding down to a QLT or gene, we start with specific SNP and extract the set of phenotypes that covary with the SNP. In this particular case, we can select rs32098314 that is located on Chr 15 at 87.988948 Mb, in the middle of the QTL peak. We compute the correlations of this SNP against the entire BXD phenome. Of course, we expect to see trait 10265 near the top of this list of covariates, but we also see the following traits that map into this same region of Chr 15: 11693: latency to enter an open quadrant of a zero maze (a test of anxiety-like behavior); 10254: DRD1 protein expression level; 10304: cocaine stereotypy; 10022: saccharin sweet taste response; 17062: serotonin concentration in striatum after treatment with the neurotoxin, MPTP; 17043: the baseline level of tyrosine hydroxylase in striatum, and finally, trait 10277: dopamine transporter level (DAT, SLC6A3). Many of these traits are biologically coherent, and they also map into the same regions of Chr 15. This provides corroboration of the DRD2 results and could also motivate a second round of studies to confirm and refine the QTL to make candidate gene analysis more practical. A logical next step to discover a gene that modulated DRD2 protein level in mouse striatum would be to double the sample size of BXDs. One way to do this would be to select those BXDs with recombinations on Chr 15 between 82 and 92 Mb. These "recombinant" strains can help refine the map. They are easy to select by mapping an SNP such as rs32098314 as if it were a phenotype over the interval (be sure to select "Haplotype Analyst"; some examples include BXD22, 38, 49, 73b, 113, 139, 186, 205).

In this second example using GN, we have defined about a 10-Mb interval of interest with respect not only to DRD2, but also as shown earlier for DRD1 and DAT. We could, in principle, combined all of these "DA-associated traits" into one summarized metatrait or eigentrait using tools in GN that enable users to synthesize groups of phenotypes and transcripts, but in this case, combining across phenotypes will not gain us much more than we already know, which is that a region of distal Chr 15 is of interest (Fig. 6.8). What we might do instead, before starting up more studies, is to evaluate the expression of genes in Chr 15 from 82 to 92 Mb. What genes have highest expression? What genes are associated with high-impact missense mutations or cis-acting eQTLs. If we have done our job well in this review, you will know how to begin answering these questions. Your first step, as usual, is to link to www.genenetwork.org, and if you get stuck, then review the first example or the even more detailed primer by Mulligan et al.[6] Happy hunting.

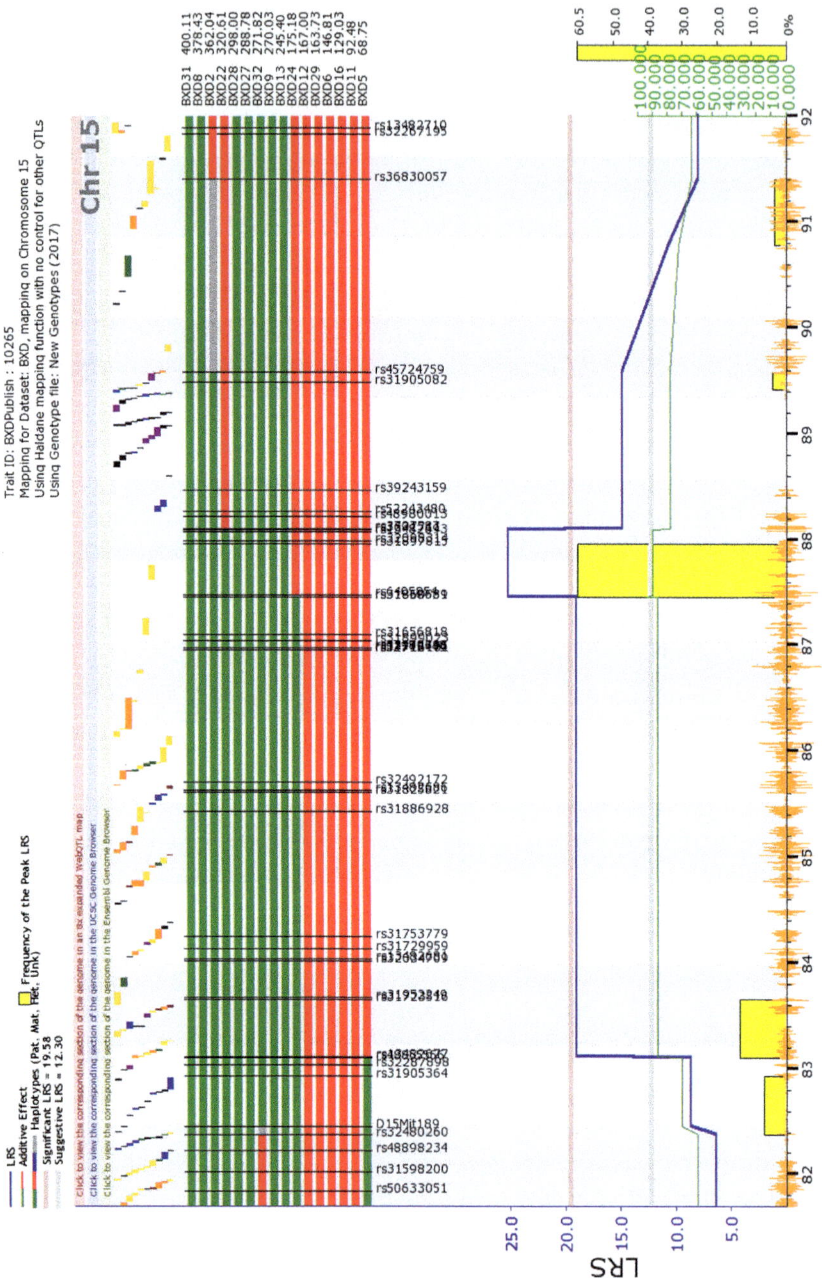


Figure 6.8 Mapping on Chr 15, trait 10265.

## CONCLUSIONS

GN, specifically the QTL analysis, enables us to analyze the genetic differences among the population. The database contains more than 5000 phenotypes obtained from the worldwide research laboratories. Using the cocaine data, we have described how to analyze the individual genetic differences in locomotor response to cocaine among the different RI strains. The program contains the data regarding dose-dependent cocaine behavior, receptor densities in the CNS, calculates the genetic correlations with graphic representation, and locates the candidate genes involved in the cocaine-related behavior and neurochemistry. We have analyzed the behavioral data, biologic data, and their correlation, and moreover, the correlation with the previously published data, called the network analysis. Combining the behavioral, neurobiological, and other related phenotypes and gene expression data, GN becomes a powerful tool for pointing to mechanisms and for hypothesis generation.

## REFERENCES

1. Peirce JL, Lu L, Gu J, Silver LM, Williams RW. A new set of BXD recombinant inbred lines from advanced intercross populations in mice. *BMC Genet.* 2004;5:7.
2. Jones BC, Tarantino LM, Rodriguez LA, et al. Quantitative-trait loci analysis of cocaine-related behaviours and neurochemistry. *Pharmacogenetics.* 1999;9:607–617.
3. Taylor BA. Recombinant inbred strains. In: Lyon ML, Searle AG, eds. *Genetic Variants and Strains of the Laboratory Mouse.* 2nd ed. Oxford: Oxford UK; 1989:773–796.
4. Gerlai R. Gene-targeting studies of mammalian behavior: is it the mutation or the background genotype? *Trends Neurosci.* 1996;19:177–181.
5. Wang X, Pandey AK, Mulligan MK, et al. Joint mouse-human phenome-wide association to test gene function and disease risk. *Nat Commun.* 2016;7:10464.
6. Mulligan MK, Mozhui K, Prins P, Williams RW. GeneNetwork: a toolbox for systems genetics. *Methods Mol Biol.* 2017;1488:75–120.
7. Rosen GD, Chesler EJ, Manly KF, Williams RW. An informatics approach to systems neurogenetics. *Methods Mol Biol.* 2007;401:287–303.
8. Williams RW, Mulligan MK. Genetic and molecular network analysis of behavior. In: *Bioinformatics of Behavior: PT 2*International Review of Neurobiology. vol. 104. 2012:135–157.
9. Philip VM, Duwuru S, Gomero B, et al. High-throughput behavioral phenotyping in the expanded panel of BXD recombinant inbred strains. *Genes Brain Behav.* 2010;9:129–159.
10. Fazzari P, Paternain AV, Valiente M, et al. Control of cortical GABA circuitry development by Nrg1 and ErbB4 signalling. *Nature.* 2010;464:1376–1380.
11. Krivosheya D, Tapia L, Levinson JN, et al. ErbB4-neuregulin signaling modulates synapse development and dendritic arborization through distinct mechanisms. *J Biol Chem.* 2008;283:32944–32956.
12. Vullhorst D, Neddens J, Karavanova I, et al. Selective expression of ErbB4 in interneurons, but not pyramidal cells, of the rodent hippocampus. *J Neurosci.* 2009;29:12255–12256.
13. Wen L, Lu YS, Zhu XH, et al. Neuregulin 1 regulates pyramidal neuron activity via ErbB4 in parvalbumin-positive interneurons. *Proc Natl Acad Sci USA.* 2010;107:1211–1216.
14. Chen YJ, Zhang M, Yin DM, et al. ErbB4 in parvalbumin-positive interneurons is critical for neuregulin 1 regulation of long-term potentiation. *Proc Natl Acad Sci USA.* 2010;107(50): 21818–21823.
15. Del Pino I, García-Frigola C, Dehorter N, et al. Erbb4 deletion from fast-spiking interneurons causes schizophrenia-like phenotypes. *Neuron.* 2013;79:1152–1168.

16. Bi LL, Sun XD, Zhang J, et al. Amygdala NRG1-ErbB4 is critical for the modulation of anxiety-like behaviors. *Neuropsychopharmacology*. 2015;40:974–986.
17. Koob GF, Pietro PS, Bloom FE. Neuroscience of addiction. *Neuron*. 1998;21:467–476.
18. Baldessarini RJ, Tarazi FI. Brain dopamine receptors: a primer on their current status, basic and clinical. *Harv Rev Psychiatry*. 1996;3:301–325.
19. Hitzemann R, Hitzemann B, Rivera S, et al. Dopamine D2 receptor binding, DRd2 expression and the number of dopamine neurons in the BXD recombinant inbred series: genetic relationships to alcohol and other drug-associated phenotypes. *Alcohol Clin Exp Res*. 2003;27:1–11.
20. Li J, Reynolds R, Pomp D, Allison DB, Yi N. Mapping interacting QTL for count phenotypes using hierarchical Poisson and binomial models: an application to reproductive traits in mice. *Genet Res (Camb)*. 2010;92:13–23.

SECTION II

# Searching for New Genes: Natural Genetic Variation

# CHAPTER 7

# Spontaneous Versus Induced Mutations: Conceptual and Methodological Considerations for the Neurobehavioral Geneticist

**Robert T. Gerlai**
University of Toronto Mississauga, Mississauga, ON, Canada

## INTRODUCTION

As science evolves, and as methods get increasingly sophisticated and data accumulate, scientists tend to become increasingly specialized.[1] The more narrowly defined specialization allows them to remain competitive experts within their field. The narrow specialization, however, may lead to reduced ability to switch from one method or concept to another. We often walk the path in front of us defined by our specific interests, the methods we have established in our laboratories, and the concepts with which we are familiar. We do not always explore whether an alternative approach may have some advantages. In this introductory chapter, I attempt to compare and contrast two distinct classes of approaches to the genetic analysis of behavior and brain function that have been utilized in separate laboratories: behavior genetic analysis based upon naturally occurring genetic variation and the other approach, which utilizes experimentally induced mutations. The chapters that follow this current one will provide ample details on how these distinct methods may be employed. They will discuss numerous methodological considerations for these approaches, and they will provide excellent experimental examples for the reasons they have been chosen. The goal of the current chapter is not this. I will not reiterate the points illuminated by the subsequent chapters. Instead, I will set the stage for the reader to consider why genetic analysis based upon naturally occurring genetic variability versus upon experimentally induced mutations may be appropriate for his or her specific purposes, and how differently these two distinct approaches may tackle questions about the genetic mechanisms and evolution of brain function and behavior.

*Molecular-Genetic and Statistical Techniques for Behavioral and Neural Research*
ISBN 978-0-12-804078-2, https://doi.org/10.1016/B978-0-12-804078-2.00007-6

## UTILIZING NATURALLY OCCURRING VARIANCE: THE OLD SCHOOL OF BEHAVIOR GENETICS?

Before the molecular revolution, behavior genetics was mainly confined to using methods that relied upon naturally occurring genetic variance.[2] Naturally occurring genetic variance here is defined loosely as any genetic variability that is not experimentally induced, i.e., using mutagens or recombinant DNA techniques. Although chemical or electromagnetic radiation-based (X-ray, for example) methods did allow scientists to experimentally induce mutations even before the molecular revolution,[3] in the absence of modern recombinant DNA techniques, the genes carrying these mutations could not be identified, so mechanistic details about the protein products of these genes could not be elucidated. Thus, this early era of behavior genetics was mostly characterized by the focus on naturally occurring genetic variation.[4] However, I must emphasize that the more seasoned history behind using naturally occurring genetic variation in the analysis of brain function and behavior does not by any means imply that this approach has no place in modern neurobehavioral genetics. It does. There are many reasons one may turn to this approach.

From an evolutionary/ecological standpoint, naturally occurring genetic variance may be adaptive, i.e., it may increase evolutionary fitness, or it may be neutral. Extensive reviews have been written on this topic from the perspective of evolutionary biology,[5–7] i.e., from the perspective of phylogenetic causation, so I will not reiterate all points. Instead, I focus on the proximate question, i.e., phenogenetic causation, of how naturally occurring mutations work in an organism, and why their analysis may allow us to understand biological mechanisms of brain function and behavior.

Unlike induced mutations, naturally occurring mutations tend not to be lethal.[8] Furthermore, often, they turn out to be only minor modulators of behavior and brain function and not crucial determinants of it. From a translational perspective, it is noteworthy that genes whose protein products are modulators and not major switches for, or determinants of, certain brain functions tend to be better targets of drug development. This is because alteration of the function of their protein products by pharmacological agents may have less strong effects (modulators), and therefore, development of drugs that have fine-tuned, as opposed to "hammer-like," effects on the brain may be easier.

If natural mutations occur with higher frequency (i.e., they are present in at least 1% of the population), they are called genetic polymorphisms, and the mutants, or genetic variants, are often regarded as "normal."[9] These are the mutations that are either neutral from an evolutionary fitness viewpoint or those that may represent an evolutionary stable strategy, e.g., maintained by frequency-dependent selection (natural selection that favors the less frequent phenotypes leading to an oscillating and/or genetically variable population), or those that under certain fluctuating environmental conditions may confer some fitness advantages. If the mutation occurs with smaller frequency, the mutant outliers may be considered abnormal. These are the mutations that are carried by genes, e.g., in the human population, whose variants we may call disease-causing or predisposing alleles. Irrespective of their frequency in a natural population, however, naturally occurring mutations "fit in" with the rest of the genome of their carriers, at least in the sense that they allow the carrier to function and reproduce.

Numerous methods have been developed to identify and utilize naturally occurring genetic variation. The oldest one is artificial selection.[10] In an artificial selection experiment, individuals exhibiting a desired phenotypical extreme are mated with each other at every generation for potentially many generations. If selection response occurs, i.e., if the offspring of such matings show a change in their phenotypical mean toward the desired direction, this change is taken as evidence of genetic variance in the starting population. The speed with which the change occurs across multiple generations allows one to judge whether genes with major or minor effects are involved, and whether the genetic variance in the starting population was due to variability at one, a few, or many loci.[11] In the past, artificial selection experiments have been used to obtain an approximate estimate of the number of variable loci, to develop selected lines comprising individuals with potentially desirable phenotypical traits, and to analyze correlated responses presumably due to genetic pleiotropy or linkage. Apart from these pieces of information or goals, very little knowledge could be gained about the biological mechanisms underlying the achieved changes. In the 21st century, however, with the help, for example, of high throughput and comprehensive gene expression analyses,[12,13] one can gain insights into the molecular mechanisms and the identity of genes underlying the selection response.[14]

Another classical behavior genetic method that was employed in the past, and one which still has excellent utility, is the comparison of inbred strains,[15,16] and specific breeding systems utilizing such strains.[17–20] Breeding of inbred strains usually starts from a genetically highly heterogeneous population and entails brother-sister mating, also known as sib-mating. After 20 generations of sib-mating, the strain is considered to be inbred, i.e., 99.9% of its loci is in a homozygous form. Inbred strains have been particularly useful in studies with mice and rats, as there are a large number of phenotypically and genetically well-characterized strains available for these species[15,21–23] (also see https://www.jax.org/jax-mice-and-services/find-and-order-jax-mice/most-popular-jax-mice-strains/ and http://www.criver.com/find-a-model?animal_type=Rats&loc=CA). Researchers using zebrafish also have a choice among genetically well-defined strains of this species[24] (also see https://zebrafish.org/fish/lineAll.php), albeit these strains possess much fewer loci in a homozygous form than inbred strains of mice or rats do. Achieving near-complete homozygosity may not require labor-intensive sib-mating, as it may also be achieved in one step, using gynogenesis.[25,26] However, the latter method has not been practical with vertebrate species other than amphibians and fish.

Irrespective of how an inbred strain is generated, the key point is that within an inbred strain, individuals are expected to be genetically identical. Due to random genetic drift resulting from sib-mating, two inbred strains are expected to exhibit differences at many genetic loci. An inbred strain may be viewed as source of a practically limitless number of individuals that have the same genome, a genome that may have characterized a putative single subject from the genetically variable starting population. From this perspective, inbreeding essentially converts individual genetic differences into strain differences. Unlike in the increasingly fashionable analysis of "individual differences,"[27–29] inbred strains thus allow the systematic dissection of genetic and environmental variance, i.e., the analysis of the contribution of genetic versus environmental effects to phenotypical differences. Inbred strains also allow the use of sophisticated statistical methods, because unlike in the analysis of individual differences, where the sample size ($n$) for a particular genome equals 1, the number of individuals tested from a given inbred strain can be as large as the experimenter wants.

Breeding systems developed using inbred strains utilize such statistical methods, essentially modified variance analyses or bivariate, covariance-based, statistics.[18–21] Examples of such breeding systems are the Mendelian cross[17,18] and the diallel cross[19,20] systems. The Mendelian cross system includes two inbred parental strains ($P_1$ and $P_2$), two reciprocal $F_1$ generations ($F_{1a} = P_1$ female $\times$ $P_2$ male; $F_{1b} = P_1$ male $\times$ $P_2$ female), at minimum one $F_2$ generation ($F_1 \times F_1$), and two backcross generations ($B_1 = P_1 \times F_1$; $B_2 = P_2 \times F_1$). The diallel cross comprises of $n$ number of inbred parental strains (with $n =< 3$) and the reciprocal $F_1$ crosses between these parental strains in all possible combinations. Phenotypical characterization of the generations and strains of these breeding systems allows the investigator to reveal information on the pattern of inheritance, the genetic architecture. Genetic architecture in this case means how genes interact with each other and how they interact with the environment. For example, these breeding systems can be used to estimate the contribution of additive genetic effects and dominance-related effects (intralocus interactions) of alleles, as well as epistatic (interlocus) interactions among alleles to genetic variance. They also allow the dissection of phenotypical variance to environmental and genetic variance components. The bivariate extension of the diallel method even enables scientists to estimate correlations between traits solely arising from genetic effects or solely due to environmental effects.[30] But why is this interesting? What can one do with such pieces of information? Is quantitative genetics, as the field was called, simply an exercise of complex statistical procedures for its own sake?

At the early stages of behavior genetics, this criticism may have been valid. But by now, it has become clear that the information scientists of this field gathered may be useful for several reasons. Without being exhaustive, here, I highlight some examples of how information about the genetic architecture gathered using quantitative genetic methods has been or can be utilized. Quantitative genetics allows one to estimate heritability, i.e., the ratio of genetic variance and total phenotypical variance (broadly defined heritability) or the ratio between genetic variance arising solely from additive genetic effects and total phenotypical variance (narrow definition of heritability). Heritability, especially the narrowly defined kind, predicts how fast artificial selection may progress. Uncovering the dominance relationships among multiple alleles, i.e., answering the question of which trait corresponding to an allele manifests in the $F_1$ generation when inbred strains are crossed, allows one to judge whether a particular trait has been under ambidirectional selection (stabilizing selection that prefers intermediate phenotypes) during the recent evolutionary past of the species, or whether the selection pressure was directional (natural selection preferring one extreme), or whether the selection was disruptive (natural selection that prefers the two extremes).

Although inbred and genetically defined outbred strains could only be generated using laboratory animals, quantitative genetics also flourished in the analysis of human behavior. Twin studies could utilize naturally occurring genetic variance and the experiments of nature. Comparison of monozygotic twins, dizygotic twins, and genetically unrelated children reared together or apart allowed scientists to dissect the contribution of genes and of the environment to phenotypical differences.[31] Finally, after the molecular revolution, classical quantitative genetic methods set the stage for subsequent molecular follow-up studies. There are two fundamentally different ways a quantitative genetic analysis of naturally occurring genetic variance may be investigated with molecular methods. One approach would be to identify the locus of the mutation, and the second approach would be to investigate gene expression differences underlying phenotypical variance.

For the identification of the locus of naturally occurring mutations, one may use linkage analysis-based positional cloning, a method developed for chemically induced mutations.[32] A conceptually similar method often employed for the identification of the locus of naturally occurring mutations is quantitative trait locus (QTL) mapping.[33] Both approaches use genetic markers and measure cosegregation of the particular mutant, or variant, phenotype with these markers. The two most frequently employed laboratory organisms for such mapping studies are the mouse and the zebrafish, species for which tens of thousands of molecular markers (with known chromosomal locations) have been developed.[34–36]

The other fundamentally different method that one can utilize to follow up on the quantitative genetic analysis of naturally occurring mutations is to analyze mRNA expression. The phenotypical differences between two distinct genotypes may be the result of a natural mutation-induced differential gene expression. Although this method may not directly lead one to the precise identification of the gene carrying the mutation, it may delineate the genes that encode a functional network of proteins that is affected by the mutation.

In sum, a quantitative genetic method that, for example, allows the estimation of the number of loci underlying observed differences between or among inbred strains may be highly useful, as it informs the investigator of what to expect. For example, if the estimated number of variable loci is low, one may expect a somewhat less complex gene expression profile difference between or among the inbred strains, and one may expect only a few major QTLs in a QTL analysis. In other words, fewer resources may need to be allocated, and faster success may be expected in a linkage analysis study aimed at identifying the gene(s) underlying the observed phenotypical variance. Similarly, potentially less expensive and less comprehensive gene expression and bioinformatics analyses may be sufficient to identify major differentially expressed genes and the functional interactions in which they are involved. On the other hand, if the estimated number of variable loci is large, or if the epistatic (interlocus interaction) effects are estimated to be high, one may expect a rather complex genetic architecture that may be more difficult to disentangle with molecular methods. In the latter case, one may be able to utilize another classical quantitative genetic approach, the use of recombinant inbred strains.

Recombinant inbred strains are a set of strains that are generated by crossing two inbred parental strains.[37] During the gametogenesis, i.e., meiotic cell division, in the $F_1$ generation of this cross, recombinant chromosomes are generated. A recombinant chromosome at this stage usually contains a single crossover, so it can be imagined as made up of two parts, one from parental strain 1 and the other from parental strain 2. Further breeding, i.e., generation of $F_2$, $F_3$, $F_n$ generation, results in increasing number of crossovers, thus a more elaborate recombination pattern between the two parental strain genomes. If one generates a number of recombinant inbred strains this way, each strain eventually will fix a particular recombinant pattern for each of its chromosomes, and in a homozygous form. Due to the random nature of the location of crossover during meiosis, each strain will have a different recombination pattern. If a large enough number of recombinant inbred strains is generated, these strains may be used as a mapping tool: cosegregation of the target phenotype with genetic markers may be easily observed without having to go through detailed linkage analysis, as used in QTL mapping.[37] And given that the recombinant inbred strains are homozygous, i.e., express no genetic variance within a strain, they represent a stable resource, and their phenotyping can accumulate crucial information over time, making the resource increasingly useful for mapping and understanding genetic correlations and underlying mechanisms too.[38] Perhaps the most comprehensive and best known recombinant inbred strain set is the B × D lines, generated from the highly genetically distinct parental inbred mouse strains, C57Bl/6 and DBA/2.[38] If, however, the number of genetic loci underlying the observed phenotypical difference between these two strains is large, the previously described recombinant strain-based mapping approach is rather complicated. To reduce complexity, congenic recombinant inbred strains may be generated by backcrossing to one of the parental strain.[39] Backcrossing, say, to C57Bl/6, will reduce the DBA/2 chromosomal regions in the resulting generation. If one develops multiple separate backcross lines, each line will possess a unique recombination pattern, so in each line, particular, but different, regions of the chromosomes will remain DBA/2-type, and this region will be smaller than in the case of recombinant inbred strains. Briefly, recombinant congenic inbred strains will reduce genetic variance and will allow one to focus on shorter chromosomal regions affecting the phenotype. This approach thus will simplify a potentially complex genetic contribution to phenotypical differences between the two parental strains, and therefore, it may facilitate the investigation of single gene effects.[39]

Another, conceptually similar method is the recombinant consomic strains.[40] Consomic strains can be generated similarly to recombinant congenic inbred strains, but in their case, the backcrossing to a particular parental inbred strain is repeated for multiple generations. Furthermore, the genotype of the strains can also be tested using molecular markers, further aiding the development of these strains. The resulting strains are similar to recombinant congenic strains, except that most of the genome of these animals is expected to carry loci of the type that are found in the strain to which the backcrossing was performed, with the exception of a very small region where loci of the other strain may be found. If a large enough number of consomic strains are developed, they allow even more precise localization of loci affecting the phenotype because the chromosomal region in which these strains differ is short and affects only a particular chromosome.[40]

In summary, there is an array of quantitative genetic methods that allow the investigator to investigate naturally occurring genetic variance. These methods may prepare the investigator for a subsequent molecular genetic analysis as to what to expect. They, for example, may allow one to reduce genetic complexity, and facilitate a simplified molecular analysis. However, quantitative genetic analyses of naturally occurring genetic variance cannot accomplish one important goal: it cannot study silent genes, loci that show no genetic variance.

## ANALYSIS OF INDUCED MUTATIONS: 21ST-CENTURY RECOMBINANT DNA TECHNOLOGY AND ITS ADVANTAGES

Lack of genetic variance at a particular locus can arise for several reasons. For example, inbreeding induced genetic drift is expected to lead to the development of substantial differences among inbred strains, but which alleles get fixed and which ones get lost during the process is a matter of chance. If only a limited number of inbred strains are developed, just by chance, one may expect several of the loci of these inbred strains to harbor the same alleles, i.e., lack of genetic variance for these loci among the strains. Similarly, not all loci may show genetic variance in the starting, genetically heterogeneous population, so inbreeding starting from such a population will result in inbred strains that are going to carry the same alleles on these loci. Genetic variance may also be eliminated in natural populations. In fact, in case of strong ambidirectional selection, i.e., if the trait has high fitness value, genetic variance is expected to be very low. In other words, genes whose protein products are essential for a particular biological function are expected to show low variance or no variance at all. For quantitative genetic approaches, such genes may remain undetected (silent). It is important to note that in heritability studies, one may confuse lack of genetic variance with lack of importance of genetic effects. Lack of genetic variance (zero heritability) is sometimes misinterpreted as proof of lack of genetic effects. In fact, lack of genetic variance often means the exact opposite of what one would assume: it means crucial importance of the gene in a biological function. The effect of such genes, and the mechanisms or biochemical interactions in which the proteins they encode are involved, can only be studied using induced mutations, i.e., with modern molecular biology methods.

But there are other advantages of inducing mutations to exploiting naturally occurring ones in the analysis of genetic mechanisms of brain function and behavior. Artificially induced mutations can be generated on an otherwise genetically homogeneous genetic background. If accomplished, this genetic homogeneity is a major advantage when the effects of the induced mutations need to be quantified or interpreted. In an ideal scenario, a homogeneous genetic background means that the only difference between the mutant and the wild-type animal is the presence versus absence of the mutation. This was the inherent assumption for many classical gene targeting,[41] or knockout, studies with mice in which the authors utilized homologous recombination in embryonic stem (ES) cells.[42,43] But the original technology as employed turned out to suffer from an issue that has become known as the hitchhiking allele problem.[44,45]

Numerous gene-targeting studies utilized ES cells for the induction of homologous recombination-based gene replacement (knockout). The selection for this event was made efficient because it was also conducted in the ES cells in the petri dish. The ES cells most often originated from a 129 substrain of mice.[46] Many of these substrains, however, suffer from major phenotypical peculiarities, including structural abnormalities in the brain of the mice (lack of or significantly reduced size corpus callosum), inactivity, and impaired learning and memory.[22,47,48] For these reasons, the germ-line transmitting chimera (in which the 129 strain—derived ES cells carrying the artificially introduced null allele made the testes, so the sperm carried the null allele) was crossed with a female from another strain, usually C57BL/6, which did not suffer from these phenotypical abnormalities.[45] Although generation of such hybrid mice solved the phenotypical peculiarity issue, it created another problem; the resulting null mutant carriers could differ from wild-type control mice for two separate reasons: (1) the presence versus absence of the null mutation (the desired difference) and (2) the presence of 129-type alleles at several loci surrounding the target locus when it carried the null allele versus the absence of these 129-type alleles at loci surrounding the target locus when it carried the wild-type (nonnull mutated) allele.[45] The problem became known in the literature as the flanking,[45] or hitchhiking, allele issue.[46] But despite its widespread acceptance as a problem, and despite the fact that some peer-reviewed journals even implemented a strict policy forcing authors to address the issue before accepting their papers,[49] publications using gene targeting often continued to ignore it. The reason for this resistance may be twofold. One, if the null mutation had a robust effect, some could argue that, relatively speaking, the effects of hitchhiking alleles would be small and thus could be ignored. This argument is not likely to be valid, given that alleles of several hundred loci surrounding the target locus would remain 129-type even after 2 years of backcrossing to C57Bl/6, and many of these loci may have genes whose protein products could potentially affect the phenotype of interest, or interact with the target gene's protein and/or biochemical processes in which this protein is involved. The second argument, however, is valid: there are not many good solutions to the hitchhiking alleles issue, and those that are available represent rather time-consuming and expensive work.[45] Luckily, however, by now, we have several recombinant DNA approaches available that may represent good alternatives to the homologous recombination in the ES cell—based method. One of these is the Clustered Regularly Interspaced Repeats (CRISPR)-Cas system.[50]

The CRISPR-Cas system, discussed in detail in two chapters of this book,[50,51] utilizes recently discovered molecular components of the bacterial immune system.[52] Essentially, this genome-editing/gene-targeting system entails generation of a short recognition sequence that is complementary to the nucleotide sequence of the target gene one wants to mutate. The short sequence is spliced in between two of the series of CRISPR obtained from the bacterial genome and bound to the Cas9 endonuclease, whose DNA cutting specificity is then determined by the engineered sequence. The thus engineered Cas9 endonuclease acts as a pair of molecular scissors that targets the specific sequence of interest. The result is double-stranded cutting and abnormal end joining of DNA around the target locus, a mutation that often leads to a shift in the reading frame and thus silences the gene of interest. Importantly, the method requires no multiple steps of breeding, so it eliminates the issue arising from the hybrid genetic background in classical gene targeting. Furthermore, and perhaps even more importantly, it allows one to generate targeted mutations in practically any species of choice, not just those for which ES cells have been developed (i.e., the mouse or the rat). Although not without some disadvantages, the CRISPR/Cas9 technology also enables one to target multiple genes with identical sequences. Thus, one can potentially affect/manipulate entire gene families with suspected overlapping functions, a feat that could not be achieved in one step with classical gene targeting, a method often criticized to suffer from the issue of compensatory or helper genes.[53]

## SUMMARY AND CONCLUSIONS

Natural genetic variance allows one to investigate the effects of those genes that are likely to have been under weak selection pressure and thus are likely to be modifiers or modulators of brain functions and behaviors. Analysis of these gene effects can shed light on the evolutionary past and the ecology of the species in question, but it can also be useful for the understanding of biological mechanisms. Analysis of naturally occurring genetic variance has been the focus of the first pioneering studies of behavior genetics, but now such studies can be followed up using modern recombinant DNA methods that enable the investigator to gain mechanistic details including the precise identity of the gene carrying the naturally occurring mutation, as well as the roles this gene may play in brain function or dysfunction. The disadvantage of relying on naturally occurring genetic variance, however, is that several loci may remain nonvariable either because of random fixation of alleles in the studied population(s) or because of strong selection pressure. The latter possibility is particularly troublesome, because it implies that analysis of naturally occurring genetic variance can mostly allow one to study genes whose effects are not very important from evolutionary and biological mechanism standpoints. To study the effects of these latter genes, one needs to employ artificially induced mutations. Gene targeting is a reverse genetic method, one of the approaches one may employ to artificially induce mutations. Classical gene targeting, although an elegant and frequently employed method successfully utilized for decades, has been known to suffer from interpretive issues. Modern genome editing tools, including the CRISPR/Cas9 system, may be an excellent alternative, one that will likely revolutionize gene targeting.

## REFERENCES


1. Kuhn TS. *The Structure of Scientific Revolutions*. 3rd ed. Chicago: University of Chicago Press; 1996. ISBN 0-226-45807-5.
2. Fuller JL, Thompson WR. *Behavior Genetics*. New York: John Wiley & Sons, Inc.; 1960:396.
3. Snell GD. The induction by X-rays of hereditary changes in mice. *Genetics*. 1935;20:545–567.
4. Fuller JL. Behavior genetics. *Annu Rev Psychol*. 1960;11:41–70.
5. Neher RA. Genetic draft, selective interference, and population genetics of rapid adaptation. *Annu Rev Ecol Evol Syst*. 2013;44:195–215.
6. Kruuk LEB, Slate JW, Alastair J. New answers for old questions: the evolutionary quantitative genetics of wild animal populations. *Annu Rev Ecol Evol Syst*. 2008;39:525–548.
7. Curtsinger JW, Service PM, Prout T. Antagonistic pleiotropy, reversal of dominance, and genetic-polymorphism. *Am Nat*. 1994;144:210–228.
8. van der Neut R. Targeted gene disruption: applications in neurobiology. *J Neurosci Methods*. 1997;71: 19–27.
9. Hedrick P. *Genetics of Populations*. Jones & Bartlett Learning; 2011:104. ISBN 978-0-7637-5737-3.
10. Hirsch J, Erlenmeyer-Kimling L. Sign of taxis as a property of the genotype. *Science*. 1961;134: 835–836.
11. Gerlai R, Csányi V. Artificial bidirectional selection for a species specific behavioural element, staccato movement, in paradise fish (*Macropodus opercularis*). *Anim Behav*. 1994;48:1293–1300.
12. Westwood T. DNA microarrays and high though put sequencing for the analysis of behavior and brain function. In: Gerlai, ed. *Molecular-Genetic and Statistical Techniques for Behavioral and Neural Research*. 1st ed. Amsterdam: Elsevier; 2018. ISBN 9780128040782.
13. Hitzemann R. Genes, behavior and next-generation RNA sequencing. In: Gerlai, ed. *Molecular-Genetic and Statistical Techniques for Behavioral and Neural Research*. 1st ed. Amsterdam: Elsevier; 2018. ISBN 9780128040782.
14. Crabbe J. Using signatures of directional selection to guide discovery. In: Gerlai, ed. *Molecular-Genetic and Statistical Techniques for Behavioral and Neural Research*. 1st ed. Amsterdam: Elsevier; 2018. ISBN 9780128040782.

15. Nguyen PV, Gerlai R. Behavioural and physiological characterization of inbred mouse strains: prospects for elucidating the molecular mechanisms of mammalian learning and memory. *Gene Brain Behav.* 2002;1:72–81.
16. Wehner JM, Radcliffe RA, Bowers BJ. Quantitative genetics and mouse behavior. *Annu Rev Neurosci.* 2001;24:845–867.
17. Schicknick H, Hoffmann HJ, Schneider R, Crusio WE. Genetic analysis of isolation-induced aggression in the mouse. III. Classical cross-breeding analysis of differences between two closely related inbred strains. *Behav Neural Biol.* 1993;59:242–248.
18. Csányi V, Gerlai R. Open field behavior and the behavior genetic analysis of the paradise fish (*Macropodus opercularis*). *J Comp Psychol.* 1988;102:326–336.
19. Gerlai R, Csányi V. Diallel genetic analysis of the elements of paradise fish's (*Macropodus opercularis*, L.) ethogram. *Acta Biol Hung.* 1989;40:67–106.
20. Crusio WE, Genthner-Grimm G, Schwegler H. A quantitative-genetic analysis of hippocampal variation in the mouse. *J Neurogenet.* 1986;3:203–214.
21. Sprott RL, Staats J. Behavioral studies using genetically defined mice. A bibliography. *Behav Genet.* 1975;5:27–82.
22. Crawley JN, Belknap JK, Collins A, et al. Behavioral phenotypes of inbred mouse strains: implications and recommendations for molecular studies. *Psychopharmacology.* 1997;132:107–124.
23. Gill 3rd TJ, Smith GJ, Wissler RW, Kunz HW. The rat as an experimental animal. *Science.* 1989;245: 269–276.
24. Johnson SL, Zon LI. Genetic backgrounds and some standard stocks and strains used in zebrafish developmental biology and genetics. *Methods Cell Biol.* 1999;60:357–359.
25. Bongers AB, Sukkel M, Gort G, Komen J, Richter CJ. Development and use of genetically uniform strains of common carp in experimental animal research. *Lab Anim.* 1998;32:349–363.
26. Pelegri F, Schulte-Merker S. A gynogenesis-based screen for maternal-effect genes in the zebrafish, *Danio rerio*. *Methods Cell Biol.* 1999;60:1–20.
27. Falco AM, Bevins RA. Individual differences in the behavioral effects of nicotine: a review of the preclinical animal literature. *Pharmacol Biochem Behav.* 2015;138:80–90.
28. Tran S, Nowicki M, Muraleetharan A, Chatterjee D, Gerlai R. Neurochemical factors underlying individual differences in locomotor activity and anxiety-like behavioral responses in zebrafish. *Prog Neuropsychopharmacol Biol Psychiatry.* 2016;65:25–33.
29. Tran S, Gerlai R. Individual differences in activity levels in zebrafish (*Danio rerio*). *Behav Brain Res.* 2013;257:224–229.
30. Gerlai R, Crusio WE. Organization of motor and posture patterns in paradise fish (*Macropodus opercularis*): environmental and genetic components of phenotypical correlation structures. *Behav Genet.* 1995; 25:385–396.
31. Grasby KL, Verweij KJH, Mosing MA, Zietsch BP, Medland SE. Estimating heritability from twin studies. *Methods Mol Biol.* 2017;1666:171–194.
32. Reijmers LG, Coats JK, Pletcher MT, Wiltshire T, Tarantino LM, Mayford M. A mutant mouse with a highly specific contextual fear-conditioning deficit found in an N-ethyl-N-nitrosourea (ENU) mutagenesis screen. *Learn Mem.* 2006;13:143–149.
33. Flint J. Analysis of quantitative trait loci that influence animal behavior. *J Neurobiol.* 2003;54:46–77.
34. Dietrich WF, Copeland NG, Gilbert DJ, Miller JC, Jenkins NA, Lander ES. Mapping the mouse genome: current status and future prospects. *Proc Natl Acad Sci USA.* 1995;92:10849–10853.
35. Knapik EW, Goodman A, Ekker M, et al. A microsatellite genetic linkage map for zebrafish (*Danio rerio*). *Nat Genet.* 1998;18:338–343.
36. Kelly PD, Chu F, Woods IG, et al. Genetic linkage mapping of zebrafish genes and ESTs. *Genome Res.* 2000;10:558–567.
37. Plomin R, McClearn GE, Gora-Maslak G, Neiderhiser JM. Use of recombinant inbred strains to detect quantitative trait loci associated with behavior. *Behav Genet.* 1991;21:99–116.
38. Adamová E, Williams RW, Jones BC. The use of recombinant inbred strains in systems genetics and functional analyses in behavioral pharmacology. In: Gerlai, ed. *Molecular-Genetic and Statistical Techniques for Behavioral and Neural Research.* 1st ed. Amsterdam: Elsevier; 2018. ISBN 9780128040782.

39. Fortin A, Diez E, Rochefort D, et al. Recombinant congenic strains derived from A/J and C57BL/6J: a tool for genetic dissection of complex traits. *Genomics*. 2001;74(1):21–35.
40. Ohno T, Hata K, Baba T, et al. Establishment of consomic strains derived from A/J and SM/J mice for genetic analysis of complex traits. *Mamm Genome*. 2012;23:764–769.
41. Capecchi MR. Altering the genome by homologous recombination. *Science*. 1989;244:1288–1292.
42. Silva AJ, Stevens CF, Tonegawa S, Wang Y. Deficient hippocampal long-term potentiation in alpha-calcium-calmodulin kinase II mutant mice. *Science*. 1992;257:201–206.
43. Silva AJ, Paylor R, Wehner JM, Tonegawa S. Impaired spatial learning in alpha-calcium-calmodulin kinase II mutant mice. *Science*. 1992;257:206–211.
44. Gerlai R. Gene-targeting studies of mammalian behavior: is it the mutation or the background genotype? *Trends Neurosci*. 1996;19:177–181.
45. Gerlai R. Gene targeting: technical confounds and potential solutions in behavioral brain research. *Behav Brain Res*. 2001;125:13–21.
46. Ledermann B. Embryonic stem cells and gene targeting. *Exp Physiol*. 2000;85:603–613.
47. Wahlsten D. Deficiency of corpus callosum varies with strain and supplier of the mice. *Brain Res*. 1982; 239:329–347.
48. Contet C, Rawlins JN, Deacon RM. A comparison of 129S2/SvHsd and C57BL/6JOlaHsd mice on a test battery assessing sensorimotor, affective and cognitive behaviours: implications for the study of genetically modified mice. *Behav Brain Res*. 2001;124:33–46.
49. Crusio WE, Goldowitz D, Holmes A, Wolfer D. Standards for the publication of mouse mutant studies. *Gene Brain Behav*. 2008;8:1–4.
50. El Khoury LY, Campbell JM, Clark KJ. The transition of zebrafish functional genetics from random mutagenesis to targeted integration. In: Gerlai, ed. *Molecular-Genetic and Statistical Techniques for Behavioral and Neural Research*. 1st ed. Amsterdam: Elsevier; 2018. ISBN 9780128040782.
51. Melgoza A, Guo S. Systematic screens in zebrafish shed light on cellular and molecular mechanisms of complex brain phenotypes. In: Gerlai, ed. *Molecular-Genetic and Statistical Techniques for Behavioral and Neural Research*. 1st ed. Amsterdam: Elsevier; 2018. ISBN 9780128040782.
52. Horvath P, Barrangou R. CRISPR/Cas, the immune system of bacteria and archaea. *Science*. 2010;327: 167–170.
53. Gerlai R. Gene targeting in neuroscience: the systemic approach. *Trends Neurosci*. 1996;19:188–189.

CHAPTER 8

# Reduced Complexity Cross Design for Behavioral Genetics

**Camron D. Bryant[1], Martin T. Ferris[2], Fernando P.M. De Villena[2], M. Imad Damaj[3], Vivek Kumar[4], Megan K. Mulligan[5]**

[1]Boston, MA, United States; [2]University of North Carolina at Chapel Hill, Chapel Hill, NC, United States; [3]Richmond, VA, United States; [4]Bar Harbor, ME, United States; [5]The University of Tennessee Health Science Center, Memphis, TN, United States

## INTRODUCTION

Mammalian model organisms, in particular rodents, are essential for scientific discovery and advances in human health and well-being. Mouse fanciers such as Miss Abby Lathrop and enterprising geneticists such as C.C. Little and W.E. Castle were responsible for generating several inbred lines of mice in the early 1900s. Inbred strains arise after strict brother-sister mating for a minimum of 20 generations,[1,2] which ultimately results in the fixation of alleles to homozygosity at a genome-wide level and yields individuals within a strain that are presumed to be genetically identical. There are now hundreds of inbred mouse lines (e.g., 211 inbred lines at The Jackson Laboratory as of early 2017) and several more transgenic lines that are commercially available. Isogenic, or inbred, lines are not exclusive to mice and exist as a genetic resource for many different organisms critical to genetic research, including rats, *Caenorhabditis elegans*, and *Drosophila melanogaster*. Access to these isogenic populations provides greater reproducibility and standardization across experiments and laboratories, and it allows for the cumulative integration of data over time. To meet scientific demand, many commercial vendors maintain large colonies of these inbred and isogenic animals. Individual laboratories also maintain large breeding colonies of inbred strains over many years. Separation and maintenance of inbred colonies at different institutions can lead to fixation of residually heterozygous variants (especially if strains were not fully inbred prior to separation) or colony-specific spontaneous mutations (a process called genetic drift) and the subsequent, and usually unintentional, creation of nonisogenic subpopulations or substrains (Fig. 8.1). A prominent example is that of the C57BL/6 (B6) substrains described later.

The most widely used inbred mouse strain in biomedical research is B6. Starting in the 1950s and continuing into the 1980s, widespread demand led to the separation and expansion of this strain by separate mouse vendors and the subsequent creation of a number of B6-derived substrains (Fig. 8.1). These strains include C57BL/6J (J), C57BL/6EiJ (E), C57BL/6NJ (NJ), C57BL/6ByJ (B), C57BL/6NCrl (N), and C57BL/6NTac (T), among several others. Note that substrains are differentiated by

*Molecular-Genetic and Statistical Techniques for Behavioral and Neural Research*
ISBN 978-0-12-804078-2, https://doi.org/10.1016/B978-0-12-804078-2.00008-8

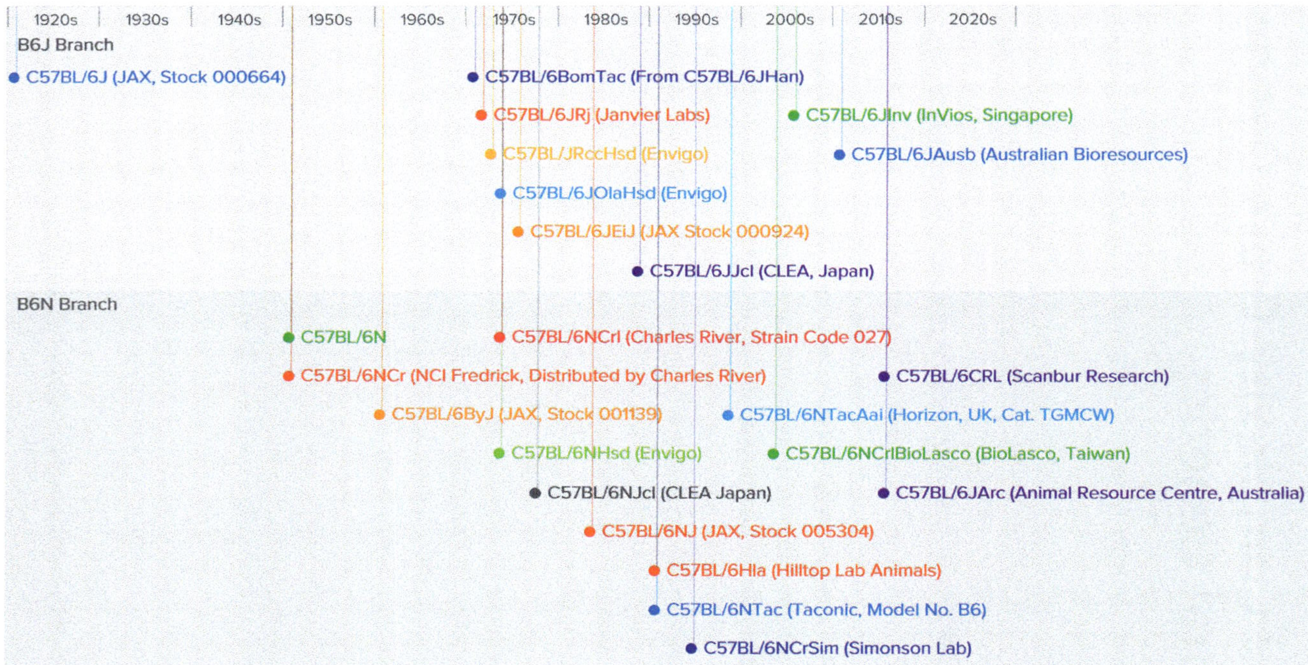


**Figure 8.1** ***C57BL/6 phylogeny.*** A timeline is provided for the separation of substrains from the B6J and B6N branches.

abbreviations that indicate their origin from different labs or commercial vendors. For example, the designations J, Ei, N, By, Crl, and Tac represent sublines from The Jackson Laboratory, E.M. Eicher, the NIH, D. Bailey, Charles River Laboratories, and Taconic Biosciences, respectively. The J substrain of B6 is widely used in the United States, was among the first inbred strains to have their genome sequenced, and serves as the reference genome for *Mus musculus*. It is also the reference mouse strain for the Allen Brain Atlas (http://mouse.brain-map.org). A second major B6 substrain is the NIH (N) strain that was separated from the J breeding colony at the 32nd generation of inbreeding in the early 1950s.

A seminal study by the Wellcome Trust Sanger Institute Mouse Genomes Project identified over 10,000 SNPs segregating between the NJ and J substrains.[3] Thus far, over 50 variants in coding regions have been validated between these substrains and are predicted to be functional (Table 8.1).[4] Of these, only a small number have been shown to have a direct impact on trait variation, including exonic deletions in the J allele of the *Nnt* gene that alter regulation of insulin[5] and a mutation in the NJ allele of the *Crb1* gene that leads to mild retinal degeneration[6] and morphological abnormalities of the retina.[4] Genomic differences are hypothesized to contribute to the wide range of phenotypic differences observed between substrains, including alcohol consumption and

**Table 8.1** Validated coding and noncoding variants

| Symbol | N/J | Chr | Mb | Private allele | Variant type | Publication |
|---|---|---|---|---|---|---|
| *Spata31* | T/C | 13 | 65023280 | N | Nonsense | 8 |
| *Bmpr2* | A/G | 1 | 59904011 | N | Missense | 8 |
| *Ecm1* | C/T | 3 | 95538799 | J | Missense | 8 |
| *Pdzk1* | G/A | 3 | 96658480 | J | Missense | 8 |
| *Sfrs18* | G/C | 4 | 21800831 | J | Missense | 8 |
| *Hp1bp3* | T/C | 4 | 137777588 | N | Missense | 8 |
| *Padi3* | G/A | 4 | 140354038 | N | Missense | 8 |
| *Casz1* | C/T | 4 | 148318468 | J | Missense | 8 |
| *Adamts3* | T/C | 5 | 90204376 | N | Missense | 8 |
| *Fras1* | C/T | 5 | 97187161 | J | Missense | 8 |
| *Myo18b* | T/C | 5 | 113191741 | N | Missense | 8 |
| *Mkrn1* | A/T | 6 | 39350455 | J | Missense | 8 |
| *Nlrp12* | C/T | 7 | 3222538 | J | Missense | 8 |
| *Herc2* | A/G | 7 | 63386662 | J | Missense | 8 |
| *Acan* | C/A | 7 | 86256240 | J | Missense | 8 |
| *Olfr577* | T/C | 7 | 110121823 | N | Missense | 8 |
| *Zp2* | A/G | 7 | 127278693 | N + Spretus | Missense | 8 |
| *Plk1* | T/C | 7 | 129311164 | N | Missense | 8 |
| *Herpud2* | G/C | 9 | 24935069 | N | Missense | 8 |

*Continued*

**Table 8.1** Validated coding and noncoding variants—cont'd

| Symbol | N/J | Chr | Mb | Private allele | Variant type | Publication |
|---|---|---|---|---|---|---|
| *Jmjd1c* | C/T | 10 | 66700922 | J | Missense | 8 |
| *Vmn2r80* | G/A | 10 | 78632222 | N | Missense | 8 |
| *Pmch* | C/T | 10 | 87554578 | N | Missense | 8 |
| *Cyfip2* | A/G | 11 | 46036117 | N | Missense | 5; 8 |
| *Stxb4* | T/C | 11 | 90341985 | N | Missense | 8 |
| *Nkapl* | G/A | 13 | 21560172 | J | Missense | 8 |
| *Ndufs6* | G/A | 13 | 73465884 | J | Missense | 8 |
| *Cmya5* | G/C | 13 | 93833534 | J | Missense | 8 |
| *Fam160b2* | T/G | 14 | 70986011 | N | Missense | 8 |
| *Adamts12* | T/G | 15 | 11266138 | N | Missense | 8 |
| *Apol11b* | C/A | 15 | 77468437 | J | Missense | 8 |
| *Adcy5* | A/G | 16 | 35291630 | N | Missense | 8 |
| *Guca1a* | C/T | 17 | 47537359 | J | Missense | 8 |
| *Armcx4* | A/C | X | 131227581 | N | Missense | 8 |
| *Tbc1d19* | G/A | 5 | 54280548 | J | Splice site | 8 |
| *Gabra2* | A/− | 5 | 71014638 | J | Splice site | M.K. Mulligan (unpublished data) |
| *Crb1* | −/G | 1 | 141133664 | N | Frameshift | 8; 13 |
| *Cilp* | −/G | 9 | 65127938 | J | Frameshift | 8 |
| *Tlk1* | −/+ | 2 | 70619835:70620080 | J | SINE Ins (intronic) | 8 |
| *Mbnl1* | +/− | 3 | 60336036:60336037 | J | Del (large; intronic) | 8 |
| *Pde4b* | −/+ | 4 | 101954274:101954395 | N | Del (intronic) | 8 |
| *Mast2* | −/+ | 4 | 116051393:116051799 | J | MaLR Ins (intronic) | 8 |
| *Chl1* | −/+ | 6 | 103669536:103676487 | J | LINE Ins (intronic) | 8 |
| *Vmn2r65* | −/+ | 7 | 92095990:92096149 | N | Del (exon) | 8 |
| *Cyp2a22* | −/+ | 7 | 27636128:27748456 | J | Ins (whole gene) | 8 |
| *Cpxm2* | −/+ | 7 | 139306094:139307981 | J | MaLR Ins (intronic) | 8 |
| *Csmd1* | +/− | 8 | 16716381:16716382 | J | Del (large; intronic) | 8 |
| *2410076I21Rik* | −/+ | 9 | 58544415:58546304 | J | MaLR Ins (intronic) | 8 |
| *Nkain2* | −/+ | 10 | 32536420:32543464 | J | LINE Ins (intronic) | 8 |
| *Rptor* | −/+ | 11 | 119560391:119566827 | J | MTA Ins (intronic) | 8 |
| *Immp2l* | −/+ | 12 | 42023964:42032747 | N | Del (intronic) | 8 |
| *Nnt* | +/− | 13 | 120164268:120164269 | J | Del (large; intronic) | 8; 12 |
| *Zfp91* | +/− | 19 | 12863187:12863188 | J | Del (large; intronic) | 8 |

*Del*, deletion; *Ins*, insertion; *J*, C57BL/6J; *N*, C57BL/6N.
Potentially functional variants known to be segregating between C57BL/6 substrains are shown here by genome location (GRCm38/mm10).

preference,[7] acute and sensitized cocaine- and amphetamine-induced locomotor activity,[8] binge eating,[9] anxiety-like behaviors in the open field, elevated plus maze, and light/dark transition test,[4,10] motor activity and coordination,[4,10,11] hot-plate thermal test,[10] acoustic startle and prepulse inhibition,[4,10] organ morphology, physiology, and metabolism,[4] depression-like behaviors in the forced swim test,[10] fear conditioning,[11] and response to stress.[12]

Forward genetic mapping studies pioneered by chapter authors Vivek Kumar and Camron Bryant have been successful in isolating additional single gene mutations with a large impact on trait variation by leveraging the reduced genetic complexity between substrains. In these studies, two inbred B6 substrains (J and NJ) were crossed to generate an isogenic $F_1$ (first filial) generation, followed by intercrossing to generate a recombinant $F_2$ population suitable for genetic mapping (Fig. 8.2). The use of this reduced complexity cross (RCC) combined with modeling predictions of deleterious variants led to the discovery of a missense mutation in *Cyfip2* in the NJ strain that modulates sensitivity to cocaine and amphetamine[8] and binge eating of sweetened palatable food.[9]

A typical RCC design (such as the ones described earlier) involves the following: (1) observation of robust trait variation between two nearly isogenic strains bred in house, (2) generation of an F2 cross between the two parental inbred strains, (3) measurement of the trait of interest in the resulting recombinant progeny, (4) genotyping of the RCC progeny using a custom marker panel, and (5) genetic mapping and possibly fine mapping to resolve quantitative trait loci (QTLs) and quantitative trait genes (QTGs) underlying trait variation. In the resulting RCC, large candidate genomic intervals are counterbalanced by a low density of segregating variants, thus greatly facilitating the pace of gene discovery. In any given animal within an F2 population, the number of recombination events per chromosome is very low (0–3). This means that the statistically defined chromosomal regions associated with trait variance (i.e., QTL) are quite large, ranging from 10 to 30 cM, and contain several hundred genes and several thousand genetic variants. However, in the RCC, these large regions are greatly offset by the low number of variants contained within the QTLs. There are orders of magnitude fewer variants to consider within each QTL interval in the RCC compared to an F2 cross, making it easier to narrow down and validate the potential list of candidate QTL genes. For example, consider a cross between J and either NJ (cross A between substrains) or DBA/2J (cross B between inbred strains). In cross A, there are roughly 20,000 segregating variants compared to ~5 million in cross B. Effectively, a QTL region in cross A will have 250 times fewer variants than a similarly sized region in cross B.

RCCs are a valuable genetic resource for studying the impact of naturally occurring mutations on complex traits; however, careful consideration is required at all stages of the RCC experimental design. It is time consuming and expensive to generate, genotype, and phenotype progeny from an F2 cross. Thus, it is important to collect prior evidence that the trait in question is strongly influenced by genetic factors and resistant to

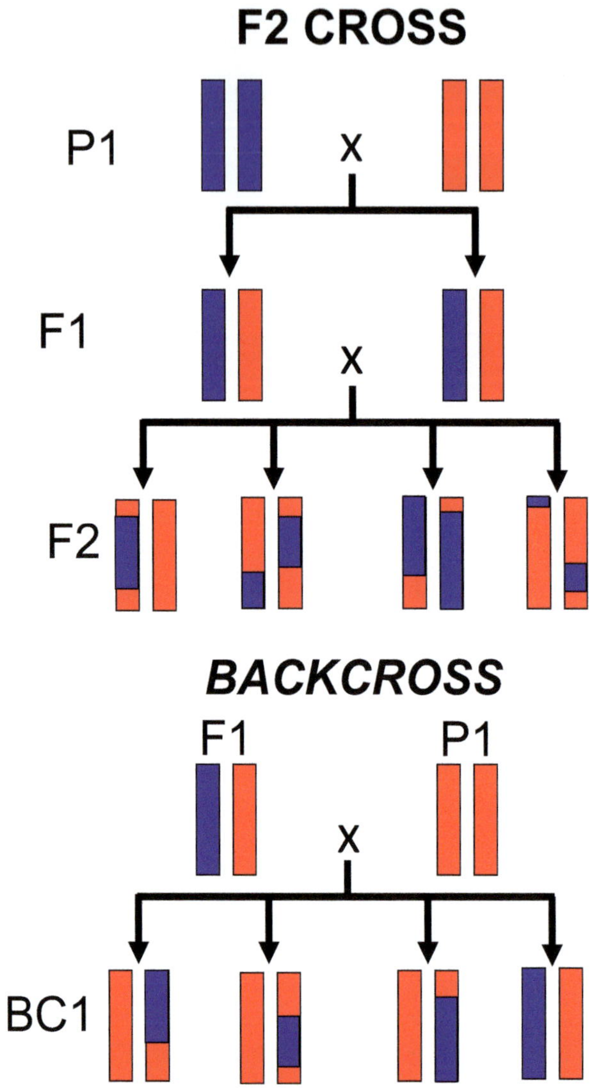


**Figure 8.2** ***F2 and backcross mapping populations.*** A forward genetic screen leverages genetically diverse populations to locate regions of the genome (QTLs) that contain variants contributing to a trait of interest. Common crosses include a backcross and F2 cross. An F2 cross involves the breeding of isogenic parental lines to create a genetically homogenous F1 population that is crossed to yield a recombinant F2 mapping population. In a backcross, F1 individuals are crossed back to an isogenic parental strain to yield a recombinant mapping population. *QTL*, quantitative trait loci.

environmental variation. It is also critical that the magnitude of the effect is large enough to detect given a reasonable number of F2 progeny (e.g., 200 as opposed to 1000) and that the depth of the marker panel is sufficient to define QTL regions. These design steps are considered in greater detail subsequently.

## TRAIT SELECTION

Environmental interactions, redundancy, pleiotropy, and epistasis both define and obfuscate the genetic study of complex behavioral traits. Environmental variation,[13] often in the form of differences in animal handlers and handling[14] and institutional variation in animal husbandry, diet, climate control, and testing procedures[15–18] can result in a lack of reproducibility in behavioral measures. For example, the sex of the researcher conducting the animal experiment can profoundly alter the behavioral and stress response of the animal. This experimenter-based variability has been demonstrated particularly well by Jeff Mogil's group, which found that rodents had increased corticosterone levels and decreased outward expressions of pain (facial grimaces) in the presence of male, but not female, investigators.[19] When selecting traits for genetic dissection, it is therefore crucial to pay close attention to the reproducibility of the behavior across different testing conditions and environments, such as the laboratory and the experimenter. Traits that are less influenced by these subtle nongenetic variables will be much easier to reproduce and quantitate. An elegant screen for behavioral and physiological differences between closely related B6 substrains was performed across four different laboratories by Michelle Simon and colleagues[4] and clearly demonstrates that many, but not all, differences between B6 substrains are robust and resistant to environmental variation.

### Cross Considerations

Both backcrosses and F2 crosses are commonly used in forward genetic screens. In contrast to the F2 cross, a backcross is generated by crossing F1 individuals back to one of the parental strains (Fig. 8.2). Unless the mode of inheritance of your favorite phenotype (YFP) is clearly known, it is generally safer to map traits using an F2 cross. A locus with a purely additive mode of inheritance explains twice as much variance in an F2 when compared to a backcross, meaning that twice as many backcross individuals will be needed to detect the QTL. If the locus is dominant and the progenitor strain with the dominant allele is used for generation of the backcross, the sample size is half that of the F2. In contrast, the sample size is doubled if the other progenitor strain is used for the backcross. In the context of reduced genetic complexity, it can be assumed that there are few variants contributing to phenotypic variation, so the additive effect of other background QTLs on YFP is minimal. An F2 tests more genotypes (AA, BB, and AB) and allows for more options to test different modes of inheritance compared to a backcross.[20,21] However, a backcross can be a more cost-efficient strategy if YFP is inherited in a dominant manner and the parental strain with the dominant allele is known (see Rosen and colleagues[22] for an example of a backcross between isogenic strains to evaluate the role of gene variants in cortical heterotopia).

## Effect Size and Heritability

The magnitude of the difference in trait expression between strains is described as the effect size and is simple to measure:

$$[\text{Mean of Group 1}] - [\text{Mean of Group 2}]/\text{SD}_{\text{pooled}}$$

where $SD_{pooled}$ is equivalent to the standard deviation of the individuals in both groups. Effect size is often used when designing behavioral experiments to determine how many individuals need to be tested to achieve statistical significance. If the effect size is small, more individuals of each genotype will need to be tested to detect a difference. In contrast, if the effect size is large, fewer individuals of each genotype will need to be tested. The effect size of a behavioral phenotype is an important consideration that must be balanced with an analysis of the sources of variation that may drive phenotypic differences between individuals.

Heritability is the proportion of the trait variance attributed to genetic factors. Trait or phenotypic variation ($V_P$) is due to the combined influence of genetic ($V_G$) and environmental ($V_E$) factors:

$$V_P = V_G + V_E$$

Heritability ($h^2$) can be estimated from data collected from inbred strains by comparing the between-strain (genetic) variance with the total variance (between- and within-strain or environmental variance)[23]:

$$h^2 = V_G/V_P$$

This can be easily calculated from an ANOVA results table by dividing the between-strain sum of squares (SS) term by the sum of the between-strain and within-strain SS terms.[24] Note that the $h^2$ values range from 0 to 1, with higher values indicating a larger genetic contribution to trait variance. A trait with high heritability has comparatively lower within-strain (environmental) and higher between-strain (genetic) variation. In general, traits with $h^2$ values of 0.2 (20% of the total phenotypic variance is attributable to genetic factors) or greater are amenable to genetic mapping. A trait with a large effect size but low heritability is not amenable to dissection using genetic crosses due to the large degree of nongenetic (e.g., environmental and/or epigenetic) variance. The ideal scenario recapitulated in an RCC is a trait with moderate to high effect size that is also highly heritable.

To assist with RCC design and increase the likelihood of QTL discovery, it is often desirable to first perform a power analysis. A power analysis considers the sources of variation contributing to YFP, the type of cross, and marker density to estimate the number of individuals needed to detect a QTL. With current genotyping technologies, markers are not usually the limiting factor (this will be discussed in more detail in the

QTL Mapping and QTG Discovery in the RCC section). Lynch and Walsh[25] estimate the number of F2 individuals needed to detect a single completely linked QTL (assuming the QTL is completely linked to a single marker and marker spacing is < 20 cM) using the phenotypic variance explained by the QTL, the desired levels of $\alpha$ and $\beta$, and the mode of inheritance of the trait (additive or dominant; Fig. 8.3):

$$n_{(\text{F2 sample size})} = \left(1 - r^2\right)/r^2 \times \left[\left(z(1 - [\alpha/2])\Big/\left(\sqrt{1 - r^2}\right)\right) + z(1 - \beta)\right]^2 \times \left[1 + \left(k^2/2\right)\right]$$

where $r^2$ is the fraction of phenotypic variation in YFP explained by the QTL, $z$ is the Z score based on the standard normal distribution, and $k$ is the dominance coefficient ($k = 0$ for completely additive and $k \pm 1$ for completely dominant). There are freely available and useful tools for cross design and power analysis, including the R/qtlDesign package,[21] which is specifically designed for power analysis in an F2 cross. Furthermore, as mentioned before, there are fewer variants segregating in an RCC and less contribution from other background variants/QTLs. Thus, $r^2$ should be higher in RCCs compared to other crosses, where multiple segregating variants can influence YFP.

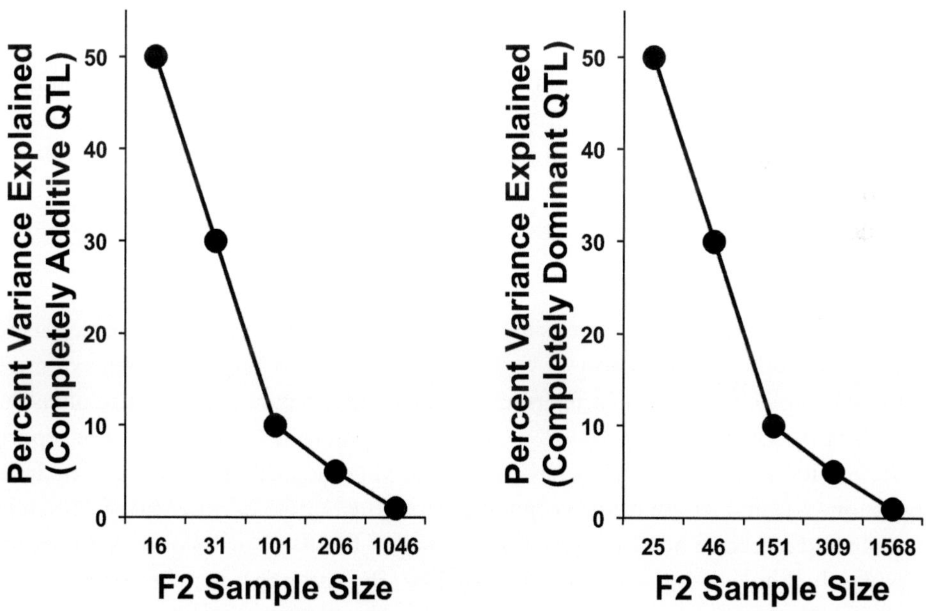


**Figure 8.3** ***Theoretical power analysis QTL mapping in an F2 cross.*** Number of F2 individuals needed to detect a QTL (in complete linkage with a single marker) with $\alpha = 0.05$ (5% probability of a false positive) and $\beta = 0.10$ (10% probability of a false negative) is shown based on variance explained for a completely additive QTL (left) and a completely dominant QTL (right). *QTL*, quantitative trait loci.

## Identification of a Strain Difference in YFP: Ruling Out Nongenetic Explanations

An inbred strain difference in YFP implicates a heritable genetic basis for your trait, provided that environmental differences are not responsible. However, there are multiple precautions that one should take to rule out nongenetic explanations before advancing to the costly, timely, and laborious process of QTL mapping. First, breeding and then phenotyping the strains in house under the same environmental conditions will ensure that temperature, diet, stress, etc., are controlled. As an example (discussed later in the A Cautionary Case Study section), failure to replicate phenotypic differences between closely related substrains of the BALB/c strain were attributed to environmental factors or other unknown, unstable factors, so QTL mapping was not conducted.[26] Due diligence would also require a second generation of breeding and phenotyping in house to fully rule out the potential for transgenerational epigenetic inheritance of changes in gene transcription that could contribute to your phenotype.[27] Second, strains should always be phenotyped within the same cohort and, ideally, within the same experimental session to minimize batch effects and circadian effects. Third, testing F1 mice simultaneously with the parental strains bred in house can support heritability for the trait, in particular if F1 mice exhibit an intermediate phenotype. This observation would provide evidence for an additive mode of inheritance, and the data can then be used to inform heritability estimates.[8] As a final point, a well-designed study should attempt to minimize, standardize, balance, or control all possible environmental factors that may influence YFP. This may also include controlling for litter effects (e.g., parity, size, and sex composition), batch effects, breeding scheme (e.g., pair or trio breeding), and in some cases, maternal effects (e.g., reciprocal cross designs or cross fostering).

## A Cautionary Case Study

BALB/c inbred mice exhibit aggressive, anxiety-like, and depression-like behaviors that are partially modulated by treatment with antidepressants and have been used extensively to study these aspects of psychiatric disease. Maintained as separate substrain colonies since 1935, BALB/cJ and BALB/cByJ are genetically polymorphic[26] and have been reported to differ in a number of behavioral and physiological phenotypes, including aggression[28–32] and length of the corpus callosum (the bundle of nerve fibers connecting the right and left brain hemispheres).[33] In particular, BALB/cJ males have been reported to exhibit higher levels of aggression than BALB/cByJ males that may be related to alterations in brain and corpus callosum morphology. Abraham Palmer's group attempted to quantitate and replicate these differences and generated $F_1$, $F_2$ (RCC), and $N_2$ (a backcross between the F1 individuals and the parent inbred strain BALB/cJ) crosses between BALB/cJ and BALB/cByJ with the goal of identifying the underlying gene variants responsible for phenotypic variation.[26] They identified differences in anxiety-like (BALB/cJ > BALB/cByJ) and depression-like behavior (BALB/cJ < BALB/cByJ) and aggression (BALB/cJ < BALB/cByJ) in the parental substrains (purchased from The Jackson Laboratory). However, their results were inconsistent with the findings of other laboratories and, importantly, they did not find a high degree of trait variation among F2 or N2 individuals. One potential explanation is that environmental differences between laboratories interacted with strain genotype to induce phenotypic differences (phenotype instability) among progenitor BALB/c substrains. Their finding of phenotypic instability in the BALB/cJ and BALB/cByJ substrains likely results from a larger contribution of nongenetic (environmental, epigenetic) as opposed to genetic factors and environmental noise associated with the behavioral measures that made replication across laboratories difficult.

## GENOTYPING STRATEGIES

The availability of sequenced genomes for many organisms and rapidly evolving platforms for high-throughput and massively multiplexed genotyping greatly facilitate genotyping and genetic mapping of YFP in an RCC. There are many different genotyping platforms available that offer varying levels of computation, throughput, and scale. Platform selection largely depends on characteristics of your cross, such as the number of individuals requiring genotypes, the number of genotypic markers per individual, and cost. For crosses derived from nearly isogenic lines where there are very few variants between parental lines, it may not be possible to locate regularly spaced markers. However, the low number of variants segregating in an RCC makes candidate variant identification feasible even in cases where there are very few markers available per chromosome. In this section, three strategies are discussed: (1) marker design and targeted high-throughput genotyping, (2) genotyping by array, and (3) genotyping by sequencing (GBS).

### Marker Design and Targeted High-Throughput Genotyping

Whole genome sequences are available for many inbred strains, and these permit selection of informative markers (SNPs or other variants) spaced at regular frequencies across the genome. In a typical F2 cross, markers are spaced every 10 to 20 centiMorgans (cM). In mice, 1 cM is on average equivalent to 2 Mb, but this depends locally on the recombination frequency within a chromosomal region. The mouse genome comprises 2.8 billion base pairs; thus, a marker panel consisting of 70−140 markers spaced every 20−40 Mb for an F2 panel is sufficient to cover the entire genome. The cM distance between each marker can be estimated using the sex-averaged MGI Mouse Map Converter (http://cgd.jax.org/mousemapconverter),[34] while keeping in mind that recombination rates vary by strain, cross, chromosome, and sex.

In a design-based approach to genotyping, it is possible to design assays that specifically target a subset of single nucleotide polymorphism (SNP) markers. When selecting this marker panel, it is important to consider the quality of the variants selected as markers. Variants called from whole genome next-generation sequencing (NGS) should have sufficient read coverage and base quality scores so they are likely to be reproducibly called using different chemistries and platforms. Ideally, these variants should not be located near insertions or deletions, structural variants, complex or repetitive DNA sequences, or regions with high guanine-cytosine (GC) content that will interfere with design of primers and amplification of the target genomic interval containing the variants. As these approaches can get quite costly, it is also a good idea to validate (ensure that the markers are indeed polymorphic and can be reliably detected on your desired platform) at least a subset of candidate markers in your panel prior to genotyping, especially if you are using a smaller marker panel.

For a small to moderate numbers of markers (24—196), multiplex high-throughput genotyping using quantitative polymerase chain reaction (qPCR) may be a cost effect solution that requires minimal preparation and computational requirements. One such platform is the Genotyping Dynamic Array Integrated Fluidic Circuits offered by Fluidigm (South San Francisco, CA, USA). This platform allows some flexibility in the number of markers assayed in each sample per genotyping run; current formats (samples × genotypes) are 192 × 24, 48 × 48, and 96 × 96. The 96 × 96 format, which includes 96 custom markers evenly spaced 15 cM apart (approximately 30 Mb; Fig. 8.4),

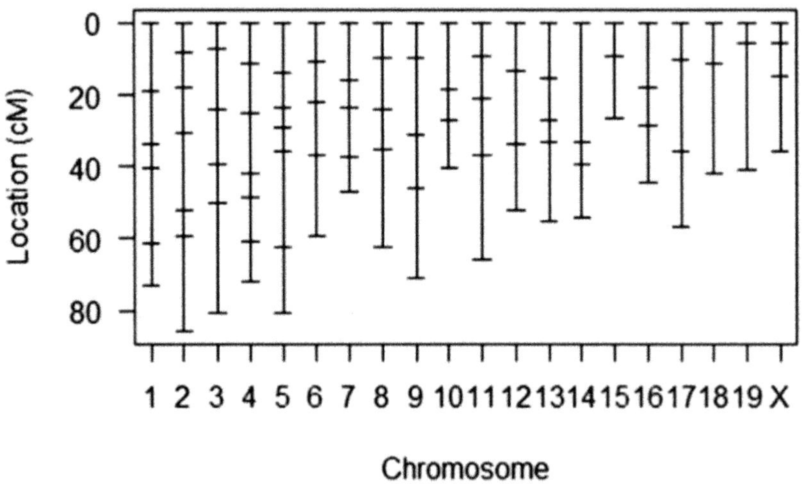


**Figure 8.4** ***Genetic map of SNP markers used in the RCC-F2 cross between C57BL/6 substrains.*** The *x*-axis indicates chromosome location. The *y*-axis indicates cM location for a given chromosome. The cM position of each SNP marker was estimated using the Mouse Map Converter containing the sex-averaged cM distances.[34] *RCC*, reduced complexity cross.

has been used previously by the authors of this chapter to map traits in an RCC derived from C57BL/6 substrains.[8,9] We selected SNP markers from the Sanger mouse variant database (https://www.sanger.ac.uk)[3,35] and from other published whole genome sequencing datasets.[4,8] Genotypes are detected on this platform using combinations of locus- and/or allele-specific primers and fluorescent probes to selectively amplify and detect all three allele combinations (AA, BB, and AB) for all markers in each sample (96 F2 samples × 96 markers equivalent to 9216 genotypes per array) following multiple cycles of PCR and imaging of the array. End-point or real-time imaging is available depending on the genotyping chemistry used, typically TaqMan assays from Applied Biosystems or SNP Type assays from Fluidigm.

Even higher throughput genotyping options are available, and this technology is expanding and changing rapidly. Only a few approaches are mentioned in detail here; this includes the Sequenom MassARRAY platform and NGS approaches. The Sequenom commercial technology uses multiplex PCR to generate allele-specific products with different masses from a targeted set of marker loci that can then be identified following MALDI-TOF mass spectrometry. A variety of targeted NGS options are also available and involve multiplexed and targeted amplification of genomic loci containing SNP markers in every sample, followed by sequencing of the resulting libraries, alignment, and variant calling to identify genotypes. Costs for these higher throughput approaches can vary greatly depending on the number of markers (and locus-specific primers) and individuals that need to be genotyped, and they can also be computationally intensive if not commercially outsourced. However, these approaches allow for massive multiplexing of samples and genotypes, and the end result is very high-throughput genotyping of dense marker panels across many individuals.

## Genotyping by Array

Microarrays represent another design-based approach to genotyping that depends on a priori knowledge of variants segregating in your cross of interest. Microarray technology was first developed in the 1980s and is now used ubiquitously for many different applications. Genotyping arrays are composed of DNA probes (short, synthesized DNA sequences) attached to a solid support to form an organized array representing different allele combinations at hundreds of thousands of SNPs. Hybridization of labeled sample DNA to complimentary probes on the microarray followed by imaging is used to detect and call genotypes. Designing a custom array can be expensive; however, a suitable commercial genotyping array may be available for your cross. For example, the latest version of the Mouse Universal Genotyping Array (MUGA; GigaMUGA, ~$100 per array) platform was designed for genotyping crosses from common inbred strains and the Collaborative Cross and Diversity Outcross populations derived from eight different progenitor mouse strains. The GigaMUGA (Illumina Infinium HD platform) also contains 96 to 391 informative markers that differentiate multiple substrains, including C57BL/6J and C57BL/6N, 129S1/SvImJ and 129S6/SvEvTac, A/J and A/WySnJ, BALB/cJ and BALB/cByJ, C3H/HeJ and C3HeB/FeJ, and DBA/2J and DBA/2DeJ substrains.[36]

## Genotyping by Sequencing

In contrast to the previous design-based genotyping platforms, this approach leverages NGS technology for simultaneous variant discovery and genotyping. This approach can be useful when genomic sequence is unavailable for your progenitor isogenic strains, as is often the case for selected lines. Not only can many informative markers be discovered by NGS, they can subsequently be used as markers for genetic mapping of YFP. In GBS, combinations of restriction enzymes are used to create a reduced representation of the genome. In other words, instead of sequencing the entire genome, the focus is limited to restriction sites where polymorphisms impacting the restriction sites, or in adjacent genomic DNA, can be detected.[37] Using this GBS approach, coverage (the number of times a region of genomic DNA is sampled) is determined by the combination of restriction enzymes used and can result in greater coverage at polymorphic regions than would otherwise result from random whole genome sequencing. It is worth mentioning that this approach should be used with caution in a typical RCC, which is expected to be segregating very few variants. A low number of variants will decrease the likelihood that a variant will be located near a restriction enzyme cut site and thus decrease the likelihood that it will be detected and included in subsequent genotyping.

A related approach involves RNA sequencing of tissue from F1 or F2 individuals. RNA-seq of the F2 can be used to identify variants (typically in coding exons) that can then be used for genotyping. Alternatively, RNA-seq of the F1 can be used to analyze allele-specific expression (whether or not one allele is associated with higher expression) to identify gene variants associated with *cis* expression QTLs (eQTLs).[38] Although these strategies permit simultaneous generation of global transcriptional profiles and genotypes, or the identification of genes regulated by local sequence variants, they are not typically recommended because the total number of coding variants segregating in RCCs is generally low, so far fewer variants will be located in exons and accessible to sequencing. For example, only ~50 validated coding variants have been identified between B6 substrains, which is not quite sufficient for QTL mapping. An additional consideration is that a larger number of reads per sample may be required, depending on transcript abundance within a tissue of interest, which can drastically increase sequencing costs.

## Tools of the Future in QTL and Genomic Studies in the RCC

Genetic mapping in RCCs has the potential to identify single nucleotide variants that drive divergent phenotypic responses between closely related strains. One of the greatest challenges for expansion and feasibility of RCC-style approaches in the near future is in the identification of genetic differences between these closely related strains, and in the development of useful, diagnostic markers between these strains. Recent work has shown that in some populations[37] GBS approaches can be used for the identification of variants associated with phenotypic differences within a cross. However, GBS approaches can be quite wasteful in the context of RCCs, as the vast majority of the genomic data generated (e.g., five times coverage of each animal's genome) will be uninformative. In contrast, the development of genotyping arrays designed to capture markers segregating between pairs of substrains can prove to be quite effective, providing the genome sequences of the substrains are known (as in the aforementioned cases of C57BL/6J and C57BL/6NJ) and that the price makes genotyping cost effective. At the time of this writing, a new iteration of the MUGA genotyping array[36] is being developed. By using whole genome sequences of a large number of mouse strains, markers have been selected to conduct genetic mapping across a large number of substrains from nine separate inbred strains groups (the 129 strain cluster, A/, BALB/, C57BL6/, C57BL10/, C3H/, DBA1/, DBA2/, FVB/, and NOD/ strain-groups). An interesting development is that as more mouse strains and substrains have whole genome sequence available, new markers can be added to this array platform, permitting the concept of RCC mapping to be applied to a wide range of strain contrasts, extending our understanding of genetic variants controlling traits.

## QTL MAPPING AND QTG DISCOVERY IN THE RCC

A variety of statistical methods and tools have been developed for QTL mapping and implemented in free software for public use. These methods are well suited for simple backcross and F2 RCC populations. R/qtl[9,39] was developed for identification of QTLs and higher order modeling. Another Web-based tool, GeneNetwork or WebQTL (GeneNetwork.org),[40] was developed for QTL mapping and to explore associations between variants, molecular traits (e.g., gene expression), and higher order phenotypes (e.g., behavior) and facilitate QTG identification. In this section, we will focus mainly on QTL analysis performed in F2 mice using the R package R/qtl. For a review of GeneNetwork tools and functions, see Ref. 41.

A variety of analytical methodologies are available in the R/qtl package, including, e.g., composite interval mapping or Haley-Knott regression (see Ref. 42 for discussion). The "scanone" function in R/qtl is used to calculate log of the odds (LOD) scores. Permutation analysis (perm = 1000) is used to establish the significance threshold for each phenotype ($P < .05$). Additive and/or interactive covariates can be added to the model (e.g., sex, treatment, batch effect). Bayesian credible intervals expanded to the nearest markers can be used to define the QTL intervals that can then be used as guide posts for fine mapping (see following).

## Fine Mapping Using F2 Recombinants

Fine mapping QTLs involves the introduction of additional recombination events within a QTL interval to iterate which subregions either capture or fail to capture the causal variants.[43] Introducing recombination events can be accomplished either via the use of an advanced intercross[44] or via the use of congenic lines.[45] The goal is to narrow the interval to the smallest number of polymorphic genes possible before the final step of gene validation via gene editing.[46] The assumption is that there is a single variant within a larger genomic region that underlies the QTL and that whittling away at the larger locus into progressively smaller subregions will eventually lead to the identification of the polymorphic genetic factor underlying YFP. However, the assumption that there is a single smaller locus/single gene/single variant is not always correct. For example, we fine mapped a QTL influencing methamphetamine behavioral sensitivity via back-crossing a chromosome substitution to generate a 10 Mb congenic capturing the QTL. However, our subsequent attempt to further dissect the region into smaller intervals failed since none of the subcongenics captured a phenotypic difference in behavior. These observations led us to conclude that localized epistasis involving more than one locus was a contributing factor to the overall QTL phenotype and frankly, led us to a dead end with regard to gene identification.[47]

Despite this example of failure, the congenic approach to fine mapping QTL intervals can still be successful in narrowing intervals and identifying QTGs.[48,49] Furthermore, the chance for localized epistasis is greatly reduced in an RCC due to very low level of genetic diversity and the expected Mendelian inheritance of most traits. Nevertheless, fine mapping QTGs from F2-derived loci using congenics can take several years or more than a decade.[49] This long timeline is in part due to the several years that it takes just to accomplish the first step of generating a congenic line that is isogenic for one of the two background strains while retaining only the polymorphic region of the donor strain within the QTL interval (typically tens of Mb in size).[50] The "speed congenic" approach involves the use of marker-assisted selection that can maximize the efficiency at eliminating heterozygosity at each generation and can shorten the timeline.[51] After the first step of generating and confirming that a congenic line is isogenic and captures a QTL, additional recombination events can then be introduced via backcrossing to further narrow the locus, which can take several more years. Clearly, faster and more efficient alternatives to fine mapping F2-derived QTLs would be a welcome improvement.

A major advantage of the RCC is the potential to eliminate the step of making congenic lines and immediately begin fine mapping F2-derived QTLs by backcrossing select F2 recombinant mice. Because the segregating background is already greater than 99.99% genetically identical and because QTL scans typically lead to the identification of a single major locus accounting for a majority of the genetic variance, the assumption is that the mixed background outside of the QTL has negligible additive or epistatic effects on trait variance, and thus, there is very little risk for background loci interfering with detection of the QTL signal. Thus, subsequent genotyping only needs to be conducted within the major QTL of interest, and genetic analysis can be restricted to polymorphisms and recombination events within the QTL interval. F2 recombinants can be backcrossed to one of the parental strains, and offspring can be genotyped and phenotyped at each generation (Fig. 8.5). Ideally, the QTL will either be dominantly or additively inherited, which would eliminate the need for the slow and laborious process of intercrossing at each and every generation following the establishment of a heterozygous F2 recombinant line (Fig. 8.5).

Additional markers within the QTL interval can be chosen from the Sanger mouse variant database (http://www.sanger.ac.uk/) and confirmed via PCR amplification and Sanger sequencing. As described in the Effect Size and Heritability section, the effect size of the genotype on trait variance should first be calculated using the F2 dataset to determine the appropriate sample size required for phenotyping the F2 recombinant lines.

## Cis-eQTL Analysis to Further Inform Positional Candidates

The RCC will lead to the identification of QTLs that contain a drastically reduced number of polymorphic genes underlying QTLs for YFP, which can potentially greatly facilitate the identification of causal genes and causal variants. However, an important consideration is that the causal variants may not lie within coding regions of protein-coding genes but instead lie within intergenic regulatory regions (e.g., within enhancers) that cannot be assigned to any particular nearby (or distant) gene. Accordingly, a systems genetic approach of combining forward genetics with complementary −omics datasets can bridge the mechanistic gap between DNA sequence and complex traits and could, e.g., identify nearby genes regulated by intergenic variants as well as downstream causal and reactive networks underlying YFP (Fig. 8.6).[52]

One intermediate phenotypic analysis at the genomic level that can aid in honing candidate QTGs is *cis*-eQTL mapping, which involves identifying genes whose transcript levels correlate with genotypes at nearby SNP markers. The assumption is that markers are in linkage disequilibrium with a causal variant that mediates both changes in gene expression and YFP (Fig. 8.6). This approach is limited by the number of tissues, cell types, and developmental time points that can be assayed, and it is not always clear which of these sets of parameter(s) are the most relevant to YFP. Furthermore, splicing QTLs (sQTLs) rather than eQTLs could comprise the molecular mechanism linking DNA variants with YFP[53]; thus, sQTL analysis could uncover genes that would not normally be detected at the level of differential gene expression (DGE),[53] and thus, a differentially

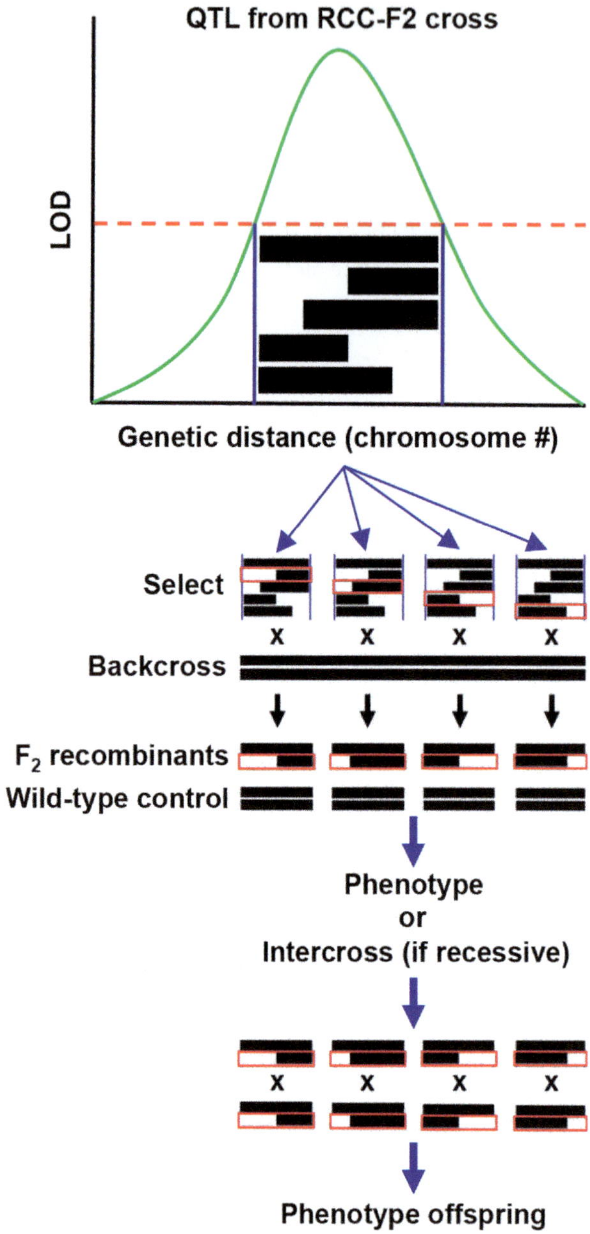


**Figure 8.5** ***Schematic for immediate, rapid fine mapping in select $F_2$ recombinants of the RCC-$F_2$ cross.*** Top panel: Genome-wide significant QTL (green trace; *red dashed line* = significance threshold; *blue vertical lines* = Bayes credible interval). Middle panel: Heterozygous, recombinant $F_2$ mice are selected for backcrossing to one of the parental RCC strains. In the case of dominant or additive mode of inheritance of the QTL, offspring are then phenotyped alongside control mice that are homozygous for the wild-type allele within the QTL interval. Bottom panel: Alternatively, in the case of recessive mode of inheritance of the QTL, heterozygous recombinant $F_2$ mice must then be intercrossed at each generation to generate offspring for phenotyping that contain two copies of the recessive allele. *QTL*, quantitative trait loci; *RCC*, reduced complexity cross.

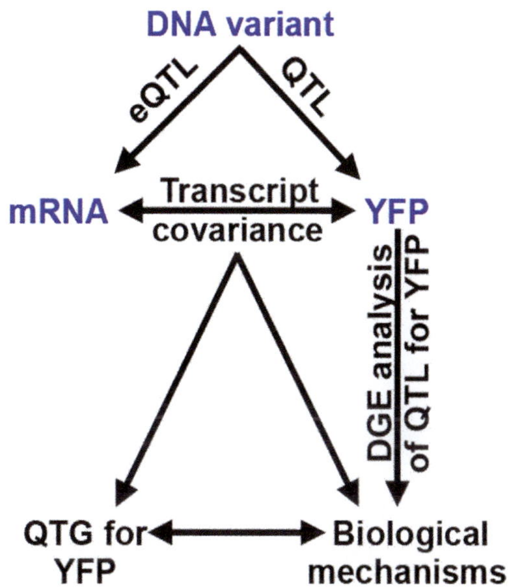


**Figure 8.6** ***Systems genetic analysis of YFP in the RCC-F$_2$ cross to identify QTGs and biologic mechanisms.*** Combining *cis*-acting expression QTL (eQTL) analysis with QTL analysis of *YFP* can aid in identifying functionally modified genes that underlie behavior. Genes containing eQTLs within the QTL interval of YFP are candidate QTGs. Differential gene expression (*DGE*) analysis as a function of QTL genotype for YFP can identify *trans*-altered gene transcripts that could comprise downstream, perturbed biologic pathways and gene networks that underlie the molecular mechanisms of YFP. *QTGs*, quantitative trait genes; *QTL*, quantitative trait loci; *RCC*, reduced complexity cross; *YFP*, your favorite phenotype.

spliced gene could comprise the QTG. Finally, it is important to consider that DNA variants could be more closely aligned with changes in other intermediate molecular phenotypes besides transcript levels that in turn, are more directly aligned to YFP. Thus, the inclusion of additional −omics approaches such as proteomics, metabolomics, and epigenomics will maximize the full utility of the systems genetics approach to understanding genetic and biologic mechanisms of YFP.[52]

## Differential Gene Expression Analysis of RCC Mice Based on Major QTL Genotype

In addition to eQTL analysis, DGE analysis can also be conducted within the same F2 RNA-seq dataset by examining the effect of the major QTL genotype on genome-wide changes in gene expression. Samples can be selectively chosen to capture the entire QTL genotype throughout the QTL confidence interval (defined by possessing the same genotype at multiple SNP markers), and then pairwise t-tests of the two genotypes (heterozygous vs. homozygous) or regression analysis of the three genotypes

(B6J/B6J, heterozygous, B6NJ/B6NJ) can be conducted for DGE analysis to obtain a list of differentially expressed genes, fold-changes in gene expression, *P*-values, and false discovery rates. This analysis can serve multiple complementary purposes to eQTL mapping, including confirming eQTLs within the YFP QTL interval and identifying genome-wide enriched biological pathways and gene networks that are specifically associated with inheritance of the QTL interval for YFP that, thus, could be relevant to understanding the biological mechanisms (Fig. 8.6).

## Genome-Wide Transcript Covariance Analysis With YFP

Correlating DNA sequence with nearby transcript abundance (eQTL analysis) can aid in identifying functionally modified (and potentially causal) genetic factors associated with YFP. An additional complementary analysis that can be conducted if both datasets are available from the same samples is covariance of YFP with gene expression, i.e., transcript covariance analysis (Fig. 8.6).[54] Genes that possess both an eQTL and a significant transcript correlation with YFP can be further prioritized as candidate QTGs. In addition, transcript covariance analysis can unveil genes that are located outside of the major QTL(s) but nevertheless are highly correlated with YFP and, thus, could contribute to the overall biologic mechanism (Fig. 8.6).

## Testing the Private Variant Hypothesis

Classical inbred strains including B6 strains were all derived from a small, common set of founders,[55] so any given two strains contain fixed segments of the genome that are identical by descent. Residually heterozygous segments that later became fixed following the separation of the B6J and B6N substrains in the 1950s can thus be assigned to a particular haplotype that is shared among multiple inbred strains. An additional source of genetic variance is spontaneous mutations that occurred after the time of separation and eventually became fixed or could still be segregating within the colony. In this case, an allele from one strain can be assigned to a particular haplotype shared among multiple inbred strains, whereas the second allele cannot be assigned to a haplotype. Private variants can be tentatively identified using the online Sanger mouse variant database (https://www.sanger.ac.uk/) containing whole genome sequences of 36 inbred strains versus the B6J reference strain by choosing B6N and other classical inbred strains with which to compare haplotypes. To narrow the list of variants to those that are tentatively private to B6J, one can focus solely on those variants that are shared between B6N and one or more classical inbred strains and where the alternate allele (B6J) is not found in any other classical inbred strains.

One situation where we found the B6J private variant hypothesis to be particularly useful was when we recently identified a QTL influencing opioid behavioral sensitivity and opioid drug withdrawal that mapped to nearly the same region as a neurobehavioral

QTL "hotspot" on distal chromosome 1 (172–178 Mb).[56] This hotspot has been mapped for several QTLs related to brain function and neuropsychiatric-related traits. A majority of the studies that were summarized in the study by Mozhui and colleagues (in particular, the behavioral studies) used B6J as one of the mapping strains. Thus, a logical and parsimonious hypothesis is that a variant private to B6J is responsible for several of these QTLs, including the QTL that we identified for multiple opioid behavioral traits (unpublished data).

We are currently testing the private variant hypothesis by prioritizing genes containing variants that are private to B6J versus several other classical inbred strains that were used to identify QTLs for phenotypes similar to opioid behavioral sensitivity and opioid drug withdrawal, including emotionality, alcohol dependency, alcohol withdrawal, pentobarbital withdrawal, basal locomotor activity, ethanol-induced locomotor activity, and haloperidol-induced catalepsy.[56] Based on the hypothesis that our QTL lies within the 172–178 Mb neurobehavioral hotspot containing a variant within a protein-coding gene that is private to B6J and whose alternate allele is shared among B6N and these other classical inbred strains, we further narrowed the already limited list of polymorphic genes from 19 to 6 genes (data not shown). We will test the private variant hypothesis using gene editing of candidate functional variants that are private to B6J on a B6N background and by editing the B6N allele on a B6J background.

## Characterizing the Biologic Function of Candidate Genes: the Role of *Cyfip2*

As mentioned in the introduction, an RCC design using forward genetic mapping in the recombinant F2 progeny derived from closely related B6 substrains led to the discovery of a missense mutation in *Cyfip2* in the NJ strain that modulates sensitivity to cocaine and amphetamine[8] and binge eating of sweetened palatable food.[9] Here, we discuss the detailed characterization of this mutation and its functional role in behavior. Since *CYFIP1* and *CYFIP2* share 88% amino acid identity, we discuss the biology of both genes here.

CYFIP1/2 have been studied in two independent contexts, as a component of fragile X mental retardation protein (FMRP) translational inhibition pathway and a component of the WAVE regulatory complex (WRC), which regulates actin polymerization. Both pathways have been extensively shown to regulate neuronal function. FMRP mutations lead to the most common inherited form of mental retardation in humans, including intellectual disability and autism spectrum disorder. The WAVE complex has been shown to regulate neural connectivity in several model organisms. CYFIP1 (p140Sra1) was initially identified as an interactor of RAC1 through biochemical affinity purification.[57] *Cyfip2* (PIR121) was originally identified as a transcriptional target that was specifically induced by p53 mutant 121F.[58] Subsequently, *Cyfip1* was isolated in a yeast two-hybrid screen using N-terminal FMRP as bait, and *Cyfip2* was shown to interact to

FMRP in a similar manner. Surprisingly, this paper reported that *Cyfip2*, but not *Cyfip1*, could also interact with FMRP orthologs, FXR1 and FXR2.[59]

CYFIP1/2 are canonical members of the WRC, which consists of WAVE (also known as WASF1), CYFIP1/2, NAP1, ABI1, and BRK1 (also known as HSPC300).[60] WRC is normally inhibited until activated by Rac GTPase, kinases, phosphatidylinositols, and WRC interacting receptor sequence (WIRS) motifs.[61,62] Upon activation, the WRC can interact with ARP2/3 complex, leading to the formation of filamentous actin from globular actin. This is mediated by the verprolin homology, central and acidic region (VCA) domain of WAVE, which is sequestered in the inactive WRC. Upon signaling, the VCA domain is disinhibited and can mediate actin polymerization (Fig. 8.7). This 400 kDa WRC complex has been crystallized and much of the mechanistic basis of this activation has been established.[63]

FMRP functions as a key translational inhibitor in the nervous system, and CYFIP1/2 have been shown to interact with FMRP in a guanosine triphosphate (GTP)-dependent manner, although the mechanism through which it regulates FMRP function remains unclear.[59,64] CYFIP1 has been shown to bind translation initiation factor eIF4E and mediate translational repression. The initial finding showed that an untranslated RNA (BC1) interacted with FMRP to increase the affinity of CYFIP1 and eIF4E.[65,66] However, the existence of the FMRP-BC1 complex has been controversial in vivo and in vitro.[67] There is good evidence of genetic interaction between *Drosophila* FMRP (dFMR1) and *Cyfip* (dCYFIP).[68]

Mouse homozygous knockouts of *Cyfip1* and *Cyfip2* are lethal during embryogenesis or soon after birth.[8,69,70] Heterozygous mice have been phenotyped and shown to have behavioral deficits. *Cyfip1* heterozygous mice have increased mGluR-dependent long-term depression (LTD) and behaviorally showed enhanced extinction of inhibitory avoidance. In rat and *Drosophila*, inactivation of CYFIP leads to decreased axonal growth, suggesting that *Cyfip* activity promotes neurite maturation.[69] Heterozygous knockout of *Cyfip2* leads to changes in cocaine responses.[8] Recently, additional behavioral characterization of *Cyfip2* knockout showed deficits in dendritic structure and plasticity leading to fragile-X like behaviors. These changes were linked to mGluR-induced expression that is regulated by *Cyfip2*.[70]

## CONCLUSION

RCCs between substrains and nearly isogenic lines are a novel forward genetic strategy that maximizes the potential to identify the underlying genetic factors and architecture contributing to trait variation. With the right traits and tools, it can be straightforward to identify genes underlying trait variation in a reasonable timeframe (e.g., less than 3 years). Underlying assumptions of an RCC are that (1) there are relatively few variants segregating between progenitor strains and recombinant F2 progeny and (2) a small

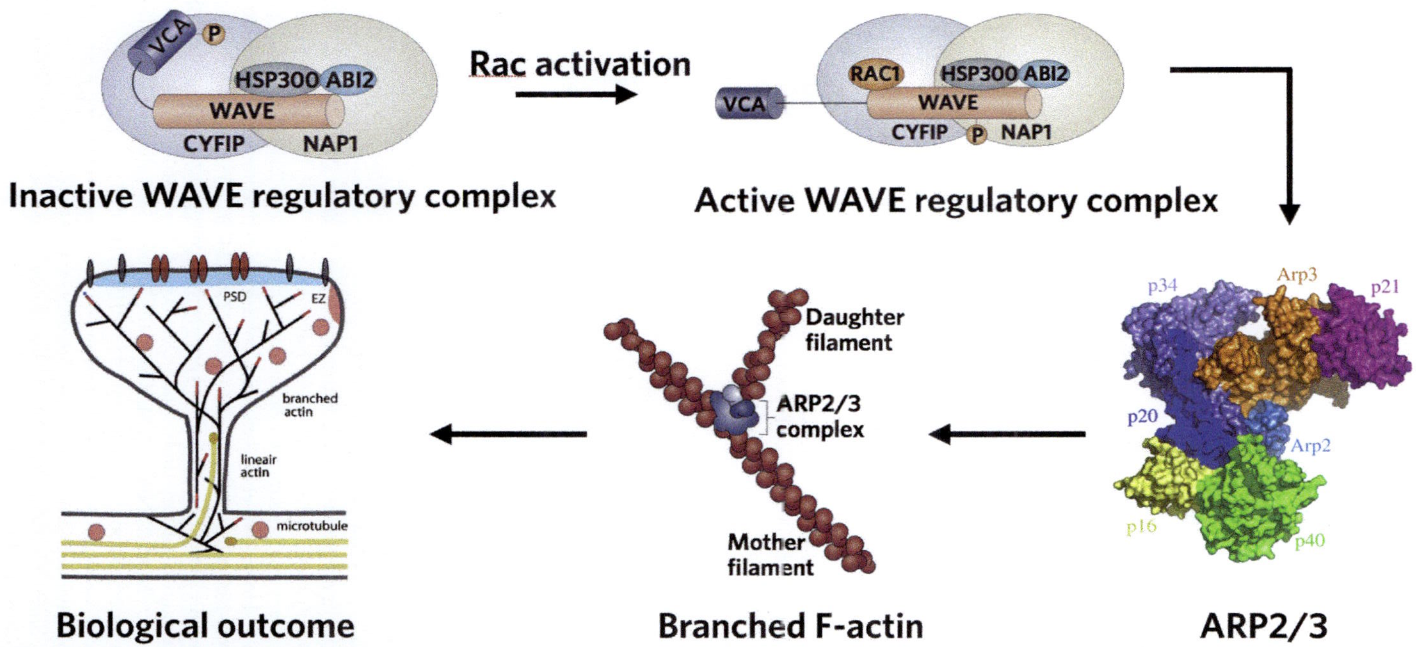


**Figure 8.7** ***Biological role of CYFIP in actin reorganization.*** Wave regulatory complex containing CYFIP1/2 is inactive, and upon Rac activation the VCA domain is disinhibited. This active complex interacts with ARP2/3 complex to mediate conversion of g-actin to f-actin. These filamentous actin structures are critical for many functions including structural changes in neurons.

number of large effect gene variants (usually acting in a dominant or strongly additive mode of inheritance) contribute to phenotypic variation in the parents and offspring. Large trait variation in the context of low background genetic diversity is expected to facilitate the identification of underlying QTL and QTGs, even in an F2 cross where the number of recombination events per chromosome is low. This assumption has held true for other RCC generated in mice[8,9] and rats.[71]

## REFERENCES


1. Green L. *Genetics and Probability in Animal Breeding Experiments: A Primer and Reference Book on Probability, Segregation, Assortment, Linkage and Mating Systems for Biomedical Scientists Who Breed and Use Genetically Defined Laboratory Animals for Research*. Oxford University Press; 1981.
2. Morse III HC. *Origins of Inbred Mice*. Elsevier Science; 2012.
3. Keane TM, Goodstadt L, Danecek P, et al. Mouse genomic variation and its effect on phenotypes and gene regulation. *Nature*. 2011;477(7364):289–294.
4. Simon MM, Greenaway S, White JK, et al. A comparative phenotypic and genomic analysis of C57BL/6J and C57BL/6N mouse strains. *Genome Biol*. 2013;14(7):R82.
5. Freeman HC, Hugill A, Dear NT, Ashcroft FM, Cox RD. Deletion of nicotinamide nucleotide transhydrogenase: a new quantitive trait locus accounting for glucose intolerance in C57BL/6J mice. *Diabetes*. 2006;55(7):2153–2156.
6. Mattapallil MJ, Wawrousek EF, Chan CC, et al. The Rd8 mutation of the Crb1 gene is present in vendor lines of C57BL/6N mice and embryonic stem cells, and confounds ocular induced mutant phenotypes. *Invest Ophthalmol Vis Sci*. 2012;53(6):2921–2927.
7. Mulligan MK, Ponomarev I, Boehm 2nd SL, et al. Alcohol trait and transcriptional genomic analysis of C57BL/6 substrains. *Genes Brain Behav*. 2008;7(6):677–689.
8. Kumar V, Kim K, Joseph C, et al. C57BL/6N mutation in cytoplasmic FMRP interacting protein 2 regulates cocaine response. *Science*. 2013;342(6165):1508–1512.
9. Kirkpatrick SL, Goldberg LR, Yazdani N, et al. Cytoplasmic FMR1-interacting protein 2 is a major genetic factor underlying binge eating. *Biol Psychiatry*. 2017;81(9):757–769.
10. Matsuo N, Takao K, Nakanishi K, Yamasaki N, Tanda K, Miyakawa T. Behavioral profiles of three C57BL/6 substrains. *Front Behav Neurosci*. 2010;4:29.
11. Bryant CD, Zhang NN, Sokoloff G, et al. Behavioral differences among C57BL/6 substrains: implications for transgenic and knockout studies. *J Neurogenet*. 2008;22(4):315–331.
12. Sturm M, Becker A, Schroeder A, Bilkei-Gorzo A, Zimmer A. Effect of chronic corticosterone application on depression-like behavior in C57BL/6N and C57BL/6J mice. *Genes Brain Behav*. 2015;14(3):292–300.
13. Crabbe JC, Wahlsten D, Dudek BC. Genetics of mouse behavior: interactions with laboratory environment. *Science*. 1999;284(5420):1670–1672.
14. Hurst JL, West RS. Taming anxiety in laboratory mice. *Nat Methods*. 2010;7(10):825–826.
15. Lewejohann L, Reinhard C, Schrewe A, et al. Environmental bias? Effects of housing conditions, laboratory environment and experimenter on behavioral tests. *Genes Brain Behav*. 2006;5(1):64–72.
16. Mandillo S, Tucci V, Holter SM, et al. Reliability, robustness, and reproducibility in mouse behavioral phenotyping: a cross-laboratory study. *Physiol Genom*. 2008;34(3):243–255.
17. Richter SH, Garner JP, Zipser B, et al. Effect of population heterogenization on the reproducibility of mouse behavior: a multi-laboratory study. *PLoS One*. 2011;6(1):e16461.
18. Wolfer DP, Litvin O, Morf S, Nitsch RM, Lipp HP, Wurbel H. Laboratory animal welfare: cage enrichment and mouse behaviour. *Nature*. 2004;432(7019):821–822.
19. Sorge RE, Martin LJ, Isbester KA, et al. Olfactory exposure to males, including men, causes stress and related analgesia in rodents. *Nat Methods*. 2014;11(6):629–632.
20. Flint J, Valdar W, Shifman S, Mott R. Strategies for mapping and cloning quantitative trait genes in rodents. *Nat Rev Genet*. 2005;6(4):271–286.

21. Sen S, Satagopan JM, Broman KW, Churchill GA. R/qtlDesign: inbred line cross experimental design. *Mamm Genome*. 2007;18(2):87–93.
22. Rosen GD, Azoulay NG, Griffin EG, et al. Bilateral subcortical heterotopia with partial callosal agenesis in a mouse mutant. *Cereb Cortex*. 2013;23(4):859–872.
23. Falconer DS, Mackay TFC, eds. *Introduction to Quantitative Genetics*. Essex, UK: Longman; 1996.
24. Lariviere WR, Mogil JS. The genetics of pain and analgesia in laboratory animals. *Methods Mol Biol*. 2010;617:261–278.
25. Lynch M, Walsh B. *Genetics and Analysis of Quantitative Traits*. Sinauer Associates, Inc.; 1998.
26. Sittig LJ, Jeong C, Tixier E, Davis J, Barrios-Camacho CM, Palmer AA. Phenotypic instability between the near isogenic substrains BALB/cJ and BALB/cByJ. *Mamm Genome*. 2014;25(11–12):564–572.
27. Blake GE, Watson ED. Unravelling the complex mechanisms of transgenerational epigenetic inheritance. *Curr Opin Chem Biol*. 2016;33:101–107.
28. Ciaranello RD, Lipsky A, Axelrod J. Association between fighting behavior and catecholamine biosynthetic enzyme activity in two inbred mouse sublines. *Proc Natl Acad Sci USA*. 1974;71(8): 3006–3008.
29. Couppis MH, Kennedy CH, Stanwood GD. Differences in aggressive behavior and in the mesocorticolimbic DA system between A/J and BALB/cJ mice. *Synapse*. 2008;62(10):715–724.
30. Dow HC, Kreibich AS, Kaercher KA, et al. Genetic dissection of intermale aggressive behavior in BALB/cJ and A/J mice. *Genes Brain Behav*. 2011;10(1):57–68.
31. Kessler S, Elliott GR, Orenberg EK, Barchas JD. A genetic analysis of aggressive behavior in two strains of mice. *Behav Genet*. 1977;7(4):313–321.
32. Velez L, Sokoloff G, Miczek KA, Palmer AA, Dulawa SC. Differences in aggressive behavior and DNA copy number variants between BALB/cJ and BALB/cByJ substrains. *Behav Genet*. 2010;40(2):201–210.
33. Fairless AH, Dow HC, Toledo MM, et al. Low sociability is associated with reduced size of the corpus callosum in the BALB/cJ inbred mouse strain. *Brain Res*. 2008;1230:211–217.
34. Cox A, Ackert-Bicknell CL, Dumont BL, et al. A new standard genetic map for the laboratory mouse. *Genetics*. 2009;182(4):1335–1344.
35. Yalcin B, Wong K, Agam A, et al. Sequence-based characterization of structural variation in the mouse genome. *Nature*. 2011;477(7364):326–329.
36. Morgan AP, Fu CP, Kao CY, et al. The mouse universal genotyping array: from substrains to subspecies. *G3 (Bethesda)*. 2015;6(2):263–279.
37. Parker CC, Gopalakrishnan S, Carbonetto P, et al. Genome-wide association study of behavioral, physiological and gene expression traits in outbred CFW mice. *Nat Genet*. 2016;48(8):919–926.
38. Castel SE, Levy-Moonshine A, Mohammadi P, Banks E, Lappalainen T. Tools and best practices for data processing in allelic expression analysis. *Genome Biol*. 2015;16:195.
39. Broman KW, Wu H, Sen S, Churchill GA. R/qtl: QTL mapping in experimental crosses. *Bioinformatics*. 2003;19(7):889–890.
40. Chesler EJ, Lu L, Wang J, Williams RW, Manly KF. WebQTL: rapid exploratory analysis of gene expression and genetic networks for brain and behavior. *Nat Neurosci*. 2004;7(5):485–486.
41. Mulligan MK, Mozhui K, Prins P, Williams RW. GeneNetwork: a toolbox for systems genetics. *Methods Mol Biol*. 2017;1488:75–120.
42. Broman KW, Sen S. *A Guide to QTL Mapping with R/qtl*. Springer; 2009.
43. Iraqi F. Fine mapping of quantitative trait loci using advanced intercross lines of mice and positional cloning of the corresponding genes. *Exp Lung Res*. 2000;26(8):641–649.
44. Gonzales NM, Palmer AA. Fine-mapping QTLs in advanced intercross lines and other outbred populations. *Mamm Genome*. 2014;25(7–8):271–292.
45. Denmark DL, Milner LC, Buck KJ. Interval-specific congenic animals for high-resolution quantitative trait loci mapping. *Alcohol Res Health*. 2008;31(3):266–269.
46. Turner TL. Fine-mapping natural alleles: quantitative complementation to the rescue. *Mol Ecol*. 2014; 23(10):2377–2382.
47. Bryant CD, Kole LA, Guido MA, Cheng R, Palmer AA. Methamphetamine-induced conditioned place preference in LG/J and SM/J mouse strains and an F45/F46 advanced intercross line. *Front Genet*. 2012;3:126.

48. Bryant CD, Parker CC, Zhou L, et al. Csnk1e is a genetic regulator of sensitivity to psychostimulants and opioids. *Neuropsychopharmacology*. 2012;37(4):1026–1035.
49. Yazdani N, Parker CC, Shen Y, et al. Hnrnph1 is a quantitative trait gene for methamphetamine sensitivity. *PLoS Genet*. 2015;11(12):e1005713.
50. Iakoubova OA, Olsson CL, Dains KM, et al. Genome-tagged mice (GTM): two sets of genome-wide congenic strains. *Genomics*. 2001;74(1):89–104.
51. Wong GT. Speed congenics: applications for transgenic and knock-out mouse strains. *Neuropeptides*. 2002;36(2–3):230–236.
52. Civelek M, Lusis AJ. Systems genetics approaches to understand complex traits. *Nat Rev Genet*. 2014; 15(1):34–48.
53. Li YI, van de Geijn B, Raj A, et al. RNA splicing is a primary link between genetic variation and disease. *Science*. 2016;352(6285):600–604.
54. Young EE, Bryant CD, Lee SE, et al. Systems genetic and pharmacological analysis identifies candidate genes underlying mechanosensation in the von Frey test. *Genes Brain Behav*. 2016;15(6):604–615.
55. Yang H, Wang JR, Didion JP, et al. Subspecific origin and haplotype diversity in the laboratory mouse. *Nat Genet*. 2011;43(7):648–655.
56. Mozhui K, Ciobanu DC, Schikorski T, Wang X, Lu L, Williams RW. Dissection of a QTL hotspot on mouse distal chromosome 1 that modulates neurobehavioral phenotypes and gene expression. *PLoS Genet*. 2008;4(11):e1000260.
57. Kobayashi K, Kuroda S, Fukata M, et al. p140Sra-1 (specifically Rac1-associated protein) is a novel specific target for Rac1 small GTPase. *J Biol Chem*. 1998;273(1):291–295.
58. Saller E, Tom E, Brunori M, et al. Increased apoptosis induction by 121F mutant p53. *EMBO J*. 1999; 18(16):4424–4437.
59. Schenck A, Bardoni B, Moro A, Bagni C, Mandel JL. A highly conserved protein family interacting with the fragile X mental retardation protein (FMRP) and displaying selective interactions with FMRP-related proteins FXR1P and FXR2P. *Proc Natl Acad Sci USA*. 2001;98(15):8844–8849.
60. Eden S, Rohatgi R, Podtelejnikov AV, Mann M, Kirschner MW. Mechanism of regulation of WAVE1-induced actin nucleation by Rac1 and Nck. *Nature*. 2002;418(6899):790–793.
61. Chen B, Brinkmann K, Chen Z, et al. The WAVE regulatory complex links diverse receptors to the actin cytoskeleton. *Cell*. 2014;156(1–2):195–207.
62. Ismail AM, Padrick SB, Chen B, Umetani J, Rosen MK. The WAVE regulatory complex is inhibited. *Nat Struct Mol Biol*. 2009;16(5):561–563.
63. Chen Z, Borek D, Padrick SB, et al. Structure and control of the actin regulatory WAVE complex. *Nature*. 2010;468(7323):533–538.
64. Abekhoukh S, Bardoni B. CYFIP family proteins between autism and intellectual disability: links with Fragile X syndrome. *Front Cell Neurosci*. 2014;8:81.
65. Napoli I, Mercaldo V, Boyl PP, et al. The fragile X syndrome protein represses activity-dependent translation through CYFIP1, a new 4E-BP. *Cell*. 2008;134(6):1042–1054.
66. Zalfa F, Adinolfi S, Napoli I, et al. Fragile X mental retardation protein (FMRP) binds specifically to the brain cytoplasmic RNAs BC1/BC200 via a novel RNA-binding motif. *J Biol Chem*. 2005;280(39): 33403–33410.
67. Iacoangeli A, Rozhdestvensky TS, Dolzhanskaya N, et al. On BC1 RNA and the fragile X mental retardation protein. *Proc Natl Acad Sci USA*. 2008;105(2):734–739.
68. Schenck A, Bardoni B, Langmann C, Harden N, Mandel JL, Giangrande A. CYFIP/Sra-1 controls neuronal connectivity in *Drosophila* and links the Rac1 GTPase pathway to the fragile X protein. *Neuron*. 2003;38(6):887–898.
69. Bozdagi O, Sakurai T, Dorr N, Pilorge M, Takahashi N, Buxbaum JD. Haploinsufficiency of Cyfip1 produces fragile X-like phenotypes in mice. *PLoS One*. 2012;7(8):e42422.
70. Han K, Chen H, Gennarino VA, Richman R, Lu HC, Zoghbi HY. Fragile X-like behaviors and abnormal cortical dendritic spines in cytoplasmic FMR1-interacting protein 2-mutant mice. *Hum Mol Genet*. 2015;24(7):1813–1823.
71. Zhou Z, Karlsson C, Liang T, et al. Loss of metabotropic glutamate receptor 2 escalates alcohol consumption. *Proc Natl Acad Sci USA*. 2013;110(42):16963–16968.

CHAPTER 9

# Collaborative Cross as the Next-Generation Mouse Genetic Reference Population Designed for Dissecting Complex Traits

**Hanifa J. Abu Toamih Atamni, Fuad A. Iraqi**
Tel Aviv University, Tel Aviv, Israel

## BACKGROUND

Genetic studies of complex diseases with complex etiologies under controlled and standardized environmental conditions are unfeasible in humans because of assorted limitations. The major limitations are due to the difficulty in performing controlled environmental challenges and the fact that susceptibility by itself is a "complex trait," suggested to be controlled by the cumulative effects and interactions of multiple environmental factors and genetic components. In crosses between genetically characterized strains of mice, chromosomal regions controlling the genetic variance of complex traits can be mapped as quantitative trait loci (QTLs) using experimental populations maintained under common environmental conditions.[1–3] Furthermore, comparative mapping reveals that a multitude of murine genes have recognized human gene homologues, emphasizing the relevance of QTL mapping and gene identification in a mouse model for the realization of complex diseases in humans. Once a QTL or corresponding genes are mapped in the mouse genome, then translational genetic analysis can be extended successfully to the human genome. It is recognized that the utilization of mouse models is tightly established in research studies of human medicine.[4–6] Through QTL mapping studies in mice and other species, accumulated evidence for genetic loci controlling a wide variety of complex traits has emerged; and yet, albeit with some significant exceptions, the majority of the genes within the mapped genomic loci remain undisclosed. Consequently, the realization of the necessity of a new mouse model population allocated for genetic mapping and understanding of the complex traits of complex human diseases arose. Therefrom, broad and profound panel discussions of the genetics community proposed the Collaborative Cross (CC) mouse model design, to provide a new mouse resource dedicated to high-resolution mapping of complex traits, placing extra emphasis on relevance to extensive human health traits.[7,8] The CC mouse model reference genetic resource (CC lines) comprises a unique set of approximately 350 recombinant inbred lines (RILs) created from full reciprocal matings of eight divergent

*Molecular-Genetic and Statistical Techniques for Behavioral and Neural Research*
ISBN 978-0-12-804078-2, https://doi.org/10.1016/B978-0-12-804078-2.00009-X

mouse strains: A/J, C57BL/6J, 129S1/SvImJ, NOD/LtJ, NZO/HiLtJ, CAST/Ei, PWK/PhJ, and WSB/EiJ. Intercrosses during the breeding process were performed with controlled randomization to minimize and disperse large linkage disequilibrium blocks, as well as recombining the natural genetic variation that exists in these inbred strains, aiming for a unique and inexhaustible resource of RI strains exhibiting extensive phenotypic and genotypic diversity.[9] By way of illustration, Fig. 9.1 presents the development scheme of an individual CC line with a mosaic genotype structure following a series of inbreeding generations.

Computer simulations propose that the 100 RILs being developed at our lab will allow fine mapping of QTLs explaining a total of 5% of the RIL phenotypic variations (equivalent to 2.5% of F2 variation) to an average mean location error of 3.5 cM.[10] Originally, all CC mice were genotyped with a mouse diversity array (MDA) consisting of 620,000 single-nucleotide polymorphism (SNP) markers.[11] Subsequently, all SNPs with heterozygous or missing genotypes in the eight CC founders, or that were not in

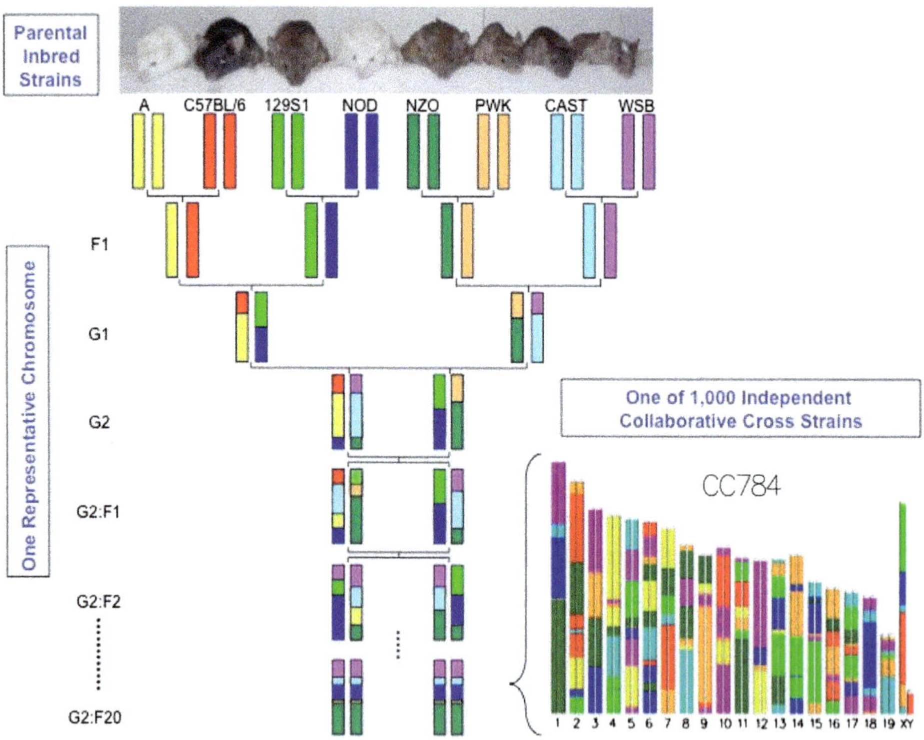


**Figure 9.1** ***Single Collaborative Cross (CC) line (CC784) funnel breeding scheme using eight founder strains.*** Each CC line originates from an independent breeding funnel of the eight CC founders so that every recombination site in the CC population is uniquely generated. In this example, a CC line is produced from the breeding funnel and labeled with laboratory code.[81]

common between the arrays, were filtered out, leaving 170,935 SNPs. The SNPs were mapped onto Build 37 of the mouse genome. Next, using the HAPPY HMM software,[12] a descent distribution probability was computed for each of the 170,935 SNP intervals, which were reduced to 8533 intervals by averaging the matrices in groups of $n = 20$ consecutive SNPs. This reduction minimized the effects of genotyping errors to improve analysis speed quality. Additionally, mean heterozygosity was computed across each window of the 20 SNPs. An illustration of the genomic reconstruction of two CC lines following genotyping with MDA and using HAPPY software[12] is presented in Fig. 9.2. Later on, at an advanced generation, the CC lines were regenotyped with a new custom-designed 7500-SNP array, named the mouse universal genotype array (MUGA), providing the genomic architecture of the CC lines.[39] Afterward, five advanced generations, all CC lines, were regenotyped with MegaMUGA. Eventually, merge analysis was performed, merging the genotypes of the three SNP arrays into a single genotype file to be used for QTL mapping.[14]

The eight CC founder strains included three wild-derived strains, CAST/EiJ, PWK/PhJ, and WSB/EiJ, representing the subspecies *Mus musculus castaneus*, *Mus musculus musculus*, and *Mus musculus domesticus*, respectively. These three wild-derived strains contribute a large number of sequence variants not segregating among classical strains descended from *M. musculus domesticus*, whereas most classical strains differ from the reference C57BL/6J at about 4 million SNPs, though PWK and CAST each differ at about 17 million SNPs, and WSB differs at 6 million.[15] Accordingly, the QTL mapping using the CC mouse model achieved success, revealing novel QTLs that involve contrasts between the wild-derived strains. A pilot experiment presenting this success included

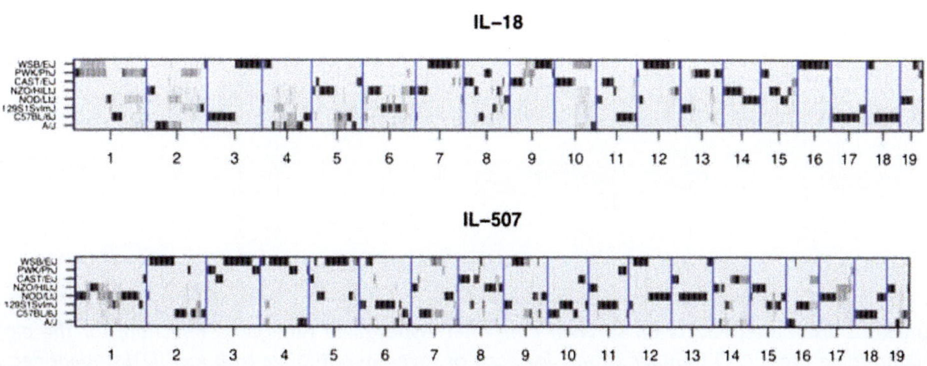


**Figure 9.2** ***Reconstructions of the genomes of representative Collaborative Cross (CC) lines IL-18 and IL-507 from the hidden Markov model implemented by the HAPPY program.*** The *x* axis shows the 19 autosomes, the *y* axis shows the eight parental founder strains of the CC lines. The *black horizontal bands* represent genetically fixed loci contributed by only one founder, while *gray bands* represent loci with residual heterozygosis.[13]

fine-mapped QTLs that were associated with survival in response to *Aspergillus fumigatus* (AF) infection challenge, in which we mapped eight QTLs, five involving contrasts with wild-derived strains, a result that would not have been achieved in a cross between classical laboratory inbred strains (Fig. 9.3). These early studies confirmed that phenotyping a relatively modest number of CC lines (around 70 lines), with sufficient replication, will enable mapping of QTLs to a resolution of about 1 Mb.[10] In another study using CC lines, we demonstrated that the main determinant of gut microbiome diversity was contributed by the genetic makeup of the CC lines,[16] manifesting the power of the CC mouse model in dissecting environmental effects on host response and performance.

Moreover, candidate genes located within the significant QTL genomic interval can be refined by incorporating variation data from the genotypic data of the CC founders, available from the Sanger Mouse Genomes Project,[15] and restricting attention to variants whose differences across the founders are consistent with the pattern of action of the QTL.[17] Overall, our published studies examine and approve the special efficacy of the CC lines for mapping QTLs to a resolution of about 1 Mb via phenotyping a relatively modest number of approximately 70 CC lines. Table 9.1 presents the heritability and coefficient of genetic variation underlying the mapped complex traits.[13,14,18–32]

Nowadays, mouse genetic resource populations (GRPs) are frequently used for complex traits and biological systems studies, in both medical and life sciences applications, because genotyping is required only once. It is described as a "genotype once, phenotype many times" paradigm, in which the same genotype will be used for replicate individuals, allowing optimal designs of case/control and gene-by-environment.[33]

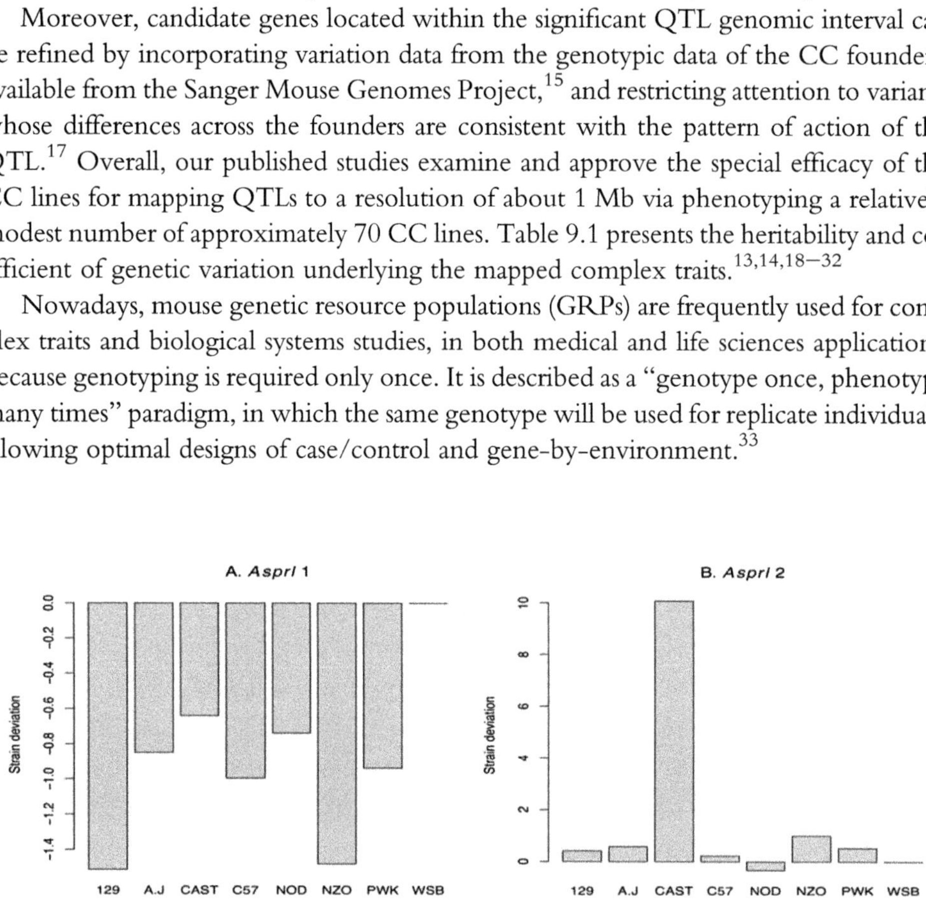


**Figure 9.3** ***Estimated effects on survival time after*** **Aspergillus fumigatus** ***infection, for the eight Collaborative Cross (CC) founder strains for each of seven quantitative trait loci (QTLs) identified in a screen of 66 CC lines.*** Effects are shown as deviations relative to WSB/EiJ, which is arbitrarily assigned the trait effect of 0. The *x* axis of each plot shows the founder strains; the *y* axis shows the estimated effect on survival for the haplotype of the corresponding CC founder. QTLs A and B involve a contrast between a wild-derived strain (CAST, PWK, or WSB) and the other founders.[13]

**Table 9.1** Estimated heritability of a variety of phenotypic complex traits assessed using the Collaborative Cross mouse model

| Topic | Challenge | Trait | $H^2$ |
|---|---|---|---|
| Infectious diseases | *Aspergillus fumigatus* | Survival time (days) | 0.78 |
| | *Klebsiella pneumonaie* | Survival time (days) | 0.45 |
| | *Porphyromonas gingivalis* and *Fusobacterium nucleatum* | CBV | 0.4 |
| | | RBV | 0.4 |
| | | LBV | 0.2 |
| | *Pseudomonas aeruginosa* | Survival time (days) | 0.54 |
| | | Body weight loss(g) | 0.28 |
| | Lipopolysaccharide | Survival time (h) | 0.19 |
| Tumorigenesis | CC-C57BL/6 lines carrying $Apc^{Min/+}$ mutation | Total polyps number | 0.64 |
| | | %Packed cell volume | 0.51 |
| | | %Spleen weight | 0.53 |
| Type 2 diabetes | High-fat diet (42%) | ♀IPGTT total AUC | 0.43 |
| | | ♂IPGTT total AUC | 0.37 |
| Obesity | High-fat diet (42%) | ♀Body weight gain | 0.53 |
| | | ♂Body weight gain | 0.38 |
| Body composition | DEXA scan | Body %fat | 0.44 |
| | | Bone mineral density ($g/cm^2$) | 0.00 |
| Immune system | FACS analysis | T cells | 0.34 |
| | | B cells | 0.30 |
| | | Macrophages | 0.16 |
| Trabecular bone microarchitecture | Microtomography scan | BY/TV | 0.63 |
| | | Tb.N | 0.71 |
| | | Tb.Th | 0.56 |
| | | Conn.D | 0.67 |

*AUC*, area under curve; *BV/TV*, trabecular bone volume fraction; *CBV*, control bone volume; *Conn.D*, connectivity density; *DEXA*, dual-energy X-ray absorptiometry; *FACS*, fluorescence-activated cell sorting; $H^2$, broad-sense heritability; *IPGTT*, intraperitoneal glucose tolerance test; *RBV*, residual bone volume; *Tb.N*, trabecular bone number; *Tb.Th*, trabecular bone thickness.

In this chapter, we present a series of our multidisciplinary studies, published and unpublished, using the CC mouse population for research on complex traits. The studies map the genetic components contributing to the phenotypic variation observed among certain populations, either in baseline biological/physiological variations or at the level of host response to environmental challenges that induce infectious or chronic diseases. Studies of infectious diseases in our lab scanned independently the CC lines' response to various known pathogens, including AF,[13] *Klebsiella pneumoniae* (KP),[18] coinfection with *Porphyromonas gingivalis* and *Fusobacterium nucleatum*,[34] and *Pseudomonas aeruginosa*,[35] and host response to microbial toxins [lipopolysaccharide (LPS)/lipoteichoic acid (LTA)] for a study of sepsis. Our studies of complex chronic diseases included diet-induced type 2 diabetes (T2D) and metabolic syndrome,[36,37] development of colon cancer,[38] and autism. Additionally, the CC populations were scanned for genetic factors underlying basic phenotypic variation in the baseline state without disease in the background; this included influencing immune response cell lineages in peripheral blood and body composition traits using dual-energy X-ray absorptiometry (DEXA). Understanding the host susceptibility to all of these complex traits is important because they all affect systems under the same genetic background.

### Research-Based Implementations of the Collaborative Cross Mouse Model

As of this writing, the CC mouse model has been assessed in many fields of scientific study, whether in the study of communicable diseases (infectious diseases) or noncommunicable diseases or for mapping genetic components of baseline body features. A majority of these multidisciplinary studies were accomplished in our lab, while additional studies were accomplished in collaboration with partners. For the most part published reports are presented in this chapter, but also presented are other ongoing studies. Based on simulation studies,[10] the CC lines (RILs) used in these projects (both females and males) should allow successful mapping of a QTL explaining a total of 5% of the RIL phenotypic variation (equivalent to 2.5% of F2 variation) to an average mean location error of 3.5 cM or less. At this time, about 70 CC lines are under development at our lab at the Small Animal Facility of Tel-Aviv University, between generations G25 and G53 of inbreeding by full-sib mating[2,39]; CC lines for the various studies were selected randomly based on breeding availability in each weaning cycle, which is held every 3 weeks by our research team at Tel-Aviv University. Interestingly, owing to genetic variations between the CC lines, the breeding rate and number/sex of litters in each cycle might vary. Therefore, the CC lines were assessed in the various projects based on the litters' availability, while we make our best efforts to scan as many of the CC lines as possible with representation of both sexes.

## COMMUNICABLE DISEASES

### Mapping Host Susceptibility to *Aspergillus fumigatus* Pathogen

Aspergillosis is considered a serious propagating and invasive disease in humans, inflicting severe damage to the kidneys, liver, spleen, brain, heart, and other organs.[40,41] Susceptibility to infectious pulmonary aspergillosis is particularly higher among immunocompromised individuals (impaired immune system), including neutropenia patients, patients treated with corticosteroids or other immunosuppressive therapies, and patients with late-stage human immunodeficiency viral infection. However, significant variations were observed between individuals in their responses to AF pathogen, showing different levels of susceptibility or resistance and different patterns of disease progress and severity, suggesting that host response is a complex trait controlled by the genetic background of the host. In our previous study, we reported the mapping of host susceptibility to AF by phenotyping 371 mice from 66 CC lines for susceptibility to infection by AF. For the infection challenge, AF strain 293 (AF293) was obtained by growth on Sabouraud dextrose agar (SDA; Difco Laboratories) containing chloramphenicol (Sigma Chemicals). Conidia were harvested in 0.2% (v/v) Tween 80 (Sigma Chemicals), resuspended in triple-distilled water, and counted with a hemocytometer. Conidia numbers were confirmed by colony counts of SDA plate dilutions after 48 h incubation at 37°C. On the day of the challenge, mice were weighed before infection (time 0) and then inoculated intravenously via the lateral tail vein using a 27-gauge syringe. Each mouse was challenged with $10^7$ conidia of freshly harvested AF293 with a final volume of 0.2 mL in saline medium. This dose should be sufficient to ensure that severe infection occurs regardless of the weight or sex of the animal. The challenge period started on the day of infection (day 1) and lasted for 10 days, and mouse survival was monitored daily. The body weight (BW) of the mice was recorded before and throughout the challenge period at various time points (day 0, 5, 10). Our findings demonstrated that survival time postinfection varied significantly between CC lines ($P < .05$) within a broad range of 4–28 days of survival, as presented in Fig. 9.4. The calculated broad-sense heritability ($H^2$) for the host response, i.e., days of survival, was 0.18, while QTL mapping based on survival analysis and ancestral haplotype reconstruction of the CC genomes identified genome-wide significant QTLs on chromosomes 2, 3, 8, 10 (two QTLs), 15, and 18. The QTL mapping resolution varied between 2 and 11.6 Mb. Furthermore, variation data analysis using the genomes of CC founder strains refines these QTLs, further suggesting several candidate genes.[13] To our knowledge, this was the first report on mapping susceptibility loci for invasive aspergillosis in immune-competent mice. Subsequently, the study was extended to further CC lines, assessing AF susceptibility and also performing hepatic gene expression analysis (RNA sequencing) using deep sequence technology for in-depth analysis at the level of actual gene expression variations during disease. Liver samples for RNA sequencing in this project were collected from each CC line (two mice per line) at different time points during the challenge (0, 5, and 10 days postinfection), harvested in liquid nitrogen, and stored in a −150°C freezer. RNA was extracted using an RNeasy Plus Universal Mini Kit (Qiagen; Cat. No. 73404). Briefly, tissue samples were homogenized in QIAzol lysis reagent. After addition of gDNA Eliminator solution and chloroform, the homogenate was separated into aqueous and organic phases by centrifugation. RNA partitioned to the upper, aqueous phase was then collected. Thereafter, the RNA was purified using RNeasy spin columns. RNA sequencing was performed by the CRG Institute, a node of the European Sequence and Genotyping Institute in Barcelona, Spain. Meanwhile, the hepatic gene expression analysis is ongoing as of this writing and is expected to be published soon after completion.

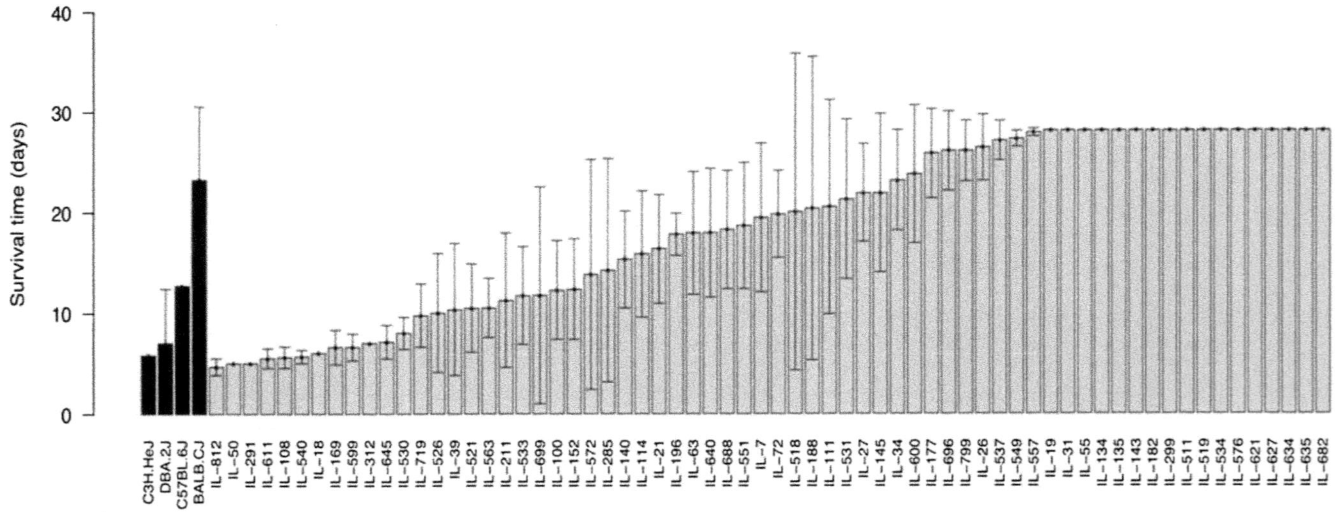


**Figure 9.4** ***Mean survival time (days) of various Collaborative Cross (CC) lines in response to* Aspergillus fumigatus *infection.*** The *x* axis represents the CC lines, while the *y* axis represents the mean survival days (±SEM). Full details of the analysis are presented in Durrant et al.[13]

## Mapping Host Susceptibility to *Klebsiella pneumoniae* Pathogen

KP is a considerable pulmonary pathogen leading to severe pneumonia and sepsis, mainly in immunocompromised patients. Host susceptibility to KP was successfully mapped in our lab, using the CC lines, through intraperitoneal (IP) injection of KP as an infectious challenge and measurement of survival time postinfection at various time points. The KP strain used in this study, K2, was originally isolated from human blood (further details in Ref. 18). K2 was grown for 18 h at 37°C in Luria broth (LB), consisting of 10% Bacto Tryptone (Difco), 5% yeast extract, and 5% NaCl (pH 7.5), and thereafter inoculated into LB for an extra 4 h. Moreover, K2 concentration was assessed by comparing the spectrophotometric absorbance at 595 nm with a standard curve. Bacterial dilution was performed using sterilized endotoxin-free saline providing $10^4$ colony-forming units (CFU) in a final volume of 0.2 mL. Overall, 328 mice generated from 73 CC lines were infected intraperitoneally with $10^4$ CFU/mL K2. The response of the mice postinfection was monitored daily for a period of 15 days, by recording the phenotypes of BW and survival. In Vered et al.[18] we describe the study findings, in which changes in BW were not significantly different between the CC lines during the infection period, whereas mean survival time varied significantly ($P < .05$) between the CC lines (Fig. 9.5A), ranging between 2 and 15 days of survival postinfection. Importantly, mice that survived the seventh day postinfection eventually survived the complete challenge period up to day 15, suggesting the possibility of a critical time window during the disease progress. The observed wide variations in host response to KP demonstrate the high hereditary component of the host response to be controlled by genes. In fact, calculation of the $H^2$ confirms that, with an $H^2$ of 0.45. Moreover, variances of the mean survival time at different time points between days 1 and 2 were successfully mapped, using only 48 CC lines. Mapping results revealed at least three QTLs associated with host susceptibility to KP as presented in Fig. 9.5B, named KP-resistant locus 1, 2, and 3 (*Kprl1*, *Kprl2*, and *Kprl3*), and located on chromosomes 4, 8, and 18, respectively. The mapped QTLs were time-specific, suggesting the idea of disease progress-related "players" (genes) changing at different stages of the body response (Chr. 4, day 2, and Chrs. 8 and 18, day 8). These achievements emphasize the power of the CC population to enable fine mapping with high resolution of narrow genomic intervals (<0.6 Mb). Estimations of a founder's effect revealed that wild-derived strains were the main contributors of the observed genetic variance leading to phenotypic variance in host susceptibility. Furthermore, the search for candidate genes located within the significant QTLs using gene browsing confirmed their relevance to this complex trait of KP susceptibility. The QTL significant genomic intervals include known protein-coding genes, such as on *Kprl1* QTL, the candidate genes *Ikbkap*, inhibitor of κ light polypeptide gene enhancer in B cells, a kinase complex-associated protein, and *Actl7a* and *Actl7b*, actin-like 7a and 7b, which are involved in cell adhesion and cytoskeleton structure (phagocytosis and eliminating

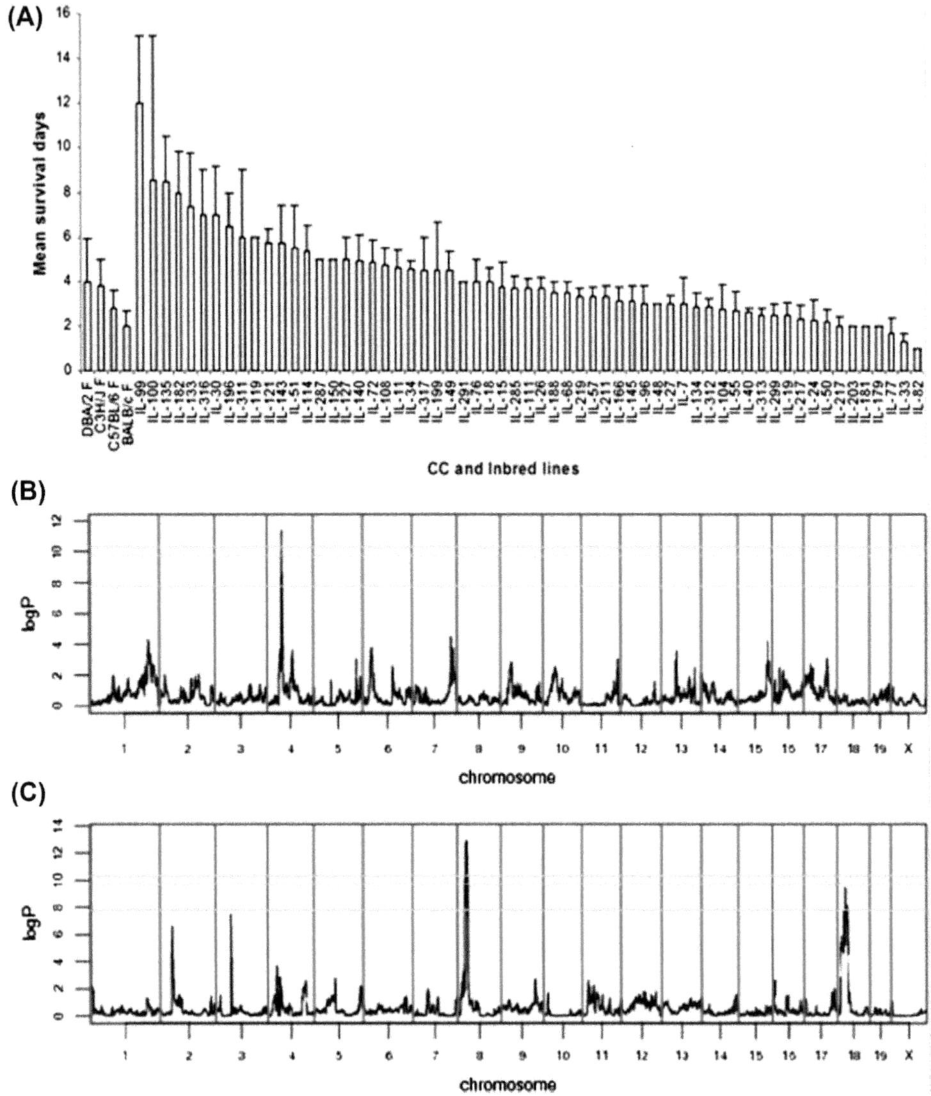


**Figure 9.5** Genetic dissection of host response to *Klebsiella pneumoniae* (KP) infection. (A) Phenotypic profile of mean survival time (days) of the Collaborative Cross (CC) lines and four inbred strains after infection with KP. The *x* axis represents the CC lines and the *y* axis represents the mean survival time in days (±SEM). (B and C) The quantitative trait locus (QTL) mapping results using the phenotypic and genotypic data of the CC lines. In total the results revealed three significant QTLs associated with survival time after infection with KP, on day 2, one QTL on chromosome 4 named *Kprl1* (in B), and on day 8, two QTLs named *Kprl2* and *Kprl3* on chromosomes 8 and 18, respectively (in C). The *x* axis shows the 19 chromosomes and the *y* axis the logP of the linkage analysis.[18]

bacterial infection). The *Kprl1* QTL also contains *Ctnnal1*, catenin α-like 1, a gene that plays a central role in cell adhesion processes and captures bacterial molecules to eliminate them.

This was the novel report of mapping multilocus KP host susceptibility in immune-competent mice, supporting the genetic complexity of host response to KP infection. These QTLs were KP-specific genomic regions that absolutely differ from AF QTLs.

## Mapping Host Susceptibility to Oral Bacterial Mixed Infection

Periodontitis is considered to be one of the most prevalent chronic inflammatory diseases in humans, and is defined as periodontal infection accompanied by the destruction of tooth-supporting tissues that eventually leads to tooth loss. The complex process is characterized by degradation of the periodontal ligaments, formation of periodontal pockets, and alveolar bone resorption.[42] The pathogenesis of periodontitis requires pathogenic periodontal bacterial accumulation as a subgingival biofilm to encompass the inflammatory response stimulation of the host gingiva, whereby excessive sustained response may lead to chronic inflammation.[43] Accumulated evidence of studies asserts a significant genetic component contributing to host susceptibility to chronic periodontitis.[44–46]

Accordingly, we have used the CC lines and an oral mixed infection system (of the two anaerobic gram negative bacteria *P. gingivalis* and *F. nucleatum*) as was previously described by Polak,[47] to ascertain and characterize the genetic component involved in the development of periodontitis. The measured phenotype for periodontitis progress in our study was residual alveolar bone volume ($mm^3$) postinfection. The bone volume trait was measured using the accurate micro-computed tomography (CT) technique, providing evaluation of the ligament loss around the host teeth.[48] We assessed 272 mice (103 females and 169 males) from 23 CC lines challenged with *P. gingivalis* and *F. nucleatum* infection as a trigger for periodontitis development in susceptible CC lines. The experiment consisted of two study groups: (1) control—placebo infection treated with phosphate-buffered-saline (PBS) without bacteria (control bone volume; CBV) and (2) challenge—infection with oral mixed bacteria (*P. gingivalis* and *F. nucleatum*). Both groups were represented in each CC line. To allow a controlled infection, prior to initiation of the experiment, the normal oral flora of the mice was suppressed by treating the mice with the antibiotic sulfamethoxazole (Resprim) 10/500 mL in water ad libitum for 10 days followed by a 3-day washout (antibiotic-free period). Thereafter, the mice were superinfected with 400 μL of a mixture of *P. gingivalis* ($10^9$/mL) and *F. nucleatum* ($10^9$/mL) in PBS with 2% carboxymethylcellulose by gavage three times at 2-day intervals.[45,47] The bacterial strains *P. gingivalis* 381 and *F. nucleatum* 1594 were grown in peptone yeast extract containing hemin and vitamin K in an anaerobic chamber, with 85% $N_2$, 5% $H_2$, and 10% $CO_2$, for 48 h, followed by three washes in PBS. Bacterial concentration was determined spectrophotometrically, standardized to

OD 650 nm = 0.1 for *P. gingivalis*, corresponding to $10^{10}$ bacteria/mL, and OD 660 nm = 0.26 for *F. nucleatum*, corresponding to $10^9$ bacteria/mL.[49] All the CC mice were sacrificed and dissected 42 days after the third infection. Mice were anesthetized with an injection of a xylazine:ketamine solution (1:3 ratio, distilled in sterile saline), at 0.2 mL solution per 25 g BW. Thereafter, the jaws (maxillae and mandible) were harvested from each mouse and saved in tubes with 300 μL RNA Save solution. Then the alveolar bone loss was measured using the micro-CT technique, whereby the maxillary hemijaws were analyzed by compact fan-beam-type CT (μCT 40, Scanco Medical, Brüttisellen, Switzerland). The samples were placed in a cylindrical sample holder, and approximately 200 microtomographic slices at increments of 12 μm were acquired for each jaw. Image segmentation of bone, dentin, enamel, and pulp was obtained by applying a manually selected threshold for all the specimens. The alveolar bone loss phenotype for each CC line postinfection was calculated as the difference in residual bone volume (RBV) between the experimental groups. Consequently, the study findings reported highly significant variations ($P < .05$) in bone loss phenotype between the CC lines in response to infectious challenge (Fig. 9.6), unique results that were successfully published in Shusterman et al.[34] The CC lines showed multiple response patterns clustered in subgroups of susceptible, resistant, and intermediate, based on their bone volume

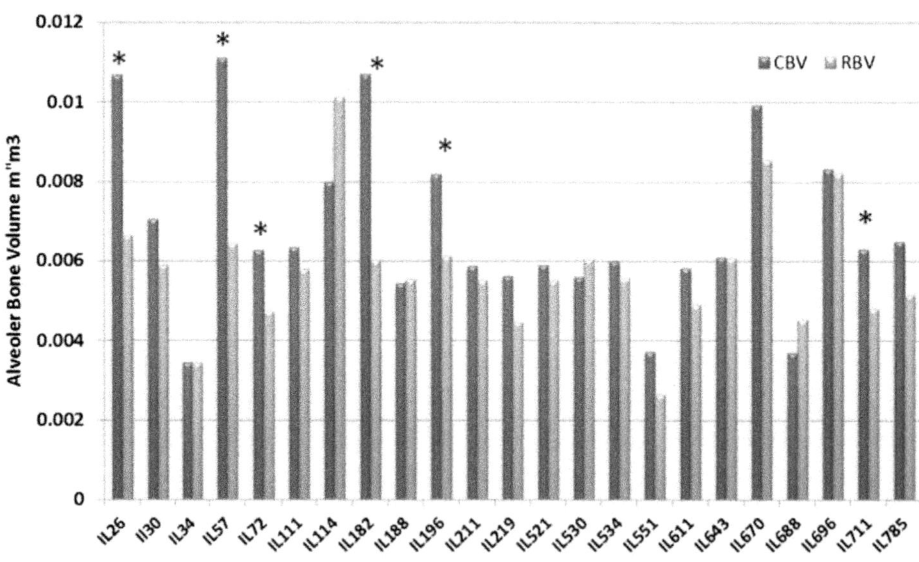


**Figure 9.6** ***Collaborative Cross (CC) lines alveolar bone volume changes in response to oral mixed infection.*** *Dark gray columns* represent control bone volume (CBV) and *light gray columns* represents residual bone volume (RBV). Of the total 23 CC lines, six (Il26, IL57, IL72, IL182, IL196, and IL711) show a significant bone loss ($P < .05$) and are considered to be susceptible to the infection (labeled with *asterisk*). The *x* axis represents the CC lines, the *y* axis represents the alveolar bone volume in cubic millimeter units (m"m$^3$).[34]

changes postinfection. Overall, 25% of the assessed CC lines were susceptible to infection as assessed by postinfection bone loss (IL26, IL57, IL72, IL182, IL196, and IL711), and the remaining CC lines were either resistant or moderately susceptible (no significant change in bone volume). The estimated $H^2$ calculated for the measured phenotypes was 0.4 for both CBV and RBV and 0.2 for loss of bone volume (LBV).

We hereby affirm the existence of strong genetic components controlling the host response to the development of periodontitis. Our findings reconfirm the advantageous utilization of the CC mouse model lines for QTL mapping of susceptibility to periodontal infection. Hence, this study was successfully extended to assess ~100 CC lines for periodontitis as described above, followed by mapped QTLs associated with host susceptibility to periodontitis, which will be published soon.

## Mapping Host Susceptibility to *Pseudomonas aeruginosa*

*P. aeruginosa* is one the major leading resistant pathogens in hospitals and health care systems worldwide (10%–15% of cases) to be listed among the World Health Organization's leading resistant pathogens with greatest public health concern. The primary concern is for intensive care unit (ICU) patients with a higher risk of development of severe response to infection with *P. aeruginosa* leading to ventilator-associated pneumonia and sepsis.[50] Extensive variations have been reported among populations in host response to *P. aeruginosa* infection, particularly high-risk individuals and cystic fibrosis patients. Hitherto, multiple studies have been carried out in attempts to map the *P. aeruginosa* genotypic components contributing to the observed phenotypic variations in disease progress and pathogenesis.[51–54] Furthermore, approaches have also been suggested for dissecting the host genetic background, which might play a crucial role in contributing to the phenotypic variations observed in response severity.[55] Thus, multiple studies based on murine models have been performed, challenging different mouse strains with *P. aeruginosa* infection, to confirm repeatedly significant variation in the phenotypic response postinfection between the strains.[56,57]

The next study was done using the CC mouse population, aiming for high-resolution fine mapping of these genomic regions. During this study, 92 mice (50 males, 42 females) generated from 17 CC lines were assessed by *P. aeruginosa* infection challenge (intratracheal injection), and subsequently the host phenotypic responses of survival time (days) and BW (gram) changes were recorded for 7 days postinfection. This experiment was done using the *P. aeruginosa* clinical strain AA2, grown in trypticase soy broth at 37°C. The infection was performed under anesthesia by intratracheal injection of $10^6$ CFU, cannulated into the lungs as described in Lorè et al.[35] Postinfection, the survival time and BW of the mice were monitored daily over 1 week; then the surviving mice were euthanized and tested for CFU in blood, liver, and lung. Study findings were published in Lorè et al.[35] reporting significant variations between the CC lines in their survival time

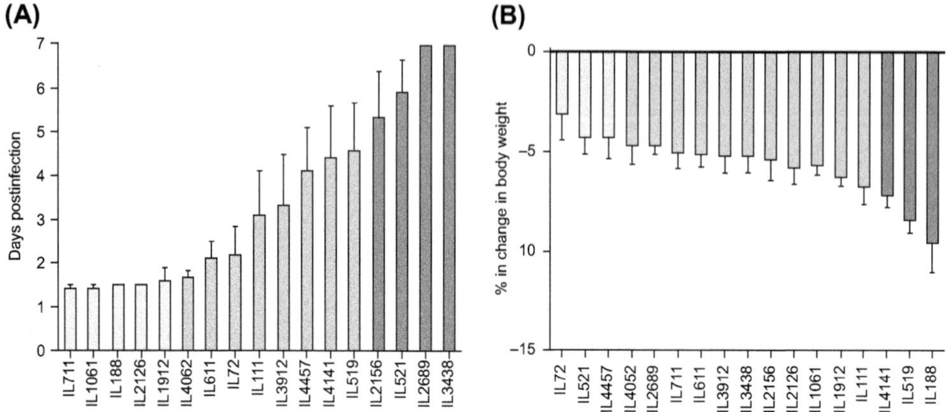


**Figure 9.7** Collaborative Cross (CC) lines' response to *Pseudomonas aeruginosa* airway infection. (A) Mean survival time in days (±SEM). (B) Percentage change in body weight (BW) following infection. The *x* axis represents the CC lines, the *y* axis represents the mean survival time in days (±SEM) in (A) and percentage change in BW in (B).

(Fig. 9.7A), ranging from high susceptibility (lethal response: survival 1.5 days) to complete resistance (100% survival). Additionally, significant variations were observed between the CC lines in their postinfection BW changes, whereby several CC lines significantly decreased in BW, while others showed BW recovery 5 days postinfection. Fig. 9.7B presents mean survival time and percentage change in BW on day 1 postinfection. Once again, in this study we emphasize the strong role of genetic components controlling host response despite strictly controlled environmental conditions. Furthermore, calculations of estimated $H^2$ introduce solid evidence for high heritability of both postinfection traits, survival time, with $H^2 = 0.54$, and BW change, with $H^2 = 0.28$. Accordingly, a number of CC lines were extended to enable successful QTL mapping for these traits, which will be published soon.

## Mapping Host Susceptibility to Bacterial Toxin-Induced Sepsis

Sepsis is a complex immune reaction leading to severe health complications triggered by infectious disease, most likely bacterial infection.[58] Typical treatment for sepsis consists of a combination of antibiotic medications with aggressive operative intervention and supportive treatment for the body systems in the ICU. However, elevated morbidity and mortality from sepsis are reported despite the continual advances in antibiotic therapies and medical support.[59] The pathogenesis of sepsis is investigated in multiple directions; the first aspect focuses on dissecting the microbial genomic architecture contributing to the disease, while the second aspect focuses on dissecting the host genomic architecture contributing to sepsis development and progress. Focusing on the second aspect, various studies suggest host septic response to be triggered by microbial toxins, particularly the

LPS components in gram negative bacteria and LTA in gram positive bacteria.[60,61] Actually, extensive variations are observed among populations in sepsis prevalence and progress, implying a major role for strong genetic components in controlling the development of microbial toxin-induced sepsis. Hence, mapping these genetic components is essential for the prevention and curing of sepsis and for the development of personalized therapies to combat sepsis deterioration.

For this purpose, two separate projects using the CC lines as a genetic reference and LPS or LTA injections as an environmental challenge for host exposure to microbial toxins are ongoing in our lab. The LPS project comprises 296 mice from 16 CC lines receiving LPS IP injection (dissolved in PBS). Subsequently the phenotypic traits of the host response are monitored for 72 h, measured as survival time (hours) and body temperature (°C) and weight (gram) changes. Mice were injected intraperitoneally with LPS from *Escherichia coli* (0111:B4) at a dose of 15 mg/kg mouse BW dissolved in PBS. The volume of the injection was set for 100 μL using a 1-mL 25-gauge syringe. Following the infection, BW was monitored using a digital balance, and rectal body temperature was monitored using an EcoScan Temp 4 thermometer with special probe designed for small animals, inserted 0.5 cm inside the rectum. Both measures were recorded at the following time points: 0, 2, 4, 6, 24, 30, 48, 54, and 72 h postinjection.

Preliminary results indicate (Fig. 9.8A) significantly high variations between the CC lines in survival time, in that several CC lines showed extreme LPS susceptibility and died about 30 h postinfection, whereas other CC lines showed LPS resistance, surviving the overall challenge for 72 h postinfection. Moreover, two-way analysis of variance (ANOVA) for sex cross line interactions revealed highly significant sex effects ($P < .05$), in which for the majority of the CC lines, females were apparently LPS resistant to a greater extent than males of the same CC line (Fig. 9.8B). As for the host thermal response, the major response during the first 30 h post-LPS infection was hypothermia for most of the CC lines; thereafter the CC lines varied in their thermal dynamics, whether hyperthermia or hypothermia (Fig. 9.8C). In a similar way, significant variations were observed between the CC lines in the dynamics of BW changes postinfection (Fig. 9.8D).

The aforementioned findings show the significant variations in host response to toxic microbial LPS despite the shared homogeneous environments, suggesting the involvement of strong genetic components of multiple body systems. These projects are being extended, assessing further CC lines to enrich the phenotypic reference, which will enable fine QTL mapping of the different traits. Similar experiments are ongoing as of this writing, with LTA challenge as well.

## NON-COMMUNICABLE DISEASES MAPPING HOST SUSCEPTIBILITY TO INTESTINAL CANCER

Familial adenomatous polyposis (FAP) syndrome is considered a rare dominant genetic disease, comprising one form of hereditary colorectal cancer.[62–66] The FAP syndrome

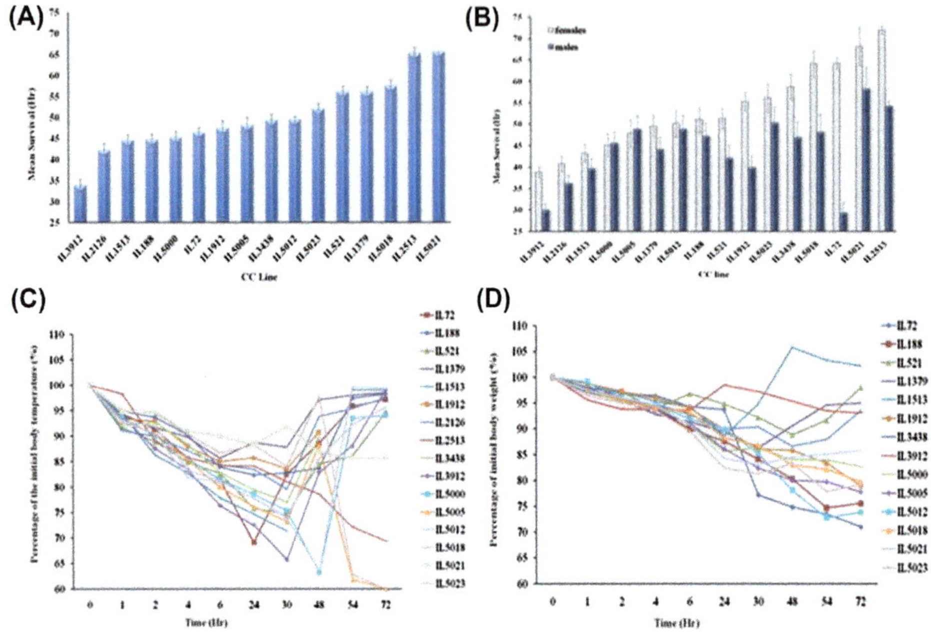


**Figure 9.8** Collaborative Cross (CC) lines' septic response to the microbial toxic component lipopolysaccharide (LPS). (A) CC lines mean (±SEM) survival time (h) profile for the 72 h after infection. (B) The data split by sex in each CC line showing a significant sex effect, *white columns* for females, *blue columns* for males. In (A) and (B), *x* axis represents the CC lines, *y* axis represents the mean survival time in hours (±SEM). (C and D) Multiple patterns of temperature (C) and body weight (BW) (D) changes during the 72 h postinfection. Changes in temperature and BW were calculated as a percentage of the initial value. *x* axis represents time after LPS injection (h), *y* axis represents percentage of the initial body temperature (C) or weight (D).

is characterized by multiple intestinal polyps and known to be caused by a functional mutation in the adenomatous polyposis coli (*Apc*) gene. Hitherto, FAP syndrome studies have relied on the knockout mouse model named $Apc^{Min+/-}$, which mimics polyp formation in humans.[67–71] The $Min^{+/-}$ mouse model has been examined for identification of *Apc* gene modifiers influencing the *Apc* gene function. Nowadays, various genomic regions are mapped in linkage with *Apc* gene modifiers (MOM 1–10); however, owing to the limitation of low mapping resolution, the candidate genes affecting *Apc* gene function within these loci were obscure.[72–77] To overcome these limitations, we proposed using the CC mouse as a powerful model for high-resolution mapping of QTLs in linkage with complex traits, with outstanding resolution to the level of individual candidate genes with top priority on the mapped QTL. For this purpose, we generated an F1 mouse population of CC-C57BL/6 lines carrying the $Apc^{Min/+}$ mutation, by crossing male mice of C57BL/6J-$Apc^{Min+/-}$ with

CC females. Overall, the study consisted of 27 F1 CC-C57BL/6 lines carrying the $Apc^{Min/+}$ mutation. At the age of 4 weeks, tail DNA of the cross litters (F1 mice) was extracted using the NaOH boiling method. Thereafter, the polymerase chain reaction was applied to confirm whether cross litters were *Apc* mutant gene carriers or wild type as described in Dorman et al.[38] F1 heterozygous carriers of the mutated *Apc* gene were maintained at an animal facility for an experimental period of 5 months, on a 12-h light/dark cycle. BW (g) was recorded biweekly during the experimental period. The mice were then killed; packed cell volume was recorded as an indicator for anemia, and internal organs were collected for further examination (spleen and intestines). Furthermore, spleen weight was recorded as an indicator for inflammation, while intestines were divided into different intestinal regions: small intestines and colon. Thereafter, the intestines were characterized by length (cm), weight (g), and polyp count using the binuclear method after overnight fixation in 10% neutral buffered formalin and staining by 0.02% methylene blue. Study findings were published in Dorman et al.[38] reporting extensive significant variations for all the phenotypic parameters, including total number of polyps (Fig. 9.9), between the CC-C57BL/6 lines carrying the $Apc^{Min/+}$ mutation, despite sharing fixed environmental conditions. These significant variations were suggested to be mainly contributed by the CC lines' genetic

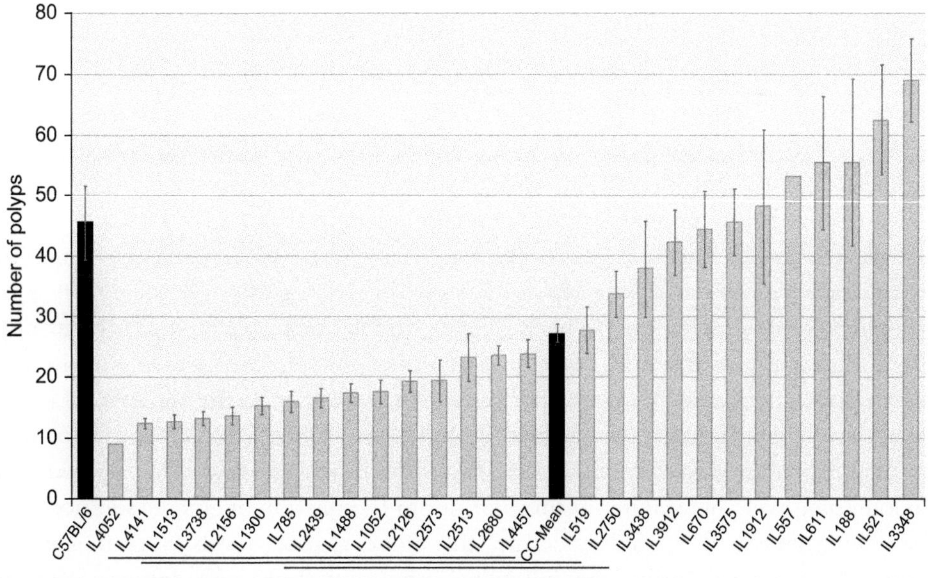


**Figure 9.9** Average number of polyps (±SEM) in 27 F1 CC-C57BL/6-*Min* lines ($n = 1-16$ mice/line). *x* axis represents Collaborative Cross (CC) lines and the C57BL/6 strain, which is an $Apc^{Min/+}$ mutation carrier (first *black column* on the left), *y* axis represents number of polyps.[38]

backgrounds. $H^2$ estimation confirms this statement, reporting a high $H^2$ value of 0.64 for the total number of polyps. Dorman et al.[38] presents secondary effects of the polyp formation, including BW changes, anemia and splenomegaly (enlarged spleen). Estimated Pearson correlations revealed a significant negative correlation between the number of polyps and packed cell volume ($r = -0.84$, $P < .01$); in other words, the more polyps (intestinal cancer), the lower the packed cell volume (anemia). On the other hand, a positive significant correlation ($r = 0.74$, $P < .01$) was reported between the number of polyps and the spleen adjusted weight (% weight), indicating that more polyps (intestinal cancer) related to larger spleens (splenomegaly). Linkage analysis for modifiers for *Apc* genes and their ability to induce intestinal cancer will be mapped and published soon.

## COLLABORATIVE CROSS LINES FOR TYPE 2 DIABETES STUDY

The prevalence of T2D in the past decades has increased to epidemic proportions—150 million in the year 2000—and is expected to double by the year 2025. Several lines of evidence provide support for the role of genetic variation in the pathogenesis of T2D and insulin resistance. In this context, diabetes can be considered a complex disease, for which the experimental challenge is the Western diet and CC lines may be used as the genetic reference. The publication of our study by Abu Toamih Atamni et al.[36,78] introduces the powerful utility of a CC line population for understanding the genetic basis of high-fat (42% fat) diet-induced T2D (HFD-induced T2D). The study cohort consisted of 683 mice generated from 21 CC lines. The mice were divided into two experimental groups: (1) control—maintained on a standard chow rodent diet and (2) challenge group—maintained on an HFD. Each CC line was represented in both groups. The dietary challenge started at the age of 8 weeks and lasted for 12 weeks, during which BW (g) and size (length, waist circumference) were assessed biweekly. At the terminal time point, an intraperitoneal glucose tolerance test (IPGTT) was performed to evaluate the diabetogenic status of the mice in response to the dietary challenge, and the results were compared between CC lines within each group (genetic effect) as well as between groups (diet effect). The GTT measures the clearance of an intraperitoneally injected glucose load from the body and is used to detect disturbances in glucose metabolism that can be linked to diabetes or metabolic syndrome. At the end of the 12-week dietary challenge, and after all the aforementioned biweekly measurements were taken, the mice were fasted for 6 h (6:00−12:00 a.m.) with free access to water. Fasted blood glucose levels were determined at time 0, prior to IP injection of glucose solution (2.5 g glucose/kg mouse BW). Subsequently, the blood glucose levels were measured at various time points during the following 3 h (time 0, 15, 30, 60, 90, 120, and 180 min after glucose injection). Thereafter the mice were returned back to their cages with food and free access to water for overnight recovery from the IPGTT procedure and sacrificed the next day. Published findings in Abu Toamih Atamni et al.[36,78] present an extensive variation in all the measured phenotypes between the CC lines; importantly, body diabetogenic and weight response to the HFD challenge was significantly influenced by the host genetic background, with a strong sex effect in each CC line, indicating that males and females from the same CC line experience the disease differently. CC lines showed multiple patterns for glucose clearance (Fig. 9.10A–D), in

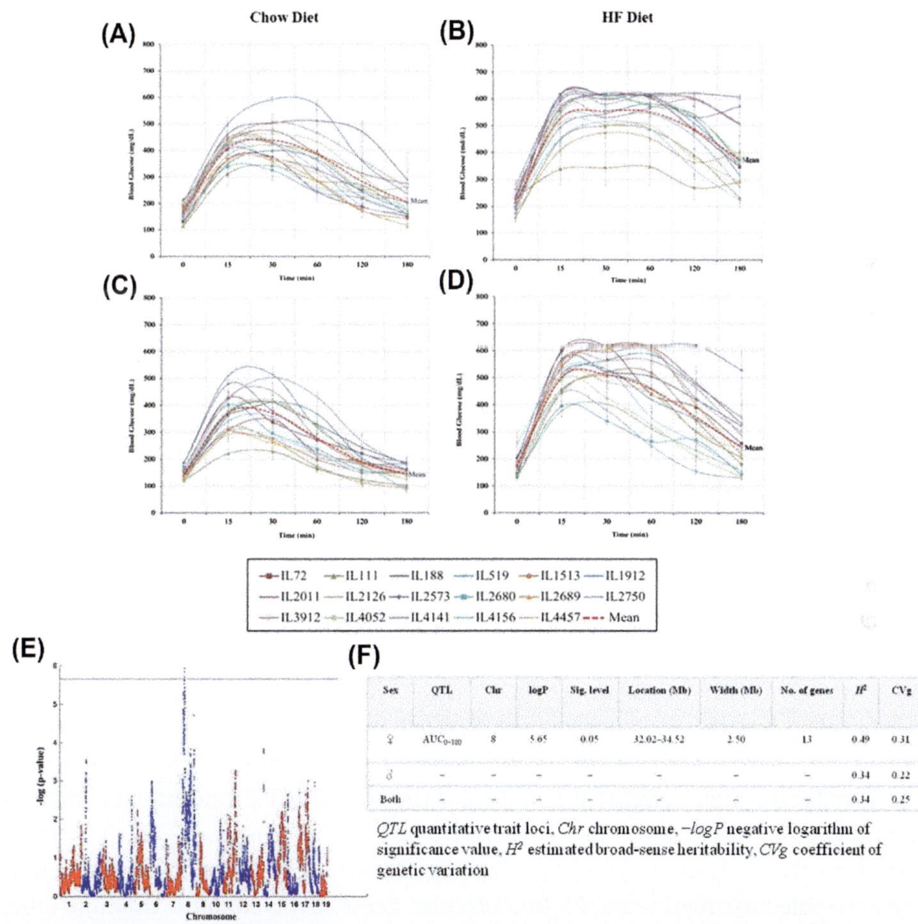


| Sex | QTL | Chr | logP | Sig. level | Location (Mb) | Width (Mb) | No. of genes | $H^2$ | CVg |
|---|---|---|---|---|---|---|---|---|---|
| ♀ | $AUC_{0-180}$ | 8 | 5.65 | 0.05 | 32.02-34.52 | 2.50 | 13 | 0.49 | 0.31 |
| ♂ | – | – | – | – | – | – | – | 0.34 | 0.22 |
| Both | – | – | – | – | – | – | – | 0.34 | 0.25 |

*QTL* quantitative trait loci, *Chr* chromosome, $-logP$ negative logarithm of significance value, $H^2$ estimated broad-sense heritability, *CVg* coefficient of genetic variation

**Figure 9.10** Kinetics of glucose clearance for 17 Collaborative Cross (CC) lines (split by sex) as assessed for 180 min by intraperitoneal glucose tolerance test after 12 weeks on chow diet (18% fat) and on high-fat diet (HFD, 42% fat). The *dashed line* shows average glucose level (mg/dL). (A) Males, chow; (B) males, HFD; (C) females, chow; (D) females, HFD.[36,78] (E and F) Significant quantitative trait loci (QTLs) mapping for females' total area under the curve of the glucose clearance trait ($AUC_{0-180}$) on chromosome 8 and positions of the QTLs associated with impaired glucose tolerance in response to 12 weeks HF dietary challenge in CC lines, respectively.

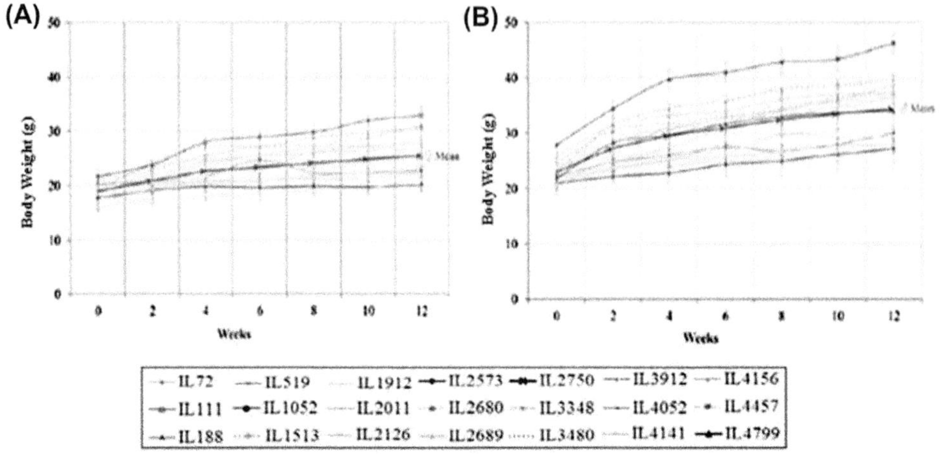


**Figure 9.11** Kinetics of body weight (g). Means (±SEM) of (A) females and (B) males of 21 Collaborative Cross lines during 12 weeks on high-fat dietary challenge (HFD, 42% fat).[36,78]

which different outcomes were observed, including resistance, susceptibility, and moderate intermediate response. Some CC lines were highly susceptible to consumption of the HFD, showing impaired glucose tolerance compared to mice from control groups of the same CC line. Similar findings were seen at the level of BW changes, showing multiple patterns, some toward continuous gain, some toward moderate gain, and some even showed high BW resistance, keeping their BW balanced despite HFD consumption (Fig. 9.11).

The bottom line of this comprehensive study is that "one size does not fit all" in the context of diets; the genetic background of the host plays a critical role in body response to HFD or standard diet. Therefore, discovering the candidate genes involved in susceptibility or resistance will bring us closer to personalized diets for prevention and treatment of obesity and T2D development.

QTLs that determine susceptibility to HFD challenge, as expressed in BW and total area under the curve of overall IPGTT phenotypes, were identified. The QTL of the total area under the curve (Fig. 9.10E and F) of the overall GTT was published, and was significant only for females.[37] The published female sex-dependent QTL (2.5 Mb) was mapped on chromosome 8 (32–34.5 Mb) (Fig. 9.10E). Gene browsing of the genomic interval revealed the reported QTL for BW/size, genes involved in the immune system, and two main protein-coding genes involved in glucose homeostasis, *Mboat4* and *Leprotl1*. In conclusion, genetic factors controlling glucose tolerance differ between sexes, requiring independent studies for females and males as regards T2D prevention and therapy. Furthermore, precise QTL mapping for individual time points during the IPGTT and BW accumulation is ongoing as of this writing and will be analyzed and published soon.

# COLLABORATIVE CROSS LINES FOR FUNDAMENTAL HOST BODY TRAITS

## Dissecting Genetic Background of Body Composition

In this study, we aim to determine the genetic factors underlying body composition using the CC population and the power of the DEXA scan technique. The following phenotypes were recorded by DEXA scan: (1) bone tissue features—bone area ($cm^2$), bone mineral content, and bone mineral density (BMD)—and (2) soft tissues—fat (g), lean (g), and percentage fat (%fat). Using the phenotypic and genotypic data of the CC lines will enable a genome-wide scan analysis for identifying QTLs associated with these traits. Additionally, body length (BL; distance in cm between nose and anus), waist circumference, BW, and body mass index ($BW/BL^2$) were measured for each mouse, which will also be used as phenotypes for determining variation between the CC lines. On average, 4–10, 18- to 20-week-old, mice of both sexes from each of 15 randomly used CC lines, which were maintained on a standard rodent diet, were sacrificed and analyzed. The DEXA scans were applied using the Lunar PIXImus densitometer (GE Medical Systems) with an image area of 80 mm × 65 mm, revealing data output of the body composition at the level of bone and soft tissues. Quality control and calibration for the PIXImus densitometer were performed prior to each experiment following the manufacturer's instructions and using the quality control phantom with known values (BMD = 0.0609 g/$cm^2$ and %fat = 11.8%). Fig. 9.12A–D shows the profiles of the tested traits of different CC lines using the DEXA scan. The statistical analysis of the phenotypic results showed significant ($P < .05$) variations between the CC lines. A sex effect was also observed with the various tested traits.

## Genetic Background of Immune Response

Mapping QTLs affecting the immune response cell lineages in the peripheral blood, and subsequently identifying the genes underlying these QTLs, can lead to better understanding of the host response to infectious diseases. For this purpose, we initiated a study aimed at mapping the genes that control the production of various subsets of cells that participate in the host defense. Here, we present the immune profile data on 15 CC mouse lines by quantifying the amount of peripheral blood T cells, B cells, and macrophages of mature mice (three to five mice per line, 10 weeks of age). The immunophenotyping was performed using flow cytometry analysis using fluorescence-activated cell sorting technology. Blood samples were collected from the orbital venous plexus of mice. Blood samples from individual mice were incubated with specific antibodies, including anti-CD3e (T cells), anti-CD19 (B cells), and anti-CD11b (macrophages) (Fig. 9.13). Our study revealed marked differences in the relative proportions of cells in the different CC mouse lines. The relative proportions varied between 13.5% and 43% for B lymphocytes and between 6.5% and 36% for T lymphocytes. As for macrophages, their

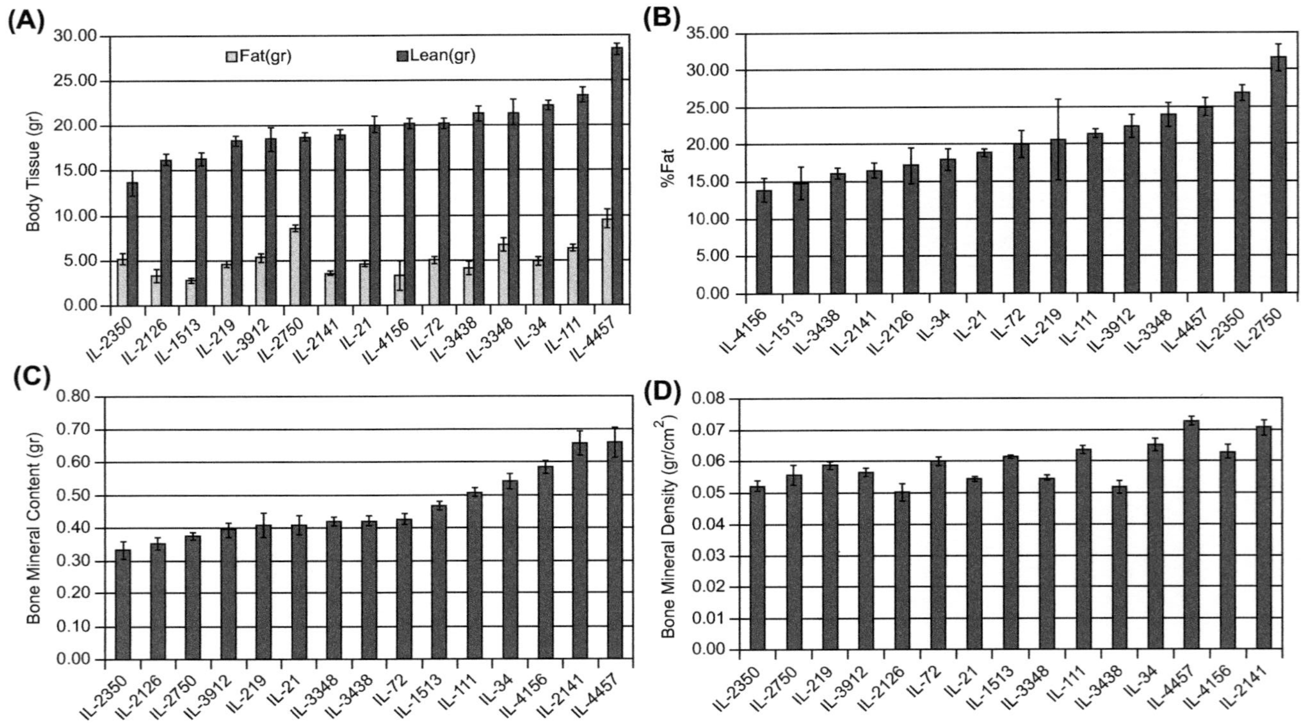


**Figure 9.12** Phenotypic profiles of the body composition complex traits assessed by dual-energy X-ray absorptiometry scan of various Collaborative Cross (CC) lines. *x* axis represents the CC lines, *y* axis represents phenotypic trait value. (A) Body tissue weight in grams, (B) percentage fat, (C) bone mineral content in grams, (D) bone mineral density in $g/cm^2$.

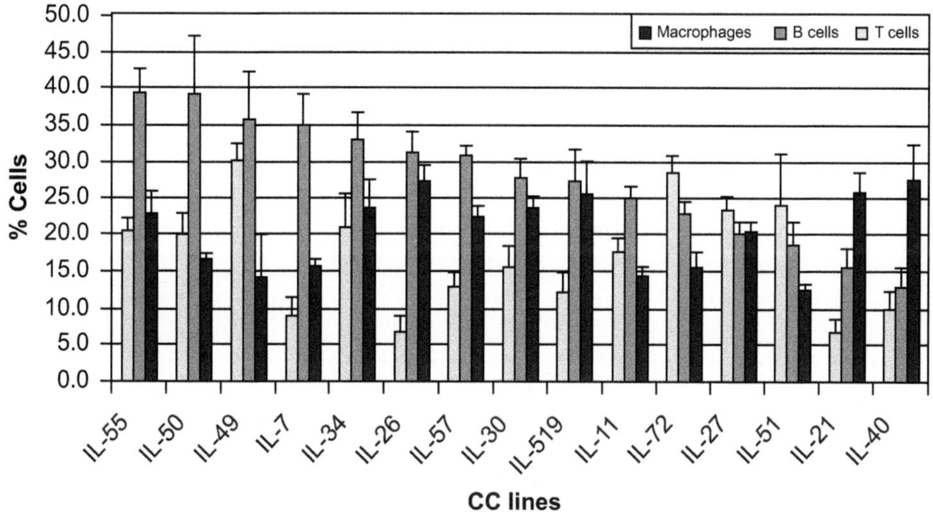


**Figure 9.13** ***Percentage of T, B, and macrophage immune cells in 15 naïve Collaborative Cross (CC) mouse lines.*** The *x* axis represents the CC lines, the *y* axis represents the percentage, including standard errors, of the three tested immune cell lines in 8- to 9-week-old CC mice. The percentages of T, B, and macrophage immune cells in peripheral blood were assessed by fluorescence-activated cell sorting analysis.

proportion value range was 10.6%—39% in the CC lines. Our results are compatible with the results already published for different commercially available inbred mouse strains. Thus, these results form the basis for extending the study of more CC lines, and using phenotypic data from 63 lines, we were able to map QTLs associated with the immune profile cells (unpublished data).

## Mapping QTLs for Trabecular Bone Microarchitecture

The microarchitecture of trabecular bone structure is considered a complex trait controlled by multiple genetic factors. It is known to differ among populations, resulting in phenotypic variations in development of osteoporosis and bone fractures (mechanic strength). Trabecular bone microstructure variations can lead to fracture risk variations between individuals despite their similar BMD, indicating that additional bone features, such as trabecular bone microstructure, determine bone strength.[79] In an attempt to characterize the genetic background underlying these features, Levy et al.[14] scanned the femoral distal metaphysic of 160 mice generated from 31 CC lines for trabecular traits measured by micro-CT analysis. The measured trabecular traits comprised trabecular bone volume fraction (BV/TV), number (Tb.N), thickness (Tb.Th), and connectivity density (Conn.D). Using a relatively small number of CC lines, Levy et al.[14] succeeded in mapping six QTLs, with four loci on chromosomes 2 and 7 for BV/TV and Tb.Th (two loci for each trait), and two additional loci on chromosomes 8 and 14 for Tb.N and Conn.D, respectively (Fig. 9.14) (1% false discovery rate). Genome browsing of the mapped QTLs suggests the new genes with relevance to skeletal biology were further validated. $H^2$ values of these traits confirmed their high genetic component, ranging from 0.6 to 0.7. In this study, Levy et al.[14] demonstrated for the first time significant genetic linkage between multiple genetic loci and trabecular microstructural features, suggesting new candidate genes to be further validated for their contribution to microstructure trabecular bone traits.

## Mapping QTLs for Hepatic Weight and Fattiness

Hepatic features are complex traits, known to be controlled by polygenic factors that differ between and within populations. Dissecting the genetic architecture controlling these variations will facilitate the study of multiple hepatic disorders, including nonalcoholic fatty liver disease and primary/secondary hepatomegaly. In this study, we aimed to map QTLs associated with phenotypic variations of liver weight (g) and fattiness (%fat) in CC lines. For this purpose, 506 mice from 39 CC lines (both sexes) were assessed at the age of 20 weeks for liver weight (g), measured manually by electronic balance, and for fattiness, measured by DEXA scanner. Data analysis revealed that liver weight (Fig. 9.15A and B) and fattiness (Fig. 9.16A and B) had significant variations ($P < .05$) between CC lines, with significant sex differences. QTL linkage analysis was then performed using the phenotypic and genotypic data of the CC lines. QTL mapping was significant only for females. A liver weight QTL was mapped on chromosome 8 (88.61−93.38 Mb), and three suggestive QTLs were mapped at chromosomes 4, 12, and 13 (Fig. 9.15C and D), named *LWL1*−*LWL4* (liver weight loci 1−4). In-depth examinations of the significant genomic interval of *LWL1* for candidate genes revealed multiple genes involved in different body systems and pathways. Importantly, the *Ces1a* to *Ces1h* family plays a central role in lipid metabolism and xenobiotic clearance, and *IrxB* (Iroquois homeobox genes complex B) plays major roles in growth and development. The *FTO* gene is related to obesity and metabolic syndrome, *Mmp2* gene has a key role in tissue growth and regulation, and *Slc6a2* gene is involved in growth, development, body size, metabolism, and homeostasis. As for the liver fattiness, QTLs were mapped on chromosomes 17 (3 Mb) and 18 (2 Mb) (Fig. 9.16C), and less significantly on chromosome 4 (2 Mb), named *Flal1*−*Flal3* (fatty liver accumulation loci 1−3), respectively. A gene browser search of the significant genomic interval of *Flal1* revealed the candidate genes *Ccr6*, which is involved in inflammatory/immunological responses and *Tagap*, which is involved in T cell activation during autoimmune diseases; *Flal2* candidate genes were *Nedd41*, which plays an important role in insulin regulation pathways, and *Akt* and *Sgk*, which are involved in the insulin signaling pathway. Additionally, the estimation of a founder effect on chromosome 17 for liver fattiness showed major

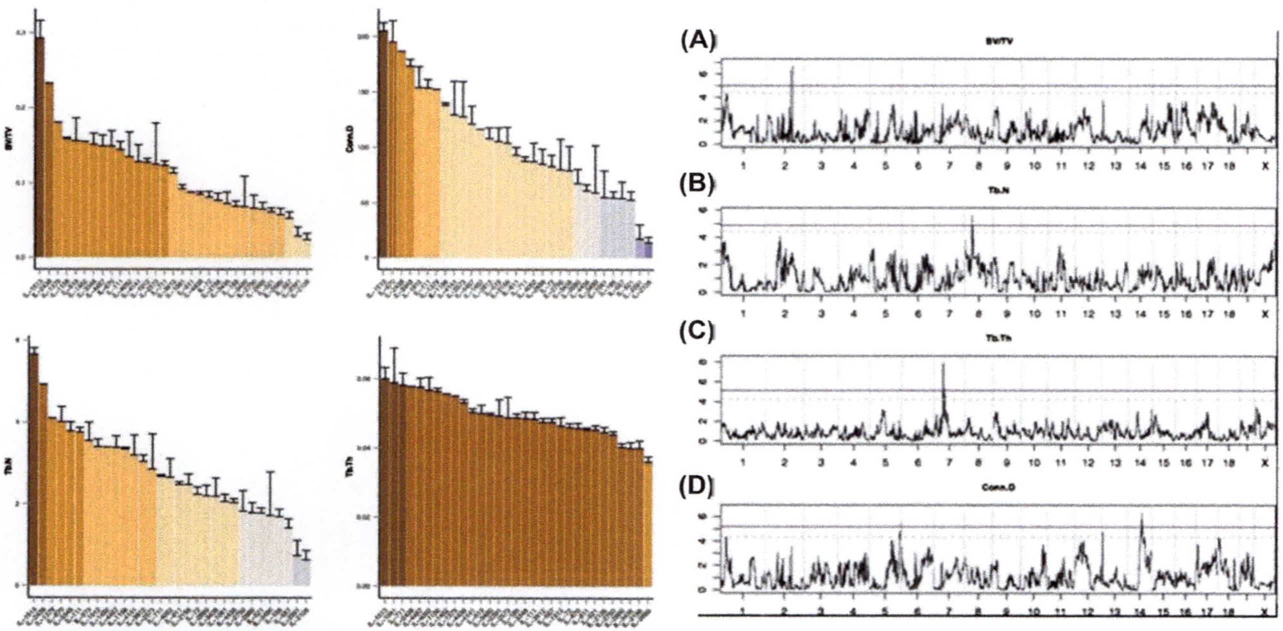


**Figure 9.14** ***Quantitative trait locus (QTL) haplotype mapping of trabecular bone traits using the Collaborative Cross (CC) lines.*** Left: Phenotypic profile of the trabecular bone features BV/TV, Tb.N, Tb.Th, and Conn.D. *x* axis represents CC lines and *y* axis represents the phenotypic trait value. Right: QTL mapping results revealing six significant QTLs using the phenotypic and genotypic data of the CC lines and HAPPY software. *BV/RV*, trabecular bone volume fraction; *Conn.D*, connectivity density; *Tb.N*, trabecular bone number; *Tb.Th*, trabecular bone thickness.[14]

(A) (B)

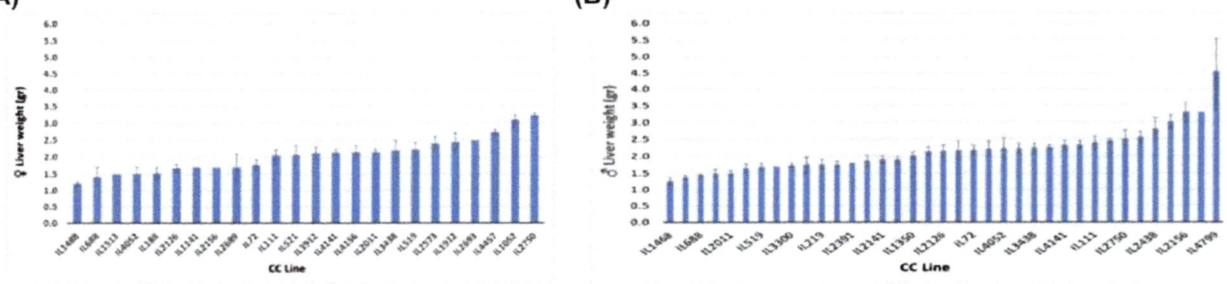

(C)

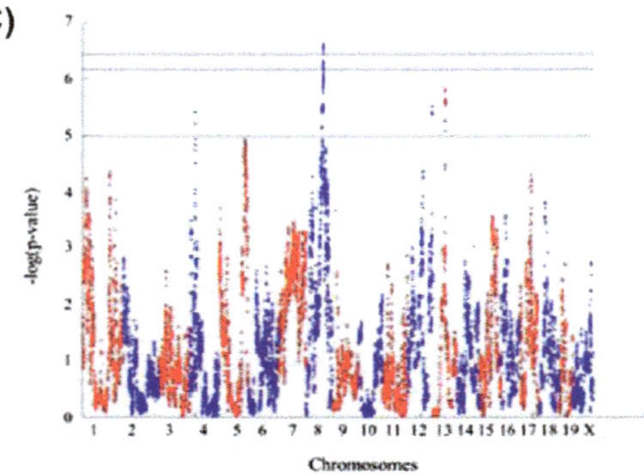

(D)

| QTL | Chr | logP | Sig. | 50% CI (Mb) | | 90% CI (Mb) | | 95% CI (Mb) | |
|---|---|---|---|---|---|---|---|---|---|
| | | | | Position | Width | Position | Width | Position | Width |
| *LWL1* | 8 | 6.62 | 0.05 | 88.61-93.38 | 4.77 | 88.61-93.38 | 4.77 | 88.61-93.38 | 4.77 |
| *LWL2* | 4 | 5.33 | 0.50 | 29.39 - 30.02 | 0.63 | - | - | - | - |
| *LWL3* | 12 | 5.51 | 0.50 | 115.59 - 117.82 | 2.23 | - | - | - | - |
| *LWL4* | 13 | 5.78 | 0.50 | 62.94 - 65.15 | 2.21 | - | - | - | - |

**Figure 9.15** ***Genetic mapping of the quantitative trait loci (QTLs) underlying baseline liver weight of various Collaborative Cross (CC) lines at age of 20 weeks of age in a naïve environment.*** (A and B) Females' and males' liver weight (g) profiles, respectively. (C and D) QTL mapping for females' liver weight trait and genomic locations of the QTLs, respectively.

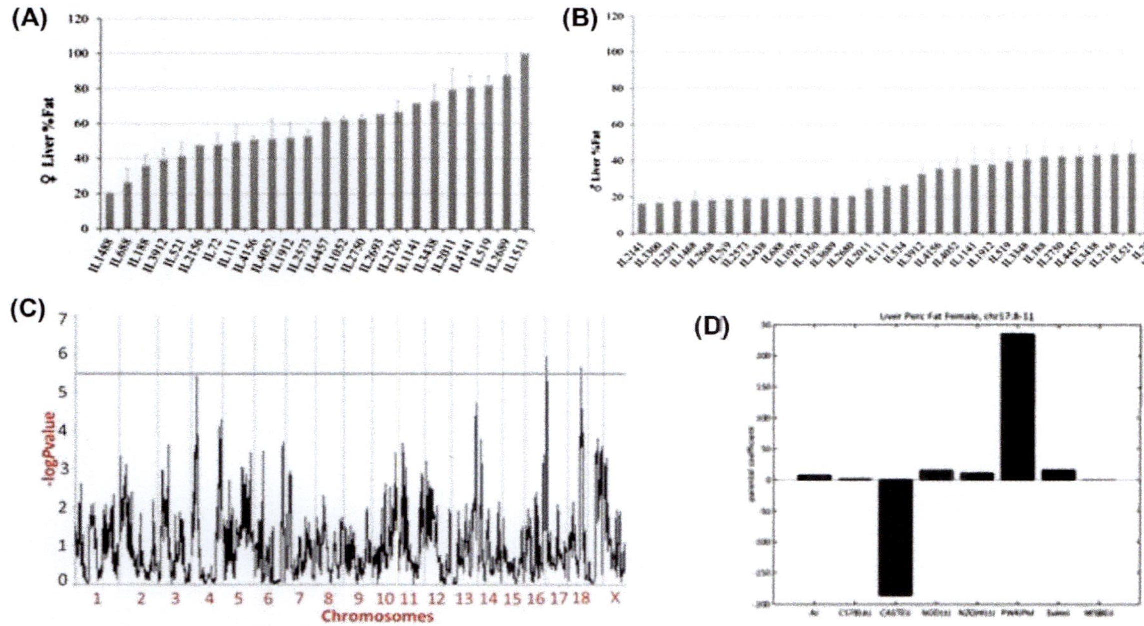


**Figure 9.16** ***Genetic mapping of the quantitative trait loci (QTLs) underlying baseline liver fattiness (%fat) of various Collaborative Cross lines at age of 20 weeks in a naïve environment.*** (A and B) Females' and males' liver fattiness profiles, respectively. (C and D) QTL mapping for females' liver fattiness and estimation of founder effect, respectively.[36,78]

effects contributed by the wild-derived strain PWK (Fig. 9.16D). Our findings demonstrate, for the first time, the utilization of the CC lines for mapping QTLs associated with baseline liver weight and fattiness, which differ between the sexes. Mapping these hepatic genetic components in health will contribute to the study of hepatic abnormalities (growth retardation/hepatomegaly/non-alcoholic fatty liver disease). Further details of the QTL mapping and data analysis are available in our publication Atamni et al.[78]

## COLLABORATIVE CROSS LINES FOR NEXT-GENERATION PHARMACOGENETICS

Pharmacogenetics is the study of genetic components controlling the host response to therapeutic agents that might lead to observed phenotypic variations among populations and sexes in drug disposition, including adverse effects. Mapping the genetic architecture involved in drug metabolic pathways will empower our knowledge toward further advances in personalized medicine. Utilization of the CC mouse model will be ideal for this purpose, enabling real-time evaluation of gene expression variations between the CC lines sharing joint environmental conditions. Herein are presented the unique findings of a published joint study of our lab with collaborators[80] based on data analysis of baseline hepatic gene expression of CC lines under naïve environmental conditions.

### Mapping Genetic Components Controlling Hepatic Drug Metabolism

This study demonstrates the utilization of the CC lines' powerful phenotypic diversity for characterization of their genetic diversity in regard to drug, chemical, and xenobiotic disposition processes, including detoxification reactions. For this purpose, hepatic gene expression profiles of 55 male mice at the age of 8–10 weeks, generated from 29 CC lines from a naïve environment (baseline), were estimated using a next-generation RNA sequencing approach. Mice were sacrificed and livers collected into sterilized (DNAse/RNAse free) tubes and subsequently stored in liquid nitrogen (−196°C). Next, RNA extraction was performed using the RNeasy Plus kit (Cat. No. 73404; Qiagen). Thereafter, RNA quality was assessed by the 2100 BioAnalyzer (Agilent), approving samples with RNA integrity number above 7. The next step was preparation of the RNA-Seq libraries using the TruSeq Stranded mRNA library preparation kit (Illumina), followed by library pooling and sequencing as described in Nachshon et al.[80] Thereafter, linkage analysis was performed between hepatic gene expression levels and data collection of metabolic biotransformation pathways, aiming for mapping of expression QTLs (eQTLs) associated with drug disposition. Eventually, linkage analysis for proximal eQTLs revealed successful construction of the hepatic eQTL–drug connectivity across the CC line populations (Table 9.2). In this study, Nachshon et al.[80] emphasizes the significant impact of genetic variations between CC lines on drug metabolism, proposing as well the interesting cross-talk existing between drug metabolism and lipid

**Table 9.2** A summary of proximal eQTLs underlying the biotransformation of drugs in the CC mouse population

| Reaction | Enzyme class | Drug metabolizing enzymes (DMEs) | eQTL (Mb) |
|---|---|---|---|
| Functionalization: oxidation | Cytochrome P450 | *Cyp2a22* | chr7:28 |
| | | *Cyp2c40** | chr19:40 |
| | | *Cyp2c44* | chr19:40 |
| | | *Cyp2d11* | chr15:84 |
| | | *Cyp2d12* | chr15:86 |
| | | *Cyp3a13* | chr5:138 |
| | | *Cyp3a16* | chr5:147 |
| | | *Cyp3a25*** | chr5:147 |
| | Alcohol dehydrogenase | *Adhfe1** | chr1:12 |
| | Aldehyde dehydrogenase | *Aldh8a1* | chr10:22 |
| | | *Aldh16a1* | chr7:50 |
| | Thiol-disulfide oxidoreductases | *Glrx2*** | chr1:146 |
| | Flavin-containing monooxygenase (FMO) | *Fmo1* | chr1:160 |
| Functionalization: reduction | Aldo-keto reductase | *Akr1c13* | chr13:3 |
| | | *Akrlc19* | chr13:5 |
| Functionalization: hydrolysis | Esterase | *Ces1g* | chr8:99 |
| | | *Ces2h* | chr8:112 |
| | | *Ces3a** | chr8:108 |
| | | *Siae* | chr9:40 |
| | | *Sulf2* | chr2:167 |
| | Epoxide hydrolase | *Ephx2* | chr14:62 |
| | Dihydropyrimidinase | *Dpys** | chr15:39 |
| | Glucuronidase | *Gusb** | chr5:130 |
| | Glyoxalase | *Glo1*** | chr17:32 |
| Conjugation | UDP-glucuronosyl transferase | *Ugt1a6a** | chr1:92 |
| | | *Ugtla6b*** | chr1: 90*,92 |
| | | *Ugtla10* | chrl:92 |
| | | *Ugt3a1* | chrl5:11 |
| | | *Ugt3a2* | chr15:9 |
| | Methyl transferase | *Tpmt*** | chr13:51 |
| | *N*-acetyl transferase | *Nat8* | chr6:86 |
| | Amino acid transferase | *Ccbl1** | chr2:30 |
| | | *Ccbl2** | chr3:143 |
| | Glutathione S-transferase | *Mgst3* | chr1:170 |
| | | *Gsta2** | chr9:82 |
| | | *Gstm6* | chr3:105 |
| | | *Gstz1* | chr12:90 |
| Transport | ATP-binding cassette | *Abcc6* | chr7:51 |
| | Solute-linked carrier | *Slco1a1* | chr6:142 |
| Transcription factors | Nuclear receptors | *Nr1i3 (CAR)** | chr1:174 |

For each drug disposition reaction (column 1) shown are various enzyme classes (column 2), eQTL-associated enzymes in each class (column 3) and the position of their proximal eQTLs (column 4). The expression of each enzyme is either associated with an isoform-ratio eQTL (asterisk), total eQTL (no asterisk) or both (two asterisks). Overall, we found 30 total eQTLs and 15 isoform-ratio eQTLs that underlie the expression of 40 different drug disposition enzymes.[80] *CC*, Collaborative Cross; *eQTLs*, expression quantitative trait loci.

metabolism to be further investigated. Research findings have produced supporting evidence for the power of CC line utilization in the characterization of the genomic architecture controlling drug, chemical, and xenobiotic disposition pathways. Genomic mapping of these traits is expected to promote our knowledge base further toward development of personalized medicine. Table 9.2 shows an eQTL summary as presented in Nachson et al.[80]

## CONCLUSIONS

The CC is a next-generation mouse genetic reference population designed by the mouse genetics community for high-resolution mapping of genetic factors of relevance to human health, such as susceptibility to infectious and chronic diseases and response to medical interventions. However, it is a very flexible resource with very wide potential application to complex traits of interest for human medicine and livestock agriculture.

The CC will eventually consist of a set of about 350 RILs, which are currently available for study. These lines were generated by reciprocal crosses between eight founder lines, including five relatively unrelated classical laboratory strains and three lines derived from recent wild accessions (two *Mus* subspecies and one *M. musculus domesticus*). The CC resource contains an extremely wide genetic diversity relative to existing mouse genetic resources and has already demonstrated unprecedented power for high-resolution QTL mapping. Genotypes and all generated phenotypes will be stored centrally and made publicly available, making the CC a "genotype once, phenotype many times" resource. The phenotypic distribution of CC lines is also consistent with the presence of very large variation among them in quantitative traits as well.

The phenotyping of CC lines for the identification of QTLs related to obesity, diabetes, metabolism, or microbiome interaction traits has clearly demonstrated the enormous value of the mouse GRP resources. The published and unpublished results have highlighted the usefulness of this mouse GRP to identify complex genetic traits, which may lead to translation to human diseases. A number of studies have shown that the knowledge gained in the mouse experimental systems can be applied to humans. This research will continue to make major discoveries in biomedical research and, thus, contribute to a better understanding and treatment of human and livestock diseases.

The advantage of using the CC mouse population for modifier mapping is the numerous genetic variants segregating in the population (there are over 36 million SNPs), and the relatively high level of recombination events compared to other mouse RIL sets (4.4 million SNPs segregate between the founders). The three wild-derived founders of the CC represent different mouse subspecies, *M. musculus castaneus*, *M. musculus musculus*, and *M. musculus domesticus*, and contribute many sequence variants not segregating among classical laboratory strains descended from *M. musculus domesticus*. Interestingly, based on recent publications, it was shown that the mapped QTLs in CC mice tend to map contrast alleles between wild-derived strains and laboratory strains,[28,29,39] and stimulation of QTL analysis in the CC population showed that resolution of mapping will be less than 1 Mb, which is more powerful than any suggested classical approach of QTL studies using F2 backcrosses, advanced intercross lines,[82] outbred populations, or consomic, congenic, and RILs, in which two parental strains that differ phenotypically and genetically are used.

## REFERENCES


1. Iraqi F, Clapcot S, Kuman P, et al. Fine mapping of trypanosomiasis resistance QTLs in mice using advanced intercross lines. *Mamm Genome*. 2000;11:645–648.
2. Iraqi F, Churchill G, Mott R. The collaborative cross, developing a resource for mammalian systems genetics: a status report of the Wellcome Trust cohort. *Mamm Genome*. 2008;19:379–381.
3. Welsh CE, Miller RD, Manly KF, et al. Status and access to the collaborative cross population. *Mamm Genome*. 2012;23:706–712.
4. Klopocki E, Hennig BP, Dathe K, et al. Deletion and point mutations of PTHLH cause brachydactyly type E. *Am J Hum Genet*. 2010;86(3):434–439.
5. Rankinen T, Zuberi A, Chagnon YC, et al. The human obesity gene map: the 2005 update. *Obesity (Silver Spring)*. 2006;14(4):529–644. Review.
6. Wicker LS, Clark J, Fraser HI, et al. Type 1 diabetes genes and pathways shared by humans and NOD mice. *J Autoimmun*. 2005;(25 suppl):29–33.
7. Threadgill DW, Hunt KW, Williams RW. Genetic dissection of complex and quantitative traits: from fantasy to reality via a community effort. *Mamm Genome*. 2002;16:344–355.
8. Churchill G, Airey D, Allayee H, et al. The collaborative cross, a community resource for the genetic analysis of complex traits. *Nat Genet*. 2004;36:1133–1137.
9. Roberts A, Villena FP, Wang W, McMillan L, Threadgill DW. The polymorphism architecture of mouse genetic resources elucidated using genome-wide resequencing data. *Mamm Genome*. 2007; 18(6):473–481.
10. Valdar W, Flint J, Mott R. Simulating the collaborative cross: power of quantitative trait loci detection and mapping resolution in large sets of recombinant inbred strains of mice. *Genetics*. 2006;172(3): 1783–1797.
11. Yang H, Ding Y, Hutchins LN, et al. A customized and versatile high-density genotyping array for the mouse. *Br J Pharmacol*. 2009;6(9):663–666.
12. Mott R, Talbot CJ, Turri MG, Collins AC, Flint J. A method for fine mapping quantitative trait loci in outbred animal stocks. *Proc Natl Acad Sci USA*. 2000;23(7):12649–12654.
13. Durrant C, Tayem H, Yalcin B, et al. Collaborative cross mice and their power to map host susceptibility to *Aspergillus fumigatus* infection. *Genome Res*. 2011;21:1239–1248.
14. Levy R, Mott RF, Iraqi FA, Gabet Y. Collaborative cross mice in a genetic association study reveal new candidate genes for bone microarchitecture. *BMC Genom*. 2015;16(1):1013.
15. Keane TM, Goodstadt L, Danecek P, et al. Mouse genomic variation and its effect on phenotypes and gene regulation. *Nature*. 2011;14(477):289–294.
16. Kovacs A, Ben-Jacob N, Tayem H, et al. Genotype is a stronger determinant than sex of the mouse gut microbiota. *Microb Ecol*. 2011;61(2):423–428.
17. Yalcin B, Flint J, Mott R. Using progenitor strain information to identify quantitative trait nucleotides in outbred mice. *Genetics*. 2005;171:673–681.
18. Vered K, Durrant C, Mott R, Iraqi FA. Susceptibility to *Klebsiella pneumonaie* infection in collaborative cross mice is a complex trait controlled by at least three loci acting at different time points. *BMC Genom*. 2014;15:865.

19. Aylor DL, Valdar W, Foulds-Mathes W, et al. Genetic analysis of complex traits in the emerging collaborative cross. *Genome Res.* 2011;21:1213–1222.
20. Philip VM, Sokoloff G, Ackert-Bicknell CL, et al. Genetic analysis in the collaborative cross breeding population. *Genome Res.* 2011;21:1223–1238.
21. Kelada SN, Aylor DL, Peck BC, et al. Genetic analysis of hematological parameters in incipient lines of the collaborative cross. *G3.* 2012;2(2):157–165.
22. Xiong H, Morrison J, Ferris MT, et al. Genomic profiling of collaborative cross founder mice infected with respiratory viruses reveals novel transcripts and infection-related strain-specific gene and isoform expression. *G3.* 2014;4(8):1429–1444.
23. Ram R, Mehta M, Balmer L, Gatti DM, Morahan G. Rapid identification of major-effect genes using the collaborative cross. *Genetics.* 2014;198(1):75–86.
24. Gralinski LE, Ferris MT, Aylor DL, et al. Genome wide identification of SARS-CoV susceptibility loci using the collaborative cross. *PLoS Genet.* 2015;11(10):e1005504.
25. Rogala AR, Morgan AP, Christensen AM, et al. The collaborative cross as a resource for modeling human disease: CC011/Unc, a new mouse model for spontaneous colitis. *Mamm Genome.* 2014; 25(3–4):95–108.
26. Phillippi J, Xie Y, Miller DR, et al. Using the emerging collaborative cross to probe the immune system. *Genes Immun.* 2014;15(1):38–46.
27. Ferris MT, Aylor DL, Bottomly D, et al. Modeling host genetic regulation of influenza pathogenesis in the collaborative cross. *PLoS Pathog.* 2013;9(2):e1003196.
28. Thaisz J, Tsaih SW, Feng M, et al. Genetic analysis of albuminuria in collaborative cross and multiple mouse intercross populations. *Am J Physiol Renal Physiol.* 2012;303(7):F972–F981.
29. Bottomly D, Ferris MT, Aicher LD, et al. Expression quantitative trait loci for extreme host response to influenza a in pre-collaborative cross mice. *G3.* 2012;2(2):213–221.
30. Mathes WF, Aylor DL, Miller DR, et al. Architecture of energy balance traits in emerging lines of the collaborative cross. *Am J Physiol Endocrinol Metab.* 2011;300(6):E1124–E1134.
31. Gelinas R, Chesler EJ, Vasconcelos D, et al. A genetic approach to the prediction of drug side effects: bleomycin induces concordant phenotypes in mice of the collaborative cross. *Pharmgenomics Pers Med.* 2011;4:35–45.
32. Zombeck JA, Deyoung EK, Brzezinska WJ, Rhodes JS. Selective breeding for increased home cage physical activity in collaborative cross and Hsd:ICR mice. *Behav Genet.* 2011;41(4):571–582.
33. Broman KW. The genomes of recombinant inbred lines. *Genetics.* 2005;169(2):1133–1146.
34. Shusterman A, Salymah Y, Nashef A, et al. Genotype is an important determinant factor of host susceptibility to periodontitis in the collaborative cross and inbred mouse populations. *BMC Genet.* 2013; 14:68–79.
35. Lorè NI, Iraqi FA, Bragonzi A. Host genetic diversity influences the severity of *Pseudomonas aeruginosa* pneumonia in the collaborative cross mice. *BMC Genet.* 2015;16:106.
36. Abu-Toamih Atamni HJ, Mott R, Soller M, Iraqi FA. High-fat-diet induced development of increased fasting glucose levels and impaired response to intraperitoneal glucose challenge in the collaborative cross mouse genetic reference population. *BMC Genet.* 2016;5(17):10.
37. Abu-Toamih Atamni HJ, Ziner Y, Mott R, et al. Glucose tolerance female-specific QTL mapped in collaborative cross mice. *Mamm Genome.* 2017;28(1–2):20–30.
38. Dorman A, Baer D, Tomlinson I, Mott R, Iraqi FA. Genetic analysis of intestinal polyp development in collaborative cross mice carrying the Apc (Min/+) mutation. *BMC Genet.* 2016;19(17):46.
39. Iraqi FA, Mahajne M, Salaymah A, et al. The genome architecture of the collaborative cross mouse genetic reference population. *Genetics.* 2012;190(2):389–401.
40. Latgé JP. *Aspergillus fumigatus* and aspergillosis. *Clin Microbiol Rev.* 1999;12(2):310–350.
41. Soubani AO, Chandrasekar PH. The clinical spectrum of pulmonary aspergillosis. *Chest.* 2002;121(6): 1988–1999.
42. Williams RC. Periodontal disease. *N Engl J Med.* 1990;322(6):373–382. Review.
43. Wilson M. Biological activities of lipopolysaccharides from oral bacteria and their relevance to the pathogenesis of chronic periodontitis. *Sci Prog.* 1995;78(Pt 1):19–34.

44. Baker PJ, Dixon M, Roopenian DC. Genetic control of susceptibility to *Porphyromonas gingivalis*-induced alveolar bone loss in mice. *Infect Immun*. 2000;68:5864–5868.
45. Baker PJ, Evans RT, Roopenian DC. Oral infection with *Porphyromonas gingivalis* and induced alveolar bone loss in immunocompetent and severe combined immunodeficient mice. *Arch Oral Biol*. 1994;39:1035–1040.
46. Baker PJ, Roopenian DC. Genetic susceptibility to chronic periodontal disease. *Microbes Infect*. 2002;4:1157–1167.
47. Polak D, Wilensky A, Shapira L, et al. Mouse model of experimental periodontitis induced by *Porphyromonas gingivalis/Fusobacterium nucleatum* infection: bone loss and host response. *J Clin Periodontol*. 2009;36(5):406–410.
48. Wilensky A, Gabet Y, Yumoto H, Houri-Haddad Y, Shapira L. Three-dimensional quantification of alveolar bone loss in *Porphyromonas gingivalis*-infected mice using micro-computed tomography. *J Periodontol*. 2005;76(8):1282–1286.
49. Kolenbrander PE, Andersen RN. Inhibition of coaggregation between *Fusobacterium nucleatum* and *Porphyromonas* (Bacteroides) *gingivalis* by lactose and related sugars. *Infect Immun*. 1989;57:3204–3209.
50. Gellatly SL, Hancock RE. *Pseudomonas aeruginosa*: new insights into pathogenesis and host defenses. *Pathol Discov*. 2013;67(3):159–173.
51. Bianconi I, Milani A, Cigana C, et al. Positive signature-tagged mutagenesis in *Pseudomonas aeruginosa*: tracking patho-adaptive mutations promoting airways chronic infection. *PLoS Pathog*. 2011;7(2):e1001270.
52. Cigana C, Curcurù L, Leone MR, et al. *Pseudomonas aeruginosa* exploits lipid A and muropeptides modification as a strategy to lower innate immunity during cystic fibrosis lung infection. *PLoS One*. 2009;4(12):e8439.
53. Bragonzi A, Paroni M, Nonis A, et al. *Pseudomonas aeruginosa* microevolution during cystic fibrosis lung infection establishes clones with adapted virulence. *Am J Respir Crit Care Med*. 2009;180(2):138–145.
54. Nguyen D, Singh PK. Evolving stealth: genetic adaptation of *Pseudomonas aeruginosa* during cystic fibrosis infections. *Proc Natl Acad Sci USA*. 2006;103(22):8305–8306.
55. Weiler CA, Drumm ML. Genetic influences on cystic fibrosis lung disease severity. *Front Pharmacol*. 2013;4:40.
56. Bragonzi A. Murine models of acute and chronic lung infection with cystic fibrosis pathogens. *Int J Med Microbiol*. 2010;300(8):584–593.
57. De Simone M, Spagnuolo L, Lorè NI, et al. Host genetic background influences the response to the opportunistic *Pseudomonas aeruginosa* infection altering cell-mediated immunity and bacterial replication. *PLoS One*. 2014;9(9):e106873.
58. Levy MM, Fink MP, Marshall JC, et al. 2001 CCM/ESICM/ACCP/ATS/SIS International Sepsis Definitions Conference. *Crit Care Med*. 2003;31(4):1250–1256.
59. Friedman G, Silva E, Vincent JL. Has the mortality of septic shock changed with time. *Crit Care Med*. 1998;26(12):2078–2086.
60. Van der Poll T, Opal SM. Host-pathogen interactions in sepsis. *Lancet Infect Dis*. 2008;8(1):32–43.
61. Mattsson E, Verhage L, Rollof J, et al. Peptidoglycan and teichoic acid from *Staphylococcus epidermidis* stimulate human monocytes to release tumour necrosis factor-alpha, interleukin-1 beta and interleukin-6. *FEMS Immunol Med Microbiol*. 1993;7(3):281–287.
62. Moser AR, Mattes EM, Dove WF, et al. Min, a mutation in the murine Apcgene, predisposes to mammary carcinomas and focal alveolar hyperplasias. *Proc Natl Acad Sci USA*. 1993;90:8977–8981.
63. Rubinfeld B, Souza B, Albert I. Association of the APC gene product with betacatenin. *Science*. 1993;262:1731–1734.
64. Sheng JQ, Cui WJ, Fu L, et al. APC gene mutation in Chinese familial adenomatous polyposis patients. *World J Gastroenterol*. 2010;16:1522–1526.
65. Silverman KA, Koratkar RA, Siracusa LD, Buchberg AM. Exclusion of Madh2, Madh4, and Madh7 as candidates for the modifier of Min 2 (Mom2) locus. *Mamm Genome Res*. 2003;14:119–129.
66. Suraweera N, Haines J, McCart A, et al. Genetic determinants modulate susceptibility to pregnancy-associated tumourigenesis in a recombinant line of Min mice. *Hum Mol Genet*. 2006;15:3429–3435.

67. Joslyn G, Carlson M, Thliveris A, et al. Identification of deletion mutations and three new genes at the familial polyposis. *Cell*. 1991;66:601–613.
68. Kinzler KW, Nilbert MC, Su LK, et al. Identification of FAP locus genes from chromosome 5q21. *Science*. 1991;253:661–665.
69. Dietrich WF, Lander ES, Smith JS, Moser AR, Gould KA. Genetic identification of Mom-1, a major modifier locus affecting Min-induced intestinal neoplasia in the mouse. *Cell*. 1993;75:631–639.
70. Harada N, Tamai Y, Ishikawa T, et al. Intestinal polyposis in mice with a dominant stable mutation of the beta-catenin gene. *EMBO J*. 1999;18:5931–5942.
71. Hirohashi S, Kanai Y. Cell adhesion system and human cancer morphogenesis. *Cancer Sci*. 2003;94: 575–581.
72. Lawrence KN, Shedlovsky A, Biehl BS, et al. Identification of Mom7, a novel modifier of $APC^{min/+}$ on mouse chromosome 18. *Genetics*. 2007;176:1237–1244.
73. Cormier RT, Bilger AJ, Lillich RB, Halberg KH. The Mom1AKR intestinal tumor resistance region consists of Pla2g2a and a locus distal to D4Mit64. *Oncogene*. 2000;19:3182–3192.
74. Morin PJ, Sparks AB, Korinek V, et al. Activation of β-catenin-Tcf signaling in colon cancer by mutations in β-catenin or APC. *Science*. 1997;275:1787–1790.
75. Moser AR, Dove WF, Roth KA, Gordon JI. The Min (multiple intestinal neoplasia) mutation: its effect on gut epithelial cell differentiation and interaction with a modifier system. *J Cell Biol*. 1992;716: 1517–1526.
76. Oikarineny SI, Cleveland AG, Cork KM, et al. Genetic mapping of Mom5, a novel modifier of ApcMin-induced intestinal tumorigenesis. *Carcinogenesis*. 2009;30:1591–1596.
77. Silverman KA, Koratkar RA, Siracusa LD, Buchberg AM. Identification of the modifier of Min 2 (Mom2) locus, a new mutation that influences Apc-induced intestinal neoplasia. *Genome Res*. 2002; 12:88–97.
78. Abu-Toamih Atamni HJ, Botzman M, Mott R, et al. Mapping liver fat female-dependent quantitative trait loci in collaborative cross mice. *Mamm Genome*. 2016;27(11–12):565–573.
79. Chappard D, Baslé MF, Legrand E, Audran M. New laboratory tools in the assessment of bone quality. *Osteoporos Int*. 2011;22(8):2225–2240.
80. Nachshon A, Abu-Toamih Atamni HJ, Steuerman Y, et al. Dissecting the effect of genetic variation on the hepatic expression of drug disposition genes across the collaborative cross mouse strains. *Front Genet*. 2016;7:172.
81. Threadgill DW, Churchill GA. Ten years of the collaborative cross. *G3 (Bethesda)*. 2012;2(2):153–156.
82. Iraqi F. Fine mapping of quantitative trait loci using advanced intercross lines of mice and positional cloning of the corresponding genes. *J Exp Lung Res*. 2000;26:641–649.

# CHAPTER 10

# Discovery of Novel Genes and Other Lineage-Specific Features Through Comparative Genomics

**Claudio V. Mello[1], Peter V. Lovell[1], Morgan Wirthlin[1,2]**
[1]Oregon Health and Sciences University, Portland, OR, United States; [2]Carnegie Mellon University, Pittsburgh, PA, United States

## INTRODUCTION

The genetic basis of species- and lineage-specific traits is a fundamental but still elusive question in evolutionary genomics. The recent sharp decreases in sequencing costs and the broad application of next-generation sequencing methods have led to an explosive growth in the number of available genomes from a broad range of organisms, including microorganisms, plants, insects, and vertebrates. Such advances have led to many efforts to identify genomic features that are unique to a given species or lineage. Indeed, comparative genomics studies have identified many genes that lack orthologs outside a given species or lineage of interest (e.g., Refs. 1−6). These novel genes, which are often referred to as "orphan" or "taxonomically restricted genes" based upon their placement within a phylogeny, are thought to arise through events that may include, but are not limited to, duplications, truncations, elongations, fusions, and translocations, and they may result from various mechanisms including nonallelic homologous recombination, transposable element activity, or gene mutations. The resulting duplicated genes may undergo subsequent divergence that in some cases allows the novel paralog to take on a different functional role than the parent gene. Genes can also be created de novo when a previously untranscribed region of DNA becomes transcribed due to mutations in *cis*-regulatory sequences that result in the expression of a novel RNA transcript, which may over time result in the evolution of a novel gene.[7] Regardless of mechanism, several efforts have attempted to find potential associations between novel genes and the unique traits exhibited by a given species or lineage. This includes searching for novel genes related to structural traits, such as the loss of teeth[8] or gain of feathers in birds,[9] homeothermy, or formation of the placenta in mammals,[10] as well as those related to behavioral traits such as social behavior in bees[11] or vocal learning in songbirds (e.g., Refs. 12−15).

Due to a range of factors, the task of conclusively establishing the unique occurrence of a novel gene in a given species or lineage is particularly challenging. Current methods for identifying novel genes primarily rely on detecting species differences in sets of computationally predicted orthologs (e.g., Compara,[16] CAFE,[17] OrthoMCL,[18]

*Molecular-Genetic and Statistical Techniques for Behavioral and Neural Research*
ISBN 978-0-12-804078-2, https://doi.org/10.1016/B978-0-12-804078-2.00010-6

INPARANOID[5]). These methods depend on the accuracy and sensitivity of in silico or ab initio gene model prediction and annotation algorithms (e.g., Ensembl's[6] and NCBI's Gnomon[19] automatic genome annotation pipelines). This leads to two major problems that prevent the accurate detection of novel genes. First, these algorithms generally do not confirm gene orthology by systematic synteny verification. This can result in gene annotations where a gene identified as an ortholog actually represents a paralogous gene within the same gene family.[12,14,20] Second, while in silico gene prediction algorithms work reasonably well on high-quality genome assemblies (e.g., human or mouse), they fail to reliably and accurately predict gene models in genomes that are incomplete (i.e., contain very frequent and/or large sequence gaps), as is usually the case with genomes assembled primarily from short-read sequencing (<300 bp; e.g., next-generation sequencing technologies like Illumina or SOLiD). Accurate gene annotation is particularly challenging in regions containing high G/C content and/or repetitive elements. Such regions frequently occur in proximity to regulatory promoter regions, although coding sequences can also be affected. While newer methods, like PacBio sequencing and assembly technology, appear to be better suited to handle these difficult regions than previous sequencing and assembly platforms,[21] such methods are still costly and not as widely applied. The presence of genomic gaps greatly limits the ability of both ab initio— and in silico—based gene prediction pipelines (e.g., Ensembl, Gnomon) to accurately predict complete open reading frames (ORFs) of protein-coding genes and thus correctly discriminate them from loci associated with noncoding transcripts. As a result of these and other limitations intrinsic to in silico prediction pipelines, many orthologs that are actually present in a given genome fail to be identified or, conversely, end up being incorrectly cataloged as pseudogenes, noncoding transcripts, or novel genes.[12] These issues are major contributing factors to the limited sets of 1-to-1 orthologs in databases such as Ensembl, UniGene, RefSeq, etc. In other cases, in silico prediction algorithms make multiple partial predictions for the same gene, which results in incorrect conclusions about gene duplications and/or family expansions.[12,14,20] These various categories of errors are further propagated when the computational annotation pipelines used to identify sets of orthologous or paralogous genes are based solely on alignments of gene annotations from other species, which may themselves have only been predicted through such pipelines. These serious errors highlight the continued need for direct manual verification of gene annotations, as well as the further development of computational pipelines that incorporate gene synteny analysis to accurately establish orthology in annotated gene sets.

In our own studies, we have employed comparative genomics and extensive manual and computational annotation strategies to identify genes associated with lineage-specific traits. Our efforts have focused on finding novel genomic features associated with vocal learning (the ability to imitate the vocalizations of a tutor through sensorimotor learning) in passerines, and lifespan determination in parrots and other long-lived birds. To identify novel features associated with vocal learning, we initially compared sets of gene predictions derived from the genomes of the zebra finch[22] (a vocal learning songbird) with those predicted in chicken[23] (a vocal nonlearning galliform), the two best-assembled and annotated avian genomes. This effort required curating several hundred Ensembl gene predictions annotated as novel zebra finch protein-coding genes (i.e., no orthologs in other species) by applying pairwise alignments using UCSC's BLAT alignment toolkit[24,25] in these two species, and using comparative synteny to establish orthology of the identified loci. To further test for an association with vocal learning, we verified the presence of candidate novel genes in other songbird and nonsongbird avian genomes, as well as in nonavian vertebrate outgroups. This extensive analysis was made possible by the >50 avian genomes that became available in 2014 as a result of the collaborative avian phylogenomics initiative.[26]

Our comparative analysis revealed a set of 10 de novo novel or paralogous loci that are present in all songbirds, but absent in all nonlearner sister taxa, including suboscines.[12] As an example, *YTHDC2L1* is a songbird-specific duplication (Fig. 10.1A) that shares 97% identity with its putative parent gene, the RNA-binding helicase *YTHDC2*, but lacks domains like the RNA-binding YTH. This apparent truncation suggests a possible novel "dominant negative"-like regulatory role for the paralog. *YTHDC2L1* is part of a family expansion and is uniquely expressed in LMAN, a vocal control nucleus critical for vocal learning and plasticity (Fig. 10.1B, left), whereas the parent gene *YTHDC2* is undetectable in the brain (Fig. 10.1B, right). As another example, *TMRA* (a.k.a. Transmembrane protein of the Robust nucleus of the Arcopallium) is a de novo novel gene found solely in songbirds, with no orthologs in other bird groups, and it is selectively expressed in RA (Fig. 10.1C), a vocal control nucleus critical for the production of learned vocalizations. *TMRA*'s predicted transmembrane and cell surface recognition domains (Fig. 10.1D) suggest possible roles in cell-cell or cell-substrate interactions. In another study, a similar curation pipeline applied to the entire set of potassium channel genes ($n = 107$) in the zebra finch genome discovered six previously undescribed paralogs,[14] including several enriched in song nuclei. Thus, the genomes of zebra finches and other vocal learning songbirds contain novel genes, including paralogs whose function diverged from the parent gene and became linked to songbird neurobiology. These findings are consistent with the overarching hypothesis that the emergence of novel genes in songbirds coevolved with the trait of vocal learning.

As illustrated schematically in Fig. 10.2, in some cases a given novel gene might represent a novel paralog that apparently resulted from a duplication at a site of a chromosomal rearrangement (Fig. 10.2, *case 1*: compare synteny of chicken Gene A with A-like copies in finch). In other cases, the novel gene is present at a new locus, and there are no apparent indications as to the mechanism of the duplication (Fig. 10.2, *case 2*: compare A and A-like in finch with A in chicken). We note that in both scenarios, an evaluation of synteny (which requires alignments to the genome) was critical to clearly establish

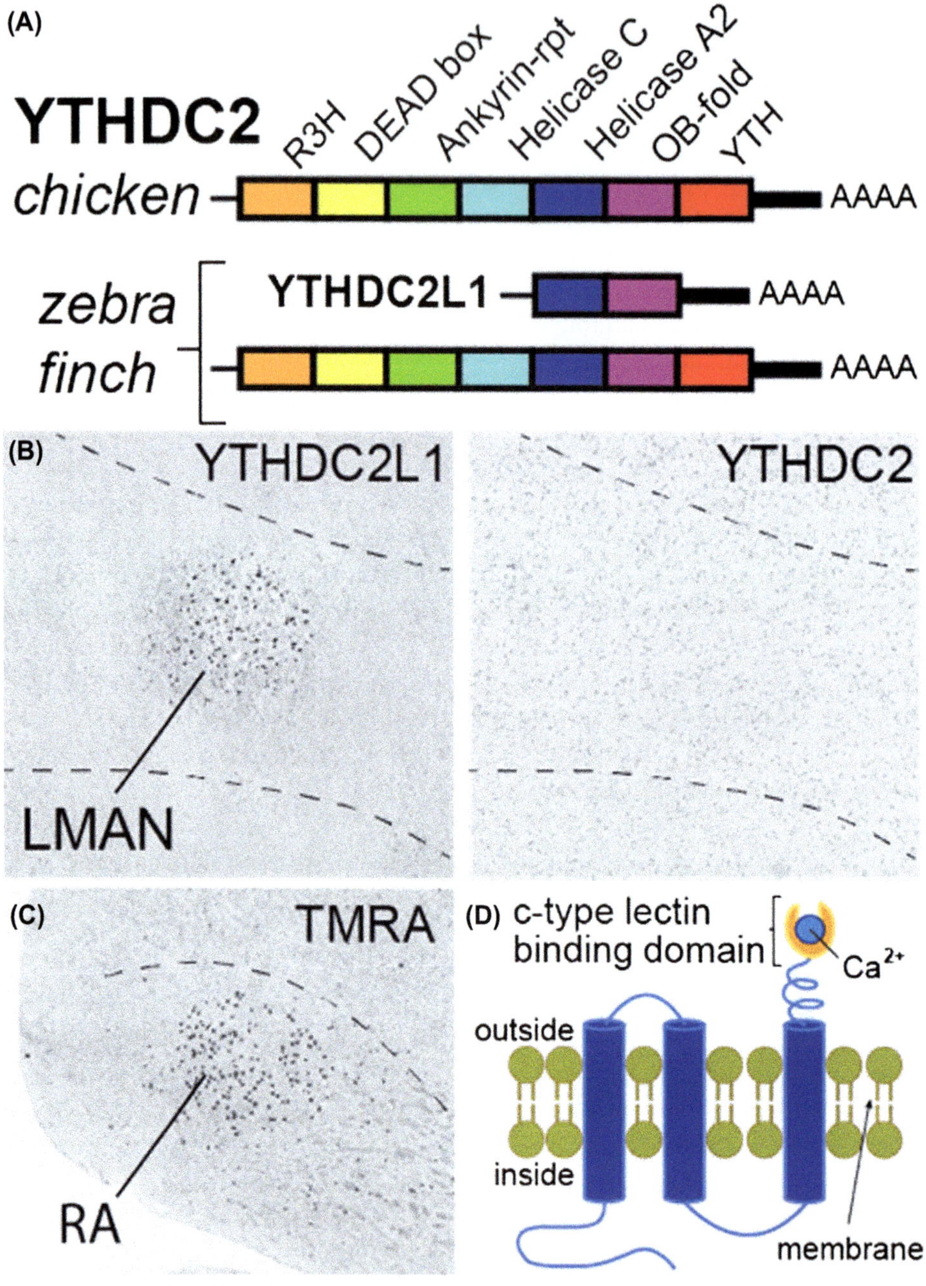


**Figure 10.1** Examples of novel genes in songbirds. (A) *YTHDC2L1* is unique to songbirds and lacks key functional domains typical of this gene family. (B) In situ hybridization reveals that *YTHDC2L1*, but not its parent gene *YTHDC2*, is selectively expressed in vocal nucleus LMAN. (C) *TMRA* expression is highly enriched in vocal-motor nucleus RA. (D) Topology of *TMRA*'s predicted peptide suggests membrane localization and roles in cell-to-cell signaling. *TMRA*, transmembrane protein of the robust nucleus of the arcopallium. Adapted from Wirthlin et al. (2014); Figure 4 (BioMed Central).

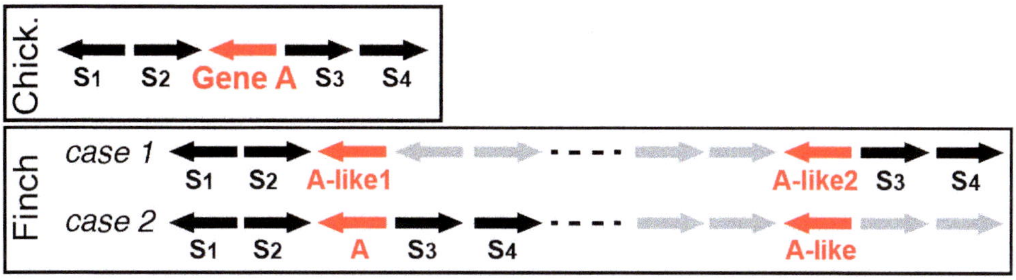


**Figure 10.2** Analysis of genes (e.g., Gene A) in comparative syntenic context (i.e., syntenic genes S1–S4) can reveal finch duplications associated with sites of chromosomal rearrangements (*case 1*: A-like 1 and 2) or other unknown mechanisms (*case 2*: A-like).

orthologous/paralogous relationships of members of a given gene set of interest. In fact, due to the frequent occurrence of paralogous genes and the high conservation of shared domains across different members of the same gene family, we learned that conclusions about orthology are often not reliable if based only on reciprocal alignment criteria without synteny verification.

Our efforts also revealed that a large proportion of gene predictions annotated as novel in zebra finches actually consist of artifacts due to gaps, assembly issues, and partial or failed in silico predictions. We found, for example, numerous instances of multiple partial predictions of a single gene, with one of the partial models misannotated as a novel duplication of that gene (e.g., *PLXNA4*; Fig. 10.3), or as an altogether novel gene. Such cases could be readily identified and corrected by aligning the partial finch gene predictions and experimentally derived expressed sequences (e.g., ESTs, mRNAs), as well as by aligning gene predictions from species with more thoroughly annotated genes, as in our case of chicken predictions aligned to the finch genome. In this regard, we have found UCSC's genome browser and the BLAT alignment toolset to be particularly helpful and user-friendly.

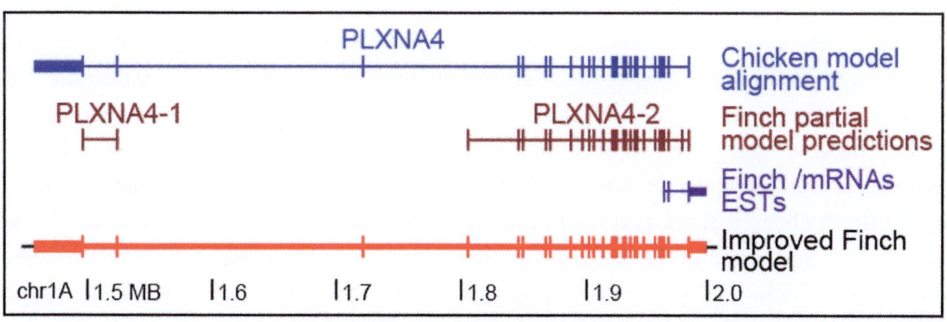


**Figure 10.3** A duplicate gene misannotation (for *PLXNA4*) due to partial finch models, detected by BLAT alignments of these models together with the full chicken gene and finch ESTs; notice the identification of the 5′ and 3′ untranslated regions (UTRs) and third exon, absent in the finch predictions.

By closely evaluating the strengths, outcomes, and difficulties of our comparative genomics efforts, we have devised a conceptual framework and pipeline for generating high-confidence sets of novel genes in a given species or lineage of interest, representing genes with no orthologs in other species or lineages. Although we have successfully employed this strategy for discovering novel genes in specific avian genomes, we believe our approach should be of general applicability.

## STRATEGIES FOR IDENTIFYING LINEAGE-SPECIFIC EXPANSIONS AND DE NOVO GENES

The key element of our strategy for novel gene discovery consists of performing comparative alignments of the candidate gene (or gene set) of interest to the genomes of both a target species of interest and a contrast species, followed by careful verification of the outcomes of these alignments, to confirm true novel gene cases and exclude false positives and/or artifacts. This overall process can be schematically divided into three steps: (1) identification of candidate novel loci in the target genome; (2) classification of the candidate novel loci as orthologous or novel genes (unique to the target genome); and (3) evaluation of novel gene predictions to confirm findings and exclude artifacts.

### Target and Contrast Species Considerations

The target species consists of a representative of the lineage that expresses the trait of interest and for which a high-quality genome is available. The contrast species, on the other hand, consists of a representative species of a lineage that lacks the trait, but is also as close as possible phylogenetically to the target species. Furthermore, since the gene predictions of the contrast species are used to identify orthologous genes in the target species, the contrast species should also have a high-quality genome, preferably with a well-annotated set of gene predictions. A close phylogenetic distance between the target and contrast species minimizes the chance that possible differences in genomic features are due to the evolutionary distance between species rather than the presence/absence of a given trait of interest. However, the need for high-quality genomes can also substantially limit options. In our own studies, we initially analyzed the genomes of zebra finch versus chicken, since these represented the highest-quality and best-annotated avian genome assemblies available. Ideally, species from more closely related avian groups to songbirds, such as a sister taxon among the suboscines (also vocal nonlearners; e.g., Golden-collared manakin[27]), would represent a much more stringent contrast to zebra finches than chicken. However, in this case, all available suboscine genomes are currently substantially more incomplete, which would create several other obstacles to the comparative analyses discussed subsequently. Nonetheless, after conducting our initial screening of zebra finch versus chicken, the inclusion of multiple genomes from other relevant avian species, even in cases where their genomes and gene predictions were incomplete, greatly facilitated the tasks of confirming or disproving the initial candidates.

## Identifying Lineage-Specific Expansions and Paralogs

Gene duplication events, resulting from chromosomal rearrangements, represent a major mechanism by which new genetic material, including genes, can be generated during evolution. Therefore, a logical starting point for finding novel genes in a lineage or species is to conduct comparative analyses of large or expanded gene families. As an example, our analysis of the full complement of potassium channel—related genes in zebra finch revealed six paralogs that are absent in mammals but variously present in different vertebrate and avian lineages. In our experience, the gene predictions available in Ensembl or NCBI's RefSeq databases represent reasonable starting points. In contrast, in silico algorithms that attempt to predict ORFs based on sequence features alone, such as Genscan or Augustus,[28,29] represent less ideal starting points. These tools were developed for, and trained on, model species genomes such as mouse and human. In practice, they are highly prone to predicting false positive and partial models in species genomes more distantly related to these model organisms, likely due to the differences in fundamental sequence properties across vertebrate taxa. ESTs/mRNAs can be informative, but additional care must be taken to confirm annotations, and many of these transcripts are incomplete or lack coding domains, as discussed later.

The implementation of steps (1) and (2) in our novel gene search strategy, as well as possible resulting scenarios and their interpretation, are illustrated diagrammatically in Fig. 10.4. After a set of predicted gene models (e.g., from Ensembl, RefSeq) and associated ESTs/mRNAs, if available, have been retrieved, they are aligned to both the target and contrast species genomes using BLAT. This step is followed by a close verification of the alignment outcomes. Whenever possible, we highly recommend using UCSC's genomic browser. However, due to the limited availability of genomes on that browser, we suggest using BLAT for local genomic alignments, which can be done by uploading the BLAT toolset and genomes of interest to private servers and applying the method in a customized manner to those genomes.

To illustrate our strategy for discriminating gene predictions that represent orthologs versus paralogs, in *case 1* (Fig. 10.4, top row, Step 1), a gene prediction of interest from the target species (target model A) and its known ortholog in the contrast species (contrast model A) are depicted as aligning with comparably high scores to both the primary locus (left), and also a secondary locus (right). While the primary alignment is present in both the target and contrast genomes (consisting of a locus flanked by the same syntenic genes in both species), the secondary alignment is absent in the contrast genome, and present, but not predicted in the target genome. Under this scenario, we would conclude that the secondary alignment likely represents a novel paralog of gene A (Fig. 10.4, top row, Step

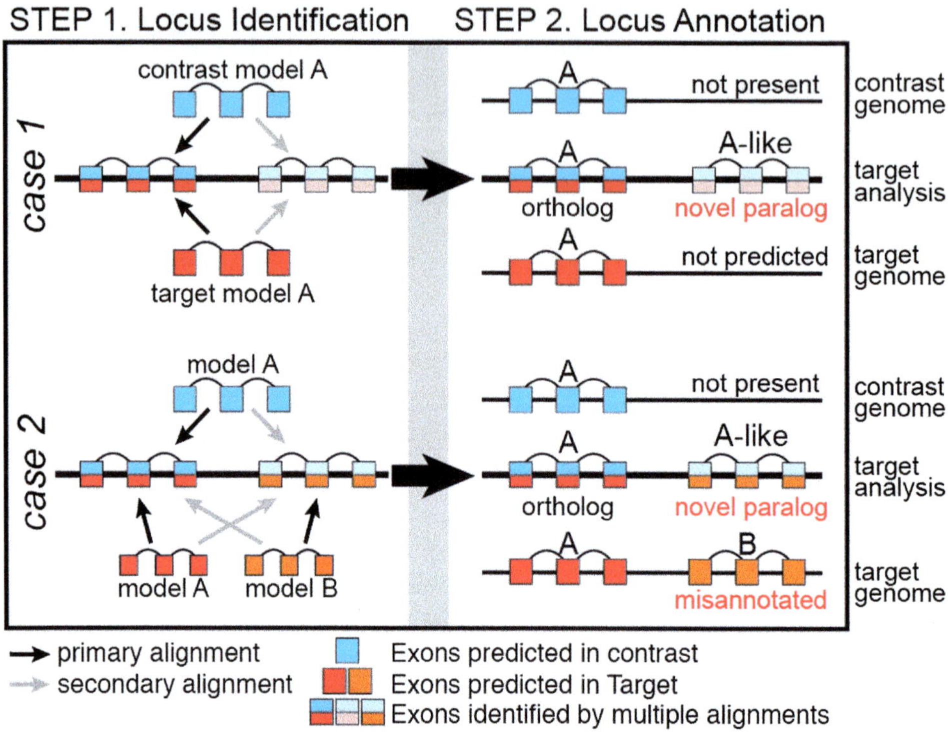


**Figure 10.4** Detection of novel loci in a target versus contrast species comparison. Step 1: Alignments of target and contrast models reveal possible loci in target genome; Step 2: Comparative analysis identifies orthologs and novel paralogs. *Case 1*: locus present, but not predicted in the target is identified by secondary alignments of models A (left), and annotated as novel A-like (right); *Case 2*: reciprocal cross-alignments of models A and B (left) reveals that the latter is a misannotated novel A-like paralog (previously unrelated to A).

2). As a possible variant of this scenario, multiple high-scoring alignments to the target species genome could occur, which would point to an expansion of that gene family (similar to the *YTHDC2* case in songbirds[12]).

Step 3 of our gene search strategy consists of further evaluating the identified candidate species- or lineage-specific novel gene predictions to better understand their structure and possible origins, as well as to exclude false positive predictions. Most critically, a close examination of the corresponding syntenic region in the contrast species is required to determine that the lack of the secondary alignment is not due to a local genomic sequence gap in that species. To confirm putative novel gene predictions and to better understand the evolutionary history of the novel gene (or gene family expansion), one should also verify that it is present in other members of the lineage containing the target species, but not in the outgroup lineage containing the contrast species. A confirmation that the novel gene is present in other species of the target lineage helps strengthen the correlation between the presence of that novel gene and the occurrence of the trait of interest, but it also further discards the alternate hypothesis that the identified novel feature is specific to the target species and not shared by other members of that lineage. Furthermore, examination of outgroups to both lineages is key to determining whether the novel gene represents a gain in the target lineage or a possible loss in the contrast lineage. In either case, close examination of the region syntenic to the novel locus in both lineages would also help evaluate whether that gene duplication might be associated with a chromosomal rearrangement (similar to *case 1* in Fig. 10.2), or whether other unknown mechanisms are more likely to have been involved (similar to *case 2* in Fig. 10.2).

Under a different scenario (illustrated by *case 2* in Fig. 10.4, bottom row), there are predictions for two closely related genes (models A and B) in the target species, but only one of them (model A) is present in the contrast species. This case may have arisen after an initial search (e.g., *case 1* in Fig. 10.4) revealed a secondary locus that was associated with an unannotated model. In an attempt to annotate this locus, a second round of alignments would be conducted. The paralogous relationship between models A and B is established by assessing two criteria: first, that the models represent the reciprocal first- and second-best alignments to one another in the target species. Second, in the contrast species genome, both models should have their first-best alignment to the locus corresponding to the syntenic location of model A from the target species, with no alignment to the locus corresponding to the syntenic location of model B in the target species. As in *case 1*, a close examination is required to determine whether the lack of alignment in the contrast species is not due to a local genomic sequence gap in that species. If this possibility can be discarded, then the conclusion would be that target species model B represents a novel paralog of gene A, which following convention should be renamed "Gene A-like." We note that in cases of this type, the misannotation of model B masked its paralogous relationship to model A. In our experience, this is most often the consequence of incompleteness of one or both predicted models.

It is worthwhile pointing out that the genomic alignments discussed earlier can also improve and, in some cases, extend existing gene models in the target species, thereby correcting problems such as partial models and unpredicted exons (see *PLXNA4* example in Fig. 10.3). Such improvements by themselves can often clarify misannotations, filtering out false positives and thus helping to streamline the set of candidate novel genes. The likelihood of correcting such problems is highest when the gene predictions for the orthologous genes in the contrast species are complete and when high coverages of expression data from ESTs or RNA-Seq analyses are available for the relevant tissues in the target species.

A significant question that arises in both scenarios discussed before is why the novel paralog was not predicted in the target species. Although a clear answer is not always available, it is our experience that the existing in silico prediction algorithms are highly sensitive to gaps where the start codon of the gene has been lost, or that significantly truncate ORFs. It is thus possible that the lack of a gene prediction at the site of the novel paralog in the target species is due to a genomic gap that partially deleted the corresponding ORF. Fortunately, BLAT nucleotide alignments do not depend on correct ORF predictions and provide an excellent and highly sensitive tool for detecting the presence of that novel gene, even in the case of partial models or when predictive algorithms completely failed to predict a given gene. An alternative possibility, however, is that there are no gaps in the region containing the novel paralog. In such cases, it is more likely that the secondary locus represents a pseudogene that was similarly not detected by the prediction algorithms. In either case, examining the alignment of experimentally derived evidence of transcription—including ESTs, sequenced RNAs, or RNA-Seq data from relevant tissues in the target species—can be extremely helpful. Evidence of specific alignment of transcriptional data to a given candidate novel locus would help establish the latter as a functional paralog, even though it might appear truncated due to the genomic sequence gap. Conversely, completeness of the local genomic sequence coupled with a lack of locus-specific alignments of transcriptional data would support the alternative possibility that the novel gene copy is a nonfunctional duplication or pseudogene.

## IDENTIFYING DE NOVO GENES

The scenarios discussed in the preceding paragraphs consist of situations where a clear parent ortholog for the gene of interest is present in the contrast species and the secondary locus represents a novel paralog in the target species. An alternate scenario occurs when an ortholog for the putative novel gene of interest cannot be identified in any other genome or expression databases, including the contrast species and other species from related lineages or outgroups. In most such cases, these gene model predictions are annotated as "unknown" or "novel." In addition, for many species large collections of expressed sequence tags (ESTs) and/or sequenced mRNAs may exist that cannot be annotated by aligning to large databases of published nucleotide sequences, such as NCBI's NR/NT nucleotide database. This provides a second source of candidate novel genes. In these cases, the application of our BLAT alignment–based search pipeline described previously would result in a similar outcome to that shown for *case 1* in Fig. 10.4, except that the primary locus of alignment would only be present in the target species. As for the interpretation of outcomes, detection of one or multiple high-scoring alignments in the genome of the target species (or of the target species as well as other members of a target lineage), but no significant hits in the contrast species (or lineage), would indicate that the target model of interest represents a de novo novel gene (or de novo gene family) in the target species (or lineage). We emphasize that it is critical for conclusively determining if a putative novel gene of interest is present in the contrast species. There may be cases where the gene is actually present in the contrast species but there are no gene predictions at that site (likely due to genomic gaps that truncated the corresponding

ORF), as well as cases where there are no local sequence gaps in the contrast species but the orthologous locus has become pseudogenized. In both cases, a high-scoring alignment of a candidate novel gene sequence at the correct syntenic location in the contrast species would clearly establish the candidate novel gene as a false positive.

## HIGH-THROUGHPUT STRATEGIES FOR IDENTIFYING NOVEL GENES

Although our general strategy for identifying lineage-specific expansions and de novo genes has been discussed thus far as a series of stepwise alignments of gene predictions and transcriptional data to target and contrast species genomes and subsequent analyses of synteny, one could also employ a more high-throughput automated approach. The goal in this case would be to define all putative gene loci in the target genome that are not present in the contrast genome and subsequently categorize predicted models as orthologs, paralogs, or novel genes. These categorizations would be a result of comparative syntenic analysis of alignments conducted in parallel in both the target and contrast species, and they could be scaled to include multiple genomes from target and contrast lineages of interest. This can be accomplished by initially retrieving all the gene predictions from both the target and contrast species genomes, as well as all available EST/mRNA sequences from both species, as these may include transcripts from existing, but nonpredicted novel loci. These complete sets of target and contrast predicted gene models and transcriptional data are then aligned to the target genome, to identify all possible gene loci in that genome (Fig. 10.4, Step 1). It is critical in this process to optimize alignment parameters, so the true structure of novel loci is recovered. For example, if aligning gene model sequences derived from chicken onto finch is required, then one should set alignment parameters to be more permissive for a cross-species analysis. This can be accomplished by using a translated BLAT server with the "minScore" and "minIdentity" parameters set to zero, the stepSize parameter lowered to 2 or 3, and the repMatch parameter set to a large value (we have used 1,000,000). In contrast, when aligning transcriptional data or sequence models derived from a species back onto that species, then one should set parameters to be stricter, for instance, using a standard BLAT server with default parameters. This is followed by combining all overlapping alignments into a single set of loci, which can be accomplished using genome alignment file manipulation tools such as bedtools[30] or BEDOPS.[31]

After generating the complete set of candidate gene models for a given assembly, the next step is to implement a "locus-based approach" to annotate the target species models as orthologous, paralogous, and novel relative to the set of gene models from a contrast species or set of species representing an outgroup lineage (Fig. 10.4, Step 2). This strategy, described previously by our group,[12] represents a fundamentally different strategy from traditional alignment-based approaches, where models are annotated by taking into account their alignment scores, following a BLAST alignment of the models to some reference set of nucleotide sequences. These approaches fail to take synteny or genomic location into account. In contrast, the locus-based annotation approach described here takes into account the locations of the complete gene model set in the target species of interest and the aligned locations of all gene models from a well-annotated contrast species onto the target species, and only the single, highest-scoring alignments of each model are preserved (i.e., the "primary alignments" in Fig. 10.4). Genes within the target species' complete gene model set that overlap with a "best hit" from the contrast species alignments are classified as orthologs, which can be scored with increasing confidence based on the preservation of synteny. The remaining gene models in the complete target species gene model set are then classified as putative novel genes. For this remaining set, a second comparison is made to the aligned locations of all gene models from a well-annotated contrast species onto the target species, but this time all alignments are included, whether they represent highest-scoring hits, or second-best hits, etc. Gene models that were not already annotated as orthologs, but which overlap a contrast species locus, and which lack preservation of synteny with adjacent gene loci, are then classified as novel paralogs, and they are given gene names that reflect this parent-child gene relationship as described previously. The remaining gene models, which could not be classified as orthologs nor novel paralogs, and which appear in gapless regions in the matching syntenic location in the contrast species genome, are classified as de novo novel genes.

It bears repeating that the incorporation of synteny analysis into this algorithmic annotation is critical for correctly establishing orthology, particularly when a gene duplication has occurred, and the original parent gene has subsequently become pseudogenized. In such cases, the novel paralog in the target species would have a higher alignment score to the parent ortholog from the contrast species, as pseudogenes are prone to gradual sequence drift due to the loss of stabilizing selection (for an example of this phenomenon, see the discussion of the avian *YTHDC2* expansion in Ref. 12). Thus, annotation algorithms that favor the preservation of synteny over alignment scores alone in orthology determination perform with greater accuracy than those that fail to take this chromosomal context into account.

Besides greatly facilitating the examination of a large number of candidate novel genes, another significant advantage of this high-throughput automated approach is that paralogous relationships across unannotated or misannotated genes (such as those illustrated in Fig. 10.4, *case 2*) can be readily identified in the alignments. Such cases are much harder to detect when performing alignments of individual models, since paralogous relationships cannot be defined a priori in the case of unannotated, misannotated, or partially predicted gene models. In our own studies, we have taken preliminary steps to implement several custom scripts for the steps described before in our novel gene search strategy, including both the "Locus Identification Algorithm," which combines BLAT,[24] BEDOPS,[31] and custom scripts,[12] and a "Locus Classification Algorithm." We note, however, that it is particularly difficult to implement effective automated algorithm for synteny verification. Some suspicious annotations may require manual synteny verification, especially when a candidate novel gene query has multiple similarly high-scoring hits to one or more genomes. We also acknowledge that synteny verification may be particularly difficult to establish computationally for incomplete genome assemblies; such cases may require further manual efforts or sequence/assembly improvements. It is also critical to develop effective filters to classify and exclude candidate novel loci that may represent false positives (e.g., pseudogenes, noncoding RNAs) or several types of genomic artifacts, including repetitive elements, assembly artifacts, and allelic variants. Previous efforts have identified genomic signatures (e.g., lack of ORF, loss of intron-exon structure), which will influence filtering algorithms using a training set of positive and negative controls from previous studies.

## PREDICTING THE FUNCTION(S) OF NOVEL GENES

Once novel genes have been identified and confirmed in the target species or lineage, an important follow up in the case of protein-coding genes is a characterization of the ORF and associated domains. Novel genes and paralogous pairs can take on divergent functions by adding or losing protein-coding domains, or through an accumulation of *indels* and/or nonsynonymous substitutions. While there are numerous ways to perform such characterization, we recommend retrieving the predicted protein sequences and analyzing them using Interproscan 5,[32] PSI-BLAST,[33] and CD-search[34] to identify and, whenever possible, classify conserved domains. This analysis will help determine if members of paralogous pairs or gene sets differ by the presence/absence of specific domains, indels, or nonconservative substitutions, suggesting divergence in gene function.[12,14]

While the universe of possible protein-coding domains is extensive, certain domain types are strong predictors of possible gene functions. For example, the prediction of transmembrane domains can link a protein to the plasma membrane, DNA-binding motifs can link proteins to transcription factor activity, and specific catalytic domains (e.g., kinase, phosphatase, hydrolase, proteinase) could indicate enzymatic activity. The differential presence of such domains in a paralogous pair would be very helpful to reveal possible functional divergences (e.g., *YTHDC2L1* vs. *YTHDC2* in Fig. 10.1A). The search for conserved domains is particularly important for understanding the possible function of de novo novel genes, since inferences about function cannot be derived from the sequence of a known ortholog. In either case, a limitation of this analysis is that incomplete models may not include the full complement of functional domains. Thus, it is critical to incorporate to the gene prediction any additional sequences revealed in the target genome by BLAT alignments of other gene predictions or expression databases and use the longest possible predicted protein sequence. We also note that while the databases of conserved protein domains are quite extensive, there is no guarantee that their search will successfully assign putative function to a given novel gene. In such cases, we recommend conducting general BLAST searches of the entire universe of available sequences, as the occurrence of short, conserved domains can provide preliminary insights into possible function.[35]

Another important follow-up in attempting to link the occurrence of a given novel gene or paralog to a specific phenotype is to demonstrate that the gene is expressed in the relevant tissue(s). For behavioral phenotypes, this involves demonstrating expression in the brain circuits associated with the behavior, a task that is greatly facilitated if those circuits are known. While expression databases[36–38] and RNA-Seq data[22,39] can provide valuable resources for a first pass verification of whether novel genes or paralogs and/or their parent genes are expressed in the brain, in situ hybridization is an invaluable tool, as it reveals regional patterns of mRNA expression on histological sections with cellular resolution, thus defining brain distribution with high accuracy.

In our own studies, we have found that novel paralogous pairs often differ significantly in their tissue distributions and expression patterns, supporting a differentiation of function after the gene duplication event. We have also taken advantage of the fact that the pathways for vocal learning and production have been identified and extensively characterized in songbirds. These pathways consist of a set of discrete, interconnected brain nuclei that control and integrate the vocal production organ (syrinx) with respiratory function to copy a tutor's song with a high degree of precision. These brain structures, as well as the trait of vocal learning they support, are uniquely present in vocal learners, such as songbirds, and absent in their vocal nonlearner relatives. Using in situ hybridization, we have shown that some novel paralogs or de novo genes in songbirds are selectively expressed in one or multiple of the cortical-like and/or basal ganglia areas within the vocal control circuitry (Fig. 10.1B and C), suggesting a likely association with the regulation of vocal behavior. Future analyses of these confirmed song system markers will be geared toward combining in situs with tract-tracing or utilizing double labeling methods to reveal cell type specificity, in an attempt to link these genes to more specific circuit components. Overall, the ability to perform detailed expression analysis proved critical because songbirds differ from other avian groups by a broad range of variables besides vocal learning. Accordingly, the expression of some identified novel songbird genes turned out not to be associated with the vocal system. In some cases, an enrichment was observed in the ventricular region, a site of known adult neurogenesis, suggesting a possible association with adult neuronal replacement. In other cases, the expression was undetectable in the brain, suggesting a primary function related to other organs or tissues.

On a technical note, an accurate assessment of the differential expression of a novel paralog requires using highly specific molecular probes capable of differentiating between the duplicated and the parent genes. In this regard, we recommend utilizing 3′-UTR biased riboprobes, which tend to be more locus-specific compared to probes derived from coding regions and can be easily derived from clones present in typical brain EST/cDNA collections. Alternatively, this differential expression analysis may require generating paralog-specific probes through polymerase chain reaction (PCR) cloning. We also recommend using an optimized digoxigenin-based nonradioactive protocol that can be customized for detection of transcripts expressed at a broad range of levels in the brain. This well-established protocol has been extensively utilized for examining >500 brain-expressed transcripts in zebra finch, with high-resolution images displaying the distribution of brain-expressed genes at the subcellular level having been made available online as a molecular atlas of gene expression in the zebra finch brain (ZEBrA; www.zebrafinchatlas.org).[40]

## CONCLUSIONS

Advancements in genomic sequencing technologies and an ever-expanding library of genomes have delivered on the promise of allowing researchers to identify genomic features associated with unique behavioral phenotypes expressed by individual organisms or lineages. Here, we provide a strategy for identifying novel gene family expansions (paralogs) and de novo genes that may be associated with unique structural, physiological, and/or behavioral traits (e.g., vocal learning) that are unique to a species or lineage. While functional demonstrations of novelty will ultimately require gene manipulations, or pharmacological intervention, our approach provides a first pass computational approach that can be applied to reliably identify and classify candidate novel genes in a species of interest.

## ACKNOWLEDGMENTS

Funded by grants from the NIDCD (DC014432) and NIGMS (GM092842).

## REFERENCES

1. Yang L, Zou M, Fu B, He S. Genome-wide identification, characterization, and expression analysis of lineage-specific genes within zebrafish. *BMC Genom.* 2013;14:65.
2. Zhang G, Wang H, Shi J, et al. Identification and characterization of insect-specific proteins by genome data analysis. *BMC Genom.* 2007;8:93.
3. Domazet-Loso T, Tautz D. An evolutionary analysis of orphan genes in *Drosophila. Genome Res.* 2003; 13(10):2213–2219.
4. Lin H, Moghe G, Ouyang S, et al. Comparative analyses reveal distinct sets of lineage-specific genes within *Arabidopsis thaliana. BMC Evol Biol.* 2010;10:41.
5. Sonnhammer EL, Ostlund G. InParanoid 8: orthology analysis between 273 proteomes, mostly eukaryotic. *Nucleic Acids Res.* 2015;43(Database issue):D234–D239.
6. Aken BL, Ayling S, Barrell D, et al. The Ensembl gene annotation system. *Database (Oxford).* 2016; 2016.
7. Knowles DG, McLysaght A. Recent de novo origin of human protein-coding genes. *Genome Res.* 2009;19(10):1752–1759.

8. Meredith RW, Zhang G, Gilbert MT, Jarvis ED, Springer MS. Evidence for a single loss of mineralized teeth in the common avian ancestor. *Science*. 2014;346(6215):1254390.
9. Lowe CB, Clarke JA, Baker AJ, Haussler D, Edwards SV. Feather development genes and associated regulatory innovation predate the origin of Dinosauria. *Mol Biol Evol*. 2015;32(1):23–28.
10. Knox K, Baker JC. Genomic evolution of the placenta using co-option and duplication and divergence. *Genome Res*. 2008;18(5):695–705.
11. Johnson BR, Tsutsui ND. Taxonomically restricted genes are associated with the evolution of sociality in the honey bee. *BMC Genom*. 2011;12:164.
12. Wirthlin M, Lovell PV, Jarvis ED, Mello CV. Comparative genomics reveals molecular features unique to the songbird lineage. *BMC Genom*. 2014;15:1082.
13. Zhang G, Li C, Li Q, et al. Comparative genomics reveals insights into avian genome evolution and adaptation. *Science*. 2014;346(6215):1311–1320.
14. Lovell PV, Carleton JB, Mello CV. Genomics analysis of potassium channel genes in songbirds reveals molecular specializations of brain circuits for the maintenance and production of learned vocalizations. *BMC Genom*. 2013;14(1):470.
15. Lovell PV, Clayton DF, Replogle KL, Mello CV. Birdsong "transcriptomics": neurochemical specializations of the oscine song system. *PLoS One*. 2008;3(10):e3440.
16. Vilella AJ, Severin J, Ureta-Vidal A, Heng L, Durbin R, Birney E. EnsemblCompara GeneTrees: complete, duplication-aware phylogenetic trees in vertebrates. *Genome Res*. 2009;19(2):327–335.
17. Han MV, Thomas GW, Lugo-Martinez J, Hahn MW. Estimating gene gain and loss rates in the presence of error in genome assembly and annotation using CAFE 3. *Mol Biol Evol*. 2013;30(8):1987–1997.
18. Li L, Stoeckert Jr CJ, Roos DS. OrthoMCL: identification of ortholog groups for eukaryotic genomes. *Genome Res*. 2003;13(9):2178–2189.
19. Gnomon. http://www.ncbi.nlm.nih.gov/projects/genome/guide/gnomon.shtml.
20. Lovell PV, Wirthlin M, Wilhelm L, et al. Conserved syntenic clusters of protein coding genes are missing in birds. *Genome Biol*. 2014;15(12):565.
21. Quail MA, Smith M, Coupland P, et al. A tale of three next generation sequencing platforms: comparison of Ion Torrent, Pacific Biosciences and Illumina MiSeq sequencers. *BMC Genom*. 2012;13:341.
22. Warren WC, Clayton DF, Ellegren H, et al. The genome of a songbird. *Nature*. 2010;464(7289): 757–762.
23. Hillier LW, Miller W, Birney E, Warren WC, Consortium, atmotCG. Sequence and comparative analysis of the chicken genome provide unique perspectives on vertebrate evolution. *Nature*. 2004; 432(7018):695–716.
24. Kent WJ. BLAT—the BLAST-like alignment tool. *Genome Res*. 2002;12(4):656–664.
25. Kuhn RM, Karolchik D, Zweig AS, et al. The UCSC genome browser database: update 2009. *Nucleic Acids Res*. 2009;37(Database issue):D755–D761.
26. Jarvis ED, Mirarab S, Aberer AJ, et al. Whole-genome analyses resolve early branches in the tree of life of modern birds. *Science*. 2014;346(6215):1320–1331.
27. Gilbert MP, Jarvis ED, Li B, Schlinger BA. Genomic data of the golden-collared manakin (*Manacus vitellinus*). *GigaSci Database*. 2014. https://doi.org/10.5524/101010.
28. Burge C, Karlin S. Prediction of complete gene structures in human genomic DNA. *J Mol Biol*. 1997; 268(1):78–94.
29. Stanke M, Waack S. Gene prediction with a hidden Markov model and a new intron submodel. *Bioinformatics*. 2003;19(suppl 2):ii215–ii225.
30. Quinlan AR, Hall IM. BEDTools: a flexible suite of utilities for comparing genomic features. *Bioinformatics*. 2010;26(6):841–842.
31. Neph S, Kuehn MS, Reynolds AP, et al. BEDOPS: high-performance genomic feature operations. *Bioinformatics*. 2012;28(14):1919–1920.
32. Jones P, Binns D, Chang HY, et al. InterProScan 5: genome-scale protein function classification. *Bioinformatics*. 2014;30(9):1236–1240.
33. Liu T, Zheng X, Wang J. Prediction of protein structural class for low-similarity sequences using support vector machine and PSI-BLAST profile. *Biochimie*. 2010;92(10):1330–1334.

34. Marchler-Bauer A, Lu S, Anderson JB, et al. CDD: a conserved domain database for the functional annotation of proteins. *Nucleic Acids Res*. 2011;39(Database issue):D225–D229.
35. Wirthlin M. *Evolutionary Neurogenomic Approaches Provide Insight into the Molecular Basis of Vocal Learning*. Department of Behavioral Neuroscience, OHSU; 2016.
36. Replogle KL, Arnold AP, Ball GF, et al. The Songbird Neurogenomics (SoNG) Initiative: community-based tools and strategies for study of brain gene function and evolution. *BMC Genom*. 2008;9(1):131.
37. Wada K, Howard JT, McConnell P, et al. A molecular neuroethological approach for identifying and characterizing a cascade of behaviorally regulated genes. *Proc Natl Acad Sci USA*. 2006;103(41): 15212–15217.
38. Li X, Wang XJ, Tannenhauser J, et al. Genomic resources for songbird research and their use in characterizing gene expression during brain development. *Proc Natl Acad Sci USA*. 2007;104(16): 6834–6839.
39. Whitney O, Pfenning AR, Howard JT, et al. Core and region-enriched networks of behaviorally regulated genes and the singing genome. *Science*. 2014;346:1256780. American Association for the Advancement of Science.
40. Carleton JB, Lovell PV, McHugh A, Marzulla T, Horback KL, Mello CV. An optimized protocol for high-throughput in situ hybridization of zebra finch brain. *Cold Spring Harb Protoc*. 2014;2014(12): 1249–1258.

CHAPTER 11

# Using Signatures of Directional Selection to Guide Discovery

**John C. Crabbe**
Oregon Health & Science University, and VA Portland Health Care System, Portland, OR, United States

## INTRODUCTION

Selective breeding has been practiced since ancient times to improve crops and domestic animals. The 21st-century genomics and informatics revolutions have increased the power of the method when applied to complex traits, and they have deepened our understanding of the genomic and population genetic consequences of artificial selection. At its core, the method has not changed, and its practice mimics observed processes underlying natural selection. Individuals in a population are assessed for a phenotype of interest. When individual variation is found, extremely high-scoring (or extremely low-scoring) individuals are chosen for mating to produce the next generation. To the extent that the underlying phenotypic differences reflect genetic influences, this process when iterated across multiple generations will increase (or decrease) the population mean phenotypic values at a rate proportionate to the degree of genetic influence. This is accomplished by alterations in the frequencies of alleles at those genes affecting the selected trait. By mating like-scoring pairs, alleles favoring the trait at those genes become less and less polymorphic and under most circumstances ultimately are fixed homozygous for the favoring allele. At all trait-irrelevant genes, polymorphisms are retained at population equilibrium proportions, given a host of limitations.[1,2] Practice of the method thus demands careful and reliable assessment of a valid phenotype and careful control over as many nongenetic (environmental) influences as feasible, and its success is greatly enhanced by use of very large numbers of subjects. Among the alternatives, population genetic statistical models developed by Falconer[1] and Lynch and Walsh[2] are frequently used as the basis for quantifying the results and differentiating among the types of genetic and environmental influences at work.

In neuroscience, one of the oldest and most widely known applications of selective breeding was Robert Tryon's attempt to breed Maze-Bright and Maze-Dull rats,[3] seeking evidence for the idea of "general intelligence," or "g."[4] Fig. 11.1 shows the results of this study. Thorough explorations of the power of selective breeding were provided by Jerry Hirsch's early experiments with selecting *Drosophila melanogaster* for phototaxis or negative geotaxis [e.g., Refs. 5,6]. In these and subsequent selection experiments to date, the goal was generally to use the lines of animals with exaggerated phenotypic

*Molecular-Genetic and Statistical Techniques for Behavioral and Neural Research*
ISBN 978-0-12-804078-2, https://doi.org/10.1016/B978-0-12-804078-2.00011-8

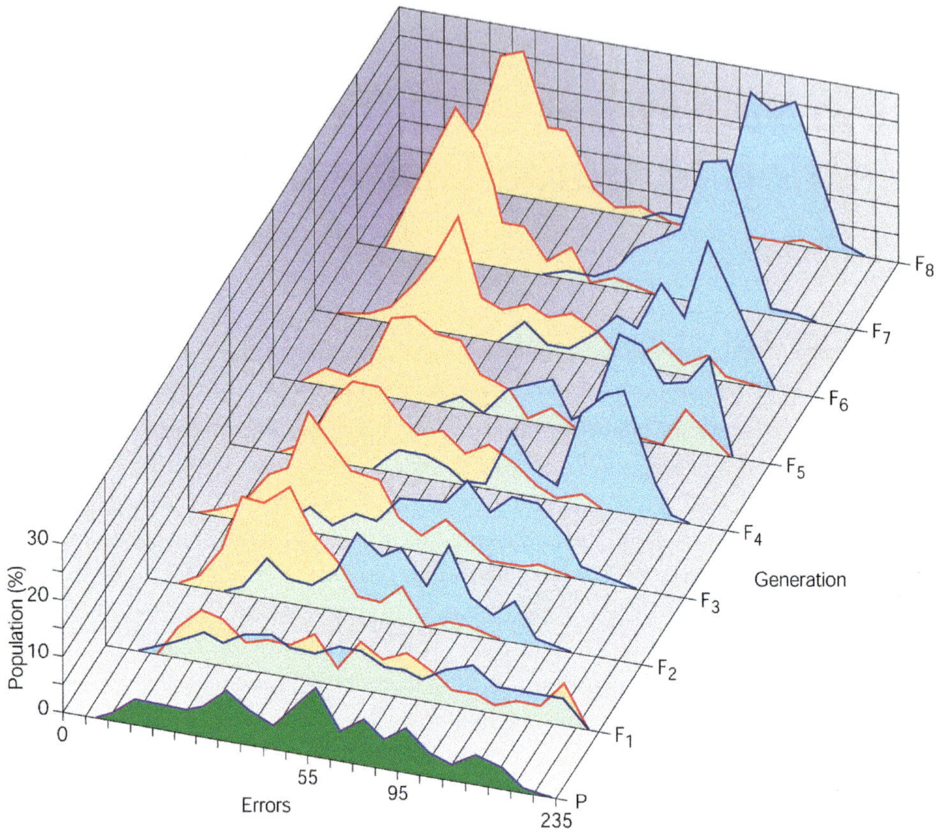


**Figure 11.1** Depiction of Robert Tryon's classic selective breeding experiment with rats. The frequency distributions [Population (%): *z* axis] of individual differences in numbers of errors made during training are plotted (*x* axis) for successive generations (F1–F8: *y* axis). Starting with the progenitor generation (P), rats were selected for mating pairs to create a "Maze-Bright" (yellow) and a "Maze-Dull" (blue) line; animals to breed the P population were selected without regard for their scores (green). From F1 on, the three populations were reproductively isolated from one another. *(From Ref. 77.)*

characteristics to explore the underlying neurobiologic changes that had been created. Because selection (choice of breeders) had been directed solely at a specific trait (often behaviorally defined), these accompanying changes in the populations were, technically, "correlated responses to selection" and could be attributed to pleiotropic effects of some of the same genes affecting the selected trait.

A few examples include murine locomotor activity,[7,8] body fat content,[9] aggressive behavior,[10,11] and thermoregulatory nest building.[12] Rat lines have been bred to be hypertensive,[13] multiple times for good versus poor shuttle box avoidance learning,[14,15] anxiety-like behavior,[16] response to novelty,[17] and many other traits. Results detailing correlated responses to selection for some of the older of these studies can be found in a series of review articles (Experientia Vol. 47, 1991).

However, the field of addiction research has made the most extensive use of this technique by breeding rodents for their relative sensitivity to the behavioral effects of several drugs of abuse.[18] Within the broad addiction field, the majority of selection experiments have been conducted for alcohol-related traits: primarily, the willingness of rodents to ingest solutions of laboratory ethanol in preference to water. The early adoption of the technique can be traced to the pioneering studies of Jorge Mardones at the University of Chile to create UChB high-preferring rats in the 1950s,[19] followed by Kalervo Eriksson's Finnish alcohol accepting rats[20] in the 1960s and T.-K. Li's preferring (P) rats in Indianapolis in the 1970s.[21] Reviews of correlated behavioral-neurobiologic responses to those selections appear regularly, and much has been learned about the neural substrates of risk for addiction-related behaviors. References for many specific reviews of these lines have been collated previously.[22,23]

## SELECTIVE BREEDING: THE BASICS

There are two basic advantages of selective breeding. With a reasonably heritable trait, an animal line can be produced with a very extreme phenotype for use in mechanistic studies. Usually, a low-scoring alternative line is produced in parallel. Because the altered phenotypes (and thus the difference between the divergently selected lines) are due mostly to changes in genetics, other traits that diverge between the lines must necessarily have done so due to the influence of the genes originally selected for. The ability to identify and study such correlated responses to selection (evidence of gene pleiotropy) offers the second major advantage. There are also some disadvantages to the method. For rodent species, it is a very expensive undertaking. The resulting selected lines are not completely inbred, so the influence of other genes may also affect experiments. If replicate lines are bred (see later), they will almost certainly diverge phenotypically, but even when they resemble the original pair of selected lines, this is very likely because of the influence of some different combinations of genes. Note that this can also be construed as an advantage.

To initiate selective breeding requires several starting conditions. First and foremost, the phenotype selected for (and against) needs to be well defined. Simpler phenotypes will be more tractable than more complex ones. That said, in agriculture, "index selection" is frequently practiced: here the selected phenotype is a complex quantitative derivative of multiple measures. And, the desired trait may simply not be measurable in a very simple way. A second requirement is that the trait is heritable. The presence of individual differences among animals, or even among families of animals, is not a guarantee that those differences can be traced to the additive genetic variance necessary for successful selection. These issues are discussed in detail in more technical works.[1,2] Finally, the best practice is to start with a genetically defined, replicable foundation population. To maximize the genetic variability, a systematically intercrossed population of several (or many) inbred strains is a good choice, such as the fairly simple heterogeneous stock HS/Npt mice,[24] the much more genetically diverse HS/CC collaborative cross mice,[25] or the DO diversity outbred stock.[26] Much simpler, but still replicable starting populations such as the F2 cross of two inbred lines can also be used and may provide advantages in some situations.[27,28]

The process of mating like-scoring parents ensures that each parent has a higher than normal probability of having a "favorable" allele at any gene influencing the trait. *Ipso facto*, their offspring will have a higher likelihood of inheriting two such alleles. As generations accrue, the probability of individuals in the emerging selected line having that trait-influencing allele increases, until at some point, it likely becomes invariant, or fixed, in all individuals in the population. This is called inbreeding. However, the goal of *selective* breeding is to maintain genetic diversity as long as possible, while selectively promoting inbreeding for only the trait-relevant genes. This is because inbreeding at all genes generally reduces reproductive fitness and viability. Breeding populations in laboratories are necessarily relatively small (in evolutionary terms), so it is necessary to maintain as many animals as feasible and to avoid inbreeding in general as much as possible: this is usually accomplished by forbidding sibling matings (because siblings share 50% of their alleles by descent), or even cousin matings.

Setting aside complicated methods of selection used in agriculture, there are two basic types of selection. Individual selection chooses the highest-scoring animals in a generation and mates them, ignoring from which families they come. This method usually produces a response to selection more quickly, and it is by far the simplest. However, some families inevitably contribute more offspring to the highest-scoring pool than others, so the general rate of inbreeding is increased. The alternative, within-family selection, is often used. Here, each family contributes only two offspring, the highest-scoring male and female from that family. It is often combined with a rotational mating scheme that delays inbreeding for several generations by systematically preventing brother-sister matings. It is most effective where sibling correlations are high and the underlying heritability is high, and the effective breeding population size is about twice that of individual section.[1]

The tradeoff between accumulating unavoidable inbreeding at genes unrelated to the selected trait and systematically (in)breeding at trait-relevant genes is why it is important to study putative genetically correlated traits fairly early in the selection experiment. The more generations that selection progresses, the more like it is that irrelevant genes will become homozygously fixed in different configurations in the oppositely selected lines. This can lead to "false-positive" genetic correlates. The best way to avoid this trap is to produce duplicate selected lines. If the same correlates are found in multiple replicates of the selection, they are more likely really to reflect pleiotropic influence [for discussion of these issues, see Ref. 29]. This is why a genetically reproducible but genetically variable starting population is useful. It is also very helpful to maintain nonselected control lines (see Fig. 11.1). Such lines are started from the same foundation stock and are maintained in the same number of families, and they are mated according to the same rules as the selected lines, except that the phenotypic value is ignored. Thus, for within-family selection, one male and one female are chosen at random as breeders for the control line. The response of the control lines across generations can reveal the systematic influences of environmental factors (e.g., seasons) and be used to estimate the stochastic variability in the average trait value from generation to generation.[8]

## GENETIC AND GENOMIC CONSEQUENCES OF SELECTION: QTL AND EXPRESSION ARRAYS

As modern molecular genetic techniques matured, studies with selected lines began to be able to survey the consequences of selection on specific genes, starting with changes in gene-encoding sequence (i.e., alleles). Trait-to-chromosomal locus mapping of gene's effects had been pursued from very early days, mostly in cases where single-gene, or Mendelian, traits would produce an obvious change in phenotype within a generation or two. Dominant and recessive mutations' effects were thus related to a growing genome map by systematically producing recombinations through targeted matings [for a clear example of this using *Drosophila*, see Ref. 30]. Most behavioral traits quickly proved to be quantitative, and starting in the 1980s the earliest quantitative trait loci (QTL) were reported. These were loci of influence, as the term QTL suggests, because the crudity of early techniques could not resolve the genomic area of influence to a region containing only a single gene. Each trait, however, proved to be influenced by numerous QTL. When we first reviewed the available QTL data for drugs of abuse, we provisionally identified about 170 QTLs for 20 different responses to ethanol, 11 responses to methamphetamine or cocaine, 5 responses to morphine, and 1 to nitrous oxide. The number of putative QTL reflected our "false-positive" criteria, which were generous, and the recent emergence of microsatellite markers to augment the available (and sparse) morphologic and histologic chromosomal maps and those based on restriction fragment length polymorphisms.[18] We now know that each QTL typically will contribute to a small proportion of genotypic and phenotypic variance, and a similar review of the same phenotypic data, if performed today with similar stringency, would likely find hundreds to thousands of QTL signals for these drug traits. Needless to say, QTL mapping did not lead to immediate discovery of genes of powerful influence that could guide scientists to novel therapeutic targets for disease-related genes. However, it remains an integral part of gene-finding strategies when QTL data are aggregated with other information.

With the advent of microarrays, explorations of gene expression became feasible for multiple genes simultaneously. Knowing gene sequence variations associated with disease is not sufficient for complete understanding, as most genes are expressed only some of the time, and expression patterns may vary with location even within a tissue. As microarrays became affordable and the knowledge of the rough sequence of the whole genome became more and more clearly articulated, studies proliferated where tissue from selectively bred rodent lines were compared to establish the genome-wide differences in gene expression between risk-prone and risk-resistant lines, as well as lines studied before and after a treatment or disease process. Studies in the addictions have proliferated along these lines as well, and the reader is referred to reviews.[31–34]

Both QTL mapping and global gene expression studies, however, were limited in two ways. First, because they were hypothesis-free and exploratory (of all the genes out there, which ones are important?), each involved performing very large numbers of gene-by-gene tests of association between genotype (or gene expression level in terms of RNA) and phenotype, which in turn required strict statistical corrections for multiple comparisons. The industry standard for such corrections is quite stringent.[35] Often, therefore, the statistically significant result of such a query would yield only a handful of significant genes or gene variants, and these could be genes of unknown function. The second problem was that such gene lists could be extremely heterogeneous with respect to the functions of their target proteins. The analog of this problem in human genetics is the Genome-Wide Association Study (GWAS) of disease X, which tends to find hardly any significant associations with modest numbers of subjects. Hence, the current trend is to aggregate extremely large numbers (>100,000 for common diseases) of cases and controls. Such GWAS studies can often then report many significant gene effects, each of vanishingly small likely individual importance.

## GENETIC AND GENOMIC CONSEQUENCES OF SELECTION: BEYOND GENE LISTS

Recent studies have used the strong genetic differences produced by directional selection as starting material. They have applied what might be thought of as informatics-driven filters to try to focus on individual genes whose manipulation will affect the selected trait. A clear example comes from Zhifeng Zhou and David Goldman's work with P and non-preferring (NP) rats.[36] P rats have been bred for many generations to have high preference for 10% alcohol versus water, and the NP rat line was bred for low preference. They first sequenced the exomes of six individual rats from each selected line, finding >120,000 single nucleotide polymorphisms (SNPs). About 20% of these SNPs were homozygous and consistently different between P and NP rats, suggesting that they represented the signature of response to selection pressure for high versus low drinking. Alternatively, given the relatively small population sizes necessarily involved in producing the selected lines, these SNPs could represent accidental inbreeding. To distinguish between these alternatives, they mapped the differential SNPs and found numerous relatively large haplotype blocks. About 1000 such blocks of SNPs could be mapped to known functional genes.

Clearly, it was necessary to further reduce the number of potential targets. Two of the segregating SNPs revealed stop codons in known genes, and 31 others showed exomic sequence differences predicted to adversely affect protein function. At this point, Zhou and Goldman turned back to behavior. They used an F2 segregating population of P and NP rats obtained by intercrossing inbred variants of P and NP lines. They tested hundreds of F2 rats for alcohol preference and performed a standard QTL linkage analysis to identify four variants in three genomic regions linked to preference variation. Comparing the QTL results to the exome sequencing SNP patterns consistently identified a stop codon in the *Grm2* gene (*Grm2* *407) (see Fig. 11.2): this gene encodes the metabotropic glutamate receptor 2, and the stop codon variant in P rats leads to widely deficient receptor function. With a clear target gene of interest, these investigators amassed a wealth of evidence that consistently implicated the mGluR2 receptor in P rats, including absence of receptor expression and impaired mGluR-mediated synaptic depression. They demonstrated elevated alcohol consumption in mice with a null mutation for *Grm2*. Overall, this paper shows how the power of directional selection to influence the genome can yield fruitful results. Three additional possible genes of interest were discussed in the paper, but these have not yet been validated.

A similar conceptual approach was also applied to ethanol preference drinking, also in rats. These investigators took advantage of the other rat preference lines selected at Indiana University, the High Alcohol Drinking and Low Alcohol Drinking rats.[37] HAD and LAD rats were bred for high versus low preference in the same way P and NP rats were.[38] This population offered three major advantages from a genetic perspective. First, the lines were derived from a genetically heterogeneous intercross of eight inbred rat strains and

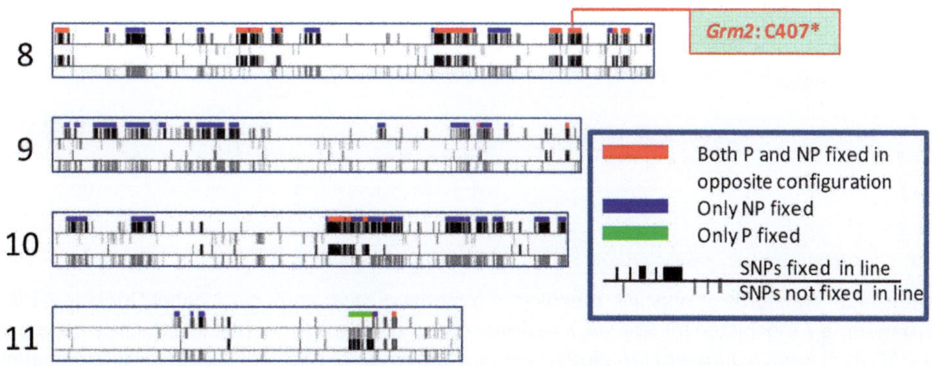


**Figure 11.2** Exome sequencing of preferring (P) and nonpreferring (NP) rats identified the *mGluR2* gene as important for ethanol preference drinking. SNPs are shown for chromosomes 8, 9, 10, and 11. *(From Zhou Z, Karlsson C, Liang T, et al. Loss of metabotropic glutamate receptor 2 escalates alcohol consumption.* Proc Natl Acad Sci USA. *2013;110:16963–16968 with permission.)*

thus presented high genetic diversity at the outset.[39] Rats from this segregating population were also available to serve as controls. Second, two independent populations of HAD rats, and two of LAD rats, were developed. Because laboratory populations are always relatively small (from a population genetic perspective), many differences between divergently selected lines emerge during the course of selection due to accidental fixation of gene variants (i.e., loss of genetic variation, or inbreeding). The occurrence of identical gene or chromosomal regional fixation during selection in two completely independent pairs of lines, but not in the nonselected controls, is thus relatively unlikely; this improves the detection of signatures of selection versus noise.[1,2,29]

Many details of the genomics analyses in this experiment differed from the analysis of P versus NP rats. Signatures of selection in the HAD/LAD populations were found for nearly 1000 genes, and most were located within a single gene. Within those genes, very few (four) were found in exonic regions: most were in promoter or intronic regions (see Fig. 11.3). Functional overrepresentation analyses suggested an important role for genes involved with synaptic transmission, memory, and reward pathways and included those coding for several ion channels and excitatory neurotransmitter receptors.[37] Unlike the NIAAA group, these investigators did not pursue individual candidate genes and attempt to provide further confirmatory evidence.

Our group has also taken advantage of the existence of replicated selectively bred lines.[40] A binge has been defined by the National Institute on Alcohol Abuse and Alcoholism of the US National Institutes of Health as a period of temporally focused drinking that leads to a blood ethanol concentration (BEC) > 80 mg%, or 0.8 mg/mL.[41] Binge drinking is a risk for development of an alcohol use disorder, and most alcohol abusers binge drink. Binge drinking is also a strong predictor of medical diagnosis and has deleterious health consequences.[42] Prevalence of binge drinking is increasing in the United

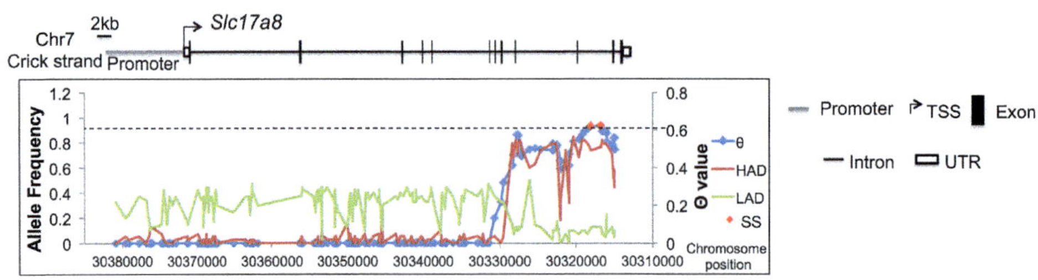


**Figure 11.3** Signatures of selective breeding in genome of High and Low Alcohol Drinking rat lines. Chromosome 7 is depicted for *Slc17a8*, a vesicular glutamate transporter (see inset). Allele frequencies for HAD (red) and LAD (green) are plotted versus chromosomal position. Excessive differentiation in genomic architecture between lines across both replicates, termed signatures of selection (SS), are plotted in red based on intraclass correlation values (θ), plotted in blue. *(Adapted from Lo CL, Lossie AC, Liang T, et al. High resolution genomic scans reveal genetic architecture controlling alcohol preference in bidirectionally selected rat model.* PLoS Genet. *2016;12:e1006178 with permission.)*

States,[43,44] and it is highly prevalent in both veterans and active military duty personnel. Alcohol use disorder is comorbid with many other psychiatric conditions: in these populations, posttraumatic stress disorder is also a frequent diagnosis.[45] To develop an animal model of binge-like drinking, we explored several alternatives with the goal of achieving a simple behavioral assay for binge-like drinking in the mouse. Following earlier work in the area,[46] we developed the drinking in the dark (DID) assay, where mice consume enough alcohol in 2–4 h to reach intoxicating BECs.[47] The basic paradigm we used was to substitute 20% ethanol for water for a limited period each day, during the early hours of the circadian dark cycle, as this is when rodents consume much of their daily food and fluid. We have determined the optimal time after "lights off" to start access,[47] and the optimal duration of access to result in elevated blood alcohol levels.[48] Consumption of the ethanol solutions remains relatively consistent across 12 days. When we examined panels of multiple inbred strains in the DID procedure, we found that the trait was reliable upon retest and significantly heritable.[49,50]

We subsequently selectively bred high DID (High Drinking in the Dark [HDID]-1 and HDID-2) mouse lines for high BECs after a 4-h DID session; these mice drink to the point of behavioral intoxication and reach blood levels that average about 200 mg/mL [Refs. 48,51; see Fig. 11.4]. Behavioral characterization of HDID mice has revealed that HDID mice exhibit behavioral impairment after drinking, withdrawal after a single binge drinking session, and escalate their intake in response to induction of successive cycles of dependence.[48,50] Notably, HDID mice do not exhibit altered tastant preference or alcohol clearance rates.[52,53] One clear limitation of the DID model is that ethanol is not offered as a choice versus water, and when it is, ethanol intake and blood alcohol levels are somewhat lower.[48,53] This selection has one or two unusual features. The first is that we bred for a pharmacological endpoint (blood alcohol level after drinking) rather than for increased intake. As Fig. 11.4 shows, animals did nonetheless show elevated drinking across generations, which was expected. However, they achieve higher blood levels by patterning their drinking differently. HDID-1 mice show larger (longer) bouts of sustained drinking, while HDID-2 mice show more frequent, smaller bouts.[54] Second, given that the HS/Npt foundation population shows very low blood levels (and intake: see data at Generations S0 and S14 in Fig. 11.4), we elected not to develop parallel lines for low blood levels after DID.

In an initial attempt to explore the genomic structure of response to intense selection, we compared patterns of gene expression in ventral striatum tissue from 48 naïve, male mice from all three genotypes.[40] We compared HDID-1 mice from the 22nd selected generation, HDID-2 mice from the 15th selected generation, and HS/Npt unselected controls. Using Illumina WG 8.2 arrays, we analyzed SNP variation in 3683 markers: analyses were carried out marker by marker and with Weighted Gene Coexpression Network Analysis [WGCNA: Refs. 55,56]. For both QTL analyses and network analyses, we predicted that genetic variability across animals would be greatest in the

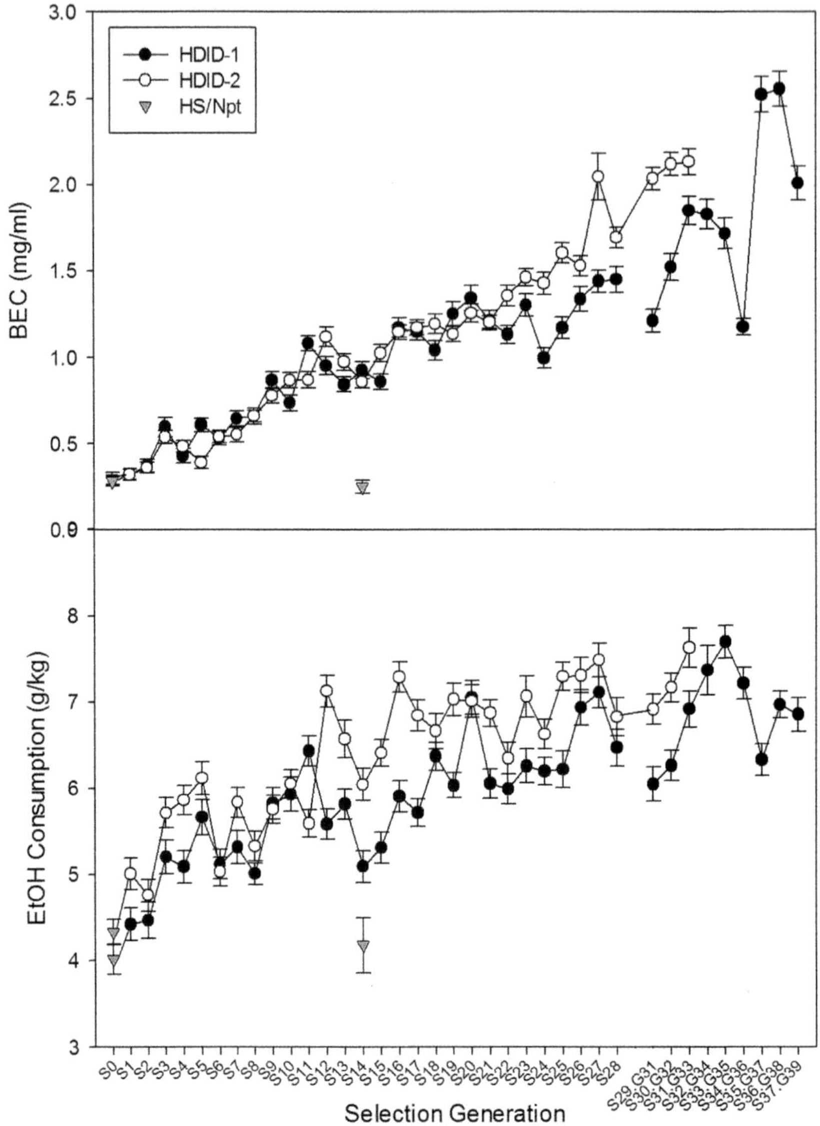


**Figure 11.4** Upper panel: Response to unidirectional selective breeding for blood ethanol concentration (BEC) in mice. High Drinking in the Dark (HDID) mice were offered 20% ethanol in place of water for 4 h starting 3 h into their circadian dark session. Each data point represents the mean ± standard error BEC at the end of drinking for that generation's population of about 100–150 mice (approximately half males and half females). HDID-1 mice (*closed symbols*) have been selected for 37 generations. The gap between S28 and S29 for HDID-1 represents 2 generations where selection was relaxed. Selection of HDID-2 mice (*open symbols*) was initiated 2 years later and is at the 31st selected generation. Response lines are plotted versus selected generation. Lower panel: Consumption in g ethanol/kg body weight is also shown, although selection was based entirely on BEC. HS/Npt data are shown at the outset (generation S0) and for 14th generations later. This genetically segregating population served as the foundation population for both lines and has never been directionally selected. For details, see Refs. 48,51.

unselected HS mice, which we found to be true. We predicted that the HDID-2 mice would show differences from HS, and that the HDID-1 mice would show even larger changes given their greater response to selection at that point. Of the more than 9000 transcripts (more than 7000 unique genes) surveyed, there were more genes differentially expressed between HDID-1 and HS than HDID-2 and HS, and 94 transcripts differed from HS in both lines, with the same directionality.

One interesting finding from this study is shown in Fig. 11.5. The WGCNA identified 21 modules, each representing a number of coexpressed genes. Of these, four modules were strikingly and consistently affected by selection. In some instances, the coherence of the module was increased by increased selection (i.e., HDID-1 > HDID-2 > HS), while for others, the opposite was true (see Fig. 11.5). The overall signature of selection indicated that intramodule coherence was more meaningfully responsive to selection (i.e., consistent across the two replicates) than the specific differences in expression of individual genes.[40]

## GENETIC AND GENOMIC CONSEQUENCES OF SELECTION: BEYOND RATS AND MICE

Although most studies seeking to elucidate gene functional variation underlying a trait employ either rats or mice, some investigators have elected to incorporate other types of data to converge on a short list of potential genes of interest. Elissa Chesler's group has developed GeneWeaver, a database of >60,000 curated functional genomic data sets that includes both QTL and gene expression data, searchable in many ways including annotation by Mammalian Phenotype Ontology terms.[57] They used this tool to seek candidate genes that might affect the well-known negative genetic association between ethanol preference drinking and ethanol withdrawal seizure severity in mice.[58,59] Through a series of hierarchic queries, they discovered common representation of a single gene, *Ap3m*, across multiple data sets (see Fig. 11.6). We had divergently selected lines of mice to be alcohol Withdrawal Seizure—Prone (WSP-1, WSP-2) or —Resistant (WSR-1, WSR-2) for 25 generations,[60] and inbred versions of WSP-2 and WSR-1 had been used to identify QTL associated with alcohol withdrawal severity.[61] Analysis of sequence variation in the iWSP-2 and iWSR-1 lines was shown to converge with prior QTL data for the *Ap3m2* gene in a segregating region of the genome; however, no sequence polymorphisms were detected in the coding regions. Expression differences were found between iWSP-2 and iWSR-1. One upstream SNP, and H3K4me3 site, showed strain differences in methylation. Together, these analyses showed how cross-species functional genomics data bases could be related to lead to a functional variant underlying two ethanol-response traits.[62]

We are currently collaborating with the Zhou-Goldman laboratory to characterize withdrawal behaviors in WSP and WSR mice. We developed a battery of withdrawal

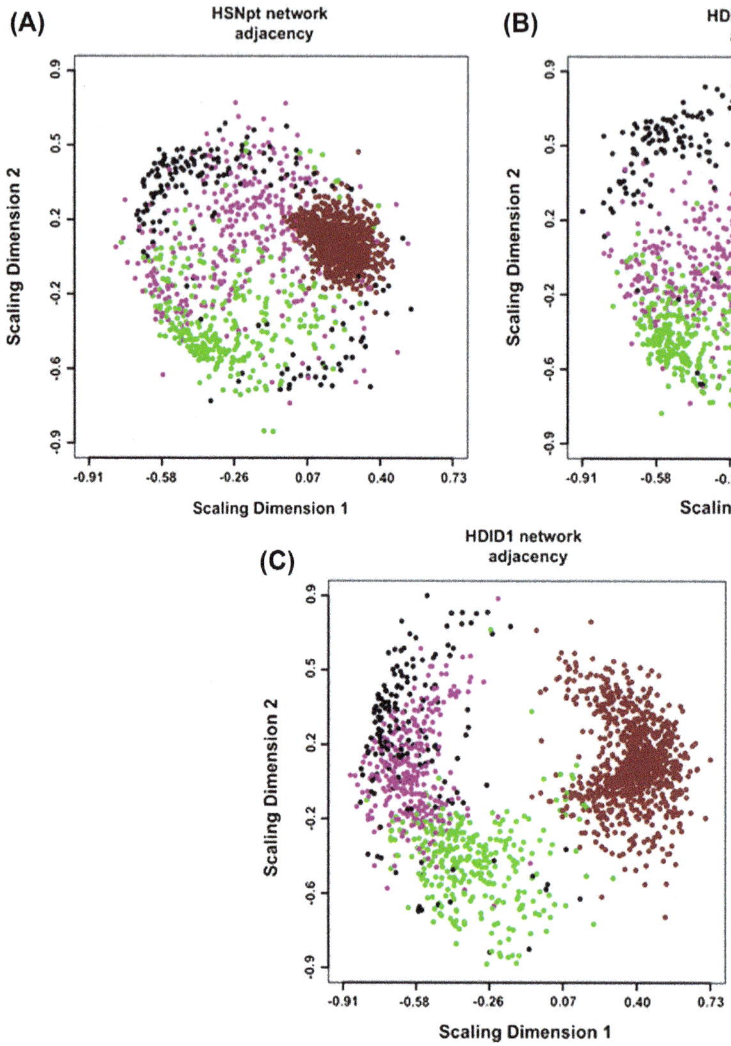


**Figure 11.5** Multidimensional scaling plots of the coexpression networks in (A) heterogeneous stock (HS/Npt), (B) High Drinking in the Dark (HDID)-2, and (C) HDID-1 datasets. For visual clarity, only the four modules most consistently affected by selection ("black," "magenta," "dark red," and "green") are depicted. Each *dot* represents a transcript, with colors corresponding to module assignments. The distances between points correspond to network adjacency. The figure illustrates (1) the modularity of the networks, with similar colors clustered together, and (2) the effect of selection on the network structure, with HDID-2 and HDID-1 successively diverging more from the original HS/Npt network structure. In particular, the "dark-red" module appears have become more dispersed, while the "magenta" module appears to have become more compacted in the selection networks. *(From Iancu OD, Overbeck D, Darakjian P, et al. High Drinking in the Dark selected lines and brain gene coexpression networks.* Alcohol Clin Exp Res. *2013;37:1295–1303 with permission.)*

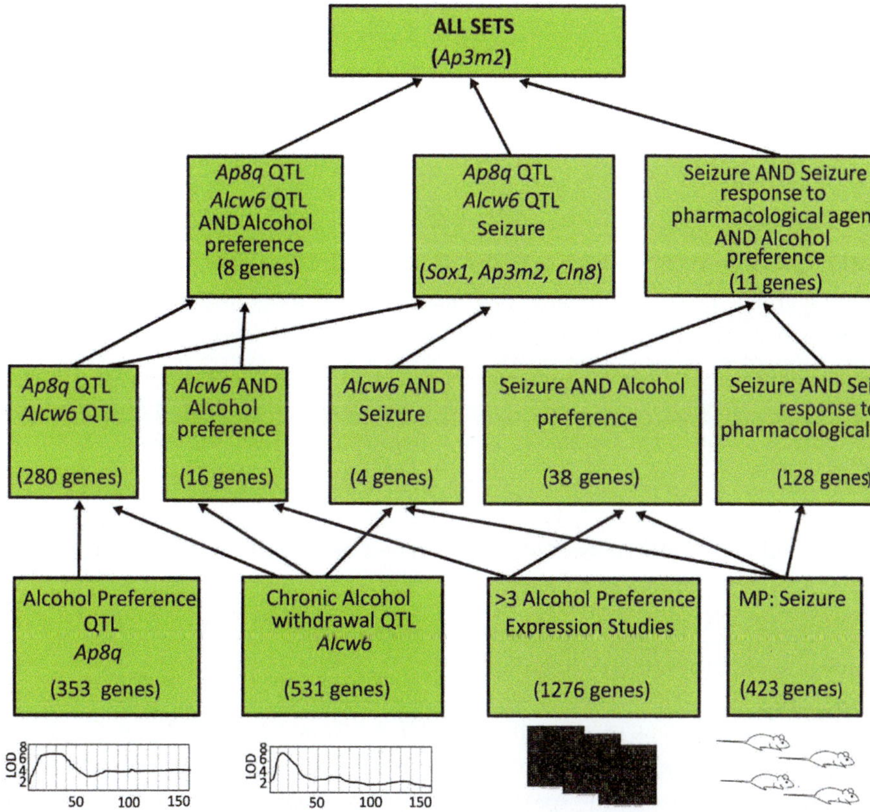


**Figure 11.6** The hierarchic similarity for data from GeneWeaver for alcohol preference– and alcohol withdrawal–related gene sets. The analysis starts at the bottom of the chart with aggregated individual gene sets for alcohol preference QTL and withdrawal QTL, differentially expressed genes related to preference, and murine seizure-related genes. Each successive level from the bottom to the top of the chart represents increasingly higher order intersections among these gene sets, with the most highly connected genes at the top. *(From Bubier JA, Jay JJ, Baker CL, et al. Identification of a QTL in* Mus musculus *for alcohol preference, withdrawal, and Ap3m2 expression using integrative functional genomics and precision genetics.* Genetics. *2014;197:1377–1393 with permission.)*

tests examining behavioral symptoms from multiple domains that could be measured over 4 days in mice after administration of ethanol in vapor for 72 continuous hours. Withdrawal produced hypothermia, greater neuroexcitability (seizures and tremor), anxiety-like behaviors using some apparatus (such as reduced transitions between light and dark compartments of a maze), anhedonia (reduced sucrose preference), Straub tail, backward walking, and reductions in activity. Withdrawing mice did not differ from controls in thermal pain sensitivity, hyperreactivity to handling, or anxiety-like emergence behaviors in other apparatus. Using these data, we constructed a battery of withdrawal tests. Interestingly, we found that individual differences in severity of withdrawal among different tests were weakly correlated at best.[63]

Tissues from WSP-1, WSP-2, WSR-1, WSR-2, WSC-1, and WSC-2 female mice were initially analyzed. This identified 1983 SNPs that were significantly different between WSP-1 and WSR-1 mice and significantly different between WSP-2 and WSR-2 mice ($P < E-8$), but not different between WSC-1 and WSC-2 mice ($P > .05$). Many SNPs were known to have deleterious effects on protein structure, and therefore may have functional significance in ethanol withdrawal severity. We also tested more than 600 WSP-2 X WSR-1 F2 intercross mice using the withdrawal battery, and tissue from those mice is being analyzed as well. Using approximately 400 convergent markers as well as variants that are predicted to be functional from among those identified in the parental lines, we are screening the exomes of the F2 mice. The exome sequencing will be followed by genotyping verification of variants of greatest interest, followed by verification using genetic segregation with withdrawal phenotype using a QTL map already generated using from the inbred F2 cross, iWSP-2 X iWSR-1.[61]

Another approach combined data from multiple species. This recently published study started with a GWAS analysis of >700 related cases with an alcohol dependence diagnosis and >1700 unscreened population controls: all were from Ireland.[64] The initial analysis identified 14 SNPs significantly associated with alcohol dependence: 13 of them were within the chromosome 2 collagen gene, *COL6A3*. These 14 SNPs and 262 others less strikingly associated were analyzed in a replication sample of >15,000 subjects drawn from four additional studies. Although the *COL6A3* association did not significantly replicate, the authors pursued functional analyses of this gene and three others using mouse models and the nematode species, *Caenorhabditis elegans*, and in *D. melanogaster*. Genes orthologous to *COL6A3* were manipulated in these species, and species-specific responses to ethanol were investigated. Traits in the three model species ranged from ethanol sensitivity to tolerance to withdrawal. One of the tools used by the group was GeneWeaver, which also allowed incorporation of data from rats and nonhuman primates. Overall, the investigators identified several genes of potential relevance for alcohol dependence risk in humans.[64]

Another example is drawn from a project exploring risk for methamphetamine abuse in mice that recently discovered an intriguing polymorphism exerting a very large effect. Methamphetamine High-Drinking (MAHDR) and Low-Drinking (MALDR) mouse lines were bred by selection from an F2 intercross of C57BL/6J × DBA/2J inbred strains.[28] MAHDR mice drink large amounts of methamphetamine, while MALDR mice avoid the amphetamine solution nearly entirely. One interesting feature of this project is that response to selection was extremely rapid, perhaps due to the somewhat simpler genetics of an F2 background as opposed to an eight-strain intercross, but also because there is a major gene effect accounting for a large proportion of genetic variance. These mice are not continued under directional selection beyond five generations. Instead, a new selection for exactly the same trait is initiated: four consecutive pairs of lines have responded nearly identically, allowing the Phillips lab to amass a large amount of data on correlated responses to selection over the years.

A QTL analysis was performed that compared the MADR lines from the second selected generation and the fifth generation. In addition, QTL data were collected for replicates 1 and 2 for the fourth selected generation.[65] The initial QTL analysis showed a large signal on proximal chromosome 10, near *Oprm1*, the gene encoding the mu opioid receptor. In an ensuing network analysis, when it was added to the most significant network, results suggested that *Oprm1* is significantly regulated by genes that have a primary role. A subsequent survey of differential sensitivity to a variety of opiate effects in MAHDR and MALDR mice revealed several significant differences, and a drug that acts as a partial agonist at the mu opioid receptor was able to reduce methamphetamine intake in the MAHDR mice.[66,67] However, higher resolution gene mapping within the chromosome 10 QTL interval eventually ruled out the *Oprm1* gene as a source of genetic risk for methamphetamine intake (Eastwood et al., submitted).

Another gene in the chromosome 10 interval showed persistent association with drinking. The Trace amine-associated receptor 1 gene *Taar1* encodes a G protein–coupled receptor stimulated by methamphetamine and neurotransmitter precursors and metabolites, including tyramine and beta-phenylethylamine. TAAR1 modulates dopamine activity, and it may participate in many physiological and behavioral responses, including predisposition to drug abuse [for review, see Ref. 68]. *Taar1* mapped within the confidence interval of the chromosome 10 QTL controlling a majority of methamphetamine drinking.[69] The DBA/2J inbred strain (one of the founders of the MAHDR/MALDR lines and a high methamphetamine drinking strain) was found to possess a unique variant of the gene that codes for a nonfunctional receptor. Using a cell culture assay, it was determined that receptors from DBA/2J mice (but not C57BL/6) were insensitive to methamphetamine stimulation of cyclic AMP in vitro.[69] Next, constitutive knockouts of *Taar1* were compared with wild-type (low drinking) C57BL/6N mice and displayed high methamphetamine drinking and reduced sensitivity to methamphetamine-induced hypothermia and a methamphetamine-conditioned taste aversion,[69] phenotypes consistent with the differences found between the MADR lines. Furthermore, the MADR lines were found to differ in the frequency of this *Taar1* allele, with all individuals of the MAHDR line, but no members of the MALDR lines, exhibiting homozygosity.

Subsequently, Aaron Janowsky's and Tamara Phillips' groups showed that TAAR1 from DBA/2J (but not C57BL/6) lost sensitivity to in vitro stimulation of cyclic AMP by multiple agonists.[68] They also pursued the SNP variant by examining the published sequence from multiple standard inbred strains.[68] They also extracted DNA from tissue samples from many of the BXD recombinant inbred strains, which (like MAHDR/MALDR) were developed from F2 crosses of C57BL/6 × DBA/2J, and sequenced for the *Taar1* SNP. They expected to find a number of strains with C57BL/6 markers spanning the *Taar1* locus and a number with the variant DBA/2J SNP. To their surprise, all of the 57 BXD RI strains derived in three distinct waves of intercrosses (approximately in 1969, 1991, and 1998) had the C57BL/6 *Taar1* allele. However, 20 of 44 BXD RI strains that were derived beginning in 2008 had the DBA/2J *Taar1* allele. All sets of the BXD RIs used the Jackson Laboratory DBA/2J mice as the crossing partner. This means that the SNP variant in the now-existing DBA/2J mouse occurred in the Jackson Laboratory founder colony after 1998. Finally, they explored human *TAAR1* receptor variants in vitro and discovered a range of functional responses. This intriguing story may well have important implications for understanding human individual differences in susceptibility to abuse methamphetamine and potentially enable identification of personalized therapeutics that target an individual's particular *TAAR1* genotype.[68]

A final example of cross-species, cross-database integration is represented by a collaborative effort between our laboratory and those of Adron Harris and Dayne Mayfield at the University of Texas (Ferguson, Ozburn et al., in press),[70] supported by the National Institute on Alcohol Abuse and Alcoholism's Integrative Neuroscience Initiative on Alcoholism-Neuroimmune consortium. We are attempting to use transcriptome-based drug discovery to identify potential novel drug therapies that could be used to treat binge-like alcohol drinking. The project takes advantage of the HDID mice discussed earlier. The Mayfield and Harris laboratories began by comparing gene expression signatures from eight different brain regions in naïve male HDID-1 mice, selected for high binge-like drinking blood alcohol levels, and nonselected HS/Npt mice. They used multiple computational methods to compare the differences in gene expression between the genotypes, assuming that this difference represented a genomic signature for high genetic risk for binge drinking. They then queried the Library of Integrated Cellular Signatures L1000 database. This in vitro/informatics resource contains gene expression signatures produced by applying thousands of compounds to over 100 cell lines and is continually growing. It has been used to predict candidate drugs with high likelihood of antagonizing the gene expression signatures for many complex diseases including many cancers and inflammatory bowel disease.[71,72] The reasoning underlying the query strategies was that an interesting drug should present an overall gene expression profile that mirrored the difference between HDID-1 and HS, i.e., it would be expected to reverse that difference in HDID-1 mice.

Two drugs emerged as top novel candidates. One was a Bruton's tyrosine kinase inhibitor, terreic acid, that had previously been shown to participate in ischemic responses in rodents.[73] The other was the mixed dopamine-serotonin receptor agonist pergolide, which had been used in early clinical trials with humans to treat cocaine dependence in a study that included a subset of alcoholic patients.[74] We obtained these two compounds and tested them for efficacy to reduce DID and blood alcohol levels in naïve HDID-1 mice. We found that both drugs showed dose-dependent efficacy.

This ongoing effort has therefore provided the first evidence for the use of transcriptome-based drug discovery to target a trait related to the addictions (indeed, for any psychiatric disease). However, this project is in its infancy. Neither of these compounds would have been targeted by any logical analysis based on known mechanisms of action of ethanol, so in this respect the approach is succeeding. However, we already knew that pergolide had been withdrawn from the human pharmacopeia in the United States due to adverse cardiac side effects.[74] We elected to test it because it was inexpensive, readily crossed the blood brain barrier, and affected plausible biologic systems important for reward processes (e.g., dopamine, serotonin). Most drugs nominated to reduce drinking fail in the clinic despite having shown efficacy in previous tests with laboratory animal models.[75] Therefore, we pursued the specificity of the effects to reduce ethanol drinking by testing each compound for its tendency to reduce drinking plain water and saccharin-water solutions. Low doses of terreic acid appeared to reduce binge-like ethanol drinking selectively. We saw decreased water intake only after the highest dose tested, and terreic acid did not reduce saccharin intake, so it was not simply reducing thirst motivation or blunting the effects of any rewarding stimulus. The behavioral profile for pergolide was somewhat less selective, as it also reduced water intake at all doses tested, as well as saccharin intake. However, neither drug reduced locomotor activity. Thus, the effects on ethanol intake cannot be attributed to sedation or malaise for either drug. We are currently exploring terreic acid and related derivatives further but have abandoned further studies related to pergolide (Ferguson, Ozburn et al., in press).[70]

## UTILITY OF SELECTION SIGNATURES IN THE FUTURE

It is quite hard to predict the future in neuroscience. Seventeen years ago, Robert Plomin and I attempted to predict how molecular genetics would come to affect our understanding of behavior. We got a few things right, but we were breathtakingly modest in assessing how deep the transformation would be.[76] Nonetheless, I hope the examples I have discussed have indicated some ways to use the advantages of selective breeding in the future. The great strength of selective breeding has always been its capacity to sculpt the genome in ways that create very large phenotypic differences. Since it is by definition only considered ethical to do so in laboratory model species, it leads directly to the ability to explore how that has been accomplished at whatever level of reduction the investigator chooses to apply. Now that genomics and informatics are affordable (as compared to even 5 years ago, and especially 10–15 years ago), selective breeding is no longer a blinded process: we are finally able to see what we have done. Our ability to interpret the results and understand what we have done will continue to develop apace as well, given the radical increases in statistical power and informatics approaches available. I am confident that selection will sustain a valuable place in the array of biologic tools for understanding the relationships between that most fundamental molecular level of analysis, DNA, and its most complex end product, behavior.

## ACKNOWLEDGMENTS

Preparation of this MS was supported by grants AA013519, AA020245, and AA010760 from the NIH; 101BX000313 from the US Department of Veterans Affairs; and by the family of John Andrews. Thanks to Mark Rutledge–Gorman for help with the manuscript.

## REFERENCES

1. Falconer DS, Mackay TFC. *Introduction to Quantitative Genetics*. 4th ed. Harlow, England: Longman; 1996.
2. Lynch M, Walsh B. *Genetics and Analysis of Quantitative Traits*. Sunderland, MD: Sinauer Associates; 1998.
3. Tryon RC. Genetic differences in maze learning in rats. In: *39th Yearbook of the National Society for Social Studies in Education: Part I*. Bloomington, IL: Public Schools Publishing Company; 1940:111–119.
4. Spearman C. "General intelligence," objectively determined and measured. *Am J Psychol*. 1904;15: 201–292.
5. Hirsch J, Boudreau JC. Studies in experimental behavior genetics. I. The heritability of phototaxis in a population of *Drosophila melanogaster*. *J Comp Physiol Psychol*. 1958;51:647–651.
6. Hirsch J, Ksander G. Studies in experimental behavior genetics. V. Negative geotaxis and further chromosome analyses in *Drosophila melanogaster*. *J Comp Physiol Psychol*. 1969;67:118–122.
7. Halcomb RA, Hegmann JP, DeFries JC. Open-field behavior in mice: a diallel analysis of selected lines. *Behav Genet*. 1975;5(3):217–231.
8. Swallow JG, Carter PA, Garland Jr T. Artificial selection for increased wheel-running behavior in house mice. *Behav Genet*. 1998;28:227–237.
9. Eisen EJ, Benyon LS, Douglas JA. Long-term restricted index selection in mice designed to change fat content without changing body size. *Theor Appl Genet*. 1995;91:340–345.
10. Hyde JS, Sawyer TF. Selection for agonistic behavior in wild female mice. *Behav Genet*. 1980;10: 349–359.
11. Lagerspetz KM, Lagerspetz KY. Changes in the aggressiveness of mice resulting from selective breeding, learning and social isolation. *Scand J Psychol*. 1971;12:241–248.
12. Bult A, Lynch CB. Multiple selection responses in house mice bidirectionally selected for thermoregulatory nest-building behavior: crosses of replicate lines. *Behav Genet*. 1996;23(4):439–446.
13. Okamoto K, Aoki K. Development of a strain of spontaneously hypertensive rats. *Jpn Circ J*. 1963;27: 282–293.
14. Brush FR, Froelich JC, Sakellaris PC. Genetic selection for avoidance behavior in the rat. *Behav Genet*. 1979;9:309–316.
15. Bignami G. Selection for high rates and low rates of avoidance conditioning in the rat. *Anim Behav*. 1965;13:221–227.
16. Liebsch G, Montkowski A, Holsboer F, Landgraf R. Behavioural profiles of two Wistar rat lines selectively bred for high or low anxiety-related behaviour. *Behav Brain Res*. 1998;94:301–310.
17. Stead JD, Clinton S, Neal C, et al. Selective breeding for divergence in novelty-seeking traits: heritability and enrichment in spontaneous anxiety-related behaviors. *Behav Genet*. 2006;36:697–712.
18. Crabbe JC, Belknap JK, Buck KJ. Genetic animal models of alcohol and drug abuse. *Science*. 1994;264: 1715–1723.
19. Mardones J, Segovia-Riquelme N. Thirty-two years of selection of rats by ethanol preference: UChA and UChB strains. *Neurobehav Toxicol Teratol*. 1983;5:171–178.
20. Eriksson K. Rat strains specially selected for their voluntary alcohol consumption. *Ann Med Exp Biol Fenniae*. 1971;49:67–72.
21. Lumeng L, Hawkins TD, Li T-K. New strains of rats with alcohol preference and non-preference. In: Thurman RG, Williamson JR, Drott H, Chance B, eds. *Alcohol and Aldehyde Metabolizing Systems*. New York: Academic Press; 1977:537–544.

22. Crabbe JC, Phillips TJ. Genetic animal models. In: Sher KJ, ed. *The Oxford Handbook of Substance Use Disorders*. Oxford, UK: Oxford University Press; 2014.
23. Crabbe JC. Alcohol use disorders. In: Pietropaulo S, Sluyter F, Crusio WE, eds. *Genetic Mouse Models of Neurobehavioral Disorders*. Cambridge: Cambridge University Press; 2014:293–302; vol. II.
24. Hitzemann B, Dains K, Kanes S, Hitzemann R. Further studies on the relationship between dopamine cell density and haloperidol-induced catalepsy. *J Pharmacol Exp Ther*. 1994;271(2):969–976.
25. Churchill GA, Airey DC, Allayee H, et al. The Collaborative Cross, a community resource for the genetic analysis of complex traits. *Nat Genet*. 2004;36:1133–1137.
26. Chesler EJ, Gatti DM, Morgan AP, et al. Diversity outbred mice at 21: maintaining allelic variation in the face of selection. *G3 (Bethesda)*. 2016;6:3893–3902.
27. DeFries JC, Hegmann JP. Genetic analysis of open-field behavior. In: Elliot RM, MacCorquodale K, Lindzey G, Clark KE, eds. *Contributions to Behavior-Genetic Analysis: The Mouse as a Prototype*. New York: Appleton-Century-Crofts, Deucational Division/Meredith Corporation; 1970:23–56.
28. Wheeler JM, Reed C, Burkhart-Kasch S, et al. Genetically correlated effects of selective breeding for high and low methamphetamine consumption. *Gene Brain Behav*. 2009;8:758–771.
29. Crabbe JC, Phillips TJ, Kosobud A, Belknap JK. Estimation of genetic correlation: interpretation of experiments using selectively bred and inbred animals. *Alcohol Clin Exp Res*. 1990;14(2):141–151.
30. Dudai Y, Jan YN, Byers D, Quinn WG, Benzer S. dunce, a mutant of Drosophila deficient in learning. *Proc Natl Acad Sci USA*. 1976;73:1684–1688.
31. Warden AS, Mayfield RD. Gene expression profiling in the human alcoholic brain. *Neuropharmacology*. 2017;122:161–174.
32. Zhou Z, Enoch MA, Goldman D. Gene expression in the addicted brain. *Int Rev Neurobiol*. 2014;116: 251–273.
33. Edenberg HJ, Foroud T. The complex genetics of alcoholism. In: Noronha ABC, Cui C, Harris RA, Crabbe JC, eds. *Neurobiology of Alcohol Dependence*. San Diego: Academic Press; 2014:539–550.
34. Saba LM, Flink SC, Vanderlinden LA, et al. The sequenced rat brain transcriptome – its use in identifying networks predisposing alcohol consumption. *FEBS J*. 2015;282:3556–3578.
35. Benjamini Y, Hochberg Y. Controlling the false discovery rate: a practical and powerful approach to multiple testing. *J Roy Stat Soc B*. 1995;57:289–300.
36. Zhou Z, Karlsson C, Liang T, et al. Loss of metabotropic glutamate receptor 2 escalates alcohol consumption. *Proc Natl Acad Sci USA*. 2013;110:16963–16968.
37. Lo CL, Lossie AC, Liang T, et al. High resolution genomic scans reveal genetic architecture controlling alcohol preference in bidirectionally selected rat model. *PLoS Genet*. 2016;12:e1006178.
38. Murphy JM, Stewart RB, Bell RL, et al. Phenotypic and genotypic characterization of the Indiana University rat lines selectively bred for high and low alcohol preference. *Behav Genet*. 2002;32: 363–388.
39. Hansen C, Spuhler K. Development of the National Institutes of Health genetically heterogeneous rat stock. *Alcohol Clin Exp Res*. 1984;8(5):477–479.
40. Iancu OD, Overbeck D, Darakjian P, et al. High drinking in the dark selected lines and brain gene coexpression networks. *Alcohol Clin Exp Res*. 2013;37:1295–1303.
41. NIAAA council approves definition of binge drinking. *NIF Newsl*. 2004:3.
42. Dawson DA, Li TK, Grant BF. A prospective study of risk drinking: at risk for what? *Drug Alcohol Depend*. 2008;95:62–72.
43. Han BH, Moore AA, Sherman S, Keyes KM, Palamar JJ. Demographic trends of binge alcohol use and alcohol use disorders among older adults in the United States, 2005–2014. *Drug Alcohol Depend*. 2017; 170:198–207.
44. Esser MB, Hedden SL, Kanny D, Brewer RD, Gfroerer JC, Naimi TS. Prevalence of alcohol dependence among US adult drinkers, 2009–2011. *Prev Chronic Dis*. 2014;11:E206.
45. Norman SB, Myers US, Wilkins KC, et al. Review of biological mechanisms and pharmacological treatments of comorbid PTSD and substance use disorder. *Neuropharmacology*. 2012;62:542–551.
46. Sharpe AL, Tsivkovskaia NO, Ryabinin AE. Ataxia and c-Fos expression in mice drinking ethanol in a limited access session. *Alcohol Clin Exp Res*. 2005;29:1419–1426.

47. Rhodes JS, Best K, Belknap JK, Finn DA, Crabbe JC. Evaluation of a simple model of ethanol drinking to intoxication in C57BL/6J mice. *Physiol Behav.* 2005;84:53–63.
48. Crabbe JC, Metten P, Belknap JK, et al. Progress in a replicated selection for elevated blood ethanol concentrations in HDID mice. *Gene Brain Behav.* 2014;13:236–246.
49. Rhodes JS, Ford MM, Yu C-H, et al. Mouse inbred strain differences in ethanol drinking to intoxication. *Gene Brain Behav.* 2007;6:1–18.
50. Crabbe JC, Metten P, Huang LC, et al. Ethanol withdrawal-associated drinking and drinking in the dark: common and discrete genetic contributions. *Addiction Genet.* 2012;1:3–11.
51. Crabbe JC, Metten P, Rhodes JS, et al. A line of mice selected for high blood ethanol concentrations shows drinking in the dark to intoxication. *Biol Psychiatr.* 2009;65:662–670.
52. Crabbe JC, Spence SE, Brown LL, Metten P. Alcohol preference drinking in a mouse line selectively bred for high drinking in the dark. *Alcohol.* 2011;45:427–440.
53. Barkley-Levenson AM, Crabbe JC. High Drinking in the Dark mice: a genetic model of drinking to intoxication. *Alcohol.* 2014;48:217–223.
54. Barkley-Levenson AM, Crabbe JC. Ethanol drinking microstructure of a High Drinking in the Dark selected mouse line. *Alcohol Clin Exp Res.* 2012;36:1330–1339.
55. Zhang B, Horvath S. A general framework for weighted gene co-expression network analysis. *Stat Appl Genet Mol Biol.* 2005;4 [Article 17].
56. Langfelder P, Horvath S. WGCNA: an R package for weighted correlation network analysis. *BMC Bioinf.* 2008;9:559.
57. Baker EJ, Jay JJ, Bubier JA, Langston MA, Chesler EJ. GeneWeaver: a web-based system for integrative functional genomics. *Nucleic Acids Res.* 2012;40:D1067–D1076.
58. Metten P, Phillips TJ, Crabbe JC, et al. High genetic susceptibility to ethanol withdrawal predicts low ethanol consumption. *Mamm Genome.* 1998;9:983–990.
59. Metten P, Iancu OD, Spence SE, et al. Dual-trait selection for ethanol consumption and withdrawal: genetic and transcriptional network effects. *Alcohol Clin Exp Res.* 2014;38:2915–2924.
60. Crabbe JC, Kosobud A, Young ER, Tam BR, McSwigan JD. Bidirectional selection for susceptibility to ethanol withdrawal seizures in *Mus musculus*. *Behav Genet.* 1985;15:521–536.
61. Bergeson SE, Warren RK, Crabbe JC, Metten P, Erwin VG, Belknap JK. Chromosomal loci influencing chronic alcohol withdrawal severity. *Mamm Genome.* 2003;14:454–463.
62. Bubier JA, Jay JJ, Baker CL, et al. Identification of a QTL in *Mus musculus* for alcohol preference, withdrawal, and *Ap3m2* expression using integrative functional genomics and precision genetics. *Genetics.* 2014;197:1377–1393.
63. Metten P, Schlumbohm JS, Huang LC, et al. An alcohol withdrawal test battery measuring multiple behavioral symptoms in mice. *Alcohol*, in press.
64. Adkins AE, Hack LM, Bigdeli TB, et al. Genomewide association study of alcohol dependence identifies risk loci altering ethanol-response behaviors in model organisms. *Alcohol Clin Exp Res.* 2017;41:911–928.
65. Belknap JK, McWeeney S, Reed C, et al. Genetic factors involved in risk for methamphetamine intake and sensitization. *Mamm Genome.* 2013;24:446–458.
66. Eastwood EC, Barkley-Levenson AM, Phillips TJ. Methamphetamine drinking microstructure in mice bred to drink high or low amounts of methamphetamine. *Behav Brain Res.* 2014;272:111–120.
67. Eastwood EC, Phillips TJ. Opioid sensitivity in mice selectively bred to consume or not consume methamphetamine. *Addict Biol.* 2014;19:370–379.
68. Shi X, Walter NA, Harkness JH, et al. Genetic polymorphisms affect mouse and human trace amine-associated receptor 1 function. *PLoS One.* 2016;11:e0152581.
69. Harkness JH, Shi X, Janowsky A, Phillips TJ. Trace amine-associated receptor 1 regulation of methamphetamine intake and related traits. *Neuropsychopharmacology.* 2015;40:2175–2184.
70. Ferguson LB, Ozburn AR, Ponomarev I, et al. Genome-wide expression profiles drive discovery of novel compounds that reduce binge drinking in mice, *Neuropsychopharmacology*, in press.
71. Dudley JT, Sirota M, Shenoy M, et al. Computational repositioning of the anticonvulsant topiramate for inflammatory bowel disease. *Sci Transl Med.* 2011;3, 96ra76.

72. San Lucas FA, Fowler J, Chang K, Kopetz S, Vilar E, Scheet P. Cancer in silico drug discovery: a systems biology tool for identifying candidate drugs to target specific molecular tumor subtypes. *Mol Canc Ther.* 2014;13:3230–3240.
73. Ito M, Shichita T, Okada M, et al. Bruton's tyrosine kinase is essential for NLRP3 inflammasome activation and contributes to ischaemic brain injury. *Nat Commun.* 2015;6:7360.
74. Malcolm R, Herron J, Sutherland SE, Brady KT. Adverse outcomes in a controlled trial of pergolide for cocaine dependence. *J Addict Dis.* 2001;20:81–92.
75. Egli M. Can experimental paradigms and animal models be used to discover clinically effective medications for alcoholism? *Addict Biol.* 2005;10:309–319.
76. Plomin R, Crabbe J. DNA. *Psychol Bull.* 2000;126:806–828.
77. Plomin R. The genetics of g in human and mouse. *Nat Rev Neurosci.* 2001;2:136–141.

SECTION III

# Discovery of New Genes and New Functions of Genes Using Gene Expression Analyses

CHAPTER 12

# Using Transcriptomics to Study Behavior

**J.T. Westwood**
University of Toronto, Mississauga, ON, Canada

## INTRODUCTION

During the lifetime of a multicellular animal, it will experience or perform a number of behaviors related to feeding, reproduction, nurturing of offspring, avoidance of danger, responses to changes in their environment, and social interactions, just to name a few. In addition, there are a number of conditions or diseases that can affect the well-being and overall behavior of individuals. There are good examples of specific genes that play a role in some of the aforementioned behaviors. For example, the *egr1* gene, which codes for a transcription factor and thus can regulate the transcription of downstream genes, has been shown to be transcriptionally induced in response to social interaction and plays a role in zebra finch song recognition during courtship, male dominance in male cichlid fish, and mothering behavior in rats (see Robinson et al.[1] for review). Some other examples include the *period* gene in *Drosophila* male courtship and the foraging gene in *Drosophila* and honeybee feeding behavior.[1]

The fact that different alleles of a gene can affect behaviors of an animal in the long run, and that the products of these genes are often transcription factors, implies that the changes in gene expression (i.e., changes in gene transcription rates) likely underlie and/or modulate the behavior. Therefore, it is logical to study gene expression changes that may underlie factors that influence behavior and brain function. Examples of factors that influence behavior and brain function include (but are not limited to) are: social interactions; different alleles of the same gene; and changes in the environment.

There are some well-studied models for specific behaviors that have been developed in insects. For some of these behaviors, gene expression studies have been conducted to gain insight into the molecular mechanisms that underlie the behavior. For example, honey bees have been a model for studying many kinds of behaviors (or factors affecting them) including, but not limited to differences in the behavior of different members of bee "society" (e.g., male drones, different workers such as nurses, guards, and foragers, and the queen), aggressive behavior in response to intruders, and pheromone and hormone manipulations that affect bee maturation and behavior. All of the aforementioned models have been examined in gene expression studies, and hundreds of reproducible

*Molecular-Genetic and Statistical Techniques for Behavioral and Neural Research*
ISBN 978-0-12-804078-2, https://doi.org/10.1016/B978-0-12-804078-2.00012-X

changes in gene expression have been identified and could be linked to specific developmental and behavioral states (for a review, see Zayed and Robinson).[2]

Numerous genetic studies and mutational screens have been carried out to find genes that affect specific behaviors in *Drosophila* as well. For example, screens to find genes affecting circadian rhythm behavior have led to the discovery of the *clock* and *cycle* genes, which were later found to code for transcription factors, as well as genes like *period* and *timeless* that code for proteins that regulate the stability of *clock* and *cycle*.[3] Gene expression studies with *clock* mutants and wild-type flies uncovered 134 Clock-regulated genes that are involved in a variety of processes.[4] Another example in *Drosophila* is the *foraging* gene that encodes a cGMP-dependent protein kinase. Different natural alleles of *for* show different locomotion during larval feeding. Mutations in *dgcalpha1*, which encodes a soluble guanylyl cyclase subunit, increase both PKG activity and foraging locomotion. DNA microarray studies on this mutant in different genetic *for* backgrounds identified many genes that are differentially transcribed, and interestingly, relatively few are affected in both backgrounds. Several differentially expressed genes were found to exhibit enhanced or suppressed expression in a background-dependent manner.[5]

Gene expression analysis can also be a useful tool in the study of conditions that affect behavior such as alcoholism and drug addiction. Zebrafish that were chronically exposed to alcohol showed gene expression changes in more than 1900 genes, including a number of Cytochrome P450 family genes, ion channel genes, and solute carrier family members that provide insights on the effects of alcohol on physiology and how it might be affecting behavior.[6] For example, the solute carrier genes code for transmembrane proteins that play key roles in the transport of small molecules, including neurotransmitters across vesicular and plasma membranes, a class of molecules whose function was suspected to play roles in alcoholism but never conclusively proven before.[6]

An area of intense current scientific research is the effect of chromatin state on gene transcription with posttranslational modifications of histone proteins and DNA methylation underlying the epigenetic mechanisms that regulate gene expression. There are several examples of epigenetic alterations of behavior. In *Drosophila*, a mutation in *EHMT/G9a*, a gene that encodes a histone methyltransferase that methylates histone 3 at lysine 9 (H3K9), alters several behaviors including larval locomotor behavior, nonassociative learning, and courtship memory.[7] DNA microarray studies with the *EHMT/G9a* mutation revealed a number of genes that were differentially expressed.[7] In summary, gene expression changes may be induced by a variety of factors, and these changes may influence a broad range of behaviors that can be studied in several laboratory species.

The goal of this chapter is to provide conceptual and methodological information to researchers who are considering performing transcriptomics to investigate molecular mechanisms of brain function and/or behavior. While detailed step-by-step protocols are often useful, they can also be limited and dependent upon the exact organism and tissue being studied, as well as upon which exact experimental platform and equipment is being used. There are numerous ways to obtain transcriptome data, and in many instances, a researcher will choose an experimental platform based on what equipment and/or services are available at local (or distant) genomic service centers. Therefore, rather than providing several different highly detailed protocols, this chapter primarily focuses on general design guidelines for gene expression studies with an emphasis on RNA isolation and cDNA production, as these are often the starting material that is given to genomic centers to generate experimental data. For the sake of simplicity, in this chapter, changes in gene expression specifically refer to changes in mRNA transcript levels.

## METHODOLOGY

It is important to consider the many factors that can affect the quality and type of transcriptomic data in the design of an experiment. From choosing the appropriate source for the biologic samples, to generating high-quality RNA (and whether it is total or polyA RNA), to deciding which transcriptomic platform to employ and deciding upon an analysis pipeline. Some of the aforementioned factors have been addressed in detail elsewhere,[8,9] but many have not, so they will be reviewed here.

### Experiment Design and RNA Preparation

#### *General Experimental Design Considerations*

There are many possible schemes or designs that can be employed in gene expression projects as well as factors to consider in their design. One of the most important factors is to decide how many replicates of each sample group are needed for the investigator to be able identify a sufficient number of statistically significant differentially expressed genes between groups. Results must be reproducible in any experimental system for the researcher to be confident in their validity, which in most cases means achieving statistical significance. Gene expression experiments are no exception, despite the early trend in the microarray field of using arbitrary fold-change cutoffs (e.g., twofold) to select differentially expressed genes. Although having three independent replicates is a minimum requirement for any statistical analysis, depending on the type of experiment and the nature of the biologic samples themselves, more than three replicates may be needed. If the biologic material is fairly homogeneous, for example, tissue culture cells grown in vitro at the same time or several insects of the same sex from a genetically identical population, and the experimental treatment or condition is fairly pronounced (e.g., exposure to a drug with strong physiological effects), then three replicates are likely to be sufficient. However, if the biologic samples are more heterogeneous (e.g., different individual animals that are not siblings, mixed populations of cells of different cell types, or individual cells), and the experimental treatment or condition is more subtle (e.g., age of an animal), then more than three replicates will likely be needed. Because gene expression experiments are fairly expensive, it is important to consider the total number of replicates early, so if the total number of samples for a project is limited by budgetary constraints, then it is wiser to have fewer experimental groups with more replicates than to have more groups with only three replicates each. Often, one does not know at the start of a project how many replicates are needed until the first set of data has been analyzed. If time permits, pilot experiments with only two experimental groups (e.g., treatment vs. control) with three replicates can be informative. Similarly, if one is able to make additional biologic replicates of the sample groups (e.g., 4, 5, or 6 replicates) when the initial experiment is being done, then if it turns out that three replicates are not sufficient to give very many differentially expressed genes, the additional replicates can be processed and added to the data set. The advantages of adding biologic replicates during the initial experiment is that variables such as time of year or having a different population of cells or individuals is minimized, as is the time delay in completing the project. More information on the number of replicates needed for gene expression studies can be found in Conesa et al.[9]

Another important experimental design consideration is what biologic material one should select for RNA extraction. In the analysis of behavior, the obvious choice is the brain if the study organism is large enough to extract the tissue, as is the case for most vertebrates, from zebrafish to rat. In the case of some insects, one may use whole heads as the starting material. In mammals and other larger vertebrates, the whole brain is most likely too large of a tissue given the known idiosyncrasies of functions of particular brain areas. Therefore, analysis of the transcriptome of specific, well-defined brain regions may be required. That is, if the gene expression changes underlying the behavior are only occurring in one part of the brain, using whole-brain extract may make such changes undetectable in total brain RNA. A variety of approaches can be used to isolate RNA from specific very small areas of the brain and/or specific neurons in the brain. These include tissue dissection using laser capture microscopy; the homogenization of tissue and sorting of cells with cell-specific fluorescently (e.g. green fluorescent protein) labeled marker proteins followed by fluorescence-activated cell sorting; or the homogenization of tissue and immunoprecipitation of cells using antibodies to unique cell surface markers (immunopanning) (see Okaty et al. for review).[10] When RNA is obtained from a relatively small number of cells, the RNA (and/or resulting cDNA) often has to be amplified to have enough material to either perform microarray experiments or RNA sequencing. The fact that a smaller number of cells is used in the original sample and that the material is being amplified typically leads to more variation in the data from one sample to another. Therefore, more replicates of a sample type will likely be needed to obtain statistically significant gene expression changes.

### *Choice of Experimental Platform to Generate Transcriptomic Data*

This is often the first choice that is made when deciding upon a study because it will usually affect all of the subsequent aspects of the study design and methodology. For

**Table 12.1** Differences between the DNA microarrays and high-throughput sequencing in transcriptome experiments

| | **DNA microarrays** | **High-throughput sequencing** |
|---|---|---|
| Species that experiment can be done | Limited to model organisms | Any (but analysis is easier with model organisms) |
| RNA type needed for experiment | Total or polyA | polyA (or rRNA depleted) |
| Time to process RNA samples prior to experiment | Moderate (for total RNA) | Relatively long |
| Time and complexity to prepare samples before hybridization or sequencing | Moderate | Typically long and complex (many steps) |
| Cost of preparing/labeling each sample | Moderate | Relatively high |
| Cost of array or sequencing | Moderate | Moderate to high |
| Time to obtain data | Moderate | Moderate to long |
| Time (and ease) of analyzing | Moderate | Moderate to long (and more difficult) |
| Ability to detect rare transcripts | Relatively low | Moderate to high |

generating genome-wide transcriptomic data, there are two main technology platforms: DNA microarrays or high-throughput sequencing.

Once the cost of high-throughput sequencing became more affordable and the centers offering this type of sequencing services became more available, i.e., since about 2012, the vast majority of transcriptome profiling has been done using high-throughput DNA sequencing. While there are numerous advantages and disadvantages for each approach, one of the main disadvantages of high-throughput sequencing was the relatively high cost, but it has come down in recent years. A comparison of some of the key differences between the DNA microarray and RNA sequencing platforms is shown in Table 12.1.

Before discussing the two main experimental platforms for obtaining the transcriptome of a given sample, both approaches require obtaining high-quality RNA, a topic that will be addressed here first.

## *RNA Type and Isolation*

A very important factor contributing to a successful transcriptome analysis experiment is to begin with high-quality RNA. For DNA microarray experiments, either total RNA or mRNA may be used, but for high-throughput sequencing, either mRNA- or rRNA-depleted RNA must be used; otherwise a majority of the sequence reads will be rRNA.

### Total RNA

Total RNA may be isolated using a variety of methods, but those most commonly employed rely on homogenization of the cells, tissue, or whole organisms in the presence of a highly denaturing guanidine isothiocyanate (GITC)-containing buffer, which immediately inactivates RNases and ensures isolation of intact RNA. Such buffers can be combined with phenol:chloroform for a single-solution approach to the isolation and purification of total RNA as originally described by Chomczynski and Sacchi.[11] Several commercially available reagents such as Invitrogen/ThermoFisher TRIzol are examples of this approach. Other commercial available RNA isolation kits combine GITC-containing buffers with purification columns or membranes, such as Zymo Direct-zol, Qiagen's RNeasy, and Ambion/ThermoFisher Trizol Plus purification kits. Our experience has led us to rely on TRIzol to extract RNA from most samples. Whereas different purification columns are required depending on the mass or type of sample tissue, the TRIzol method is robust and scalable, and samples may also be homogenized directly in the reagent, which eliminates the need to use additional RNase-protective buffers that may interfere with the extraction. In some instances, the use of total RNA may lead to poor or no reverse transcription of the template RNA and/or to poor incorporation of modified nucleotides and/or fluorescent background on the hybridized array due to contaminants present in the RNA preparation. In such cases, an additional cleanup of the total RNA using column-based purification kits can be helpful (e.g., Zymo RNA clean and concentrator or Ambion/ThermoFisher Megaclear).

There are instances where the collection of tissue may be time consuming (e.g., the dissection of specific parts of the brain) and care must be taken to reduce RNA degradation prior to the addition of GITC-containing buffer. In these cases, it is advisable to place the tissue in RNAlater (Ambion/ThermoFisher) for either a short term (4°C) or long term (−20°C) prior to RNA isolation.

### mRNA- and rRNA-Subtracted RNA

As mentioned, total RNA is generally not suitable for high-throughput sequencing, so one must either purify mRNA or mRNA plus long noncoding (nc) RNA prior to the creation of DNA libraries for sequencing.

**mRNA Isolation** There are a few different ways to purify mRNA (i.e., polyA RNA), but the ones that employ oligo dT linked to paramagnetic beads are among the most efficient and popular. In all cases, total RNA is purified first (see earlier), and it is then mixed with the beads to bind to the polyA RNA. Two commercially available beads are Magnetic mRNA isolation beads (NEB catalog number (cat. no.). S1550S) and the Dynabeads mRNA purification kit (Invitrogen/ThermoFisher cat. no. 61006). To use these beads, one must have magnetic stands that can hold 1.5-mL microfuge tubes (e.g., Promega/ThermoFisher cat. no. PR-Z5342. NEB cat. no. S1509S, or Qiagen cat. no. 36912) or in the case of processing a large number of samples, 96-well magnetic plates (e.g., Alpaqua cat. no. 96S Super Magnet Plate).

**rRNA Subtracted RNA** In cases where it also is desirable to have long ncRNA data in addition to mRNA data, then one can prepare RNA samples that are depleted of ribosomal RNA. Ribosomal RNA is removed using beads that have single-stranded oligonucleotide sequences that can hybridize to rRNA, so the RNA that does not bind to the beads is mostly mRNA and large ncRNA. The preferred beads for preparing rRNA-depleted RNA are the Ribo-Zero rRNA Removal Kit (Illumina cat. no. MRZH116) or the Ribo-Zero Gold kit (Illumina cat. no. MRZG126), which also removes mitochondrial RNA. While these beads have antisense oligonucleotides specific to mammalian rRNA sequences, they also work quite well for removing most animal rRNAs.

### RNA Quantification and Quality Control

It is essential that the RNA be quantified and checked for overall integrity (i.e., absence of degradation). RNA can be quantified and checked for protein contamination using a UV spectrophotometer (e.g., ThermoFisher Nanodrop). For a more accurate quantification, especially when there is only a very small amount of RNA, fluorometric dyes can be used (i.e., Quant-iT RiboGreen dye, Molecular Probes/ThermoFisher cat. nos. R11490 or R11491) in conjunction with a fluorometric plate reader or a single-sample fluorometer (e.g., ThermoFisher Quant-iT).

The integrity of total RNA samples can be assayed using devices such as the Agilent Bioanalyzer or TapeStation. These are microfluidic capillary gel electrophoresis instruments that can quickly separate RNA (or DNA or protein) samples, and the rRNA peaks can be observed. Degraded RNA samples have rRNA peaks that are not of the expected amplitude and/or show spreading. RNA integrity can also be examined using nondenaturing RNA gels, but this method requires higher amounts of RNA.

More details on RNA quantification and quality control can be found in Appendix A.

### RNA Storage and Shipping

RNA samples are typically snap frozen in liquid nitrogen and stored in a −80°C freezer until they are processed further. When using the services of genomic centers, samples are usually shipped on dry ice. Sometimes, this is not practical, especially when the samples have to be shipped overseas. In those cases, products like RNAstable (Biomatrica cat. no. 93221-001) allow RNA to be lyophilized onto a stable matrix, shipped at ambient temperature, and then to be reconstituted with nuclease-free water to be processed further.

After one has obtained high-quality RNA, the processing of those RNAs will largely depend upon which experimental platform one will be using to obtain the transcriptome data. Therefore, the remaining sections are divided into the two platforms addressed in this chapter: DNA microarrays and high-throughput sequencing.

## DNA Microarrays

With the advent of sequencing the entire genomes of humans and other model organisms, a variety of high-throughput approaches were developed in the late 1990s and early 21st century to study entire transcriptomes. One of these is DNA microarray technology. Early versions of microarray production involved the spotting of dsDNA fragments or ssDNA oligonucleotides onto specially coated glass microscope slides, and these were often produced by academic labs and genomic centers using robotic printers.[12,13] Several commercial microarrays were developed, including but not limited to Affymetrix (GeneChips), Illumina (GEX bead arrays), Agilent arrays, and Nimblegen/Roche arrays. The Nimblegen platform no longer exists, but the other three do. Each of the still-existing platforms attach or synthesize thousands of single-stranded DNA oligonucleotides with different sequences to a substrate. Briefly, Affymetrix arrays synthesize 20 mer oligonucleotides using photoactivatable chemistry on a transparent quartz surface, Illumina uses more traditional synthesis of 50—70 mers onto silica beads, and Agilent synthesizes 60 mers at specific locations on a coated microscope slide using inkjet technology (see Bumgarner[14] for more information). Each of these commercial array types have their advantages and disadvantages that can vary depending on the specific application for which the arrays are used (e.g., single nucleotide polymorphism analysis vs. transcriptome analysis). Both the Affymetrix and Agilent have predesigned arrays for many model organisms whose genomes have been sequenced, while Illumina Bead transcriptome arrays are mainly limited to human, mouse, and rat. One distinct advantage of the Agilent arrays is that it is possible and economical to make custom arrays.

### *cDNA Synthesis*

Once high-quality RNA is obtained, one typically makes fluorescently labeled, double-stranded cDNA that can be hybridized to the microarray and quantified. Some manufacturers' experimental workflow (e.g., Affymetrix, Agilent) makes labeled, amplified RNA instead of cDNA. However, the production of cDNA does provides added advantages, including that the final labeled products are likely to be more stable, unused cDNA can be labeled again if the initial labeling reaction failed or hybridizations did not work, and unused cDNAs can be used for verification experiments such as qRT PCR.

A detailed protocol for cDNA preparation is provided in Appendix A.

### *Labeling cDNAs*

cDNAs can be fluorescently labeled during cDNA synthesis using direct (e.g., incorporation of Cy dye-labeled nucleotides[15]) or indirect protocols (e.g., incorporation of amino allyl dUTP followed by conjugation to Alexa fluor reactive dyes).[16] Alternatively, cDNAs can be labeled afterward through the use of Cy dye-labeled random nonamer primers and the Klenow fragment of DNA polymerase.[17] This latter approach has the added advantages of no inhibition or "dye bias" in cDNA synthesis, cDNAs becoming amplified during the labeling protocol by DNA polymerase, and that the initial unused cDNAs are unlabeled and therefore can be used again for other experiments if needed.

We have used this latter direct labeling of cDNAs with Cy dye-labeled nonamers successfully for Nimblegen array-based transcriptome analyses[6,18–21] as well as RNA immunoprecipitation analyses.[22–24]

A detailed protocol for labeling ds cDNA with Cy dye-labeled random nonamers is provided in Appendix B.

### Array Hybridization and Scanning

Each commercial array manufacturer typically has specialized equipment and protocols for carrying out the hybridization of the labeled samples to each microarray. Most protocols require the hybridization of a single labeled cDNA (or aRNA) to the array, with or without labeled control RNA samples. Once the array has been hybridized, washed, scanned, and the signals quantified, replicate samples of sample type can be used to compare to replicate samples of a different sample type (e.g., mutant vs. control). We have found that for long oligonucleotide-based arrays (e.g., Nimblegen and Agilent), samples can be labeled with either Cy3- or Cy5-labeled nonamers and two samples with different Cy labels mixed and cohybridized to the same array at the same time. During array scanning, a quantification of each Cy dye signal can be obtained. Thus, one can use half the number of arrays to perform an experiment with no noticeable change in the outcome of the experimental data.

While there have been a few different microarray scanner manufacturers over the years, since 2005 the scanners that have been primarily made are Affymetrix/Thermo-Fisher GeneChip scanners, Illumina iScan bead array scanners, Agilent SureScan scanners, and Axon/Molecular Devices GenePix scanners. The latter two can be used to scan almost any microarray that is of $1 \times 3''$ standard microscope slide size, while the GeneChip and bead array scanners are specific for a specified type of array. Each array manufacturer provides detailed protocols on how their arrays should be scanned.

Scanned arrays typically generate large image (e.g., TIFF) files from which the fluorescent signals from specific locations on the array can be quantified using specialized software. Quantification software is made by each scanner manufacturer (e.g., GenePix Pro for GenepPix scanners) and/or array manufacturer (Nimblescan for Nimblegen arrays), but free quantification software (e.g., TIGR Spotfinder) is also available.

### Analysis of Microarray Data

The quantified image data are exported in a tabular format from one of the aforementioned software programs and are ready to be analyzed. There are many things that need to be considered during analysis: background subtraction, filtering of low signals, flagging of aberrant features, and data normalization. These topics are discussed in detail by Neal and Westwood.[8] A number of software packages have the ability to perform many of the aforementioned processes automatically using "default" parameters, but it is always advisable to inspect image files prior to quantification and further analysis, in case further steps need to be taken during analysis.

There are also several software packages that enable one to perform postscanning normalization as well as statistical analyses and the clustering of genes with similar expression patterns (i.e., visualization tools). These include commercial software such as GeneSpring (Agilent), Spotfire (TIBCO), Acuity (Axon/Molecular Devices), and ArrayStar (DNASTAR), as well as open-source software such as TIGR MIDAS and MeV (www.tm4.org), BASE (http://base.thep.lu.se/), and Bioconductor and R (www.bioconductor.org). Further information on the microarray analysis, and in particular statistical analysis of microarray data, can be found in Neal and Westwood.[8]

Additional information on software that can aid in the visualization of microarray data as well as help interpret the functional meaning of differentially expressed genes (e.g., gene ontology [GO], pathway and network analysis) can be found in Visualization and Pathway Tools section.

## High-Throughput Sequencing

High-throughput sequencing, also known as massively parallel sequencing and next-generation sequencing, as the name implies, sequences tens of thousands to billions of DNA molecules at the same time. The first commercially available high-throughput sequencers appeared in about 2005, and they were made by 454 (later Roche), based on a pyrosequencing technology, and Solexa (now Illumina) based on sequencing by synthesis technology. There are a few other companies that also make high-throughput sequencers, but the majority of these were never produced in high numbers because they were quite expensive, and they are mainly used for their unique characteristics (e.g., long sequence read by Pacific Biosciences sequencers that facilitate assembly of entire genomes). More recently, companies such as Ion Torrent (Thermo-Fisher) and Nanopore make sequencers that are increasingly popular because they are relatively inexpensive to buy and operate. However, the vast majority of high-throughput DNA sequencing that is currently being done worldwide is performed on Illumina sequencers, so the subtopics in this section are focused on data generated by these sequencers.

In previous sections, we discussed RNA preparation. The guidelines provided in those sections apply for preparing samples for high-throughput sequencing as well. For Illumina transcriptome sequencing, mRNA (or rRNA-depleted RNA) is converted to cDNA libraries of a specific nucleotide size range that will be sequenced. For many researchers, the DNA library preparation (see later) and sequencing will be done by a genomics center, so all they will need to do is to prepare high-quality RNA in sufficient quantity and deliver it to the center.

### DNA Library Preparation

Most DNA libraries for Illumina sequencing are prepared from kits that provide almost all of the necessary reagents to make the libraries. Kits are made by a number of manufacturers, including Illumina, New England Biolabs, and Kapa (Roche). For those researchers who will be making the libraries themselves, one can either convert their mRNA to cDNA first (e.g., using a modified version of the protocol in Appendix A), which can be used to make the library, or use a kit where the starting material is mRNA. In the past few years, most of the kits that make libraries from mRNA are "stranded" kits that preserve the identity of the sense strand. Examples of such kits are the Illumina TruSeq stranded mRNA kit, the NEBNext Ultra II Directional RNA Library Prep Kit for Illumina kit, and KAPA (Roche) Stranded mRNA-Seq kit. KAPA also makes HyperPlus kits for making libraries from DNA (or cDNA) in a much shorter period of time than traditional kits.

If a researcher is planning on making libraries using a kit, one item that is generally not provided by a kit is the purification columns or paramagnetic beads required for binding DNA products at various steps of the library preparation. Most kit manufacturers recommend using AMPure beads (Agencort, Beckman Coulter) for DNA cleanup. Significant cost savings can be achieved using Sera-Mag Magnetic Speed-beads (SPRI beads) (GE Life Sciences cat. no. 24152105050250) and by following the protocols outlined by Rohland and Reich[25] and DeAngelis and coworkers.[26]

For researchers who are planning to make large numbers of libraries, significant cost savings can also be achieved using protocols that employ individually purchased enzymes and reagents as opposed to kits (e.g., Wang et al.[25,27]).

### RNAseq Analysis

The analysis of RNAseq data is an extensive topic on its own, and only a brief overview will be provided here. A detailed review of many of the current pipelines and publicly available software for RNAseq data analysis can be found in Conesa et al.[9] Some online resources for information and/or tutorials on publicly available software, as well as sources for where the software can be run, include the Galaxy Project (https://galaxyproject.org/tutorials/rb_rnaseq/),[28] Bioconductor RNAseq workflow (https://www.bioconductor.org/help/workflows/rnaseqGene/),[29] and GenePattern (http://software.broadinstitute.org/cancer/software/genepattern/rna-seq-analysis).[30]

There are also several commercial software packages that generally rely on publicly available programs for most stages of the analysis but provide an easier to use interface and superior visualization tools. Several also have tools for things like GO term, pathway, and network analysis to get more insight when one obtains lists of differentially expressed genes. Most of these software packages are subscription based, and they can be expensive. Some of these include A.I.R. (https://transcriptomics.cloud/), ArrayStar (https://www.dnastar.com/t-sub-products-genomics-arraystar.aspx), Avadis (http://www.strand-ngs.

com/), BaseSpace apps for RNAseq analysis (Illumina, https://www.illumina.com/informatics/sequencing-data-analysis/rna.html), CLC Workbench (https://www.qiagenbioinformatics.com/solutions/gene-expression/), and Partek Genomics Suite and Pathway (http://www.partek.com/).

### Quality Control Checkpoints

Poor RNA sample quality or library preparation can lead to the presence of a large number of adaptor sequences, overrepresented k-mers, duplicated reads, PCR artifacts, and other sequencing errors in the data. Samples that have a large number of these errors and artifacts should be excluded from further analysis. Programs such as FastQC (Andrews, http://www.bioinformatics.babraham.ac.uk/projects/fastqc/) can perform analysis of the raw sequence reads.

In addition to quality control of raw reads, which is typically done, other quality control analysis can also be performed after read alignment and quantification of transcript numbers with the goal to again filter out samples that have large numbers of anomalies.[9] Once individual samples pass quality control, they can be compared to each other for overall sample reproducibility. Reproducibility among replicates should be generally high (e.g., Spearman $R^2 > 0.9$), but the cutoff for acceptable reproduction will largely depend upon the heterogeneity of the initial biologic samples themselves. As mentioned in the section on Experimental Design, samples with poorer reproducibility will require more replicates to obtain statistically significant differences in gene expression.

### Transcript Identification

The next step in the analysis pipeline is to align the raw sequence reads. This is greatly facilitated when there is a reference genome for the organism being studied. Two commonly used programs for aligning transcript sequences are TopHat[31,32] and STAR.[33]

### Transcript Quantification

Once the sequence reads have been aligned, they then need to be quantified to estimate gene and transcript expression. RPKM (reads per kilobase of exon model per million reads), FPKM (fragments per kilobase of exon model per million mapped reads), and TPM (transcripts per million) are the most commonly reported formats for reporting RNAseq gene expression. One of the most highly used programs for quantifying transcripts is Cufflinks,[34] which estimates transcript expression from mappers such as TopHat using an expectation-maximization approach. This approach takes into account biases such as the nonuniform read distribution along a gene.

### Differential Gene Expression Analysis

Differential gene expression (DGE) analysis requires that gene expression values be compared between sample group types. RPKM and FPKM normalize the most important factor for comparing samples-sequencing depth. Normalizing approaches that are based simply on total or effective counts tend to perform poorly when samples have highly and/or differentially expressed genes, but programs like TMM[35] and DESeq[36] can minimize the effects of such "abnormally" expressed genes. Different programs that perform DGE analysis use varied approaches for normalizing data, so they perform differently depending on the type of experimental data and number of sample replicates. LimmaVoom[37] performs well under many circumstances.[38–40] Two of the most commonly used programs, DESeq and edgeR, perform similarly in ranking differentially expressed genes, but DESeq has been shown to have relatively conservative false discovery rates (FDRs), whereas edgeR FDRs are relatively liberal.[38,41]

### *Visualization and Pathway Tools*

Visualization tools can be broken into different categories with two of the main ones being viewing normalized transcript abundance along annotated genomes and viewing DGE patterns in a graphic format (e.g., heat maps). Transcript abundance can be viewed using genome browsers such as the UCSC browser,[42] integrative genomics viewer,[43] and RNAseqViewer.[44] Graphic representations of DGEs can be achieved with programs like MeV and QuickRNAseq.[45] As previously mentioned, many of the commercially available software packages have excellent tools for visualizing DGEs.

Once one has obtained lists of differentially expressed genes, the next step usually involves visualizing the results in way that is easier to understand (e.g., heat maps), as well as trying to interpret the biologic implications of the DGEs through GO function, biochemical or molecular pathway, and/or network analysis. Again, there are a variety of publicly available software applications for doing this. This is usually accomplished by comparing a specific list(s) of DGEs, based on statistical criteria, to the rest of the genome for functions that are overrepresented. Alternatively, the entire list of differentially expressed genes can be used, and a gene set enrichment analysis may be performed. Many of these tools were originally developed for microarray and other large genomic (e.g., proteomic) datasets. Tools that specifically take into account some of the biases that are inherent to RNAseq data include GOseq,[46] Gene Set Variation Analysis (GSVA),[47] and SeqGSEA.[48]

## SUMMARY

In this chapter, some key aspects of gene expression experimental design and RNA preparation were addressed. Comparisons outlining some of the differences between obtaining transcriptomic data using DNA microarrays versus high-throughput sequencing were made, and workflows for both of these technology platforms were presented. Since most current gene expression experiments involve obtaining data using high-throughput sequencing, often the researcher needs only to prepare the RNA samples and provide them to a genomic service center. Therefore, the success of a gene expression study will largely depend upon the original design of the experiment, including the source of tissue or even specific cells from which the RNA was obtained and having enough replicate samples to reach statistically significant results. Other factors that will be key to a successful study include isolation of high-quality RNA, having a robust and appropriate analysis pipeline, and the use of software tools to identify the underlying biologic and molecular processes and pathways that are being revealed by the changes in gene expression being observed.

## ACKNOWLEDGMENTS

I would like to thank past and present members of the Canadian *Drosophila* Microarray Centre (CDMC) and Westwood Lab including Anna Soltyk, Scott Neal, Sarah Gonsalves, Kaigo Mo and Zak Razak for their contributions to various gene expression studies. I would also like to thank Hang Noh Lee and Brian Oliver (NIDDK, NIH, Bethesda, MD, USA) for their insights and advice on RNAseq. The gene expression studies performed in or in collaboration with the Westwood Lab was supported by an NSERC Discovery grant to J.T.W. The CDMC was originally supported by a multiuser maintenance and NET grant from CIHR and an equipment and facility grant from NSERC.

## REFERENCES

1. Robinson GE, Fernald RD, Clayton DF. Genes and social behavior. *Science*. 2008;322(5903):896–900. https://doi.org/10.1126/science.1159277.
2. Zayed A, Robinson GE. Understanding the relationship between brain gene expression and social behavior: lessons from the honey bee. *Annu Rev Genet*. 2012;46:591–615. https://doi.org/10.1146/annurev-genet-110711-155517.
3. Peschel N, Helfrich-Förster C. Setting the clock—by nature: circadian rhythm in the fruit fly *Drosophila melanogaster*. *FEBS Lett*. 2011;585(10):1435–1442. https://doi.org/10.1016/j.febslet.2011.02.028.
4. McDonald MJ, Rosbash M. Microarray analysis and organization of circadian gene expression in *Drosophila*. *Cell*. 2001;107(5):567–578.
5. Riedl CAL, Neal SJ, Robichon A, Westwood JT, Sokolowski MB. *Drosophila* soluble guanylyl cyclase mutants exhibit increased foraging locomotion: behavioral and genomic investigations. *Behav Genet*. 2005;35(3):231–244. https://doi.org/10.1007/s10519-005-3216-1.
6. Pan Y, Kaiguo M, Razak Z, Westwood JT, Gerlai R. Chronic alcohol exposure induced gene expression changes in the zebrafish brain. *Behav Brain Res*. 2011;216(1):66–76. https://doi.org/10.1016/j.bbr.2010.07.017.
7. Kramer JM, Kochinke K, Oortveld MAW, et al. Epigenetic regulation of learning and memory by *Drosophila* EHMT/G9a. *PLoS Biol*. 2011;9(1):e1000569. https://doi.org/10.1371/journal.pbio.1000569.
8. Neal SJ, Westwood JT. Optimizing experiment and analysis parameters for spotted microarrays. *Methods Enzymol*. 2006;410:203–221. https://doi.org/10.1016/S0076-6879(06)10010-5.
9. Conesa A, Madrigal P, Tarazona S, et al. A survey of best practices for RNA-seq data analysis. *Genome Biol*. 2016;17(1):13. https://doi.org/10.1186/s13059-016-0881-8.
10. Okaty BW, Sugino K, Nelson SB. Cell type-specific transcriptomics in the brain. *J Neurosci*. 2011;31(19):6939–6943. https://doi.org/10.1523/JNEUROSCI.0626-11.2011.
11. Chomczynski P, Sacchi N. Single-step method of RNA isolation by acid guanidinium thiocyanate-phenol-chloroform extraction. *Anal Biochem*. 1987;162(1):156–159. https://doi.org/10.1016/0003-2697(87)90021-2.
12. Schena M, Shalon D, Davis RW, Brown PO. Quantitative monitoring of gene expression patterns with a complementary DNA microarray. *Science*. 1995;270(5235):467–470.

13. DeRisi J, Penland L, Brown PO, et al. Use of a cDNA microarray to analyse gene expression patterns in human cancer. *Nat Genet.* 1996;14(4):457–460. https://doi.org/10.1038/ng1296-457.
14. Bumgarner R. Overview of DNA microarrays: types, applications, and their future. *Curr Protoc Mol Biol.* 2013. https://doi.org/10.1002/0471142727.mb2201s101 [Chapter 22: Units 22.1–22.1.11].
15. Neal SJ, Gibson ML, So AK-C, Westwood JT. Construction of a cDNA-based microarray for *Drosophila melanogaster*: a comparison of gene transcription profiles from SL2 and Kc167 cells. *Genome.* 2003;46(5):879–892. https://doi.org/10.1139/g03-056.
16. Semotok JL, Westwood JT, Goldman AL, Cooperstock RL, Lipshitz HD. Measuring mRNA stability during early *Drosophila* embryogenesis. *Methods Enzymol.* 2008;448:299–334. https://doi.org/10.1016/S0076-6879(08)02616-5.
17. Ouellet M, Adams PD, Keasling JD, Mukhopadhyay A. A rapid and inexpensive labeling method for microarray gene expression analysis. *BMC Biotechnol.* 2009;9(1):97. https://doi.org/10.1186/1472-6750-9-97.
18. Gonsalves SE, Moses AM, Razak Z, Robert F, Westwood JT. Whole-genome analysis reveals that active heat shock factor binding sites are mostly associated with non-heat shock genes in *Drosophila melanogaster*. *PLoS One.* 2011;6(1):e15934.
19. Le Masson F, Razak Z, Kaigo M, et al. Identification of heat shock factor 1 molecular and cellular targets during embryonic and adult female meiosis. *Mol Cell Biol.* 2011;31(16):3410–3423. https://doi.org/10.1128/MCB.05237-11.
20. Zhang J, Marshall KE, Westwood JT, Clark MS, Sinclair BJ. Divergent transcriptomic responses to repeated and single cold exposures in *Drosophila melanogaster*. *J Exp Biol.* 2011;214(Pt 23): 4021–4029. https://doi.org/10.1242/jeb.059535.
21. Siddiqui NU, Li X, Luo H, et al. Genome-wide analysis of the maternal-to-zygotic transition in *Drosophila* primordial germ cells. *Genome Biol.* 2012;13(2):R11. https://doi.org/10.1186/gb-2012-13-2-r11.
22. Laver JD, Li X, Ancevicius K, et al. Genome-wide analysis of Staufen-associated mRNAs identifies secondary structures that confer target specificity. *Nucleic Acids Res.* 2013;41(20):9438–9460. https://doi.org/10.1093/nar/gkt702.
23. Laver JD, Li X, Ray D, et al. Brain tumor is a sequence-specific RNA-binding protein that directs maternal mRNA clearance during the *Drosophila* maternal-to-zygotic transition. *Genome Biol.* 2015; 16(1). https://doi.org/10.1186/s13059-015-0659-4.
24. Chen L, Dumelie JG, Li X, et al. Global regulation of mRNA translation and stability in the early *Drosophila* embryo by the Smaug RNA-binding protein. *Genome Biol.* 2014;15(1). https://doi.org/10.1186/gb-2014-15-1-r4.
25. Rohland N, Reich D. Cost-effective, high-throughput DNA sequencing libraries for multiplexed target capture. *Genome Res.* 2012;22(5):939–946. https://doi.org/10.1101/gr.128124.111.
26. DeAngelis MM, Wang DG, Hawkins TL. Solid-phase reversible immobilization for the isolation of PCR products. *Nucleic Acids Res.* 1995;23(22):4742–4743.
27. Wang L, Si Y, Dedow LK, Shao Y, Liu P, Brutnell TP. A low-cost library construction protocol and data analysis pipeline for Illumina-based strand-specific multiplex RNA-seq. Hudson ME, ed. *PLoS One.* 2011;6(10):e26426. https://doi.org/10.1371/journal.pone.0026426.
28. Afgan E, Baker D, van den Beek M, et al. The Galaxy platform for accessible, reproducible and collaborative biomedical analyses: 2016 update. *Nucleic Acids Res.* 2016;44(W1):W3–W10. https://doi.org/10.1093/nar/gkw343.
29. Love MI, Anders S, Kim V, Huber W. RNA-Seq workflow: gene-level exploratory analysis and differential expression. *F1000 Res.* 2016;4:1070. https://doi.org/10.12688/f1000research.7035.2.
30. Reich M, Liefeld T, Gould J, Lerner J, Tamayo P, Mesirov JP. GenePattern 2.0. *Nat Genet.* 2006;38(5): 500–501. https://doi.org/10.1038/ng0506-500.
31. Trapnell C, Pachter L, Salzberg SL. TopHat: discovering splice junctions with RNA-Seq. *Bioinformatics.* 2009;25(9):1105–1111.
32. Kim D, Pertea G, Trapnell C, Pimentel H, Kelley R, Salzberg SL. TopHat2: accurate alignment of transcriptomes in the presence of insertions, deletions and gene fusions. *Genome Biol.* 2013;14(4): R36. https://doi.org/10.1186/gb-2013-14-4-r36.

33. Dobin A, Davis CA, Schlesinger F, et al. STAR: ultrafast universal RNA-seq aligner. *Bioinformatics*. 2013;29(1):15–21. https://doi.org/10.1093/bioinformatics/bts635.
34. Trapnell C, Williams BA, Pertea G, et al. Transcript assembly and quantification by RNA-Seq reveals unannotated transcripts and isoform switching during cell differentiation. *Nat Biotechnol*. 2010;28(5):511–515.
35. Robinson MD, Oshlack A. A scaling normalization method for differential expression analysis of RNA-seq data. *Genome Biol*. 2010;11(3):R25. https://doi.org/10.1186/gb-2010-11-3-r25.
36. Anders S, Huber W. Differential expression analysis for sequence count data. *Genome Biol*. 2010;11(10): R106. https://doi.org/10.1186/gb-2010-11-10-r106.
37. Law CW, Chen Y, Shi W, Smyth GK. Voom: precision weights unlock linear model analysis tools for RNA-seq read counts. *Genome Biol*. 2014;15(2):R29. https://doi.org/10.1186/gb-2014-15-2-r29.
38. Rapaport F, Khanin R, Liang Y, et al. Comprehensive evaluation of differential gene expression analysis methods for RNA-seq data. *Genome Biol*. 2013;14(9):R95. https://doi.org/10.1186/gb-2013-14-9-r95.
39. Soneson C, Delorenzi M. A comparison of methods for differential expression analysis of RNA-seq data. *BMC Bioinform*. 2013;14:91. https://doi.org/10.1186/1471-2105-14-91.
40. Seyednasrollah F, Laiho A, Elo LL. Comparison of software packages for detecting differential expression in RNA-seq studies. *Brief Bioinform*. 2015;16(1):59–70. https://doi.org/10.1093/bib/bbt086.
41. Robles JA, Qureshi SE, Stephen SJ, Wilson SR, Burden CJ, Taylor JM. Efficient experimental design and analysis strategies for the detection of differential expression using RNA-sequencing. *BMC Genom*. 2012;13:484. https://doi.org/10.1186/1471-2164-13-484.
42. Kent WJ, Sugnet CW, Furey TS, et al. The human genome browser at UCSC. *Genome Res*. 2002; 12(6):996–1006. https://doi.org/10.1101/gr.229102.
43. Thorvaldsdottir H, Robinson JT, Mesirov JP. Integrative genomics viewer (IGV): high-performance genomics data visualization and exploration. *Brief Bioinform*. 2013;14(2):178–192. https://doi.org/10.1093/bib/bbs017.
44. Roge X, Zhang X. RNAseqViewer: visualization tool for RNA-seq data. *Bioinformatics*. 2014;30(6): 891–892. https://doi.org/10.1093/bioinformatics/btt649.
45. Zhao S, Xi L, Quan J, et al. Quick RNASeq lifts large-scale RNA-seq data analyses to the next level of automation and interactive visualization. *BMC Genom*. 2016;17:39. https://doi.org/10.1186/s12864-015-2356-9.
46. Young MD, Wakefield MJ, Smyth GK, Oshlack A. Gene ontology analysis for RNA-seq: accounting for selection bias. *Genome Biol*. 2010;11(2):R14. https://doi.org/10.1186/gb-2010-11-2-r14.
47. Hanzelmann S, Castelo R, Guinney J. GSVA: Gene set variation analysis for microarray and RNA-seq data. *BMC Bioinform*. 2013;14:7. https://doi.org/10.1186/1471-2105-14-7.
48. Wang X, Cairns MJ. Gene set enrichment analysis of RNA-Seq data: integrating differential expression and splicing. *BMC Bioinform*. 2013;14(suppl 5):S16. https://doi.org/10.1186/1471-2105-14-S5-S16.

## APPENDIX A: PREPARATION OF cDNAs

Adapted from Roche Nimblegen Gene Expression Analysis Guide (version 5.1) and Ouellet and coworkers.[17] Note that the major modification is to make half-size reactions throughout the protocols.

### Materials

- Anchored oligo dT Primer (500 ng/μL, Ambion/ThermoFisher cat. no. AB1247)
- Random Hexamers (100 μM (0.2 μg/μL), Invitrogen/ThermoFisher cat. no. SO142)
- RNase A Solution (4 mg/mL Promega cat. no. A7973)
- SuperScript II Double-Stranded cDNA Synthesis Kit (Invitrogen/ThermoFisher cat. no. 11917-010)

### *Step 1. Spectrophotometric Quality Control of RNA*

Prior to synthesizing cDNA, verify that the RNA samples are of sufficient purity, typically with a Nanodrop spectrophotometer (ThermoFisher).

1. Quantitate each RNA sample according to the following formula:

   RNA concentration (μg/mL) = A260 × 40 × dilution factor.

   RNA samples must have a concentration ≥ 1.0 μg/μL.
2. Verify all samples meet the following requirements: A260/A280 ≥ 1.8; A260/A230 ≥ 1.8.

### *Step 2. Bioanalyzer/Gel QC of RNA*

Verify RNA samples are of sufficient molecular weight. This can be done using an Agilent 2100 Bioanalyzer (preferred especially for smaller samples) or using a denaturing agarose gel.

1. Transfer 250 ng total RNA or 250 ng polyA + RNA to a sterile microcentrifuge tube. Store the remainder of your sample on ice or at −80°C.
2. Analyze samples using the Agilent Bioanalyzer and RNA 6000 Nano Kit.
3. Compare the Bioanalyzer traces to published electropherogram images in the Agilent Bioanalyzer manual or to samples known to be intact. Degraded samples appear as significantly lower intensity traces with the main peak area shifted to the left and typically exhibit much more noise in the trace.

   Samples exhibiting degradation should not be used for cDNA synthesis because there is a high risk for obtaining poor results.

### *Step 3. First Strand cDNA Synthesis*

Use the Invitrogen (ThermoFisher) SuperScript II Double-Stranded cDNA Synthesis Kit to synthesize double-stranded cDNA.

1. Thaw and maintain the following components on ice. Combine components in a 0.2-mL tube on ice according to the following tables.

   Note: prepare a primer mix solution by mixing one part random hexamers to three parts anchored oligo dT primer.

| | Total RNA | PolyA + RNA (mRNA) amount |
|---|---|---|
| RNA amount | 5 μg | 0.5 μg |
| Oligo dT/random primer mix | 1 μL | 1 μL |
| DEPC water | To volume | To volume |
| Total | 5.5 μL | 5.5 μL |

2. Heat sample(s) to 70°C for 10 min in a thermocycler. Briefly spin tubes in a microcentrifuge and place them on ice for 5 min.

3. Add the following to each sample tube. (You can use a master mix when preparing multiple samples.)

| Component | Volume (μL) |
|---|---|
| Reaction volume from step 2 | 5.5 |
| 5X first strand buffer | 2 |
| 0.1 M DTT | 1 |
| 5 mM dNTP Mix | 1 |
| Total | 9.5 |

4. Mix gently (avoid vortexing). Briefly spin the tube(s) in a microcentrifuge.
5. Place sample(s) in a thermocycler set at +42°C for 2 min.
6. Add 1 μL of SuperScript II RT and mix gently (avoid vortexing).
7. Incubate the sample(s) at +42°C for 60 min.
8. STOP POINT: Briefly spin the tube(s) in a microcentrifuge. Place the sample(s) on ice until the second strand synthesis. You can store the sample(s) overnight at −20°C.

### *Step 4. Second Strand cDNA Synthesis*

1. Add the following components to the first strand reaction(s) in the indicated order; you can use a master mix. Keep tube(s) on ice or in a PCR tube chiller rack.

| Component | Volume (μL) |
|---|---|
| Reaction volume from Step 3.8 | 10.5 |
| DEPC water | 45 |
| 5X second strand buffer | 15 |
| 10 mM dNTP mix | 1.5 |
| 10 U/μL DNA ligase | 0.5 |
| 10 U/μL DNA polymerase I | 2 |
| 2 U/μL RNase H | 0.5 |
| Total | 75 |

2. Mix gently (avoid vortexing). Briefly spin the tube(s) in a microcentrifuge. Incubate at +16°C for 2 h.
3. Add 1 μL of 5 U/μL T4 DNA polymerase to each reaction. Incubate at +16°C for an additional 5 min. Do not allow the reaction temperature to exceed +16°C during this step.
4. STOP POINT: Place the sample(s) on ice or in a PCR tube chiller rack, and add 5 μL of 0.5 M EDTA. You can store samples overnight at −15 to −25°C.

### *Step 5. RNase A Cleanup*

1. Add 1 μL of 2 mg/mL RNase A solution to the tubes from Step 4.4. Use caution when working with RNase A. Change gloves after use. Use RNaseZap (Thermo-Fisher) to clean work area surfaces.

2. Mix gently (avoid vortexing). Briefly spin the tube(s) in a microcentrifuge.
3. Incubate sample(s) at +37°C for 10 min.
4. During incubation, centrifuge Phase Lock tube(s) (ThermoFisher) at 12,000x g for 2 min. Label one Phase Lock tube and two 1.5-mL centrifuge tubes for each sample with sample names.
5. Add 82 μL of phenol:chloroform:isoamyl alcohol (25:24:1) to one set of 1.5-mL centrifuge tubes.
6. Transfer the sample(s) to the tube(s) containing phenol:chloroform:isoamyl alcohol. Vortex well.
7. Transfer samples with the phenol:chloroform:isoamyl alcohol to Phase Lock tubes.
8. Centrifuge at 12,000x g for 5 min.
9. Transfer the upper, aqueous layer to a clean, labeled 1.5-mL tube.

### Step 6. cDNA Precipitation

1. Add 8 μL (0.1 volume of Step 5.9) of 7.5 M ammonium acetate to the samples. Mix by repeated inversion. Briefly spin the tube(s) in a microcentrifuge.
2. Add 3.5 μL of 5 mg/mL glycogen to the samples. Mix by repeated inversion. Briefly spin the tube(s) in a microcentrifuge.
3. Add 163 μL (2 volumes of Step 5.9) of ice-cold absolute ethanol to the samples. Mix by repeated inversion.
4. Centrifuge at 12,000x g for 20 min.
5. Remove supernatant. Take care not to disturb the pellet.
6. Add 250 μL of ice-cold 80% ethanol (v/v). Mix by repeated inversion.
7. Centrifuge tubes at 12,000x g for 5 min.
8. Remove supernatant. Take care not to disturb the pellet.
9. Repeat steps 6–8.
10. Dry the pellet in a DNA vacuum concentrator.
11. Rehydrate samples with 10 μL of sterile, nuclease-free water.

### Step 7. Spectrophotometric QC of cDNA

1. Quantify each cDNA sample according to the following formula:
   cDNA concentration (μg/mL) = A260 × 50 × dilution factor.
2. Verify that all samples meet the following requirements: concentration ≥100 ng/μL, A260/A280 ≥ 1.8, A260/A230 ≥ 1.8.

### Step 8. Bioanalyzer/Gel QC of cDNA

1. Transfer 250 ng cDNA to a sterile microcentrifuge tube. Store the remainder of your sample on ice or at −15 to −25°C.
2. Analyze the samples using the Agilent Bioanalyzer and RNA 6000 Nano Kit.

3. Compare the Bioanalyzer traces to the traces displayed next. Verify that all samples meet the following requirement for acceptance:
   Median size ≥400 bp when compared to a DNA ladder.
   It looks similar to the examples of good cDNA sample traces displayed next.
   If using an agarose gel, compare the gel images to the Bioanalyzer's electropherogram images.

## APPENDIX B: cDNA LABELING

Adapted from Nimblegen Gene Expression Analysis Guide (version 2). Note that the main modification is to perform half-size reactions.

### Materials

- 5′ Cy3- and Cy5-labeled Random Nonamers (9 mer "Wobble") TriLink Biotechnologies
  (8 O.D. or 50 O.D. sizes)
- Klenow Fragment (3′→5′ exo-) 50 U/μL (NEB cat. no. M0212M)

1. Prepare random 9 mer buffer

| | |
|---|---|
| Nuclease-free water | 8.6 mL |
| 1M tris HCl, pH 7.4 | 1.25 mL |
| 1M MgCl2 | 125 μL |
| β-Mercaptoethanol | 17.5 μL |
| Total | 10 mL |

2. Dilute Cy3- and Cy5-labeled 9 mers to 1 O.D./42 μL random 9 mer buffer. Aliquot to 20 μL individual reaction volumes in 0.2-mL thin-walled PCR tubes and store at −20°C; protected from light.
3. Assemble the following components in separate 0.2-mL thin-walled PCR tubes:

| | Amount |
|---|---|
| cDNA | 0.5 μg |
| Cy3 or Cy5-labeled 9 mers | 20 μL |
| Nuclease-free water | To volume |
| Total | 40 μL |

4. Heat denature samples in a thermocycler at 98°C for 10 min. Quick-chill in an ice-water bath for 10 min.
   Important: quick-chilling after denaturation is critical for high-efficiency labeling.

5. Prepare 10 μL of the following dNTP/Klenow master mix for each sample prepared in Step 4:

   Important: keep all reagents and dNTP/Klenow master mix on ice. Do not vortex after addition of Klenow.

| 5X dNTP/Klenow master mix | Recipe per sample (μL) |
|---|---|
| 10 mM dNTP mix | 5 |
| Nuclease-free water | 4 |
| Klenow (50 U/μL) | 1 |
| Total | 10 |

6. Add 10 μL of the dNTP/Klenow master mix to each of the denatured samples.

| | Volume (μL) |
|---|---|
| Reaction volume from step 4 | 40 |
| 5X dNTP/Klenow master mix | 10 |
| Total | 50 |

7. Mix well by pipetting up and down 10 times. (Important: do not vortex after addition of Klenow.)
8. Quick-spin to force contents to bottom of the tube.
9. Incubate for 2 h at 37°C in a thermocycler protected from light.
10. Stop the reaction by addition of 5 μL 0.5M EDTA to each tube. Total volume = 55 μL.
11. Add 5.8 μL 5M NaCl to each tube. Total volume = 60.8 μL
12. Vortex briefly, spin, and transfer the entire contents to a 1.5-mL tube containing 55 μL isopropanol. Total volume = 115.8 μL
13. Vortex well. Incubate for 10 min at room temperature; protected from light.
14. Centrifuge at 12,000x g for 10 min. Remove supernatant with a pipette. Pellet should be pink for Cy3-labeled samples and light blue for Cy5-labeled samples.
15. Rinse pellet with 500 μL 80% ice-cold ethanol. Dislodge pellet from tube wall.
16. Centrifuge at 12,000x g for 2 min. Remove supernatant with a pipette.
17. Dry contents in a SpeedVac on low heat until dry (~5 min), protected from light.
18. STOP POINT: Proceed to Step 19, or store labeled samples at −20°C, protected from light, until ready to continue experiment.
19. Spin tubes briefly prior to opening. Rehydrate pellets in 12.5 μL nuclease-free water.
20. Vortex for 30 s and quick-spin to collect contents in bottom of the tube. Continue to vortex or let sit at room temperature, protected from light, until the pellet is completely rehydrated, then vortex again and quick-spin.
21. Quantify each sample using the following formula: concentration (μg/mL) = A260 × 50 × dilution factor.

22. The amount of Cy3- or Cy5-labeled cDNA sample required for each hybridization will be dependent on which array type is being used. For example, 4-plex Nimblegen or Agilent arrays would require 2 μg per sample.
23. STOP POINT: Dry contents in a SpeedVac on low heat, protected from light.
    Store samples at −80 or −20°C.

CHAPTER 13

# Genes, Behavior, and Next-Generation Sequencing: The First 10 Years

**Robert Hitzemann[1,2], Priscila Darakjian[1], Denesa Oberbeck[1], Christina Zheng[1], Nikki Walter[1,3], Ovidiu D. Iancu[1], Robert Searles[1], Christina A. Harrington[1], Shannon McWeeney[1,3]**

[1]Oregon Health & Science University, Portland, OR, United States; [2]Veterans Affairs Medical Center, Portland, OR, United States; [3]Oregon National Primate Research Center, Portland, OR, United States

## INTRODUCTION

Mortazavi et al.[1] appear to have been the first to sequence a brain transcriptome (C57BL/6J mouse whole brain); they compared the data with microarray results from the same sample and found good concordance ($r = 0.83$). Although the adoption of RNA-Seq in the brain-behavior context was initially slow, by 2012, 45 publications reported using the technique. We reviewed the RNA-Seq brain-behavior literature, emphasizing the differences and similarities between sequencing the brain transcriptome compared with microarray-based results.[2] Our review was updated early in 2014 as part of a volume on sequencing and the brain.[3] As sequencing the brain transcriptome approaches its 10th year, it is appropriate to revisit the literature, to focus on the progress that has been made and to speculate on what the future will hold. In 2016, there were 181 publications reporting on some aspect of sequencing the brain transcriptome, and 2017 is on pace to produce >200 publications. Of course, there are thousands more RNA-Seq publications, e.g., in cancer research, that have helped refine the technique, provided new analysis tools, and have helped set the parameters for experimental design. Conesa et al.[4] have provided an excellent survey of best practices for RNA-Seq experimental design and data analysis. Importantly, these authors emphasize that no one design or analysis pipeline is best for all experiments. Further, the brain presents special problems, given that for any given behavior or condition, multiple brain regions and cell types are likely to be involved. In addition, the brain transcriptome is highly enriched in alternative transcripts and noncoding RNAs that require deep sequencing for adequate analyses (see, e.g., Ref. 5). Finally, although our emphasis here is on four species (mouse, rat, macaque, and human), we recognize that RNA-Seq in the brain-behavior context is being implemented with a wide range of species, e.g., flies, worms, fish, and bees, each of which may require a special sequencing perspective.

This chapter focuses on the historical context of using RNA-Seq to analyze the brain transcriptome, on examples of current uses, and on some of the emerging technologies

*Molecular-Genetic and Statistical Techniques for Behavioral and Neural Research*
ISBN 978-0-12-804078-2, https://doi.org/10.1016/B978-0-12-804078-2.00013-1

such as single cell sequencing and the sequencing of circular RNAs (circRNAs). We argue that the historical context is important since many of the early problems with RNA-Seq remain, e.g., the oversampling of highly expressed and multiexon genes. In addition, the question of whether RNA-Seq is superior to microarray-based approaches remains unsettled for many applications.

## THE EARLY YEARS: MICROARRAYS TO RNA-SEQ

To appreciate RNA-Seq in the brain-behavior context, one needs to begin with the use of microarrays. Cirelli and Tononi[6] were among the first to report genome-wide brain gene expression profiling associated with a behavioral phenotype; both mRNA differential display and cDNA arrays were used to examine the effects of sleep deprivation on rat prefrontal cortex gene expression. Sandberg et al.[7] used Affymetrix microarrays to detect differences in brain gene expression between two inbred mouse strains (C57BL/6J [B6] and 129SvEv [129; now 129S6/SvEvTac]). Importantly, these authors observed that some differentially expressed (DE) genes were found in chromosomal regions with known behavioral quantitative trait loci (QTLs). For example, *Kcnj9*, which encodes for GIRK3, an inwardly rectifying potassium channel, was DE (higher expression in the 129 strain) and is located on distal chromosome 1 in a region where QTLs had been identified for locomotor activity, alcohol and pentobarbital withdrawal, open-field emotionality, and certain aspects of fear-conditioned behavior (see Ref. 7). This alignment of global brain gene expression data and behavioral QTLs was reported in numerous publications and discussed in multiple symposia and reviews (e.g., Refs. 8–15). This association gained further support as the focus turned to genes whose expression appeared to be regulated by a factor or factors within the behavioral QTL interval. Web tools were developed to facilitate integrating behavioral and brain microarray data (e.g., www.genenetwork.org and http://phenogen.ucdenver.edu/PhenoGen/index.jsp). This integration was successful in detecting several candidate quantitative trait genes for behavioral phenotypes (see, e.g., Refs. 13,14,16–18).

The alignment of DE genes with a behavioral phenotype was further explored using a variety of secondary analyses, including examination of DE gene clustering within known gene ontology categories[19] or association with known protein-protein interaction networks.[20,21] DE genes were also grouped on the basis of common transcription factors and other regulatory elements (e.g., Ref. 22). In addition to DE genes, microarrays also facilitated gene coexpression-based analyses, such as the weighted gene coexpression network analysis (WGCNA).[23,24] Bayesian approaches were also used to infer network structure.[25] The rationale behind these approaches is that coexpressed genes frequently code for interacting proteins. This has led to new insights into protein function(s) and in some cases to the discovery of protein function.[26] Further, some evidence supported the view that coexpression networks had significant translational value (see, e.g., Refs. 26,27). Coexpression analysis was used to analyze differences in functional brain organization between nonhuman primates and humans,[28] regional differences in the functional organization of the human brain,[29] and the molecular pathology of autism.[30] Iancu et al.[31,32] showed that this approach can be used to detect gene modules affected by behavioral selection (haloperidol response) and that the gene modules affected were relatively constant across a wide breadth of genetic diversity.

Despite the successes of the microarray-based approaches, just a few of which are noted before, there was a growing realization that the strategy was not without problems, especially when working with complex crosses of mice and rats. *First*, differences in brain gene expression among genetically unique individuals or lines selected for distinct behavioral traits were generally small; reported differences of 15%–25% were not uncommon. To some extent, these small variations occurred because the hybridization isotherms for oligonucleotide arrays were frequently not linear due to probe saturation.[33] The solution to this problem, running multiple dilutions of the same sample, was and never has been a standard practice.

A *second* problem with oligonucleotide arrays was the effect of single nucleotide polymorphisms (SNPs).[34–38] Rodent oligonucleotide arrays are upon the sequence of the C57BL/6 (B6) mouse and Brown-Norway rat. Walter et al.[37] used next-generation sequencing (NGS) to address the SNP problem, building upon the repeated observation that, when comparing gene expression in the B6 and DBA/2J (D2) inbred mouse strains (or crosses and selected lines formed from these strains) and after masking for known SNPs in the D2 strain, there remained an excess of genes showing higher expression in the B6 strain. Similarly, this was also observed in the case of cis-eQTLs showing higher expression associated with the B6 allele (see Refs. 17,34,36). The two possible explanations for these observations were (1) gene expression was actually higher in the B6 strain or (2) there were many uncharacterized D2 SNPs, which led to decreased binding of D2-derived target on probes containing the SNP locale. Preliminary direct sequencing and quantitative polymerase chain reaction (PCR) data pointed to missing SNPs. NGS was used to analyze a 3-Mbp region of Chr 1 (171.5–174.5 Mbp) that was enriched in a number of behavioral QTLs (see earlier) and transcripts DE between the B6 and D2 strains. B6 and D2 bacterial artificial chromosome (BAC) clones tiled across the region were sequenced using the short-read Illumina IIx and ABI SOLiD 2 platforms. The results obtained (30–100X coverage) illustrated that there were 160% more SNPs in the region than previously reported.[37] The integration of these SNPs to the array mask markedly reduced the disparity in DE genes between the B6 and D2 strains.

The SNP problem for oligonucleotide arrays increases almost exponentially when working with complex crosses such as the recombinant inbred (RI) collaborative cross (CC) mouse strains, the heterogeneous stock-collaborative cross (HS-CC), the closely related diversity outbred (DO) population, and the rat HS/NIH.[31,39–44] These complex crosses allow one to survey a great range of genetic diversity for genomic features associated with the phenotypes of interest. However, while RNA-Seq would seem to be to solution to the SNP effect for such crosses, the RNA-Seq solution is not entirely perfect (see, e.g., Ref. 45).

A *third* set of problems that affected the performance, especially of the early oligonucleotide microarrays, were the annotation and summarization issues associated with predefined reporters/probes (e.g., Refs. 46,47). Interestingly, on some arrays, a significant number of the represented transcripts were actually long noncoding (lnc) RNAs (see, e.g., Ref. 48). With the third and fourth generation of oligonucleotide arrays, e.g., the Affymetrix Clariom D array, the annotation issues have significantly improved.

Finally, the *fourth* problem with the oligonucleotide arrays was that 3′UTR-orientated microarrays provided relatively little information about alternative splicing, which is particularly high in brain.[1,49–51] Although the Affymetrix Mouse 1.0 Exon ST array collected data on alternative splicing, when used to detect *differential* alternative splicing, the array was particularly sensitive to the "SNP effect" due to the smaller number of probes per probe set.[52]

There is another piece of microarray history that is relevant to the development and acceptance of RNA-Seq: several analysis controversies that emerged with the introduction of microarrays. In 1999, *Nature Genetics* devoted an entire issue (volume 21, January) to microarrays. Cautionary concerns were raised around issues of data analysis.[53] Microarray experiments, at the time, were generally expensive, limiting sample sizes. Small sample sizes and thousands of independent observations per sample were seen as a prescription for statistical disaster. Initial attempts to deal with this problem frequently involved using a nonstatistical threshold for a meaningful difference, e.g., a twofold difference in expression. This approach frequently worked well in some applications, e.g., when comparing cancerous and noncancerous tissue; however, this approach was destined not to work well in brain, where differences in expression among experimental groups were generally smaller (see earlier). Initially, journal reviewers, editors, and study sections panned microarray experiments as being "fishing expeditions," with no clear hypothesis. The idea of discovery science as a valuable strategy was a minority opinion. On the other hand, microarray data confirmed what was being learned from genome-wide association study (GWAS), namely that most phenotypes were associated with small effects from a large number of genes. For example, Mulligan et al.,[17] in the first meta-microarray analysis for any behavioral trait, found from the analysis of 105 microarray samples that ethanol two-bottle choice was associated with the differential expression of >3000 genes.

## THE EARLY YEARS: DEVELOPMENT OF RNA-SEQ

Most of the current RNA-Seq is done on one of the Illumina platforms. However, this was not always the case, and there are several excellent reviews of the first- and second-generation NGS platforms (e.g., Refs. 54–57). Table 1 in Metzker[56] provides a historically useful comparison of the available NGS platforms based on instrument cost, complexity, and method of library preparation, sequencing, and detection method, read length, read depth, run time, and cost per sequenced base. Sample and library preparation, read depth, and read length remain today as key features in designing a RNA-Seq experiment. In addition to the various Illumina platforms, commercially available instruments today (see http://www.molecularecologist.com/next-gen-table-4-2016/) include the Pacific Biosciences Sequel System and PacBio RS II, both of which use the proprietary SMRT sequencing technology. The Sequel System is a high-throughput instrument suitable for projects generating whole genome de novo assemblies; the RS II is suitable for the whole genome sequencing of smaller genomes and targeted sequencing of DNA and RNA. There are three different Ion Torrent platforms (PGM, Proton, and S5/S5xl) that are generally of lower cost per base pair than Illumina. To our knowledge, there are no reported uses of PacBio or Ion Torrent in a RNA-Seq brain/behavior context. For the reader interested in updates on the chemistry associated with each of today's commercially available instruments, there are excellent tutorials available at the manufacturer's websites. Finally, we recommend reviewing three publications[58–60] that address the causes of and solutions for dealing with biases in large-scale RNA-Seq data, including biases that occur across laboratories. In the following paragraphs, we focus on some of the parameters we consider to be particularly critical.

The first and perhaps the most important step of a RNA-Seq experiment is the same as that for a microarray experiment: the isolation of high-quality RNA. Although both RNA-Seq and microarrays can be used on fragmented RNA, such as that found in formalin-fixed-paraffin-embedded samples, the biases present in such samples for genome-wide sequencing are difficult to assess. However, short-reads (<50 bp) are likely to be more suitable for degraded RNA.

RNA quality is routinely examined on an instrument such as the Agilent Bio-Analyzer; an RNA integrity number (RIN) of $\geq 8$ is generally considered high quality. Unfortunately, for brain samples, the amount of beginning tissue may be very small, and obtaining a reliable RIN number may be difficult. Even within very discrete brain regions, there are multiple cell types, and some experiments need to focus on a specific subset of cells or even a single cell. These experiments require significant sample amplification. Improvements in library preparation and instrument performance have substantially reduced the amount of nonamplified RNA required per sample. For example, 50 ng of RNA is now sufficient for the Illumina platform. Chen et al.[61] appear to be the first to couple laser capture microdissection (LCM) and RNA-Seq to examine brain gene expression. They examined rat GABAergic neurons projecting from the nucleus accumbens to the ventral pallidum. Cells were labeled using the retrograde tracer, Fluorogold. Approximately 1500 cells were labeled and isolated by LCM in each of four animals; this in turn produced ~4 ng of RNA per animal, and the average RIN was

8.1. Samples were independently amplified for microarray and RNA-Seq; for genes detected on both platforms, the correlation for gene expression was ~0.7. Not surprisingly, the correlation was better for the highly expressed genes. We have routinely used LCM to isolate discrete regions from the mouse brain, e.g., the central nucleus of the amygdala or the nucleus accumbens core and shell (see Ref. 62) and have found the technique quite reliable. However, we note that LCM requires processing of the tissue, including several dehydration steps, which in turn can affect RNA quality.

Chen et al.[61] (earlier in chapter) did not deplete the samples of ribosomal RNA (rRNA); the amplification procedure used selectively enriched mRNA. However, for samples beginning with larger amounts of starting material, amplification is not used, and the question of how to address the rRNA issue (rRNA makes up ~95% of the RNA in most samples) must be addressed. Most RNA-Seq experiments have selected for poly(A)-RNA using some variation of an oligo-d(T) hybridization pull-down, similar to what is used for microarray experiments. This approach nicely enriches for mRNA but will miss many ncRNAs and RNAs without a poly(A) tail.[63] RNA-Seq is well suited to look for these RNAs, many of which may be poorly annotated and may not be found by microarray platforms. As noted by Guennewig and Cooper,[64] noncoding RNAs are associated with chromatin modification, transcriptional regulation, alternative splicing, RNA editing, and translation; further, noncoding RNAs appear to have a significant role in neuropsychiatric and neurodegenerative diseases. Products such as Ribo−Zero Gold (Illumina) can be used to deplete ribosomal RNA and prepare samples for the sequencing of both coding and noncoding RNAs. There is, however, a price to be paid for using ribosome-depleted samples if read density is kept constant. The number of coding reads will be significantly decreased. Cui et al.,[65] using a first-generation ribosome depletion product, compared the read profile of RNA-Seq of ribosome-depleted (rdRNA) and poly(A)-selected (mRNA) samples; the starting total RNA was extracted from BALB/c mouse whole brain. The authors found (on a percent basis) that there were marked read distribution differences between samples. The percentage of known exon reads was twice as high in the mRNA sample (60%), while the percentage of both intronic and intergenic reads was twice as high in the rdRNA sample (25% and 44%, respectively). Both samples detected reads in essentially the same population of RefSeq defined genes, i.e., there was not a substantial read bias. Our experience (RNA-Seq of the mouse ventral striatum) with rdRNA and mRNA is similar, i.e., the relative proportions among coding genes remain the same. However, we have found that for the rdRNA samples, there is a large (four- to fivefold) increase in the number of multimapping sequences, with the percentage of uniquely mapping sequences decreasing from 90% to about 60%. Analytical strategies for dissecting the multimapping reads have been developed.[66,67] Illustrating the potential benefit of quantifying ncRNAs, we found that in our rdRNA samples a noncoding RNA (NONMMUG007939.2) with no known function accounted for 1.75% of the uniquely mapped reads.

Based on data collected in our laboratory and elsewhere,[1,4,62,68–70] it appears that 20–40 million reads are generally adequate for most estimates of gene expression and for constructing gene coexpression networks. However, if the goal is to quantitatively measure expression at the exon level and detect novel transcripts, then the read density must be increased significantly, perhaps by an order of magnitude (see Refs. 71,72). A related issue is whether one should use single-end or paired-end sequencing, given that paired-end sequencing facilitates the detection of novel transcripts. Given the heterogeneous nature of the brain and especially if one is comparing data from multiple brain regions and or cell types, the possibility of detecting novel transcripts and splice junctions would seem to be reasonably high, thus arguing that the paired-end approach would be preferred. To our knowledge, no one has sequenced to great depth (e.g., 1010 paired-end reads per sample) across brain regions in a very well-annotated genome (e.g., the B6 mouse) and determined if there are rare and regionally specific transcripts undetectable at a lower read density. However, Lee et al.[72] suggest that rare and functionally important transcripts may indeed be present and only detected at very high read density. One could ask a relatively similar question when comparing transcripts across a panel of inbred mouse or rat strains. Zheng et al.[73] used paired-end sequencing ($200 \times 10^7$ reads/sample) to compare the striatal splice junction profiles of the eight inbred strains that were used to form the CC[39]; the strains encompass 90% of *Mus musculus* genetic diversity.[74] The three wild-derived strains (PWK/PhJ, CAST/EiJ, and WSB/EiJ) had the most unique splice junctions, but these were rare (0.1%–0.3%).

RNA-Seq remains expensive, and it seems unlikely that costs will substantially decrease. High costs mean that sample sizes are frequently reduced and/or replication experiments are not performed. As reviewers of RNA-Seq manuscripts, we frequently find ourselves asking for a power analysis. The problem here is that RNA-Seq data are digital, the dynamic range is large, and the coefficient of variation for each of the thousands of genes being measured differs markedly. Thus, the usual analytic strategies for power analysis do not apply. Hart et al.[75], Conesa et al.,[4] and Wu and Wu[76] have provided some guidance on this issue. Importantly, the power to detect DE genes depends on read depth, which in turn may be influenced by whether or not one is interested in detecting alternatively spliced transcripts or other factors. When constructing coexpression networks, we have empirically determined the sample sizes necessary to generate reproducible modules (clusters of correlated or coexpressed genes)[77]. Obviously, this approach is not entirely objective since the experimenter can set the number of genes included in the network construction, the minimum module size, and other clustering parameters. We found that in datasets with ~30 samples, a read density of ~20 million reads per sample, where ~7000 genes were included in the network, the majority of genes clustered in modules. Importantly, a majority of these modules were reproducible to a significant extent across species and brain regions.[78] A related issue is the extent to which network results are stable when the number of samples is in the range of

10–20. Here, we found that RNA-Seq compares favorably with Illumina and Affymetrix microarray platforms.[32]

One might have predicted that the microarray experience would have laid the groundwork for the acceptance of NGS and RNA-Seq. However, with the introduction of the 454 sequencer[79] the argument was made that the data sets were so large that only one of the established genome centers would have the necessary bioinformatics expertise. But as the technology improved, so did the analytic approaches, so by 2007/2008, RNA-Seq data appeared from several different laboratories.[1,68,80–82] Workflows emerged that addressed the measurement not only of DE genes but also differential alternative splicing and the detection of novel transcripts.[68,73,78] Bullard et al.[83] examined a number of statistical issues associated with using RNA-Seq to detect DE genes. Similar to Marioni et al.,[68] they found that most sources of technical variation had only small effects on detecting DE transcripts. The most significant effect on DE transcripts was data normalization. Bullard et al.[83] concluded that their "main novel finding is the extent to which normalization affects differential expression results: sensitivity varies more between normalization procedures than between test statistics…we propose scaling gene counts by a quantile of the gene count distribution (the upper-quartile)."

RNA-Seq data have some unique properties that affect the strategies for data analysis.[4,84] *First,* unlike microarray data where the output is fluorescence intensity (more or less a continuous measure), the output from an RNA-Seq experiment is digital. For the microarray but not the RNA-Seq experiment, familiar statistics such as a *t*-test or ANOVA are appropriate. Robinson et al.[85] proposed the use of the **e**mpirical analysis of **d**igital **g**ene **e**xpression in R (edgeR), a variant of a procedure used to analyze SAGE data. edgeR models count data using an overdispersed Poisson model and use an empirical Bayes' procedure to moderate the degree of overdispersion across genes; the overdispersion reflects the biologic variation among samples.[85] An early implementation of edgeR to detect striatal DE between two mouse strains is found in Bottomly et al.[69]

*Second,* RNA-Seq data are biased in several important ways. First, the majority of the counts are produced by a small number (<10% of the total) of very highly expressed genes. As noted previously, a single transcript can provide nearly 2% of the total read counts. Thus, many genes of interest may have only moderate to low counts. Also, for genes with equal levels of expression, the long genes will be overrepresented, distorting the relative expression among genes. Similarly, within a given gene, long exons are overrepresented. Normalization and weighting algorithms can be used to address these issues, but they in turn may introduce new biases.

*Third,* RNA-Seq provides a substantial amount of data with very low read counts, which will be quite variable and thus, regardless of the analytic strategy, make detecting DE genes difficult. This variability is not necessarily rectified by increasing the read counts two- or threefold. We routinely set the threshold for inclusion in the data analysis as

1 read per million reads or approximately 30 reads/gene in a typical multiplexed experiment. This threshold passes through a large number of noncoding RNAs that typically have a relatively low expression.

*Fourth,* and as noted before, some RNA-Seq data may include a large number of multimapping multireads. The multireads arise predominantly from conserved domains in paralogous genes and from repeats.[63] Mortazavi et al.[1] found that, in the mouse brain, 76% of the 25-bp transcriptome sequence segments uniquely mapped; 6% mapped 2–10 times in the genome; and the remainder mapped more than 10 times. Depending on the gene model used and assuming a high-read density, ignoring the multireads may only have a minimal effect on detecting DE genes.

*Fifth,* RNA-Seq collects data across splice junctions that (1) are ignored by some alignment tools and/or (2) may be unknown. While there are <25,000 known protein-coding genes in the mammalian genome, the number of gene-related transcripts may well be several times higher.[51,86] Given the heterogeneous nature of brain tissue, the complexity problem is significantly amplified. Tools are available that detect splice junctions and will estimate the minimum number of gene isoforms that account for the observed data.[4,87–92] Iancu et al.[78] developed a novel method to look at alternative splicing. Genes were represented as vectors/lists of exon counts, and distance measures sensitive to exon inclusion rates quantified differences across samples. For all gene pairs, distance matrices were correlated across samples, resulting in cosplicing or cotranscriptional network matrices. It was observed that networks including cosplicing information were scale-free and distinct from coexpression. In the networks capturing cosplicing a novel set of hubs were detected with unique characteristics distinguishing them from coexpression hubs: heavy representation in neurobiologic functional pathways, strong overlap with markers of neurons and neuroglia, and long coding lengths.

## RECENT PROGRESS IN RNA-SEQ, THE BRAIN, AND BEHAVIOR

We briefly discuss here some recent publications from our laboratories and others that illustrate current uses of RNA-Seq in the brain-behavior context. There is no intent to be comprehensive. Given the rate that new publications are appearing, this would be impossible. Rather, we hope that the reader will sense that the field is maturing and looks very different from what it did 10 years ago. In addition, we focus on some areas that are not entirely new but which are emerging as key brain-behavior applications of RNA-Seq.

Historically, selective breeding for behavioral traits in mice and rats has utilized as the founder populations simple $F_2$ intercrosses or HS that captured only a small portion of the available genetic diversity (see, e.g., Ref. 74). The question naturally arose as to whether or not more genetically diverse populations would yield new insights into the relevant behavioral genetics. The CC, a panel of RI lines, was established with this question in mind.[39,44] As an adjunct to the CC, two outbred populations were created: the HS-CC and the DO.[31,40] Selective breeding from these populations and the subsequent transcriptome analyses are very well suited to the RNA-Seq paradigm, given the underlying genetic complexity. Colville et al.[62] selectively bred for ethanol two-bottle choice from the HS-CC; after four generations the High and Low preference lines were significantly different both genotypically and phenotypically. The nucleus accumbens shell of ethanol naïve $S_4$ mice was interrogated using RNA-Seq; $N = 24$/line and the consensus module approach[31] were used. Gene networks were constructed using the WGCNA assessing both coexpression and cosplicing. Selection targeted one of the network coexpression modules (greenyellow) that was significantly enriched in genes associated with receptor signaling activity including *Chrna7, Grin2a, Htr2a,* and *Oprd1*. Connectivity in the module as measured by changes in the hub nodes was significantly reduced in the Low preference line. Analysis of the cosplicing network data revealed a significant effect of selection on a large cluster of *Ras*GTPase binding genes including *Cdkl5, Cyfip1, Ndrg1, Sod1*, and *Stxbp5*. These data in part supported the earlier observation of Mulligan et al.[17] that preference was linked to *Ras/Mapk* pathways.

Beginning with microarray data and increasing dramatically with RNA-Seq data sets, most of the results in the typical RNA-Seq publication are found in the "Supplemental Material" section of that publication. This is, of course, not peculiar to studies of the brain transcriptome. Further, nearly all journals require that the raw data be deposited into a publicly available database such as GEO (https://www.ncbi.nlm.nih.gov/geo/). So, it is easy to argue that the interested and *very dedicated* reader has ample opportunities for detailed data review. However, one must wonder if this is really the best approach for data dissemination. Here we simply paraphrase the results section from one of our recent publications, to illustrate some areas of potential ambiguity. Hitzemann et al.[93] focused on the observation that even within a highly selected line for ethanol binge consumption, there was a more than fourfold variation in the blood ethanol concentration (BEC) at the end of the 4-h drinking-in-the dark (DID) trial. RNA-Seq analysis was performed on 34 animals (striatum) taken 3 weeks after the DID trial. 14,899 genes met the criteria for inclusion in the analysis (see Ref. 93, Supplemental Table 4). (*It seems unlikely that many readers will check to see if genes that should be included are here and/or in the right proportion for a striatal sample*). Pearson correlations were used to assess the relationships between gene expression and the BEC. Four hundred and seventy-five genes met the criteria for inclusion in the gene set enrichment analysis (see Ref. 93, Supplemental Table 4) This group of genes, denoted as the "affected gene set" were characterized using a gene set enrichment analysis strategy. Both the GO Process and Component categories contained features with significant enrichment. Key Process features that were enriched included modulation of synaptic transmission (FDR $< 2 \times 10^{-5}$), signal release (FDR $< 5 \times 10^{-5}$), and calcium ion regulated exocytosis (FDR $< 8 \times 10^{-5}$). A complete listing of the enriched features and the associated genes is found in Supplemental Table 5 of Hitzemann et al.[93]

*(Reviewing this table is important, since someone may recognize a gene or gene set that has been previously associated with excessive ethanol consumption and/or other aspects of alcohol and drug abuse.)* The genes associated with the modulation of synaptic function include six receptors: *Adora1, Adrb1, Cckbr, Grik5, Grin1*, and *Htr2a*. The enrichment in Component features includes synapse (FDR $< 5 \times 10^{-10}$), neuron part (FDR $< 8 \times 10^{-8}$), and postsynaptic density (FDR $< 6 \times 10^{-5}$). A complete listing of the enriched features and the associated genes is found in Supplemental Table 5.[93] The genes associated with the postsynaptic density category include *Camk2a, b & c, Mapk10, Rims1, Syn1*, and *Syngap1*. *(A general comment here concerns the secondary analyses; we use EnrichR, GeneMania and GOrilla. Other tools are available. Tools such a GOrilla,[94] which allow one to designate the background dataset against which enrichment is measured, are preferred.)*

The number of RNA-Seq/brain/behavior publications have dramatically increased over the past 4–5 years. Behavior here is broadly defined as both normal and abnormal behavior, including neuropathological conditions. Here, we note some recent favorites, illustrating the breadth of the publications. He et al.[95] have used RNA-Seq to compare gene expression in the neocortex of humans, chimpanzees, and macaques. The authors compared cortical layers and the adjacent white matter using unsupervised sectioning followed by sequencing. The authors found that despite the conservation of histological features across species, the human neocortex has undergone substantial reorganization affecting more than 5% of the transcriptome. Hosseinn-nezhad et al.[96] have reported on the transcriptome profiling of cerebrospinal fluid (CSF) taken from subjects with Parkinson disease and matched healthy controls. Very little CSF (300 μL) was required for the analysis. Genes showing differential expression and confirmed with real-time PCR included *CCR3, SSTR3, PTPRC SCN9, SCN9* antisense and two lnc RNAs (*AC79630* & *UC001lva.4*). Importantly, these data suggest that RNA-Seq of the CSF could provide useful biomarkers for disease progression. Pantazatos et al.[97] used RNA-Seq (Broadman area 9) to investigate the transcriptomes of human subjects separated into several groups; controls, major depressive suicides and major depressive nonsuicides. Gene ontology annotation of the DE genes suggested lower expression of genes involved in oligodendrocyte differentiation, regulation of glutamate neurotransmission, and oxytocin receptor expression in both suicide and depression. Continuing the focus on noncoding RNAs, Zhong et al.[98] found that in the mouse, traumatic brain injury was associated with the upregulation of 667 lncRNAs and the downregulation of 156 lncRNAs. Mills et al.[99] have identified an intergenic lncRNA (*LINC00507*) that is expressed in a cortex-specific manner in nonhuman primates and humans; further, the expression appears to be age-specific. Kroeze et al.[100] have identified 137 intergenic lncRNAs that are differentially regulated during development in the medial prefrontal cortex; these lncRNAs may be important to our understanding of the mechanisms associated with neurodevelopmental and neuropsychiatric disorders. We have found (unpublished observation) that the gene showing the largest change (>200-fold increase) in the selection for high DID is a small nuclear RNA, *Gm22513*. As more of these data become available, it follows that the systematic manipulation of these noncoding RNAs will be required to infer function.

There are many different types of noncoding RNAs: intergenic, intronic, antisense, and small nuclear to just name a few. The functions for the majority of these noncodes remain unknown. However, perhaps the most enigmatic on the noncoding RNAs are the circRNAs.[101–103] The existence of individual circRNAs has been known for more than 40 years. These RNAs were largely considered to be splicing errors with no known function. This view dramatically changed when Salzman et al.[104] using RNA-Seq found that circRNAs were the dominant isoform for hundreds of human genes. The detection of circRNAs in eukaryotic cells remains a work in progress (see Ref. 103). As noted by Chen and Schuman[101] "CircRNAs derived from introns and exons represent two different RNA types with distinct biogenesis mechanisms and likely also exhibit different biological functions. Whereas the intron-derived circRNAs originate from intron lariats that escape debranching and degradation, accumulate in nucleus, and appear to function by regulating the expression of their host gene (e.g., Ref. 105), exon-derived circRNAs are formed by back-splicing (i.e., the covalent joining of a downstream splice donor site with an upstream splice acceptor site from the same gene locus); exon-derived circRNAs are pre-dominantly localized in cytoplasm and likely function via a variety of mechanisms."[106] In the brain/behavior context, circRNAs appear to be important as they are more highly expressed in neuronal tissue. Deep RNA-Seq of several tissues has revealed that in brain "a significantly greater fraction of reads are circRNA junctional reads, a greater number of genes are hosts to circRNAs, and there is a greater number of tissue-specific circRNA hosts."[107] Indeed, 20% of the protein-coding genes in brain produce circRNAs.[101] Despite their ubiquitous presence, the functions of the circRNAs remain largely unknown; however, there is more than suggestive evidence to suggest that they serve as miRNA sponges (e.g., Ref. 108).

Most brain RNA-Seq data has been obtained using hand- or LCM-dissected tissue; the samples often contain multiple nuclei and almost always multiple cell types. However, greater precision (from the tissue perspective) is now possible given the significant decrease in the amount of unamplified RNA required for analysis and given the improvements in single-cell RNA amplification. The technology has now sufficiently improved, so it is possible to assess coexpression patterns among individually sequenced cells.[109] In brain a specialized group of mRNAs, ncRNAs, and miRNAs are shuttled to synapses where they are used for synaptic maintenance and for the regulation of synaptic plasticity. Synaptosomes and even growth cones can be isolated by differential centrifugation (see, e.g., Ref. 110); synaptoneurosomes are isolated somewhat differently[111] but have similar properties. It is important to remember that these synaptic preparations will have variable levels of contamination from other cellular elements; these should be routinely assessed using marker enzymes and electron microscopy. Chen et al.[112] have used RNA-Seq to examine the effects of aging on the mouse hippocampal transcriptome. Aging-related changes were detected in the expression of genes associated with neurite guidance, synaptosomal physiology, and RNA splicing. The authors also discovered alterations in the expression of thousands of novel, unannotated lincRNAs. Of related interest, Ma and Chang[113] have observed in the human neocortex that while overall the expression of lncRNAs is low compared to mRNAs, in some cells the expression of lncRNAs is quite high, suggesting that the noncodes may provide an additional strategy for characterizing individual cells.

Under ideal circumstances, it would be possible to know with some surety (1) that the connectivity among multiple brain regions was associated with the behavioral phenotype of interest, (2) that it was possible to isolate the various cell types in these regions without compromising their transcriptome integrity, (3) that individual cell transcriptomes would be sequenced in sufficient density to detect coexpression and cosplicing networks within and across specific cell types, and (4) the transcriptome profiles would be predictive of protein abundance and/or functional interactions at the cellular level. Progress is being made across all of these issues, but perhaps most important are the data integrating functional and transcriptome connectivity (e.g., Refs. 114,115). It is assumed that this integration will be enhanced by determining the specific transcriptome signature of those cells within the region of interest that is most closely aligned with function. Importantly, RNA-Seq of single cells has increased from 41 publications in 2012 to 238 in 2016 and is on track to exceed 400 publications in 2017. Skene and Grant[114] have identified vulnerable cell types in major brain disorders using single-cell transcriptomes and a novel analysis strategy (weighted cell type enrichment). Using their approach, they concluded that genetic susceptibility primarily affected microglia in Alzheimer disease and multiple sclerosis and was shared between interneurons and pyramidal neurons in autism and schizophrenia, while intellectual disabilities and epilepsy were attributable to a range of cell types, with the strongest enrichment in interneurons. One problem that remains with single-cell analysis is the processing of the tissue, generally for fluorescence-activated cell sorting (FACS). Processing effects on the transcriptome will be considerable. Krishnaswami et al.[116] have suggested an alternative approach that focuses on the processing of postmortem tissue. These authors have developed a procedure for the capture of single nuclei. The authors note that the method also allows investigation of biologic features unique to nuclei, such as enrichment of certain transcripts and precursors of some noncoding RNAs.

Although not widely used in the brain-behavior context, RiboTag profiling is well suited to analyzing the transcriptome of specific cell types in vivo, i.e., the processing artifacts noted before are largely absent. Sanz et al.[117] described the creation of the RiboTag mouse line. The line "carries an Rpl22 allele with a floxed wild-type C-terminal exon followed by an identical C-terminal exon that has three copies of the hemagglutinin (HA) epitope inserted before the stop codon. When the RiboTag mouse is crossed to a cell-type-specific Cre recombinase-expressing mouse, Cre recombinase activates the expression of epitope-tagged ribosomal protein RPL22HA, which is incorporated into actively translating polyribosomes. Immunoprecipitation of poly-somes with a monoclonal antibody against HA yields ribosome-associated mRNA transcripts from specific cell types." There appear to be 19 RiboTag publications, nine of which focus on some aspect of the brain transcriptome.[117–124] These reports include targeting the transcriptome of retinal ganglion cell axons[123] and the transcriptome of astrocytes.[124]

## CONCLUSIONS

The first 10 years of RNA-Seq from the brain/behavior perspective have witnessed some remarkable advances in technology and analytical strategies. In addition to providing new tools to look at gene expression and alternative splicing, RNA-Seq allows one to look at the noncoding RNAs that may have little or no annotation but are functionally important. In addition, RNA-Seq is well suited to detailed brain mapping, given the single-cell resolution in many applications. The next 10 years should be even more remarkable.

## ACKNOWLEDGMENTS

Sponsored in part by AA 10760 and AA 13484.

## REFERENCES

1. Mortazavi A, Williams BA, McCue K, Schaeffer L, Wold B. Mapping and quantifying mammalian transcriptomes by RNA-Seq. *Br J Pharmacol.* 2008;5(7):621–628. https://doi.org/10.1038/nmeth.1226.
2. Hitzemann R, Bottomly D, Darakjian P, et al. Genes, behavior and next-generation RNA sequencing. *Gene Brain Behav.* 2013;12(1):1–12. https://doi.org/10.1111/gbb.12007.
3. Hitzemann R, Darakjian P, Walter N, Iancu OD, Searles R, McWeeney S. Introduction to sequencing the brain transcriptome. *Int Rev Neurobiol.* 2014;116:1–19. https://doi.org/10.1016/B978-0-12-801105-8.00001-1.
4. Conesa A, Madrigal P, Tarazona S, et al. A survey of best practices for RNA-seq data analysis. *Genome Biol.* 2016;17(1). https://doi.org/10.1186/s13059-016-0881-8.
5. Veneziano D, Nigita G, Ferro A. Computational approaches for the analysis of ncRNA through deep sequencing techniques. *Front Bioeng Biotechnol.* 2015;3. https://doi.org/10.3389/fbioe.2015.00077.
6. Cirelli C, Tononi G. Differences in brain gene expression between sleep and waking as revealed by mRNA differential display and cDNA microarray technology. *J Sleep Res.* 1999;8(suppl 1):44–52.
7. Sandberg R, Yasuda R, Pankratz DG, et al. Regional and strain-specific gene expression mapping in the adult mouse brain. *Proc Natl Acad Sci USA.* 2000;97(20):11038–11043.
8. Hoffman PL, Miles M, Edenberg HJ, et al. Gene expression in brain: a window on ethanol dependence, neuroadaptation, and preference. *Alcohol Clin Exp Res.* 2003;27(2):155–168. https://doi.org/10.1097/01.ALC.0000060101.89334.11.
9. Bergeson SE, Berman AE, Dodd PR, et al. Expression profiling in alcoholism research. *Alcohol Clin Exp Res.* 2005;29(6):1066–1073. https://doi.org/10.1097/01.ALC.0000171043.29384.3E.
10. McBride WJ, Kerns RT, Rodd ZA, et al. Alcohol effects on central nervous system gene expression in genetic animal models. *Alcohol Clin Exp Res.* 2005;29(2):167–175.
11. Matthews DB, Bhave SV, Belknap JK, et al. Complex genetics of interactions of alcohol and CNS function and behavior. *Alcohol Clin Exp Res.* 2005;29(9):1706–1719.

12. Sikela JM, Maclaren EJ, Kim Y, et al. DNA microarray and proteomic strategies for understanding alcohol action. *Alcohol Clin Exp Res*. 2006;30(4):700–708. https://doi.org/10.1111/j.1530-0277.2006.00081.x.
13. Tabakoff B, Saba L, Printz M, et al. Genetical genomic determinants of alcohol consumption in rats and humans. *BMC Biol*. 2009;7(70). https://doi.org/10.1186/1741-7007-7-70.
14. Saba LM, Bennett B, Hoffman PL, et al. A systems genetic analysis of alcohol drinking by mice, rats and men: influence of brain GABAergic transmission. *Neuropharmacology*. 2011;60(7–8):1269–1280. https://doi.org/10.1016/j.neuropharm.2010.12.019.
15. Farris SP, Miles MF. Ethanol modulation of gene networks: implications for alcoholism. *Neurobiol Dis*. 2012;45(1):115–121. https://doi.org/10.1016/j.nbd.2011.04.013.
16. Hitzemann R, Reed C, Malmanger B, et al. On the integration of alcohol-related quantitative trait loci and gene expression analyses. *Alcohol Clin Exp Res*. 2004;28(10):1437–1448.
17. Mulligan MK, Ponomarev I, Hitzemann RJ, et al. Toward understanding the genetics of alcohol drinking through transcriptome meta-analysis. *Proc Nat Acad Sci USA*. 2006;103(16):6368–6373. https://doi.org/10.1073/pnas.0510188103.
18. Hofstetter JR, Hitzemann RJ, Belknap JK, Walter NAR, McWeeney SK, Mayeda AR. Characterization of the quantitative trait locus for haloperidol-induced catalepsy on distal mouse chromosome 1. *Gene Brain Behav*. 2008;7(2):214–223. https://doi.org/10.1111/j.1601-183X.2007.00340.x.
19. Pavlidis P, Qin J, Arango V, Mann JJ, Sibille E. Using the gene ontology for microarray data mining: a comparison of methods and application to age effects in human prefrontal cortex. *Neurochem Res*. 2004;29(6):1213–1222.
20. Bebek G, Yang J. PathFinder: mining signal transduction pathway segments from protein-protein interaction networks. *BMC Bioinf*. 2007;8(335). https://doi.org/10.1186/1471-2105-8-335.
21. Feng G, Shaw P, Rosen ST, Lin SM, Kibbe WA. Using the bioconductor GeneAnswers package to interpret gene lists. *Meth Mol Biol*. 2012;802:101–112. https://doi.org/10.1007/978-1-61779-400-1_7.
22. Vadigepalli R, Chakravarthula P, Zak DE, Schwaber JS, Gonye GE. PAINT: a promoter analysis and interaction network generation tool for gene regulatory network identification. *OMICS A J Integr Biol*. 2003;7(3):235–252. https://doi.org/10.1089/153623103322452378.
23. Zhang B, Horvath S. A general framework for weighted gene co-expression network analysis. *Stat Appl Genet Mol Biol*. 2005;4 [Article 17]. https://doi.org/10.2202/1544-6115.1128.
24. Horvath S, Zhang B, Carlson M, et al. Analysis of oncogenic signaling networks in glioblastoma identifies ASPM as a molecular target. *Proc Nat Acad Sci USA*. 2006;103(46):17402–17407. https://doi.org/10.1073/pnas.0608396103.
25. Friedman N, Linial M, Nachman I, Pe'er D. Using Bayesian networks to analyze expression data. *J Comput Biol*. 2004;7(3–4):601–620. https://doi.org/10.1089/106652700750050961.
26. Zhao W, Langfelder P, Fuller T, Dong J, Li A, Hovarth S. Weighted gene coexpression network analysis: state of the art. *J Biopharm Stat*. 2010;20(2):281–300. https://doi.org/10.1080/10543400903572753.
27. Oti M, van Reeuwijk J, Huynen MA, Brunner HG. Conserved co-expression for candidate disease gene prioritization. *BMC Bioinf*. 2008;9(208). https://doi.org/10.1186/1471-2105-9-208.
28. Oldham MC, Horvath S, Geschwind DH. Conservation and evolution of gene coexpression networks in human and chimpanzee brains. *Proc Nat Acad Sci USA*. 2006;103(47):17973–17978. https://doi.org/10.1073/pnas.0605938103.
29. Oldham MC, Konopka G, Iwamoto K, et al. Functional organization of the transcriptome in human brain. *Nat Neurosci*. 2008;11(11):1271–1282. https://doi.org/10.1038/nn.2207.
30. Voineagu I, Wang X, Johnston P, et al. Transcriptomic analysis of autistic brain reveals convergent molecular pathology. *Nature*. 2011;474(7351):380–384. https://doi.org/10.1038/nature10110.
31. Iancu OD, Darakjian P, Walter NAR, et al. Genetic diversity and striatal gene networks: focus on the heterogeneous stock-collaborative cross (HS-CC) mouse. *BMC Genom*. 2010;11(585). https://doi.org/10.1186/1471-2164-11-585.
32. Iancu OD, Darakjian P, Malmanger B, Walter NAR, McWeeney S, Hitzemann R. Gene networks and haloperidol-induced catalepsy. *Gene Brain Behav*. 2012;11(1):29–37. https://doi.org/10.1111/j.1601-183X.2011.00736.x.

33. Pozhitkov AE, Boube I, Brouwer MH, Noble PA. Beyond Affymetrix arrays: expanding the set of known hybridization isotherms and observing pre-wash signal intensities. *Nucleic Acids Res.* 2010; 38(5):e28. https://doi.org/10.1093/nar/gkp1122.
34. Peirce JL, Li H, Wang J, et al. How replicable are mRNA expression QTL? *Mamm Gen.* 2006;17(6): 643–656. https://doi.org/10.1007/s00335-005-0187-8.
35. Sliwerska E, Meng F, Speed TP, et al. SNPs on chips: the hidden genetic code in expression arrays. *Biol Psychiatr.* 2007;61(1):13–16. https://doi.org/10.1016/j.biopsych.2006.01.023.
36. Walter NAR, McWeeney SK, Peters ST, Belknap JK, Hitzemann R, Buck KJ. SNPs matter: impact on detection of differential expression. *Br J Pharmacol.* 2007;4(9):679–680. https://doi.org/10.1038/nmeth0907-679.
37. Walter NAR, Bottomly D, Laderas T, et al. High throughput sequencing in mice: a platform comparison identifies a preponderance of cryptic SNPs. *BMC Genom.* 2009;10(379). https://doi.org/10.1186/1471-2164-10-379.
38. Duan F, Pauley MA, Spindel ER, Zhang L, Norgren RB. Large scale analysis of positional effects of single-base mismatches on microarray gene expression data. *BioData Min.* 2010;3(1):2. https://doi.org/10.1186/1756-0381-3-2.
39. Churchill GA, Airey DC, Allayee H, et al. The Collaborative Cross, a community resource for the genetic analysis of complex traits. *Nat Genet.* 2004;36(11):1133–1137. https://doi.org/10.1038/ng1104-1133.
40. Svenson KL, Gatti DM, Valdar W, et al. High-resolution genetic mapping using the Mouse Diversity outbred population. *Genetics.* 2012;190(2):437–447. https://doi.org/10.1534/genetics.111.132597.
41. Hansen C, Spuhler K. Development of the National Institutes of Health genetically heterogeneous rat stock. *Alcohol Clin Exp Res.* 1984;8(5):477–479.
42. Solberg Woods LC, Holl KL, Oreper D, Xie Y, Tsaih S-W, Valdar W. Fine-mapping diabetes-related traits, including insulin resistance, in heterogeneous stock rats. *Physiol Genom.* 2012;44(21): 1013–1026. https://doi.org/10.1152/physiolgenomics.00040.2012.
43. Keane TM, Goodstadt L, Danecek P, et al. Mouse genomic variation and its effect on phenotypes and gene regulation. *Nature.* 2011;477(7364):289–294. https://doi.org/10.1038/nature10413.
44. Morgan AP, Welsh CE. Informatics resources for the Collaborative Cross and related mouse populations. *Mamm Gen.* 2015;26(9–10):521–539. https://doi.org/10.1007/s00335-015-9581-z.
45. Green R, Wilkins C, Ferris MT, Gale M. RNA-seq in the collaborative cross. *Meth Mol Biol.* 2017; 1488:251–263. https://doi.org/10.1007/978-1-4939-6427-7_11.
46. Allison DB, Cui X, Page GP, Sabripour M. Microarray data analysis: from disarray to consolidation and consensus. *Nat Rev Genet.* 2006;7(1):55–65. https://doi.org/10.1038/nrg1749.
47. Lu J, Lee JC, Salit ML, Cam MC. Transcript-based redefinition of grouped oligonucleotide probe sets using AceView: high-resolution annotation for microarrays. *BMC Bioinf.* 2007;8(108). https://doi.org/10.1186/1471-2105-8-108.
48. Liao Q, Liu C, Yuan X, et al. Large-scale prediction of long non-coding RNA functions in a coding-non-coding gene co-expression network. *Nucleic Acids Res.* 2011;39(9):3864–3878. https://doi.org/10.1093/nar/gkq1348.
49. Licatalosi DD, Darnell RB. Splicing regulation in neurologic disease. *Neuron.* 2006;52(1):93–101. https://doi.org/10.1016/j.neuron.2006.09.017.
50. Li Q, Lee J-A, Black DL. Neuronal regulation of alternative pre-mRNA splicing. *Nat Rev Neurosci.* 2007;8(11):819–831. https://doi.org/10.1038/nrn2237.
51. Johnson MB, Kawasawa YI, Mason CE, et al. Functional and evolutionary insights into human brain development through global transcriptome analysis. *Neuron.* 2009;62(4):494–509. https://doi.org/10.1016/j.neuron.2009.03.027.
52. Laderas TG, Walter NAR, Mooney M, et al. Computational detection of alternative exon usage. *Front Neurosci.* 2011;5(69). https://doi.org/10.3389/fnins.2011.00069.
53. Lander ES. Array of hope. *Nat Genet.* 1999;21(1 suppl):3–4. https://doi.org/10.1038/4427.
54. Mardis ER. The impact of next-generation sequencing technology on genetics. *Trends Genet.* 2008; 24(3):133–141. https://doi.org/10.1016/j.tig.2007.12.007.

55. Mardis ER. A decade's perspective on DNA sequencing technology. *Nature*. 2011;470(7333): 198–203. https://doi.org/10.1038/nature09796.
56. Metzker ML. Sequencing technologies - the next generation. *Nat Rev Genet*. 2010;11(1):31–46. https://doi.org/10.1038/nrg2626.
57. Martin JA, Wang Z. Next-generation transcriptome assembly. *Nat Rev Genet*. 2011;12(10):671–682. https://doi.org/10.1038/nrg3068.
58. Li S, Łabaj PP, Zumbo P, et al. Detecting and correcting systematic variation in large-scale RNA sequencing data. *Nat Biotechnol*. 2014;32(9):888–895. https://doi.org/10.1038/nbt.3000.
59. Li S, Tighe SW, Nicolet CM, et al. Multi-platform assessment of transcriptome profiling using RNA-seq in the ABRF next-generation sequencing study. *Nat Biotechnol*. 2014;32(9):915–925. https://doi.org/10.1038/nbt.2972.
60. SEQC/MAQC-III Consortium. A comprehensive assessment of RNA-seq accuracy, reproducibility and information content by the Sequencing Quality Control Consortium. *Nat Biotechnol*. 2014;32(9): 903–914. https://doi.org/10.1038/nbt.2957.
61. Chen H, Liu Z, Gong S, et al. Genome-wide gene expression profiling of nucleus accumbens neurons projecting to ventral pallidum using both microarray and transcriptome sequencing. *Front Neurosci*. 2011;5(98). https://doi.org/10.3389/fnins.2011.00098.
62. Colville AM, Iancu OD, Oberbeck DL, et al. Effects of selection for ethanol preference on gene expression in the nucleus accumbens of HS-CC mice. *Gene Brain Behav*; 2017. https://doi.org/10.1111/gbb.12367.
63. Costa V, Angelini C, De Feis I, Ciccodicola A. Uncovering the complexity of transcriptomes with RNA-Seq. *J Biomed Biotechnol*. 2010;2010(853916). https://doi.org/10.1155/2010/853916.
64. Guennewig B, Cooper AA. The central role of noncoding RNA in the brain. *Int Rev Neurobiol*. 2014; 116:153–194. https://doi.org/10.1016/B978-0-12-801105-8.00007-2.
65. Cui P, Lin Q, Ding F, et al. A comparison between ribo-minus RNA-sequencing and polyA-selected RNA-sequencing. *Genomics*. 2010;96(5):259–265. https://doi.org/10.1016/j.ygeno.2010.07.010.
66. Storvall H, Ramsköld D, Sandberg R. Efficient and comprehensive representation of uniqueness for next-generation sequencing by minimum unique length analyses. *PLoS One*. 2013;8(1):e53822. https://doi.org/10.1371/journal.pone.0053822.
67. Dao P, Numanagić I, Lin Y-Y, et al. ORMAN: optimal resolution of ambiguous RNA-Seq multi-mappings in the presence of novel isoforms. *Bioinformatics*. 2014;30(5):644–651. https://doi.org/10.1093/bioinformatics/btt591.
68. Marioni JC, Mason CE, Mane SM, Stephens M, Gilad Y. RNA-seq: an assessment of technical reproducibility and comparison with gene expression arrays. *Genome Res*. 2008;18(9):1509–1517. https://doi.org/10.1101/gr.079558.108.
69. Bottomly D, Walter NAR, Hunter JE, et al. Evaluating gene expression in C57BL/6J and DBA/2J mouse striatum using RNA-Seq and microarrays. *PLoS One*. 2011;6(3):e17820. https://doi.org/10.1371/journal.pone.0017820.
70. Iancu OD, Colville A, Walter NAR, et al. On the relationships in rhesus macaques between chronic ethanol consumption and the brain transcriptome. *Addiction Biol*; 2017. https://doi.org/10.1111/adb.12501.
71. Łabaj PP, Leparc GG, Linggi BE, Markillie LM, Wiley HS, Kreil DP. Characterization and improvement of RNA-Seq precision in quantitative transcript expression profiling. *Bioinformatics*. 2011;27(13): i383–391. https://doi.org/10.1093/bioinformatics/btr247.
72. Lee C, Mayfield RD, Harris RA. Altered gamma-aminobutyric acid type B receptor subunit 1 splicing in alcoholics. *Biol Psychiatr*. 2014;75(10):765–773. https://doi.org/10.1016/j.biopsych.2013.08.028.
73. Zheng CL, Wilmot B, Walter NA, et al. Splicing landscape of the eight collaborative cross founder strains. *BMC Genom*. 2015;16(52). https://doi.org/10.1186/s12864-015-1267-0.
74. Roberts A, Pardo-Manuel de Villena F, Wang W, McMillan L, Threadgill DW. The polymorphism architecture of mouse genetic resources elucidated using genome-wide resequencing data: implications for QTL discovery and systems genetics. *Mamm Gen*. 2007;18(6–7):473–481. https://doi.org/10.1007/s00335-007-9045-1.

75. Hart SN, Therneau TM, Zhang Y, Poland GA, Kocher J-P. Calculating sample size estimates for RNA sequencing data. *J Comput Biol.* 2013;20(12):970–978. http://doi.org/10.1089/cmb.2012.0283.
76. Wu Z, Wu H. Experimental design and power calculation for RNA-seq experiments. In: Mathé E, Davis S, eds. *Statistical Genomics.* vol. 1418. New York, NY: Springer; 2016:379–390. https://doi.org/10.1007/978-1-4939-3578-9_18.
77. Langfelder P, Zhang B, Horvath S. Defining clusters from a hierarchical cluster tree: the Dynamic Tree Cut package for R. *Bioinformatics.* 2008;24(5):719–720. https://doi.org/10.1093/bioinformatics/btm563.
78. Iancu OD, Colville A, Oberbeck D, Darakjian P, McWeeney SK, Hitzemann R. Cosplicing network analysis of mammalian brain RNA-Seq data utilizing WGCNA and Mantel correlations. *Front Genet.* 2015;6(174). https://doi.org/10.3389/fgene.2015.00174.
79. Margulies M, Egholm M, Altman WE, et al. Genome sequencing in microfabricated high-density picolitre reactors. *Nature.* 2005;437(7057):376–380. https://doi.org/10.1038/nature03959.
80. Weber APM, Weber KL, Carr K, Wilkerson C, Ohlrogge JB. Sampling the Arabidopsis transcriptome with massively parallel pyrosequencing. *Plant Physiol.* 2007;144(1):32–42. https://doi.org/10.1104/pp.107.096677.
81. Sugarbaker DJ, Richards WG, Gordon GJ, et al. Transcriptome sequencing of malignant pleural mesothelioma tumors. *Proc Nat Acad Sci USA.* 2008;105(9):3521–3526. https://doi.org/10.1073/pnas.0712399105.
82. Torres TT, Metta M, Ottenwälder B, Schlötterer C. Gene expression profiling by massively parallel sequencing. *Genome Res.* 2008;18(1):172–177. https://doi.org/10.1101/gr.6984908.
83. Bullard JH, Purdom E, Hansen KD, Dudoit S. Evaluation of statistical methods for normalization and differential expression in mRNA-Seq experiments. *BMC Bioinf.* 2010;11(94). https://doi.org/10.1186/1471-2105-11-94.
84. Garber M, Grabherr MG, Guttman M, Trapnell C. Computational methods for transcriptome annotation and quantification using RNA-seq. *Br J Pharmacol.* 2011;8(6):469–477. https://doi.org/10.1038/nmeth.1613.
85. Robinson MD, McCarthy DJ, Smyth GK. edgeR: A bioconductor package for differential expression analysis of digital gene expression data. *Bioinformatics.* 2010;26(1):139–140. https://doi.org/10.1093/bioinformatics/btp616.
86. Pan Q, Shai O, Lee LJ, Frey BJ, Blencowe BJ. Deep surveying of alternative splicing complexity in the human transcriptome by high-throughput sequencing. *Nat Genet.* 2008;40(12):1413–1415. https://doi.org/10.1038/ng.259.
87. Trapnell C, Pachter L, Salzberg SL. TopHat: discovering splice junctions with RNA-Seq. *Bioinformatics.* 2009;25(9):1105–1111. https://doi.org/10.1093/bioinformatics/btp120.
88. Trapnell C, Williams BA, Pertea G, et al. Transcript assembly and quantification by RNA-Seq reveals unannotated transcripts and isoform switching during cell differentiation. *Nat Biotechnol.* 2010;28(5):511–515. https://doi.org/10.1038/nbt.1621.
89. Trapnell C, Roberts A, Goff L, et al. Differential gene and transcript expression analysis of RNA-seq experiments with TopHat and Cufflinks. *Nat Protoc.* 2012;7(3):562–578. https://doi.org/10.1038/nprot.2012.016.
90. Guttman M, Garber M, Levin JZ, et al. Ab initio reconstruction of cell type-specific transcriptomes in mouse reveals the conserved multi-exonic structure of lincRNAs. *Nat Biotechnol.* 2010;28(5):503–510. https://doi.org/10.1038/nbt.1633.
91. Katz Y, Wang ET, Airoldi EM, Burge CB. Analysis and design of RNA sequencing experiments for identifying isoform regulation. *Br J Pharmacol.* 2010;7(12):1009–1015. https://doi.org/10.1038/nmeth.1528.
92. Dobin A, Davis CA, Schlesinger F, et al. STAR: ultrafast universal RNA-seq aligner. *Bioinformatics.* 2013;29(1):15–21. https://doi.org/10.1093/bioinformatics/bts635.
93. Hitzemann R, Oberbeck D, Iancu O, et al. Alignment of the transcriptome with individual variation in animals selectively bred for High Drinking-In-the-Dark (HDID). *Alcohol.* 2017;60:115–120. https://doi.org/10.1016/j.alcohol.2017.02.176.

94. Eden E, Navon R, Steinfeld I, Lipson D, Yakhini Z. GOrilla: a tool for discovery and visualization of enriched GO terms in ranked gene lists. *BMC Bioinf.* 2009;10(48). https://doi.org/10.1186/1471-2105-10-48.
95. He Z, Han D, Efimova O, et al. Comprehensive transcriptome analysis of neocortical layers in humans, chimpanzees and macaques. *Nat Neurosci.* 2017;20(6):886–895. https://doi.org/10.1038/nn.4548.
96. Hossein-nezhad A, Fatemi RP, Ahmad R, et al. Transcriptomic profiling of extracellular RNAs present in cerebrospinal fluid identifies differentially expressed transcripts in Parkinson's disease. *J Parkinson's Dis.* 2016;6(1):109–117. https://doi.org/10.3233/JPD-150737.
97. Pantazatos SP, Huang Y-Y, Rosoklija GB, Dwork AJ, Arango V, Mann JJ. Whole-transcriptome brain expression and exon-usage profiling in major depression and suicide: evidence for altered glial, endothelial and ATPase activity. *Mol Psychiatr.* 2017;22(5):760–773. https://doi.org/10.1038/mp.2016.130.
98. Zhong J, Jiang L, Cheng C, et al. Altered expression of long non-coding RNA and mRNA in mouse cortex after traumatic brain injury. *Brain Res.* 2016;1646:589–600. https://doi.org/10.1016/j.brainres.2016.07.002.
99. Mills JD, Ward M, Chen BJ, Iyer AM, Aronica E, Janitz M. LINC00507 is specifically expressed in the primate cortex and has age-dependent expression patterns. *J Mol Neurosci.* 2016;59(4):431–439. https://doi.org/10.1007/s12031-016-0745-4.
100. Kroeze Y, Oti M, van Beusekom E, et al. Transcriptome analysis identifies multifaceted regulatory mechanisms dictating a genetic switch from neuronal network establishment to maintenance during postnatal prefrontal cortex development. *Cerebr Cortex*; 2017. https://doi.org/10.1093/cercor/bhw407.
101. Chen W, Schuman E. Circular RNAs in brain and other tissues: a functional enigma. *Trends Neurosci.* 2016;39(9):597–604. https://doi.org/10.1016/j.tins.2016.06.006.
102. Cortés-López M, Miura P. Emerging functions of circular RNAs. *Yale J Biol Med.* 2016;89(4):527–537.
103. Szabo L, Salzman J. Detecting circular RNAs: bioinformatic and experimental challenges. *Nat Rev Genet.* 2016;17(11):679–692. https://doi.org/10.1038/nrg.2016.114.
104. Salzman J, Gawad C, Wang PL, Lacayo N, Brown PO. Circular RNAs are the predominant transcript isoform from hundreds of human genes in diverse cell types. *PLoS One.* 2012;7(2):e30733. https://doi.org/10.1371/journal.pone.0030733.
105. Li J, Yang J, Zhou P, et al. Circular RNAs in cancer: novel insights into origins, properties, functions and implications. *Am J Cancer Res.* 2015;5(2):472–480.
106. Hentze MW, Preiss T. Circular RNAs: splicing's enigma variations. *EMBO J.* 2013;32(7):923–925. https://doi.org/10.1038/emboj.2013.53.
107. You X, Vlatkovic I, Babic A, et al. Neural circular RNAs are derived from synaptic genes and regulated by development and plasticity. *Nat Neurosci.* 2015;18(4):603–610. http://doi.org/10.1038/nn.3975.
108. Zheng Q, Bao C, Guo W, et al. Circular RNA profiling reveals an abundant circHIPK3 that regulates cell growth by sponging multiple miRNAs. *Nat Commun.* 2016;7(11215). http://doi.org/10.1038/ncomms11215.
109. Crow M, Paul A, Ballouz S, Huang ZJ, Gillis J. Exploiting single-cell expression to characterize co-expression replicability. *Genome Biol.* 2016;17(1). https://doi.org/10.1186/s13059-016-0964-6.
110. Hitzemann RJ, Loh HH. Influence of pentobarbital on synaptic plasma membrane protein synthesis. *Life Sci.* 1977;20(1):35–41.
111. Hollingsworth EB, McNeal ET, Burton JL, Williams RJ, Daly JW, Creveling CR. Biochemical characterization of a filtered synaptoneurosome preparation from Guinea pig cerebral cortex: cyclic adenosine 3':5'-monophosphate-generating systems, receptors, and enzymes. *J Neurosci.* 1985;5(8):2240–2253.
112. Chen BJ, Ueberham U, Mills JD, et al. RNA sequencing reveals pronounced changes in the noncoding transcriptome of aging synaptosomes. *Neurobiol Aging.* 2017;56:67–77. https://doi.org/10.1016/j.neurobiolaging.2017.04.005.

113. Ma Q, Chang HY. Single-cell profiling of lncRNAs in the developing human brain. *Genome Biol.* 2016;17(1). https://doi.org/10.1186/s13059-016-0933-0.
114. Skene NG, Grant SGN. Identification of vulnerable cell types in major brain disorders using single cell transcriptomes and expression weighted cell type enrichment. *Front Neurosci.* 2016;10. https://doi.org/10.3389/fnins.2016.00016.
115. Richiardi J, Altmann A, Milazzo A-C, et al. Correlated gene expression supports synchronous activity in brain networks. *Science.* 2015;348(6240):1241–1244. http://doi.org/10.1126/science.1255905.
116. Krishnaswami SR, Grindberg RV, Novotny M, et al. Using single nuclei for RNA-seq to capture the transcriptome of postmortem neurons. *Nat Protoc.* 2016;11(3):499–524. https://doi.org/10.1038/nprot.2016.015.
117. Sanz E, Yang L, Su T, Morris DR, McKnight GS, Amieux PS. Cell-type-specific isolation of ribosome-associated mRNA from complex tissues. *Proc Natl Acad Sci USA.* 2009;106(33):13939–13944.
118. Brackett DM, Qing F, Amieux PS, Sellers DL, Horner PJ, Morris DR. FMR1 transcript isoforms: association with polyribosomes; regional and developmental expression in mouse brain. *PLoS One.* 2013;8(3):e58296. https://doi.org/10.1371/journal.pone.0058296.
119. Gonzalez C, Sims JS, Hornstein N, et al. Ribosome profiling reveals a cell-type-specific translational landscape in brain tumors. *J Neurosci.* 2014;34(33):10924–10936. https://doi.org/10.1523/JNEUROSCI.0084-14.2014.
120. Puighermanal E, Cutando L, Boubaker-Vitre J, et al. Anatomical and molecular characterization of dopamine D1 receptor-expressing neurons of the mouse CA1 dorsal hippocampus. *Brain Struct Funct.* 2017;222(4):1897–1911. https://doi.org/10.1007/s00429-016-1314-x.
121. Kim T, Lim C-S, Kaang B-K. Cell type-specific gene expression profiling in brain tissue: comparison between TRAP, LCM and RNA-seq. *BMB Rep.* 2015;48(7):388–394. https://doi.org/10.5483/BMBRep.2015.48.7.218.
122. Chandra R, Francis TC, Konkalmatt P, et al. Opposing role for Egr3 in nucleus accumbens cell subtypes in cocaine action. *J Neurosci.* 2015;35(20):7927–7937. https://doi.org/10.1523/JNEUROSCI.0548-15.2015.
123. Shigeoka T, Jung H, Jung J, et al. Dynamic axonal translation in developing and mature visual circuits. *Cell.* 2016;166(1):181–192. https://doi.org/10.1016/j.cell.2016.05.029.
124. Srinivasan R, Lu T-Y, Chai H, et al. New transgenic mouse lines for selectively targeting astrocytes and studying calcium signals in astrocyte processes in situ and in vivo. *Neuron.* 2016;92(6):1181–1195. https://doi.org/10.1016/j.neuron.2016.11.030.

CHAPTER 14

# Abnormal Social Behaviors and Dysfunction of Autism-Related Genes Associated With Daily Agonistic Interactions in Mice

**Natalia N. Kudryavtseva, Irina L. Kovalenko, Dmitry A. Smagin, Anna G. Galyamina, Vladimir N. Babenko**
Institute of Cytology and Genetics, Siberian Branch of Russian Academy of Sciences, Novosibirsk, Russia

## INTRODUCTION

The ability of people to communicate with each other is a necessary component of social behavior, and it is important for normal development of humans who live in a community. Decline in sociability may be the result of a negative social environment, or it may be the result of other environmental or genetic factors underlying the development of affective and neurological disorders. A pronounced impairment in communication is found in autism, which is a neurodevelopment disorder with an onset at an early age.[1] Autism is characterized by impaired socialization, low communication, and restricted and/or repetitive behaviors.[2] Autism may also be accompanied by other neurobiologic symptoms.[3] Moreover, there is a growing consensus that autism is a complex syndrome with a range of severity and phenotypical features forming what we call the autism spectrum disorder (ASD). Underlying ASD may be several mechanisms acting simultaneously during early development[1,4,5] influenced by numerous environmental factors, for example, prenatal stress.[6] ASD risk factors are thought to include unbalanced diet, exposure to toxins, infectious diseases, stress, and a multitude of other factors.[7,8] Furthermore, the symptoms of ASDs may overlap with those of other diseases.[2] ASD has been found to be comorbid with several disorders, including epilepsy, schizophrenia, attention deficit hyperactivity disorder (ADHD), depression, and anxiety.[9–12]

To date, experimental studies have been focused on the identification of adequate models to investigate the mechanisms and potential treatments of impaired social behaviors. Animal studies have demonstrated that social impairment in rodents may be produced by prenatal stress,[13,14] early handling,[15] maternal separation,[16] and early social isolation.[17,18] Several knockouts or mutant mice, such as 4E-BP2, Fmr1, Mecp2, Ube3a, Nf1, Pten, Tsc1/Tsc2, and Shank, exhibited low sociability compared with wild-type animals.[19–23] Among the strains of mice that exhibit low communication, the BTBR T+tf/J, BALB/cJ, C57BL/6J, A/J, and 129S1/SvImJ mouse lines have also

*Molecular-Genetic and Statistical Techniques for Behavioral and Neural Research*
ISBN 978-0-12-804078-2, https://doi.org/10.1016/B978-0-12-804078-2.00014-3

been considered models of autism.[24–28] Thus, both environmental as well as genetic factors may lead to the impairment of social behaviors.

Social behavior of laboratory animals (mainly rodents) is typically evaluated as the behavioral response to a conspecific in the home cage or in a neutral test apparatus.[25,27,29] Reduced response to the tester or stimulus subject as compared to control is considered a sign of impairment, an effect that may be utilized for investigating the mechanisms of decreased social behavior as well as its treatment. Knockout mice and wild type controls were compared using this method, but other changes in social behaviors as well as in repetitive behaviors have also been quantified.[25,26,28] In addition to impaired social behavioral responses, reduced activity, impaired memory of previously seen conspecific subjects, as well as increased anxiety may also be detected. Restricted (repetitive) behaviors (circling, digging, and jumping), increased motor activity (spontaneous activity, exploration), and increased self-involved behaviors (self-grooming, scratching) may also accompany impaired social behavior of animals.[25,26,28]

Our studies have previously demonstrated that male mice with long-term, daily repeated experience of aggression accompanied by wins or chronic social defeats exhibited disturbances in social and motivation-related behaviors, as well as showing pronounced anxiety, aggressiveness, a depression-like state, and stereotypic behaviors.[30–36] The initial findings indicated that these mice—winners and losers—suffered from impaired communication and sociability.[32,37] The current chapter presents a detailed analysis of social behaviors of male mice that have been living in a socially hostile environment. Our goal is to identify behavioral changes induced by this environmental treatment that other authors[25,26,28] have considered autistic spectrum symptoms similar to those seen in humans.

Our hypothesis is that autistic symptoms may develop in adult mice in a hostile environment in our behavioral paradigm. We also suggest that changes in social behaviors may involve changes in the expression of genes associated with autism in the brain. To investigate this hypothesis, we used RNA-Seq analysis of whole transcriptome and analyzed gene expression in the midbrain raphe nuclei (MRN), ventral tegmental area (VTA), dorsal striatum (STR), hippocampus (HPC), and hypothalamus (HPT) of the winners and losers. Our analysis focused on identification of changes in the expression of genes that have been found to be strongly associated with autism according to multiple databases (OMIM, http://omim.org/; MalaCards, http://www.malacards.org; http://autism.mindspec.org/autdb/Welcome.do). Thus, we have chosen to analyze the following 27 genes: *Tph2*, *Maoa*, *Slc6a4*, *Htr7*, *Shank2* (*auts17*), *Shank3*, *Mecp2* (*autsx3*), *Ube3a*, *Gabrb3*, *Cntn3*, *Reln*, *Pten*, *Ctnnd2*, *Nrxn1*, *Nrxn2*, *Nlgn1*, *Nlgn2*, *Nlgn3* (*autsx1*), *Pcdh10*, *Cadps2*, *Foxp2*, *Fmr1*, *Auts2*, *En2*, *Arx*, *Oxtr*, and *Ptchd1* (*autsx4*). Most of these genes are related to signaling and neurotransmission.

# MATERIALS AND METHODS

## Animals

Adult male mice of the C57BL/6 strain were obtained from Animal Breeding Facility, Branch of Institute of Bioorganic Chemistry of the Russian Academy of Sciences (ABF BIBCh, RAS) (Pushchino, Moscow region). The animals were housed under standard conditions (12:12 h light/dark regime, switch-on at 8:00 a.m.; food (pellets) and water available ad libitum). Mice were weaned at 1 month of age and housed in groups of 8−10 in plastic cages (36 × 23 × 12 cm). Experimental animals were 10−12 weeks old. All procedures were in compliance with the European Communities Council Directive 210/63/EU on September 22, 2010. The study was approved by Scientific Council N 9 of the Institute of Cytology and Genetics SB RAS of March 24, 2010, N 613.

## Induction of Positive and Negative Social Experience in Male Mice

Prolonged positive and negative social experience (wins and defeats) in male mice was induced by daily agonistic interactions.[38,39] Pairs of weight-matched animals were each placed in a steel cage (14 × 28 × 10 cm) bisected by a perforated transparent partition allowing the animals to see, hear, and smell each other, but preventing physical contact. The animals were left undisturbed for 2−3 days to adapt to their new housing conditions. Every afternoon (14:00−17:00 p.m.), the cage lid was replaced by a transparent one, and 5 min later (the period necessary for individual's activation), the partition was removed for 10 min to encourage agonistic interactions. The superiority of one of the males was firmly established within 2−3 encounters with the same opponent. The superior male (winner) would be attacking, biting, and chasing the other, who would be displaying only defensive behavior (sideways postures, upright postures, withdrawal, lying on the back, or freezing). As a rule, aggressive interactions between males are discontinued by lowering the partition if the sustained attacks have lasted 3 min (in some cases less), thereby preventing the damage of defeated mice. Each defeated mouse (loser) was exposed to the same winner for 3 days. Subsequently, each loser was placed, once a day after the fight, in an unfamiliar cage with an unfamiliar winner behind the partition. Each winning mouse (winners, aggressors) remained in its original cage. This procedure was performed once a day for 20 days and yielded an equal number of winners and losers.

In the behavioral study, the following three groups of animals were used: (1) Losers were groups of chronically defeated mice after 10 and 20 days of agonistic interactions (Los-10 and Los-20); (2) Winners were groups of chronically winning mice that exhibited daily aggression during 10 and 20 days (Win-10 and Win-20) of agonistic interactions; and (3) Controls were mice without a consecutive experience of agonistic interactions.[38,39]

20-day winners and 20-day losers with the most expressed behavioral phenotypes of the winners and losers were selected for the transcriptome analysis. Different mice were used for the behavioral study and transcriptome analysis. All mice were simultaneously decapitated, including 20-day winners and losers, 24 h after the last agonistic interaction. Control animals were also analyzed simultaneously. The brain regions were dissected by the same experimenter according to the Allen Mouse Brain Atlas (http://mouse.brain-map.org/static/atlas). All biologic samples were placed in RNAlater solution (Life Technologies, USA) and were stored at −70°C until sequencing.

The brain regions were selected for the analysis based on their functions and localization of neurons of neurotransmitter systems. These areas were as follows: the MRN, which comprises a multifunctional brain region and contains the majority of serotonergic neuronal bodies; the VTA, which contains the bodies of dopaminergic neurons, is widely implicated in the natural reward circuitry of the brain, and is important in cognition, motivation, drug addiction, and emotions related to several psychiatric disorders; the dorsal STR, which is responsible for the regulation of motor activity and stereotypical behaviors and is also potentially involved in various processes; the HPC, which belongs to the limbic system, is essential for memory consolidation and storage, and plays important roles in neurogenesis and emotional mechanisms; and the HPT, which regulates the stress reaction and many other physiological functions.

## BEHAVIORAL ANALYSIS

### Partition Test

The partition test can estimate behavioral reactivity of mice to a conspecific behind a transparent perforated partition dividing the experimental cage into equal parts.[29] The number of approaches to the partition made by the mouse and the total time spent near it (moving near the partition, smelling and touching it with one or two paws, clutching and hanging, putting noses into the holes, or gnawing the holes) were scored in this study during a 5-min recording session as indices of reacting to the partner in the neighboring compartment of the home cage. The time experimental males showed sideways position or "turning away" near the partition was not included in the total session time.

The experimental procedure was as follows. Two mice (a winner and a loser) were kept for about 1 day together in a home cage with a partition (familiar partner). Then, the steel cover of the cage was replaced by a transparent one. The behaviors were videotaped for 5 min.

### Social Interaction Test

The social interaction test was developed to quantitatively assess social behavior in mice shown toward a standard tester, i.e., a group-housed unfamiliar subject of the same strain, weight, and age. Every tester was used once. The tested experimental mouse and standard tester were simultaneously placed in a neutral territory, i.e., in a clean cage, measuring 36 × 23 × 12 cm in size, in opposite corners of the cage. The cage did not have litter and fodder, which may distract the mouse from the partner.

During a 10-min test, the following behavioral parameters were recorded: (1) *Avoidance behavior*, avoidance of the approaching partner, flight, or freeze in response to his approach, which included freezing in the upright posture; (2) *Approach behavior* directed toward the unfamiliar partner: approaching, sniffing, and following the opponent; (3) *Attack*, attacking, biting, or chasing the opponent; (4) *Aggressive grooming* (mounting the unfamiliar opponent's back, holding the opponent down, and licking and nibbling at the scruff of the neck); the opponent is wholly immobilized or may stretch out the neck and freeze; (5) *Rearing,* an upright position with forepaws in the air, a form of exploratory activity; (6) *Immobility*, the animal sits in the corner or near the wall; (7) *Locomotor activity* of animals in the cage without attention to the partner; (8) *Self-grooming*, body care activities, such as fur licking and head or body washing, a form of displacement activity. The total time and number of events, as well as the percent of animals that demonstrated particular behaviors during the test were measured. After each encounter, the cage was washed several times with water and was dried with a clean tissue paper.

The animals were taken to a test room and remained there for 5 min, so they could adapt to the new environment. Animal behavior during the test was video-recorded, and the videos were later replayed and processed.

## RNA-Seq

The collected brain samples were sequenced at ZAO Genoanalytica (www.genoanalytica.ru, Moscow, Russia). mRNA was extracted using a Dynabeads mRNA Purification Kit (Ambion, Thermo Fisher Scientific, Waltham, MA, USA). cDNA libraries were constructed using the NEBNext mRNA Library PrepReagent Set for Illumina (New England Biolabs, Ipswich, MA, USA) following the manufacturer's protocol and were subjected to Illumina sequencing. More than 20 million reads were obtained for each sample. The resulting "fastq" format files were used to align all reads to the GRCm38.p3 reference genome using the TopHat aligner.[40] DAVID Bioinformatics Resources 6.7 (http://david.abcc.ncifcrf.gov) was used for the description of differentially expressed gene ontology. The Cufflinks program was used to estimate gene expression levels in fragments per kilobase of transcript per million mapped reads (FPKM) and subsequently to identify differentially expressed genes. Each brain region was represented by three samples each for the control, aggressive, and defeated animals. Only annotated gene sequences were used in the analysis. Genes were considered differentially expressed when the probability of the null hypothesis (lack of difference) was smaller than 5%, i.e., $P \leq .05$. $q < 0.05$ (the false discovery rate (FDR)-adjusted *P* value after Benjamini-Hochberg correction for multiple testing) was also taken into consideration.

We have previously conducted studies of gene expression in males in similar experiments using RT-PCR method with larger samples for each compared experimental group, i.e., winners, losers, and control ($n > 10$ animals). The direction and extent of changes in expressions of the *Tph2*, *Slc6a4*, *Bdnf*, *Creb1*, and *Gapdh* genes in the MRN of males compared with the control produced by two methods, including RT-PCR[41,42] and RNA-Seq, were found generally consistent, thus verifying transcriptome analyses of the data provided by the ZAO Genoanalitika (http://genoanalytica.ru, Moscow).

Analysis of the correlation between microarrays and RNA-Seq detection of transcripts in the same tissue sample to explore the reproducibility between the techniques showed that the results obtained by RNA-Seq and microarrays were highly reproducible.[43] Therefore, we can be sure that the data obtained by the RNA-Seq method reflects the real changes that arise in the brain under the influence of environmental impact.

The Human Gene Database (http://www.genecards.org/); GeneMania database (http://www.genemania.org), Online Mendelian Inheritance in Man Database (OMIM, http://omim.org/); Human Disease database (MalaCards, http://www.malacards.org); Autism database (http://autism.mindspec.org/autdb/Welcome.do), and Human Biological Pathway Unification database (http://pathcards.genecards.org/) were used for the description and analysis of the data obtained.

## Selection of Target Genes

In the analysis of the whole transcriptome using RNA-Seq in five brain regions of male mice with chronic social defeats or wins in daily agonistic interactions, we identified changes in the expression of numerous genes, which are considered candidates for autism according to those previously described in the Autism database (http://autism.mindspec.org/autdb/Welcome.do). Notably, a substantial portion of the candidate genes corresponds to neurospecific genes involved in synaptic transmission. This chapter is focused on the analysis of genes that are strongly associated with the development of autism or ASDs in humans and are also highly neurospecific. These genes included the following: serotonergic genes: *Tph2*, *Maoa*, *Slc6a4*, and *Htr7*; genes of cell adhesions and synaptic proteins: *Reln*, *Pten*, *Ctnnd2*, *Nrxn1*, *Nrxn2*, *Nlgn1*, *Nlgn2*, *Nlgn3* (*autsx1*), *Shank2* (*auts17*), and *Shank3*; genes that code receptors: *Gabrb3*, *Htr7*, *Oxtr*, and *Pcdh10*; genes involved in brain development: *Mecp2* (*autsx3*), *Arx*, *Auts2*, *En2*, *Foxp2*, *Fmr1*, and *Ptchd1* (*autsx4*); transcription factors: *Foxp2*; translation repressors: *Fmr1*; and factor of protein degradation: *Ube3a* gene. We built the interaction network using a String's association network resource (http://string-db.org).[44] It demonstrates interactions between proteins that are coded by most of our genes related to the signal neurotransmission networks

(Fig. 14.1), which include serotonergic genes and genes of cell adhesions and synaptic proteins. The String database located all serotonergic genes closely related to each other and the *Gabrb3* gene (Fig. 14.1). A significant association has previously been demonstrated between *Gabrb3* and serotonergic genes in the regulation of schizophrenia symptoms, such as delusions, hallucinations, mania, depression, and negative symptoms.[45,46] Furthermore, we analyzed the expression of the *Bdnf* gene, which encodes a protein involved in neural plasticity and, as a consequence, in many psychiatric disorders, particularly autism and ASDs.[47,48]

## Statistical Analysis

Statistical analysis of the behavioral data was performed using one-way ANOVA with factor "groups" (Controls, Win-10, Win-20, Los-10, and Los-20) and one-way ANOVA with factor "experience" (Controls, Win-10, and Win-20 and Controls, Los-10, and Los-20), followed by Bonferroni *post hoc* $P < .05$ or, for a number of repetitive behaviors, the LSD test. The data are reported as the mean ± SEM. Statistical analyses were performed using a two-way ANOVA, followed by a Bonferroni's *post hoc* $P < .05$. A Chi-square analysis was used to identify significant differences in the percent of animals that exhibited attacks and aggressive grooming. Statistical significance was accepted when the probability of the null hypothesis was less than 5%, i.e., $P \leq .05$. Sample sizes were $n = 10$ mice in the control group and $n = 11$ mice in the Win-10, Win-20, Los-10, and Los-20 groups, respectively.

For the transcriptome data, a principal component analysis (PCA) was conducted using the XLStat software package (www.xlstat.com). PCA was based on Pearson product moment correlation matrix calculated on the FPKM value profiles of 27 analyzed genes. Agglomerative hierarchic clustering (AHC) and multidimensional scaling (MDS) were performed on the same data with the XLStat software package. We also used a Pearson correlation as a similarity metric for the AHC/MDS analysis. The agglomeration method comprised an unweighted pair-group average. MDS was performed by minimization of the Kruskal stress statistic, with a random initial configuration, five repetitions, and stop conditions as follows: convergence = 0.00001/iterations = 500.

The identification of alternatively spliced events in the RNA-Seq data was performed with rMATs software.[49] Refseq v. 10.0 was used as a template for mapping the reads and alternative event annotations. Only exon skipping events were considered.

# RESULTS

## Social Interactions and Partition Tests

One-way ANOVA revealed a significant influence of the factor "groups" on all parameters of behaviors studied (number of events or/and total time) for avoidance, approach,

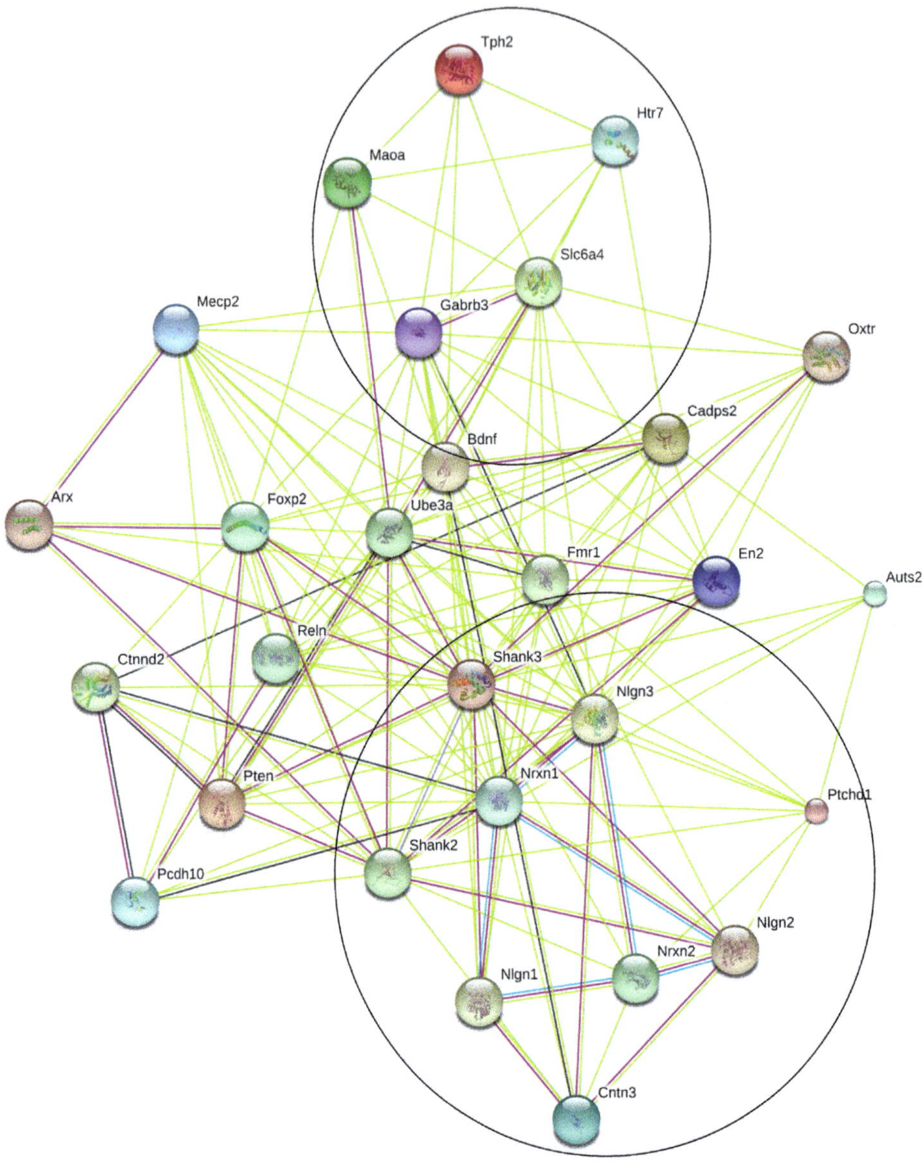


**Figure 14.1** Gene association network of 27 target genes based on the String text mining tool (http://string-db.org). Synaptic genes (low right) and the serotonergic (top center) system are encircled. All genes were connected on the first layer with no mediator second layer proteins. The majority of the associations are based on publication cooccurrence (*salad lines*[43]); however, a significant number of the interactions is experimentally (*red lines*) confirmed. The center node corresponds to the *Shank3* gene, which maintains multiple experimentally confirmed interactions with other genes.

**Table 14.1** One-way ANOVA analysis of behavioral data

| Tested behaviors | Factor "groups" | Factor "experience" in the losers | Factor "experience" in the winners |
|---|---|---|---|
| ***Avoidance behavior*** | | | |
| Number of events | F (4, 48) = 23.64; $P < .001$ | F (2, 28) = 15.51; $P < .001$ | NS |
| Total time | F (4, 48) = 33.07; $P < .001$ | F (2, 28) = 24.24; $P < .001$ | NS |
| ***Approach behavior*** | | | |
| Number of events | F (4, 48) = 16.05; $P < .001$ | F (2, 28) = 29.52; $P < .001$ | NS |
| Total time | F (4, 47) = 13.21; $P < .001$ | F (2, 27) = 30.92; $P < .001$ | F (2, 27) = 10.01; $P < .001$ |
| ***Rearing behavior*** | | | |
| Number of events | F (4, 48) = 8.44; $P < .001$ | F (2, 28) = 21.40; $P < .001$ | F (2, 28) = 5.45; $P < .01$ |
| Total time | F (4, 48) = 12.44; $P < .001$ | F (2, 28) = 26.19; $P < .001$ | F (2, 28) = 10.01; $P < .001$ |
| ***Immobility*** | | | |
| Total time | F (4, 48) = 78.71; $P < .001$ | F (2, 28) = 47.02; $P < .001$ | F (2, 28) = 16.42; $P < .001$ |
| ***Locomotor activity*** | | | |
| Total time | F (4, 48) = 21.86; $P < .001$ | F (2, 28) = 7.81; $P < .002$ | F (2, 28) = 17.59; $P < .001$ |
| ***Self-grooming*** | | | |
| Total time | F (4, 47) = 2,77; $P < .038$ | F (2, 28) = 4.75; $P < .017$ | F (2, 27) = 3.42; $P < .047$ |
| ***Partition behaviors*** | | | |
| Number of approaches | F (4, 47) = 7.47; $P < .001$ | F (2, 28) = 9.84; $P < .001$ | NS |
| Total time | F (4, 47) = 40.18; $P < .001$ | F (2, 28) = 34.30; $P < .001$ | F (2, 27) = 7.79; $P < .002$ |

and rearing behaviors, immobility, locomotor activity, self-grooming, and the behaviors in the partition test (Table 14.1). Additionally, one-way ANOVA opened a significant influence of the factor "experience" on all behaviors in the losers. Exceptions in the winners were avoidance behavior (both parameters) and number of approaches and partition behaviors (Table 14.1). Following Bonferroni, *post hoc* test or, for a number of repetitive behaviors, the LSD test is presented on Figs. 14.2 and 14.3.

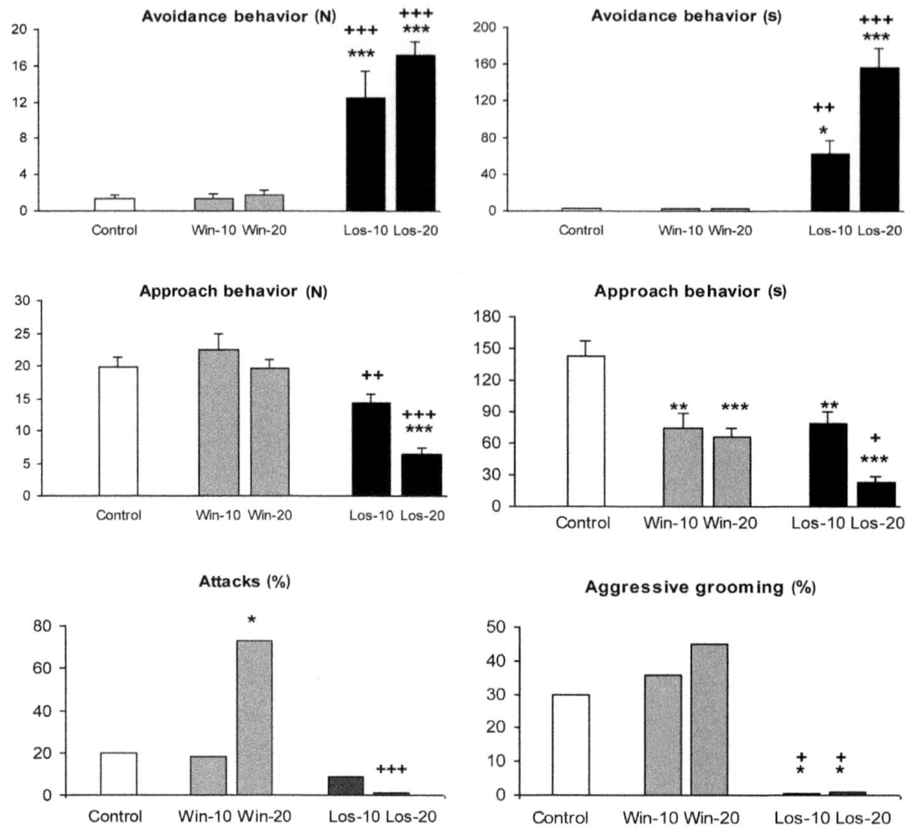


**Figure 14.2** Social behaviors of the controls, winners, and losers in the social interaction test. Los-10, Los-20 and Win-10, Win-20 male mice with experiences of social defeats and victories, respectively, with 10 and 20 days of agonistic interactions. *, $P < .05$; **, $P < .01$; ***, $P < .001$ versus controls; +, $P < .05$; ++, $P < .01$; +++, $P < .001$ versus winners of respective social experience. Significant differences between the groups of Win-10 versus Win-20 and Los-10 versus Los-20 are described in the Results section.

In the social interactions test, the Win-20 exhibited short aggression time (18.4 s/test on the average), and only one winner showed about 3 min of attacks and chasing of a loser in total. Seventy percent of Win-20 (8/11) and 20% of the controls (2/10) attacked the tester mice. One loser out of 11 Los-10 (1/11) and no Los-20 (0/11) attacked the tester mice. There were no differences between proportion of the controls (3/10), Win-10 (4/11), and Win-20 (5/11) exhibiting aggressive grooming (Fig. 14.2). The losers never showed aggressive grooming toward the unfamiliar opponent (0/11). Chi-square analysis revealed differences in the number of animals exhibiting attacks between the controls and Win-20 ($P = .016$) and between the Los-20 and Win-20 ($P = .001$). Differences in the number of animals showing aggressive grooming were

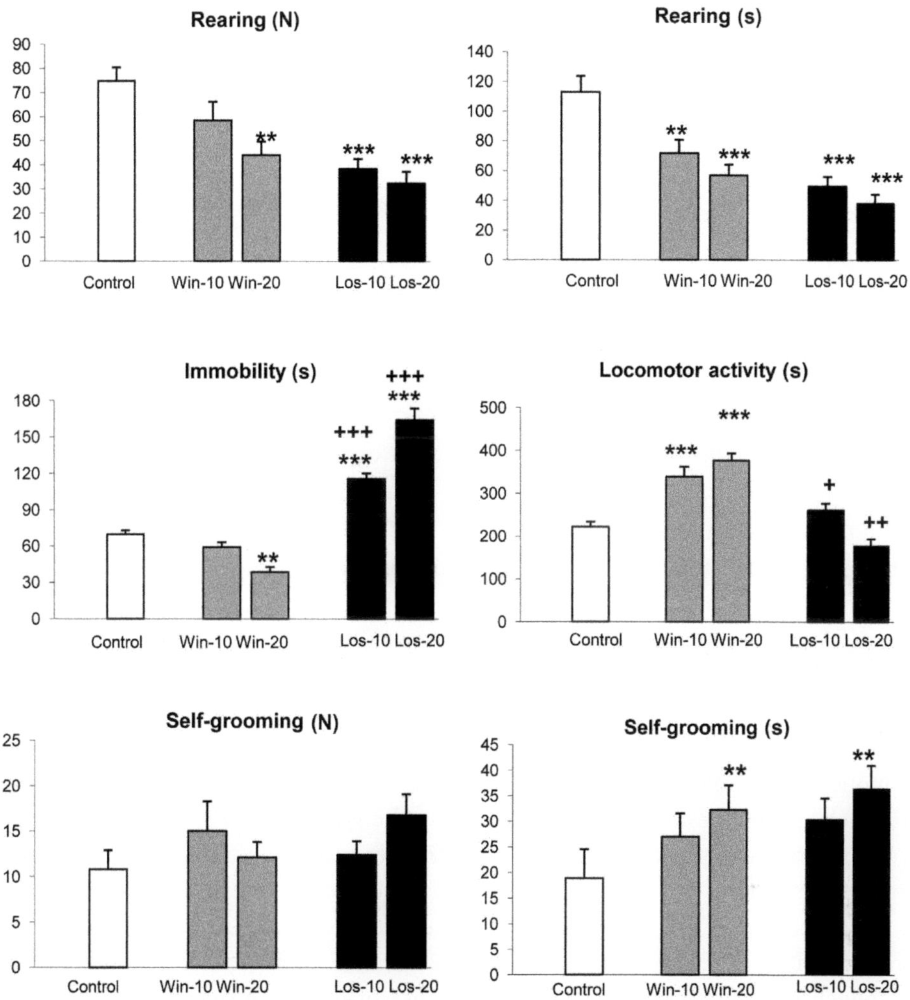


**Figure 14.3** Individual behaviors of the controls, winners, and losers in the social interaction test. Los-10, Los-20 and Win-10, Win-20 male mice with experiences of social defeats and victories, respectively, with 10 and 20 days of agonistic interactions. *, $P < .05$; **, $P < .01$; ***, $P < .001$ versus controls; +, $P < .05$; ++, $P < .01$; +++, $P < .001$ versus winners of respective social experience. Significant differences between the groups of Win-10 versus Win-20 and Los-10 versus- Los-20 are described in the Results section.

revealed between the controls and the Los-10 and Los-20 (for both $P < .05$), between the Los-10 and Win-10 ($P = .027$), and between the Los-20 and Win-20 ($P = .011$).

Fig. 14.4 shows behaviors of the controls, Win-20, and Los-20 expressed as a percentage of total time for locomotor activity, avoidance, approach behavior, rearing, and immobility in the social interactions test. In the neutral territory, the controls exhibited mainly exploration: locomotor activity and rearing time was about 60% of total testing

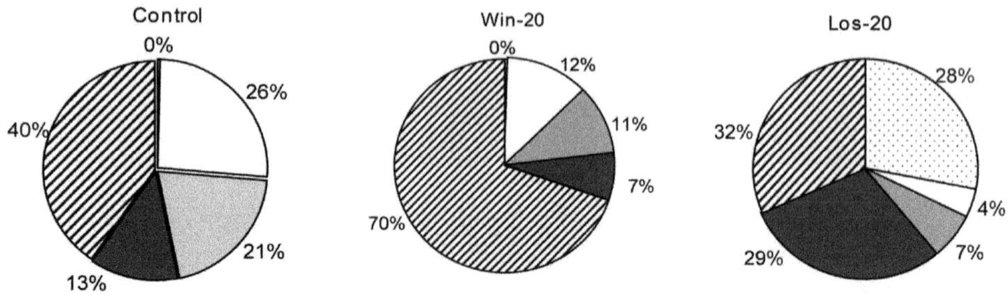


**Figure 14.4** Behaviors of the controls, Win-20, and Los-20 expressed as a percentage of the total time for the demonstration of locomotor activity, avoidance, approach behavior, rearing, and immobility during the social interaction test. ▨, locomotor activity; ⬚, avoidance; □, approach behavior; ▩, rearing; ■, immobility.

time. They displayed interest for the unfamiliar partner: the controls approached the unfamiliar male and spent a substantial amount of time sniffing it. Control mice never demonstrated avoidance behavior.

For most of the 10-min test, i.e., for approximately 5–6 min, the winners randomly moved in the cage and did not pay attention to the partner. Approaching the stranger, they attacked or displayed aggressive grooming. Winners very seldom demonstrated avoidance behavior toward a forthcoming unfamiliar partner. They approached him; however, the total time of following the opponent was decreased because the winners, instead, attacked or exhibited aggressive grooming behavior. The total time spent with locomotion was significantly increased compared with that in the losers. The total duration of self-grooming episodes was increased in the winners.

The losers actively avoided an amicable stranger or demonstrated freezing behavior about 58% of testing time. They rarely approached the unfamiliar opponent or sniffed it as the control mice did. They also sat in the corner or near a wall of the cage, watched the opponent, and never displayed aggressive behavior or aggressive grooming toward him, which may be considered passive avoidance of social contact with other conspecifics. The total time of self-grooming behavior was increased in the losers.

Decreased exploratory activity was identified both in the winners and losers: the number and total duration of rearing events were reduced compared to the control. An ANOVA indicated the influence of social experience duration (days) on the expression of most behavioral parameters in both social groups. The severity of behavioral changes was greater in mice with 20-day agonistic interactions.

In the partition test, one-way ANOVA revealed a significant influence of the factor "groups" on the number of approaches and total time parameters (Fig. 14.5; Table 14.1). Compared with the controls, the number of approaches was decreased in the Los-10 and Los-20, and significant differences were shown in comparison of the Los-10 versus Win-10 and the Los-20 versus Win-20. Total time spent near the partition differs significantly

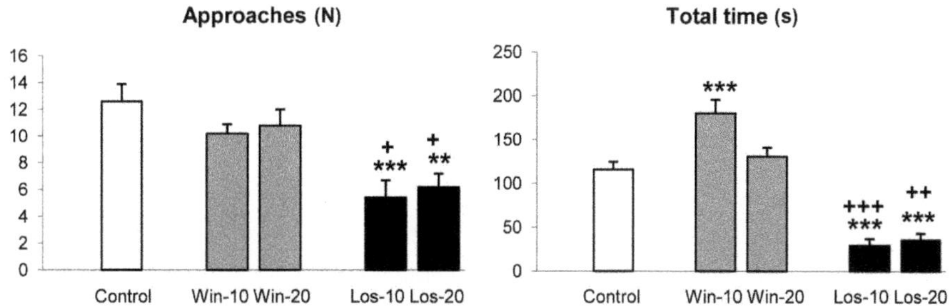


**Figure 14.5** Behaviors of the controls, winners, and losers in the partition test. Los-10, Los-20 and Win-10, Win-20 male mice with experiences of social defeats and victories, respectively, with 10 and 20 days of agonistic interactions. **, $P < .01$; ***, $P < .001$ versus controls; +, $P < .05$; +++, $P < .001$ versus winners of respective social experience.

in comparison of the control versus Los-10 and Los-20 and versus the Win-10, the Los-10 versus Win-10, and in Los-20 versus Win-20. One-way ANOVA revealed a significant influence of the factor "experience" in the losers on number of approaches and total time and in the winners on total time (Table 14.1). The losers exhibited low interest in the partner in the home cage as follows: the number of approaches and total time spent near the partition as a reaction to the partner in the neighboring compartment were significantly decreased.

## RNA-Seq Study

Analyzing whole transcriptome data of the MRN, VTA, STR, HPC, and the HPT, we found similar and differing changes in the expression of autistic genes in the winners and losers.

In the HPT (Fig. 14.6A and B; Table 14.2), autism-associated genes decreased their expression both in the winners and losers compared to control. These genes included *Maoa, Gabrb3, Fmr1, Ube3a, Pten, Mecp2, Ptchd1,* and *Nrxn1.* In both social groups, expression of the *Shank3* gene was increased. In comparison with the controls, in the winners, expression of genes was decreased in the following genes: *Reln, Foxp2,* and *Arx* genes. In the losers, expression of the *Nrxn2, Nlgn2, Nlgn3, Auts2* and *Pcdh10* genes was increased, and expression of the *Nlgn1* and *Cntn3* genes was decreased. In total, 12 autistic genes in the winners and 16 genes in the losers changed their expression under agonistic interactions.

In the MRN, expression of the *Tph2, Slc6a4, Foxp2, Reln,* and *Oxtr* genes is decreased in both social groups (Fig. 14.7; Table 14.2). Additionally, in the winners, expression of the *Ube3a, Fmr1, Nrxn1, Cadps2,* and *Cntn3* genes increased. There were no specific changes in this brain region in the losers. In total, 10 genes in the winners and 5 genes in the losers changed their expression in the MRN under agonistic interactions.

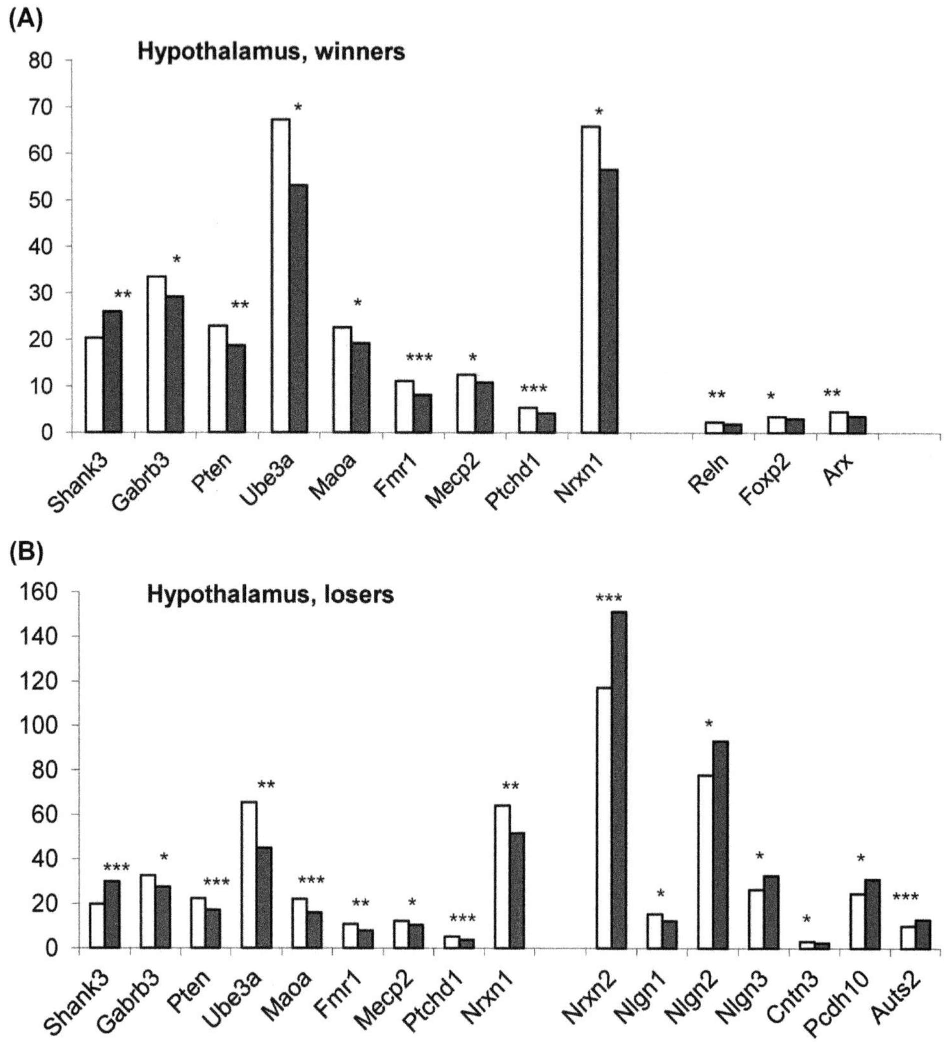


**Figure 14.6** Differentially expressed autistic genes in the hypothalamus of the winners (A) and losers (B) under long agonistic interactions. The Cufflinks program was used to estimate the gene expression levels in FPKM units. The autistic gene expression levels are presented in the control (*white columns*) and the winners and losers (*black columns*). *, $P < .05$; **, $P < .01$; ***, $P < .001$ versus controls.

In the HPC, reduced expression of the *Shank3*, *Auts2*, *Nrxn2*, and *Ctnnd2* genes was revealed in the winners. The *Shank2*, *Nlgn2*, *Pcdh10*, *Reln*, and *Arx* genes increased their expression in the losers (Fig. 14.8A; Table 14.2). In total, nine autistic genes (four genes in the winners and five genes in the losers) changed their expression in this brain region under agonistic interactions.

**Table 14.2** Differentially expressed autism-related genes in different brain regions in male mice with positive (winners) and negative (losers) social experience in daily agonistic interactions

| Genes | Annotation | Winners | Losers | Exon skipping events |
|---|---|---|---|---|
| *Tph2* | tryptophan hydroxylase 2 | ▼▼▼▼* MRN<br>▲▲▲▲* VTA | ▼▼▼▼* MRN<br>▲▲▲▲* VTA | 0 |
| *Maoa* | monoamine oxidase A | ▼ HPT | ▼▼▼▼* HPT | 1 |
| *Slc6a4* | solute carrier family 6 (serotonin transporter), member 4 | ▼▼▼▼* MRN<br>▲▲▲▲* VTA | ▼▼▼▼* MRN<br>▲▲▲▲* VTA | 0 |
| *Htr7* | 5-hydroxytryptamine (serotonin) receptor 7 | | ▲▲▲▲* STR | 1 |
| *Shank2* | SH3 And Multiple Ankyrin Repeat Domains 2 (auts17) | | ▲ HPC | 2 |
| *Shank3* | SH3/ankyrin domain gene 3 | ▲▲ HPT<br>▼▼ HPC | ▲▲▲* HPT | 3 |
| *Mecp2* | Methyl-CpG-Binding Protein 2 (autx3) | ▼ HPT | ▼ HPT<br>▲ VTA | 1 |
| *Ube3a* | ubiquitin protein ligase E3A | ▼ HPT<br>▲ MRN | ▼▼* HPT | 2 |
| *Gabrb3* | gamma-aminobutyric acid (GABA) A receptor, subunit beta 3 (*auts4*) | ▼ HPT | ▼ HPT | 0 |
| *Cntn3* | contactin 3 | ▲ MRN | ▼ HPT | 0 |
| *Reln* | reelin | ▼▼▼▼* MRN<br>▼▼ HPT | ▼▼▼▼* MRN<br>▲ HPC<br>▼ STR | 1 |
| *Pten* | phosphatase and tensin homolog | ▼▼* HPT<br>▼▼ STR | ▼▼▼* HPT | 1 |
| *Ctnnd2* | catenin (cadherin associated protein), delta 2 | ▼ HPC | | 1 |
| *Nrxn1* | neurexin I | ▼ HPT<br>▲ MRN | ▼ HPT | 7 |
| *Nrxn2* | neurexin 2 | ▼▼ HPC | ▲▲▲* HPT | 11 |
| *Nlgn1* | neuroligin1 | | ▼ HPT | 1 |
| *Nlgn2* | neuroligin2 | | ▲ HPT<br>▲ HPC | 1 |
| *Nlgn3* | neuroligin3 (autsx1) | | ▲ HPT | 3 |
| *Pcdh10* | protocadherin 10 | | ▲ HPT<br>▲▲ HPC | 0 |
| *Cadps2* | Ca2+-dependent activator protein for secretion 2 | ▲ MRN | ▲▲ VTA | 3 |
| *Foxp2* | forkhead box P2 | ▼▼▼▼* MRN<br>▼ HPT | ▼▼▼▼* MRN<br>▲▲▲▲* VTA | 0 |
| *Fmr1* | fragile X mental retardation syndrome 1 homolog | ▼▼▼* HPT<br>▲ MRN | ▼▼* HPT | 1 |
| *Auts2* | autism susceptibility gene 2 protein | ▼ HPC | ▲▲▲* HPT | 3 |
| *En2* | engrailed 2 (auts1) | | ▲▲ STR<br>▲▲▲▲* VTA | 0 |
| *Arx* | aristaless related homeobox | ▼▼ HPT | ▲▲▲▲* HPC<br>▼ STR | 0 |
| *Oxtr* | oxytocin receptor | ▼▼ MRN | ▼▼ MRN | 0 |
| *Ptchd1* | patched domain containing 1 | ▼▼▼* HPT<br>▼▼ STR | ▼▼▼▼* HPT<br>▲ VTA | 0 |

Note: ▼ - downregulation; ▲ - upregulation genes expression in comparison with the controls; ▲ - $p \le 0.05$; ▲▲ - $P \le 0.01$; ▲ ▲▲ - $P \le 0.001$; ▲▲▲▲ - $P \le 0.0001$; * - $q < 0.05$. MRN – midbrain raphe nuclei; HPC – hippocampus; STR – striatum; VTA – ventral tegmental area; HPT – hypothalamus. Gray colour – changes of genes expression common for both - the winners and losers.

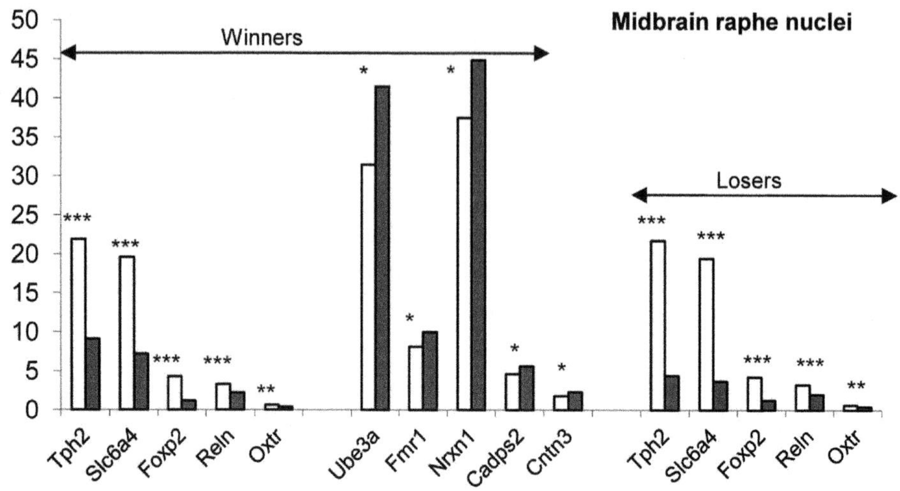


**Figure 14.7** Differentially expressed autistic genes in the midbrain raphe nuclei of the winners and losers under long agonistic interactions. The Cufflinks program was used to estimate the gene expression levels in FPKM units. The autistic gene expression levels are presented in the control (*white columns*) and the winners and losers (*black columns*). *, $P < .05$; **, $P < .01$; ***, $P < .001$ versus controls.

In the VTA, increased expression of the *Tph2* and *Slc6a4* genes was found for both social groups. The *Mecp2*, *Ptchd1*, *Foxp2*, *En2*, and *Cadps2* genes increased their expression only in the losers (Fig. 14.8B; Table 14.2).

In the STR, the *Pten* and *Ptchd1* genes in the winners and the *Reln* and *Arx* genes in the losers decreased the expression under repeated agonistic interactions (Fig. 14.8C; Table 14.2). In the losers, expression of the *Htr7* and *En2* genes was increased.

Expression of the *Bdnf* gene was increased in the STR of the losers, and it was decreased in the MRN in the winners and losers.

## Concordant and Discordant Changes in Gene Expression in the Winners and Losers

In the HPT of the winners and losers, the expression of the *Maoa*, *Mecp2*, *Ube3a*, *Gabrb3*, *Cntn3*, *Pten*, *Nrxn1*, *Fmr1*, and *Ptchd1* genes was downregulated, and the *Shank3* gene was upregulated. In the MRN of both social groups, the *Tph2*, *Slc6a4*, *Reln*, *Foxp2*, and *Oxtr* genes were downregulated. In the VTA, the *Tph2* and *Slc6a4* genes were upregulated.

Specific changes in the winners included the downregulation of the *Reln*, *Foxp2* and *Arx* genes in the HPT, the *Shank3*, *Auts2*, *Ctnnd2*, and *Nrxn2* genes in the HPC, and the *Pten* and *Ptchd1* genes in the STR. In contrast, the *Ube3a*, *Fmr1*, *Nrxn1*, *Cadps2*, and *Cntn3* genes were upregulated in the MRN of the winners.

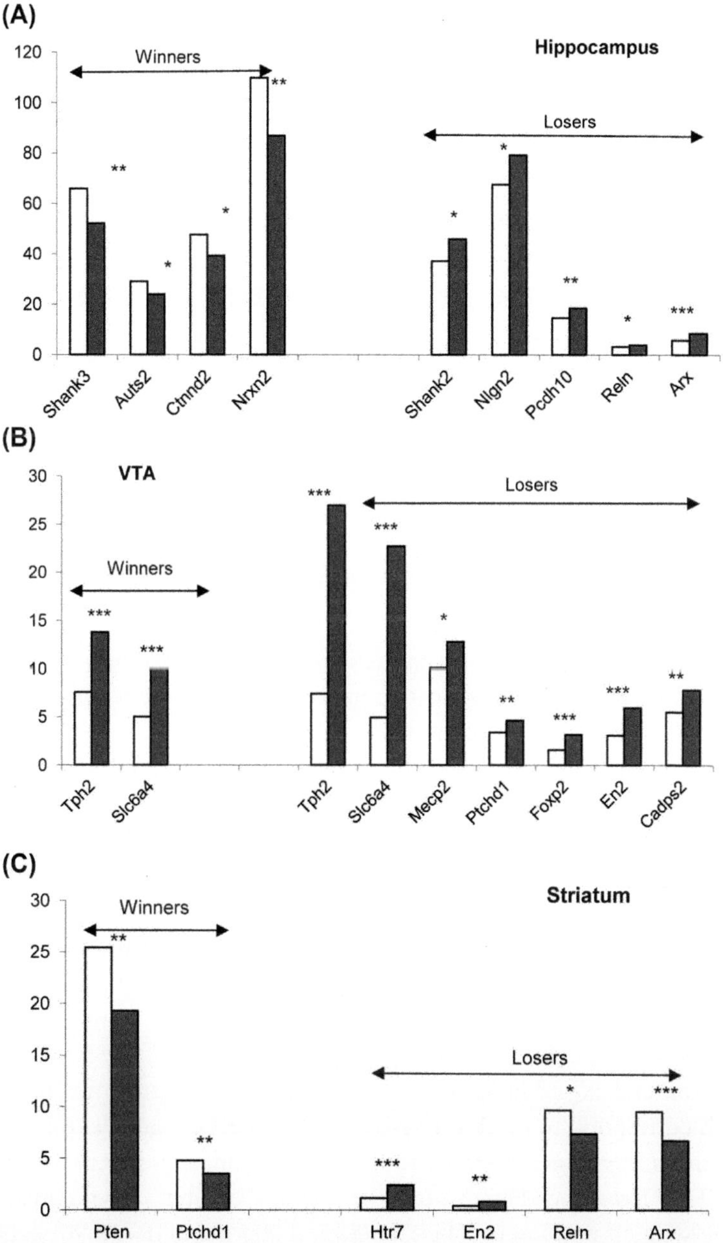


**Figure 14.8** Differentially expressed autistic genes in the hippocampus (A), VTA (B), and striatum (C) of the winners and losers under long agonistic interactions. The Cufflinks program was used to estimate the gene expression levels in FPKM units. The autistic gene expression levels are presented in the control (*white columns*) and the winners and losers (*black columns*). *, $P < .05$; **, $P < .01$; ***, $P < .001$ versus controls. *VTA*, ventral tegmental area.

Specific changes in the losers included the upregulation of the *Shank2*, *Nlgn2*, *Pcdh10*, *Reln*, and *Arx* genes in the HPC; the *Mecp2*, *Ptchd1*, *Foxp2*, *En2*, and *Cadps2* genes in the VTA; the *Nrxn2*, *Nlgn2*, *Nlgn3*, *Pcdh10*, and *Auts2* genes in the HPT; and the *En2* and *Htr7* genes in the STR. An opposite downregulation was identified for the *Ube3a* and *Nlgn1* genes in the HPT and the *Reln* and *Arx* genes in the STR of the losers.

## Principal Component Analysis of the Expression of Autism-Associated Genes

To assess the degree of brain region-specific expression of genes of interest, we performed PCA based on the covariation of 27 genes using the expression profiles of 45 samples, which comprised RNA-Seq FPKM data for five brain regions of nine mice. Ovals correspond to brain regions. We observed compact clustering of samples in the HPT, STR, and HPC based on gene expression profiles (Fig. 14.9, encircled), whereas samples of the MRN and VTA were not distinct. Compact clustering of samples underscores distinct pattern of considered genes expression in each brain region.

## Agglomerative Hierarchic Clustering of Autistic Coding Transcripts and Identification of Outliers

We applied AHC to the initial sample of 27 genes across 45 samples, which comprised RNA-Seq FPKM data for five brain regions of nine mice based on their expression profiles. Fig. 14.10 presents the dendrogram of gene clustering.

An AHC analysis elucidates closely correlated genes, which thus strengthens the confidence of concordant differential expressions of a gene pair based on their highly correlated expression profiles across 45 samples (Fig. 14.10). Thus, we reaffirm the significant differential expression given correlated pairs of genes and independent tests using the Cufflinks program. For example, the *Tph2* and *Slc6a4* genes both exhibit downregulation in the MRN and upregulation in the VTA of the winners and losers while maintaining the coexpression correlation value across 45 samples, $r = 0.99$; $df = 43$; $P < 3.7E\text{-}39$. The *Maoa* and *Ube3a* genes ($r = 0.93$) and the *Cntn3* and *Mecp2* genes ($r = 0.58$) exhibit downregulation in the HPT of the winners and losers. Both the *Nrxn2* and *Nlgn2* genes ($r = 0.90$) exhibit upregulation in the HPT of the losers (which correspond to inhibitory synapses[50]). The *Oxtr* gene exhibits downregulation and the *Cadps2* gene exhibits upregulation in the MRN of the winners ($r = 0.85$). The *Reln* and *Foxp2* genes ($r = 0.89$) exhibit downregulation in the MRN of the winners and losers. The *Ctnnd2* and *Arx* genes ($r = 0.96$) exhibit downregulation in the HPT and HPC in the winners. The *Shank3* and *Auts2* ($r = 0.62$) genes exhibit downregulation in the HPC of the winners and upregulation in the HPT of the losers. The *Shank2* gene exhibits upregulation in the HPC and the *Nlgn3* gene ($r = 0.87$) exhibits upregulation in the HPT of the losers. Last, the *Gabrb3* and *Nlgn1* ($r = 0.89$) genes exhibit downregulation in the HPT in the losers. All correlation values

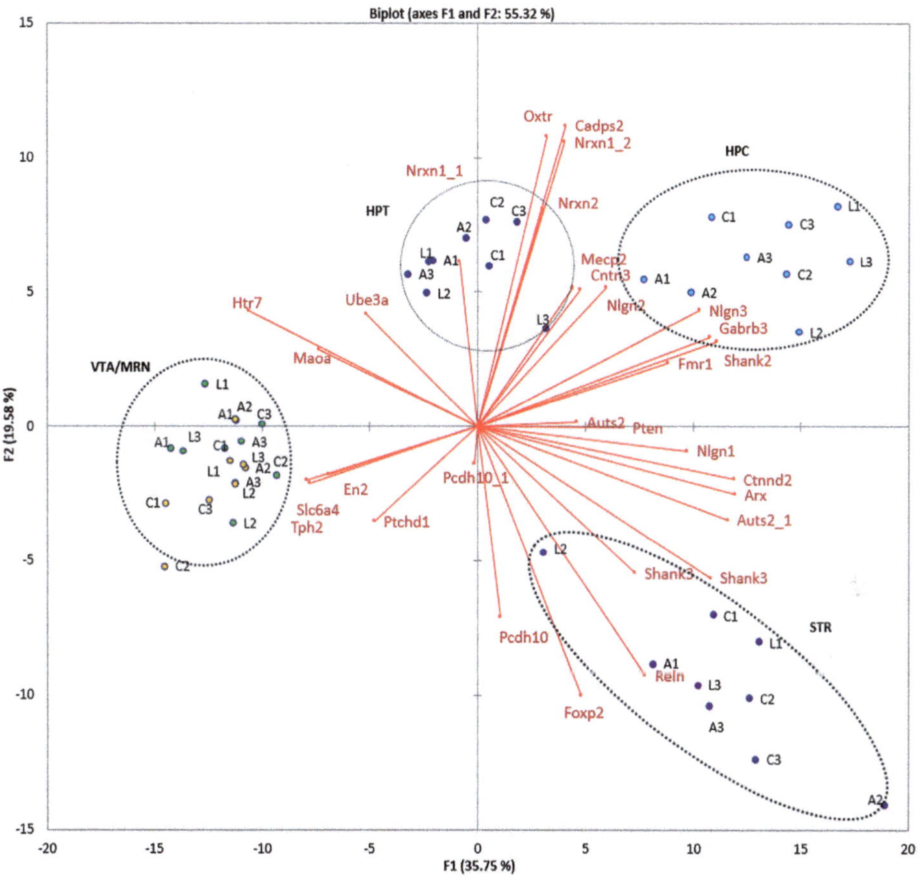


**Figure 14.9** Principal component analysis plot based on covariation of 27 autistic genes using the expression profiles of 45 samples, which comprised RNA-Seq FPKM data for five brain regions of nine mice. Ovals correspond to brain regions. *A (A1, A2, A3)*, aggressive mice; *C (C1, C2, C3)*, control; *HPC*, hippocampus; *HPT*, hypothalamus; *L (L1, L2, L3)*, defeated mice, losers; *MRN*, midbrain raphe nuclei; *STR*, striatum; *VTA*, ventral tegmental area. Distinct clustering of three brain regions occurred, whereas the MRN and VTA brain compartments were not distinct.

presented are significant with at least $P<0.05$ ($P\_value(r = 0.293)=0.05$ here and below).

## Multidimensional Scaling Analysis of 27 Neurospecific Target Genes and Comparison With the Strings Gene Association Network

We reconstructed the MDS plot based on the initial sample of 27 genes across 45 samples, which comprised RNA-Seq FPKM data for five brain regions of nine mice based on their expression profiles. Fig. 14.11 presents the gene relations based on the Pearson

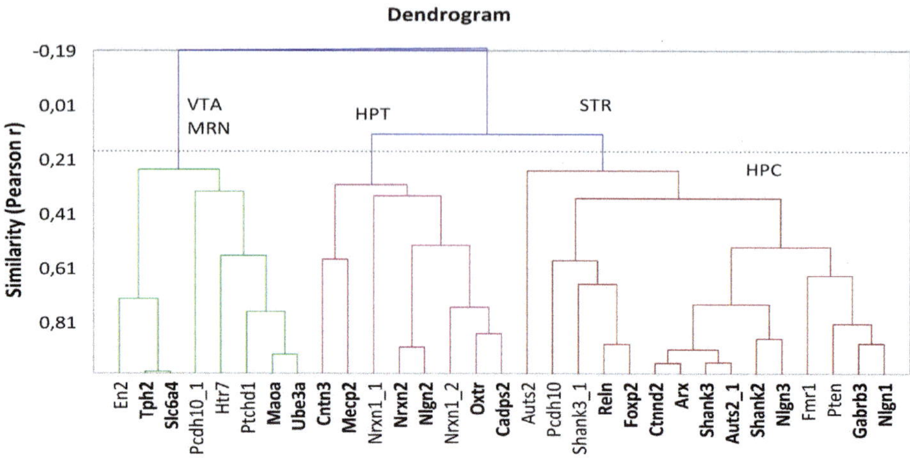


**Figure 14.10** Agglomerative hierarchic clustering tree for 27 autistic genes across 45 samples, which comprised RNA-Seq FPKM data for five brain regions of nine mice. The genes elucidated as differentially expressed compared with the control, aggressive mice and defeated mice are marked in *bold* type script. The similarity ordinate corresponds to the Pearson correlation coefficient (df = 44). Several clusters specific for the brain regions were elucidated. *HPC*, hippocampus; *HPT*, hypothalamus; *MRN*, midbrain raphe nuclei. Based on the expression, we determined the clusters as HPC-altered (typically down); STR-altered; HPT-altered; and MRN- and VTA-altered cluster.

pairwise correlation coefficient values. We identified clustering of synapse versus serotonergic genes as indicated in the Strings gene network (Fig. 14.1). Intriguingly, in the comparison of the Strings network and MDS plot, we identified discrepancies as a result of the high negative correlation values, which thus misplaced the genes in the MDS plot while maintaining close relations. For example, the *Gabrb3* gene exhibits a distinct negative correlation with the serotonergic genes cluster including *Maoa* (r = −0.25), *Tph2* (r = −0.53), *Slc6a4* (r = −0.5), and *Htr7* (r = −0.66) based on our RNA-Seq data. Thus, it is positioned distantly from these four genes in Fig. 14.11 in contrast to Fig. 14.1. This finding underscores that the Strings network (Fig. 14.1) does not elaborate on gene relations being negative or positive.

## Alternative Splicing

Seventeen of 27 autism-related genes express alternatively spliced isoforms detected in our data (Fig. 14.12; Table 14.1). The major splice variability is featured by synapse-specialized genes as previously reported.[50] The largest number of isoforms observed in our data attributes to the *Nrxn1,2,3* genes family (eight transcripts; Table 14.2; Fig. 14.12). Notably, the isoforms are distinctly expressed in particular brain regions as deduced from the PC analysis for the *Nrxn* and *Nlgn* gene families (Fig. 14.12).

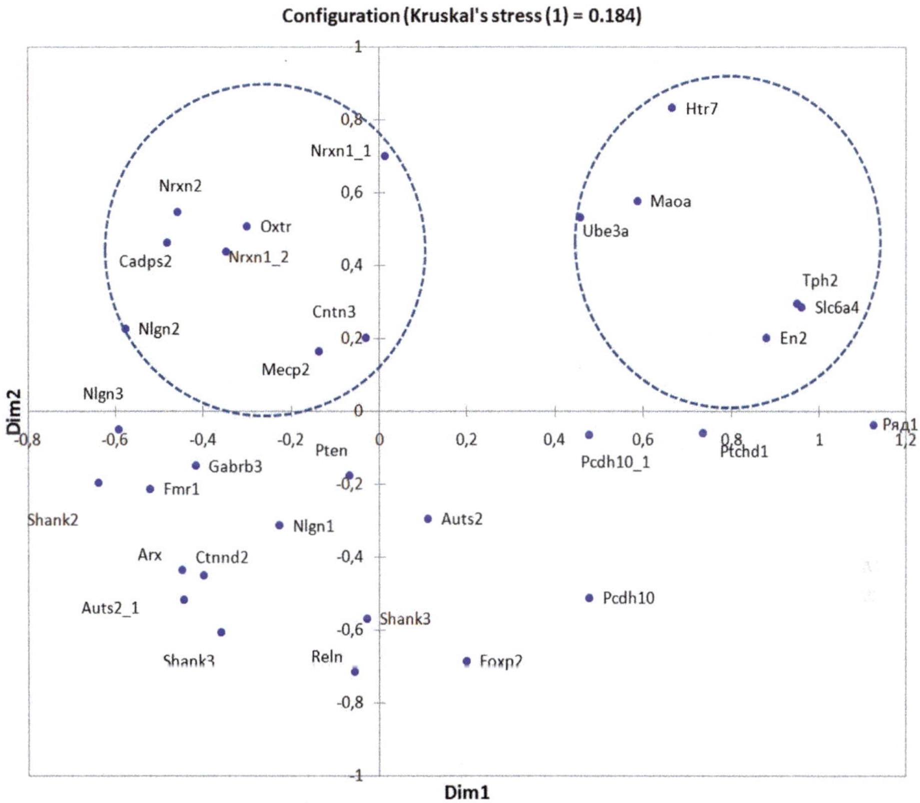


**Figure 14.11** Multidimensional scaling plot of RNA-Seq covariation data. *Left circle* corresponds to synaptic genes, and *right circle* partially corresponds to the serotonergic system. Note the overall consistency and some discrepancies with corresponding clusters in Fig. 14.1 due to negative expression correlation for a range of genes observed in our data.

# DISCUSSION

## Abnormal Social Behaviors Associated With Autistic Spectrum Symptoms Induced by Chronic Agonistic Interactions in Male Mice

In our previous experiments, we noticed that repeated agonistic interactions induce the development of psychopathologic states, which are accompanied by disturbances in social behaviors in male mice. Abnormal forms of behaviors, which are never demonstrated by intact mice, resembled symptoms that other authors cited in experimental studies as autism-like symptoms in knockout mice, which were recognized as adequate models of autism. To confirm these behavioral observations, we carried out detailed behavioral analyses in male mice in the situation of a social interaction test.

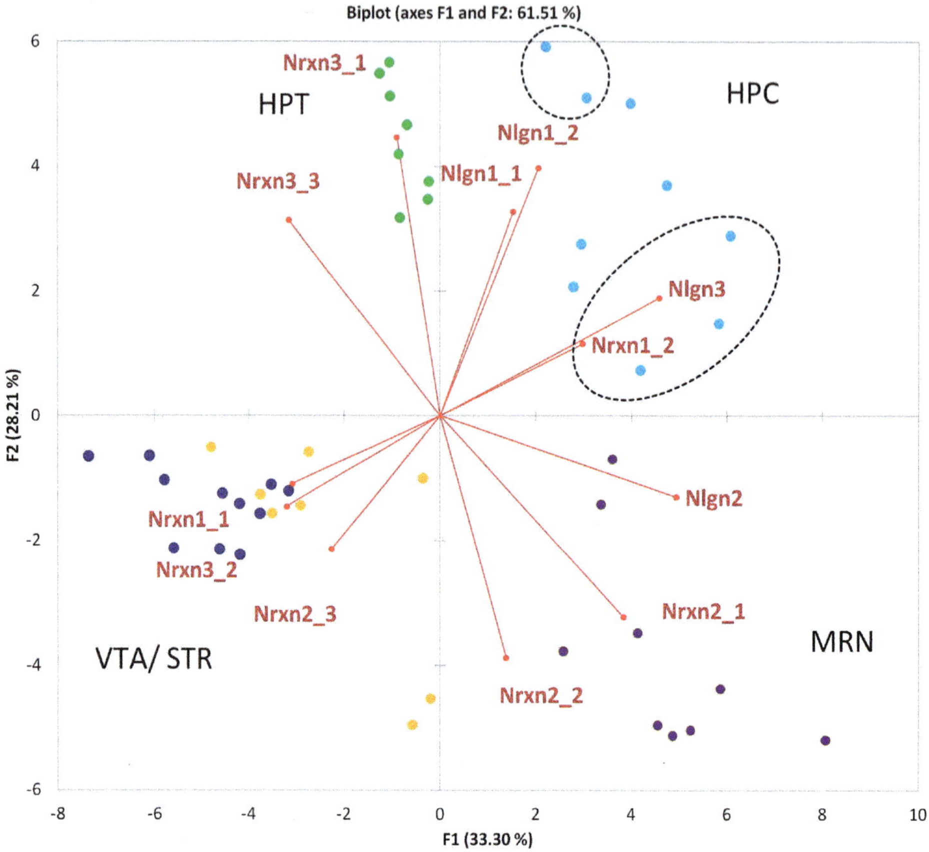


**Figure 14.12** Biplot of principal component analysis based on *Nrxn1,2,3* and *Nlgn1,2* genes transcripts expression variation observed in RNA-Seq data for the control, loser, and aggressive mice (45 samples total, 9 mice, and 5 brain regions). Nlgn1 and Nrxn2 gene isoforms are highly correlated, implying isofunctional transcript alterations within these genes. In contrast, Nrxn1_1 transcript expression rate (HPC region) is oppositely correlated to isoform Nrx1_2 (VTA/STR regions), manifesting brain region isoform specific expression in our data. In the hippocampus (HPC) loser mice (encircled lower three points) manifest increased Nrxn1_2/Nlgn1 isoforms expression, while aggressive mice (encircled upper two points) manifest preferential Nlgn1 expression. Nlgn2 is highly correlated with both isoforms of Nrxn2. Both (Nlgn2/Nrxn2) manifest twofold higher expression than any other gene from this family. Midbrain raphe nuclei (MRN) region manifests threefold higher expression of these genes. Brain region denotations: *green* – hypothalamus (HPT); *light blue* – hippocampus (HPC); *violet* – MRN; *dark blue* – ventral tegmental area (VTA); *orange* – striatum (STR). Isoforms transcript denotations: (*Nlgn1_1*:NM_001163387; *Nlgn1_2*:NM_138666; *Nlgn2*:NM_198862; *Nlgn3*: NM_172932; *Nrxn1_1*:NM_020252; *Nrxn1_2*:NM_177284; *Nrxn2_1*:NM_001205234; *Nrxn2_2*: NM_001205235; *Nrxn2_3*:NM_020253; *Nrxn3_1*:NM_001198587; *Nrxn3_2*:NM_001252074; *Nrxn3_3*: NM_172544).

This study demonstrates that the behaviors of male mice with alternative types of social behavior, the winners and losers, significantly differed in the social interaction test on neutral territory compared with the control animals. Mice with repeated experiences of social defeats demonstrate immobility or avoidance behavior toward an amicable stranger or freezing behavior, which may be considered passive avoidance of social contact with other conspecifics. Similar to previous studies,[29,33,35] the losers also exhibited low interest in the partner in the home cage. A low response to a stranger may be interpreted as low communication and an avoidance of social interactions.

Winners demonstrated strong aggressiveness in the social interaction test. Increased reaction to the partners in the neighboring compartment in the partition test has previously been connected with aggressive motivation because behavioral activity measured prior to an agonistic interaction has been positively correlated with the total time of attacks, which winners demonstrated after the partition was removed.[29,34] Chaotic movement in the cage without attention to the partner may be considered a sign of attention deficit and hyperactivity symptoms. Self-grooming episodes, which we initially considered a replacement activity, were increased in the winners and losers and may be considered an increase of repetitive behaviors developing under daily agonistic interactions.

Low exploratory activity was identified in both social groups. However, in the winners, the exploration time decreased as a result of increased locomotion. In the losers, the exploration time decreased as a result of increased immobility. The different behavioral motivations may also be considered for the decreased approaching behavior time, which may reflect a deficit in communication or abnormal social behavior.

Careful consideration of other behavioral results enabled us to note some intriguing similarities in the social behaviors and psychoemotional state of our experimental mice with several symptoms that have been considered autistic-like behaviors in animal models of autism.[25,27,28,51] Impairments in social recognition were common in the winners and losers. Experienced winners lose the ability to differentiate sex, age, and social status of conspecifics and demonstrate elevated irritability and aggression toward opponents even in low-provoking situations.[34] Winners differentiated males from receptive females substantially less well than the controls did, and they equally attacked other males that demonstrated full submission or active defense,[31,52] as well as young males.[53] We suggest that this behavior may be considered low empathic and aberrant social behavior. In addition to self-grooming behavior, the winners demonstrated different forms of other stereotypic repetitive behaviors, such as jumps, rotations, back circles, head shakes, and rigid tail, as demonstrated by previous studies.[34,55] However, the expression of stereotypical behavior may be different in the winners of different mouse strains.

The losers exhibited avoidance behavior and immobility toward any opponent. They differentiated familiar and unfamiliar conspecifics in the partition test significantly less well than the controls did, and they exhibited full indifference, helplessness, and absence of interest to novelty.[30,33,54]

Enhanced anxiety, which did not disappear 2 weeks after stopping agonistic interactions,[56,57] was demonstrated by winners and losers. Notably, according to different reports, anxiety disorders are present in 11%–84% of children with autistic symptoms.[58–61] Furthermore, the losers showed signs of increased anxiety and fear, and the winners exhibited signs of increased anxiety and aggression.[30,33,36] Similar to ASD patients,[62] both the losers and winners have been found to exhibit general reward dysfunction. For example, they decreased sucrose solution consumption in a free two-bottle choice paradigm.[34,63] Thus, numerous behavioral changes induced by repeated and long-term exposure to agonistic encounters in mice resemble some of the symptoms associated with autism (Table 14.3).

Could the changes in communication and sociability of the experimental mice, as well as their altered social behaviors and psychoemotional state be regarded as ASD-like symptoms? According to DSM-V,[2] the diagnostic criteria of autism include the following triad of symptoms: impaired social interactions, low communication, and repetitive behaviors. These basic symptoms may be accompanied by restricted interests and altered activity, atypical motivated behavior,[62] anxiety disorders,[59–61] enhanced aggressiveness and irritability, and attention/deficit hyperactivity.[11,64] Moreover, the social behaviors of chronically defeated and aggressive mice are essentially indicative of all symptoms of ASDs (Table 14.3), which have been used in models of autism, such as in animals of specific strains, transgenic, or knockout mice.[25,27,28,51] We consider our findings as a first step for the study of the possible disturbances of social behaviors under chronic agonistic interactions with use of additional behavioral tests that could clarify what exactly is happening to their animals. However, these results indicate that autism-like symptoms may develop under influence of social environment in adult animals. Other studies have also demonstrated social impairment in rodents under prenatal stress, early handling, maternal separation, and early social isolation.[13–18]

However, the preceding conclusion is not accepted by the majority of researchers working in the field of autism modeling. This is because most agree that autism and ASDs in general are developmental disorders, resulting from intrautero environmental or genetic effect that first manifests during early childhood. This is a reasonable argument against our repeated agonistic encounter-induced mouse model. However, we point out that our analysis of the expression of genes known to be involved in the mechanisms of autism do suggest that our model may be appropriate for the analysis of this human disorder. We identified numerous changes in the expression of recognized autism genes in our mice exhibiting abnormal social behaviors. Thus,

**Table 14.3** Autism spectrum disorder symptoms in humans, their animal analogs in gene modified animals and in the winners and losers

| Symptoms in humans[2] | Symptoms in animal models[25–28] | Winners[31,34,55,56]; these data | Losers[30,32,33,63]; these data |
|---|---|---|---|
| ***Main symptoms*** | | | |
| Low communication | Reduced or low sociability | Reduced or low sociability | Reduced or low sociability<br>Avoidance of social contact |
| Impaired socialization | Impaired social recognition<br>Cognitive deficit | Impaired social recognition<br>Cognitive deficit | Impaired social recognition<br>Cognitive deficit |
| Restricted and/or repetitive behavior, seizures, stereotypies, dyskinesia | Increased repeated motor activities (spontaneous activity, exploration, circling, digging, jumping) and increased self-involved behaviors (self-grooming, scratching, circling behavior) | Increased repetitive self-grooming, jumping, circling, jerk, digging, movement back, rotations, head shakes etc. | Increased repetitive self-grooming |
| ***Associated symptoms*** | | | |
| Attention/deficit | | Attention/deficit | Social indifference |
| Hyperactivity | | Hyperactivity, hyperexcitation | Hypoactivity (freezing, immobility, helplessness, depressiveness) |
| Hyperresponsiveness<br>Hyporesponsiveness | | Hyperresponsiveness to sensory stimuli | Hyporesponsiveness to sensory stimuli |
| Anxiety, depression | Increased anxiety-like behavior | Increased anxiety | Increased anxiety, depression |
| Aggression, irritability | | Aggression, irritability | Social avoidance |
| Exploration deficit | Exploration deficit | Exploration deficit | Exploration deficit |
| Reward dysfunction | | Reward dysfunction | Reward dysfunction |

environmental social manipulations in adults may induce the autistic-like behavior that is accompanied by alteration in expression rate of autism-related genes.

## Changes in Autism-Related Gene Expression in Brain Regions of Male Mice With Abnormal Social Behaviors Induced by Repeated Agonistic Interactions

Similar changes in the expression of several genes found in the winners and losers, as well as the changes of gene expression specific for loser or winner mice, as well as specific to particular brain regions, have been identified in this study (Table 14.2).

The *Tph2*, *Maoa*, *Slc6a4*, and *Htr7* genes are associated with CNS serotonergic activity and, according to several databases, are involved in numerous psychiatric disorders. Using RT-PCR, we previously identified decreased expression of the *Tph2*, *Maoa*, and *Slc6a4* genes in the MRN of the winners[42] and losers.[41] The RNA-Seq transcriptome analysis data we present here confirmed our previous findings for the *Tph2* and *Slc6a4* genes. We also identified decreased *Maoa* gene expression in the HPT and increased expression of the *Tph2* and *Slc6a4* genes in the VTA of the winners and losers. *Htr7* gene expression was also found increased in the STR of the losers.

Alterations in the peripheral and central indices of serotonin production, storage, and signaling have been associated with autism.[65,66] Brain-derived neurotrophic factor (BDNF) has been shown to promote the survival and differentiation of serotonergic neurons.[67] In accordance with these findings, involvement of BDNF in symptoms of ASDs has been demonstrated in animal models,[68–70] as well as in autistic patients.[71–73] In the current and previous studies with mice,[41,42] decreased *Bdnf* gene expression was identified in MRN in the losers and winners. Changes in the expression of these genes in mice exposed to social stress may underlie reduced or low sociability, exploration deficits, reward dysfunction, and anxiety.

The *Shank* genes encode proteins of a family of synaptic proteins, which may function as molecular scaffolds in the postsynaptic density of excitatory synapses. Diseases associated with *Shank2* (*Auts17*) gene, found upregulated in the HPC of the loser mice of the current study, include ASDs. Reduced *Shank3* expression is a major causative factor in several neurological symptoms.[74,75]

In the winners and losers, *Shank3* gene overactivation was identified in the HPT. On the other hand, reduced expression of this gene was found in the HPC of the winners, a change that was associated with lower levels of social recognition and exploration, and increased repetitive behaviors and aggression. Thus, synthesis of Shank family synaptic proteins is disturbed in the HPT and HPC in male mice with abnormal social behaviors.

The *Ube3a* gene encodes an E3 ubiquitin-protein ligase, which is a component of the ubiquitin protein degradation system. The expression of rodent *Ube3a* transcripts is associated with changes in the expression of the *Gabrb3* (*auts4*) gene, which encodes a protein that serves as the receptor for gamma-aminobutyric acid. We identified increased expression of the *Ube3a* gene in the MRN of the winners and decreased expression of the *Ube3a* and *Gabrb3* (*auts4*) genes in the HPT in both the winners and losers.

The *Reln* and *Pten* genes, as well as the *Ctnnd2*, *Nrxn1*, *Nrxn2*, *Nlgn1*, *Nlgn2*, and *Nlgn3* genes, encode proteins that are involved in cell adhesion processes, which enable cells to interact and attach to a surface, substrate, or another cell. In both the winners and the losers, strongly decreased expression of the *Reln* gene in the MRN and the *Pten* gene in the HPT was identified. The winners exhibited decreased expression of the *Reln* gene in the HPT, the *Pten* gene in the STR, and the *Ctnnd2* and *Nrxn2* genes in the HPC, as well as increased expression of the *Nrxn1* gene in the MRN. The losers exhibited increased expression of the *Nrxn2*, *Nlgn2*, and *Nlgn3* genes in the HPT and the *Nlgn2* gene in the HPC, as well as decreased expression of the *Nrxn1* and *Nlgn1* genes in the HPT, the *Reln* gene in the STR, and increased expression in the HPC. Increased expression of the *Pcdh10* gene, a potential calcium-dependent cell-adhesion protein, was identified in the HPT and HPC of the losers. Increased expression of the *Cadps2* gene, which encodes a member of the calcium-dependent activator of secretion protein family, was also demonstrated in the MRN of the winners and the VTA of the losers. The *Bdnf* gene has been shown to be involved in biochemical interactions associated with proteins encoded by the *Cadps2*, *Reln*, *Ctnnd2*, and *Nlgn1* genes.

Dysfunction of several genes is strongly suspected as a genetic cause of ASDs. The *Mecp2* (*autsx3*) gene plays an essential role in mammalian development, and the MeCP2 protein binds to forms of DNA that have been methylated. Loss of function as well as overexpression of the *Mecp2* gene may both lead to disease.[76]

We identified increased expression of the *Mecp2* gene in the VTA of the losers and decreased expression in the HPT in both the winners and losers.

The *Fmr1* gene encodes the fragile X Mental Retardation Protein 1, which binds RNA and is associated with polysomes. Loss of function of the *Fmr1* gene as a result of microdeletion is responsible for the clinical phenotype of fragile X syndrome (FXS).[77] In our study, hypoexpression of the *Fmr1* gene was identified in the HPT of both social groups and overexpression in the MRN of the winners. Animals in both social groups demonstrated cognitive dysfunctions, social anxiety, and stereotypic behaviors. The winners also exhibited aggression and hyperactivity, attention deficit behavior, and increased tendency to attack conspecifics, as well as irritability. Similar to children with FXS, aggressive mice exhibit hypersensitivity to visual, auditory, tactile, pain, and olfactory stimuli (behavioral observations[39]). The losers exhibited social withdrawal behaviors, including avoidance, indifference, and depression. Interestingly, a reduction in *Bdnf* expression in Fmr1 knockout mice has been found to worsen cognitive deficits and to improve hyperactivity and sensorimotor deficits.[68]

Decreased expression of the *Arx* gene in the HPT of the winners and increased expression in the HPC and decreased expression in the STR of the losers were identified in animals with abnormal social behaviors in the present study. Mental retardation is associated with mutations of the *Arx* gene, which has been described in several patients with autism or autistic features, and it is also known to cause a diverse spectrum of other diseases that includes cognitive impairment and epilepsy.

We identified downregulation of the *Foxp2* gene in the MRN of the winners and losers, the HPT of the winners, and overexpression in the VTA of the losers. The *Auts2*, Autism Susceptibility Candidate 2 gene, is thought to be a transcription factor required for normal brain development. It has been implicated in neurodevelopment and as a candidate gene for numerous neurological disorders that result from *Auts2* gene deficiency. *Auts2* gene expression was decreased in the HPC of the winners and increased in the HPT of the losers.

*Ptchd1* (*autsx4*) gene diseases, which are associated with PTCHD1 protein, include autism x-linked 4 and intellectual disability. There was decreased expression in the HPT in both social groups. Decreased expression was specific only for the winners in the STR, with increased expression in the VTA of the losers. The *Oxtr* gene, which codes the oxytocin receptor, is significantly associated with ASD. Decreased expression of this gene was identified in the MRN of the winners and losers. The *En2* (*auts1*) gene is a Homeobox-containing gene and is thought to have a role in controlling development and contributes to as many as 40% of ASD cases, approximately twice the prevalence in the general population.[78] Increased expression of this gene was identified in the STR and VTA.

The changes in the expression of numerous autism-related genes in several specific brain regions are noteworthy. In the HPC, all autism-related genes we studied that changed their expression differed in their change between loser and winner mice. Moreover, all genes decreased expression in the winners and increased expression in the losers. Previously, most of the various *Rps* and *Rpl* genes were found upregulated in the winners[79] and downregulated in the losers[80] in this brain region, and they were different, with the exception of one *Rps39* gene. The pattern of these differences may be associated with activation of neurogenesis identified in the winners[81] and decreased cell division and neurogenesis in the losers.[82–85] The HPC is responsible for cognitive function, memory consolidation, and storage, and the gene expression changes we detected we suspect may lead to differential cognitive processes in the winners and losers.

The VTA, which is widely implicated in the brain's natural reward circuitry and in emotions, exhibited increased expression of all genes in the loser and winner mice. Our previous behavioral data demonstrated reward dysfunction in both winners and losers.[34,63] In the STR, which is responsible for the regulation of motor activity and stereotypical behaviors, we identified decreased expression of most genes, which may be associated with the disturbances in motor activity. We suspect altered expression of these genes may underlie stereotypic behaviors identified in both winner and loser mice of the current study. In the MRN, a multifunctional brain region that contains the majority of serotonergic neuronal cell bodies, we identified decreased expression of genes in the winners and losers and an upregulation of specific genes in the winners only. In the HPT, decreased gene expression was identified in the winners and losers, as well as upregulation of specific genes in the losers.

Autism is considered a multifactorial disorder that involves many genes that only predispose the patient to, but do not determine the development of autism. Several of these genes, for example, *Shank2* (*Auts17*), *Mecp2* (*autsx3*), *Ptchd1* (*autsx4*), *En2* (*auts1*), *Nlgn3* (*Autsx1*), and *Gabrb3*(*auts4*), which are strongly associated with autism, changed their expression in the winners and losers exposed to multiple repeated agonistic interactions. Considering all autism-related genes investigated, we suggest that the social environment may influence their expression, thereby forming symptoms similar to those of human ASDs.

## Alternative Splicing Expression Profiling

Alternative splicing of mRNA has been shown to contribute to ASDs.[86,87] For example, aberrant mRNA expression of specific genes, including the *Shank3*, *Nrxn1*, *Nlgn3*, and *Nlgn4X*, as well as others, have been identified in ASDs.[88–90]

In addition, a range of mutations located within splice sites or splicing factor binding sites of autism-related genes[91] lead to splicing alterations. Thus, we may assume that even in the absence of deleterious mutations, a stress-induced misregulation of alternatively spliced transcripts may influence the complex homeostasis of mRNA isoforms in brain tissues.[92] The highest emergence of isoforms in the 27 selected gene samples was identified in synapse specialized genes as previously reported and provides flexible neuronal communication.[93] Thus, we assume that alterations in the splicing-specific isoform expression profiles may lead to abnormal neurotransmitter responses that involve ASD associated genes, such as neurexins, neuroligins, *Shank1-3*, and *Pcdh10*. It should be noted that isoforms often coordinately express, particularly in genes associated with synaptic functions.[92,93] Moreover, we identified the coordinated expression of *Nrxn2*- and *Nlgn2*-specific isoforms in the MRN and HPT (Fig. 14.12). Importantly, the losers exhibited the most intense expression of synaptic transmission genes, which indicates that the signal of inhibitory synapses is increased. Thus, we emphasize the importance of paying attention to isoform coexpression profiling. In summary, our current findings demonstrate that autism-related genes may be involved in disturbances of social behaviors induced by chronic agonistic interactions in adult mice.

# CONCLUSION

As previously indicated, ASDs include substantially broader phenotypical abnormalities with the less severe effects than those characteristic of full-fledged autism. The ASD diagnosis covers patients with symptoms who do not meet the full criteria for autism or other disorders. Moreover, according to the APA, in the DSM-V, various affective and neurological disorders, such as depression, anxiety, schizophrenia, and ADHD, as well as many somatic disorders, may be accompanied by impairments in social behaviors.[2] Investigation of networks of genes identified both similar and unique sets underlying the impairment of social behaviors and bipolar disorder.[94] Comorbidity between these two disease clusters has been attributed to the genes and biochemical pathways common to both disorders. It may be assumed that common genes may determine the disturbances in social behaviors similar to "autistic-like symptoms" under the psychopathology of aggressive behavior similar to the psychosis, as demonstrated in the winners with chronic experience of aggression[34] or an anxiety and depression-like state in defeated mice, which develop under chronic social defeat stress in animals.[30,33,35,95] We suggest that abnormal social behaviors and changes in gene expression are a result of comorbidity of these pathological states in male mice.

We assessed the RNA-Seq data consistency in the target genes using two approaches. First, we built up the PC variation plot and identified brain region-specific compact clustering, which thus confirmed the consistent expression pattern of the 45 samples (Fig. 14.9). Second, we visualized the Pearson pairwise correlation matrix via MDS Plot (Fig. 14.11). We also identified segregation of the synapse genes and serotonergic genes, as well as other compact gene clusters, which supports functional covariation of genes, compared with the String gene association network (Fig. 14.1) based on external publications, databases, and experimental data. Thus, we assume that the RNA-Seq data adequately represents functioning of neurospecific genes in the 45 samples considered. We also observed differential splicing profiles in *Nrxn1* isoforms specific for defeated mice, implying some synaptic plasticity alterations. We also observed consistent *Htr7* splicing deviation.

Our behavioral approach mimicked many signs of impaired social behaviors in male mice, as well as changes in autism-associated gene expression, which may be useful for investigating brain molecular changes that may determine the subsequent psychoemotional state, cognitive dysfunction, and disturbances in social behaviors. Other approaches for the study mechanisms of autistic-like behaviors help to understand the role of certain genes, for example, with use of knockouts or mutant mice, which exhibited low sociability compared with wild-type animals.[19–23] Mice of some strains that exhibit low communication[24–28] have also been considered models of autism. This approach may be useful for the study of genetic predisposition to the development of autism under different effects, namely stress, development of some disorders, as well as negative social environment. We also studied the behavior of CBA/Lac mice[37] with repeated experience of aggression or social defeats in daily agonistic interactions with accent on searches of associations with autistic symptoms compared with C57BL/6J animals. It has been shown that male mice of both strains with alternative social behaviors demonstrated the changes in social behaviors; however, the expression of some behavioral forms was different. Thus, both environment and genetic factors may lead to the impairment of social behaviors. Obviously, multifold increase of the cases with autistic spectrum symptoms over the last years[96] cannot be attributed exclusively to genetic mutations or heredity. Environmental contribution to the development of autistic symptoms has to be considered.

It has been shown that changes in gene expression may also be caused in adults by different environments and may persist for a long time. Evidence has been obtained in experiments on the molecular consequences of chronic agonistic interactions. The persistence of changed social behaviors[57] and changed expression of many genes has been demonstrated for the losers[41,95] and winners.[97] Changes in the function of "specific autism genes" or genes of the neurotransmitter systems, which are involved in communication mechanisms and/or associated behavior patterns, may appear as a hereditary impairment and may explain the epigenetic nature of autism mechanisms discussed by researchers.[7,98–100] These novel insights open broader perspectives for the experimental studies of the molecular mechanisms of ASDs and the identification of therapeutic approaches for pharmacological correction.

## ABBREVIATIONS

**AHC** Agglomerative hierarchic clustering
**ASDs** Autism spectrum disorders
**FPKM** Fragments per kilobase of transcript per million mapped reads
**HPC** Hippocampus
**HPT** Hypothalamus
**Los-10, Los-20** Groups of chronically defeated mice after 10 or 20 days of agonistic interactions
**MDS** Multidimensional scaling
**MRN** Midbrain raphe nuclei
**PC** Principal components
**STR** Striatum
**VTA** Ventral tegmental area
**Win-10, Win-20** Groups of chronically victorious mice demonstrating daily aggression, during 10 or 20 days

## FUNDING

This work was funded by Russian Science Foundation, no. 14-15-00063; Russian Foundation for Basic Research, no. 16-04-00905A.

## AUTHOR CONTRIBUTIONS

I.L.K. substantially contributed to behavioral data acquisition and analysis of transcriptome data; I.L.K., D.A.S., and A.G.G. contributed to acquisition of animals with alternative social behaviors and brain materials; V.N.B. implemented bioinformatics assessing of RNA-Seq data, analyzed and interpreted data, and wrote the main manuscript text. N.N.K. has made substantial contributions to conception and design, analysis, and interpretation of data, and wrote the main manuscript text. All authors have been involved in drafting the manuscript, have given final approval of the version to be published, and agree to be accountable for all aspects of the work in ensuring that questions related to the accuracy or integrity of any part of the work are appropriately investigated and resolved. N.N.K. and V.N.B. equally contributed to the paper.

## ACKNOWLEDGMENTS

We thank N.N. Lang for technical assistance and Dr. V.A. Naprimerov for organizational support on the behavioral experiment. We are infinitely grateful to Prof. R. Gerlai for very helpful and productive advice, discussion, comments, and suggestions that we took into account working with this chapter.

## REFERENCES

1. Happé F, Ronald A. The 'fractionable autism triad': a review of evidence from behavioural, genetic, cognitive and neural research. *Neuropsychol Rev.* 2008;18(4):287–304.
2. APA. *Diagnostic and Statistical Manual of Mental Disorders.* 5th ed. Washington, DC: American Psychiatric Publishing; 2013.
3. Neuhaus E, Beauchaine TP, Bernier R. Neurobiological correlates of social functioning in autism. *Clin Psychol Rev.* 2010;30(6):733–748.
4. Arndt TL, Stodgell CJ, Rodier PM. The teratology of autism. *Int J Dev Neurosci.* 2005;23(2–3):189–199.
5. Kolevzon A, Gross R, Reichenberg A. Prenatal and perinatal risk factors for autism. *Arch Pediatr Adolesc Med.* 2007;161(4):326–333.
6. Kinney DK, Munir KM, Crowley DJ, Miller AM. Prenatal stress and risk for autism. *Neurosci Biobehav Rev.* 2008;32(8):1519–1532.
7. Iwata K, Matsuzaki H, Takei N, Manabe T, Mori N. Animal models of autism: an epigenetic and environmental viewpoint. *J Cent Nerv Syst Dis.* 2010;2:37–44.
8. Newschaffer CJ, Croen LA, Daniels J, et al. The epidemiology of autism spectrum disorders. *Annu Rev Publ Health.* 2007;28:235–258.
9. Croen LA, Zerbo O, Qian Y, Massolo ML. Psychiatric and medical conditions among adults with ASD. In: *International Meeting for Autism Research, May 14–17, 2014, Atlanta, GA*; 2014. https://imfar.confex.com/imfar/2014/webprogram/Paper17783.html.
10. Levy SE, Mandell DS, Schultz RT. Autism. *Lancet.* 2009;374(9701):1627–1638.
11. O'Dwyer L, Tanner C, van Dongen EV, et al. Brain volumetric correlates of autism spectrum disorder symptoms in attention deficit/hyperactivity disorder. *PLoS One.* 2014;9(6):e101130.
12. Weiss LA. Autism genetics: emerging data from genome-wide copy number and single nucleotide polymorphism scans. *Expert Rev Mol Diagn.* 2009;9(8):795–803.
13. Shalev U, Feldon J, Weiner I. Gender- and age-dependent differences in latent inhibition following pre-weaning non-handling: implications for a neurodevelopmental animal model of schizophrenia. *Int J Dev Neurosci.* 1998;16:279–288.
14. Shalev U, Weiner I. Gender-dependent differences in latent inhibition following prenatal stress and corticosterone administration. *Behav Brain Res.* 2001;126:57–63.
15. Peters SL, Gray JA, Joseph MH. Pre-weaning non-handling of rats disrupts latent inhibition in males, and results in persisting sex- and area-dependent increases in dopamine and serotonin turnover. *Behav Pharmacol.* 1991;12:215–223.

16. Ellenbroek BA, Geyer MA, Cools AR. The behavior of APO-SUS rats in animal models with construct validity for schizophrenia. *J Neurosci.* 1995;15:7604–7611.
17. Wilkinson LS, Killcross SS, Humby T, Hall FS, Geyer MA, Robbins TW. Social isolation in the rat produces developmentally specific deficits in prepulse inhibition of the acoustic startle response without disrupting latent inhibition. *Neuropsychopharmacology.* 1994;10(1):61–72.
18. Roubertoux PL, Guillot PV, Mortaud S, et al. Attack behaviors in mice: from factorial structure to quantitative trait loci mapping. *Eur J Pharmacol.* 2005;526(1–3):172–185.
19. Bernardet M, Crusio WE. Fmr1 KO mice as a possible model of autistic features. *Sci World J.* 2006;6: 1164–1176.
20. Gkogkas CG, Khoutorsky A, Ran I, et al. Autism-related deficits via dysregulated eIF4E-dependent translational control. *Nature.* 2013;493(7432):371–377.
21. Provenzano G, Zunino G, Genovesi S, Sgady P, Bozzi Y. Mutant mouse models of autism spectrum disorders. *Dis Markers.* 2012;33(5):225–239.
22. Shinoda Y, Sadakata T, Furuichi T. Animal models of autism spectrum disorder (ASD): a synaptic-level approach to autistic-like behavior in mice. *Exp Anim.* 2013;62(2):71–78 [Review].
23. Swanberg SE, Nagarajan RP, Peddada S, Yasui DH, LaSalle JM. Reciprocal co-regulation of EGR2 and MECP2 is disrupted in Rett syndrome and autism. *Hum Mol Genet.* 2009;18(3):525–534.
24. Ey E, Leblond CS, Bourgeron T. Behavioral profiles of mouse models for autism spectrum disorders. *Autism Res.* 2011;4(1):5–16.
25. Crawley JN. Translational animal models of autism and neurodevelopmental disorders. *Dialogues Clin Neurosci.* 2012;14(3):293–305.
26. Lewis MH, Tanimura Y, Lee LW, Bodfish JW. Animal models of restricted repetitive behavior in autism. *Behav Brain Res.* 2007;176(1):66–74.
27. Moy SS, Nadler JJ, Young NB, et al. Social approach in genetically engineered mouse lines relevant to autism. *Genes Brain Behav.* 2009;8(2):129–142.
28. Tordjman S, Drapier D, Bonnot O, et al. Animal models relevant to schizophrenia and autism: validity and limitations. *Behav Genet.* 2007;37(1):61–78.
29. Kudryavtseva NN. Use of the "partition" test in behavioral and pharmacological experiments. *Neurosci Behav Physiol.* 2003;33(5):461–471.
30. Avgustinovich DF, Alekseyenko OV, Bakshtanovskaia IV, et al. Dynamic changes of brain serotonergic and dopaminergic activities during development of anxious depression: experimental study. *Usp Fiziol Nauk.* 2004;35(4):19–40 [Russian].
31. Bondar NP, Kudriavtseva NN. Impaired social recognition in male mice with repeated experience of aggression. *Zh Vyssh Nerv Deiat Im I P Pavlova.* 2005;55(3):401–407.
32. Kovalenko IL, Kudryavtseva NN. Development of autistic spectrum symptoms under chronic social defeat stress in anxious male mice: effects of diazepam. *Psychopharmacol Biol Narcol.* 2010;10(1–2): 2624–2635 [Russian].
33. Kudryavtseva NN, Avgustinovich DF. Behavioral and physiological markers of experimental depression induced by social conflicts (DISC). *Aggress Behav.* 1998;24:271–286.
34. Kudryavtseva NN. The psychopathology of repeated aggression: a neurobiological aspect. In: Morgan JP, ed. *Perspectives on the Psychology of Aggression.* NY: Nova Science Publichers, Inc.; 2006: 35–64 [chapter 2].
35. Kudryavtseva NN, Bakshtanovskaya IV, Koryakina LA. Social model of depression in mice of C57BL/6J strain. *Pharmacol Biochem Behav.* 1991;38(2):315–320.
36. Kudryavtseva NN, Bondar NP, Avgustinovich DF. Association between repeated experience of aggression and anxiety in male mice. *Behav Brain Res.* 2002;133(1):83–93.
37. Kovalenko IL, Kudryavtseva NN. Alteration of social behaviors in male mice of CBA/Lac strain under agonistic interactions. *Zh Vyssh Nerv Deiat Im IP Pavlova.* 2015;65(4):486–497 [Russian].
38. Kudryavtseva NN. The sensory contact model for the study of aggressive and submissive behaviors in male mice. *Aggress Behav.* 1991;17(5):285–291.

39. Kudryavtseva NN, Smagin DA, Kovalenko IL, Vishnivetskaya GB. Repeated positive fighting experience in male inbred mice. *Nat Prot.* 2014;9(11):2705–2717.
40. Trapnell C, Hendrickson DG, Sauvageau M, Goff L, Rinn JL, Pachter L. Differential analysis of gene regulation at transcript resolution with RNA-seq. *Nat Biotechnol.* 2013;31(1):46–53.
41. Boyarskikh UA, Bondar NP, Filipenko ML, Kudryavtseva NN. Downregulation of serotonergic genes expression in the raphe nuclei of midbrain under chronic social defeat stress in male mice. *Mol Neurobiol.* 2013;48(1):13–21.
42. Smagin DA, Boyarskikh UA, Bondar NP, Filipenko ML, Kudryavtseva NN. Reduction of serotonergic gene expression in the midbrain raphe nuclei under positive fighting experience [Special Issue on Gene Expression] *Adv Biosci Biotechnol.* 2013;4(10B):36–44.
43. Chen L, Sun F, Yang X, et al. Correlation between RNA-Seq and microarrays results using TCGA data. *Gene.* 2017. pii: S0378-1119(17)30579–6.
44. Jensen LJ, Kuhn M, Stark M, et al. STRING 8—a global view on proteins and their functional interactions in 630 organisms. *Nucleic Acids Res.* 2009;37(Database issue):D412–D416. https://doi.org/10.1093/nar/gkn760.
45. Bergen SE, Fanous AH, Walsh D, O'Neill FA, Kendler KS. Polymorphisms in SLC6A4, PAH, GABRB3, and MAOB and modification of psychotic disorder features. *Schizophr Res.* 2009; 109(1–3):94–97.
46. Sun J, Jayathilake K, Zhao Z, Meltzer HY. Investigating association of four gene regions (GABRB3, MAOB, PAH, and SLC6A4) with five symptoms in schizophrenia. *Psychiatry Res.* 2012;198(2): 202–206.
47. Meng WD, Sun SJ, Yang J, Chu RX, Tu W, Liu Q. Elevated serum brain-derived neurotrophic factor (BDNF) but not BDNF gene Val66Met polymorphism is associated with autism spectrum disorders. *Mol Neurobiol.* 2017;54(2):1167–1172.
48. Halepoto DM, Bashir S, Zeina R, Al-Ayadhi LY. Correlation between hedgehog (Hh) protein family and brain-derived neurotrophic factor (BDNF) in autism spectrum disorder (ASD). *J Coll Physicians Surg Pak.* 2015;12:882–885.
49. Chen S, Park JW, Lu ZX, et al. rMATS: robust and flexible detection of differential alternative splicing from replicate RNA-Seq data. *Proc Natl Acad Sci USA.* 2014;111(51):E5593–E5601.
50. Iijima Y, Behr K, Iijima T, Biemans B, Bischofberger J, Scheiffele P. Distinct defects in synaptic differentiation of neocortical neurons in response to prenatal valproate exposure. *Sci Rep.* 2016;6:27400. https://doi.org/10.1038/srep27400.
51. Brodkin ES. BALB/c mice: low sociability and other phenotypes that may be relevant to autism. *Behav Brain Res.* 2007;176(1):53–65.
52. Kudryavtseva NN, Amstislavskaya TG, Kucheryavy S. Effects of repeated aggressive encounters on approach to a female and plasma testosterone in male mice. *Horm Behav.* 2004;45(2):103–107.
53. Kovalenko IL, Galyamina AG, Smagin DA, Michurina TV, Kudryavtseva NN, Enikolopov G. Extended effect of chronic social defeat stress in childhood on behaviors in adulthood. *PLoS One.* 2014;9(3):e91762.
54. Lipina TV, Kudriavtseva NN. The study of explorative behavior in CBA/Lac male mice under influence of positive and negative social experience. *Zh Vyssh Nerv Deiat Im IP Pavlova.* 2008;8(2):194–201 [Russian].
55. Vishnivetskaia GB, Avgustinovich DF, Kudriavtseva NN. Development of movement disorders in DBA/2J male mice under repeated experience of aggression. *Zh Vyssh Nerv Deiat Im I P Pavlova.* 2013;63(2):235–245 [Russian].
56. Kudryavtseva NN, Smagin DA, Bondar NP. Modeling fighting deprivation effect in mouse repeated aggression paradigm. *Prog Neuropsychopharmacol Biol Psychiatry.* 2011;35(6):1472–1478.
57. Avgustinovich DF, Kovalenko IL, Kudryavtseva NN. The animal model of anxious depression: persistence of behavior pathology. *Neurosci Behav Physiol.* 2005;35(9):917–924.
58. Amaral DG, Corbett BA. The amygdala, autism and anxiety. In: Bock G, Goode J, eds. *Autism: Neural Basis and Treatment Possibilities: Novartis Foundation Symposium 251.* Novartis Foundation; 2003:177–197.

59. Gillott A, Furniss F, Walter A. Anxiety in high-functioning children with autism. *Autism*. 2001;5: 277–286.
60. Muris P, Steerneman P, Merckelbach H, Holdrinet I, Meesters C. Comorbid anxiety symptoms in children with pervasive developmental disorders. *J Anxiety Disord*. 1998;2:387–393.
61. White SW, Oswald D, Ollendick T, Scahill L. Anxiety in children and adolescents with autism spectrum disorders. *Clin Psychol Rev*. 2009;29(3):216–229.
62. Kohls G, Schulte-Rüther M, Nehrkorn B, et al. Reward system dysfunction in autism spectrum disorders. *Soc Cogn Affect Neurosci*. 2013;8(5):565–572.
63. Bondar NP, Kovalenko IL, Avgustinovich DF, Smagin DA, Kudryavtseva NN. Anhedonia in the shadow of chronic social defeat stress, or when the experimental context matters. *Open Behav Sci J*. 2009;3:17–27.
64. Doyle CA, McDougle CJ. Pharmacologic treatments for the behavioral symptoms associated with autism spectrum disorders across the lifespan. *Dialog Clin Neurosci*. 2012;14(3):263–279.
65. Harrington RA, Lee LC, Crum RM, Zimmerman AW, Hertz-Picciotto I. Serotonin hypothesis of autism: implications for selective serotonin reuptake inhibitor use during pregnancy. *Autism Res*. 2013;6(3):149–168.
66. Veenstra-Vanderweele J, Jessen TN, Thompson BJ, et al. Modeling rare gene variation to gain insight into the oldest biomarker in autism: construction of the serotonin transporter Gly56Ala knock-in mouse. *J Neurodev Disord*. 2009;1:158–171.
67. Martinowich K, Lu B. Interaction between BDNF and serotonin: role in mood disorders. *Neuropsychopharmacology*. 2008;33:73–83.
68. Uutela M, Lindholm J, Louhivuori V, et al. Reduction of BDNF expression in Fmr1 knockout mice worsens cognitive deficits but improves hyperactivity and sensorimotor deficits. *Genes Brain Behav*. 2012;11(5):513–523.
69. Sadakata T, Shinoda Y, Oka M, et al. Reduced axonal localization of a Caps2 splice variant impairs axonal release of BDNF and causes autistic-like behavior in mice. *Proc Natl Acad Sci USA*. 2013;09(51): 21104–21109.
70. Scattoni ML, Martire A, Cartocci G, Ferrante A, Ricceri L. Reduced social interaction, behavioural flexibility and BDNF signaling in the BTBR T+ tf/J strain, a mouse model of autism. *Behav Brain Res*. 2013;251:35–40.
71. Han JC, Thurm A, Golden WC, et al. Association of brain-derived neurotrophic factor (BDNF) haploinsufficiency with lower adaptive behaviour and reduced cognitive functioning in WAGR/11p13 deletion syndrome. *Cortex*. 2013;49(10):2700–2710.
72. Garcia KL, Yu G, Nicolini C, et al. Altered balance of proteolytic isoforms of pro-brain-derived neurotrophic factor in autism. *J Neuropathol Exp Neurol*. 2012;71(4):289–297.
73. Das UN. Autism as a disorder of deficiency of brain-derived neurotrophic factor and altered metabolism of polyunsaturated fatty acids. *Nutrition*. 2013;29(10):1175–1185.
74. Durand CM, Betancur C, Boeckers TM, et al. Mutations in the gene encoding the synaptic scaffolding protein SHANK3 are associated with autism spectrum disorders. *Nat Genet*. 2007;39(1):25–27.
75. Moessner R, Marshall CR, Sutcliffe JS, et al. Contribution of SHANK3 mutations to autism spectrum disorder. *Am J Hum Genet*. 2007;81(6):1289–1297.
76. Nagarajan RP, Hogart AR, Gwye Y, Martin MR, LaSalle JM. Reduced MeCP2 expression is frequent in autism frontal cortex and correlates with aberrant MECP2 promoter methylation. *Epigenetics*. 2006;1(4):e1–e11.
77. Wohrle D, Kotzot D, Hirst MC, et al. A microdeletion of less than 250 kb, including the proximal part of the FMR-I gene and the fragile-X site, in a male with the clinical phenotype of fragile-X syndrome. *Am J Hum Genet*. 1992;51:299–306.
78. Benayed R, Gharani N, Rossman I, et al. Support for the homeobox transcription factor gene ENGRAILED 2 as an autism spectrum disorder susceptibility locus. *Am J Hum Genet*. 2005;77(5): 851–868.
79. Smagin DA, Kovalenko IL, Galyamina AG, Orlov YL, Babenko VN, Kudryavtseva NN. Heterogeneity of brain ribosomal genes expression following repeated experience of aggression in male mice as revealed by RNA-Seq. *Mol Neurobiol*. 2018. https://doi.org/10.1007/s12035-016-0327-z.

80. Smagin DA, Kovalenko IL, Galyamina AG, Bragin AO, Orlov YL, Kudryavtseva NN. Dysfunction in ribosomal gene expression in the hypotalamus and hippocampus following chronic social defeat stress in male mice as revealed by RNA-seq. *Neural Plast.* 2016, 3289187. https://doi.org/10.1155/2016/3289187.
81. Smagin DA, Park J-H, Michurina TV, et al. Altered hippocampal neurogenesis and amygdalar neuronal activity in adult mice with repeated experience of aggression. *Front Neurosci.* 2015;9:443. https://doi.org/10.3389/fnins.2015.00443.
82. Buwalda B, Kole MH, Veenema AH, et al. Long-term effects of social stress on brain and behavior: a focus on hippocampal functioning. *Neurosci Biobehav Rev.* 2005;29(1):83–97.
83. Ferragud A, Haro A, Sylvain A, Velázquez-Sánchez C, Hernández-Rabaza V, Canales JJ. Enhanced habit-based learning and decreased neurogenesis in the adult hippocampus in a murine model of chronic social stress. *Behav Brain Res.* 2010;210(1):134–139.
84. Lagace DC, Donovan MH, DeCarolis NA, et al. Adult hippocampal neurogenesis is functionally important for stress-induced social avoidance. *Proc Natl Acad Sci USA.* 2010;107(9):4436–4441.
85. Van Bokhoven P, Oomen CA, Hoogendijk WJ, Smit AB, Lucassen PJ, Spijker S. Reduction in hippocampal neurogenesis after social defeat is long-lasting and responsive to late antidepressant treatment. *Eur J Neurosci.* 2011;33(10):1833–1840.
86. Cieply B, Carstens RP. Functional roles of alternative splicing factors in human disease. *Wiley Interdiscip Rev RNA.* 2015;6:311–326. https://doi.org/10.1002/wrna.1276.
87. Smith RM, Sadee W. Synaptic signaling and aberrant RNA splicing in autism spectrum disorders. *Front Synap Neurosci.* 2011;3:1. https://doi.org/10.3389/fnsyn.2011.00001.
88. Reissner C, Runkel F, Missler M. Neurexins. *Genome Biol.* 2013;14:213.
89. Wang X, Xu Q, Bey AL, Lee Y, Jiang Y-H. Transcriptional and functional complexity of Shank3 provides a molecular framework to understand the phenotypic heterogeneity of SHANK3 causing autism and Shank3 mutant mice. *Mol Autism.* 2014;5:30. http://www.molecularautism.com/content/5/1/30.
90. Zhu L, Wang X, Li X-L, et al. Epigenetic dysregulation of SHANK3 in brain tissues from individuals with autism spectrum disorders. *Hum Mol Genet.* 2014;23(6):1563–1578.
91. Takata A, Ionita-Laza I, Gogos JA, Xu B, Karayiorgou M. De Novo synonymous mutations in regulatory elements contribute to the genetic etiology of autism and schizophrenia. *Neuron.* 2016; 89(5):940–947.
92. Iancu OD, Colville A, Oberbeck D, Darakjian P, McWeeney SK, Hitzemann R. Cosplicing network analysis of mammalian brain RNA-Seq data utilizing WGCNA and Mantel correlations. *Front Genet.* 2015;6:174.
93. Yap K, Xiao Y, Friedman BA, Je HS, Makeyev EV. Polarizing the neuron through sustained co-expression of alternatively spliced isoforms. *Cell Rep.* 2016;15(6):1316–1328.
94. Ragunath PK, Chitra R, Mohammad S, Abhinand PA. A systems biological study on the comorbidity of autism spectrum disorders and bipolar disorder. *Bioinformation.* 2011;7(3):102–106.
95. Berton O, McClung CA, Dileone RJ, et al. Essential role of BDNF in the mesolimbic dopamine pathway in social defeat stress. *Science.* 2006;311:864–868.
96. Wingate M, Kirby RS, Pettygrove S, et al. Prevalence of autism spectrum disorder among children aged 8 years – autism and developmental disabilities monitoring network, 11 sites, United States, 2010. *MMWR Surveill Summ.* 2014;63(2):1–21.
97. Bondar NP, Boyarskikh UA, Kovalenko IL, Filipenko ML, Kudryavtseva NN. Molecular implications of repeated aggression: Th, Dat1, Snca and Bdnf gene expression in the VTA of victorious male mice. *PLoS One.* 2009;4(1):e4190.
98. Abrahams BS, Geschwind DH. Advances in autism genetics: on the threshold of a new neurobiology. *Nat Rev Genet.* 2008;9(5):341–355.
99. Cook EH. Genetics of autism. *Child Adolesc Psychiatr Clin N Am.* 2001;10:333–350.
100. Rapin I, Tuchman RF. Autism: definition, neurobiology, screening, diagnosis. *Pediatr Clin North Am.* 2008;55(5):1129–1146.

# CHAPTER 15

# Epigenetic Mechanisms of Learning and Memory

**Gilda Stefanelli[1], Brandon J. Walters[2], Firyal Ramzan[1], Klotilda Narkaj[3], Cindy Tao[1], Iva B. Zovkic[1]**
[1]University of Toronto Mississauga, Mississauga, ON, Canada; [2]The Hospital for Sick Children, Toronto, ON, Canada; [3]University of Toronto, Toronto, ON, Canada

## INTRODUCTION: MEMORY, GENE REGULATION, AND EPIGENETICS

Memory formation is a multistage process dependent on a complex set of molecular events that shift as memories transition from recently acquired to remote and long-lasting.[1] For memories to become stable over time, they must undergo cellular consolidation, which typically occurs within hours of a learning event and requires changes in protein synthesis and gene expression.[2] Studies on brain injury patients have implicated the hippocampus as a crucial brain region for this initial phase of consolidation,[3–5] and lesion studies in rodents have, similarly, implicated the hippocampus as a key site for consolidating spatial memories.[6–8]

The cellular consolidation phase is transient, wherein manipulations that block memory during the first few hours after learning (e.g., protein synthesis and transcriptional inhibitors; see next section) are not effective at blocking memory when administered approximately 6 h after learning (reviewed in Ref. 9). However, ongoing molecular changes are required to maintain long-lasting memories that persist over years in humans and weeks to months in rodents. In contrast to the primary role of the hippocampus during the cellular consolidation stage, remote memories are associated with increased gene expression in the medial prefrontal cortex (mPFC),[10] and lesions or molecular manipulations of the mPFC lead to selective deficits in remote memory.[11,12] Given the shift in brain regions that appear to be primarily involved in memory at different time points, the transition of memory dependence from the hippocampus to the cortex has been termed systems consolidation.[2] Systems consolidation occurs over longer time scales compared to cellular consolidation, with rodent studies typically measuring remote memories approximately 30 days after training.[1]

A breakthrough in understanding the molecular basis of consolidation was the discovery that neurons increase levels of protein synthesis in response to a variety of stimuli.[13] Neuronal dendrites and synapses contain transcripts (mRNAs), ribosomes, and other proteins belonging to the translation apparatus, which become active after neuronal stimulation to provide a rapid local supply of protein for synaptic functions.[14] In addition,

*Molecular-Genetic and Statistical Techniques for Behavioral and Neural Research*
ISBN 978-0-12-804078-2, https://doi.org/10.1016/B978-0-12-804078-2.00015-5

neuronal stimulation and the consequent calcium ($Ca^{2+}$) influx also triggers gene transcription,[15,16] providing a reservoir for de novo protein synthesis. Gene transcription and protein synthesis play a crucial role in memory formation.[17] Pharmacological inhibition of either translation or transcription pathways impairs memory in mice and disrupts long-term potentiation (LTP), a cellular correlate of learning and memory.[18–20] The gene expression and protein synthesis waves triggered by neuronal stimulation are time-dependent, returning to baseline within hours.[15] The transient nature of key molecular events in the face of long-lasting memories led to the initial proposal that epigenetic mechanisms may serve as stable markers that would persist for the duration of the memory trace.[21,22]

Epigenetic mechanisms (see Epigenetic Mechanisms section) are involved in regulating gene expression and have the potential to persist over long periods of time, thus providing a candidate mechanism to support life-long memory retention. Numerous studies in the last decade have demonstrated that epigenetic mechanisms, particularly DNA methylation and posttranslational modifications (PTMs) of histones, are dynamically regulated in the brain in response to external stimuli, and they are key regulators of learning and memory across species.[23–26] These findings initiated a subfield of epigenetics called neuroepigenetics (also referred to as behavioral epigenetics), which studies how epigenetic mechanisms dynamically regulate the neural epigenome in experience-dependent ways.[27,28]

The term epigenetics classically refers to mitotically heritable modifications of DNA that alter gene expression without affecting the underlying DNA sequence.[25] This definition is based on the role of epigenetic marks, particularly DNA methylation (see DNA Methylation: The Code for Transcription Repression section), in maintaining "cell fate" during development, whereby genetically identical cells give rise to phenotypically different cells of the body. Given the persistence of cell fate across many rounds of cell division during the life span of an organism, the self-perpetuating nature of epigenetic marks has been referred to as "epigenetic memory" to describe how a cell "remembers" its identity.[29] This capacity for persistence and self-perpetuation was the motivating force for investigating epigenetic marks as a candidate mechanism for maintaining long-lasting memories in the brain.[21,30] Indeed, the discovery that learning-induced epigenetic regulation in the brain is highly dynamic and reversible[21] has challenged the dogma of the immutable epigenetic state that was based on epigenetic maintenance of cellular identity. It is now accepted that many dynamic epigenetic changes occur in mature cells that continuously adapt to environmental stimuli, such as immune cells producing antibodies to different pathogens[31] or skeletal muscle adapting to training.[32]

## EPIGENETIC MECHANISMS

In the cell nucleus, stretches of approximately 147 bp of DNA are packaged around a histone octamer containing two dimers of histones H2A-H2B and a tetramer of

H3–H4 to form nucleosomes (Fig. 15.1A). Repeating nucleosomes are connected through the linker histone H1 to form the building blocks of chromatin[33] (Fig. 15.1A). Within the chromatin landscape, epigenetic mechanisms most commonly studied are DNA methylation (Fig. 15.2) and PTMs of histone tails (Fig. 15.1B) that regulate transcription in different ways. The organization of chromatin into nucleosomes leads to the compaction of genes, thus creating a mechanism by which modifications of histone proteins can regulate gene accessibility to modify gene expression (see Histone Posttranslational Modifications: The Epigenetic Code Regulating Cell Function section). Specifically, highly compacted nucleosomes form heterochromatin and are associated with transcriptional repression, whereas an open nucleosome structure forms euchromatin and is associated with transcriptional activation (Fig. 15.3B). Genes can

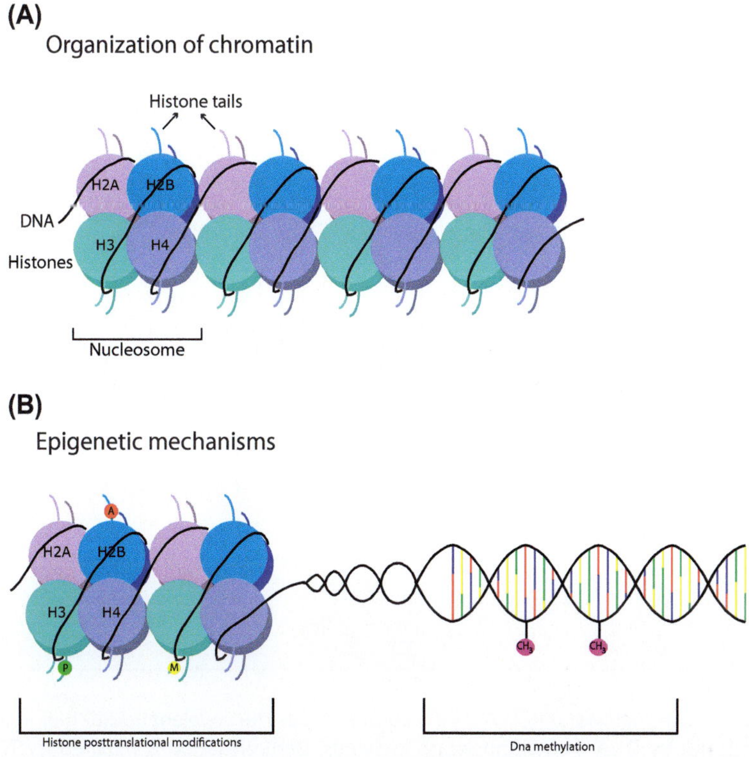


**Figure 15.1** (A) *Organization of chromatin structure*: Nucleosomes are composed of a stretch of 147 bp of DNA wrapped around and octamer of four histones (H2A, H2B, H3, H4), with tails protruding from each histone. Repeats of nucleosomes form chromatin. (B) *Epigenetic mechanisms*: epigenetic mechanisms generally affect histone tails and DNA. Histone tails can be posttranslationally modified to regulate transcription (see Fig. 15.3 and Histone Posttranslational Modifications: The Epigenetic Code Regulating Cell Function section), while CpG nucleotides can be methylated to repress transcription (see Fig. 15.2 and DNA Methylation: The Code for Transcription Repression section).

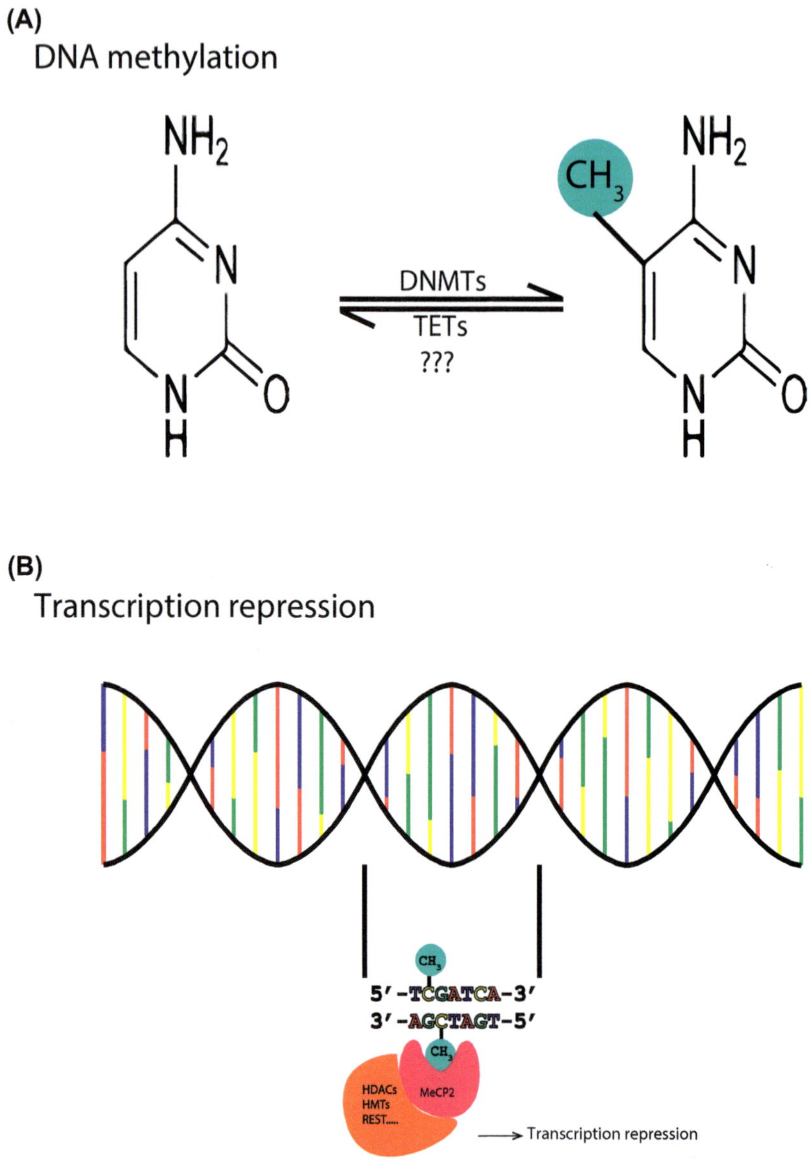


**Figure 15.2** (A) *DNA methylation*: DNA methylation involves the covalent addition of a methyl group ($CH_3$) to cytosines by DNA methyltransferase enzymes. Demethylation is initiated by TET1 proteins that mediate the conversion of 5mC to 5hmC. (B) *Transcription repression*: proteins with a methyl-binding domain can bind to methylated cytosines and recruit other modifying enzymes that alter gene expression.

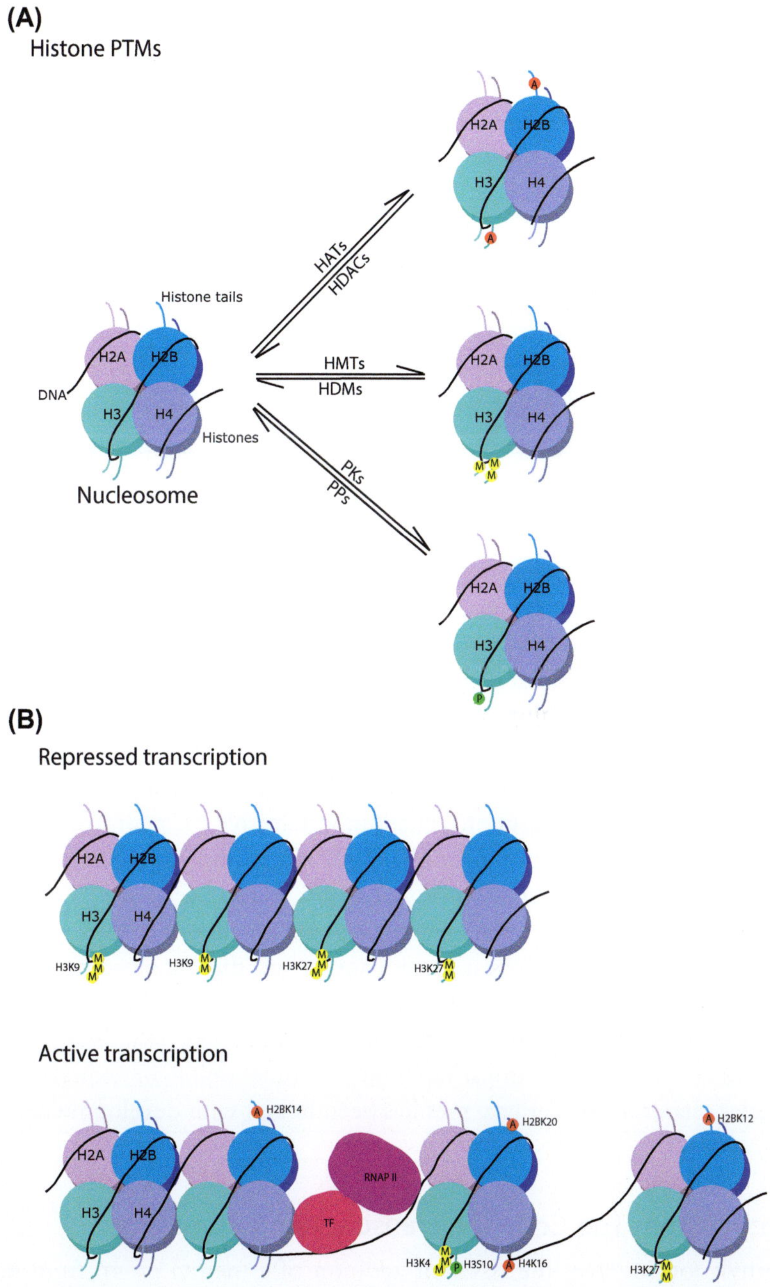


**Figure 15.3** (A) *Histone PTMs*: specific writers and erasers catalyze the dynamic addition and removal of epigenetic marks on nucleosomes. (B) The simultaneous presence of different modifications on histones determines active and repressed chromatin state. *Repressed transcription*: The presence of H3K9me3 and/or H3K27 (and others not reported in the figure) is associated with transcriptional repression and chromatin compaction. *Active transcription*: Acetylation at various marks (e.g., H2BK14, H2BK20, H2BK12, H4K16) and/or phosphorylation (e.g., at H3 serine 10) is associated with the recruitment of TFs, RNApol, and an open chromatin state that favor transcription.

**Table 15.1** Epigenetic machinery: writers, erasers, and readers

| Protein | Examples | Role/function |
|---|---|---|
| Writer enzymes | DNA methyltransferase, histone acetyltransferase, histone methyltransferase | Catalyze addition of epigenetic marks on DNA/histones |
| Eraser enzymes | Histone deacetylase, histone demethylase, histone lysine demethylase; ten-eleven translocation proteins (TET1, TET2, TET3; catalyze conversion of 5mC to 5hmC) | Remove epigenetic marks |
| Reader domains | Methyl-binding protein, bromodomains, tudor domains, chromodomains | Recognize and recruited to specific epigenetic marks |

A summary of the most commonly examples of writer and eraser enzymes and reader domains involved in the process of epigenetic modification.

also exist in a poised or bivalent state, in which promoter-proximal nucleosomes display markers of both activating and repressing histone modifications[34] (see Histone Posttranslational Modifications: The Epigenetic Code Regulating Cell Function section) to allow for rapid stimulus-induced activation or silencing of gene expression.[35] Poised states were originally identified in the context of precise gene expression patterns in development, but approximately 40% of poised states are retained in terminally differentiated neurons,[36] indicating that poised genes fulfill an important role for neuronal function.

Epigenetic markers and their functions are established and implemented through actions of various enzymes that act as *writers*, *erasers*, and *readers* (see Table 15.1). Writers are enzymes that catalyze the addition of epigenetic marks on either histone tails or the DNA itself. These changes are not necessarily permanent, and thus erasers consist of enzymes that remove these marks. Typically, these epigenetic marks exert their effects through interactions with *reader* proteins that recognize individual epigenetic marks and initiate further downstream effects.[37] Impairment in the functioning or expression level of various writer, eraser, or reader enzymes has been linked with developmental disorders, as well as cancer.[38–40]

## DNA Methylation: The Code for Transcription Repression

DNA methylation involves the covalent addition of a methyl group from the donor *S*-adenosyl-methionine to the fifth carbon of a cytosine residue adjacent to a guanine (CpG) to form 5-methylcytosine (5mC)[41,42] (Fig. 15.2A). DNA methylation is largely considered to be a repressive regulator of gene activity, based on its ability to interfere with the deposition of transcriptional machinery, and the recruitment of transcriptional repressors that promote a closed chromatin state[42] (Fig. 15.2B). Once deposited, methylation requires high levels of energy to be reversed.[43] As a result, methylation was considered to be a very stable mark that allows for persistent changes to chromatin and long-term regulation of cellular function.[44,45] Strikingly, the process was recently reported to be dynamic across a variety of cell types, including muscle,[46] breast cancer cells,[47] and neurons,[48] thus challenging this belief.

DNA methylation is catalyzed by a family of enzymes called DNA methyltransferases (DNMTs), classified as methylation *writers* (Table 15.1). Two different categories of DMNTs are responsible for maintenance and de novo DNA methylation.[48] DNMT1 acts to maintain methylation marks on newly synthesized daughter strands by copying the DNA methylation pattern on the parental DNA strand, allowing for self-perpetuation of the methylation state throughout cell divisions.[49] In contrast, DNMT3a and DNMT3b target unmodified DNA to establish novel methylation marks, and thus are called de novo methylatransferases.[48] DNMTs are heavily involved in embryonic and early life development, so germ line knockout of either DNMT1, DNMT3A, or 3b causes embryonic lethality early during development.[50] As such, research on DNA methylation has mainly focused on its role in regulating cell division and early differentiation, as well as disorders of these processes, such as cancer or other developmental disorders.[51,52] It was recently shown, however, that postmitotic neurons in the mature brain continue to express high levels of DNMTs, and that DNA methylation continues to play a role in transcription regulation, synaptic plasticity, and memory formation in the adult central nervous system.[53,54]

In addition to representing a steric obstacle for transcriptional machinery, DNA methylation also exerts repressive effects by recruiting methyl-binding proteins (MBPs), the *readers* of DNA methylation (Table 15.1). MBPs bind to methylated DNA and interact with repressor complexes, such as histone deacetylases (HDACs) (discussed in Histone Posttranslational Modifications: The Epigenetic Code Regulating Cell Function section)[55,56] to negatively influence transcription (Fig. 15.2B). As readers of DNA methylation, MBPs provide a direct link between DNA methylation, transcriptional repression, and chromatin remodeling.[42,57] All MBPs have the methyl-binding domain (MBD) in common, which allows them to bind methylated sites. Mutation in the methyl-CpG-binding protein 2 (MeCP2) is the cause of the X-linked neurodevelopmental disorder Rett syndrome,[58] highlighting the importance of all the components of epigenetic machinery, from writers and erasers to readers, during development.

In a process similar to DNA methylation, the recently discovered hydroxy-methylation adds a hydroxy-methyl group to cytosine to form 5-hydroxy-methylcytosine (5-hmC).[59] This modified base is not present in cancer cell lines and exists at high levels within neurons compared to other tissues.[60] In contrast to methylation, hydroxy-methylation is often associated with transcriptional activation.[61] Although the primary role for hydroxy-methylation is to act as an intermediate step in DNA demethylation,[62] some evidence suggests that 5-hmC may be an epigenetic regulator in its own right[63] (see The DNA methylation code section).

## Histone Posttranslational Modifications: The Epigenetic Code Regulating Cell Function

Histones are highly alkaline proteins with high affinity for DNA. They are classified into two major classes: core histones (H2A, H2B, H3, and H4) and linker histones (H1/H5), which work together to package DNA into nucleosomes (Fig. 15.3). Histones participate in epigenetic regulation largely through PTMs. The N- and C-terminal histone tails extend outside the nucleosome core and are a target of various PTMs, including acetylation, methylation, and phosphorylation[64] (Table 15.2 and Fig. 15.3A). The two major functions of histone PTMs are to alter gene accessibility by modifying histone-DNA interactions and to recruit nonhistone proteins[65] (Fig. 15.3B). Both mechanisms affect a multitude of cellular processes, including regulation of gene expression and chromatin compaction (Table 15.2).

Distinct PTMs affect gene expression and chromatin compaction in different ways. Although some modifications have been linked specifically to transcriptional activation (e.g., tri-methylation of histone 3 at lysine 4; H3K4me3)[66] or repression (e.g., H3K27me3),[67] histone PTMs generally exert combinatorial effects on gene expression. This observation forms the basis of the histone code hypothesis, which posits that the combination of different modifications can lead to either gene activation or repression.[68,69] Indeed, certain combinations of PTMs tend to cooccupy the same nucleosome.[70] For example, although H3K4me3 and H3K27me3 are individually associated

**Table 15.2** Histone posttranslational modifications (PTMs) and their associated functions

| PTM | Residues affected | Role/function |
|---|---|---|
| Methylation | Lysines K-me1/me2/me3 | Transcription activation/ repression, DNA repair |
| | Arginines R-me1/me2/me3 | Transcription activation/ repression |
| Acetylation | Lysines K-ac | Transcription activation, DNA repair, replication, condensation |
| Phosphorylation | Serines S-ph, Threonine T-ph | Transcription activation, DNA repair, condensation |
| Ubiquitylation | Lysines K-ub | Transcription activation/ repression, DNA repair |
| Sumoylation | Lysines K-su | Transcription repression |

A summary of the most commonly studied histone PTMs and cellular functions in which they play a role. Methylation, acetylation, and phosphorylation are discussed in the chapter. For a detailed discussion of ubiquitylation, sumoylation, and other PTMs, see Kouzarides.[65]

with transcriptional activation and repression, respectively, their cooccurrence is associated with poised genes.[34] Additionally, the outcome of a specific histone modification may depend on its location in the gene. For example, di-/tri-methylation at the promoter of H3K9 is associated with transcriptional repression, whereas recent evidence suggests that tri-methylation at the coding region may be associated with activation.[65,71] Thus, the same PTM does not always generate the same transcriptional outcome, and the combinatorial effects of various modifications add complexity to gene regulation.

Several enzymes regulate the deposition and removal of PTMs on histone tails (Table 15.1 and Fig. 15.3A), and their disruption has been linked with severe developmental disorders, dementia, and cancer.[72] Histones can undergo a plethora of PTMs, but the most commonly studied are acetylation, methylation, and phosphorylation (Fig. 15.3A). Histone acetylation/deacetylation involves the addition or removal of an acetyl group on lysine (K) residues on histone tails.[73] This process is catalyzed by histone acetyl-transferase and HDAC enzymes, which can add or remove acetyl groups, representing writers and erasers, respectively (Table 15.1 and Fig. 15.3A). In general, acetylation is linked with transcriptional facilitation by reducing histone-DNA interactions and by recruiting chromatin-remodeling proteins and RNA polymerase II[74] to initiate transcription (Fig. 15.3B). Histone phosphorylation occurs mainly on serine (S) and threonine (T) residues and is mediated by a number of nuclear kinases and phosphatases[23,75] (Fig. 15.3A). As with acetylation, phosphorylation is primarily associated with transcriptional activation[76] (Fig. 15.3B).

Conversely, histone methylation, the addition of a methyl group to lysine (K) and/or arginine (R) residues, can result in either activation or repression of transcription (Fig. 15.3B). For example, methylation of lysine 4 on histone H3 (H3K4me3) is associated with transcriptional activation, whereas methylation of lysine 9 (H3K9me) is associated with repression.[66,67] In addition, different levels of methylation can occur on the same residue, including mono, di-, and tri-methylation, each of which affects transcriptional activation/repression differently.[77] Histone methyl-transferases (HMTs) and demethylases (HDMs) are involved in specific deposition/removal of methyl groups on histone tails[78,79] (Table 15.1 and Fig. 15.3A). A number of HMTs and HDMs have been characterized thus far and some have been linked with memory formation, including Mll (mixed leukemia lineage 1 methyltransferase), an H3K4-specific HMT that is particularly important for hippocampal function.[80] Similarly, G9a, an H3K9-specific HMT, has been implicated in drug addiction.[81]

## Histone Variants: Novel Epigenetic Regulators of Memory

Histone variants are nonallelic counterparts that replace core histones in nucleosomes. Histones and their variants are largely conserved throughout species, and several variants have been identified for histones H2A, H2B, and H3, whereas limited evidence exists for

**Table 15.3** Histones and histone variants

| Canonical histones | Histone variants | Function/role |
|---|---|---|
| H2A | H2Ax | DNA repair, remodeling sex chromosomes |
| | H2Az | Transcription, reduced nucleosome stability |
| | Other numerous variants | |
| H2B | Numerous variants exist but are understudied in mammals | |
| H3 | H3.3 | Activated genes and regulatory elements |
| | cenH3/CENP-A | In centromeres; DNA repair |
| H4 | Largely understudied; limited evidence of variants | |

A summary of the most commonly studied histone variants and cellular functions in which they have been implicated to play a role. H2Ax, H2Az, H3.3, and cenH3/CENP-A are discussed in greater detail in the text.

the presence of H4 variants in mammals[82] (see Table 15.3). In contrast to canonical histones, histone variants are replication-independent and, as such, they can be synthesized and incorporated into nucleosomes regardless of cell division.[83] In contrast, canonical histones are replication-dependent and are only expressed/incorporated during the synthesis phase of the cell cycle, when DNA is undergoing replication.[84–86] The fact that variants are expressed ubiquitously throughout the cell cycle suggests a unique role in postmitotic cells, such as neurons.

Whereas histone variant exchange implies a replacement of a canonical histone with a variant, histone turnover involves the replacement of older histone proteins with identical newer proteins.[84,87] Initially, histone turnover was found to be fairly slow, with a half-life of 117 days for histones in the liver.[88] However, recent studies show that the half-life of histone variants H3.3 and H2A.Z in neurons is approximately 3.5 days under basal conditions and 0.7 days under induced conditions,[87] suggesting that histones can be rapidly replaced in the brain. Both histone variant exchange and histone turnover have been implicated in neuronal function and plasticity,[87,89] and their roles in memory formation will be discussed in Histone Variants: New Regulators of Memory section.

## EPIGENETIC MECHANISMS REGULATING MEMORY

### Histone Posttranslational Modifications: Dynamic Regulators of Memory

The first evidence of dynamic regulation of histone PTMs by neuronal stimulation was reported in Aplysia in 2002.[90] The authors investigated the effects of different stimuli on LTP and long-term depression (LTD), two cellular models of learning. As with memory, LTP is associated with increased gene expression that supports strengthening of synaptic transmission, and LTD is associated with decreased gene expression and is usually followed by a weakening of synaptic transmission (for a review on LTP and LTD, see Ref. 91). In *Aplysia*, simultaneous stimulation of a sensory neuron with an excitatory and an inhibitory neurotransmitter blocked LTP in favor of LTD through a mechanism involving HDAC5. Specifically, the inhibitory stimulus recruited HDAC5 and led to histone deacetylation on promoters of early-response genes, which was associated with transcriptional repression.[90] Numerous studies have since reported that in vitro neuronal stimulation and induction of LTP or LTD triggers changes in histone PTMs,[75,92] firmly implicating this process in neuronal plasticity. Accordingly, a range of memory tasks alter PTMs of different histones in vivo across brain regions and at different time points of memory formation.[23,24,80,93,94]

### Cellular Consolidation: Dynamic and Transient Histone Modifications

Many studies of epigenetic regulation in memory have utilized contextual fear conditioning (CFC), an associative learning paradigm in which animals learn to associate a novel context with the delivery of an aversive stimulus in the form of a mild foot shock.[95] Memory formation is indexed by freezing behavior upon reexposure to the context in the absence of shock. Given the ability of rodents to acquire fear conditioning in a single training session, this paradigm is especially useful for investigating time-dependent changes in memory formation.

In accordance with a transient role of the hippocampus in memory, the first study to directly link learning with histone PTMs reported a transient increase in global levels of hippocampal histone acetylation 1 h after CFC, which returned to basal levels by 24 h.[24] The pharmacological enhancement of histone acetylation with HDAC inhibitors (see Manipulating Histone Modifications section) during the cellular consolidation window resulted in enhanced memory recall at 24 h,[24] suggesting that increased histone acetylation is functionally relevant during memory formation. Since then, CFC has been associated with changes in hippocampal levels of H3K4me3, H3K9me2,[80,93] H3S10P,[23] and acetylation at a number of lysine residues on histones H3 and H4.[24,96]

Histone changes were also identified in other hippocampus-dependent tasks, including the Morris water maze. This test is an open-field procedure designed to test spatial memory.[97] Spatial learning is a form of associative learning involving multiple spatial cues instead of a single pair. In this test, animals are placed in a pool filled with an opaque liquid and are motivated to find a submerged platform that allows them to escape the distress of swimming. The animal learns to associate multiple external cues as guiding markers for finding the submerged platform. Learning is assessed as the latency to reach the platform over repeated training sessions. In this paradigm, learning is gradual, so the latency and distance traveled decrease as the number of trials increases, indicating the occurrence of learning.[97] Three days of consecutive water maze training resulted in increased acetylation of H4K12 and multiple residues of histone H2B.[98,99] Histone changes were measured at 3 days because it represented the time point at which most animals showed learning. Consistent with a transient role of the hippocampus in memory, histone PTMs in this region typically return to basal levels within a few hours of training.

Memory formation requires a balance between memory activators and suppressors at the level of transcription (i.e., memory suppressors are inhibited and plasticity-promoting genes are activated by learning) and epigenetics.[100] Accordingly, memory formation is also associated with increased levels of histone PTMs that are correlated with gene repression. For instance, CFC induced increased levels of H3K9me2, a mark of repressive chromatin, in area CA1 of the hippocampus and the entorhinal cortex (EC) 1 h after training,[101,102] with distinct functional outcomes in each region. Inhibiting G9a/GLP, a H3K9-specific dimethyltransferase in the EC enhanced fear memory, whereas G9a/GLP inhibition in area CA1 impaired fear memory at 24 h, indicating that the same histone modification can exert region-specific outcomes on memory.[102] Further, intra-EC inhibition of G9a/GLP also resulted in enhanced H3K9me2 levels in area CA1 of the hippocampus, indicating crosstalk between the two areas.[102]

These data suggest that memory formation involves changes in histone marks associated with both positive and negative transcriptional outcomes. However, the link between learning-induced changes in histone marks and gene expression is not clear. Recent work with next-generation sequencing (NGS) techniques (see High-Throughput Next-Generation Sequencing: From Candidate Genes to Genome-Wide Analysis section) found that fear conditioning induces global changes in various histone PTMs in the hippocampus, but these PTMs are poorly correlated with training-induced changes in gene expression.[103] This discrepancy between stimulus-induced changes in histone PTMs and transcription may be attributed to experimental design, such as the choice of time point, or it may reflect an alternative role for histone PTMs in response to stimuli. For example, activity-induced changes in histone PTMs may "prepare" genes to respond to additional stimuli, or for memory reactivation, rather than regulating transcription directly. Further research will elucidate the precise role of histone PTMs at the level of gene regulation.

### *Systems Consolidation: Stable and Long-Lasting Histone Modifications in the Cortex Regulate Long-Term Memory Storage*

A distinct landscape of histone PTMs has been observed during different stages of systems consolidation across different brain regions. One study investigated changes in several histone PTMs (H3S10P, H3K14, H4K5ac, and H3K36me3) in the hippocampus and prefrontal cortex during different phases of novel object memory formation, ranging from habituation to short-term, recent, and remote memory.[104] In contrast to tasks based on aversive stimuli, novel object recognition (NOR) assesses recognition memory upon exposure to a novel and a familiar object.[105] After an initial familiarization session with two objects, the animal is reexposed to a familiar object alongside a novel object and allowed to explore each freely. This test exploits rodents' preference for novel objects, whereby memory formation is measured as the difference between the amount of time spent exploring a novel compared to a familiar object.[105] Mice were exposed to three unfamiliar objects (habituation) and memory formation and the molecular analyses were carried out 10 min (short-term memory), 24 h (recent memory), or 7 days (remote memory) after training. All four PTMs were modified in the hippocampus during the initial stages of training (habituation and 10 min) and returned to baseline during recent (24 h) and remote (7 days) memory phases,[104] in accordance with previous observations. In contrast, the four PTMs appeared in the prefrontal cortex (PFC) for the first time during the recent memory phase (24 h) and persisted for at least 7 days after training (remote memory), suggesting that changes in the PFC are delayed relative to the hippocampus. However, prolonging the habituation period induced a more robust increase in the four PTMs in the PFC at 7 days and enhanced remote memory at this time point,[104] indicating that despite the delayed occurrence in the cortex, histone PTMs in the PFC are nevertheless sensitive to variation in memory strength during the initial learning stages. This study indicates that changes in histone PTMs in the PFC may occur at greater delays, but they are more persistent compared to rapid and transient changes in the hippocampus.

Cortical changes in histone acetylation have also been implicated in regulating recent memory formation in the object location task. Inhibiting cortical histone acetylation by overexpressing a dominant negative form of histone acetyltransferase *CBP* (*CREB binding protein*) in mPFC resulted in decreased *c-Fos* expression in the mPFC and impaired recent memory in the object location task.[106] This test is a variation of the NOR in which a familiar object is placed in a different position during the test compared to the habituation phase, thereby adding a spatial component to the task. Control animals show a preference for the novel location 5 min and 24 h after training, whereas *CBP* dominant negative mice only show a location preference 5 min after training, indicating a selective deficit in recent long-term, but not short-term memory.[106] These studies highlight the importance of cortical histone PTMs in regulating both recent and remote memory formation.

Despite extensive lesion and molecular evidence to support cortical maintenance of remote memory,[9] mechanisms that underlie systems consolidation are still poorly understood. An elegant study investigated the role of histone modifications in the shift of memory dependence from the hippocampus to the cortex using the social transmission of food preference paradigm.[107] In this task, rats learn about the safety of a novel food by smelling the breath of littermates who were exposed to this food.[108] The authors showed that early changes in cortical histone acetylation were essential for remote memory. They observed increased H3 acetylation in the orbitofrontal cortex (OFC) within 1 h of training[107] and blocking OFC histone acetylation before training selectively impaired remote memory at 30 days.[107] These findings led to the hypothesis that the formation of remote memories involves an early tagging of cortical neurons through a process that involves histone acetylation.[107] Overall, these data suggest that epigenetic changes have time-dependent effects on memory formation in the hippocampus and the cortex. Despite the temporally distinct roles hypothesized for the two brain regions by the systems consolidation theory,[2] both early and delayed histone changes in the cortex appear to be essential for establishing long-lasting memories.

In addition to histone changes in the hippocampus during the cellular consolidation window, memory recall is also associated with a new wave of histone PTMs after recall. Reexposure of animals to the same context 24 h after CFC training (i.e., during recall) induced increased global H3 phosphorylation and acetylation in area CA1 of the hippocampus.[109] Although the role of recall-induced changes is less clear, these changes may be essential for reconsolidation of previously consolidated memories.[109]

### *Manipulating Histone Modifications*

Histone PTMs have been heavily studied for their ability to modify memory formation, with most studies focusing on the pharmacological manipulation of HDAC inhibitors (HDACi). This class of drugs includes sodium butyrate (NaB) and trichostatin (TSA), which inhibit different classes of HDACs and result in global histone hyperacetylation. HDACi administration increases fear memory and hippocampal LTP, which has been attributed to increased levels of acetylated H3 in the hippocampus.[24] The memory-potentiating effects of HDACi have led to their widespread use in animal models of cognitive decline. For example, HDACi treatment has been used to reverse learning deficits in aged mice,[96] Alzheimer models,[110] and in a mouse model of neuronal atrophy.[94]

Although such findings have led to speculation about therapeutic targeting of histones for memory disorders, linking HDACi effects to a specific mechanism for memory improvement is difficult. In addition to modifying histone acetylation, HDACs may also exert effects on other histone modifications. For example, the HDACi NaB increased H3K4me3 and decreased H3K9me2 in response to fear conditioning, an effect that likely reflects crosstalk between different histone modifications.[93] HDACs also act on numerous nonhistone substrates, including transcription factors (TFs), transcriptional regulators, chaperones, and structural proteins.[111] Phylogenetic analyses show that HDACs preceded the evolution of histones, indicating that their main targets might be nonhistone proteins.[112] Moreover, many HDACi have a broad spectrum of action, resulting in functional inhibition of many different classes of HDACs.

Genetic studies have been utilized to tease apart the roles of specific subclasses of HDACs in regulating memory. For instance, HDAC2 deficiency facilitates memory formation and increases spine density, and overexpression of HDAC2, but not HDAC1, results in impaired memory and decreased spine density.[113] In addition, brain-specific conditional knockout of *Hdac4*, but not *Hdac5*, impaired fear memory and LTP,[114] and *Hdac3* knockdown in the hippocampus enhanced NOR memory 24 h and 7 days after training.[115] These studies suggest that different classes of HDACs exert different effects on memory and point to a need to investigate downstream targets of each HDAC class during learning. Moreover, some data suggest that different HDAC classes may have brain region-specific expression patterns, pointing to potential regional differences in their effects.[116]

Besides HDACs, other writers and erasers of histone PTMs have also been directly implicated in memory formation and are thus potential targets for therapeutic intervention. For instance, knockdown of the neuro-specific H3K20 demethylase *Lsd1n* impairs stimulus-dependent gene expression and spatial learning.[117] Identifying how specific enzymes involved in the deposition/removal of different PTMs interact with one another to affect memory and gene activity will be crucial for understanding the epigenetic code that fine tunes memories.

## DNA Methylation: A Dynamic Process for Long-Term Memory Storage

A role for dynamic changes in DNA methylation in memory formation was first identified in a seminal paper by Miller and Sweatt in 2007.[21] They found a training-induced increase in the expression of de novo DNMTs 3a and 3b in area CA1. Concomitantly, they reported first evidence for dynamic changes in methylation at the promoters of memory-related genes.[21] Specifically, fear conditioning resulted in increased methylation and decreased expression of the memory suppressor gene *protein phosphatase 1* (*PP1*), as well as demethylation and increased expression of the plasticity gene *reelin*.[21] Accordingly, administration of a DNMT inhibitor after training impaired fear memory and blocked the *PP1* downregulation typically observed with training. Learning-induced changes in hippocampal DNA methylation have since been reported for a variety of memory tests and genes.[21,118–120]

As for histone PTMs, changes in DNA methylation observed during the cellular consolidation stage in the hippocampus tend to be transient and disappear within 24 h of training. However, epigenetic changes in the cortex persist in association with remote memory. The first study to report stable changes in cortical DNA methylation showed that *reelin* is rapidly hypermethylated 1 h after training, and increased methylation persisted for 7 days, whereas an increase in *calcineurin* methylation was not found until 1 day after training and persisted for at least 30 days.[12] Increased *calcineurin* methylation was associated with decreased gene expression, and pharmacological inhibition of DNA methylation in the mPFC at ~30 days impaired fear memory and increased *calcineurin* expression,[12] suggesting that DNA methylation regulates remote memory through effects on gene expression in the cortex.

More recently, Halder and colleagues[103] carried out a genome-wide study of learning-induced changes in DNA methylation in the hippocampus and the cortex. Consistent with previous work, fear conditioning-induced changes in hippocampal

DNA methylation were transient, with genes showing both increased and decreased levels of methylation 1 h, but not 4 weeks after training. In contrast, DNA methylation in the anterior cingulate cortex (ACC) was evident both at 1 h and 4 weeks after training. Indeed, cortical changes in differentially methylated genes comparing naïve mice to mice exposed to context and shock (6250 differentially methylated genes) were much more extensive compared to changes in area CA1 of the hippocampus (1206 differentially methylated genes). However, cortical changes were less specific to the context-shock association compared to hippocampal changes at 1 h. Whereas ~3000 differentially methylated regions were detected in area CA1 when comparing mice exposed to context and shock compared to context only, no such changes were found in the anterior cingulate 1 h after training.

Consistent with temporal shifts in memory dependence from the hippocampus to the cortex, remote memory was associated with 1223 differentially methylated genes in the cortex of mice exposed to context-shock compared to naïve mice. This number represents a dramatic reduction in the number of methylated genes compared to 1 h after training, suggesting that a smaller set of genes in the ACC are involved in maintaining the memory trace than are involved in processing the initial training stimulus. When the comparison was limited to mice exposed to context only versus context-shock (i.e., representing only associative learning), the number of differentially methylated genes was reduced to 154. This study validates and extends data from candidate gene studies,[12,21] demonstrating that a subset of genes that are altered during initial training exhibit persistent changes in methylation that correlate with memory maintenance. Although it is not clear what distinguishes genes that are persistently methylated from those that are not, Halder and colleagues[103] found that differential methylation tended to occur on neuron-specific genes.

### *Role of DNA Demethylation in Memory*

Studies of learning-induced DNA methylation have established correlational evidence for both DNA methylation and DNA demethylation in memory formation. However, experimental manipulations of DNA methylation have been limited to DNMTs because a mechanism for DNA demethylation had not been identified. Recent work has identified a demethylation pathway involving the enzyme ten-eleven translocation methylcytosine dioxygenase 1 (TET1).[59] The active demethylation pathway is initiated by the enzyme TET1, which converts 5mC to 5hmC (5-hydroxy-methylcytosine) through hydroxylation.[121] Next, the APOBEC enzymatic complex (activation-induced cytidine deaminase/apolipoprotein B mRNA-editing enzyme complex) deaminates the modified base to form 5-hydroxy-methyluracil (5hmU). Finally, 5hmU undergoes glycolysis-dependent excision and repair, culminating in cytosine replacement and thus DNA demethylation.[121]

Recent evidence suggests that this pathway represents a way for neurons to achieve active demethylation and to cycle between methylation states as a way to regulate activity of memory-related genes.[122] Accumulating data now implicate ten-eleven translocations (TET) in regulating memory formation. Endogenous TET1 is expressed in neurons throughout the hippocampus, and transcript levels are downregulated following neuronal activity.[122] The overexpression of the catalytic domain of TET1 in CA1 induced decreased levels of 5mC in favor of 5hmC at baseline, resulting in the upregulation of several memory-related genes.[122] In addition, overexpressing the catalytic domain of TET1 resulted in impaired memory recall 24 h after fear conditioning training,[122] suggesting that the ability of TET1 to convert 5mC to 5hmC may mediate memory impairment. However, memory impairment was also observed with overexpression of a TET1 catalytic inactive isoform, indicating that TET1 also has additional uncharacterized functions.[122]

A crucial role for TET1 in memory was also reported with genetic manipulations of TET1. *Tet1* knockout mice exhibit decreased levels of 5hmC throughout the brain and display altered expression of memory-related genes.[123] *Tet1* KO animals also show upregulation of genes related to DNA demethylation, such as *Tet2* and *Tet3*,[123] which have redundant functions to *Tet1*, as well as *Gadd45b* and *APOBEC1* (see later), indicating that some compensatory mechanisms might follow *Tet1* ablation. *Tet1* knockout mice showed enhanced recent and remote memory measured 24 h and 30 days after fear conditioning.[123] The same behavioral results were obtained following virally mediated knockdown of TET1 in the dorsal hippocampus, indicating that the observed phenotype is caused by lack of the protein in the mature brain and not induced by any developmental alteration that might follow *Tet1* germline ablation.[123]

In contrast to enhanced fear memory observed by Kumar and colleagues,[123] Rudenko and colleagues[124] found that *Tet1* knockout mice display impaired fear extinction. The reason for this discrepancy is not clear, but it may relate to different functions of TET1 in different types of memory. These studies implicate DNA hydroxy-methylation as a regulator of memory formation and warrant further research into understanding how 5hmC affects memory.

In addition to enzymatic demethylation by TET1, a family of nonenzymatic proteins, growth arrest and DNA damage 45 (Gadd45) function to recruit enzymatic machinery responsible for base and nucleotide-excision repair in the process of DNA demethylation.[125,126] Gadd45 are stress-inducible proteins, implicated in diverse processes, such as cell cycle control, apoptosis, and DNA repair.[126] The exact mechanism by which these proteins promote DNA demethylation remains unclear, but it is proposed that their involvement in base and nucleotide excision repair is a potential candidate,[48] whereby methylated cytosines are excised and replaced by unmethylated counterparts during strand break repair.[126] In neurons, Gadd45 knockout results in impaired DNA demethylation and impaired activation of memory-related genes such as *Bdnf* and *Fgf1*,[125] but genome-wide studies showing a link between Gadd45 and DNA demethylation are lacking.

The role of Gadd45 in hippocampus-dependent memory formation is unclear. Fear conditioning has been linked with upregulation of specific members of the Gadd45 family in the hippocampus, including *Gadd45b* and *Gadd45g*, but not *Gadd45a*,[127,128] suggesting a potential gene-specific role in memory. *Gadd45b* knockout mice exhibited enhanced hippocampus-dependent fear and spatial memory, as well as enhanced LTP,[128] suggesting that *Gadd45b* may have a negative effect on memory. In contrast, another study found that *Gadd45b* knockout mice exhibit impaired contextual fear memory,[127] indicating a need for further research into the potential role of *Gadd45* in memory formation and DNA demethylation.

### *The DNA Methylation Code*

Discovery of dynamic DNA methylation has led scientists to speculate the existence of a DNA methylation code that, together with the histone code, can regulate memory formation and maintenance.[129] DNA methylation has distinct downstream effects that vary based on several factors, including the type of modification, the proteins it interacts with, and the location of the methyl mark. Whereas DNA methylation in promoters is traditionally linked to gene repression, the conversion of 5mC to 5hmC by TET proteins is linked primarily with transcriptional activation.[60,61] In addition, DNA methylation can produce distinct effects on gene expression by recruiting MBPs[42] that bind different types of methyl marks, including hemi-methylated, fully methylated, and hydroxyl-methylated DNA[21,130] to create a multitude of potentially functional interactions. MBPs themselves recruit other chromatin-remodeling complexes and histone-modifying proteins to link DNA methylation to histone modifications and gene regulatoin.[55,131] Although most of these interactions tend to be repressive, evidence indicates that the MBP MeCP2 can also recruit the transcriptional activator CREB1 to positively regulate gene expression.[132] Moreover, MBPs interact with histone-modifying enzymes, such as HDACs, to reduce levels of histone acetylation and influence gene expression.[55,131] Increased understanding of DNA methylation function at different regulatory regions (see NGS High-Throughput Next-Generation Sequencing: From Candidate Genes to Genome-Wide Analysis section) and in relation to its different interaction partners will be essential for understanding the complex contribution of DNA methylation to gene regulation and memory formation and maintenance.

## Histone Variants: New Regulators of Memory

Since neuroepigenetics was established as a field, most research focused on the role of histone PTMs and DNA methylation in memory formation. In the first in vivo study of histone variants, we recently identified H2A.Z, a variant of histone H2A, as a novel regulator of memory formation and maintenance in mice.[89] We showed that fear conditioning resulted in the removal of H2A.Z from the first nucleosome downstream (+1 nucleosome) of the transcription start site (TSS) and a concomitant increase in the expression of several memory-related genes in the hippocampus.[89] As with other epigenetic modifications, H2A.Z removal in the hippocampus was transient, returning to basal levels by 2 h after training.[89] In the mPFC, H2A.Z eviction during the cellular consolidation window (2 h after training) was also transient, whereas H2A.Z incorporation into the first nucleosome upstream (−1 nucleosome) of the TSS emerged 7 days after training and returned to baseline by 30 days. These data suggest that H2A.Z binding is subject to dynamic stimulus-induced regulation, but that it also has the capacity to support ongoing processing required for systems consolidation.

In both the hippocampus and the mPFC, H2A.Z depletion enhanced recent and remote memory, implicating H2A.Z as a novel memory suppressor.[89] However, the mechanism by which H2A.Z impacts memory is not clear. H2A.Z depletion only impacted the expression of approximately 400 genes at baseline and enhanced the learning-induced expression of only a small fraction of the memory-related genes that were studied. Even though a direct link between H2A.Z eviction and its impact on gene expression is still lacking, this study unravels new epigenetic mechanisms that so far had been neglected in neurons and may contribute to better understanding of the molecular mechanisms that underlie memory formation and maintenance.

Histone H3.3 is another variant that was recently implicated in memory formation. Maze and colleagues discovered that H3.3 accumulates throughout the lifespan in rodents and in humans, reaching almost saturating levels by adolescence.[87] This saturation is attributed to the fact that H3.3 is the only replication-independent variant of H3, allowing for its exclusive synthesis in postmitotic neurons. Despite this accumulation, H3.3 remains highly dynamic, with rapid activity-induced turnover occurring in response to neuronal stimulation. This turnover relies on proteasomal degradation of existing H3.3 and is associated with late events of transcription that take place after neuronal stimulation.[87] The authors disrupted turnover using a protein knockdown approach and showed that knockdown of H3.3 reduces the number of both excitatory and inhibitory synapses in neurons. In addition, blocking H3.3 turnover in the hippocampus impaired memory formation, suggesting that dynamic removal and incorporation of H3.3 is essential for consolidation. This study highlights the importance of histone turnover in memory formation and indicates that H3.3 dynamics are crucial for normal neuronal function.

## Transgenerational Epigenetics: Can Memories Be Inherited?

Transgenerational epigenetics refers to the hypothesis that epigenetic traits could be passed down through generations to influence behavior, similarly to genetic information. This hypothesis emerged following observations of similar and recurrent behaviors across generations. Indeed, exposing ancestral female mice (Familial 0) to a stressor resulted in future generations (i.e., F1, F2) exhibiting similar maternal care phenotype as the mother that experienced the stressor directly. In other words, offspring that received high levels of licking and grooming from a nonstressed mother also became mothers that engaged in high levels of licking and grooming with their own pups. In contrast, the offspring that received low levels of licking and grooming as a result of maternal stress exposure also provided low levels of licking and grooming to their own pups.[133–135] This behavioral inheritance of the maternal care phenotype was associated with similar patterns of methylation of the glucocorticoid and estrogen receptor genes in the stress-exposed mothers and their offspring.[133–135] These studies represent the first example of inheritance of epigenetic traits that can influence behavior, but it is important to note that the transmission of epigenetic traits was established in response to early life experience (i.e., type of maternal care), rather than genetic inheritance transmitted through the gametes.

More recently, several studies proposed that epigenetic traits established as a consequence of cue-specific ancestral memory formation may be passed down through the gametes, resulting in the inheritance of memories in future generations. One such study exploited odor memory as the learning cue, since this form of learning generally does not affect maternal care phenotypes. Exposing pregnant mice to "mint" or "cherry" odor resulted in the birth of offspring exhibiting the same odor preference as their mother.[136] Although this study demonstrates the transmission of preferences due to events that occurred before birth, the pups were nevertheless "present" during the experience, given that the learning experience occurred during pregnancy.

Additional studies used olfactory fear conditioning to train mice to associate a particular smell with an adverse situation prior to conception.[137] Olfactory fear conditioning is a cue-specific fear conditioning paradigm in which the administration of a foot-shock is paired with an odor, in this case acetophenone. F0 male mice were trained before mating and behavior was assessed in the F1 generation. F1 males fathered by fear-conditioned F0 males showed an odor adversity for acetophenone that was not seen in F1 males fathered by control F0 males.[137] Strikingly, F2 males also retained the aversive odor memory, indicating that this trait may indeed be passed down through the gametes. In addition, in vitro fertilization using sperm coming from F0 conditioned males and cross fostering studies produced offspring exhibiting the same odor adversity for acetophenone, suggesting that the memory was being passed through sperm and not by social transmission.[137] Both F0 and F1 males had increased levels of M71 (an olfactory receptor) expressing olfactory sensory neurons in the olfactory bulb compared to controls.[137] In addition, epigenetic studies on genomic DNA in the sperm of the conditioned F0 males and their F1 offspring showed hypomethylation of the M71 promoter, without influencing the pattern of histone PTMs, suggesting that changes in methylation may account for increased transcription and expression of the M71 receptor.[137] The study of transgenerational inheritance of epigenetic traits is still in its infancy, and much more research is required before we can conclude whether epigenetic inheritance of memories is actually possible. Nevertheless, these studies provide the first example of how distinct epigenetic traits have been passed through multiple generations of mice exhibiting the same behavior and point to epigenetic traits as possible inheritable material together with genomic DNA information.

## EPITRANSCRIPTOMICS: THE NEW HORIZON OF "RNA EPIGENETICS"

Memory formation requires both gene expression and protein synthesis. Although we have focused mostly on epigenetic regulation of gene expression, the control of gene translation is also vital for learning and memory. Recent characterization of the epitranscriptome, or "RNA epigenetics," offers a new method to regulate the fate of mRNA, and recent implication of this modification in learning and memory necessitates its discussion here.

On the chemical level, DNA and RNA are nearly identical, making it no surprise that both exhibit similar modifications, with the exception of histones on mRNA. In fact, RNA contains many additional modifications that have not been described in DNA.[138] Unlike DNA, the most prevalent modification to RNA is N6-methyladenine (methylation of adenine at the 6′ position, abbreviated as m6A), which is the most prevalent internal modification in eukaryotic mRNA.[139]

RNA that contains m6A is often referred to simply as RNA methylation due to the prevalence of this modification, even though the more traditional 5mC is also present within RNA. Every type of RNA can be methylated (from tRNA to snRNA), and methylation is present in RNA across living organisms, ranging from *Escherichia coli* to humans.[140,141] In humans, m6A levels are high in the brain,[142] giving additional impetus to the idea that these modifications play a role in learning and memory. Although knowledge of posttranscriptional modifications of RNA spans decades, recent studies demonstrated that these RNA modifications are highly dynamic and readily reversible, rather than static and irreversible.[143]

m6A is established by two methyltransferases (methyltransferase like 3, METTL3, and methyltransferase like 14, METTL14), often referred to as m6A *writers*.[144] Knocking down METTL3 or METTL14 in mouse or human cells reduces levels of m6A-containing mRNA[145] and the deletion or mutation of either writer interferes with differentiation of human or mouse embryonic stem cells.[146,147] Opposing the action of m6A *writers* are the demethylase *erasers*, including fat mass and obesity-associated protein (FTO)[148] and α-ketoglutarate-dependent dioxygenase alkB homolog 5 (ALKBH5).[149] Silencing or overexpression of FTO or ALKBH5 increases[149,150] or decreases levels of m6A-containing mRNA,[149,151] respectively. The strong association between an *FTO* polymorphism and obesity risk observed in a genome-wide association study in humans[152] increased research interest in FTO. Although the link between these polymorphisms and *FTO* is still being debated,[153,154] it has been demonstrated that manipulating FTO specifically in the mouse brain results in lean or obese phenotype.[155–157] Given the enrichment of m6A in the brain and its dynamic and reversible nature, RNA methylation is well suited to play an important role in memory formation.

## Functional Outcomes of mRNA Methylation

Methylated mRNA can regulate mRNA localization and promote mRNA degradation. Initial studies examining the function of RNA methylation in vitro observed an inverse correlation between the amount of RNA methylation on a transcript and its half-life,[158] implying that RNA methylation can promote transcript degradation. This finding was extended by data showing that RNA methylation can be recognized by the m6A-*reader* protein YTHDF2, which changes the localization of the associated transcript from the translatable pool to a processing body (P-body), where RNA degradation can occur.[147] As with DNA methylation, RNA methylation itself is not thought to have much intrinsic function; instead, it influences cellular outcomes through interactions with such reader proteins.

## Methylated mRNA Can Promote Translation in a Stimulus-Dependent Manner

mRNA can be methylated throughout the transcript, but the 5′ untranslated region (UTR) has less methylation than other regions.[151,159] Authors studying this phenomenon discovered that, in vitro, cells respond to certain stimuli (in this case, heat shock stress) using mRNA methylation. In these experiments, heat shock caused nuclear localization of the reader protein YTHDF2, and while in the nucleus, YTHDF2 bound to and protected certain methylated transcripts from demethylation by FTO.[160] This function was very different than the previously described role of YTHDF2 in promoting mRNA degradation, since preserving 5′ UTR methylation resulted in preferential translation of these transcripts.[160] This finding is noteworthy because it showed that mRNA methylation can promote translation using the same reader protein that can promote degradation.[147] The major difference between the two outcomes was the region of the mRNA that was methylated, giving reader proteins the ability to modulate the translational outcome based on region-specific methylation. Further work examining other readers of mRNA methylation showed that the reader protein YTHDF1 could directly promote translation of the bound transcript by actively interacting with the translation machinery.[147] These studies established that both the region of the transcript that is modified, and the protein that reads the modification, are crucial for regulating effects of m6A on translation.

## The Role of mRNA Methylation in Learning and Memory

Three studies have recently implicated mRNA methylation in learning and memory. The first focused on the effects of FTO depletion on dopaminergic pathways of the midbrain.[151] The authors discovered that FTO depletion attenuated cocaine-induced locomotor activity and D2-mediated neuronal firing. Using NGS from a pull-down of m6A-containing mRNA (MeRIP-seq), the authors probed which mRNAs were differentially methylated between *wt* and *FTO-null* mice. The data showed that FTO-targeted transcripts were enriched for synaptic function and dopaminergic signaling,[151] implicating FTO as a regulator of synaptic plasticity.

RNA methylation and FTO were directly implicated in memory formation through actions in the cortex[161] and the hippocampus.[162] In the cortex, fear conditioning increased the amount of methylated mRNA, and MeRIP-seq revealed that a large proportion of the differentially methylated transcripts were involved in synaptic function. Using neuronal culture, the authors showed that increased methylation changed the half-lives of the modified mRNA. In accordance with data from nonneuronal studies, some transcripts showed an increased and others showed a decreased lifespan. FTO was directly implicated in memory by demonstrating that FTO depletion improved memory for cued fear conditioning,[161] suggesting that higher levels of mRNA methylation are associated with improved memory.

In a separate study, one of us[162] examined how methylated mRNA affected memory formation within the hippocampus. We showed that CFC transiently increased the level of methylated mRNA and decreased the expression of *FTO* in area CA1 of the hippocampus. Previous studies in nonneuronal cells had only reported FTO localization in the nucleus.[151] To determine whether learning-induced changes in FTO were specific to a particular cellular compartment, we characterized FTO expression separately in nuclear and dendritic/synaptic regions. We found that FTO was preferentially depleted in dendritic and synaptic regions after fear conditioning, raising important questions about the potential synaptic role of mRNA demethylation. As with FTO depletion in the cortex, we showed that FTO depletion in area CA1 improved memory for CFC, suggesting that FTO-mediated mRNA demethylation may have a negative role in memory formation. Although the individual findings in these studies may be preliminary, the consistent trend is that FTO-mediated mRNA methylation and demethylation is regulated by fear conditioning, and that promoting mRNA methylation through FTO depletion promotes memory formation.

The mechanism by which mRNA methylation regulates memory is not yet known. In vitro experiments indicate that mRNA methylation can control nearly every stage of mRNAs, from splicing to translation.[147,158,160,163,164] In the context of memory formation, m6A function has only been characterized in relation to transcript stability[161] and further research is needed to investigate other functional outcomes of m6A. An exciting idea that emerged from our studies is that neurons can locally control mRNA demethylation at synapses via FTO, raising the possibility that mRNA methylation plays a vital role in how specific synapses respond to memory formation. This is an especially important point because memory formation has classically been associated with synapse-specific potentiation, which raises a fundamental question in epigenetic research: How can epigenetic changes in the nucleus result in synapse-specific changes in protein synthesis? Although this area of research is in its infancy, recent work on RNA methylation suggests that preferential regulation of FTO within synapses may locally regulate translation, suggesting that RNA and DNA epigenetics work together to regulate memory formation.

## INTEGRATING THE EPIGENETIC CODE TO FORM AND RECALL MEMORIES

Even though histone PTMs, histone variants, and DNA methylation are typically studied separately, these modifications do not occur independently and do in fact influence one another. Multiple epigenetic marks cooccupy a single locus and recruit many other factors that together exert effects on gene expression.[129] Indeed, evidence suggests that DNA methylation and histone PTMs influence and integrate with each other to finely tune gene expression. This integration occurs at several levels, including upstream signaling cascades that ultimately regulate the establishment of both marks. For example, DNA methylation and histone acetylation are both regulated in part by the *N*-methyl-D-aspartate (NMDA) pathway.[100] NMDA receptors respond to the excitatory neurotransmitter glutamate by activating a signaling cascade that reaches the nucleus and triggers many downstream events, including H3 acetylation and phosphorylation in the hippocampus.[24,165] Similarly, blocking the NMDA receptor impairs DNA methylation dynamics in the hippocampus,[118] suggesting that NMDA is a common upstream regulator of DNA methylation and histone PTMs. In addition, enzymes that modify one epigenetic mark have been shown to impact other marks. For example, HDAC inhibitors are able to induce DNA demethylation,[166] whereas DNMT inhibitors can block H3 acetylation triggered by CFC.[119] Finally, MeCP2 is able to recruit HDACs and HMTs to methylated promoters, thus linking DNA methylation to a range of histone modifications.[55,131]

Similarly, H2A.Z has an antagonistic relationship with DNA methylation, whereby DNA methylation displaces histone H2A.Z binding,[167] so one mark can alter the likelihood of another mark at the same locus. This is likely to have wide-ranging effects on the entire nucleosome, since distinct histone variants tend to preferentially cooccur with specific PTMs on neighboring histones,[168] adding an additional layer of complexity to gene regulation. Understanding how epigenetic mechanisms integrate to regulate gene transcription will allow us to unravel how the transcription pathways involved in gene expression during memory formation are regulated. The clarification of these intricate events will aid the development of new drugs to treat memory-related disorders.

## HIGH-THROUGHPUT NEXT-GENERATION SEQUENCING: FROM CANDIDATE GENES TO GENOME-WIDE ANALYSIS

Tools used to measure epigenetic changes are continually refined, but they are generally based on variations of a few key strategies (for a detailed review of epigenetic tools, see Ref. 169). Chromatin immunoprecipitation (ChIP) is commonly used for measuring histone PTMs and DNA methylation at specific genomic loci. Classic ChIP[170] is used to measure changes in histone PTMs, whereas methylated DNA immunoprecipitation (MeDIP)[171] is a more recent variation used to isolate DNA enriched for methylated regions. Both approaches utilize specific antibodies with high affinity for chromatin proteins (e.g., TF, histones, histone specific PTMs, methylated cytosine) to isolate the segments of DNA associated with these proteins or modifications. Another commonly used approach for measuring DNA methylation is methyl-CpG-binding domain (MBD) capture, in which methylated DNA is isolated using beads coated in MBD polypeptide.[172] In all cases, isolated DNA can be measured using specific primers through real-time PCR (RT-PCR)[170] or with next-generation sequencing. Comparing the amount of material isolated in different conditions (e.g., stimulated vs. unstimulated neurons) provides an indication of how various treatments influence the epigenetic modification of interest. Bisulfite sequencing is another method for measuring DNA methylation, but at a single cytosine resolution. It uses bisulfite as a reagent to convert cytosines to uracil, leaving methyl-cytosines unchanged to allow for identification of methylated and unmethylated cytosine residues using classical sequencing.[173]

Initial studies of epigenetic changes in the brain utilized a candidate gene strategy to quantify learning-induced changes in epigenetic marks. With this approach, epigenetic changes are studied on a small set of genes that have an established role in memory formation. Although this strategy is highly effective for demonstrating that particular marks are dynamically regulated by learning (specific findings are discussed in Epigenetic Mechanisms Regulating Memory section), it has limited utility for identifying mechanistic links between epigenetic changes, transcription, and memory for several reasons. Specifically, candidate genes often belong to a similar gene class, such as immediate early genes, and as such, modifications on these genes may not be representative of genes in general and may provide a skewed understanding of epigenetic influence on transcription. In addition, these strategies are also limited to specific gene regions around which qPCR primers are designed. These regions are selected a priori based on a known functional relevance, thus limiting most studies to promoters, CpG islands, and regions proximal to the TSS. Although candidate-gene studies provide crucial information about epigenetic function in memory, understanding the complexity of epigenetic regulation also requires a genome-wide research strategy.

Increased affordability and more widespread use of high-throughput NGS revolutionized the field of neuroepigenetics. NGS methods allow for the sequencing of millions of molecules at once at high resolution.[174] NGS methods can be coupled to ChIP (ChIP-seq), MeDIP, MBD, and bisulfite sequencing, allowing researchers to map histone binding, PTMs, and DNA methylation throughout the genome.

The primary example of how NGS represents an enormous advance for epigenetics is the ENCODE project (*Encyclopedia of DNA Elements*),[175] which was launched in 2003 to sequence and identify functional elements in the genome. ENCODE has generated maps of DNA methylation and histone PTMs in various tissues, including the brain,[175] and these maps are often used as a starting point to compare research results with the existing maps.

## Regulatory Regions: Epigenetic Regulation of Distal Elements to Regulate Transcription in the Brain

ENCODE has made crucial strides toward identifying the importance of regulatory regions in epigenetic studies. Whereas candidate gene approaches tend to focus on epigenetic changes at promoters and at regions flanking the TSS, the ENCODE project identified hundreds of thousands of enhancer sequences in the genome that regulate gene expression.[175] Enhancers are genomic regulatory elements that influence gene expression even though they can be found far from the genes they regulate. It is estimated that each cell type is regulated by 20–40 thousands enhancer sequences.[176] ChIP-seq at different developmental stages found that enhancers in the brain can exist in a poised state, characterized by H3K4, and an active state, characterized by H3K27ac.[177] Coupling ChIP-seq with RNA-seq demonstrated that H3K27ac at active enhancers indeed correlates with increased gene expression,[178] thus validating H3K27ac as a marker of active gene states. In addition, ChIP-seq showed that neuronal stimulation triggers CBP histone acetyl transferase recruitment at thousands of plasticity genes enhancers,[179] linking neuronal stimulation to enhancer epigenetic variation.

A regulatory role for enhancers in the brain was directly demonstrated in the nucleus accumbens in response cocaine,[180] whereby cocaine administration resulted in increased 5hmC at enhancers that was associated with increased cocaine-induced gene expression.[180] Epigenetic changes at enhancers also emerged as key regulators of cognitive decline. A recent study coupled RNA-seq with ChIP-seq to identify links between epigenetic and transcriptional outcomes in a mouse model of Alzheimer disease (AD).[181] AD was associated with decreased expression of genes involved in synaptic plasticity and increased expression of genes involved in immune function.[181] Genes that were upregulated in AD had increased levels of histone PTMs associated with gene expression at promoters and enhancers, whereas downregulated genes exhibited histone PTMs associated with gene repression at their regulatory elements.[181] The authors also investigated a possible correlation between AD-related genetic variants and epigenetically variable enhancers. Notably, immune system cells were enriched for both single nucleotide polymorphisms and the presence of enriched enhancers orthologues,[181] thereby linking cognitive decline in AD with previously unstudied genomic regions.[181] These studies provide examples of how NGS uncovered the importance of genomic regions that were mostly neglected when using the candidate gene approach and implicate these elements as crucial regulators of neural function.

## Stimulus-Induced Epigenetic Modifications and Transcription

Despite tremendous interest in epigenetic regulation of memory formation, the link between learning-induced epigenetic modifications and transcription remains relatively unclear. In a recent genome-wide study, Halder and colleagues[103] utilized ChIP-, MeDIP-, and RNA-sequencing to determine changes in DNA methylation and various histone PTMs in area CA1 and the ACC at distinct time points after fear conditioning (cellular consolidation at 1 h and memory maintenance at 4 weeks). Consistent with results using the candidate-gene approach, cellular consolidation was associated with changes in many histone PTMs in the hippocampus. Specifically, increased H3K4me3 and H3K9ac decreased H3K27me3 binding was observed at genic regions, whereas decreased H3K4me3 and increased H3K27me3 was observed at regulatory regions, like enhancers.

Only H3K79me3, a marker of active genes, was significantly changed in genes that also showed altered expression with fear conditioning, whereas other PTMs showed only weak correlations with gene expression, indicating that stimulus-induced changes in histone PTMs are weakly correlated with changes in transcription.[103] However, this does not mean that stimulus-induced changes in histone PTMs do not influence gene expression. Indeed, this study only analyzed PTMs in area CA1 1 h after fear conditioning, and it is possible that other time points may reveal a better correlation with gene expression. Notably, neurons stimulated with KCl showed increased H3K27ac in introns of genes whose expression is affected by the treatment.[103] Given the link between this modification with poised enhancers and promoters, histone PTMs may influence the poised state of genes rather than their learning-induced activation/repression.

The same study also investigated DNA methylation changes in CA1 and ACC 1 h and 4 weeks after fear conditioning. Region-specific changes in methylation at 1 h were detected in CA1 and ACC primarily in intergenic and intronic regions rather than in promoters, which are typically the focus of candidate-gene studies.[103] Contrary to stimulus-induced changes in histone PTMs, learning-induced changes in DNA methylation were significantly correlated with changes in gene expression in area CA1 and the ACC.[103] Nevertheless, the number of genes that exhibited both learning-induced changes in DNA methylation and gene expression at 1 h was very small, representing only 10% of all differentially expressed genes. The number increased to 49% in the ACC at 1 h and decreased again to 25% 4 weeks later in ACC.[103] This work highlights the advantages of genome-wide research strategies in uncovering links between gene expression and epigenetic modifications by providing unbiased and highly sensitive tools.

## Next-Generation Sequencing and Epitranscriptomics

Although mRNA methylation was discovered over 50 years ago, it remained unexplored for many reasons, but a major hurdle was technical limitations. Probing mRNA methylation was difficult, and the lack of a platform to pose transcript-wide questions regarding mRNA methylation impeded research on this modification. The rise of next-generation sequencing, primarily RNA-sequencing, allowed the role of mRNA methylation to be probed by combining RNA-seq with m6A-specific antibodies and immunoprecipitating the mRNA prior to sequencing (MeRIP-Seq). Prior to RNA-seq, researchers were limited to making broad statements about generalized changes in RNA methylation, but any specifics about which transcripts were targeted or where in the transcript methylation was occurring were unknown. With the discoveries that methylation of mRNA occurs in different amounts across the transcript,[151] and that each region could play a different role in regulation, the lack of RNA-seq-based methods fundamentally limited what would be learned about epitranscriptomics. Although RNA immunoprecipitation allowed mRNA methylation to be investigated, it nevertheless represents a fundamental barrier to future studies, as the amount of input RNA can be prohibitively large and prohibitive to research on smaller populations of neurons. While next-generation sequencing has revolutionized our understanding of mRNA methylation, further technological advances that eliminate the need for immunoprecipitation will usher in the next generation of discoveries on mRNA methylation.

# CONCLUSIONS

The development of neuroepigenetics as a field represented a paradigm shift for epigenetics as a whole. The discovery that DNA methylation is dynamic[21] in neuronal cells challenged the dogma of exclusively stable DNA methylation. Since then, evidence has accumulated in support of epigenetic regulation of neural plasticity. The studies discussed here strongly indicate that epigenetic mechanisms help create stable neural and behavioral outcomes out of transient events. We are only beginning to understand how different epigenetic mechanisms work together to influence gene expression and how epigenetic events in the nucleus ultimately affect memory through actions at specific synapses. Recent data suggest that the novel field of epitranscriptomics may prove crucial in bridging this gap.

Despite tremendous progress in a very short time, it is still unclear how different epigenetic modifications impact learning-induced gene expression and how epigenetic changes support the prolonged process of systems consolidation. NGS is an essential tool for advancing these research goals. The ability to integrate different NGS techniques to correlate different modifications at different time points and brain regions with changes in transcription is essential for understanding how these marks interact to influence memory formation and maintenance. The quest to understand the molecular basis of memory is further complicated by observations that neurons within a single brain region display different morphologies and electrophysiological properties,[182] and they are thus likely to vary in epigenetic marks as well. Recent development of single cell RNA and ChIP sequencing tools[183,184] will undoubtedly propel more fine-grained analysis into epigenetic modifications that differ in neurons that participate in memory formation from those that do not.[185] The application of these techniques could provide answers to key research questions and further elucidate the molecular mechanisms responsible for memory formation and maintenance.

## REFERENCES


1. Zovkic IB, Walters BJ. H2A.Z helps genes remember their history so we can remember ours. *Bioessays*. 2015;37(6):596–601. https://doi.org/10.1002/bies.201400223.
2. Frankland PW, Bontempi B. The organization of recent and remote memories. *Nat Rev Neurosci*. 2005;6(2):119–130. https://doi.org/10.1038/nrn1607.
3. Scoville WB, Milner B. Loss of recent memory after bilateral hippocampal lesions. *J Neuropsychiatry Clin Neurosci*. 1957;20(11):11–21. https://doi.org/10.1136/jnnp.20.1.11.
4. Amaral DG, Rempel-Clower NL, Zola SM, Squire LR. Three cases of enduring memory impairment after bilateral damage limited to the hippocampal formation. *J Neurosci*. 1996;16(16):5233–5255. http://eutils.ncbi.nlm.nih.gov/entrez/eutils/elink.fcgi?dbfrom=pubmed&id=8756452&retmode=ref&cmd=prlinks%5Cnpapers3://publication/uuid/A1CBCD08-E0C6-4325-9467-CB65A039B7E2.
5. Squire LR, Slater PC, Chace P. Retrograde amnesia: temporal gradient in very long term memory following electroconvulsive therapy. *Science*. 1975;187(4171):77–79.
6. Anagnostaras SG, Maren S, Fanselow MS. Temporally graded retrograde amnesia of contextual fear after hippocampal damage in rats: within-subjects examination. *J Neurosci*. 1999;19(3):1106–1114. http://hdl.handle.net/2027.42/56234.
7. Cho YH, Beracochea D, Jaffard R. Extended temporal gradient for the retrograde and anterograde amnesia produced by ibotenate entorhinal cortex lesions in mice. *J Neurosci*. 1993;13(4):1759–1766. http://www.ncbi.nlm.nih.gov/pubmed/8463849.
8. Clark RE, Broadbent NJ, Zola SM, Squire LR. Anterograde amnesia and temporally graded retrograde amnesia for a nonspatial memory task after lesions of hippocampus and subiculum. *J Neurosci*. 2002;22(11):4663–4669.
9. Walters BJ, Zovkic IB. Building up and knocking down: an emerging role for epigenetics and proteasomal degradation in systems consolidation. *Neuroscience*. 2015;300:39–52. https://doi.org/10.1016/j.neuroscience.2015.05.005.
10. Frankland PW, Bontempi B, Talton LE, Kaczmarek L, Silva AJ. The involvement of the anterior cingulate cortex in remote contextual fear memory. *Science*. 2004;304(5672):881 LP–883. http://science.sciencemag.org/content/304/5672/881.abstract.
11. Gilboa A, Winocur G, Grady CL, Hevenor SJ, Moscovitch M. Remembering our past: functional neuroanatomy of recollection of recent and very remote personal events. *Cereb Cortex*. 2004;14(11):1214–1225. https://doi.org/10.1093/cercor/bhh082.
12. Miller CA, Gavin CF, White JA, et al. Cortical DNA methylation maintains remote memory. *Nat Neurosci*. 2010;13(6):664–666. https://doi.org/10.1038/nn.2560.
13. Steward O, Levy WB. Preferential localization of polyribosomes under the base of dendritic spines in granule cells of the dentate gyrus. *J Neurosci*. 1982;2(3):284–291.
14. Aakalu G, Smith WB, Nguyen N, Jiang C, Schuman EM. Dynamic visualization of local protein synthesis in hippocampal neurons source of protein synthesis. A similar dependence on dendritic protein synthesis has been observed for meta-botropic receptor-induced LTD at Schaffer-collateral. *Neuron*. 2001;30:489–502.

15. Greenberg ME, Ziff EB, Greene LA. Stimulation of neuronal acetylcholine receptors induces rapid gene transcription. *Science*. 1986;234(4772):80–83. https://doi.org/10.2307/1697386.
16. West AE, Chen WG, Dalva MB, et al. Calcium regulation of neuronal gene expression. *Proc Natl Acad Sci USA*. 2001;98(20):11024–11031. https://doi.org/10.1073/pnas.191352298.
17. Bailey CH, Kandel ER, Si K. The persistence of long-term memory: a molecular approach to self-sustaining changes in learning-induced synaptic growth. *Neuron*. 2004;44(1):49–57. https://doi.org/10.1016/j.neuron.2004.09.017.
18. Stanton PK, Sarvey JM. Blockade of long-term potentiation in rat hippocampal CA1 region by inhibitors of protein synthesis. *J Neurosci*. 1984;4(12):3080–3088.
19. Hernandez PJ, Abel T. The role of protein synthesis in memory consolidation: progress amid decades of debate. *Neurobiol Learn Mem*. 2008;89(3):293–311. https://doi.org/10.1016/j.nlm.2007.09.010.
20. Pedreira ME, Dimant B, Maldonado H. Inhibitors of protein and RNA synthesis block context memory and long-term habituation in the crab chasmagnathus. *Pharmacol Biochem Behav*. 1996;54(3): 611–617. https://doi.org/10.1016/0091-3057(95)02206-6.
21. Miller CA, Sweatt JD. Covalent modification of DNA regulates memory formation. *Neuron*. 2007; 53(6):857–869. https://doi.org/10.1016/j.neuron.2007.02.022.
22. Crick F. Neurobiology: memory and molecular turnover. *Nature*. 1984;312(5990):101. https://doi.org/10.1038/312101a0.
23. Chwang WB, Arthur JS, Schumacher A, Sweatt JD. The nuclear kinase mitogen- and stress-activated protein kinase 1 regulates hippocampal chromatin remodeling in memory formation. *J Neurosci*. 2007; 27(46):12732–12742. https://doi.org/10.1523/JNEUROSCI.2522-07.2007.
24. Levenson JM, O'Riordan KJ, Brown KD, Trinh MA, Molfese DL, Sweatt JD. Regulation of histone acetylation during memory formation in the hippocampus. *J Biol Chem*. 2004;279(39):40545–40559. https://doi.org/10.1074/jbc.M402229200.
25. Levenson JM, Sweatt JD. Epigenetic mechanisms: a common theme in vertebrate and invertebrate memory formation. *Cell Mol Life Sci*. 2006;63(9):1009–1016. https://doi.org/10.1007/s00018-006-6026-6.
26. Levenson J, Sweatt J. Epigenetic mechanisms in memory formation. *Nat Rev Neurosci*. 2005;6(2): 108–118. https://doi.org/10.1038/nrn1604.
27. Day JJ, Sweatt JD. Epigenetic mechanisms in cognition. *Neuron*. 2011;70(5):813–829. https://doi.org/10.1016/j.neuron.2011.05.019.
28. Lester BM, Tronick E, Nestler E, et al. Behavioral epigenetics. *Ann NY Acad Sci*. 2011;1226(1): 14–33. https://doi.org/10.1111/j.1749-6632.2011.06037.x.
29. Bird A. DNA methylation patterns and epigenetic memory DNA methylation patterns and epigenetic memory. *Gene Dev*. 2002:6–21. https://doi.org/10.1101/gad.947102.
30. Crick F. Memory and molecular turnover. *Nature*. 1984;312:8–14.
31. Fernández-Morera JL, Calvanese V, Rodríguez-Rodero S, Menéndez-Torre E, Fraga MF. Epigenetic regulation of the immune system in health and disease. *Tissue Antigens*. 2010;76(6):431–439. https://doi.org/10.1111/j.1399-0039.2010.01587.x.
32. Lindholm ME, Marabita F, Gomez-Cabrero D, et al. An integrative analysis reveals coordinated reprogramming of the epigenome and the transcriptome in human skeletal muscle after training. *Epigenetics*. 2014;9(12):1557–1569. https://doi.org/10.4161/15592294.2014.982445.
33. Wolffe AP, Hayes JJ. Chromatin disruption and modification. *Nucleic Acids Res*. 1999;27(3):711–720. https://doi.org/10.1093/nar/27.3.711.
34. Bernstein BE, Mikkelsen TS, Xie X, et al. A bivalent chromatin structure marks key developmental genes in embryonic stem cells. *Cell*. 2006;125(2):315–326. https://doi.org/10.1016/j.cell.2006.02.041.
35. Vastenhouw NL, Schier AF. Bivalent histone modifications in early embryogenesis. *Curr Opin Cell Biol*. 2012;24(3):374–386. https://doi.org/10.1016/j.ceb.2012.03.009.
36. Mohn F, Weber M, Rebhan M, et al. Lineage-specific polycomb targets and de novo DNA methylation define restriction and potential of neuronal progenitors. *Mol Cell*. 2008;30(6):755–766. https://doi.org/10.1016/j.molcel.2008.05.007.
37. Musselman CA, Lalonde ME, Cote J, Kutateladze TG. Perceiving the epigenetic landscape through histone readers. *Nat Struct Mol Biol*. 2012;19(12):1218–1227. https://doi.org/10.1038/nsmb.2436.

38. Shen H, Laird PW. Interplay between the cancer genome and epigenome. *Cell.* 2013;153(1):38–55. https://doi.org/10.1016/j.cell.2013.03.008.
39. Tsai H-C, Baylin SB. Cancer epigenetics: linking basic biology to clinical medicine. *Nat Publ Gr.* 2011;2124(21):502–517. https://doi.org/10.1038/cr.2011.24.
40. Urdinguio RG, Sanchez-Mut JV, Esteller M. Epigenetic mechanisms in neurological diseases: genes, syndromes, and therapies. *Lancet Neurol.* 2009;8(11):1056–1072. https://doi.org/10.1016/S1474-4422(09)70262-5.
41. Chen L, MacMillan AM, Chang W, Ezaz-Nikpay K, Lane WS, Verdine GL. Direct identification of the active-site nucleophile in a DNA (cytosine-5)-methyltransferase. *Biochemistry.* 1991;30(46): 11018–11025. https://doi.org/10.1021/bi00110a002.
42. Klose RJ, Bird AP. Genomic DNA methylation: the mark and its mediators. *Trends Biochem Sci.* 2006; 31(2):89–97. https://doi.org/10.1016/j.tibs.2005.12.008.
43. Wolffe AP, Jones PL, Wade PA. DNA demethylation. *Proc Natl Acad Sci USA.* 1999;96(11): 5894–5896. https://doi.org/10.1073/pnas.96.11.5894.
44. Jaenisch R, Bird A. Epigenetic regulation of gene expression: how the genome integrates intrinsic and environmental signals. *Nat Genet.* 2003;(33 suppl):245–254. https://doi.org/10.1038/ng1089.
45. Jiang Y, Langley B, Lubin FD, et al. Epigenetics in the nervous system. *J Neurosci.* 2008;28(46): 11753–11759. https://doi.org/10.1523/JNEUROSCI.3797-08.2008.
46. Zykovich A, Hubbard A, Flynn JM, et al. Genome-wide DNA methylation changes with age in disease-free human skeletal muscle. *Aging Cell.* 2014;13(2):360–366. https://doi.org/10.1111/acel.12180.
47. Hon GC, Hawkins RD, Caballero OL, et al. Global DNA hypomethylation coupled to repressive chromatin domain formation and gene silencing in breast cancer. *Genome Res.* 2012;22(2): 246–258. https://doi.org/10.1101/gr.125872.111.
48. Day JJ, Sweatt JD. DNA methylation and memory formation. *Nat Neurosci.* 2010;13(11):1319–1323. https://doi.org/10.1038/nn.2666.
49. Santos KF, Mazzola TN, Carvalho HF. The prima donna of epigenetics: the regulation of gene expression by DNA methylation. *Braz J Med Biol Res.* 2005;38(10):1531–1541. https://doi.org/10.1590/S0100-879X2005001000010.
50. Li E, Bestor TH, Jaenisch R. Targeted mutation of the DNA methyltransferase gene results in embryonic lethality. *Cell.* 1992;69(6):915–926. https://doi.org/10.1016/0092-8674(92)90611-F.
51. Das PM, Singal R. DNA methylation and cancer therapy. *J Clin Oncol.* 2005;22(22). https://doi.org/10.1007/b139080.
52. Jones PA. Functions of DNA methylation: islands, start sites, gene bodies and beyond. *Nat Rev Genet.* 2012;13(7):484–492. https://doi.org/10.1038/nrg3230.
53. Barrett RM, Wood MA. Beyond transcription factors: the role of chromatin modifying enzymes in regulating transcription required for memory. *Learn Mem.* 2008;15(7):460–467. https://doi.org/10.1101/lm.917508.
54. Feng J, Zhou Y, Campbell SL, et al. DNMT1 and DNMT3A maintain DNA methylation and regulate synaptic function in adult forebrain neurons. *Nat Neurosci.* 2010;13(4):423–430. https://doi.org/10.1038/nn.2514.
55. Nan X, Ng HH, Johnson CA, et al. Transcriptional repression by the methyl-CpG-binding protein MeCP2 involves a histone deacetylase complex. *Nature.* 1998;393(6683):386–389. https://doi.org/10.1038/30764.
56. Wade PA, Gegonne A, Jones PL, Ballestar E, Aubry F, Wolffe AP. Mi-2 complex couples DNA methylation to chromatin remodelling and histone deacetylation. *Nat Genet.* 1999;23:1–5. https://doi.org/10.1038/12664.
57. Fujita N, Watanabe S, Ichimura T, et al. Methyl-CpG binding domain 1 (MBD1) interacts with the Suv39h1-HP1 heterochromatic complex for DNA methylation-based transcriptional repression. *J Biol Chem.* 2003;278(26):24132–24138. https://doi.org/10.1074/jbc.M302283200.
58. Amir RE, Van den Veyver IB, Wan M, Tran CQ, Francke U, Zoghbi HY. Rett syndrome is caused by mutations in X-linked MECP2, encoding methyl-CpG-binding protein 2. *Nat Genet.* 1999;23: 185–188. https://doi.org/10.1038/13810.

59. Tahiliani M, Koh KP, Shen Y, et al. Conversion of 5-methylcytosine to 5-hydroxymethylcytosine in mammalian DNA by MLL partner TET1. *Science*. 2009;324(5929):930–935. https://doi.org/10.1126/science.1170116.
60. Kriaucionis S, Heintz N. The nuclear DNA base 5-hydroxymethylcytosine is present in Purkinje neurons and the brain. *Science*. 2009;324(5929):929–930. https://doi.org/10.1126/science.1169786.
61. Sérandour AA, Avner S, Oger F, et al. Dynamic hydroxymethylation of deoxyribonucleic acid marks differentiation-associated enhancers. *Nucleic Acids Res*. 2012;40(17):8255–8265. https://doi.org/10.1093/nar/gks595.
62. Guo JU, Su Y, Zhong C, Ming GL, Song H. Hydroxylation of 5-methylcytosine by TET1 promotes active DNA demethylation in the adult brain. *Cell*. 2011;145(3):423–434. https://doi.org/10.1016/j.cell.2011.03.022.
63. Spruijt CG, Gnerlich F, Smits AH, et al. Dynamic readers for 5-(hydroxy)methylcytosine and its oxidized derivatives. *Cell*. 2013;152(5):1146–1159. https://doi.org/10.1016/j.cell.2013.02.004.
64. Luger K, Mäder AW, Richmond RK, Sargent DF, Richmond TJ. Crystal structure of the nucleosome core particle at 2.8 A resolution. *Nature*. 1997;389(6648):251–260. https://doi.org/10.1038/38444.
65. Kouzarides T. Chromatin modifications and their function. *Cell*. 2007;128(4):693–705. https://doi.org/10.1016/j.cell.2007.02.005.
66. Bernstein BE, Humphrey EL, Erlich RL, et al. Methylation of histone H3 Lys 4 in coding regions of active genes. *Proc Natl Acad Sci USA*. 2002;99(13):8695–8700. https://doi.org/10.1073/pnas.082249499.
67. Ringrose L, Ehret H, Paro R. Distinct contributions of histone H3 lysine 9 and 27 methylation to locus-specific stability of Polycomb complexes. *Mol Cell*. 2004;16(4):641–653. https://doi.org/10.1016/j.molcel.2004.10.015.
68. Jenuwein T, Allis CD. Translating the histone code. *Science*. 2001;293(5532):1074–1080.
69. Campos EI, Reinberg D. Histones: annotating chromatin. *Annu Rev Genet*. 2009;43(1):559–599. https://doi.org/10.1146/annurev.genet.032608.103928.
70. Wang ET, Sandberg R, Luo S, et al. Alternative isoform regulation in human tissue transcriptomes. *Nature*. 2008;456(7221):470–476. https://doi.org/10.1038/nature07509.
71. Vakoc CR, Mandat SA, Olenchock BA, Blobel GA. Histone H3 lysine 9 methylation and HP1 [gamma] are associated with transcription elongation through mammalian chromatin. *Mol Cell*. 2005;19(3):381–391. https://doi.org/10.1016/j.molcel.2005.06.011.
72. Bourtchouladze R, Lidge R, Catapano R, et al. A mouse model of Rubinstein-Taybi syndrome: defective long-term memory is ameliorated by inhibitors of phosphodiesterase 4. *Proc Natl Acad Sci USA*. 2003;100(18):10518–10522. https://doi.org/10.1073/pnas.1834280100.
73. Hebbes TR, Thorne AW, Crane-Robinson C. A direct link between core histone acetylation and transcriptionally active chromatin. *EMBO J*. 1988;7(5):1395–1402. https://doi.org/10.1007/BF00777468.
74. Mujtaba S, Zeng L, Zhou MM. Structure and acetyl-lysine recognition of the bromodomain. *Oncogene*. 2007;26(37):5521–5527. https://doi.org/10.1038/sj.onc.1210618.
75. Koshibu K, Gräff J, Mansuy IM. Nuclear protein phosphatase-1: an epigenetic regulator of fear memory and amygdala long-term potentiation. *Neuroscience*. 2011;173:30–36. https://doi.org/10.1016/j.neuroscience.2010.11.023.
76. Macdonald N, Welburn JPI, Noble MEM, et al. Molecular basis for the recognition of phosphorylated and phosphoacetylated histone H3 by 14-3-3. *Mol Cell*. 2005;20(2):199–211. https://doi.org/10.1016/j.molcel.2005.08.032.
77. Shilatifard A. Chromatin modifications by methylation and ubiquitination: implications in the regulation of gene expression. *Annu Rev Biochem*. 2006;75:243–269. https://doi.org/10.1146/annurev.biochem.75.103004.142422.
78. Shi Y, Lan F, Matson C, et al. Histone demethylation mediated by the nuclear amine oxidase homolog LSD1. *Cell*. 2004;119(7):941–953. https://doi.org/10.1016/j.cell.2004.12.012.
79. Tachibana M, Sugimoto K, Fukushima T, Shinkai Y. SET domain-containing protein, G9a, is a novel lysine-preferring mammalian histone methyltransferase with hyperactivity and specific selectivity to lysines 9 and 27 of histone H3. *J Biol Chem*. 2001;276(27):25309–25317. https://doi.org/10.1074/jbc.M101914200.

80. Kerimoglu C, Agis-Balboa RC, Kranz A, et al. Histone-methyltransferase MLL2 (KMT2B) is required for memory formation in mice. *J Neurosci.* 2013;33(8):3452–3464. https://doi.org/10.1523/JNEUROSCI.3356-12.2013.
81. Schaefer A, Sampath SC, Intrator A, et al. Control of cognition and adaptive behavior by the GLP/G9a epigenetic suppressor complex. *Neuron.* 2009;64(5):678–691. https://doi.org/10.1016/j.neuron.2009.11.019.
82. Banaszynski LA, Allis CD, Lewis PW. Histone variants in metazoan development. *Dev Cell.* 2010;19(5):662–674. https://doi.org/10.1016/j.devcel.2010.10.014.
83. Weber CM, Henikoff S. Histone variants: dynamic punctuation in transcription. *Gene Dev.* 2014;28(7):672–682. https://doi.org/10.1101/gad.238873.114.
84. Jin J, Cai Y, Li B, et al. In and out: histone variant exchange in chromatin. *Trends Biochem Sci.* 2005;30(12):680–687. https://doi.org/10.1016/j.tibs.2005.10.003.
85. Akey CW, Luger K. Histone chaperones and nucleosome assembly. *Curr Opin Struct Biol.* 2003;13(1):6–14. https://doi.org/10.1016/S0959-440X(03)00002-2.
86. Talbert PB, Henikoff S. Histone variants—ancient wrap artists of the epigenome. *Nat Rev Mol Cell Biol.* 2010;11(4):264–275. https://doi.org/10.1038/nrm2861.
87. Maze I, Wenderski W, Noh KM, et al. Critical role of histone turnover in neuronal transcription and plasticity. *Neuron.* 2015;87(1):77–94. https://doi.org/10.1016/j.neuron.2015.06.014.
88. Commerford SL, Carsten AL, Cronkite EP. Histone turnover within nonproliferating cells. *Proc Natl Acad Sci USA.* 1982;79(4):1163–1165. https://doi.org/10.1073/pnas.79.4.1163.
89. Zovkic IB, Paulukaitis BS, Day JJ, Etikala DM, Sweatt JD. Histone H2A.Z subunit exchange controls consolidation of recent and remote memory. *Nature.* 2014;515(7528):582–586. https://doi.org/10.1038/nature13707.
90. Guan Z, Giustetto M, Lomvardas S, et al. Integration of long-term-memory-related synaptic plasticity involves bidirectional regulation of gene expression and chromatin structure. *Cell.* 2002;111(4):483–493. https://doi.org/10.1016/S0092-8674(02)01074-7.
91. Malenka RC, Bear MF. LTP and LTD: an embarrassment of riches. *Neuron.* 2004;44(1):5–21. https://doi.org/10.1016/j.neuron.2004.09.012.
92. Alarcón JM, Malleret G, Touzani K, et al. Chromatin acetylation, memory, and LTP are impaired in CBP+/− mice: a model for the cognitive deficit in Rubinstein-Taybi syndrome and its amelioration. *Neuron.* 2004;42(6):947–959. https://doi.org/10.1016/j.neuron.2004.05.021.
93. Gupta K. Histone methylation regulates memory formation. *J Neurosci.* 2010;30(10):3589–3599. https://doi.org/10.1523/JNEUROSCI.3732-09.2010.
94. Fischer A, Sananbenesi F, Wang X, Dobbin MM, Tsai L-H. Recovery of learning and memory is associated with chromatin remodelling. *Nature.* 2007;447(7141):178–182. https://doi.org/10.1038/nature05772.
95. Rudy JW, Huff NC, Matus-Amat P. Understanding contextual fear conditioning: insights from a two-process model. *Neurosci Biobehav Rev.* 2004;28(7):675–685. https://doi.org/10.1016/j.neubiorev.2004.09.004.
96. Peleg S, Sananbenesi F, Zovoilis A, et al. Altered histone acetylation is associated with age-dependent memory impairment in mice. *Science.* 2010;328(5979):753–756. https://doi.org/10.1126/science.1186088.
97. Morris R. Developments of a water-maze procedure for studying spatial learning in the rat. *J Neurosci Methods.* 1984;11(1):47–60. https://doi.org/10.1016/0165-0270(84)90007-4.
98. Bousiges O, Vasconcelos AP, Neidl R, et al. Spatial memory consolidation is associated with induction of several lysine-acetyltransferase (histone acetyltransferase) expression levels and H2B/H4 acetylation-dependent transcriptional events in the rat hippocampus. *Neuropsychopharmacology.* 2010;35(13):2521–2537. https://doi.org/10.1038/npp.2010.117.
99. Dagnas M, Micheau J, Decorte L, Beracochea D, Mons N. Post-training, intrahippocampal HDAC inhibition differentially impacts neural circuits underlying spatial memory in adult and aged mice. *Hippocampus.* 2015;25(7):827–837. https://doi.org/10.1002/hipo.22406.
100. Zovkic IB, Guzman-Karlsson MC, Sweatt JD. Epigenetic regulation of memory formation and maintenance. *Learn Mem.* 2013;20(2):61–74. https://doi.org/10.1101/lm.026575.112.

101. Gupta S, Kim SY, Artis S, et al. NIH public access. *J Neurosci*. 2010;30(10):3589–3599. https://doi.org/10.1523/JNEUROSCI.3732-09.2010.HISTONE.
102. Gupta-Agarwal S, Franklin AV, Deramus T, et al. G9a/GLP histone lysine dimethyltransferase complex activity in the hippocampus and the entorhinal cortex is required for gene activation and silencing during memory consolidation. *J Neurosci*. 2012;32(16):5440–5453. https://doi.org/10.1523/JNEUROSCI.0147-12.2012.
103. Halder R, Hennion M, Vidal RO, et al. DNA methylation changes in plasticity genes accompany the formation and maintenance of memory. *Nat Neurosci*. 2016;19(1):102–110. https://doi.org/10.1038/nn.4194.
104. Gräff J, Woldemichael BT, Berchtold D, Dewarrat G, Mansuy IM. Dynamic histone marks in the hippocampus and cortex facilitate memory consolidation. *Nat Commun*. 2012;3:991. https://doi.org/10.1038/ncomms1997.
105. Antunes M, Biala G. The novel object recognition memory: neurobiology, test procedure, and its modifications. *Cogn Process*. 2012;13(2):93–110. https://doi.org/10.1007/s10339-011-0430-z.
106. Vieira PA, Korzus E. CBP-dependent memory consolidation in the prefrontal cortex supports object-location learning. *Hippocampus*. 2015;25(12):1532–1540. https://doi.org/10.1002/hipo.22473.
107. Lesburguères E, Gobbo OL, Alaux-Cantin S, Hambucken A, Trifilieff P, Bontempi B. Early tagging of cortical networks is required for the formation of enduring associative memory. *Science*. 2011;331(6019):924–928. http://science.sciencemag.org/content/331/6019/924.abstract.
108. Galef BG, Whiskin EE. Socially transmitted food preferences can be used to study long-term memory in rats. *Anim Learn Behav*. 2003;31(2):160–164. https://doi.org/10.3758/BF03195978.
109. Lubin FD, Sweatt JD. The IκB kinase regulates chromatin structure during reconsolidation of conditioned fear memories. *Neuron*. 2007;55(6):942–957. https://doi.org/10.1016/j.neuron.2007.07.039.
110. Rumbaugh G, Sillivan SE, Ozkan ED, et al. Pharmacological selectivity within class I histone deacetylases predicts effects on synaptic function and memory rescue. *Neuropsychopharmacology*. 2015;40(10):2307–2316. https://doi.org/10.1038/npp.2015.93.
111. Xu WS, Parmigiani RB, Marks PA. Histone deacetylase inhibitors: molecular mechanisms of action. *Oncogene*. 2007;26(37):5541–5552. https://doi.org/10.1038/sj.onc.1210620.
112. Gregoretti I, Lee Y-M, Goodson HV. Molecular evolution of the histone deacetylase family: functional implications of phylogenetic analysis. *J Mol Biol*. 2004;338(1):17–31. https://doi.org/10.1016/j.jmb.2004.02.006.
113. Guan J-S, Haggarty SJ, Giacometti E, et al. HDAC2 negatively regulates memory formation and synaptic plasticity. *Nature*. 2009;459(7243):55–60. https://doi.org/10.1038/nature07925.
114. Kim M-S, Akhtar MW, Adachi M, et al. An essential role for histone deacetylase 4 in synaptic plasticity and memory formation. *J Neurosci*. 2012;32(32):10879–10886. http://www.jneurosci.org/content/32/32/10879.abstract.
115. McQuown SC, Barrett RM, Matheos DP, et al. HDAC3 is a critical negative regulator of long-term memory formation. *J Neurosci*. 2011;31(2):764–774. https://doi.org/10.1523/JNEUROSCI.5052-10.2011.
116. Broide RS, Redwine JM, Aftahi N, Young W, Bloom FE, Winrow CJ. Distribution of histone deacetylases 1–11 in the rat brain. *J Mol Neurosci*. 2007;31(1):47–58. https://doi.org/10.1007/BF02686117.
117. Wang J, Telese F, Tan Y, et al. LSD1n is an H4K20 demethylase regulating memory formation via transcriptional elongation control. *Nat Neurosci*. 2015;18(9):1256–1264. https://doi.org/10.1038/nn.4069.
118. Lubin FD, Roth TL, Sweatt JD. Epigenetic regulation of BDNF gene transcription in the consolidation of fear memory. *J Neurosci*. 2008;28(42):10576–10586. https://doi.org/10.1523/JNEUROSCI.1786-08.2008.
119. Miller CA, Campbell SL, Sweatt JD. DNA methylation and histone acetylation work in concert to regulate memory formation and synaptic plasticity. *Neurobiol Learn Mem*. 2008;89(4):599–603. https://doi.org/10.1016/j.nlm.2007.07.016.
120. Levenson JM, Roth TL, Lubin FD, et al. Evidence that DNA (cytosine-5) methyltransferase regulates synaptic plasticity in the hippocampus. *J Biol Chem*. 2006;281(23):15763–15773. https://doi.org/10.1074/jbc.M511767200.

121. Bhutani N, Burns DM, Blau HM. DNA demethylation dynamics. *Cell.* 2011;146(6):866–872. https://doi.org/10.1016/j.cell.2011.08.042.
122. Kaas G, Zhong C, Eason D, et al. TET1 controls CNS 5-methylcytosine hydroxylation, active DNA demethylation, gene transcription, and memory formation. *Neuron.* 2013;79(6):1086–1093. https://doi.org/10.1016/j.neuron.2013.08.032.
123. Kumar D, Aggarwal M, Kaas GA, et al. Tet1 oxidase regulates neuronal gene transcription, active DNA hydroxymethylation, object location memory, and threat recognition memory. *Neurogenetics.* 2015;4:12–27. https://doi.org/10.1016/j.nepig.2015.10.002.
124. Rudenko A, Dawlaty M, Seo J, et al. Tet1 is critical for neuronal activity-regulated gene expression and memory extinction. *Neuron.* 2013;79(6):1109–1122. https://doi.org/10.1016/j.neuron.2013.08.003.
125. Ma DK, Guo JU, Ming GL, Song H. DNA excision repair proteins and Gadd45 as molecular players for active DNA demethylation. *Cell Cycle.* 2009;8(10):1526–1531. https://doi.org/10.4161/cc.8.10.8500.
126. Niehrs C, Schäfer A. Active DNA demethylation by Gadd45 and DNA repair. *Trends Cell Biol.* 2012; 22(4):220–227. https://doi.org/10.1016/j.tcb.2012.01.002.
127. Leach PT, Poplawski SG, Kenney JW, et al. Gadd45b knockout mice exhibit selective deficits in hippocampus-dependent long-term memory. *Learn Mem.* 2012;19(8):319–324. https://doi.org/10.1101/lm.024984.111.
128. Sultan FA, Wang J, Tront J, Liebermann DA, Sweatt JD. Genetic deletion of Gadd45b, a regulator of active DNA demethylation, enhances long-term memory and synaptic plasticity. *J Neurosci.* 2012; 32(48):17059–17066. https://doi.org/10.1523/JNEUROSCI.1747-12.2012.
129. Turner BM. Defining an epigenetic code. *Nat Cell Biol.* 2007;9(1):2–6. https://doi.org/10.1038/ncb0107-2.
130. Mellén M, Ayata P, Dewell S, Kriaucionis S, Heintz N. MeCP2 binds to 5hmC enriched within active genes and accessible chromatin in the nervous system. *Cell.* 2012;151(7):1417–1430. https://doi.org/10.1016/j.cell.2012.11.022.
131. Fuks F, Hurd PJ, Wolf D, Nan X, Bird AP, Kouzarides T. The methyl-CpG-binding protein MeCP2 links DNA methylation to histone methylation. *J Biol Chem.* 2003;278(6):4035–4040. https://doi.org/10.1074/jbc.M210256200.
132. Chahrour M, Jung SY, Shaw C, et al. MeCP2, a key contributor to neurological disease, activates and represses transcription. *Science.* 2008;320(5880):1224–1229. https://doi.org/10.1126/science.1153252.
133. Weaver ICG, Cervoni N, Champagne FA, et al. Epigenetic programming by maternal behavior. *Nat Neurosci.* 2004;7(8):847–854. https://doi.org/10.1038/nn1276.
134. Champagne FA, Meaney MJ. Stress during gestation alters postpartum maternal care and the development of the offspring in a rodent model. *Biol Psychiatry.* 2006;59(12):1227–1235. https://doi.org/10.1016/j.biopsych.2005.10.016.
135. Meaney MJ, Szyf M, Seckl JR. Epigenetic mechanisms of perinatal programming of hypothalamic-pituitary-adrenal function and health. *Trends Mol Med.* 2007;13(7):269–277. https://doi.org/10.1016/j.molmed.2007.05.003.
136. Todrank J, Heth G, Restrepo D. Effects of in utero odorant exposure on neuroanatomical development of the olfactory bulb and odour preferences. *Proc R Soc B Biol Sci.* 2011;278(1714):1949–1955. http://rspb.royalsocietypublishing.org/content/278/1714/1949.abstract.
137. Dias BG, Ressler KJ. Parental olfactory experience influences behavior and neural structure in subsequent generations. *Nat Neurosci.* 2014;17(1):89–96. https://doi.org/10.1038/nn.3594.
138. Sun WJ, Li JH, Liu S, et al. RMBase: a resource for decoding the landscape of RNA modifications from high-throughput sequencing data. *Nucleic Acids Res.* 2016;44(D1):D259–D265. https://doi.org/10.1093/nar/gkv1036.
139. Lee M, Kim B, Kim VN. Emerging roles of RNA modification: m(6)A and U-tail. *Cell.* 2014;158(5): 980–987. https://doi.org/10.1016/j.cell.2014.08.005.
140. Hoernes TP, Clementi N, Faserl K, et al. Nucleotide modifications within bacterial messenger RNAs regulate their translation and are able to rewire the genetic code. *Nucleic Acids Res.* 2015. https://doi.org/10.1093/nar/gkv1182.

141. Niu Y, Zhao X, Wu YS, Li MM, Wang XJ, Yang YG. N6-methyl-adenosine (m6A) in RNA: an old modification with a novel epigenetic function. *Dev Reprod Biol.* 2013;11(1):8–17. https://doi.org/10.1016/j.gpb.2012.12.002.
142. Meyer KD, Saletore Y, Zumbo P, Elemento O, Mason CE, Jaffrey SR. Comprehensive analysis of mRNA methylation reveals enrichment in 3′ UTRs and near stop codons. *Cell.* 2012;149(7): 1635–1646. https://doi.org/10.1016/j.cell.2012.05.003.
143. Jia G, Fu Y, He C. Reversible RNA adenosine methylation in biological regulation. *Trends Genet.* 2013;29(2):108–115. https://doi.org/10.1016/j.tig.2012.11.003.
144. Liu N, Pan T. N6-methyladenosine-encoded epitranscriptomics. *Nat Struct Mol Biol.* 2016;23(2): 98–102. https://doi.org/10.1038/nsmb.3162.
145. Liu J, Yue Y, Han D, et al. A METTL3-METTL14 complex mediates mammalian nuclear RNA N6-adenosine methylation. *Nat Chem Biol.* 2014;10(2):93–95. https://doi.org/10.1038/nchembio.1432.
146. Batista PJ, Molinie B, Wang J, et al. m(6)A RNA modification controls cell fate transition in mammalian embryonic stem cells. *Cell Stem Cell.* 2014;15:707–719. https://doi.org/10.1016/j.stem.2014.09.019.
147. Wang X, Lu Z, Gomez A, et al. N6-methyladenosine-dependent regulation of messenger RNA stability. *Nature.* 2014;505(7481):117–120. https://doi.org/10.1038/nature12730.
148. Jia G, Fu Y, Zhao X, et al. N6-methyladenosine in nuclear RNA is a major substrate of the obesity-associated FTO. *Nat Chem Biol.* 2011;7(12):885–887. https://doi.org/10.1038/nchembio.687.
149. Zheng G, Dahl JA, Niu Y, et al. ALKBH5 is a mammalian RNA demethylase that impacts RNA metabolism and mouse fertility. *Mol Cell.* 2013;49(1):18–29. https://doi.org/10.1016/j.molcel.2012.10.015.
150. Merkestein M, McTaggart JS, Lee S, et al. Changes in gene expression associated with FTO overexpression in mice. *PLoS One.* 2014;9(5):e97162. https://doi.org/10.1371/journal.pone.0097162.
151. Hess ME, Hess S, Meyer KD, et al. The fat mass and obesity associated gene (FTO) regulates activity of the dopaminergic midbrain circuitry. *Nat Neurosci.* 2013;16(8):1042–1048. https://doi.org/10.1038/nn.3449.
152. Frayling TM, Timpson NJ, Weedon MN, et al. A common variant in the FTO gene is associated with body mass index and predisposes to childhood and adult obesity. *Science.* 2007;316(5826):889–894. https://doi.org/10.1126/science.1141634.
153. Cedernaes J, Benedict C. Human obesity: FTO, IRX3, or both? *Mol Metab.* 2014;3(5):505–506. https://doi.org/10.1016/j.molmet.2014.05.003.
154. Smemo S, Tena JJ, Kim KH, et al. Obesity-associated variants within FTO form long-range functional connections with IRX3. *Nature.* 2014;507(7492):371–375. https://doi.org/10.1038/nature13138.
155. Church C, Moir L, McMurray F, et al. Overexpression of FTO leads to increased food intake and results in obesity. *Nat Genet.* 2010;42(12):1086–1092. https://doi.org/10.1038/ng.713.
156. Fischer J, Koch L, Emmerling C, et al. Inactivation of the FTO gene protects from obesity. *Nature.* 2009;458(7240):894–898. https://doi.org/10.1038/nature07848.
157. McMurray F, Church CD, Larder R, et al. Adult onset global loss of the FTO gene alters body composition and metabolism in the mouse. *PLoS Genet.* 2013;9(1):e1003166. https://doi.org/10.1371/journal.pgen.1003166.
158. Schwartz S, Mumbach MR, Jovanovic M, et al. Perturbation of m6A writers reveals two distinct classes of mRNA methylation at internal and 5′ sites. *Cell Rep.* 2014;8(1):284–296. https://doi.org/10.1016/j.celrep.2014.05.048.
159. Dominissini D, Moshitch-Moshkovitz S, Schwartz S, et al. Topology of the human and mouse m6A RNA methylomes revealed by m6A-seq. *Nature.* 2012;485(7397):201–206. https://doi.org/10.1038/nature11112.
160. Zhou J, Wan J, Gao X, Zhang X, Jaffrey SR, Qian SB. Dynamic m(6)A mRNA methylation directs translational control of heat shock response. *Nature.* 2015;526(7574):591–594. https://doi.org/10.1038/nature15377.
161. Widagdo J, Zhao Q-Y, Kempen M-J, et al. Experience-dependent accumulation of N6-methyladenosine in the prefrontal cortex is associated with memory processes in mice. *J Neurosci.* 2016;36(25):6771–6777. https://doi.org/10.1523/jneurosci.4053-15.2016.

162. Walters BJ, Mercaldo V, Gillon CJ, et al. The role of the RNA demethylase FTO (fat mass and obesity-associated) and mRNA methylation in hippocampal memory formation. *Neuropsychopharmacology*. 2017;42(7):1502–1510. https://doi.org/10.1038/npp.2017.31.
163. Haussmann IU, Bodi Z, Sanchez-Moran E, et al. m6A potentiates Sxl alternative pre-mRNA splicing for robust Drosophila sex determination. *Nature*. 2016;540(7632):301–304. https://doi.org/10.1038/nature20577.
164. Zhao X, Yang Y, Sun BF, et al. FTO-dependent demethylation of N6-methyladenosine regulates mRNA splicing and is required for adipogenesis. *Cell Res*. 2014;24(12):1403–1419. https://doi.org/10.1038/cr.2014.151.
165. Chwang WB, O'Riordan KJ, Levenson JM, Sweatt JD. ERK/MAPK regulates hippocampal histone phosphorylation following contextual fear conditioning. *Learn Mem (Cold Spring Harb NY)*. 2006; 13(3):322–328. https://doi.org/10.1101/lm.152906.
166. Cervoni N, Szyf M. Demethylase activity is directed by histone acetylation. *J Biol Chem*. 2001; 276(44):40778–40787. https://doi.org/10.1074/jbc.M103921200.
167. Zilberman D, Coleman-Derr D, Ballinger T, Henikoff S. Histone H2A.Z and DNA methylation are mutually antagonistic chromatin marks. *Nature*. 2008;456(7218):125–129. https://doi.org/10.1038/nature07324.
168. Ku M, Jaffe JD, Koche RP, et al. H2A.Z landscapes and dual modifications in pluripotent and multipotent stem cells underlie complex genome regulatory functions. *Genome Biol*. 2012;13(10):R85. https://doi.org/10.1186/gb-2012-13-10-r85.
169. Esteller M. Aberrant DNA methylation as a cancer-inducing mechanism. *Annu Rev Pharmacol Toxicol*. 2005;45:629–656. https://doi.org/10.1146/annurev.pharmtox.45.120403.095832.
170. Collas P. The current state of chromatin immunoprecipitation. *Mol Biotechnol*. 2010;45(1):87–100. https://doi.org/10.1007/s12033-009-9239-8.
171. Jacinto FV, Ballestar E, Esteller M. Methyl-DNA immunoprecipitation (MeDIP): hunting down the DNA methylome. *Biotechniques*. 2008;44(1):35–43. https://doi.org/10.2144/000112708.
172. Brinkman AB, Simmer F, Ma K, Kaan A, Zhu J, Stunnenberg HG. Whole-genome DNA methylation profiling using methylcap-seq. *Methods*. 2010;52(3):232–236. https://doi.org/10.1016/j.ymeth.2010.06.012.
173. Shapiro R, Braverman B, Louis JB, Servis RE. Nucleic acid reactivity, conformation: II. Reaction of cytosine and uracil with sodium bisulfite. *J Biol Chem*. 1973;248(11):4060–4064. http://www.jbc.org/content/248/11/4060.abstract.
174. Soon WW, Hariharan M, Snyder MP. High-throughput sequencing for biology and medicine. *Mol Syst Biol*. 2013;9(640):640. https://doi.org/10.1038/msb.2012.61.
175. Encode Consortium. An integrated encyclopedia of DNA elements in the human genome. *Nature*. 2012;489(7414):57–74. https://doi.org/10.1038/nature11247.
176. Romanoski CE, Glass CK, Stunnenberg HG, Wilson L, Almouzni G. Epigenomics: roadmap for regulation. *Nature*. 2015;518(7539):314–316. https://doi.org/10.1038/518314a.
177. Creyghton MP, Cheng AW, Welstead GG, et al. Histone H3K27ac separates active from poised enhancers and predicts developmental state. *Proc Natl Acad Sci USA*. 2010;107(50):21931–21936. https://doi.org/10.1073/pnas.1016071107.
178. Vermunt MW, Reinink P, Korving J, et al. Large-scale identification of coregulated enhancer networks in the adult human brain. *Cell Rep*. 2017;9(2):767–779. https://doi.org/10.1016/j.celrep.2014.09.023.
179. Kim T-K, Hemberg M, Gray JM, et al. Widespread transcription at neuronal activity-regulated enhancers. *Nature*. 2010;465(7295):182–187. https://doi.org/10.1038/nature09033.
180. Feng J, Shao N, Szulwach KE, et al. Role of Tet1 and 5-hydroxymethylcytosine in cocaine action. *Nat Neurosci*. 2015;18(4):536–544. https://doi.org/10.1038/nn.3976.
181. Gjoneska E, Pfenning AR, Mathys H, et al. Conserved epigenomic signals in mice and humans reveal immune basis of Alzheimer's disease. *Nature*. 2015;518(7539):365–369. https://doi.org/10.1038/nature14252.
182. Nelson SB, Sugino K, Hempel CM. The problem of neuronal cell types: a physiological genomics approach. *Trends Neurosci*. 2006;29(6):339–345. https://doi.org/10.1016/j.tins.2006.05.004.

183. Rotem A, Ram O, Shoresh N, et al. Single-cell ChIP-seq reveals cell subpopulations defined by chromatin state. *Nat Biotechnol.* 2015;33(11):1165–1172. https://doi.org/10.1038/nbt.3383.
184. Poulin J-F, Tasic B, Hjerling-Leffler J, Trimarchi JM, Awatramani R. Disentangling neural cell diversity using single-cell transcriptomics. *Nat Neurosci.* 2016;19(9):1131–1141. https://doi.org/10.1038/nn.4366.
185. Josselyn SA, Kohler S, Frankland PW. Finding the engram. *Nat Rev Neurosci.* 2015;16(9):521–534. https://doi.org/10.1038/nrn4000.

SECTION IV

# Discovery of Genes and Biological Mechanisms: Forward Genetics and Other Screening-Based Methods

CHAPTER 16

# Systematic Screens in Zebrafish Shed Light on Cellular and Molecular Mechanisms of Complex Brain Phenotypes

**Adam Melgoza, Su Guo**
University of California San Francisco, San Francisco, CA, United States

## INTRODUCTION

The brain, composed of many thousands of different cell types with intricate connections, is arguably the most complex organ. Together with the spinal cord, they constitute the central nervous system (CNS). The vertebrate CNS shares conserved anatomy, including subdivisions such as the forebrain, midbrain, hindbrain, and spinal cord. Studies employing vertebrate genetic model organisms such as zebrafish, xenopus, chick, and mice have shed light on the formation, maintenance, and function of complex brain structures.

The zebrafish is a relatively recent addition to the model organism arena.[1,2] In the 1970s, the late Dr. George Streisinger selected a pair of zebrafish (called A and B respectively) from a local pet shop because of the prolific nature and transparent embryonic and larval stages of this species. Through a series of elegant experiments, he showed that zebrafish have a diploid genome and could be bred to produce clonal populations.[3] Subsequently, the developmental stages and embryonic brain structures were characterized,[4,5] and two large-scale *N*-ethyl-*N*-nitrosourea (ENU) mutagenesis screens were performed that identified thousands of mutations affecting virtually every aspect of visible morphology of a vertebrate embryo.[6,7] Molecular genetic studies of these mutations have provided valuable insights into vertebrate development.

Despite these advances, the potential of zebrafish to uncover cellular and molecular mechanisms underlying complex brain phenotypes are just beginning to be realized. Morphology-based phenotypic screens cannot detect changes at the cellular or functional levels, and therefore tend to miss the processes that affect specific structures or cell types, or that affect brain function. In recent years, innovative screens targeting these phenotypes were devised, and they will be the subject of review in this chapter.

*Molecular-Genetic and Statistical Techniques for Behavioral and Neural Research*
ISBN 978-0-12-804078-2, https://doi.org/10.1016/B978-0-12-804078-2.00016-7

We will begin the chapter by outlining general strategies used to introduce genetic alterations in the zebrafish. Methods used to produce a library of individuals with random phenotypic variations will be delineated. We will then discuss the results from several screens, with an emphasis on assay development, screen execution, and impacts on our understanding of complex brain phenotypes and related human brain disorders.

## STRATEGIES FOR FORWARD, REVERSE, AND CHEMICAL GENETIC SCREENS

The general strategy for forward, reverse, and chemical genetic screens is outlined in Fig. 16.1. It starts with a phenotype of interest, which is then developed into an assay with sufficient robustness, sensitivity, and throughput suitable for screening. Such assay is then used to screen through a large number of individuals, with either mutagenized genomes (e.g., ENU, insertional mutagenesis, or genome-editing techniques), epigenetic alterations of gene expression (e.g., morpholino antisense oligonucleotide-based, RNAi, overexpression), or treated with small molecules (compound screening). Following the screen, phenotypic traits of interest will be uncovered, and further studies will allow the underlying molecular and cellular mechanisms to be elucidated.

## GENERATION OF A LIBRARY OF INDUCED GENETIC OR EPIGENETIC VARIATIONS

Several methods and technologies can be used to create a library of individuals with random or targeted genetic or epigenetic variations. Some of these methodologies have been broadly used in zebrafish, whereas others have so far only been feasible in cell cultures or invertebrate genetic model organisms, and may be potentially applicable to zebrafish.

### *N*-Ethyl-*N*-Nitrosourea Mutagenesis

ENU mutagenesis is a classic approach to introduce random mutations throughout the genome. As a highly potent mutagen, ENU introduces random mutations into the genome of exposed individuals. Adult male zebrafish treated with ENU at appropriate doses will harbor random mutations in their germ line. Upon mating with a wild-type female zebrafish, the next generation (F1) will carry random mutations in heterozygous states. Each F1 individual carries potentially unique mutations. These F1 individuals are then crossed with wild-type fish, producing F2 progeny heterozygous for the same mutation. In the F3 generation, homozygosity of induced mutations will be realized. F3s are used for phenotype-based screening to identify recessive loss-of-function mutations. To uncover dominant mutations, F1 or F2 generations can also be used.

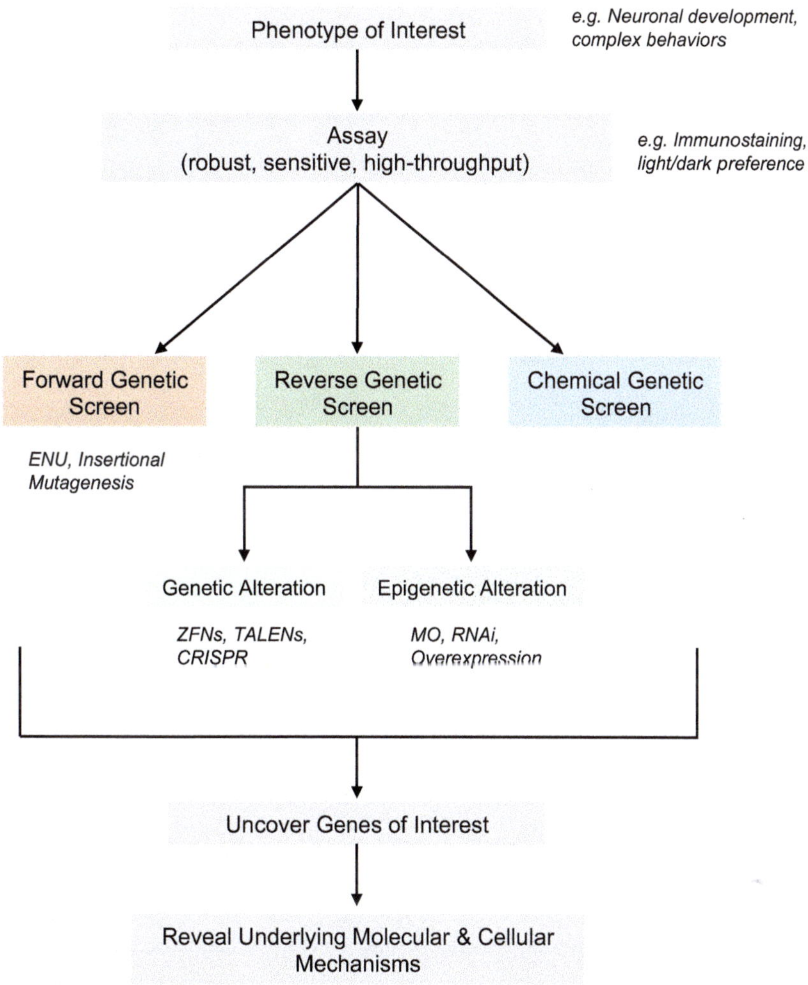


**Figure 16.1** Schematic of genetic screening. The process begins with a phenotype of interest. An assay that is robust, sensitive, and has high throughput will be chosen to test this phenotype. A variety of genetic screens can then be utilized with this assay: forward, reverse, and chemical. Forward genetic screens start by producing many random mutations. Any aberrant phenotypes are identified, then investigated to uncover the gene responsible (ENU, insertional mutagenesis). Conversely, reverse genetic screens start with specific gene targets in search of an abnormal phenotype. Reverse genetics can involve either permanent edits to the germ line (CRISPR, ZFNs, TALENs) or transient epigenetic alterations (MO, RNAi, overexpression). Small molecules are administered in chemical genetic screens. The results from these screens help to uncover genes that play a role in modulating the phenotype, providing insight on the molecular and cellular mechanisms at play.

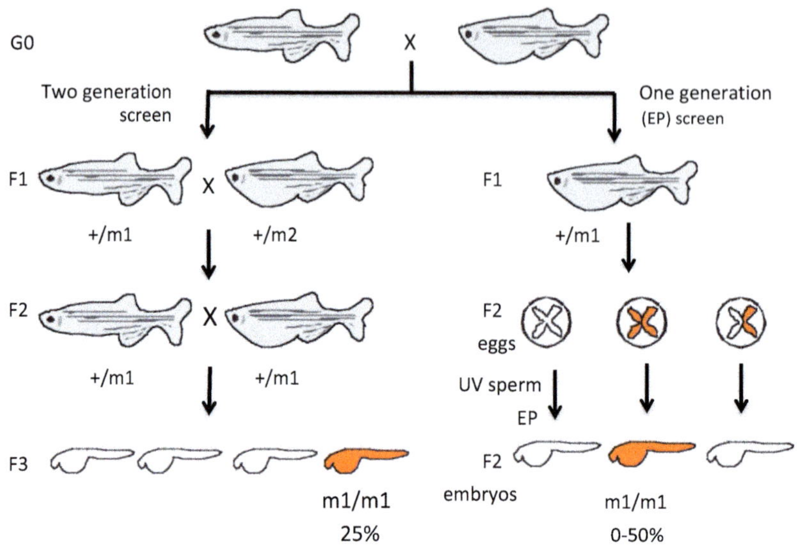


**Figure 16.2** Schematic of ENU screening. Two or one generation breeding can be carried out to screen for recessive mutant phenotypes following ENU mutagenesis. (Left) Standard two-generation breeding to obtain 25% homozygous individuals in the F3 generation. (Right) To shorten the breeding time needed, F2 oocytes can be collected from F1 females, in vitro fertilized with UV-irradiated sperm, resulting in the activation of embryonic development without contributing genetic material. Subsequently, early pressure (EP) is applied to the eggs to prevent second meiosis and enable gynogenetic development. Depending on the location of the mutated loci relative to centromeres, 0%–50% of homozygous mutants can be obtained in the F2 generation.

ENU is highly efficient in inducing mutations in the zebrafish genome.[8,9] The induced mutations are nonbiased toward specific loci. One limitation with respect to ENU-based screens is its time-consuming nature, which involves three generations of breeding. To circumvent this, early pressure-induced gynogenesis has been used to uncover homozygosity in F2 generations (Fig. 16.2).[10]

## CRISPR

In recent years, genome-editing technologies, in particular, CRISPR (clustered regularly interspaced short palindromic repeats),[11–13] have made it possible to introduce mutations in principally any genes of interest. Double-stranded DNA breaks (DSBs) can be induced in a targeted manner by delivering the bacterially derived endonuclease Cas9 and small guide RNA (sgRNA). Targeting specificity is achieved through DNA-RNA base pairing and Cas9 recognition of a short DNA sequence known as the protospacer adjacent motif. The resulting DSBs are most efficiently repaired by a mechanism known as nonhomologous end joining (NHEJ). Due to the error-prone nature of NHEJ, small deletions or insertions are introduced at the site of DSBs, thereby leading to potentially deleterious mutations. While NHEJ is typically used to create a targeted gene knockout, homology directed repair is an alternative repair mechanism that can allow for targeted knock-ins of a sequence of interest.[14] This process involves using a template to reconstruct the DSB, which can either come from the sister chromosome or be exogenously introduced.

CRISPR has been successfully applied to introduce mutations into zebrafish genes.[14–16] The procedure (Fig. 16.3) involves microinjection of the *cas9* gene product (in the form of either mRNA or protein) plus a sequence-specific sgRNA into one-cell stage zebrafish embryos. The injected embryos (F0) often carry single-allele mutations in some but not all cells (called mosaics), although in some cases with highly efficient sgRNAs and Cas9 protein delivery, widespread bi-allelic mutations can be accomplished in the injected embryos. Since the efficacy of sgRNAs vary significantly, it is important to assess the mutagenic efficiency of each sgRNA before proceeding to raise F0s for line propagation. Several methods are available, ranging from T7 endonuclease assay, surveyor assay, to restriction enzyme digestion, and sequencing. These assays are run on a pool of injected F0 embryos. Once the efficacy of sgRNAs is verified, injected F0 embryos can be raised to adulthood and used to produce F1s. Adult F1s (heterozygous for potential mutations) can then be genotyped for the presence and the nature of CRISPR-induced mutations. Further breeding of F1s can lead to recovery of homozygous mutations in the F2 generation.

An advantage of using CRISPR is the possibility to create mutations on both chromosomes (bi-allelic targeting). Also, orthologous genes may be targeted simultaneously with ease. Lastly, though CRISPR has been described earlier in the text as a method for reverse genetics, the ease with which CRISPR-based gene targeting is performed may allow this method to be used essentially as a forward genetic approach: targeting a large number of genes and searching for their phenotypic effects.

Although the efficiency of CRISPR in inducing disruptive mutations is high, due to the short nature of the sgRNAs (20 nucleotides), there can be potential off-target effects. This is generally not a major concern in zebrafish, as off-target mutations can be separated from the mutations of interest through outcross breeding. Another limitation of CRISPR-based screening is the need to breed two generations before phenotypes of homozygous mutations can be assessed. This becomes a rather tedious process if a large number of genes are to be screened. Methods to improve targeting efficiency that enable bi-allelic targeting in injected G0 embryos would circumvent these breeding needs. Another concern is a lack of phenotypes in CRISPR-induced mutations (as opposed to morpholino antisense oligonucleotides (MOs) knockdown) due to potential genetic compensation.[17] However, a lack of phenotype is not specific to the CRISPR method.

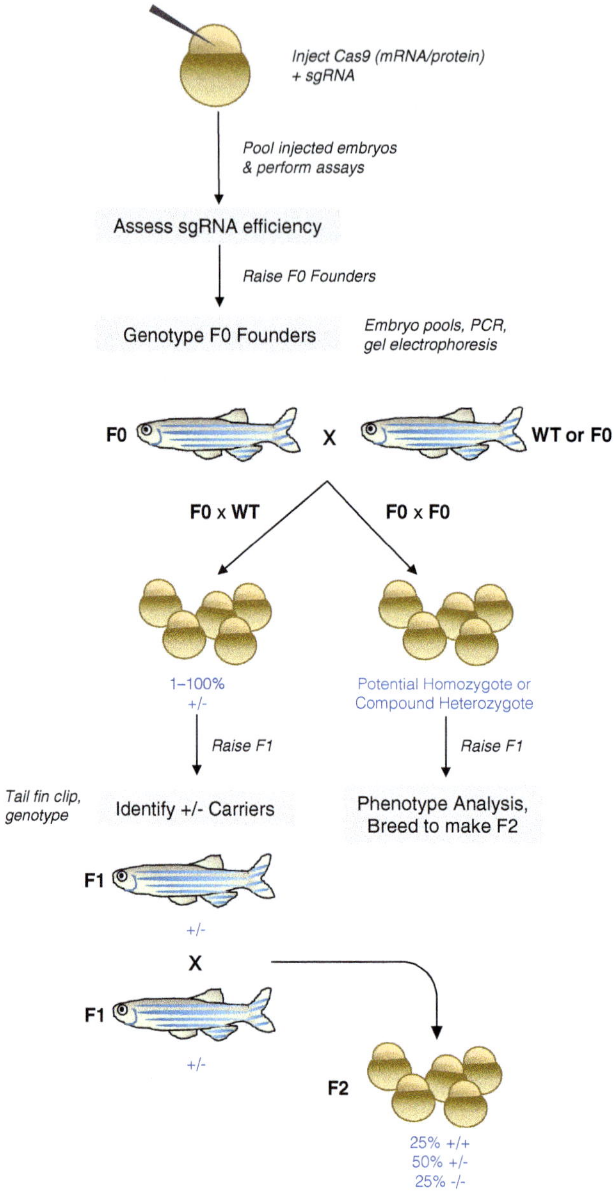


**Figure 16.3** Introducing genetic alterations via CRISPR-Cas9. Using CRISPR technology to produce gene edits in zebrafish begins with injection of both Cas9 (either mRNA or protein) and sgRNA for the gene of interest. The injected embryos are then pooled, and a variety of assays can be run to determine efficiency of the sgRNA (e.g., T7 endonuclease assay, surveyor assay, sequencing, etc.). Once mutations are confirmed, the F0 generation is raised to adulthood. As adults, they are genotyped to identify any induced genetic alterations. This is done by pooling together a clutch of their embryos, PCR, and gel electrophoresis. When F0 individuals carrying mutations are identified, they can be crossed either with a wild-type fish (outcross) or another F0 individual (incross). The outcross results

## Morpholino Antisense Oligonucleotides

The MOs are fast and effective tools for knocking down gene activity.[18] They can be designed through sequence complementarity to interfere with either mRNA translation or splicing. A key limitation with MO is its transient effect that usually lasts several days after delivery into one-cell stage embryos. Its potential off-target effects can be managed through using multiple independent MOs to target the same gene and phenotypic rescue with coexpressing the target gene that is rendered MO-resistant.

## RNA Interference (RNAi)

RNA interference (RNAi) is a powerful approach to dissect gene function.[19,20] It was originally discovered in *Caenorhabditis elegans*, where the observation of gene inactivation by both sense and antisense RNAs[21] led to the finding of dsRNA-mediated gene silencing.[22] In vertebrates, the effectiveness of RNAi was hindered by the dsRNA-induced interferon response causing nonspecific effects[23,24] until the finding of efficacious small interfering RNAs (siRNAs).[25] RNAi has proven effective in zebrafish when delivered with the microRNA (miR-30) backbone and the Gal4/UAS system,[26] which offers excellent versatility for spatiotemporal control together with amplification of gene expression level. While RNAi is highly effective when transiently expressed, stable transgenic lines so far yield moderate to weak effects, likely due to the presence of low or single copies of the transgene, resulting in insufficient knockdown. The potential off-target effects of RNAi can be ameliorated by designing multiple independent miR-shRNAs targeting the same gene of interest.

## Gene Overexpression

In addition, to interfere with gene activity, overexpression of genes can also be used to assess the impact of genetic alterations on phenotypes. This becomes particularly useful when genetic redundancy is suspected. Gene overexpression-based screens also address the sufficiency of gene activity in generating phenotypes. Successful examples of gene overexpression screens include the early pressure (EP) screens in *Drosophila*,[27] CRISPR-mediated gene activation screen in cultured cells,[28] and heat shock promoter-driven overexpression screens in zebrafish.[29] One caution with gene overexpression-induced phenotypes is that sometimes they might not be physiologically

in F1 progeny that are 1%–100% heterozygous (due to the mosaicism of the F0 founder). When this F1 generation reaches adulthood, they are genotyped to confirm individuals heterozygous for a mutation in the gene of interest. Two heterozygous fish are then crossed to produce the F2 generation, which is 25% wild type, 50% heterozygous, and 25% homozygous for the mutation. On the other hand, if the F0 fish are incrossed, progeny is produced that are either homozygous for the same mutation or compound heterozygous: containing two different mutations in the gene of interest. These F1 fish can be used for both phenotype analysis and the creation of the F2 generation.

relevant, especially when the expression level is too high and when no complementary phenotypes can be observed in loss-of-function studies.

### Bioactive Small Molecules

Another powerful way to perturb gene activity is through the use of small molecules, which can interact with gene products to either activate or inhibit their activity. Libraries of small molecules include those with known activities and those composed of diverse and novel entities. Zebrafish, in particular, small-sized embryonic and larval zebrafish that can be fit into 96-well plates, are well suited for small molecule screening since these compounds can be directly added to the medium. Small molecule screening is easy and convenient to carry out. The challenge rests in identifying biologic targets of the hit compounds and understanding their mechanisms of actions.

## POSTSCREENING: EFFORTS TO DISCOVER UNDERLYING CELLULAR AND MOLECULAR MECHANISMS

Once a hit that alters the phenotypic trait of interest is uncovered, the next step is to gain insights into the underlying biologic processes.

### Post-ENU Screening

Individual mutations identified from an ENU mutagenesis screen need to be first mapped to a chromosomal location. Such mapping is conducted by using existing genetic linkage maps, which contain many polymorphic markers.[30] A large number of individuals derived from a cross between the mutation-containing strain and a polymorphic strain will be phenotyped and genotyped to uncover linkages. Once a linkage is detected, high-resolution mapping within the region of interest can be carried out to pinpoint the gene. Such a positional cloning method has been used to successfully map and identify a number of genes.[31–34] In recent years, the cost of sequencing has been precipitously reduced, making it possible to employ whole-genome sequencing to uncover the causal mutations.[35,36] Here, pools of mutants and their wild-type siblings are collected for genomic DNA isolation and sequencing. Background polymorphisms unlinked to the mutation will be present equally in both the mutant and the sibling pools, whereas homozygosity for the mutation of interest will be unique to the mutant pool.

Once the causal mutations have been discovered, a variety of cellular molecular and biochemical techniques can be employed to understand how disruption of the particular gene might lead to the observed phenotypes.

### Post-Target-Based Screening (e.g., Employing MO, RNAi, CRISPR, or Overexpression)

Screens employing a library of agents, such as MOs, RNAi, CRISPR, or overexpression constructs, can uncover target gene(s) with ease. By nature, these libraries already have cataloged information on the target gene(s). However, it is worth noting that potential off-target effects are significant concerns with respect to these screens. Therefore, it is important to have multiple independent agents (e.g., two different shRNAs that target the same gene) for functional verification. Additionally, one also needs to bear in mind that different reagents or methods might differentially affect the activity of the same gene, resulting in different phenotypic consequences.

### Post-Small Molecule Screens

The result of a small molecule screen is a repertoire of compounds that can alter the phenotype of interest. Depending on whether the small molecules are known bioactives or novel entities, different challenges lie ahead for target identification.

For known bioactive small molecules, it is relatively easy to uncover their biologic targets. However, it is important to bear in mind that the known biologic targets might not be the ones that mediate the observed effect. To validate, one shall perform molecular genetic experiments such as knockdown or overexpression of the candidate genes and determine whether they will mimic the effect of the small molecules.

For novel small molecules, it is a considerable challenge to uncover their biologic targets. A commonly used method is affinity purification.[37] Also, the Similarity Ensemble Approach (SEA),[38] which takes an organic molecule and compares it to sets of ligands annotated to biologic activities/targets, can be used to uncover candidate biologic targets for subsequent functional validation.

## EXAMPLES OF SCREENS IN ZEBRAFISH

### ENU- and Small Molecule-Based Screens to Uncover Mechanisms of Neuronal Subtype Differentiation and Maintenance

Understanding the generation of cellular diversity is a fundamental problem that is particularly daunting in the vertebrate nervous system, which contains many thousands of different neural cell types. To gain insights into the underlying molecular and cellular mechanisms, ENU-based screens have been carried out to uncover genes and pathways that control the differentiation of dopaminergic (DA) neurons.[10] DA neurons belong to the catecholaminergic (CA) neuronal groups, and they express the tyrosine hydroxylase (*th*) gene but are devoid of the expression of dopamine beta-hydroxylase (*dbh*) gene. TH is an enzyme involved in the rate-limiting step of the dopamine synthesis pathway, whereas DBH is responsible for conversion of dopamine to noradrenaline. Being present in small numbers and discrete locations in the brain, DA neurons play important roles in movement coordination, reward, and learning and memory. The loss of substantia nigra DA neurons in humans results in Parkinson disease, the most common movement disorder and the second most common neurodegenerative disorder. After characterizing the distribution of CA neurons in developing zebrafish, an immunohistochemistry-based screen was used to find mutations that affect CA neuron number or identity. Around

700 ENU-mutagenized genomes were screened for these alterations, and five mutations were identified, causing defects in both DA and noradrenergic (NA) neurons. Subsequent molecular characterizations have revealed the identity of the disrupted genes.[32,34,39–41] Three of the five genes encode cell type-specific transcriptional regulators. The other two encode general transcription regulators, one being a component of the mediator complex that controls transcription initiation and the other being a transcription elongation regulator.[34] Similar screens have revealed additional regulators of DA neuron differentiation.[42]

A small molecule screen has uncovered compounds that regulate dopamine neuron maintenance.[43] In this screen, 5000 compounds composed of both known bioactives and novel entities were tested for their effects on DA neuron development and/or survival via high throughput immunostaining in 96-well plates. After drug administration, antibodies against TH were used to visualize DA neurons in search of aberrant neuron number or morphology. One compound named neriifolin was identified; it impaired DA neuron survival but not that of noradrenergic sympathetic neurons. Further mechanistic studies show that its biologic target, the $Na^+/K^+$ ATPase, is critical for maintaining DA neuron integrity.

## ENU- and Small Molecule-Based Screen to Study Habituation

Unbiased behavioral screening is a powerful way to reveal novel mechanisms that regulate brain function. To be suitable for a screen, the behavioral assay of interest needs to be sensitive, robust, and one that can be performed in a relatively high throughput manner. The habituation behavior in larval zebrafish has been a successful example of behavioral screening-based studies. Habituation refers to decreased sensitivity toward repeated delivery of a sensory stimulus and is a fundamental form of behavioral plasticity or nonassociative learning that occurs for virtually all behavioral responses in virtually all organisms.[44] Five-day postfertilization (dpf) larval zebrafish exhibit a highly stereotyped acoustic startle response known as the short latency C-start. Repetitive acoustic stimuli result in habituation, and the modulation of this phenomenon has been studied via small molecule[45] and ENU forward genetic screens.[46] Among 1760 known bioactive compounds screened, 11 were found to reduce startle habituation, while 19 increased it. Most targeted neurotransmitter systems were previously reported to affect mammalian startle modulation, suggesting conserved mechanisms between fish and mammals. Compounds previously unknown to be involved in regulating nonassociative learning were also uncovered. For the ENU screen, the same startle habituation assay was used to screen 405 mutagenized F2 families (corresponding to 614 genomes) and identify 14 habituation mutants. Whole-genome sequencing was used to uncover the molecular identity of two mutants, one of which encodes a vertebrate specific gene *pregnancy associated plasma protein-aa* (*pappaa*).

Further studies revealed a previously unknown role for PAPPAA-regulated IGF signaling in mediating habituation learning.

## MO-Based Screen to Uncover the Function of Candidate Genes Implicated in Autism Spectrum Disorders (ASD)

MO is a fast and convenient tool to assess gene function during development. Two studies have employed this tool, together with overexpression, to determine which gene(s) might be responsible for ASD in a copy number variation involving deletion or duplication of human chromosome 16p11.2.[47,48] In one study,[48] MOs targeting various zebrafish orthologues were individually microinjected into early embryos, followed by phenotypic evaluations of morphology and behaviors including spontaneous movement and touch response. Two genes, *adolase a* and *kinesin* family member 22, were found to give clear phenotypes when their dosage was reduced by 50%, suggesting that these genes are dosage sensors. In the other study,[47] the authors focused on the microcephaly and macrocephaly phenotypes associated with the 16p11.2 duplication and deletion respectively. By gene overexpression studies, they found that one gene named *kctd13*, when overexpressed, led to decreased brain size, whereas MO knockdown of *kctd13* led to macrocephaly. In mammalian cell cultures and brain slices, alterations of *kctd13* activity affected cell proliferation. Together, MO-based screening help decipher candidate genes involved in complex human brain disorders.

## CRISPR-Based Screen to Uncover Genes Involved in Electrical Synapse Formation

The efficiency of introducing mutations into the genome via CRISPR has been further improved for a candidate gene-based screen to uncover new regulators in electrical synapse formation.[49] By testing various concentrations of Cas9 and sgRNAs, the authors determined the appropriate doses that resulted in an optimal balance of genome editing efficiency and embryonic toxicity. Using a column- and row-based pooling strategy, 48 sgRNAs were screened, by microinjection into one-cell stage F0 embryos and imaging of electrical synapses at later stages of development in F0. Two genes with previously undescribed roles in electrical synapse formation were uncovered. Deep sequencing revealed that most target genes were successfully mutated with a frequency of 22%–85%. Although a small scale, this study showed in principle the feasibility of screening for CRISPR-induced mutations in vivo. Further optimization to reduce mosaicism and increase sgRNA mutagenic efficiency will enable larger scale screens to be carried out.

## Overexpression Screen to Identify Regulators of Sleep/Wake States

The availability of the Tol2 transposon system and Tol2 transposase[50] makes it possible to perform a primary screen of gene overexpression in the injected embryos/larvae. In a study aimed at identifying regulators of sleep/wake states,[29] a publicly available set of human open reading frames (ORF) that encode secreted proteins were cloned into the Tol2 vector under the control of a heat shock promoter. Individual plasmids were injected into one-cell stage embryos together with the transposase mRNA. For each ORF, 32 injected embryos were tested in the sleep/wake assay along with 16 matched controls injected with either hs:EGFP or an empty hs vector. The sleep/wake assay involves tracking larval zebrafish locomotor activity both pre- and post-heat shock, and parameters including number and length of sleep bouts, sleep latency at night, and changes in activity levels were quantified. After 1432 experiments that tested 1286 unique ORFs covering 1126 unique genes (29% of the human Secretome), 60 ORFs were selected to confirm in stable transgenic lines. Ten gave reproducible phenotypes as expected from the primary screen, and another two lines gave unexpected phenotypes. Among them, Neuromedin U (Nmu), when overexpressed, significantly increased locomotor activity. In situ hybridization of its orthologue in zebrafish revealed that the *nmu* gene is expressed in the hypothalamus, brainstem, and spinal cord, similar to the reported expression in mammals. Using genome-editing technologies, the authors generated mutations in the *nmu* gene, and its receptor encoding genes *nmur1a*, *1b*, and *2*. Analyses of these genetic mutations revealed complementary and at times redundant actions in arousal, activity, and body size. Further mechanistic studies uncovered the brainstem corticotropin-releasing factor neurons as a mediator of Nmu function in sleep/wake regulation.

## UNCOVERING CELLULAR AND MOLECULAR MECHANISMS THROUGH STUDYING NATURAL PHENOTYPIC VARIATIONS

Another unique resource to uncover cellular and molecular mechanisms underlying complex brain phenotypes is to study natural phenotypic variations, which already exist in the population, thus obviating the need to perform screens. Often, natural phenotypic variations represent traits that mimic complex human brain disorders.

Studies of natural phenotypic variations in model organisms such as *C. elegans*, *Drosophila*, honeybees, and mice have revealed important insights into behavioral regulation.[51–53] Similar studies have not been carried out in zebrafish, except some characterizations with behavioral lateralizations.[54]

We have recently discovered a natural behavioral variation in light/dark preference, which is associated with anxiety in larval zebrafish.[55] Preference for light or dark environment is an evolutionarily conserved phenomenon.[56,57] Adult zebrafish display a natural preference for dark, which is shown to be fear/anxiety related.[58,59] Intriguingly, larval zebrafish of around 1 week of age show opposite preference, dark aversion, which is also fear/anxiety related.[59–62] By testing 200 individual larvae derived from the wild-type laboratory strain AB, we found three phenotypic variations with respect to this behavior: a small subset of individuals displayed strong dark aversion (*sda*) with little variability across four trials, another small subset of individuals displayed variable dark aversion across the four trials (*vda*), whereas the majority showed medium level of dark aversion (*mda*). Selective breeding of these traits uncovered that sda and vda are heritable, with sda being recessive and vda being dominant to the common allele(s). The *sda* individuals also displayed heightened thigmotaxis behavior, another measure of anxiety-like behavior in larval zebrafish. Future molecular genetic studies shall reveal the cellular circuitry and molecular mechanisms underlying this natural behavioral variation.

## CONCLUSION

It is a daunting challenge to understand and treat complex brain disorders including autism, addiction, schizophrenia, anxiety, and depression. Molecular genetic studies of complex brain phenotypes employing screening in zebrafish as described in this chapter represent a promising approach that shall reveal new insights into brain development maintenance and function, and pave the way for tackling complex brain disorders in humans. The methods as described in this chapter have their own strengths and limitations. For instance, ENU screens are time consuming and labor intensive, so screening has been largely limited to phenotypes that can be very rapidly assessed. Also, mutations in small genes are underrepresented, and identifying the mutated genes requires laborious mapping and positional cloning. CRISPR screens can potentially be carried out with fewer generations to breed. Because labor is reduced, phenotypes that are relatively subtle or more time consuming to measure can be assayed. Since the target genes are already known, negative results become informative. Also, since genes that affect particular phenotypes are rare, multiple genes can be targeted in the same founder animals to increase throughput and efficiency. Together, the readers shall evaluate and decide which methods to employ based on specific interests. For all screening methods, having a robust, sensitive, and relatively high throughput assay is always an important prerequisite.

## ACKNOWLEDGMENTS

The work in authors' laboratory is supported by the US National Institute of Health (DA035680, NS095734, MH113961, DA044007, 2T32GM007175-41). We thank Michael Munchua, Hongbin Yuan, and Xingnu Zhai for excellent fish care.

## REFERENCES

1. Grunwald DJ, Eisen JS. Headwaters of the zebrafish-emergence of a new model vertebrate. *Nat Rev Genet.* 2002;3:717–724.
2. Guo S. Linking genes to brain, behavior, and neurological diseases: what can we learn from zebrafish? *Gene Brain Behav.* 2004;3:63–74.
3. Streisinger G, Walker C, Dower N, Knauber D, Singer F. Production of clones of homozygous diploid zebra fish (*brachydanio rerio*). *Nature.* 1981;291:293–296.

4. Kimmel CB. Patterning the brain of the zebrafish embryo. *Annu Rev Neurosci*. 1993;16:707–732.
5. Kimmel CB, Ballard WW, Kimmel SR, Ullmann B, Schilling TF. Stages of embryonic development of the zebrafish. *Dev Dyn*. 1995;203:253–310.
6. Haffter P, Granato M, Brand M, et al. The identification of genes with unique and essential function in the development of the zebrafish, *Danio rerio*. *Development*. 1996;123:1–36.
7. Driever W, Solnica-Krezel L, Schier AF, et al. A genetic screen for mutations affecting embryogenesis in zebrafish. *Development*. 1996;123:37–46.
8. Grunwald DJ, Streisinger G. Induction of recessive lethal and specific locus mutations in the zebrafish with ethyl nitrosourea. *Genet Res*. 1992;59:103–116.
9. Solnica-Krezel L, Schier AF, Driever W. Efficient recovery of ENU-induced mutations from the zebrafish germline. *Genetics*. 1994;136:1401–1420.
10. Guo S, Wilson SW, Cooke S, Chitnis AB, Driever W, Rosenthal A. Mutations in the zebrafish unmask shared regulatory pathways controlling the development of catecholaminergic neurons. *Dev Biol*. 1999; 208:473–487.
11. Cong L, Ran FA, Cox D, et al. Multiplex genome engineering using CRISPR/Cas systems. *Science*. 2013;339:819–823.
12. Mali P, Yang L, Esvelt KM, et al. RNA-guided human genome engineering via Cas9. *Science*. 2013; 339:823–826.
13. Jinek M, Chylinski K, Fonfara I, Hauer M, Doudna JA, Charpentier E. A programmable dual-RNA-guided DNA endonuclease in adaptive bacterial immunity. *Science*. 2012;337:816–821.
14. Li M, Zhao L, Page-McCaw PS, Chen W. Zebrafish genome engineering using the CRISPR-cas9 system. *Trends Genet*. 2016;32:815–827.
15. Hwang WY, Fu Y, Reyon D, et al. Efficient genome editing in zebrafish using a CRISPR-Cas system. *Nat Biotechnol*. 2013;31:227–229.
16. Jao LE, Wente SR, Chen W. Efficient multiplex biallelic zebrafish genome editing using a CRISPR nuclease system. *Proc Natl Acad Sci Unit States Am*. 2013;110:13904–13909.
17. Rossi A, Kontarakis Z, Gerri C, et al. Genetic compensation induced by deleterious mutations but not gene knockdowns. *Nature*. 2015;524:230–233.
18. Nasevicius A, Ekker SC. Effective targeted gene "knockdown" in zebrafish. *Nat Genet*. 2000;26: 216–220.
19. Fire AZ. Gene silencing by double-stranded RNA. *Cell Death Differ*. 2007;14:1998–2012.
20. Mello CC. Return to the RNAi world: rethinking gene expression and evolution. *Cell Death Differ*. 2007;14:2013–2020.
21. Guo S, Kemphues KJ. par-1, a gene required for establishing polarity in *C. elegans* embryos, encodes a putative ser/thr kinase that is asymmetrically distributed. *Cell*. 1995;81:611–620.
22. Fire A, Xu S, Montgomery MK, Kostas SA, Driver SE, Mello CC. Potent and specific genetic interference by double-stranded RNA in *Caenorhabditis elegans*. *Nature*. 1998;391:806–811.
23. Stark GR, Kerr IM, Williams BR, Silverman RH, Schreiber RD. How cells respond to interferons. *Annu Rev Biochem*. 1998;67:227–264.
24. Manche L, Green SR, Schmedt C, Mathews MB. Interactions between double-stranded RNA regulators and the protein kinase DAI. *Mol Cell Biol*. 1992;12:5238–5248.
25. Elbashir SM, Harborth J, Lendeckel W, Yalcin A, Weber K, Tuschl T. Duplexes of 21-nucleotide RNAs mediate RNA interference in cultured mammalian cells. *Nature*. 2001;411:494–498.
26. Dong Z, Peng J, Guo S. Stable gene silencing in zebrafish with spatiotemporally targetable RNA interference. *Genetics*. 2013;193:1065–1071.
27. Rørth P. A modular misexpression screen in Drosophila detecting tissue-specific phenotypes. *Proc Natl Acad Sci*. 1996;93:12318–12322.
28. Konermann S, Brigham MD, Trevino AE, et al. Genome-scale transcriptional activation by an engineered CRISPR-Cas9 complex. *Nature*. 2015;517:583–588.
29. Chiu CN, Rihel J, Lee DA, et al. Zebrafish genetic screen identifies Neuromedin U as a regulator of sleep/wake states. *Neuron*. 2016;89:842–856.
30. Shimoda N, Knapik EW, Ziniti J, et al. Zebrafish genetic map with 2000 microsatellite markers. *Genomics*. 1999;58:219–232.

31. Zhang J, Talbot WS, Schier AF. Positional cloning identifies zebrafish one-eyed pinhead as a permissive EGF-related ligand required during gastrulation. *Cell*. 1998;92:241–251.
32. Guo S, Brush J, Teraoka H, et al. Development of noradrenergic neurons in the zebrafish hindbrain requires BMP, FGF8, and the homeodomain protein Soulless/Phox2a. *Neuron*. 1999;24:555–566.
33. Guo S, Driever W, Rosenthal A. Mutagenesis in zebrafish: studying the brain dopamine systems. In: *Handbook of Molecular-Genetic Techniques for Brain and Behavior Research*. 1999:166–176 [chapter 2.1.8].
34. Guo S, Yamaguchi Y, Schilbach S, et al. A regulator of transcriptional elongation, which is required for vertebrate neuronal development. *Nature*. 2000;408:366–369.
35. Obholzer N, Swinburne IA, Schwab E, Nechiporuk AV, Nicolson T, Megason SG. Rapid positional cloning of zebrafish mutations by linkage and homozygosity mapping using whole-genome sequencing. *Development*. 2012;139:4380–4390.
36. Leshchiner I, Alexa K, Kelsey P, et al. Mutation mapping and identification by whole-genome sequencing. *Genome Res*. 2012;22:1541–1548.
37. Wagner BK, Schreiber SL. The power of sophisticated phenotypic screening and modern mechanism-of-action methods. *Cell Chem Biol*. 2016;23:3–9.
38. Keiser MJ, Roth BL, Armbruster BN, Ernsberger P, Irwin JJ, Shoichet BK. Relating protein pharmacology by their ligand chemistry. *Nat Biotechnol*. 2007;25:197–206.
39. Lee SA, Shen EL, Fiser A, Sali A, Guo S. The zebrafish forkhead transcription factor Foxi1 specifies epibranchial placode-derived sensory neurons. *Development*. 2003;130:2669–2679.
40. Levkowitz G, Zeller J, Sirotkin HI, et al. Zinc finger protein too few controls the development of monoaminergic neurons. *Nat Neurosci*. 2003;6:28–33.
41. Wang X, Yang N, Uno E, Roeder RG, Guo S. A subunit of the mediator complex regulates vertebrate neuronal development. *Proc Natl Acad Sci*. 2006;103:17284–17289.
42. Ryu S, Mahler J, Acampora D, et al. Orthopedia homeodomain protein is essential for diencephalic dopaminergic neuron development. *Curr Biol*. 2007;17:873–880.
43. Sun Y, Dong Z, Khodabakhsh H, Chatterjee S, Guo S. Zebrafish chemical screening reveals the impairment of dopaminergic neuronal survival by cardiac glycosides. *PLoS One*. 2012;7:e35645.
44. Thompson RF. Habituation: a history. *Neurobiol Learn Mem*. 2009;92:127–134.
45. Wolman M, Granato M. Behavioral genetics in larval zebrafish: learning from the young. *Dev Neurobiol*. 2012;72:366–372.
46. Wolman MA, Jain RA, Marsden KC, et al. A genome-wide screen identifies PAPP-AA-mediated IGFR signaling as a novel regulator of habituation learning. *Neuron*. 2015;85:1200–1211.
47. Golzio C, Willer J, Talkowski ME, et al. KCTD13 is a major driver of mirrored neuroanatomical phenotypes of the 16p11.2 copy number variant. *Nature*. 2012;485:363–367.
48. Blaker-Lee A, Gupta S, McCammon JM, De Rienzo G, Sive H. Zebrafish homologs of genes within 16p11.2, a genomic region associated with brain disorders, are active during brain development, and include two deletion dosage sensor genes. *Dis Model Mech*. 2012;5:834–851.
49. Shah AN, Davey CF, Whitebirch AC, Miller AC, Moens CB. Rapid reverse genetic screening using CRISPR in zebrafish. *Nat Methods*. 2015;12:535–540.
50. Kawakami K, Shima A, Kawakami N. Identification of a functional transposase of the Tol2 element, an Ac-like element from the Japanese medaka fish, and its transposition in the zebrafish germ lineage. *Proc Natl Acad Sci*. 2000;97:11403–11408.
51. Ben-Shahar Y, Robichon A, Sokolowski MB, Robinson GE. Influence of gene action across different time scales on behavior. *Science*. 2002;296:741–744.
52. Osborne KA, Robichon A, Burgess E, et al. Natural behavior polymorphism due to a cGMP-dependent protein kinase of Drosophila. *Science*. 1997;277:834–836.
53. de Bono M, Bargmann CI. Natural variation in a neuropeptide Y receptor homolog modifies social behavior and food response in *C. elegans*. *Cell*. 1998;94:679–689.
54. Facchin L, Argenton F, Bisazza A. Lines of *Danio rerio* selected for opposite behavioural lateralization show differences in anatomical left-right asymmetries. *Behav Brain Res*. 2009;197:157–165.
55. Wagle M, Nguyen J, Lee S, Zaitlen N, Guo S. Heritable natural variation of an anxiety-like behavior in larval zebrafish. *J Neurogenet*. 2017;31:138–148.
56. Bourin M, Hascoët M. The mouse light/dark box test. *Eur J Pharmacol*. 2003;463:55–65.

57. Gong Z, Liu J, Guo C, Zhou Y, Teng Y, Liu L. Two pairs of neurons in the central brain control Drosophila innate light preference. *Science*. 2010;330:499–502.
58. Maximino C, Marques de Brito T, Dias CA, Gouveia AJ, Morato S. Scototaxis as anxiety-like behavior in fish. *Nat Protoc*. 2010;5:209–216.
59. Lau BYB, Mathur P, Gould GG, Guo S. Identification of a brain center whose activity discriminates a choice behavior in zebrafish. *Proc Natl Acad Sci USA*. 2011;108:2581–2586.
60. Chen F, Chen S, Liu S, Zhang C, Peng G. Effects of lorazepam and WAY-200070 in larval zebrafish light/dark choice test. *Neuropharmacology*. 2015;95:226–233.
61. Steenbergen PJ, Richardson MK, Champagne DL. Patterns of avoidance behaviours in the light/dark preference test in young Juvenile zebrafish: a pharmacological study. *Behav Brain Res*. 2011;222:15–25.
62. Bai Y, Liu H, Huang B, Wagle M, Guo S. Identification of environmental stressors and validation of light preference as a measure of anxiety in larval zebrafish. *BMC Neurosci*. 2016;17:63.

# CHAPTER 17

# The Transition of Zebrafish Functional Genetics From Random Mutagenesis to Targeted Integration

**Louis Y. El Khoury[a], Jarryd M. Campbell[a], Karl J. Clark**
Mayo Clinic, Rochester, MN, United States

## INTRODUCTION

With the completion of the human genome project in 2003, it was thought that the newly read genetic sequences would bring along a clear explanation of genotype-phenotype relationships.[1] Despite knowing the sequence of the genome, the function of most genes and their contribution to human health or disease are not well understood. The surge in whole exome and whole genome sequencing is revealing a significant number of variants of unknown significance (VUS) in genes that could be associated with human disease. Moreover, the heritability of phenotypes—whether anatomic, physiologic, or behavioral—appears to be much more complicated than what was previously thought, with many diseases being influenced by multiple genes (polygenic) or epigenetic modifications that are not identified by traditional DNA sequencing. To help draw a clearer image of the role played by these genes and particular variants in diseases, functional genetic studies in model organisms are needed. Sophisticated genome engineering and gene-editing tools are rapidly being developed to enable the targeted introduction of mutations of interest that permit the analysis of their downstream effects on biologic functioning in model organisms like the zebrafish.

Alteration of genetic sequences in biologic models for the purpose of understanding the downstream effect on phenotype has been reported since the 1970s when Cohen et al.[2] transformed *Escherichia coli* with exogenous DNA to understand antibiotic resistance. In mammals, the first successful genetic modification was in 1974 when SV40 DNA was injected into mouse blastocysts to investigate the mechanisms of vertical transmission of tumor viruses.[3] The early development of methods to genetically modify the mouse is one of several reasons why this model organism is so widely used. However, the cost of maintaining mouse facilities and the complexity of the model highlight the impracticality of using only this system to understand gene function and human VUSs. On the other hand, zebrafish with their optically clear embryos, high fecundity, rapid

[a] These two authors contributed equally.

*Molecular-Genetic and Statistical Techniques for Behavioral and Neural Research*
ISBN 978-0-12-804078-2, https://doi.org/10.1016/B978-0-12-804078-2.00017-9

development, and short generation time, are an appropriate model for rapid and high-throughput screening of gene function or the impact of genetic variants. More importantly, at least 70% of human genes have at least one zebrafish orthologue that can be utilized to produce defects in a variety of organs similar to human diseases.[4] The zebrafish, as a model organism, has substantially contributed to our understanding of vertebrate biology.

This chapter will explore and highlight the advantages and limitations of various random and targeted mutagenesis tools commonly used on zebrafish models. Furthermore, it will suggest that combining highly mutagenic DNA sequences used for random mutagenesis with targeted nucleases will be an important next phase in functional genetic studies using the zebrafish model.

## ZEBRAFISH RANDOM MUTAGENESIS

Random mutagenesis encompasses a number of tools and techniques that generate mutations randomly in genomes of interest. For instance, the chemical mutagen *N*-ethyl-*N*-nitrosourea (ENU) has been a key driver of random mutagenesis in multiple model organisms, including zebrafish.[5] ENU treatment of male zebrafish can result in thousands to millions of single base pair changes throughout the zebrafish genome, which result in a high frequency of mutant phenotypes.[6] Indeed, a number of large-scale chemical mutagenesis screens produced a remarkable collection of zebrafish mutants in a breadth of biologic processes.[7,8] In spite of the high mutagenesis efficiency that ENU provides, it brings along a major limitation in one's ability to identify the single point variant responsible for a particular phenotype, simultaneously creating tens to hundreds of linked single nucleotide polymorphisms.[9]

In an effort to address the difficulties associated with identifying causal genes following chemical mutagenesis, researchers turned to various methods of insertional mutagenesis. When DNA is integrated into the genome, it can serve as a landmark to identify disrupted DNA. Although insertional mutagenesis is not as efficient as chemical mutagenesis, affected genes can be rapidly and easily cloned by various PCR methods that can identify the junction sequences between the insertional mutagen and the disrupted gene.

The development of pseudo-typed retroviruses for insertional mutagenesis[10] was an early alternative to chemical mutagenesis. Virus integration creates mutations, as discussed earlier, by directly disrupting exons or interfering with expression after integration into promoters or early introns that can disrupt transcription factors recognition sites or RNA processing. Similar to ENU, large-scale screening of retroviral insertions can identify mutations, including specific and nonspecific phenotypes.[11] In zebrafish, nonspecific mutations compromise genes necessary for cell survival and growth. On the other hand, specific mutations have resulted in abnormal pigmentation and circulation, as well as brain and body size.[6,12]

## Transposon Systems in Zebrafish

Transposons are repetitive DNA sequences capable of modifying host genomes. Transposons were discovered in the 1940s, and they make up a considerable portion of a eukaryotic genome: 50% of the human genome.[13]

Transposons are divided into two classes: class I, also called retrotransposons, are first transcribed into RNA and following reverse transcription can be reintroduced back into DNA. The new DNA copy can be inserted in the genome at a location distant from the original sequence. Unlike retrotransposons, class II transposons do not involve reverse transcription. Rather, type II or DNA transposons move via a cut-and-paste mechanism when the self-encoded transposase protein recognizes its binding sites at the termini of the DNA transposon, cuts at these sites thereby liberating the transposon, and it actively integrates the transposon to a new random location. In vertebrates, including the zebrafish, the first widely used, active DNA transposon was a Tc1-family transposon system called *Sleeping Beauty* (SB) that was reverse engineered from nonfunctional transposon sequences following the reverse engineering and elimination of accumulated mutations in transposon elements identified in teleost genomes.[14] Some of the features of SB that contributed to its popularity include its ability to efficiently and cleanly integrate DNA,[15] the production of vectors at a relatively low cost, and the reliable expression of the inserted cassette over the long term.[16] Despite advances in function, the SB transposon system failed to find a permanent place within the zebrafish research community[17] due to its delayed integration kinetics relative to other transposon systems.

SB was quickly replaced as the primary gene transfer platform by a more robust tool with rapid transposition kinetics that led to high germline integration rates and low mosaicism in somatic tissues. In the 1990s, the *Tol2* element was isolated from the Medaka *tyrosinase* locus as an autonomous mobile element, and the subsequent production of several cloning vectors has facilitated the use of this element in zebrafish.[18] An example of that would be the usage of *Tol2* mRNA in the creation of germline mutations in zebrafish embryos.[19,20] Variations in the *Tol2* vectors include a reduction in the length of the inverted terminal repeats (ITR), making expression vectors smaller by removing undefined regulatory elements such as promoter, splicing, or polyadenylation signals within the terminal repeats.[21,22] The early success of *Tol2* transposons among zebrafish researchers facilitated the creation of the zebrafish *Tol2* toolbox, which combines vectors from multiple sources including mini inverted repeat transposons and *Tol2* transposase transcription vectors. The *Tol2* transposon system became a routine genetic engineering tool in the zebrafish community thanks to these

*Tol2* kits, which included a combination of multiple plasmids and fluorescent reporter proteins, and the online "user community" platform (http://tol2kit.blogspot.com). In addition to the original Tol2Kit, there are a number of *Tol2* simple cloning vectors available, such as the pT2AL200R150G, pminiTol2, and pKTol2-SE.[17,21,22] These vectors use minimized ITRs that reduce the size of the plasmid. Choosing between the vectors comes down to suitability and the ease with which a cassette for zebrafish transgenesis can be generated.

### Genetic "Traps" via Enhanced Mutagenic Cargo

Most random DNA integrations in vertebrates, like the zebrafish, are either intergenic or within introns. This is simply due to the small percentage of the genome represented by exons. Therefore, regardless of the method of integration, researchers have a long history of trying to enhance either the signal or mutagenic capacity of integrations in or near genes. These elements were commonly named "traps," with modifiers based on what signal they were reporting. For example, an enhancer trap would not produce a reporter signal unless a nearby enhancer activated the reporter.[23,24] Since enhancers act at a distance, integration does not have to be within a gene, but nearby. Gene traps were designed to be spliced into a transcribed gene[25]; protein traps rely on splice acceptor or splice donor sequences that result in an in-frame fusion between a reporter molecule, like GFP, and the nascent peptide[26]; and poly(A) traps require splicing to downstream poly(A) signals that stabilize the resultant transcript.[27]

As pseudo-typed retroviruses and transposons were developed for efficient insertional mutagenesis in zebrafish, modifications to the cargo within the integrated DNA were developed to improve visualization of expression and/or mutagenesis potential. Initial efforts to incorporate gene trap modifications to pseudo-typed retroviruses had limited effectiveness, as inclusion of reporter molecules like green fluorescent protein and splicing signals had a dramatic effect on retroviral titer, which subsequently impacted integration efficiency. In the end, simple retroviral integrations were used for genome-wide integration screens.[12,28]

Modification to the cargo of DNA transposons, on the other hand, does not significantly alter the transgenesis efficiency. So many trapping technologies were tested and developed using transposons. Though there have been several successful traps developed and published,[29–31] two groups have produced large shared resources. Dr. Koichi Kawakami's workgroup has developed a large resource, zTrap,[22] of zebrafish strains using various *Tol2*-based enhancer and gene traps. Our lab together with Dr. Stephen Ekker's workgroup has used *Tol2* gene-break transposons to create our zfishbook resource.[17,21] The gene-breaking transposon, pGBT-RP2 or pGBT-RP8 series of vectors (Fig. 17.1), disrupt and knockdown gene expression with high efficiency. The vectors contain two trap domains: a 5′ protein trapping and a 3′ poly(A) trap cassette. The 5′ protein trapping cassette consists of a splice acceptor upstream of the red

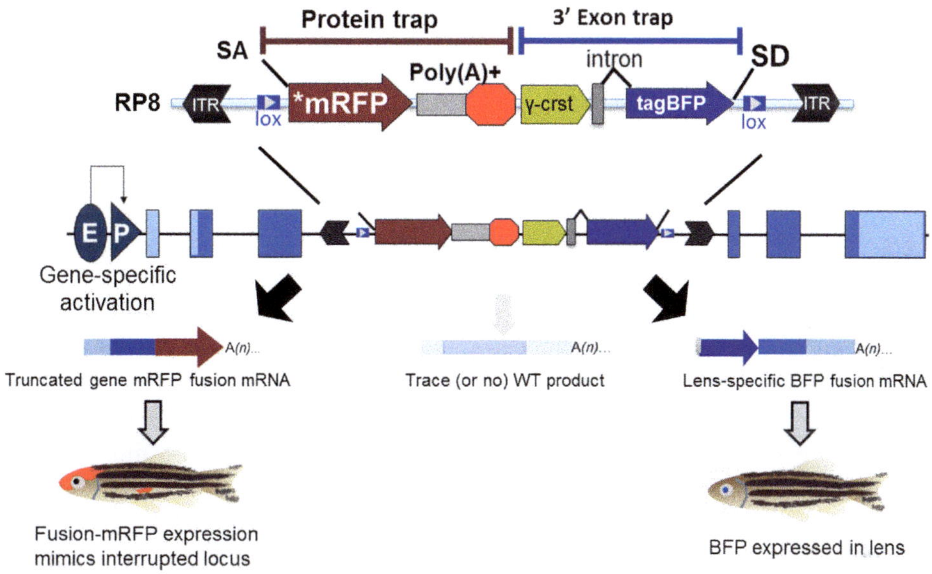


**Figure 17.1** Schematic of the RP8 gene-break transposon system composed of a protein-trap cassette with a transcriptional stop and 3′ exon trap cassette. The interrupted wild-type mRNA is either not produced or is produced in trace amounts. *ITR*, inverted terminal repeat; *SA*, splice acceptor; **mRFP*, AUG-less mRFP sequence; *poly(A)+*, polyadenylation signal with extra transcriptional terminator and putative border element; *γ-crst*, γ-crystallin promoter; *SD*, splice donor; *E*, enhancer; *P*, promoter; and *WT*, wild type.

fluorescent protein (mRFP) reporter lacking an initiation codon that thereby requires in-frame splicing to an upstream protein-coding exon. The protein trap mutates the gene, and it generates an expression profile that mimics the disrupted gene. The 3′ poly(A) trap consists of a secondary reporter. In the pGBT-RP2 series, the nearly ubiquitous b-actin promoter drives expression of GFP, whereas in the pGBT-RP8 series, the lens-specific gamma-crystallin promoter drives expression of tagBFP. In each case the reporter is followed by a downstream splice donor. For high-level reporter activation, the splice donor must splice into downstream exons that include a poly(A) signal. The 3′ poly(A) traps are useful for assessing injection efficiency during line production and are a useful alternative for identifying transgenic fish with a transposon in a transcription unit in the sense orientation regardless of in-frame mRFP fusion.

Despite the successful development of retroviral or transposon insertional mutagens, their major limitation is the fact that their integration into the genome is essentially random. Therefore, it is not practical for an individual lab to use these resources to create a mutation in their gene of interest. Instead, the community must rely on resource generators and luck to identify one of these inserts in their gene of interest, and then it may not be the ideal cargo for their biologic question. Therefore, the development of successful reverse genetic approaches targeted to a gene of interest has become the go-to approach for most zebrafish labs.

## TARGETED MODIFICATIONS

### TILLING

As the number of random mutagenic tools increased, so did the desire to acquire directed and target-specific techniques. In the year 2000, Targeting Induced Local Lesions in Genomes (TILLING) was introduced in *Arabidopsis* as the first targeted mutagenesis tool and since has spread to a variety of organisms.[32] TILLING combines random chemical mutagenesis of genomic DNA with precise genetic screening to identify point mutations in specific genes of interest. It was believed to be fast and cost friendly, and it was therefore implemented in high throughput in zebrafish.[33] The most significant problem with TILLING is the level of off-target point mutations; in an F1 generation, there is approximately one mutation for every 3−6000 base pairs.[34] Nonetheless, TILLING provided for the first time the ability to rapidly generate specific mutations in zebrafish, and these alleles are still valuable to many labs today.

### ZFNs

TILLING combined forward mutagenesis with rapid genetic screening, but the method utilizing zinc finger nucleases (ZFNs) was really the first to introduce locus-specific mutagenesis for targeted gene editing. ZFNs are artificial restriction enzymes made by the fusion of zinc finger binding domains and the nuclease domain of FokI.[35] They generally consist of an array of three to six Cys2-His2 zinc fingers, each recognizing and binding to a 3 bp segment of the DNA.[36] Each zinc finger domain binds to three consecutive nucleotides, and different zinc finger domains recognize and bind to different DNA triplets.[37] Two ZFNs are required for each locus. Once the zinc fingers bind to their respective regions, which are separated by 5−6 bp of DNA, a double-stranded break (DSB) is initiated by the FokI domain.[38]

ZFN is a very flexible system; it can be used in a variety of organisms.[39−42] Following initial synthesis and use of ZFNs, they were used in zebrafish to make targeted modifications.[43,44] These papers demonstrated that DSBs created by ZFNs in zebrafish are efficiently repaired through nonhomologous end joining (NHEJ). Imprecise repair by NHEJ often results in short indels that disrupt the reading frame of the targeted gene to create a genetic lesion. However, there are significant drawbacks to ZFNs. First, they cannot be treated as a universal targeting tool since not all DNA triplets can be recognized by zinc finger domains. Also, despite kits that made production of ZFNs more cost effective,[45] identifying the best ZFNs still required some screening as zinc fingers demonstrated neighbor effects.[46] As other gene-editing technologies became available, labs moved away from ZFNs or entered the field using alternative gene-editing technologies due to the financial and time commitments of ZFNs relative to other gene-editing methods. Despite these shortcomings, ZFNs laid the blueprint for targeted gene editing as we know it today.

## TALENs

Like ZFNs, transcription activator-like effector nucleases (TALENs) are a protein-based gene-editing tool made of a DNA-binding domain tethered to a FokI endonuclease. While ZFN recognition of DNA is complex, TALENs are simple and predictable. The DNA-binding domain of TALEs comprises multiple units of 33–35 amino acids (called TALE repeats) arranged in tandem. Two of these amino acids are variable [residues 12 and 13, called the repeat variable di-residue (RVD)] and define the nucleotide binding of a single TALE motif.[47] By knowing the sequence of the locus, you can assemble a specific TALEN pair using this one-to-one code. The biggest targeting constraint is a 5′ thymine that should be adjacent to each TALEN arm for efficient gene editing, though modified TALEs exist that can avoid even this requirement.[48]

TALENs are a great tool for introducing DSBs with the purpose of creating targeted indels in zebrafish genes. Upon their discovery, interest in TALENs spiked because they are easier and quicker to design by nonspecialist researchers than ZFNs.[49,50] TALENs can be very efficient in zebrafish, often times modifying both alleles of a targeted gene somatically.[51] This boost in efficiency also allowed zebrafish labs to create targeted knock-ins, a feat that was not demonstrated with ZFNs.[51,52] Furthermore, they have delivered more efficient gene editing on a wider scale than ZFNs and in a variety of model organisms.

The major drawbacks of TALENs are their size and to a lesser extent their availability. TALENs are significantly larger than ZFNs, which make them more difficult to deliver to many different cell types. While TALENs are much more available than ZFNs, they still require a special plasmid kit and molecular biology experience to make the final TALEN efficiently. Many labs make TALENs in-house, but purchasing TALENs is still cost-prohibitive for others. For this reason, the introduction of CRISPR/Cas9 in 2012 was a game-changer for gene editing.

## CRISPR

CRISPR/Cas9 made gene editing available to nearly any laboratory. One reason for this is that the DNA recognition motif switched from being protein-based in ZFNs and TALENs to being nucleic acid–based. The CRISPR/Cas9 system used today is made up of a single guide RNA (sgRNA) molecule, which provides the sequence specificity by Watson-Crick base pairing of RNA to the target DNA, and a Cas9 endonuclease that initiates a break in the DNA sequence. This switch from protein to RNA-targeting has allowed research labs that did not have access to ZFNs or TALENs to either synthesize the targeting domain, make gRNA with an in vitro transcription kit, or have the full sgRNA synthesized. Cas9 transcription plasmids rapidly became available followed shortly by purified Cas9 protein from commercial sources, eliminating the need for any in-house production of reagents. Today, nearly anyone that wants to try gene editing can do it using CRISPR/Cas9.

The CRISPR/Cas9 system is capable of executing a similar variety of genetic edits as ZFNs and TALENs. More so than either ZFNs or TALENs before, the CRISPR/Cas9 system was adopted by the zebrafish research community, and it is now the most commonly used gene-editing tool.[53] Like TALENs, CRISPR/Cas9 can have high activity in zebrafish and can be used both for the production of mutants via targeted introduction of indels following NHEJ repair of a double-strand break and for targeted integration using homology-directed repair (HDR).[54] One distinguishing strength of CRISPR editing is the ease of multiplexing, i.e., editing more than one locus in a single reaction.[54–57]

The fidelity and off-target effects of gene-editing tools have always been a matter of concern, and CRISPRs are not immune to that. In fact, a number of studies published shortly after CRISPR gained popularity reported notable mismatches resulting from sequence homology with genomic regions.[58,59] This led to the development of methods to reduce off-target effects such as modifying the design of the sgRNA by truncating the 3′ end or adding target-independent guanine residues to the 5′ end.[60,61] Fu et al.[62] went further to propose that the reduction of the length of the binding region can increase the specificity of sgRNA without compromising the on-target activity. It is generally believed that CRISPR/Cas9 has more off-target potential than either TALENs or ZFNs. Nonetheless, the simplicity of CRISPR has still made it the editor of choice for most labs, and improvements are continuing to be made.

## APPLICATIONS OF GENE EDITING

ZFNs, TALENs, and CRISPR/Cas9 all rely on the cellular response to DSBs for targeted modifications. Whether it is a chemical or enzymatic DSB, the cell is able to identify and respond by mounting a robust double-strand break repair (DSBR). Our DNA relies on these DSBRs to maintain genomic integrity, and we can capitalize on these repair pathways for gene editing. The most common pathways utilized are NHEJ and HDR, which includes homologous recombination (HR).[63]

## Using Nonhomologous End Joining for Site-Directed Mutagenesis

NHEJ is the major DSB repair mechanism since it takes place in both dividing and nondividing cells. NHEJ can be error prone when re-ligating the free ends, so small insertions or deletions (indels) are often made at the lesion site. Imprecise repair of the DSB via NHEJ is very efficient in zebrafish and other model systems, and this pathway is commonly used to create targeted mutations. By using a ZFN, TALEN, or CRISPR/Cas9, indels can be targeted to most places in the genome and are commonly used to make a frameshift in the coding region of a protein-coding gene. Since NHEJ is efficient in fish, many labs are capable of generating knockout lines in-house within a year of starting a project. Additionally, a subset of TALENs or CRISPR/Cas9 will cut both copies of the gene in nearly every cell of an injected animal,[51,54] which can be useful for examining potential phenotypes in injected animals without needing to make a line.

It should be noted that there are alternative uses for NHEJ. By inducing two DSBs, a large deletion can be made in-between the cut sites (Fig. 17.2). This can be used to delete entire coding regions to ensure no transcript is made.[64,65] This same method can also be used to induce chromosomal rearrangements.[66] Finally, targeted knock-ins have been demonstrated using NHEJ.[67] NHEJ-based knock-ins still have indels at both the 5′ and 3′ ends of the integration site, so more precision is sometimes required.

## Using Homology-Directed Repair for Targeted Integration

HDR is a DSBR that uses a double-stranded DNA donor that has homology to the adjacent sequences surrounding the lesion to incorporate new DNA fragments. HDR offers more precision than NHEJ and allows for seamless integration of DNA. The earliest studied HDR pathway is HR. HR uses long, double-strand DNA that is homologous to around 1 kilobase of sequence on either side of the DSB. These homologous DNA donors can come from the sister chromatid, which is used during crossover in meiosis I, or from exogenous sources of DNA like plasmids or PCR products.[68] While HR can be deployed for making exact integrations without indels, it is also inefficient.[52] In cellular models, this can be overcome with selection, like drug resistance or fluorescence. This advantage is lost when attempting to engineer a living animal like a zebrafish, so a more efficient system is required to make knock-ins feasible. It had previously been shown that HDR can be stimulated up to 1000-fold after a DSB in mouse embryonic stem cells,[69] and the boost in efficiency from ZFNs to TALENs was critical for knock-ins in fish. In this regard, Shin et al.[74] reported successful HR in zebrafish using TALENs; specifically, they found that greater

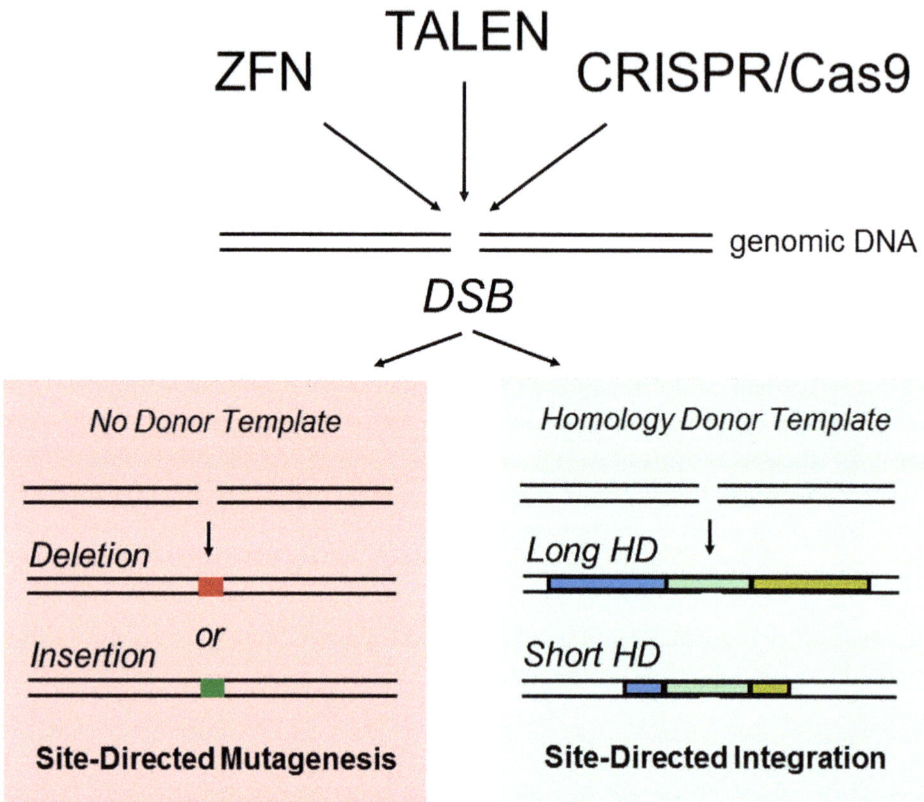


**Figure 17.2** Genomic repair and recombination events are initiated by double-strand breaks (DSBs) induced by targeted cleavage of the genome with programmable nucleases (ZFN, TALEN, CRISPR/Cas9). When template DNA is absent from directing the repair process, the nonhomologous end-joining (NHEJ) pathway heals broken ends of chromosomes in a process that often generates small deletions (*red*) and/or insertions (*green*) at the site of the lesion. When a donor template with close homology to the site of the DSB is present, the donor sequence may guide homology-directed repair (HDR), resulting in a less error prone modification of host sequences. The length of homology can be variable from very short (s) ends of 8–50 bp or traditionally long (L) homology domains of 500–2000 nt.

than 10% of injected F0 fish successfully transmitted inserts through the germline. While HR in fish has been shown possible, it is still not efficient enough to be used routinely and therefore requires further improvement to achieve desirable results.

There are alternative HDR integration methods that can be deployed that are more efficient than HR. One option is to use short, single-stranded oligodeoxynucleotides (ssODNs) that harbor small homology arms (<100 bp). This method can be used to make small integrations like point mutations or loxP sites, but it has yet to be used for DNA fragments larger than around 50 bp. While it is more efficient, it is not always precise.[51] This imprecision makes identification of a precise integration difficult, and therefore the use of ssODN-based HDR in exons is not routine. In introns, where the exact sequence outcome is not as critical, this method is still useful, for example, to integrate modified attP sites that can act as landing pads for larger DNA sequence integration via recombinases.[70]

Another integration technique is to use microhomology-mediated end joining (MMEJ) as the mechanism for knocking-in donor DNA. Like HR, MMEJ uses double-strand DNA as templates for repair. Instead of long (>1 kilobase) donor sequences, MMEJ requires short homology sequences (up to 40 bp).[71,72] Early results indicate somatic integration in up to 77% of injected animals, but the germline transmission of these integrations drops by over tenfold.[71] Although not all integrations are precise at both the 5′ and 3′ integration sites using this technique, there seems to be a higher proportion of precise integration compared with NHEJ-based or ssODN-based knock-ins. Further mechanisms and more universal protocols need to be established for MMEJ-based integration, but this technique holds potential for efficient animal knock-ins.

## Mutagenic Trap Cassettes: Novel Candidates for Targeted Integration

Before 2012, there were no published techniques for knock-ins in zebrafish. Since the rate of change in the zebrafish field has been exponential, today, most zebrafish labs will make indels in the coding region of their gene of interest to knockout a gene. However, many powerful mutagenic cassettes have been developed using random integration. For the first time, we may be able to precisely position them anywhere in the genome. For example, gene traps like RP2 or SAG[73] are more valuable than indel-based alleles because (1) knockout of the complete wild-type transcript can be reduced to below 1% of normal and eliminate many sources of *cis*-complementation like alternative transcriptional start sites, splicing, or alternative translation initiation. (2) The resulting alleles are fluorescent, allowing fish to be genotyped noninvasively, and (3) they can be conditional, either revertible back to wild type or turned off with Cre. As the germline transmission efficiency is improved, many labs will switch from making static alleles via NHEJ-based indels to integrating these mutagenic cassettes.

To push forward with this line of research, our group is using the RP2-based MMEJ integration vector to insert mutagenic cassettes in a selection of zebrafish genes. We envision that creating a DSB at a specific location using TALENs or CRISPR/Cas9 will allow the deposition of the GBT cargo at the site of the break (Fig. 17.3). In addition to the pGBT-RP2 or RP8 series shown here, there is a large assortment of mutagenic vectors available that could now be used together with custom nucleases to target any gene of interest.

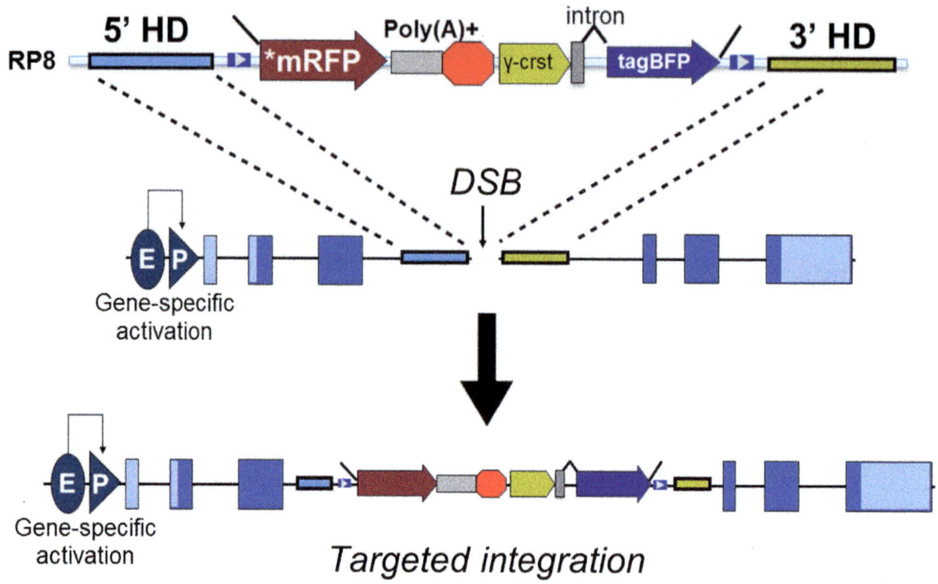


**Figure 17.3** Schematic of targeted integration of the RP8 mutagenic cassette at the site of the double-stranded break (DSB). Homology arms can be placed around the core mutagenic cassette of an insertional mutagen, in this example, RP8. Targeted integration results in the interrupted wild-type mRNA either not being produced or being produced in trace amounts. *Blue squares with grey triangles* show loxP sites flanking the mutagenic cassettes. **mRFP*, AUG-less mRFP sequence; *poly(A)+*, polyadenylation signal with extra transcriptional terminator and putative border element; *tagBFP*, blue fluorescent protein; *E*, enhancer; *P*, promoter.

## CONCLUSION

We presented in this chapter the general evolution from random to targeted mutagenesis and a variety of gene editing used for targeted mutagenesis in zebrafish. Furthermore, we illustrated the usefulness of designed mutagenic vectors, such as the pGBT-RP2 and pGBT-RP8, containing fluorescent reporter molecules that historically, through random integration, provide gene expression information, high-quality mutants, and conditional alleles. New advances in gene-editing technology will permit the coupling of designed mutagenic vectors with targeted integration into genes of interest. Zebrafish are a readily modifiable vertebrate animal model, which will be heavily relied on to optimize existing and to test novel gene-editing tools. At a time when the popularity of genetic engineering is growing exponentially, it is important to keep in mind the colossal impact this discipline will have on the understanding of the genetic basis of human diseases and on the field of gene therapy.

## REFERENCES


1. Consortium T 1000 GP. A map of human genome variation from population-scale sequencing. *Nature*. 2010;467(7319):1061–1073. https://doi.org/10.1038/nature09534.
2. Cohen SN, Chang AC, Hsu L. Nonchromosomal antibiotic resistance in bacteria: genetic transformation of *Escherichia coli* by R-factor DNA. *PNAS*. 1972;69(8):2110–2114. https://doi.org/10.1073/pnas.69.8.2110.
3. Jaenisch R, Mintz B. Simian virus 40 DNA sequences in DNA of healthy adult mice derived from preimplantation blastocysts injected with viral DNA. *Proc Natl Acad Sci USA*. 1974;71(4):1250–1254. http://www.pnas.org/content/71/4/1250.abstract.
4. Howe K, Clark MD, Torroja CF, et al. The zebrafish reference genome sequence and its relationship to the human genome. *Nature*. 2013;496(7446):498–503. https://doi.org/10.1038/nature12111.
5. Solnica-Krezel L, Schier AF, Driever W. Efficient recovery of ENU-induced mutations from the zebrafish germline. *Genetics*. 1994;136(4):1401–1420. http://www.genetics.org/content/136/4/1401.abstract.
6. Patton EE, Zon LI. The art and design of genetic screens: zebrafish. *Nat Rev Genet*. 2001;2(12): 956–966. https://doi.org/10.1038/35103567.
7. Driever W, Solnica-Krezel L, Schier AF, et al. A genetic screen for mutations affecting embryogenesis in zebrafish. *Development*. 1996;123(1):37–46. doi:9007227.
8. Haffter P, Granato M, Brand M, et al. The identification of genes with unique and essential functions in the development of the zebrafish, *Danio rerio*. *Development*. 1996;123(1). http://dev.biologists.org/content/123/1/1.short.
9. Talbot WS, Hopkins N. Zebrafish mutations and functional analysis of the vertebrate genome. *Genes Dev*. 2000;14(7):755–762. https://doi.org/10.1101/GAD.14.7.755.
10. Lin S, Gaiano N, Culp P, et al. Integration and germ-line transmission of a pseudotyped retroviral vector in zebrafish. *Science*. 1994;265(5172):666–669. https://doi.org/10.1126/science.8036514.
11. Amsterdam A, Burgess S, Golling G, et al. A large-scale insertional mutagenesis screen in zebrafish. *Genes Dev*. 1999;13(20):2713–2724. http://www.ncbi.nlm.nih.gov/pubmed/10541557.
12. Varshney GK, Lu J, Gildea DE, et al. A large-scale zebrafish gene knockout resource for the genome-wide study of gene function. *Genome Res*. 2013;23(4):727–735. https://doi.org/10.1101/gr.151464.112.
13. Cordaux R, Batzer MA. The impact of retrotransposons on human genome evolution. *Nat Rev Genet*. 2009;10(10):691–703. https://doi.org/10.1038/nrg2640.
14. Ivics Z, Hackett PB, Plasterk RH, Izsvák Z. Molecular reconstruction of Sleeping Beauty, a Tc1-like transposon from fish, and its transposition in human cells. *Cell*. 1997;91(4):501–510. http://www.ncbi.nlm.nih.gov/pubmed/9390559.
15. Izsvák Z, Stüwe EE, Fiedler D, Katzer A, Jeggo PA, Ivics Z. Healing the wounds inflicted by sleeping beauty transposition by double-strand break repair in mammalian somatic cells. *Mol Cell*. 2004;13(2): 279–290. http://www.ncbi.nlm.nih.gov/pubmed/14759372.
16. Balasubramanian S, Rajendra Y, Baldi L, Hacker DL, Wurm FM. Comparison of three transposons for the generation of highly productive recombinant CHO cell pools and cell lines. *Biotechnol Bioeng*. 2016; 113(6):1234–1243. https://doi.org/10.1002/bit.25888.
17. Clark KJ, Urban MD, Skuster KJ, Ekker SC. Transgenic zebrafish using transposable elements. *Methods Cell Biol*. 2011;104:137–149. https://doi.org/10.1016/B978-0-12-374814-0.00008-2.
18. Koga A, Suzuki M, Inagaki H, Bessho Y, Hori H. Transposable element in fish. *Nature*. 1996; 383(6595):30. https://doi.org/10.1038/383030a0.
19. Kawakami K, Shima A, Kawakami N. Identification of a functional transposase of the Tol2 element, an Ac-like element from the Japanese medaka fish, and its transposition in the zebrafish germ lineage. *Proc Natl Acad Sci USA*. 2000;97(21):11403–11408. https://doi.org/10.1073/pnas.97.21.11403.
20. Kawakami K, Shima A. Identification of the Tol2 transposase of the medaka fish *Oryzias latipes* that catalyzes excision of a nonautonomous Tol2 element in zebrafish *Danio rerio*. *Gene*. 1999;240(1): 239–244. https://doi.org/10.1016/S0378-1119(99)00444-8.
21. Balciunas D, Wangensteen KJ, Wilber A, et al. Harnessing a high cargo-capacity transposon for genetic applications in vertebrates. *PLoS Genet*. 2006;2(11):1715–1724. https://doi.org/10.1371/journal.pgen.0020169.

22. Urasaki A, Morvan G, Kawakami K. Functional dissection of the Tol2 transposable element identified the minimal cis-sequence and a highly repetitive sequence in the subterminal region essential for transposition. *Genetics*. 2006;174(2):639–649. https://doi.org/10.1534/genetics.106.060244.
23. Riesgo-Escovar J, Woodard C, Gaines P, Carlson J. Development and organization of the Drosophila olfactory system: an analysis using enhancer traps. *J Neurobiol*. 1992;23(8):947–964. https://doi.org/10.1002/neu.480230803.
24. Bayer TA, Campos-Ortega JA. A transgene containing lacZ is expressed in primary sensory neurons in zebrafish. *Development*. 1992;115(2):421–426. http://www.ncbi.nlm.nih.gov/pubmed/1425333.
25. Wurst W, Rossant J, Prideaux V, et al. A large-scale gene-trap screen for insertional mutations in developmentally regulated genes in mice. *Genetics*. 1995;139(2):889–899.
26. Morin X, Daneman R, Zavortink M, Chia W. A protein trap strategy to detect GFP-tagged proteins expressed from their endogenous loci in Drosophila. *Proc Natl Acad Sci USA*. 2001;98(26): 15050–15055. https://doi.org/10.1073/pnas.261408198.
27. Dupuy AJ, Fritz S, Largaespada DA. Transposition and gene disruption in the male germline of the mouse. *Genesis*. 2001;30(2):82–88. https://doi.org/10.1002/gene.1037.
28. Wang D, Jao L-EE, Zheng N, et al. Efficient genome-wide mutagenesis of zebrafish genes by retroviral insertions. *Proc Natl Acad Sci USA*. 2007;104(30):12428–12433. https://doi.org/10.1073/pnas.0705502104.
29. Friedrich G, Soriano P. Promoter traps in embryonic stem cells: a genetic screen to identify and mutate developmental genes in mice. *Genes Dev*. 1991;5(9):1513–1523. https://doi.org/10.1101/gad.5.9.1513.
30. Korn R, Schoor M, Neuhaus H, et al. Enhancer trap integrations in mouse embryonic stem cells give rise to staining patterns in chimaeric embryos with a high frequency and detect endogenous genes. *Mech Dev*. 1992;39(1–2):95–109. https://doi.org/10.1016/0925-4773(92)90029-J.
31. Stanford WL, Caruana G, Vallis KA, et al. Expression trapping: identification of novel genes expressed in hematopoietic and endothelial lineages by gene trapping in ES cells. *Blood*. 1998;92(12):4622–4631. https://doi.org/10.1016/S0925-4773(99)00234-8.
32. Henikoff S, McCallum CM, Comai L, Greene EA. Targeted screening for induced mutations. *Nat Biotechnol*. 2000;18(4):455–457. https://doi.org/10.1038/74542.
33. Kettleborough RNW, Busch-Nentwich EM, Harvey SA, et al. A systematic genome-wide analysis of zebrafish protein-coding gene function. *Nature*. 2013;496(7446):494–497. https://doi.org/10.1038/nature11992.
34. Moens CB, Donn TM, Wolf-Saxon ER, Ma TP. Reverse genetics in zebrafish by TILLING. *Briefings Funct Genomics Proteomics*. 2008;7(6):454–459. https://doi.org/10.1093/bfgp/eln046.
35. Kim YG, Cha J, Chandrasegaran S. Hybrid restriction enzymes: zinc finger fusions to Fok I cleavage domain. *Proc Natl Acad Sci USA*. 1996;93(3):1156–1160. https://doi.org/10.1073/pnas.93.3.1156.
36. Miller J, McLachan AD, Klug A. Repetitive zinc-binding domains in the protein transcription factor IIA from Xenopus oocytes. *EMBO J*. 1985;4(6):1609–1614.
37. Huang B, Schaeffer CJ, Li Q, Tsai MD. Splase: a new class IIS zinc-finger restriction endonuclease with specificity for Sp1 binding sites. *J Protein Chem*. 1996;15(5):481–489. http://www.ncbi.nlm.nih.gov/pubmed/8895094.
38. Cai M, Yang Y. Targeted genome editing tools for disease modeling and gene therapy. *Curr Gene Ther*. 2014;14(1):2–9. https://doi.org/10.2174/156652321402140318165450.
39. Bibikova M, Beumer K, Trautman JK, Carroll D. Enhancing gene targeting with designed zinc finger nucleases. *Science*. 2003;300(5620):764. https://doi.org/10.1126/science.1079512.
40. Geurts AM, Cost GJ, Freyvert Y, et al. Knockout rats via embryo microinjection of zinc-finger nucleases. *Science*. 2009;325(5939):433. https://doi.org/10.1126/science.1172447.
41. Watanabe M, Umeyama K, Matsunari H, et al. Knockout of exogenous EGFP gene in porcine somatic cells using zinc-finger nucleases. *Biochem Biophys Res Commun*. 2010;402(1):14–18. https://doi.org/10.1016/j.bbrc.2010.09.092.
42. Young JJ, Cherone JM, Doyon Y, et al. Efficient targeted gene disruption in the soma and germ line of the frog *Xenopus tropicalis* using engineered zinc-finger nucleases. *Proc Natl Acad Sci USA*. 2011;108(17): 7052–7057. https://doi.org/10.1073/pnas.1102030108.

43. Doyon Y, McCammon JM, Miller JC, et al. Heritable targeted gene disruption in zebrafish using designed zinc-finger nucleases. *Nat Biotechnol.* 2008;26(6):702–708. https://doi.org/10.1038/nbt1409.
44. Meng XD, Noyes MB, Zhu LHJ, Lawson ND, Wolfe SA. Targeted gene inactivation in zebrafish using engineered zinc-finger nucleases. *Nat Biotechnol.* 2008;26(6):695–701. https://doi.org/10.1038/Nbt1398.
45. Zhu C, Smith T, McNulty J, et al. Evaluation and application of modularly assembled zinc-finger nucleases in zebrafish. *Development.* 2011;138(20):4555–4564. https://doi.org/10.1242/dev.066779.
46. Isalan M. Zinc-finger nucleases: how to play two good hands. *Nat Methods.* 2011;9(1):32–34. https://doi.org/10.1038/nmeth.1805.
47. Moscou MJ, Bogdanove AJ. A simple cipher governs DNA recognition by TAL effectors. *Science.* 2009; 326(5959):1501. https://doi.org/10.1126/science.1178817.
48. Lamb BM, Mercer AC, Barbas CF. Directed evolution of the TALE N-terminal domain for recognition of all 5' bases. *Nucleic Acids Res.* 2013;41(21):9779–9785. https://doi.org/10.1093/nar/gkt754.
49. Cermak T, Doyle EL, Christian M, et al. Efficient design and assembly of custom TALEN and other TAL effector-based constructs for DNA targeting. *Nucleic Acids Res.* 2011;39(12):e82. https://doi.org/10.1093/nar/gkr218.
50. Ma AC, McNulty MS, Poshusta TL, et al. FusX: a rapid one-step transcription activator-like effector assembly system for genome science. *Hum Gene Ther.* 2016;27(6):451–463. https://doi.org/10.1089/hum.2015.172.
51. Bedell VM, Wang Y, Campbell JM, et al. In vivo genome editing using a high-efficiency TALEN system. *Nature.* 2012;491(7422):114–118. https://doi.org/10.1038/nature11537.
52. Zu Y, Tong X, Wang Z, et al. TALEN-mediated precise genome modification by homologous recombination in zebrafish. *Nat Methods.* 2013;10(4):329–331. https://doi.org/10.1038/nmeth.2374.
53. Hwang WY, Fu Y, Reyon D, et al. Efficient genome editing in zebrafish using a CRISPR-Cas system. *Nat Biotechnol.* 2013;31(3). https://doi.org/10.1038/nbt.2501.
54. Jao L-E, Wente SR, Chen W. Efficient multiplex biallelic zebrafish genome editing using a CRISPR nuclease system. *Proc Natl Acad Sci USA.* 2013;110(34):13904–13909. https://doi.org/10.1073/pnas.1308335110.
55. Cong L, Ran FA, Cox D, et al. Multiplex genome engineering using CRISPR/Cas systems. *Science.* 2013;339(6121):819–823. https://doi.org/10.1126/science.1231143.
56. Ousterout DG, Kabadi AM, Thakore PI, Majoros WH, Reddy TE, Gersbach CA. Multiplex CRISPR/Cas9-based genome editing for correction of dystrophin mutations that cause Duchenne muscular dystrophy. *Nat Commun.* 2015;6:6244. https://doi.org/10.1038/ncomms7244.
57. Sakuma T, Nishikawa A, Kume S, Chayama K, Yamamoto T. Multiplex genome engineering in human cells using all-in-one CRISPR/Cas9 vector system. *Sci Rep.* 2014;4:5400. https://doi.org/10.1038/srep05400.
58. Cradick TJ, Fine EJ, Antico CJ, Bao G. CRISPR/Cas9 systems targeting β-globin and CCR5 genes have substantial off-target activity. *Nucleic Acids Res.* 2013;41(20):9584–9592. https://doi.org/10.1093/nar/gkt714.
59. Fu Y, Foden JA, Khayter C, et al. High-frequency off-target mutagenesis induced by CRISPR-Cas nucleases in human cells. *Nat Biotechnol.* 2013;31(9):822–826. https://doi.org/10.1038/nbt.2623.
60. Cho SW, Kim S, Kim Y, et al. Analysis of off-target effects of CRISPR/Cas-derived RNA-guided endonucleases and nickases. *Genome Res.* 2014;24(1):132–141. https://doi.org/10.1101/gr.162339.113.
61. Pattanayak V, Lin S, Guilinger JP, Ma E, Doudna JA, Liu DR. High-throughput profiling of off-target DNA cleavage reveals RNA-programmed Cas9 nuclease specificity. *Nat Biotechnol.* 2013;31(9):839–843. https://doi.org/10.1038/nbt.2673.
62. Fu Y, Sander JD, Reyon D, Cascio VM, Joung JK. Improving CRISPR-Cas nuclease specificity using truncated guide RNAs. *Nat Biotechnol.* 2014;32(3):279–284. https://doi.org/10.1038/nbt.2808.
63. Ceccaldi R, Rondinelli B, D'andrea AD. Repair pathway choices and consequences at the double-strand break. *Trends Cell Biol.* 2016;26:52–64. https://doi.org/10.1016/j.tcb.2015.07.009.
64. Xiao A, Wang Z, Hu Y, et al. Chromosomal deletions and inversions mediated by TALENs and CRISPR/Cas in zebrafish. *Nucleic Acids Res.* 2013;41(14):e141. https://doi.org/10.1093/nar/gkt464.

65. Ma AC, Lee HB, Clark KJ, Ekker SC. High efficiency in vivo genome engineering with a simplified 15-RVD GoldyTALEN design. *PLoS One*. 2013;8(5). https://doi.org/10.1371/journal.pone.0065259.
66. Maddalo D, Manchado E, Concepcion CP, et al. In vivo engineering of oncogenic chromosomal rearrangements with the CRISPR/Cas9 system. *Nature*. 2014;516(7531):423–427. https://doi.org/10.1038/nature13902.
67. Auer TO, Duroure K, De Cian A, Concordet J-P, Del Bene F. Highly efficient CRISPR/Cas9-mediated knock-in in zebrafish by homology-independent DNA repair. *Genome Res*. 2014;24(1): 142–153. https://doi.org/10.1101/gr.161638.113.
68. Johnson RD, Jasin M. Sister chromatid gene conversion is a prominent double-strand break repair pathway in mammalian cells. *EMBO J*. 2000;19(13):3398–3407. https://doi.org/10.1093/emboj/19.13.3398.
69. Jasin M. Genetic manipulation of genomes with rare-cutting endonucleases. *Trends Genet*. 1996;12(6): 224–228. https://doi.org/10.1016/0168-9525(96)10019-6.
70. Mosimann C, Puller A-C, Lawson KL, Tschopp P, Amsterdam A, Zon LI. Site-directed zebrafish transgenesis into single landing sites with the phiC31 integrase system. *Dev Dyn*. 2013;242(8): 949–963. https://doi.org/10.1002/dvdy.23989.
71. Hisano Y, Sakuma T, Nakade S, et al. Precise in-frame integration of exogenous DNA mediated by CRISPR/Cas9 system in zebrafish. *Sci Rep*. 2015;5:8841. https://doi.org/10.1038/srep08841.
72. Nakade S, Tsubota T, Sakane Y, et al. Microhomology-mediated end-joining-dependent integration of donor DNA in cells and animals using TALENs and CRISPR/Cas9. *Nat Commun*. 2014;5:5560. https://doi.org/10.1038/ncomms6560.
73. Kawakami K, Takeda H, Kawakami N, Kobayashi M, Matsuda N, Mishina M. A transposon-mediated gene trap approach identifies developmentally regulated genes in zebrafish. *Dev Cell*. 2004;7(1): 133–144. https://doi.org/10.1016/j.devcel.2004.06.005.
74. Shin J, Chen J, Solnica-Krezel L. Efficient homologous recombination-mediated genome engineering in zebrafish using TALE nucleases. *Development*. 2014;141(19):3807–3818. https://doi.org/10.1242/dev.108019.

SECTION V

# Manipulating Known Genes to Understand Biological Function: Reverse Genetics

# CHAPTER 18

# Molecular Techniques Used to Explore Glutamate Receptors in Synaptic Plasticity and Memory

**Celeste Leung[1,2], Zhengping Jia[1,2]**
[1]The Hospital for Sick Children, Toronto, ON, Canada; [2]University of Toronto, Toronto, ON, Canada

## INTRODUCTION

Prior to the advent of genomic manipulation, pharmacological studies have provided most of the initial insight into the function of excitatory synapses in the mammalian central nervous system. In particular, the use of agonists and antagonists specific to glutamate receptors have established the role of α-amino-3-hydroxy-5-methyl-4-isoxazolepropionate (AMPA) and *N*-methyl-D-aspartate (NMDA) receptors in synaptic transmission and plasticity, and learning and memory.[1,2] In the past 30 years, the use of molecular genetic approaches in mice have dramatically increased our ability to specifically manipulate individual genes, allowing detailed analyses of synaptic proteins. Perhaps, one of the earliest and simplest approaches is the use of single-stranded antisense oligonucleotides and the subsequent emergence of RNA interference that allows for the knockdown of protein expression through the inhibition of gene expression and translation. However, this approach rarely eliminates the protein of interest completely, making it difficult to determine whether any residual function can give rise to unrelated processes. The introduction of pluripotent embryonic stem (ES) cells carrying targeted genomic changes has allowed for the production of mice harboring global deletions (knockout) or insertions (knock-in) of specific genes. However, the widespread knockout of a particular gene from conception may result in lethality and possible compensation that advanced the field to an approach that limits the deletion or insertion of the gene in a spatial- and temporal-dependent manner. These newer studies combined the knockout technique with site-directed mutagenesis and the bacteriophage Cre/loxP system and the tetracycline inducible transactivator system, which together have allowed for unprecedented control and precise manipulations of diverse genes of interest. With recent technical advances, innovative genome-editing tools have been developed to further circumvent the issues associated with traditional techniques involving homologous recombination and RNA interference. Specifically, clustered regularly interspaced short palindromic repeats (CRISPR) along with the expression of a Cas9 nuclease is an approach that can target and edit DNA sequences at specific locations with high

*Molecular-Genetic and Statistical Techniques for Behavioral and Neural Research*
ISBN 978-0-12-804078-2, https://doi.org/10.1016/B978-0-12-804078-2.00018-0

efficiencies. In this review, we will discuss a number of molecular-genetic techniques, including RNA interference, knockout, conditional knockout, knock-in, and CRISPR/Cas9, that have been used to study the role of synaptic proteins in the regulation of synaptic transmission and plasticity with a specific focus on two types of ionotropic glutamate receptors, AMPARs and NMDARs, in mice (Table 18.1).

AMPARs are hetero-tetrameric complexes composed of four subunits, GluA1-GluA4. Each subunit contains a large extracellular N-terminal ligand-binding domain, three full transmembrane domains, an intracellular reentrant loop, and a cytoplasmic carboxyl terminal (C-terminal) domain. Whereas GluA1 and 4 possess predominantly long C-terminal tails, GluA2 and GluA3 have relatively shorter cytoplasmic C-terminals.[3] Prior to birth and within the first postnatal week, the GluA4 subtype is predominantly expressed in excitatory hippocampal and cortical neurons.[4,5] However, GluA4 levels decline following birth, contributing to the developmental switch to the GluA2 receptor subtype. Gradually with age, hippocampal Schaffer collateral CA3-CA1 synapses are largely comprised of GluA1/2 and GluA2/3 heteromers, where GluA1/2 receptors are inserted into the synapse following activity and eventually replaced by GluA2/3 receptors over time in an activity-independent manner.[6,7] Importantly, the presence of the GluA2 subunit in AMPARs renders the channel impermeable to calcium, resulting in a linear current-voltage relationship. Thus, in the absence of GluA2, AMPARs are permeable to calcium and exhibit a strong inward rectification at positive potentials.[8] Numerous studies have shown that AMPARs play an important role in mediating fast synaptic transmission in the central nervous system and are critical for synaptic plasticity, including long-term potentiation (LTP) and depression (LTD).[2,9–12] The other key ionotropic glutamate receptor essential for the induction of synaptic plasticity is the NMDAR. NMDARs are also tetramers composed of two obligatory GluN1 subunits and two GluN2 or GluN3 subunits that determine the properties and kinetics of the receptor. The topology of each subunit is similar to that of AMPARs consisting of the N-terminal domain for binding by allosteric modulators, the agonist-binding domain (for agonists such as glycine/D-serine or glutamate), the pore domain, and the C-terminal domain that binds to various intracellular protein partners. The expression of GluN2A-D subunits varies across development (with GluN2B expression being high before birth and GluN2A expression increasing over postnatal development), which contributes to the diverse role of each subtype in synaptic plasticity. GluN2A and GluN2B are the primary subtypes found in the hippocampus and cortex.[13,14] Synaptic NMDARs in mature neurons predominantly consist of GluN1/GluN2A or GluN1/GluN2A/GluN2B heteromers.[15] NMDARs are known as coincidence detectors, as at resting potentials the channel is blocked by extracellular magnesium. It is only following presynaptic glutamate release and postsynaptic depolarization that the magnesium block is removed from NMDARs, allowing for their activation and subsequent calcium influx. Studies have shown that the activation of NMDARs is

**Table 18.1** Selected examples of genetic manipulations of AMPA and NMDA receptors

| Gene | Technique | Key phenotype | References |
|---|---|---|---|
| GluA1 | KO | Reduced NMDAR-mediated LTP | 31 |
| | KI | Reduced LTD and LTP | 50 |
| | KI | S831 A mutants showed normal LTP, but S845 A mutants showed reduced LTD | 51 |
| | KI | S831D/S845D mutants showed enhanced LTP | 53 |
| | KI | GluA1Δ7 mutants showed normal basal synaptic transmission and plasticity | 54 |
| GluA1, 2 | cKO | Decreased mEPSC frequency and amplitude | 42 |
| GluA1, 3 | cKO | Decreased mEPSC frequency and amplitude | 42 |
| GluA1-3 | RNAi | Decreased mEPSC frequency | 20 |
| | KO | Showed normal LTP and LTD | 42 |
| GluA2 | RNAi | Blocked homeostatic synaptic scaling | 23 |
| | RNAi | Influenced proper glutamate release at presynaptic terminals | 20 |
| | KO | Reduced basal synaptic transmission and enhanced LTP | 34 |
| | CRISPR | Reduced AMPAR EPSC | 61 |
| GluA2, 3 | KO | Reduced basal synaptic transmission and enhanced LTD and LTP | 35 |
| | cKO | Decreased mEPSC amplitude | 42 |
| GluA3 | KO | Enhanced LTP | 35 |
| | cKO | Decreased mEPSC frequency and amplitude. Impaired cerebellum-dependent motor learning | 43 |
| GluN1 | KO | Impaired development in the barrel cortex, ataxia, and lethal soon after birth | 36 |
| | cKO | Reduced LTP | 44 |
| | cKO | Reduced injury-induced pain transmission | 45 |
| | CRISPR | Enhanced AMPAR EPSC | 61 |
| | CRISPR | Reduced NMDAR EPSC in both mouse and rat cultures | 62 |
| GluN2A | RNAi | Showed normal LTP | 24 |
| | KO | Reduced LTP and spatial impairments | 37 |
| | KO | Enhanced the strengthening of synapses and potentiated bidirectional synaptic plasticity of the visual system | 38 |
| | KI | Reduced LTP and impaired contextual memory | 55 |

*Continued*

**Table 18.1** Selected examples of genetic manipulations of AMPA and NMDA receptors—cont'd

| Gene | Technique | Key phenotype | References |
|---|---|---|---|
| GluN2A, 2B | cKO | Enhanced synaptogenesis | 46 |
| GluN2B | RNAi | Impaired AMPAR trafficking | 22 |
| | RNAi | Influenced the induction of LTP | 24 |
| | KO | Neonatal lethality, impaired somatosensory synapse refinement, and abolished LTD | 39 |
| | cKO | Abolished LTP but can be overcome with multiple tetani. Impaired spatial learning and memory | 47, 48 |
| | KI | Perinatal lethality | 55 |
| GluN2C | KI | Impaired balance behavior | 55 |
| GluN3A | KO | Enhanced LTP. Impaired locomotor activity, increased sensitivity to pain, and enhanced recognition and spatial learning and memory | 40 |
| GluN3B | KO | Impaired motor learning and reduced novel social interactions | 41 |

required for the induction of both LTP and LTD.[16,17] This review will discuss how various genetic approaches in targeting each or a combination of the receptor subunits affect the functional and dynamic properties of the AMPAR and NMDAR in synaptic transmission and plasticity, and learning and memory (Table 18.1).

## PART 1: RNA INTERFERENCE

RNA interference (RNAi) or posttranscriptional gene silencing (PTGS) is a revolutionary tool used in diverse organisms and cell types to rapidly regulate eukaryotic gene expression in response to stress. It is a conserved biologic mechanism that targets sequence-specific gene degradation through an endogenous multiprotein complex, known as an RNAi silencing complex (RISC). The ribonucleoprotein complex acts via endonucleolytic cleavage or translational suppression followed by deadenylation and degradation through posttranscriptional gene silencing. Micro RNA (miRNA) and short interfering RNA (siRNA) are examples of two precursors of RNAi. However, siRNAs typically arise via exogenous synthetic or viral inducers of RNAi, whereas miRNA are derived endogenously from the genome. In 1998, Fire and Mello were the first to discover in a nematode worm, *Caenorhabditis elegans,* that injection of double-stranded RNA (dsRNA) activates sequence-specific inhibition of protein expression through the degradation of messenger RNA (mRNA) molecules.[18] They deduced that one strand of the dsRNA can be used as a guide for silencing specific complementary target mRNAs whose genetic code matches that of the injected RNA molecule. Hence, compared to antisense oligonucleotides that are delivered as a single strand and find their target RNA alone, siRNAs are part of a protein complex and exist as a duplex. To date, RNAi has been shown to control vital processes such as cell growth and proliferation, tissue differentiation, heterochromatin formation, and cell defense against viruses and transposons.[19] Further genetic and biochemical studies have revealed a two-step mechanism of RNAi-induced gene silencing (Fig. 18.1). In the nucleus, a 1000-nucleotide (nt) miRNA primary transcript is cleaved by RNase III family enzymes, Drosha and DiGeorge syndrome critical region gene 8 (DGCR8), into a 65-70-nt precursor miRNA prior to being shipped to the cytoplasm. It is at this step where both miRNA and exogenous siRNAs are trimmed by Dicer enzymes into a dsRNA duplex of 21-25-nt strands for loading onto Argonaute, forming a minimal RISC-loading complex. During RISC assembly, strand selection occurs simultaneously where the duplex is unwound, and the strand bound to Argonaute becomes the single-stranded guide RNA, and the other passenger strand is cleaved and eliminated. The RISC containing the guide strand becomes a probe where single-stranded RNA such as complementary mRNA bind. If there is perfect complementarity, Argonaute can cleave the targeted mRNA, which is further degraded by other endogenous nucleases and results in the loss of protein expression. However, if there is imperfect complementarity, the RISC can also induce nonendonucleolytic translational repression.

Within the realm of experimental biology, one of the simplest approaches to study the physiological function of a targeted gene is through siRNA. The transfection of chemically synthesized siRNA oligonucleotides into the cytosol can result in the "knockdown" or drastic reduction in gene expression. Hence, this particular RNAi pathway involving the exogenous delivery of siRNAs is commonly employed in diverse cell types, but the variable transfection efficiency limits its use to transient in vitro studies. Other studies have found an increased probability of incurring possible off-target effects due to the high concentration of siRNA employed to maximize gene knockdown.[20] Therefore, the off-target effects may arise when an introduced RNA contains a base sequence that can pair with and reduce the expression of multiple genes. With time, dividing cells also result in diluted siRNA concentrations rendering the long-term effects for a desired single target gene knockdown unfeasible. An alternative pathway of RNAi involves the use of short hairpin RNAs (shRNAs) via DNA vector-mediated production. The plasmid or viral vectors containing shRNAs are transcribed by RNA polymerase III or modified polymerase II promoters. Following transcription, the sequence is exported to the cytosol, where it is recognized by a variety of endogenous enzymes. Similar to miRNA and siRNA, endogenously derived shRNA binds to the target mRNA and gets incorporated into the RISC complex for target-specific degradation. The key advantage of the shRNA technique is that it may allow for stable integration of the shRNA constructs into the host genome, thus long-term knockdown of the targeted gene, which

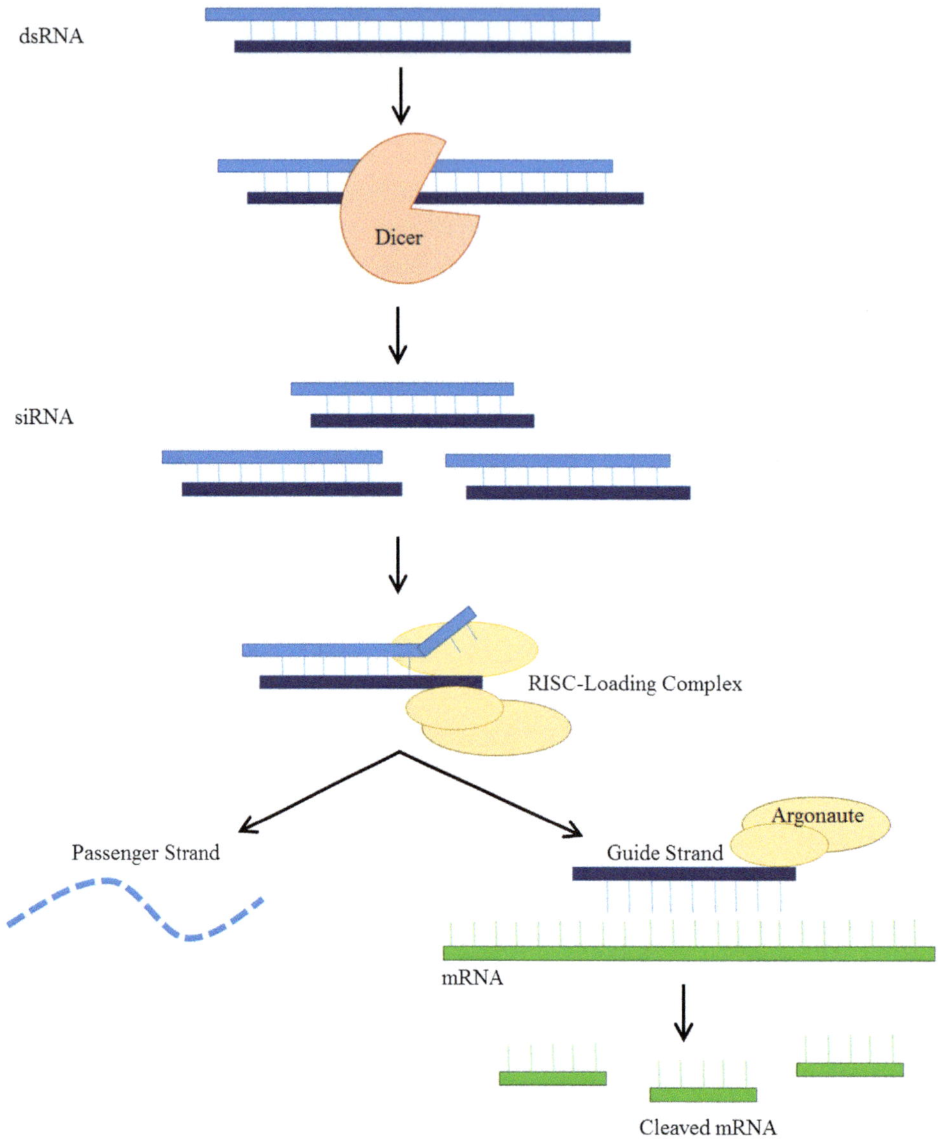


**Figure 18.1** Schematic for RNAi. For simplicity, siRNA and miRNA pathways have been combined. In the cytoplasm, dsRNA is cleaved by Dicer, and the resulting siRNA duplex is loaded onto Argonaute by the minimal RISC-loading complex, which comprises Dicer, a dsRBP protein such as TRBP, and an Argonaute protein. The guide strand in *dark blue* remains bound to Argonaute, forming the RISC. The passenger strand in *light blue* is cleaved and degraded. The RISC containing the guide strand becomes a probe where single-stranded RNA such as complementary mRNA binds (*in green*). If there is perfect complementarity, the RISC binds to the complementary target sequences and silences the targeted mRNA via Argonaute and other endogenous nucleases.

eliminates the need for multiple rounds of transfection. However, one drawback of this technique is that it is laborious and time consuming to prepare and select shRNA-positive cells by drug resistance or fluorescent markers. Although siRNA and shRNA techniques employ similar cellular mechanisms, the choice of which method to use ultimately depends on different factors such as the cell type, time demands, and the need for transient versus stable integration.[21]

For the reasons discussed before, RNAi is a useful tool to study the loss of function of targeted proteins to determine the mechanisms and molecular interactions that explain the physiological role of target genes of interest. Compared to other techniques including knockouts (KOs, to be discussed later), the major advantages are that it can be applied to multiple experimental organisms, requires less time, shortens protein turnover duration, and reduces cost to generate the mouse lines. Therefore, the RNAi strategy has been used extensively to explore the role of ionotropic glutamate receptors in synaptic plasticity. For example, Hall et al.[22] explored how the organization of ionotropic glutamate receptors changes during the developmental maturation of excitatory cortical networks. The authors created RNAi constructs to knockdown GluN2B using the pSilencer 1.0 vector to generate RNA-interfering constructs and confirmed the efficiency by recording the NMDA-evoked response, which was significantly suppressed in siRNA transfected neurons compared with nontransfected cultures. They further showed that the GluN2B NMDAR subunit negatively regulates AMPAR trafficking and AMPAR-mediated currents during a developmental period of synaptogenesis. Gainey et al.[23] examined the role of the GluA2 subunit of AMPARs in the expression of synaptic scaling in cultured rat visual cortical neurons. They employed RNAi-mediated knockdown of GluA2 using a mU6pro vector that resulted in pronounced rectification, indicating that GluA2 is greatly reduced or absent from synaptically localized AMPARs in the neurons. Furthermore, GluA2 knockdown blocked the expression of synaptic scaling but had no effect on basal synaptic transmission or the induction of LTP. These results demonstrate the importance of the GluA2 subunit for the regulated accumulation of AMPARs at the synapse during homeostatic synaptic scaling in response to chronic changes in activity. In another study, Foster et al.[24] employed RNAi plasmids targeting either GluN2A or GluN2B using a pSuper vector in organotypic hippocampal slice cultures. They demonstrate the requirement of the GluN2B cytoplasmic tail for recruiting relevant molecules in the induction of LTP. However, the GluN2A subunit was found to carry inhibitory factors for LTP in the C-terminal cytoplasmic tail and, therefore, is unnecessary for LTP. Tracy et al.[20] explored how AMPARs influence synaptic development by using plasmid-based RNAi constructs to knock down the expression of GluA1, GluA2, and GluA3 AMPAR subunits in young cultured hippocampal neurons. To confirm their model, they quantified AMPAR immunofluorescence colocalized with a presynaptic glutamatergic marker, VGLUT1 (vesicular glutamate transporter 1), and found a significant reduction (~40%) in synaptic AMPARs compared to controls. Furthermore, there was a reduction in the average miniature EPSC frequency that they attributed to the presence of postsynaptic silent synapses. To eliminate the possibility of off-target effects, all the impairments were reversed back to the control level with the co-expression of RNAi-resistant constructs of GluA1 or GluA2. Ultimately, they found that GluA RNAi effectively weakens synaptic transmission by decreasing the number of functional glutamatergic synapses containing AMPARs. Their results further show that the GluA2 N-terminal domain acts as a retrograde transsynaptic signal during synapse development to influence proper glutamate release at presynaptic terminals. Ultimately, this study elegantly conveys how RNAi-mediated knockdown of postsynaptic AMPAR expression in young neurons not only affects AMPAR-mediated synaptic currents but also causes a dramatic, corresponding decrease in NMDAR-mediated currents as well.

From these and other studies, RNAi as a method for the knockdown of genes is typically incomplete (roughly 80% or less effective), resulting in some amount of the native protein remaining.[25] Another issue of siRNAs is the activation of the cell's innate interferon response, which is a form of host-derived immunity that reacts nonspecifically to foreign genetic material. The introduction of longer dsRNA (27–29 bp) and shRNAs may affect cell viability and gene silencing of proteins other than those targeted.[26] Techniques such as avoiding the direct introduction of siRNA by stable transfection with a plasmid or conditional RNAi using a lentiviral vector (such as pSUPER) have also been developed.[27] Studies have employed controls such as scrambled RNAi sequences, but these have been found to be insufficient since the off-target effects are found to be sequence specific. Therefore, a combination of controls must be explored, such as a direct assay of interferon activation, rescue with an RNAi proof version of the target protein, and validation with a pharmacological block or dominant negative or overexpression model.

## PART 2: KNOCKOUT AND CONDITIONAL KNOCKOUT

Perhaps one of the most well-known techniques to manipulate the endogenous locus of a particular gene is the knockout strategy. However, prior to its fruition, the use of transgenic mice in elucidating the physiological functions of disease-related genes or receptors in vivo was dominant beginning in the early 1980s. To achieve transgenic expression, the pronuclear-microinjection of exogenous DNA containing the specific transgenic construct into fertilized eggs results in the random integration of the transgene into the host genome.[28] The microinjected zygotes are then surgically implanted into pseudo-pregnant recipient female mice, resulting in the birth of transgenic mice harboring inheritable traits that can be detected via biochemical analyses of transgene expression in founder mice. In this case, the specificity and strength of the transgene expression is critically dependent on the selected promoter that is also affected by the chromosomal position where the transgene is integrated.[29] Since the copy number and the site of integration of the transgene in the mouse genome is largely random and cannot be controlled, it often results in nonspecific spatial and temporal expression of the transgene; consequently, the effect of the transgene expression is difficult to interpret. To partly circumvent these issues, inducible and cell-type restricted transgene expression such as the bacterial tetracycline inducible systems have been developed. In the late 1980s, an improved alternate gene targeting approach pioneered by Mario R. Capecchi, who coined the term "knockout" (KO), employed pluripotent ES cells to target changes in the endogenous genome.[30] In this approach (Fig. 18.2), a targeting vector with the desired mutations and a drug resistance marker gene (or replacement of part of the gene by an expression cassette for a positive selection marker such as the neomycin-resistant gene, *neo*), is introduced into ES cells and undergoes homologous recombination with the endogenous wild-type allele, resulting in the incorporation of the drug resistance marker gene and the deletion or knockout of the normal wild-type copy of the gene. The correctly targeted ES cells with the altered gene and the drug resistance marker gene are selected and then transplanted into the embryos of a recipient pseudo-pregnant foster female mouse to generate chimeras. Germline transmission of the resulting offspring were selected based on the transmission of the coat color and tested for homozygosity for the targeted changes. Since the ES cells are pluripotent, they are capable of differentiating into all cell types of a chimeric mouse, resulting in the "global knockout" of a particular gene. However, two main problems exist. First, the phenotypes of global KO are difficult to attribute to a particular tissue or cell-type expressing the gene. Since the entire gene locus is disrupted in the germline cells, it is difficult to ascertain whether the phenotypes observed are due to the primary effects of the loss of the gene itself or due to the accumulation of secondary effects during development that may arise due to compensation.[29] Second, many global KOs of the CNS genes are embryonic lethal and are therefore impossible to analyze the in vivo function at an adult stage. For these reasons, this technique combined with site-directed mutagenesis and the Cre/loxP system has allowed for a more precise and/or restricted manipulation of the genes of interest.

These conditional KO (cKO) techniques have allowed gene targeting in a spatially and temporally controlled manner compared to traditional global KO models. To generate cKO mice (Fig. 18.3), the Cre/loxP system is employed, where one line of mice expresses the site-specific DNA recombinase enzyme of the bacteriophage P1, Cre, under a specific promoter that confers spatiotemporal specificity. The second line can be created via homologous recombination in ES cells and contains the locus with the gene of interest flanked by two or more recombination (loxP) sequences. The Cre recombinase enzyme specifically recognizes and targets the loxP sites within the DNA and causes recombination or DNA exchange between them. Therefore, by mating these two mouse lines, the offspring will achieve specific deletion of the target chromosomal segment in Cre-expressing tissues driven by the promoter that directs where and when the disruption of the target gene occurs in the whole body. The cKO system can be inducible because a chemical such as tetracycline or tamoxifen can be injected into the

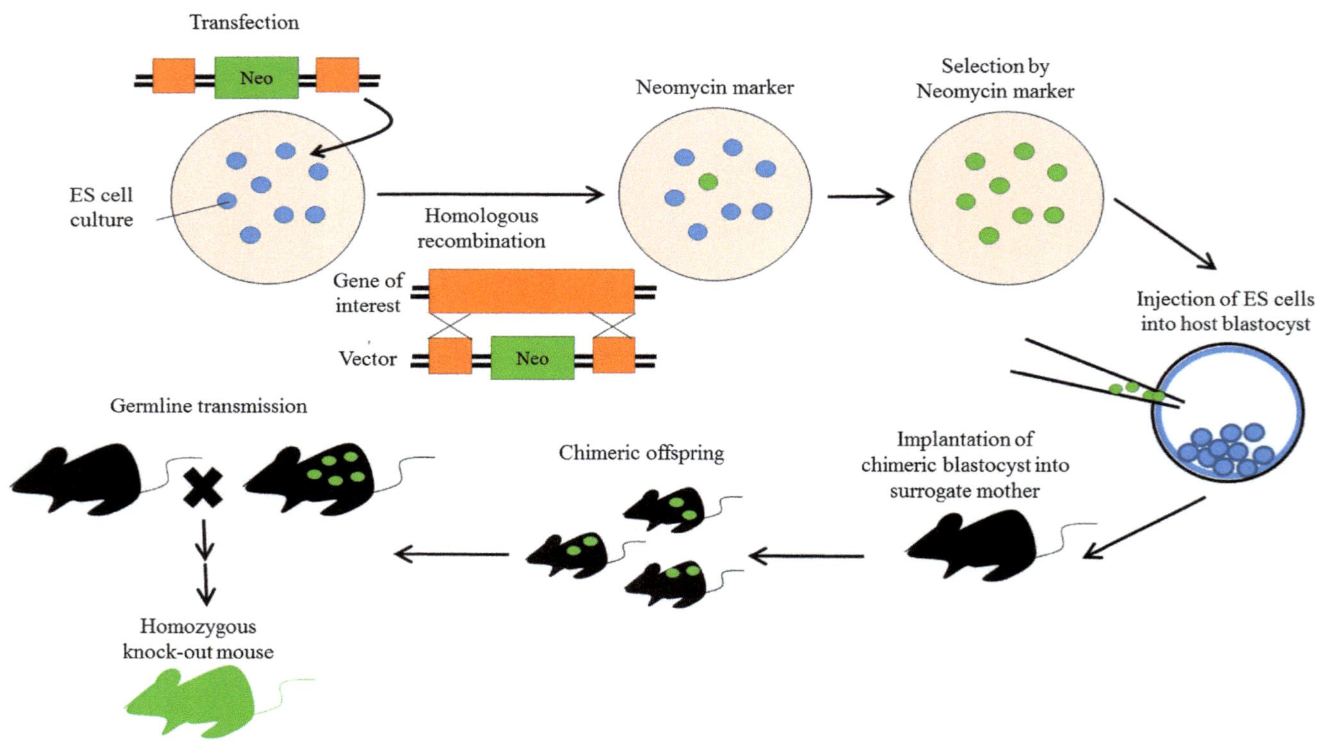


**Figure 18.2** The generation of a knockout (KO) mouse model. Embryonic stem (ES) cells are isolated from a mouse blastocyst. The desired mutations and a drug resistance marker gene (*Neo*) is introduced into ES cells via electroporation or transfection and undergoes homologous recombination with the endogenous wild-type allele. This results in the incorporation of the drug resistance marker gene and the successful recombination and deletion of the normal wild-type copy of the gene. The correctly targeted ES cells in *green*, with the drug resistance marker gene are selected. Cells in *blue* that do not integrate any of the genetic material will die as a result of poisoning with Neomycin. The ES cells are inserted into a host mouse blastocyst and are then transplanted into the embryos in the uterus of a recipient pseudo-pregnant surrogate female mouse to generate chimeras with patches on their fur. Germline transmission of the resulting offspring were selected based on the transmission of coat color and tested for homozygosity for the targeted gene.

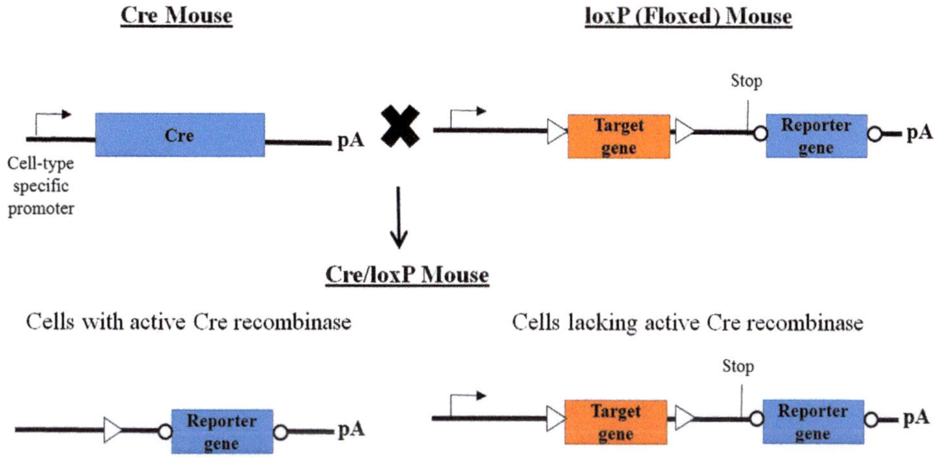


**Figure 18.3** The generation of a conditional knockout (cKO) mouse model. The Cre-loxP recombination system contains a Cre mouse with a cell-type-specific promoter to drive the expression of Cre and a floxed mouse with two loxP sites, denoted as *open triangles*, in a tandem repeat flanking the gene of interest. The Cre mouse is crossed with floxed mouse, allowing for Cre to delete the DNA sequences between the loxP sites. In the Cre/loxP offspring, cells that have an active Cre recombinase will result in a deletion event, whereas other cells lacking Cre recombinase will not catalyze recombination of the sequence. The reporter gene, which is normally flanked by two Frt sites (denoted as *open circles*), is typically removed following homologous recombination by expression of Flp-mediated recombination.

animal to initiate the activation of the Cre recombinase enzyme to knock out genes at a specific time frame or developmental period. For these reasons, both KO and cKO genetic mouse models have contributed to significant advancements in the realm of experimental biology. The following are examples of how these techniques have revolutionized our understanding of ionotropic glutamate receptors, specifically how AMPARs and NMDARs mediate synaptic transmission and plasticity and learning and memory in mice.

Zamanillo et al.[31] explored the molecular basis of LTP and the relationship between hippocampal LTP as a neural correlate of spatial learning and memory using mice lacking the GluA1 subunit of the AMPAR. GluA1 KO mice were generated through a targeting vector with parts of introns 10 and 11, and exon 11 of the GluA1 gene flanked by loxP sites and deleted upon the activation of Cre recombinase. These KO mice surprisingly exhibited standard home-cage behavior during development and normal spatial learning in the Morris water maze but impaired NMDAR-mediated LTP in CA1 pyramidal neurons that may be due to reduced functional AMPARs. Others have shown that LTD is unaffected in GluA1 KO mice.[32] Together, these findings suggest that hippocampal LTP depends on the number and subunit composition of AMPARs. Furthermore, the dichotomy between LTP and learning may be due to extrahippocampal structures important in different types of reference memory tests that involve diverse forms of episodic or working memory. In another study, Jia et al. (1996) explored the role of the GluA2 subunit in the induction of LTP. Prior in vitro studies have elucidated the function of GluA2 in inhibiting calcium influx via AMPARs.[33] Therefore, the authors wanted to explore if the subunit and calcium influx through AMPARs play a role in synaptic transmission and plasticity in mice lacking GluA2. To do this, a genomic clone was used to replace transmembrane region TM1 and the pore loop of exon 11 by a pgk-neomycin resistant cassette to completely eliminate the entire subunit.[34] GluA2KO mice exhibited reduced basal synaptic transmission and enhanced LTP but normal neuronal excitability, paired-pulse facilitation, and LTD. Interestingly, LTP can still be induced in the presence of the NMDAR inhibitor, AP5, indicating that calcium influx from GluA2-lacking receptors is sufficient to trigger LTP. Thus, GluA2 plays a crucial role in regulating both calcium influx and LTP. Meng et al.[35] also investigated the effects of GluA2 and GluA3 double KO (DKO) mice. The authors first generated GluA3 KO mice by replacing a part of the TM1 and TM2 domains of the AMPAR subunit by a pgk-neomycin resistant cassette. GluA3KO mice had normal anatomic and synaptic structures, basal synaptic response, paired-pulse facilitation, and LTD, but slightly enhanced LTP. The GluA2 and GluA3 DKO mice also showed normal gross anatomy and synaptic structures, but reduced basal synaptic transmission and enhanced LTD and LTP. Together, this data suggests that although GluA2 and GluA3 are critical for maintaining and stabilizing basal synaptic transmission, they are not absolutely required for LTP or LTD expression. These results also suggest that GluA1 is sufficient for the expression of long-lasting synaptic plasticity in vivo.

As discussed before, the induction of LTP depends on the activation of NMDARs.[16,17] The NMDAR is a tetrameric receptor comprising two obligatory GluN1 subunits and two GluN2 subunits. Li et al.[36] attempted to target the NMDAR1 gene to create a GluN1 KO mouse that lacks functional NMDARs. The authors generated GluN1 KO mice using standard homologous recombination by replacing exons 12–20, coding for the NMDAR1 gene, with a pgk-neomycin resistant cassette and confirmed the lack of the receptor function using whole-cell patch clamp recordings. GluN1 KO mice displayed abnormal development of whisker-related patterns in the barrel cortex, and most pups were unable to survive within 10–20 hours following birth due to ataxia and impaired sucking reflexes. The first successful KO study explored the deletion of the GluN2A subunit in the brain through targeting the M1–M3 putative transmembrane segments.[37] The GluN2A KO mice showed a significant reduction of the NMDAR channel current and hippocampal LTP along with spatial impairments on the Morris water maze task. In a more recent study, Cho et al.[38] showed that the genetic deletion of the GluN2A NMDAR subunit impairs experience-dependent weakening of synapses and thereby enhances the strengthening of those synapses. The mice were created by replacing the region spanning the M2 transmembrane segment of the GluN2A subunit with the neomycin resistance gene. These GluN2A-KO mice failed to exhibit deprivation-induced depression but rather exhibit precocious potentiation of nondeprived eye inputs, underscoring the bidirectional synaptic plasticity in the visual system. Conversely, GluN2B KO mice exhibited neonatal lethality, impaired whisker-related somatosensory synapse refinement, and abolished synaptic NMDA responses and LTD.[39] From these studies, the GluN2 subunit of the NMDAR plays an important role in the induction of LTP and LTD. Recent studies have also examined the role of GluN3A and GluN3B subunits of NMDARs in synaptic plasticity and learning and memory. Mohamad et al.[40] showed that the KO of GluN3A leads to impaired locomotor activity on a variety of tests, increased sensitivity to acute and subacute inflammatory pain, and enhanced recognition and spatial learning with respect to memory functions. The GluN3A KO mice had enhanced LTP that may be attributed to the increased expression and phosphorylation of $Ca^{2+}$/calmodulin-dependent kinase II (CaMKII) in the brain. The findings from this study are the first to suggest that GluN3A plays an important role in regulating proper locomotion, pain sensation, and learning and memory. Niemann et al.[41] generated viable and fertile GluN3B KO mice by deleting the first exon and replacing it with a pgk-neomycin resistant cassette. The KO mice showed an impairment in motor learning and highly increased social interaction with their home cage littermates but decreased social interaction in a novel environment. These findings recapitulate the inhibitory role of GluN3B on the functions of NMDARs. With the advancement of gene manipulation techniques, global KO models have been combined with the generation of region and time-specific KOs (cKOs) to eliminate potential side effects from traditional tools.

In a recent study, Lu et al.[42] generated the conditional knockout of GluA1, GluA2, and GluA3 subunits individually and in combination by virally expressing Cre recombinase in CA1 hippocampal cells from mice homozygous for floxed GluA alleles. The advantage of this approach is that Cre recombinase is expressed in a small subset of hippocampal neurons, so the affected neurons can be compared to adjacent wild-type neurons. This was accomplished either by infecting the hippocampus of P0–P2 mice by transcranial stereotaxic injection with a recombinant adeno-associated virus expressing Cre together with GFP (rAAV-Cre-GFP) or by biolistic transfection of a Cre-IRES-GFP construct in hippocampal slice cultures from P5–P7 mice. Following the expression of Cre-IRES-GFP in GluA1 floxed hippocampal slice cultures, there was a gradual decrease in AMPAR EPSC amplitudes that stabilized at 20% of control values, along with decreases in mEPSC frequency and amplitude, indicating that GluA1 plays a profound role in functional AMPARs (as GluA1/2 heteromers). The authors then found that the transfection of rAAV-Cre-GFP with floxed GluA2 mice resulted in a ~50% loss of AMPAR EPSCs, where half of the synapses became devoid of AMPARs. Furthermore, the other half of the synapses that were initially GluA2-containing receptors were gradually replaced by GluA2-lacking receptors. This finding suggests that two distinct populations of synapses exist based on whether they are able to recruit GluA2-lacking receptors. When the authors deleted GluA3 via rAAV-Cre-GFP infection, it resulted in a 16% decrease in AMPAR EPSCs and no changes in the rectification and the amplitude or frequency of mEPSCs. This suggests that GluA2/3 heteromers contribute modestly (~20%) to basal synaptic transmission. Next, the authors examined the consequences of the absence of two subunits through the combined deletion of GluA2 and GluA3. The combined deletion resulted in a reduction in the amplitude of AMPAR EPSC, and the rectification of the remaining AMPARs was strongly rectifying, suggesting that all receptors remaining following GluA2/3 DKO are all GluA1 homomers. The authors also found that GluA1/3 DKOs reduced the AMPAR EPSC to a greater extent than the GluA1 KO alone, and resulted in a decreased mEPSC amplitude and frequency, suggesting that the remaining current was most likely generated by synaptic GluA2 homomers that can traffic to a few synapses. The deletion of GluA1 and GluA2 resulted in a significant loss of AMPAR EPSCs, supporting the dominant role of GluA1/2 heteromers in mediating excitatory synaptic transmission onto CA1 pyramidal cells. Given that the mEPSCs generated in the GluA1/2 DKO cells were faster than those by GluA1 homomeric receptors, GluA3 receptors may assemble as homomers and traffic to synapses when GluA1 and GluA2 are both absent. Last, the authors explored triple-KO mice, which were bred by crossing single GluA1, GluA2, and GluA3 KO mice to each other. The triple KO of GluA1, GluA2, and GluA3 resulted in the complete loss of postsynaptic AMPAR-mediated responses but no changes in NMDAR-eEPSCs. Together, these findings suggest that the subunit composition of AMPARs determines not only their biophysical properties but also synaptic strength. Interestingly, these triple-KO mice were still able to express LTP and LTD, suggesting that direct modifications of AMPARs may not be the primary mechanism underlying synaptic plasticity. In a recent study by Gutierrez-Castellanos et al.[43] they explored whether the GluA3 subunit also regulates synaptic plasticity in cerebellum-dependent motor learning since GluA1 and GluA2 subunits have been previously shown to be indispensable for basal synaptic function and behavior in the CNS. They demonstrated that GluA3 KO mice displayed severe deficits in a cerebellum-dependent motor learning paradigm known as vestibulo-ocular reflex. Furthermore, the amplitude and frequency of miniature postsynaptic currents (mEPSCs) in GluA3 KO cerebellar slices were significantly reduced compared to controls. The authors then deleted GluA3 only in the Purkinje cells of the cerebellum using a Cre-loxP method. They confirmed that the open-channel probability of GluA3 in Purkinje cells is induced via cAMP-mediated EPAC (exchange proteins directly activated by cAMP) activation, which is ultimately responsible for regulating motor learning.

The disruption of the family of NMDAR subunits using the Cre-loxP recombination system in a spatiotemporal manner has provided evidence for the essential role of the NMDAR in the acquisition of synaptic plasticity and learning and memory. Tsien et al.[44] produced a mouse strain that deleted the NMDAR1 gene in the pyramidal cells of the CA1 region of the hippocampus. These mice were generated through inserting two loxP sequences into the NMDAR1 gene, one between exons 10 and 11 and the other downstream of the 3′ end of the NMDAR1 gene with a neomycin resistance gene. The mice were then crossed with Cre transgenic mice, which have the capacity to mediate Cre/loxP recombination exclusively, resulting in the deletion of the GluN1 DNA sequence between the two loxP sites in CA1 pyramidal cells. Compared to the global GluN1 KO, these cKO mice have normal development into adulthood but lack NMDAR-mediated postsynaptic currents and have impairments in LTP at CA1 synapses, along with altered spatial memory. In a more recent study, South et al.[45] similarly deleted GluN1 in the lumbar spinal cord of adult mice through the delivery of Cre recombinase using a recombinant adeno-associated virus (rAAV). The GluN1 cKO mice had normal heat or cold paw-withdrawal latencies, mechanical threshold, and motor function but experienced reduced injury-induced pain produced by intraplantar formalin. Therefore, from these findings, we can conclude that the GluN1 receptor subunit in the lumbar dorsal horn of the spinal cord is required for central sensitization and the central facilitation of pain transmission produced by peripheral injury. Gray et al.[46] examined GluN2A and GluN2B cKO mice in the hippocampus to explore the effect on AMPAR expression. The authors showed that the deletion of GluN2A or GluN2B during early postnatal development results in a period of rapid synaptogenesis, suggesting that both subunits negatively regulate synaptic AMPAR expression. While GluN2A increases synaptic strength without affecting the number of connections, GluN2B increases the number of functional synapses and plays a role in maintaining silent synapses in early development. Other studies have examined GluN2B cKO in the CA3 subfield of the hippocampus and found that it abolished NMDAR-mediated EPSCs and tetanus-induced LTP at the commissural-CA3 synapse.[47] Similarly, in mice lacking GluN2B in pyramidal neurons in the CA1 region of the hippocampus and the cortex, LTP induced by two tetani (a subsaturating form of hippocampal LTP) was impaired, but this effect could be overcome with multiple tetani.[48] Furthermore, the GluN2B KO mice exhibited decreased dendritic spine density in CA1 hippocampal neurons compared with controls and revealed deficits in corticohippocampal-mediated learning and memory on the Morris water maze test, T-maze spontaneous alternation, and Pavlovian trace fear conditioning. Together, these studies suggest that the GluN2B subunit is involved in NMDAR channel function, organization of postsynaptic macromolecular complexes, the formation of dendritic spines, and the regulation of actin cytoskeleton, which are all important for synapse development, synaptic plasticity, and ultimately, cognitive function.

From these and other studies, cKO models are an effective strategy to overcome the limits such as embryonic lethality, developmental compensation, and global effects that are associated with conventional KO models. Thus, cKO models allow the phenotypes observed to be more specifically associated with the inactivation of the particular target gene. However, cKO models require a more thorough analysis to confirm temporal and spatial gene deletions and to avoid partial expression of the gene. Furthermore, the generation of cKOs is more laborious since the deletion of the gene requires additional breeding with a tissue-specific mouse line driven by a particular promoter that limits the Cre expression in certain areas.

## PART 3: KNOCK-IN

Another strategy to study the function of endogenous genes is through the creation of knock-in (KI) mice. Compared to other gene-targeting models that involve the deletion/disruption of a particular gene, a KI mouse is generated by engineering a vector with specific mutations at a selected locus (Fig. 18.4). Thus, the mutant gene is

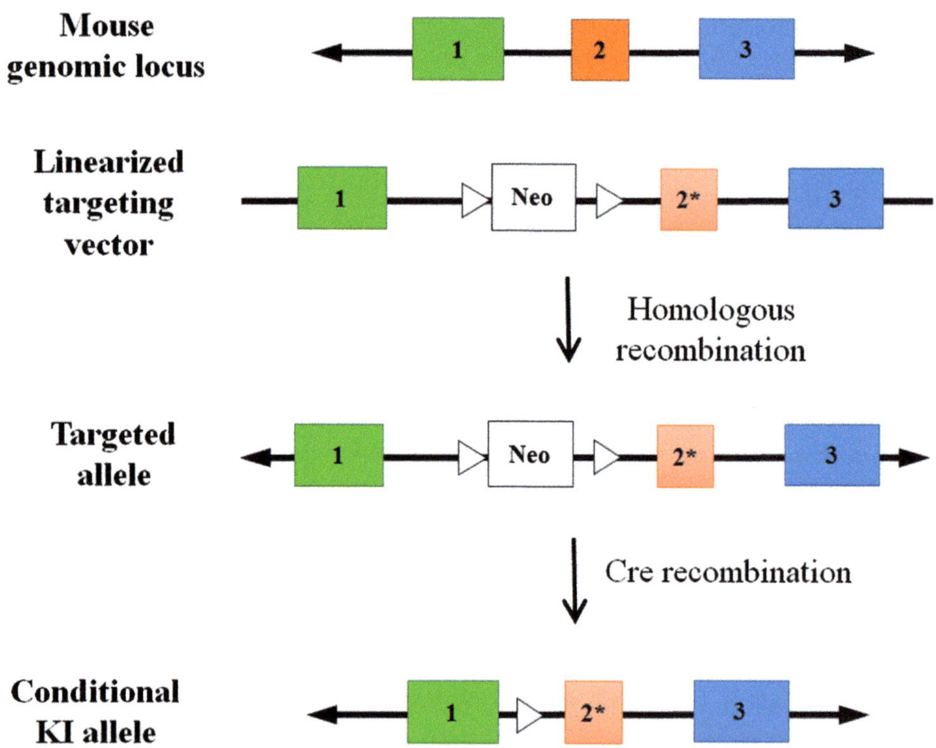


**Figure 18.4** The generation of a knock-in (KI) mouse model. A targeting vector contains a modified exon 2* and a Neo marker gene, flanked by two loxP sequences, denoted as *open triangles*. Following homologous recombination, the targeted allele contains both the Neo gene and the modified exon 2*. With Cre recombination in ES cells, a small loxP sequence (*one triangle*) remains after the excision of the Neo selection marker gene. The deletion of the marker can also be achieved by breeding the floxed mouse with a Cre-expressing mouse. The targeted gene with a modified exon 2* is now transcribed at the locus.

introduced into the mouse genome to replace an endogenous wild-type gene via homologous recombination. The procedure to generate KI mice is similar to KO mice, but the targeting vector for the KI mice has one or more mutations, but not deletions. The strategy also involves the use of the Cre/loxP system to remove the selection marker. A KI targeting construct normally contains a selection marker gene (e.g. *Neo*$^r$) flanked by two loxP sites typically in the intronic sequence and a modified exonic sequence of the gene to be targeted. Through homologous recombination, the endogenous gene is replaced by the targeted allele containing both the Neo marker and the modified gene sequence. With Cre expression either in ES cells or in vivo through breeding with a cell-type specific promoter mouse, the flanked Neo selection marker gene is removed, resulting in a targeted allele with the desired KI modification and a small insertion loxP.

In generating a successful KI mouse, the KI targeting vectors should be constructed in such a way that the modified genes acquire the transcription patterns of the endogenous gene. Roebroek et al.[49] suggest various approaches to construct KI vectors. Targeting KI vectors can be created using an exon with an inserted stop codon flanked by loxP recombination sequences to truncate the encoded protein known as a "floxed-stop" cassette approach. KI vectors can also insert, change, or delete amino acid residues or protein domains within coding sequences of an exon. In addition, KI vectors can insert or delete exons, resulting in an in frame insertion or deletion of amino acids or protein domains, to address the function of the specific portions of the protein. Furthermore, vectors can replace the complete endogenous protein-coding sequences with the coding sequences of another gene. Last, KI vectors can insert a cDNA encoding an additional open reading frame into the 3′-untranslated region preceded by an internal ribosome entry site that allows for cap-independent initiation of translation, resulting in the concomitant expression of an additional gene (e.g. Cre) along with the endogenous protein of the targeted gene.

Using the KI approach, subtle genetic changes can be made without altering the entire gene. In addition, KI allows the generation of mice carrying human mutations relevant to specific genetic disease models. Thus, compared to conventional transgenic and global KO techniques, site-specific KI mice have several benefits that include fewer effects on development or survival of the mice and more consistent levels of the mutant gene expression across generations since the gene exists in a single copy. Furthermore, the spatiotemporal expression of the targeted transgene is similar to the replaced wild-type endogenous gene, which avoids unwanted phenotypes. This allows the resulting phenotype to be attributed more definitively to the mutant protein expression of the targeted gene. However, the generation of KI models are also laborious with respect to the construction of the KI vector. Since only a limited number of mutations can be introduced within the targeting vector, choosing a crucial mutation to alter that provides the most mechanistic insight is often difficult.

Although most studies on AMPARs and NMDARs have used transgenic, KO, or cKO mouse models, there have been a number of successful examples that have employed the KI approach to introduce specific mutations into these glutamate receptors. To investigate the role of the two phosphorylation sites (S831 and S84) at the C-tail of GluA1 in synaptic plasticity, Lee et al.[50] generated KI mice where these sites were mutated to alanines to prevent their phosphorylation. A targeting vector encoding alanine substitutions at the two phosphorylation sites in exon 17 was constructed with a LoxP flanked Neo resistant marker in intron 16 in vitro. The correctly targeted ES cells were injected into C57BL/6 blastocysts and the chimeric mice carrying the mutant allele were bred and crossed with CMV-Cre mice to delete the Neo marker gene from the germline and intercrossed to produce double phosphomutant homozygous mice. The GluA1 phosphomutant mice showed normal gross anatomy and basal synaptic transmission but significantly impaired NMDAR-dependent LTD. However, only adult KI mice demonstrated reduced LTP in the CA1 region of the hippocampus. Behaviorally, both groups of KI mice exhibited normal learning but impaired spatial working memory on the Morris water maze. Together, these findings suggest that phosphorylation of the GluA1 subunit is critical for bidirectional synaptic plasticity and learning and memory in the hippocampus. More recently, the same group explored which of the two phosphorylation sites was responsible for the phenotype by generating two separate mutant mouse lines known as GluA1-S831A and GluA1-S845A.[51] The two lines of mice were generated by introducing mutations at each of the two phosphorylation sites respectively. They found that S831A mutants displayed normal LTP and LTD, whereas S845A mutants showed a specific deficit in LTD. These results suggest that the S845 site is critical for LTD expression, whereas either phosphorylation site alone may be sufficient to support LTP in the hippocampus. Other studies have also demonstrated that emotional stress can induce phosphorylation changes that are critical for the synaptic delivery of AMPARs during LTP and learning and memory.[52] In additional, Makino et al. have shown that phosphomimetic GluA1 (S831D/S845D) KI mice show enhanced LTP with a 10-Hz stimulation LTP-inducing protocol compared to wild-type mice, suggesting that the phosphorylation at S831 and S845 lowers the threshold for LTP induction and increases the probability for synaptic plasticity.[53] To investigate whether PDZ domain-containing proteins may be important in regulating the level of synaptic targeting of AMPARs during synaptic plasticity, Kim et al.[54] generated GluA1Δ7 KI mice with a truncated GluA1 carboxyl-terminal PDZ domain lacking specific residues of the GluA1 subunit. The authors targeted exon 17 coding for the C-terminus of GluA1 using a targeting vector to delete the last seven amino acids. They showed that the deletion did not affect the viability and behavior of the mice and that basal GluA1 synaptic localization, basal synaptic transmission, LTP, and LTD were all normal. These findings indicate that this domain is not required for CA1 hippocampal synaptic plasticity. In a separate set of studies, Sprengel et al.[55] investigated the physiological significance of the large intracellular C-terminal domain of the GluN2 subunit of the NMDAR. The authors altered the GluN2A, GluN2B, and GluN2C gene loci in the mouse by inserting an in-frame premature translational stop codon prior to the C-terminal, along with a Neo marker gene. Interestingly, the KI mice expressing the truncated GluN2 subunits revealed phenotypes congruent to those of the KO mice lacking the entire subunit. Specifically, the truncation of the C-terminal domain of GluN2B embryonically led to perinatal lethality, which was consistent with the full KO of GluN2B.[39] Whereas, the C-terminal truncation of GluN2A led to impaired hippocampal LTP along with memory deficits in a contextual fear-conditioning paradigm, which was similar to animals completely lacking GluN2A expression.[37] Last, the largely cerebellar-specific GluN2C receptor when C-terminally truncated resulted in a mild deficit in balance behavior on a rotarod test, which recapitulated the effects of the full genetic ablation of this subtype.[56] Hence, these data together indicate the crucial importance of the intracellular C-termini of GluN2 subunits for proper functionality of the NMDAR. The authors justify that the C-termini of GluN2 mediate the decoding and transduction of calcium signals into cellular responses entering the channel.[55] Thus, the truncation prevents the posttranslational modifications and protein interactions occurring at the C-terminal domain and blocks the physical coupling of cellular signaling components from calcium entry.

## PART 4: CRISPR

In the last decade, perhaps the most transformative and powerful genome-editing technique within biological research has been CRISPR. It is a tool that has allowed for unprecedented precision, efficiency, and adaptability, driven by an innate defense mechanism found in select bacteria and archaea. CRISPR refers to the "clustered regularly interspaced short palindromic repeats," which describes the unique DNA sequences that lie within other continuously repeating DNA sequences in bacterial genomes. As part of the bacteria's adaptive immunity, CRISPR mechanisms can recognize invading genetic material from viruses or plasmids and protect themselves through acquiring resistance against the infectious bacteriophage by integrating a genome fragment of the virus into its locus.[57] Furthermore, bacteria also employ Cas (CRISPR-associated proteins) found in close proximity to the CRISPR sequences that can precisely snip and remove invading viruses. The most well-known Cas is Cas9, an enzyme that is derived from *Streptococcus pyogenes*.[58] The type II CRISPR mechanism is unique compared to the other CRISPR systems, in that only one wild-type Cas9 nuclease is required for gene silencing (Fig. 18.5). Specifically, Cas9 binds to a short engineered RNA known as a "guide" RNA (gRNA) and a spacer sequence that defines the unique genomic target (~20 nucleotide DNA sequence in length and upstream of a protospacer adjacent motif (PAM)) to be modified through simple base-pairing principles. The PAM sequence is necessary for target binding, and the exact sequence is dependent on the species of Cas9. CRISPR

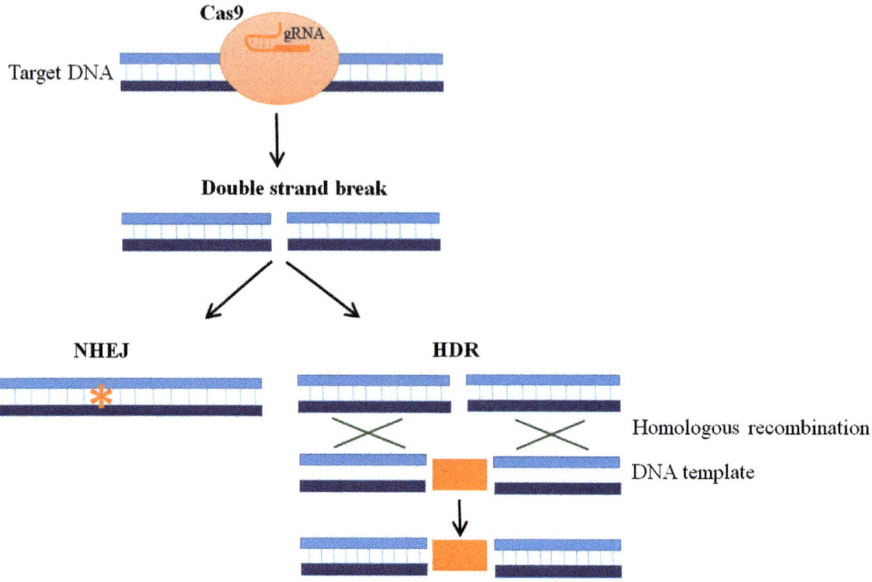


**Figure 18.5** The use of CRISPR/Cas9 for gene-editing approaches. The expression of a Cas9 enzyme and guide RNAs (gRNAs) enables the target DNA to be cut at a desired location. The gRNA consists of a 20-bp predesigned RNA sequence that binds to the target sequence of DNA and "guides" Cas9 to cut the two strands of DNA at the specified region, resulting in a double-strand break (DSB). The damaged DNA triggers repair machinery. The DSB is repaired by nonhomologous end joining (NHEJ) or homology direct repair (HDR). In NHEJ, small nucleotide insertions or deletions occur at the DSB site, which is often imperfect and results in disruptions of the open reading frame. HDR requires a DNA repair template (lacking the sequence for Cas9 to bind) that contains the desired sequence to be introduced within the DNA and additional left and right homology arms. Following homologous recombination, the target DNA with specific nucleotide changes or a large insertion can be incorporated into the target DNA.

suffers from the possibility of nonspecificity issues since the targeted recognition nucleotide sequence is relatively short and may match multiple alternative sequences in the genome. Ultimately, *S. pyogenes* Cas9 and gRNA form a riboprotein complex that undergoes a conformational change into an active DNA-binding conformation that binds to any genomic sequence marked with a PAM. Once the Cas9 complex binds, it leads to the destruction of the target DNA where two endonuclease domains, RuvC-like nuclease domain and an HNH-like nuclease domain, cleave the noncomplementary and complementary strands of the target DNA, respectively.[59] This creates a DNA double-strand break (DSB) at the genomic locus to be modified, usually 3–4 nucleotides upstream of the PAM sequence. Nuclease-induced DSBs can be repaired by nonhomologous end joining (NHEJ) or homology-directed repair pathways (HDR) via homologous recombination. The NHEJ repair pathway is an efficient but error-prone mechanism resulting in insertion or deletion mutations of variable length, whereas the

HDR pathway is less efficient but has high fidelity in the introduction of precise point mutations or insertions from single- or double-stranded DNA donor templates.

Compared to traditional KO techniques that require ES cells, CRISPR/Cas9 sequences along with the altered genes can be injected into mouse embryos where the system is able to efficiently edit both copies of the gene in one generation. CRISPR was originally employed to knock out target genes in various cell types and organisms. However, technical advances including modifications to the Cas9 enzyme have allowed the system to be adapted for gene repression or activation of specific regions of DNA.[60] CRISPR has a higher targeting efficiency and is a more cost-friendly approach compared to other widely used tools for precise genome engineering, such as zinc finger nucleases (ZFNs) and transcription activator-like effector nucleases (TALENs). These alternative techniques are able to recognize longer DNA sequences and generate permanent mutations through introducing DSBs to activate repair pathways, which theoretically allow them to have better specificity. However, ZFNs and TALENs require custom-designed protein-DNA interfaces rendering them costly and time consuming. Therefore, the process to identify a successful variation for optimization in these approaches is much more difficult than creating a RNA guide sequence for CRISPR/Cas9. In addition, with appropriate guide RNAs, CRISPR and Cas9 can be used to modify the genomes of any animal ranging from fruit flies, zebrafish, rodents, and even to humans.

Perhaps the first attempt to target glutamate receptors in excitatory hippocampal neurons using the CRISPR/Cas9 system was demonstrated by Incontro et al.[61] where they targeted the obligatory GluN1 subunit of the NMDAR and the GluA2 subunit of the AMPAR in cultured hippocampal slices. To selectively inactivate the GluN1 subunit, the authors employed a gene gun technique, where they coated gold particles with plasmids containing complementary guide GluN1 RNA, along with Cas9 into postmitotic hippocampal neurons. The design of the gRNA targeted the 5′ protein-coding region since a frameshift is likely to result in a complete loss of function. Moreover, the selected gRNAs encompassed an exon-intron junction to affect splice sites to produce a null mutation with both in-frame and out-of-frame insertions. Recordings were made at 14 days following the sparse biolistic transfection, and NMDAR-mediated excitatory postsynaptic currents were completely abolished, indicating that GluN1 was effectively deleted from the neurons. On the other hand, AMPAR EPSCs were found to increase doubly, which is consistent with the findings from GluN1 cKO mice.[44] For AMPARs, the authors designed gRNAs targeting a domain of GluA2 essential for its folding, activity, or stability. To do this, they transfected two different gRNAs and found that the transfected neurons had normal NMDAR eEPSCs but significantly reduced AMPAR eEPSCs, and the remaining AMPAR currents were all inwardly rectifying, consistent with the results from the GluA2 cKO mice.[42] Interestingly, another paper published at the same time also demonstrated that the knockdown of GluN1 via CRISPR/Cas9 in neuronal slice cultures also effectively abolished NMDAR responses. However, in this study, Straub et al.[62] injected the CRISPR constructs directly into the brain of embryonic day-15 wild-type C57BL/6J pups via in utero electroporation. To do this, a pregnant mother was anesthetized, and embryonic pups within the uterus were temporarily removed and injected with DNA containing the CRISPR-construct and soluble GFP into the left hemisphere. To direct the transfection to the hippocampus, the head of the pup was placed between paddles of tweezer electrodes and subjected to electric pulses to drive the DNA construct into mitotic progenitor cells that eventually undergo neurogenesis, migration, and differentiation to become mature neurons. Following electroporation, the pups were returned to the uterus, and the abdomen was sutured closed. In this study, the CRIPR/Cas9 target sequences were within or near the region encoding for the second, pore-forming transmembrane domain of GluN1 on exon 14. The pups at postnatal day 14 revealed no synaptic currents mediated by NMDAR in all neurons tested in the CA1 and CA3 regions of the hippocampus, which recapitulates the results of the complete GluN1 KO mice.[36] The authors also repeated their CRISPR/Cas9 experiments using the same gRNAs as discussed earlier in rat organotypic hippocampal slice cultures, through biolistic transfection in postmitotic neurons in the rat, and they found significantly reduced NMDAR currents compared to untransfected neighboring controls. Taken together, these two studies have demonstrated that CRISPR/Cas9 is an effective and efficient system to inactivate genes of interest in both in vitro and in vivo preparations. The CRISPR/Cas9 system overcomes problems associated with lethality and potential compensatory mechanisms by other subunits commonly associated with global KO models of NMDARs and AMPARs. However, several challenges still remain, including the possibility of off-target effects since this method relies on a large 20-bp recognition motif. To overcome this concern, Incontro et al.[61] designed a gRNA that encompasses an intron-exon junction of the coding unit of the particular subunit (either GluN1 or GluA2) that does not target the receptor protein encoding sequences, which allowed the cotransfection of the particular receptor cDNA to reverse the deficits observed in eEPSCs. Specifically, the cleavage site was in the exon, whereas the majority of the gRNA sequence was within the intron. Alternatively, to ensure that the rescue construct is not targeted by CRISPR, cDNA with silent mutations can also be introduced into the region recognized by the gRNA or diverse cDNA belonging to a different species. Ultimately, the future of CRISPR/Cas9 will depend on new delivery methods that extend beyond in utero electroporation and biolistic transfection, which will enable the delivery of multiple large constructs with increasingly high spatial and temporal specificity.

## CONCLUSION

In this review, we have briefly introduced a number of genetic manipulation techniques, specifically RNA interference, knockout, conditional knockout, knock-in, and

CRISPR/Cas9. For each, we have highlighted the advantages and disadvantages and briefly discussed how the use of each method has contributed to our understanding of two ionotropic glutamate receptors, NMDA and AMPA, in synaptic transmission and plasticity, and learning and memory. These genetic approaches are a glimpse of the broad range of tools currently available for the systematic investigation of the in vivo function of genes in synaptic regulation and behavior. Currently, other technologies including electrophysiological recording, fluorescent live imaging, and optogenetic and pharmacogenetic approaches are being applied in conjunction with these genetic techniques to more precisely target synapses of particular molecular and structural compositions. Therefore, the future of gene-targeting techniques still holds great promise as the development of newly refined and sophisticated models will continue to contribute to our understanding of the cellular and molecular mechanisms underlying brain function in health and disease.

## REFERENCES


1. Collingridge GL, Kehl SJ, McLennan H. Excitatory amino acids in synaptic transmission in the Schaffer collateral-commissural pathway of the rat hippocampus. *J Physiol.* 1983;334:33–46.
2. Hestrin S, Nicoll RA, Perkel DJ, Sah P. Analysis of excitatory synaptic action in pyramidal cells using whole-cell recording from rat hippocampal slices. *J Physiol.* 1990;422:203–225.
3. Sheng M, Lee SH. AMPA receptor trafficking and the control of synaptic transmission. *Cell.* 2001; 105(7):825–828.
4. Akaneya Y. Activity regulates the expression of AMPA receptor subunit GluR4 in developing visual cortex. *Eur J Neurosci.* 2007;25(6):1641–1646.
5. Zhu JJ, Esteban JA, Hayashi Y, Malinow R. Postnatal synaptic potentiation: delivery of GluR4-containing AMPA receptors by spontaneous activity. *Nat Neurosci.* 2000;3(11):1098–1106.
6. Wenthold RJ, Petralia RS, Blahos II J, Niedzielski AS. Evidence for multiple AMPA receptor complexes in hippocampal CA1/CA2 neurons. *J Neurosci.* 1996;16(6):1982–1989.
7. Shi S, Hayashi Y, Esteban JA, Malinow R. Subunit-specific rules governing AMPA receptor trafficking to synapses in hippocampal pyramidal neurons. *Cell.* 2001;105(3):331–343.
8. Jonas P, Burnashev N. Molecular mechanisms controlling calcium entry through AMPA-type glutamate receptor channels. *Neuron.* 1995;15(5):987–990.
9. Collingridge GL, Isaac JT, Wang YT. Receptor trafficking and synaptic plasticity. *Nat Rev Neurosci.* 2004;5(12):952–962.
10. Malinow R, Malenka RC. AMPA receptor trafficking and synaptic plasticity. *Annu Rev Neurosci.* 2002; 25:103–126.
11. Bredt DS, Nicoll RA. AMPA receptor trafficking at excitatory synapses. *Neuron.* 2003;40(2):361–379.
12. Kessels HW, Malinow R. Synaptic AMPA receptor plasticity and behavior. *Neuron.* 2009;61(3): 340–350.
13. Paoletti P, Bellone C, Zhou Q. NMDA receptor subunit diversity: impact on receptor properties, synaptic plasticity and disease. *Nat Rev Neurosci.* 2013;14(6):383–400.
14. Sans N, Petralia RS, Wang YX, Blahos 2nd J, Hell JW, Wenthold RJ. A developmental change in NMDA receptor-associated proteins at hippocampal synapses. *J Neurosci.* 2000;20(3):1260–1271.
15. Petralia RS, Al-Hallaq RA, Wenthold RJ. Trafficking and targeting of NMDA receptors. In: Van Dongen AM, ed. *Biology of the NMDA Receptor. Boca Raton (FL).* 2009.
16. Sobczyk A, Scheuss V, Svoboda K. NMDA receptor subunit-dependent [$Ca^{2+}$] signaling in individual hippocampal dendritic spines. *J Neurosci.* 2005;25(26):6037–6046.

17. Luscher C, Malenka RC. NMDA receptor-dependent long-term potentiation and long-term depression (LTP/LTD). *Cold Spring Harb Perspect Biol.* 2012;4(6).
18. Fire A, Xu S, Montgomery MK, Kostas SA, Driver SE, Mello CC. Potent and specific genetic interference by double-stranded RNA in *Caenorhabditis elegans*. *Nature.* 1998;391(6669):806–811.
19. Wilson RC, Doudna JA. Molecular mechanisms of RNA interference. *Annu Rev Biophys.* 2013;42: 217–239.
20. Tracy TE, Yan JJ, Chen L. Acute knockdown of AMPA receptors reveals a trans-synaptic signal for presynaptic maturation. *EMBO J.* 2011;30(8):1577–1592.
21. Moore CB, Guthrie EH, Huang MT, Taxman DJ. Short hairpin RNA (shRNA): design, delivery, and assessment of gene knockdown. *Methods Mol Biol.* 2010;629:141–158.
22. Hall BJ, Ripley B, Ghosh A. NR2B signaling regulates the development of synaptic AMPA receptor current. *J Neurosci.* 2007;27(49):13446–13456.
23. Gainey MA, Hurvitz-Wolff JR, Lambo ME, Turrigiano GG. Synaptic scaling requires the GluR2 subunit of the AMPA receptor. *J Neurosci.* 2009;29(20):6479–6489.
24. Foster KA, McLaughlin N, Edbauer D, et al. Distinct roles of NR2A and NR2B cytoplasmic tails in long-term potentiation. *J Neurosci.* 2010;30(7):2676–2685.
25. Sigoillot FD, King RW. Vigilance and validation: keys to success in RNAi screening. *ACS Chem Biol.* 2011;6(1):47–60.
26. Reynolds A, Anderson EM, Vermeulen A, et al. Induction of the interferon response by siRNA is cell type- and duplex length-dependent. *RNA.* 2006;12(6):988–993.
27. Brummelkamp TR, Bernards R, Agami R. A system for stable expression of short interfering RNAs in mammalian cells. *Science.* 2002;296(5567):550–553.
28. Leung C, Jia Z. Mouse genetic models of human brain disorders. *Front Genet.* 2016;7:40.
29. Matsuda I, Aiba A. Receptor knock-out and knock-in strategies. *Methods Mol Biol.* 2004;259:379–390.
30. Capecchi MR. Targeted gene replacement. *Sci Am.* 1994;270(3):52–59.
31. Zamanillo D, Sprengel R, Hvalby O, et al. Importance of AMPA receptors for hippocampal synaptic plasticity but not for spatial learning. *Science.* 1999;284(5421):1805–1811.
32. Selcher JC, Xu W, Hanson JE, Malenka RC, Madison DV. Glutamate receptor subunit GluA1 is necessary for long-term potentiation and synapse unsilencing, but not long-term depression in mouse hippocampus. *Brain Res.* 2012;1435:8–14.
33. Geiger JR, Melcher T, Koh DS, et al. Relative abundance of subunit mRNAs determines gating and $Ca^{2+}$ permeability of AMPA receptors in principal neurons and interneurons in rat CNS. *Neuron.* 1995; 15(1):193–204.
34. Jia Z, Agopyan N, Miu P, et al. Enhanced LTP in mice deficient in the AMPA receptor GluR2. *Neuron.* 1996;17(5):945–956.
35. Meng Y, Zhang Y, Jia Z. Synaptic transmission and plasticity in the absence of AMPA glutamate receptor GluR2 and GluR3. *Neuron.* 2003;39(1):163–176.
36. Li Y, Erzurumlu RS, Chen C, Jhaveri S, Tonegawa S. Whisker-related neuronal patterns fail to develop in the trigeminal brainstem nuclei of NMDAR1 knockout mice. *Cell.* 1994;76(3):427–437.
37. Sakimura K, Kutsuwada T, Ito I, et al. Reduced hippocampal LTP and spatial learning in mice lacking NMDA receptor epsilon 1 subunit. *Nature.* 1995;373(6510):151–155.
38. Cho KK, Khibnik L, Philpot BD, Bear MF. The ratio of NR2A/B NMDA receptor subunits determines the qualities of ocular dominance plasticity in visual cortex. *Proc Natl Acad Sci USA.* 2009; 106(13):5377–5382.
39. Kutsuwada T, Sakimura K, Manabe T, et al. Impairment of suckling response, trigeminal neuronal pattern formation, and hippocampal LTD in NMDA receptor epsilon 2 subunit mutant mice. *Neuron.* 1996;16(2):333–344.
40. Mohamad O, Song M, Wei L, Yu SP. Regulatory roles of the NMDA receptor GluN3A subunit in locomotion, pain perception and cognitive functions in adult mice. *J Physiol.* 2013;591(1):149–168.
41. Niemann S, Kanki H, Fukui Y, et al. Genetic ablation of NMDA receptor subunit NR3B in mouse reveals motoneuronal and nonmotoneuronal phenotypes. *Eur J Neurosci.* 2007;26(6):1407–1420.
42. Lu W, Shi Y, Jackson AC, et al. Subunit composition of synaptic AMPA receptors revealed by a single-cell genetic approach. *Neuron.* 2009;62(2):254–268.

43. Gutierrez-Castellanos N, Da Silva-Matos CM, Zhou K, et al. Motor learning requires Purkinje cell synaptic potentiation through activation of AMPA-receptor subunit GluA3. *Neuron*. 2017;93(2): 409–424.
44. Tsien JZ, Huerta PT, Tonegawa S. The essential role of hippocampal CA1 NMDA receptor-dependent synaptic plasticity in spatial memory. *Cell*. 1996;87(7):1327–1338.
45. South SM, Kohno T, Kaspar BK, et al. A conditional deletion of the NR1 subunit of the NMDA receptor in adult spinal cord dorsal horn reduces NMDA currents and injury-induced pain. *J Neurosci*. 2003;23(12):5031–5040.
46. Gray JA, Shi Y, Usui H, During MJ, Sakimura K, Nicoll RA. Distinct modes of AMPA receptor suppression at developing synapses by GluN2A and GluN2B: single-cell NMDA receptor subunit deletion in vivo. *Neuron*. 2011;71(6):1085–1101.
47. Akashi K, Kakizaki T, Kamiya H, et al. NMDA receptor GluN2B (GluR epsilon 2/NR2B) subunit is crucial for channel function, postsynaptic macromolecular organization, and actin cytoskeleton at hippocampal CA3 synapses. *J Neurosci*. 2009;29(35):10869–10882.
48. Brigman JL, Wright T, Talani G, et al. Loss of GluN2B-containing NMDA receptors in CA1 hippocampus and cortex impairs long-term depression, reduces dendritic spine density, and disrupts learning. *J Neurosci*. 2010;30(13):4590–4600.
49. Roebroek AJ, Wu X, Bram RJ. Knockin approaches. *Methods Mol Biol*. 2003;209:187–200.
50. Lee HK, Takamiya K, Han JS, et al. Phosphorylation of the AMPA receptor GluR1 subunit is required for synaptic plasticity and retention of spatial memory. *Cell*. 2003;112(5):631–643.
51. Lee HK, Takamiya K, He K, Song L, Huganir RL. Specific roles of AMPA receptor subunit GluR1 (GluA1) phosphorylation sites in regulating synaptic plasticity in the CA1 region of hippocampus. *J Neurophysiol*. 2010;103(1):479–489.
52. Hu H, Real E, Takamiya K, et al. Emotion enhances learning via norepinephrine regulation of AMPA-receptor trafficking. *Cell*. 2007;131(1):160–173.
53. Makino Y, Johnson RC, Yu Y, Takamiya K, Huganir RL. Enhanced synaptic plasticity in mice with phosphomimetic mutation of the GluA1 AMPA receptor. *Proc Natl Acad Sci USA*. 2011;108(20): 8450–8455.
54. Kim CH, Takamiya K, Petralia RS, et al. Persistent hippocampal CA1 LTP in mice lacking the C-terminal PDZ ligand of GluR1. *Nat Neurosci*. 2005;8(8):985–987.
55. Sprengel R, Suchanek B, Amico C, et al. Importance of the intracellular domain of NR2 subunits for NMDA receptor function in vivo. *Cell*. 1998;92(2):279–289.
56. Kadotani H, Hirano T, Masugi M, et al. Motor discoordination results from combined gene disruption of the NMDA receptor NR2A and NR2C subunits, but not from single disruption of the NR2A or NR2C subunit. *J Neurosci*. 1996;16(24):7859–7867.
57. Barrangou R, Fremaux C, Deveau H, et al. CRISPR provides acquired resistance against viruses in prokaryotes. *Science*. 2007;315(5819):1709–1712.
58. Deltcheva E, Chylinski K, Sharma CM, et al. CRISPR RNA maturation by trans-encoded small RNA and host factor RNase III. *Nature*. 2011;471(7340):602–607.
59. Jinek M, Chylinski K, Fonfara I, Hauer M, Doudna JA, Charpentier E. A programmable dual-RNA-guided DNA endonuclease in adaptive bacterial immunity. *Science*. 2012;337(6096):816–821.
60. Qi LS, Larson MH, Gilbert LA, et al. Repurposing CRISPR as an RNA-guided platform for sequence-specific control of gene expression. *Cell*. 2013;152(5):1173–1183.
61. Incontro S, Asensio CS, Edwards RH, Nicoll RA. Efficient, complete deletion of synaptic proteins using CRISPR. *Neuron*. 2014;83(5):1051–1057.
62. Straub C, Granger AJ, Saulnier JL, Sabatini BL. CRISPR/Cas9-mediated gene knock-down in post-mitotic neurons. *PLoS One*. 2014;9(8):e105584.

CHAPTER 19

# Using Herpes Simplex Virus Type 1-Based Amplicon Vectors for Neuroscience Research and Gene Therapy of Neurologic Diseases

**Maria V. Baez[1], Alejandra I. Aguirre[1], Alberto L. Epstein[2], Diana A. Jerusalinsky[1]**
[1]IBCN, University of Buenos Aires and CONICET, Ciudad de Buenos Aires, Argentina; [2]UMR INSERM U1179 – Université de Versailles Saint Quentin en Yvelines, UFR des Sciences de la Santé "Simone Veil", Montigny le Bretonneux, France

## INTRODUCTION

Amplicon vectors are particles identical to wild-type herpes simplex virus type 1 (HSV-1) from the structural as well as from the host-range points of view, but bearing a concatemeric form of a DNA plasmid, the amplicon plasmid, instead of the viral genome. HSV-1 amplicon vectors are extremely interesting and suitable tools as gene transfer vehicles due to the very high capacity to host foreign DNA, a feature that makes them unique. These vectors are suited to either express/overexpress a particular gene, involving its regulatory sequences if necessary, or to efficiently knockdown a target gene with the transgene consisting of either a high number of repeats of an antisense (AS) or an interference RNA (RNAi).[1] Although HSV-1-based vectors are particularly suited to infect neurons, being strongly neurotropic, particularly in vivo, they also share many useful features of the wild-type HSV-1,[2] as the ability to infect a broad range of dividing and nondividing host cells. HSV-1 amplicon vectors allow efficient infection of many neuronal types, with relatively long-term transgene expression, without evident side effects.[3] This is mainly due to the fact that amplicon vectors are devoid of viral genes and remain episomal, without integration into the host genome.

In the following sections, we summarize a brief description of the biology, structure, and genome of the HSV-1 to facilitate understanding of the biology, construction, production, and use of amplicon vectors.

## HERPES SIMPLEX VIRUS TYPE 1

### Brief Introduction to Herpes Simplex Virus Type 1 Biology

HSV-1 is a widespread, enveloped, neurotropic human pathogen, infecting ~70% of people worldwide, whose lifestyle is based on a long-term dual interaction with the infected host. After initial infection and lytic replication at the periphery, generally at

*Molecular-Genetic and Statistical Techniques for Behavioral and Neural Research*
ISBN 978-0-12-804078-2, https://doi.org/10.1016/B978-0-12-804078-2.00019-2

the epithelial cells of the oral mucosa where it produces a mild gingivostomatitis (the primary infection), the virus particles enter the sensory neurons that innervate the epithelia and travel in retrograde direction along the axons, reaching the neuronal nucleus in the sensory ganglia. Following delivery into the neuronal nucleus through the nuclear membrane pore complex, the viral genome ordinarily remains in a latent state for long periods. The virus is able to return to the epithelial cells by periodic reactivation; there, it produces secondary lytic infections resulting in mild illness symptoms, such as cold sores (secondary infections or recurrences). A short introduction to selected molecular aspects of the biologic cycle of HSV-1 will help understanding how HSV-1-derived vectors are generated and used. For a more comprehensive study of HSV-1 structure and molecular biology, see Roizman, Knipe and Whitley.[4]

## The Herpes Simplex Virus Type 1 Particle

HSV-1 is a 220-nm diameter, enveloped, double-stranded DNA virus, whose genome is surrounded by three concentric layers. As shown in Fig. 19.1, the main components in the mature virion are (1) the double-stranded (ds) DNA virus genome, (2) an icosadeltahedral capsid of 162 capsomeres, (3) the tegument, which is a protein layer between the capsid and the envelope that plays critical roles in virion morphogenesis and helps HSV-1 to control the expression machinery of the host cell early after infection, and (4) a lipidic envelope from cell origin with embedded viral glycoproteins and other membrane proteins, most of which are involved in receptor-mediated cellular entry.[5]

## The Herpes Simplex Virus Type 1 Genome

The linear 153-kbp ds DNA virus genome (Fig. 19.1) is densely packaged as a core within the capsid[6] and is devoid of histones.[7] This genome is arranged as long (UL) and short (US) unique segments of 126 and 26 kbp, respectively (Fig. 19.1, right). The UL and US segments are flanked by repeated sequences named TRL, IRL, IRS, and TRS (for terminal repeat L, internal repeat L, internal repeat S, and terminal repeat S).[8]

The HSV-1 genome has been fully sequenced.[9–11] There are at least 84 viral genes, which may be divided into essential and nonessential genes, depending on whether their expression is necessary or not for viral growth in cell culture. The virus genome also encodes several microRNAs (miRNAs), which play regulatory roles, mainly during establishment and/or reactivation from latency. HSV-1 genes do not contain intron sequences, except those encoding ICP0, ICP22, ICP47, UL15, and LAT. Nonessential genes often encode functions that are important for virus—host interaction in vivo, such as to counteract host innate and immune reactions, to allow replication in nondividing cells, or to shut down host protein synthesis.

Since approximately half of the viral genes are dispensable for HSV-1 replication in cultured cells, they could be replaced by exogenous genetic material, which has been the premise for the development of HSV-1-based vectors for gene therapy/transfer.[12] At variance, most genes involved in virus entry, DNA replication, capsid assembly, and DNA packaging are essential to produce viral particles.

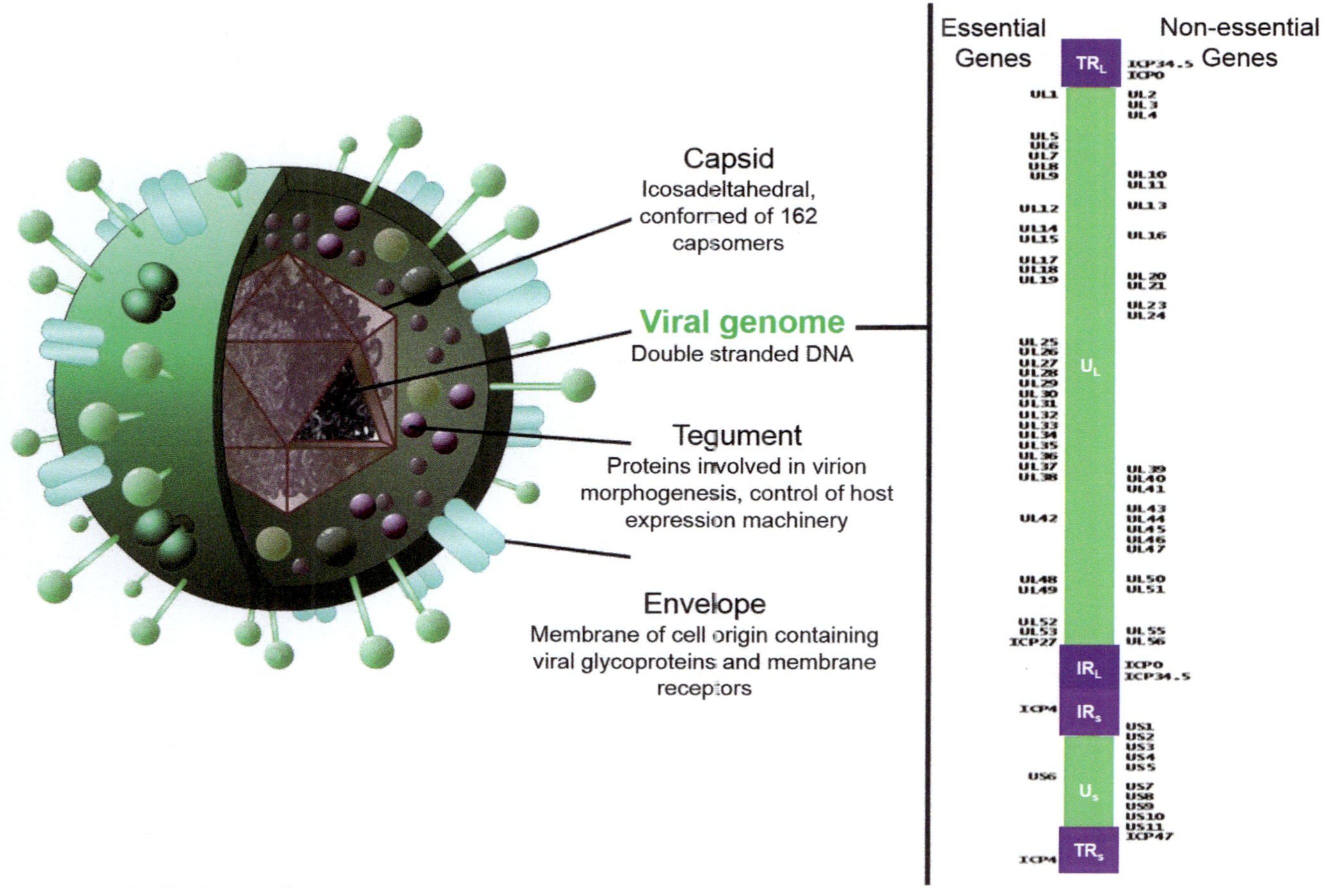


**Figure 19.1** ***Herpes simplex virus type 1 (HSV-1) structure.*** HSV-1 is composed by a lipoprotein envelope of cell origin, containing glycoproteins and other membrane proteins of viral origin, mostly involved in host cell–virus interaction. Below this envelope, there is a protein layer called the tegument, containing, among many others, proteins that activate immediate-early virus transcription and inhibit host cell translation. The viral genome is located within the capsid, an icosahedral protein structure. The viral genome (right) is a 153-kpb double-stranded DNA helix composed by a large (UL) and a short (US) fragment, separated by inverted repeats. About 50% of the virus genes are nonessential for virus multiplication in cultured cells.

The HSV-1 genome contains two *cis*-acting sequences that are essential for virus replication: the origins of viral DNA replication (ori) and the signals for cleavage and packaging of the newly synthesized DNA (pac or *a*). HSV-1 harbors three origins of replication, with two located within the repeated segments surrounding US (oriS) and one in the UL segment (oriL).[13] The oriS is located within the shared promoter regulatory region between the divergently transcribed immediate-early genes *ICP4* and *ICP22/ICP47*, whereas oriL lies in the regulatory region between two divergently transcribed early genes that are essential for viral DNA replication: the major DNA-binding protein and the HSV-1 DNA polymerase.[14] Mutant viruses lacking either oriL or both oriS copies are replication competent, indicating that all replication origins are functionally competent.[8]

The signals for cleavage and packaging of the HSV-1 genome are located within the *pac* sequences, generally found in tandem repeats at both ends of the genome, as well as at the L/S junction.[15,16] Those specific signals, termed pac1 and pac2, are located within the Ub and Uc regions, respectively.[16–18] The virus genome replicates using a combination of homologous recombination and rolling circle amplification.[19]

## BRIEF INTRODUCTION TO HERPES SIMPLEX VIRUS TYPE 1-BASED VECTORS

The improvement of ad hoc methods for gene transfer, particularly to eukaryotic cells, is a subject of increasing interest. To achieve an efficient delivery of genetic material in mammalian cells has been a major objective of molecular and cellular biology, gene therapy, and vaccine development during the last 30 years. In this line, viral-derived vectors are the most promising gene transfer tools since viruses naturally evolved as molecular gene carriers to specific tissues/organisms.

HSV-1 is a naturally epithelial-neurotropic virus, highly adapted to the host nervous system. Therefore, neurons are target cells for HSV-1-derived vectors. Coding for more than 84 genes, HSV-1 is a complex virus that can be engineered to design viral vectors for fundamental research and gene therapy of neurologic disorders, as well as for gene therapy of cancer and vaccines development.

The following features make it especially useful and particularly interesting as a vector (Fig. 19.2). (1) Although it is essentially neurotropic, under experimental conditions, HSV-1 can infect a broad cell type range because its cellular receptors are widely distributed; it infects and replicates in cells from different mammals, facilitating preclinical evaluation. (2) HSV-1 can infect dividing and nondividing cells. (3) The diseases that HSV-1 can cause are mostly nonsevere. (4) Transgene capacity is larger than in any other viral vector, allowing the delivery of multiple or very large transgenes (approximately half of the genome is nonessential in cultured cells and can be deleted without impairing vector production). In particular, HSV-1-derived amplicon vectors are devoid of almost all the viral genome (i.e., 153-kbp DNA), allowing to incorporate extremely high amounts of

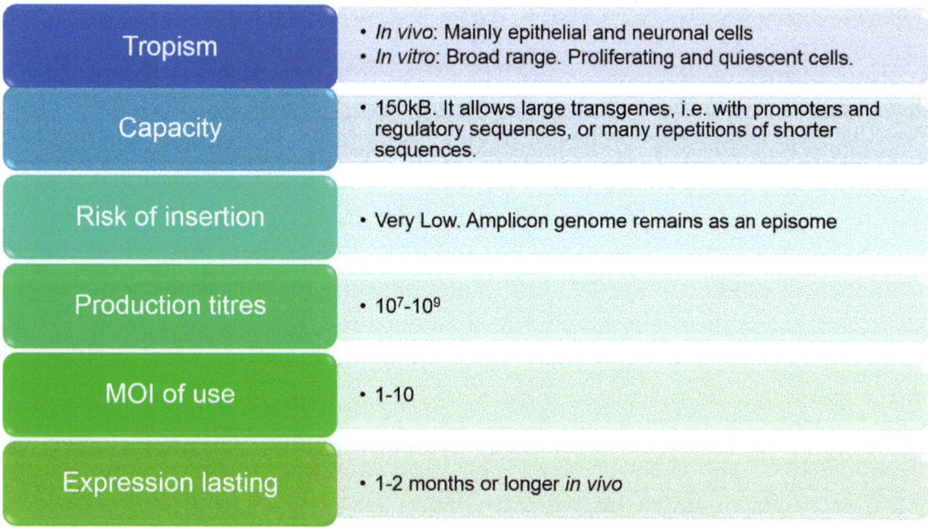


**Figure 19.2 Main advantages of amplicon vectors.**

exogenous DNA. (5) Safe replication-defective HSV-1 vectors may be prepared to high titres. (6) HSV-1 is able to establish a latent state in peripheral neurons, which may be useful, for instance, for long-term expression of therapeutic transgenes. (7) The viral genome remains episomal, therefore reducing the risk of insertional mutagenesis (Fig. 19.2).

To exploit the versatility of HSV-1 for gene transfer, three types of HSV-1 vectors have been developed: (1) attenuated replication-competent recombinant vectors, (2) replication-defective recombinant vectors, and (3) amplicon vectors.

1. Attenuated replication-competent HSV-1 vectors can replicate only in certain cell types and tissues in vivo due to deletion of some specific nonessential viral genes. Such vectors have been typically used for the development of tumor therapies and are usually referred to as oncolytic HSV-1 vectors. However, other uses of attenuated vectors are possible, i.e., delivery of genes to peripheral ganglia following vector infection at peripheral tissues.
2. Replication-defective recombinant HSV-1 vectors are devoid of one or more viral genes that are essential for lytic replication (such as the immediate-early genes ICP4 and ICP27). However, they retain the ability to establish latency.[20,21] Their production requires complementing cell lines expressing the deleted viral genes. These vectors can be used either as genetic vaccines or for gene therapy of neurologic disorders.
3. Amplicon vectors are fully defective vectors that lack all virus genes, so they require a helper system to be produced. Since this chapter will be exclusively focused on amplicon vectors, we present a more detailed description of the biology of these vectors in the following sections.

## HERPES SIMPLEX VIRUS TYPE 1-BASED AMPLICON VECTORS

Amplicon vectors, or amplicons,[22] are HSV-1 particles identical to wild-type HSV-1 from the structural (Fig. 19.3), immunologic, and host-range points of view, but carrying a concatemeric form of a DNA plasmid, the amplicon plasmid, instead of the viral genome. An amplicon plasmid (Fig. 19.3, right) is a standard *Escherichia coli* plasmid carrying one origin of viral DNA replication (generally oriS) and one packaging signal (pac) from HSV-1, in addition to a reporter gene and a multiple cloning site, where the transgene sequences can be cloned.[13,22] One major advantage of amplicons as gene transfer tools is the fact that they carry no viral genes and, consequently, do not induce synthesis of viral proteins. Therefore, these vectors are fully nontoxic for the infected cells and nonpathogenic for the inoculated organisms. Furthermore, the absence of viral genes in the amplicon genome strongly reduces the risk of reactivation, complementation, or recombination with latent or resident HSV-1 genomes.

The versatility of amplicons stems from the fact that during their production the amplicon genome will replicate, like HSV-1, via a rolling circle-like mechanism, generating long concatemers composed of tandem repeats of the amplicon plasmid.[23] Since HSV-1 particles will always package around 150-kbp DNA, corresponding to the size of the viral genome, the number of repeats that a particular amplicon vector will carry and deliver will depend on the size of the original amplicon plasmid. Therefore, an amplicon plasmid of around 5 kbp will be repeated some 30 times into the amplicon vector, while a very large amplicon plasmid carrying a 150-kbp genomic locus will generate

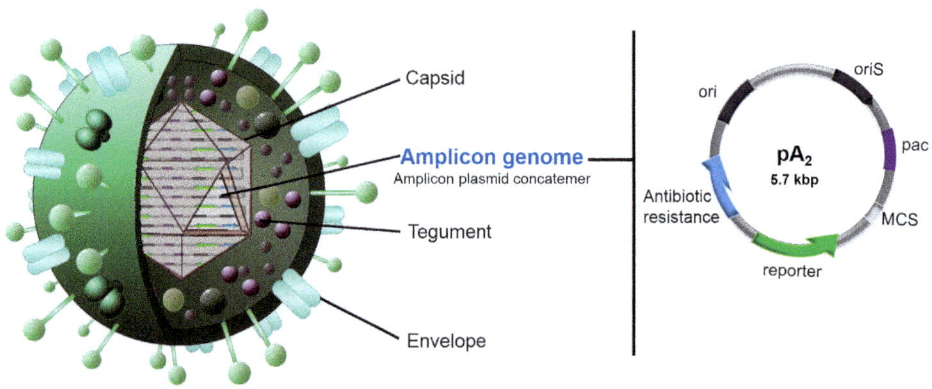


**Figure 19.3** ***Amplicon vector and amplicon plasmid structure.*** Amplicon vector particles are identical to herpes simplex virus type 1 (HSV-1) particles, but the vector genome is a concatemer derived from the amplicon plasmid (pA). The amplicon plasmid (right) contains bacterial sequences (one *Escherichia coli* replication origin and one antibiotic resistance gene) and HSV-1 regulatory sequences (one origin of virus DNA replication (oriS) and one packaging g signal (*pac*)). In addition, this plasmid contains, a multiple cloning site (MCS), where the transgene can be inserted, and a reporter gene (*green box*), generally the green fluorescent protein or β-galactosidase (β-gal).

amplicon vectors carrying a single repeat of this sequence. Hence, a second major benefit that arises from the lack of viral genes in the amplicon plasmid is that most of the 150-kbp capacity of the HSV-1 particle can be used to accommodate very large pieces of foreign DNA. That is, undoubtedly, the most outstanding property of amplicons, as there is no other viral vector system available displaying the capacity to deliver such a large amount of foreign DNA to the nuclear environment of mammalian cells.

## Preparation of Nontoxic Amplicon Vector Stocks

Since HSV-1-amplicon vectors carry no viral genes, they are replication-defective and depend on helper functions for production. The helper genome should provide all viral functions required to replicate and package the amplicon genome (including replication, structural, and DNA packaging proteins), but it should lack packaging signals to avoid packaging of the helper genome itself. It is critical that amplicon stocks that will be used for gene transfer and gene therapy do not contain contaminating helper particles, which can be toxic and induce immune responses. In spite of the difficulties to generate such helper-free amplicon vectors, two different methodologies have been developed, which allow generating amplicon stocks either completely free of helper particles or only faintly contaminated with fully defective helper particles. These two methodologies are developed in the following paragraphs and shown in Figs. 19.4 and 19.5. (The corresponding protocols are fully developed in Ref. Fraefel et al.[24])

## Production of Amplicon Vectors by Co-Transfecting Amplicon Plasmids With Helper Defective HSV-1 Genomes (Fig. 19.4)

Helper functions can be provided by replication-competent, though packaging-defective, HSV-1 genomes cloned as set of cosmids[25] or bacterial artificial chromosome (BAC).[26] Following transfection into mammalian cells, sets of cosmids that overlap and represent the entire HSV-1 genome can form circular replication-competent viral genomes via homologous recombination. These reconstituted viral genomes give rise to infectious virus progeny. Similarly, BACs that contain the entire HSV-1 genome also produce infectious virus progeny by transfecting eukaryotic cells. If the viral DNA packaging/cleavage (pac) signals are deleted from the HSV-1 cosmids or HSV-1 BACs, reconstituted virus genomes are defective for packaging; however, in the absence of the pac signals, these genomes can still provide all helper functions required for the replication and packaging of co-transfected amplicon DNA. The resulting amplicon vector stocks are essentially free of helper virus contamination, but they are produced in rather limited amounts (titre generally lower than $10^7$ transducing particles/ml). To improve safety, in the latest version of this strategy,[26] the helper genome carried by the BAC lacks a gene encoding one essential virus function (generally ICP27), where its length is oversized, to further avoiding packaging. Amplicon plasmids are therefore

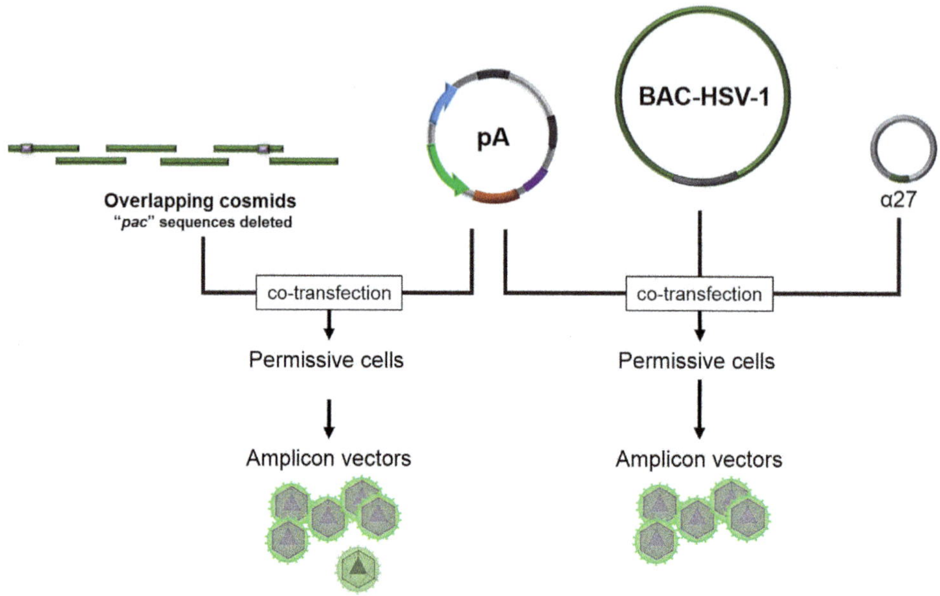


**Figure 19.4** ***Production of amplicon vectors by co-transfecting amplicon plasmid with helper defective herpes simplex virus type 1 (HSV-1) genomes.*** Left: Amplicon vectors can be obtained by co-transfecting eukaryotic cells with an amplicon plasmid and a set of overlapping cosmids carrying the entire HSV-1 genome, except *pac* and *IE4* gene. Co-transfection takes place in permissive cells carrying the IE4 genes. After replication and packaging, amplicon vectors are collected from culture supernatant. A disadvantage of this method is that some recombinant viruses could contaminate amplicon stocks. Right: An improved method that reduces contamination is achieved by replacing the cosmids by a bacterial artificial chromosome (BAC, bacterial sequences) carrying the entire HSV-1 genome (BAC-HSV-1), except sequences encoding ICP27 and *pac*. Co-transfection takes place in cells complementing the virus ICP27 protein. In addition, a plasmid expressing ICP27 can be co-transfected.

replicated and packaged in a cell line complementing the lacking virus function, and often co-transfected with a plasmid expressing this function (Fig. 19.4).

## Production of Amplicon Vectors With the Cre/LoxP1 Site-Specific Recombination System (Fig. 19.5)

Alternatively, amplicon vector stocks containing only very low amounts of otherwise defective helper virus can be prepared using a system based on the deletion of the *a* signals from the HSV-1 genome; this is achieved by Cre/loxP1-based site-specific recombination, to inhibit packaging of the helper genome in the cells that are producing the amplicons.[27] This helper virus, named HSV-1-LaLΔJ, carries a unique and ectopic *a* signal, flanked by two lox-P1 sites in parallel orientation. Therefore, these particles are Cre-sensitive and cannot be packaged in Cre-expressing cells due to the deletion of the

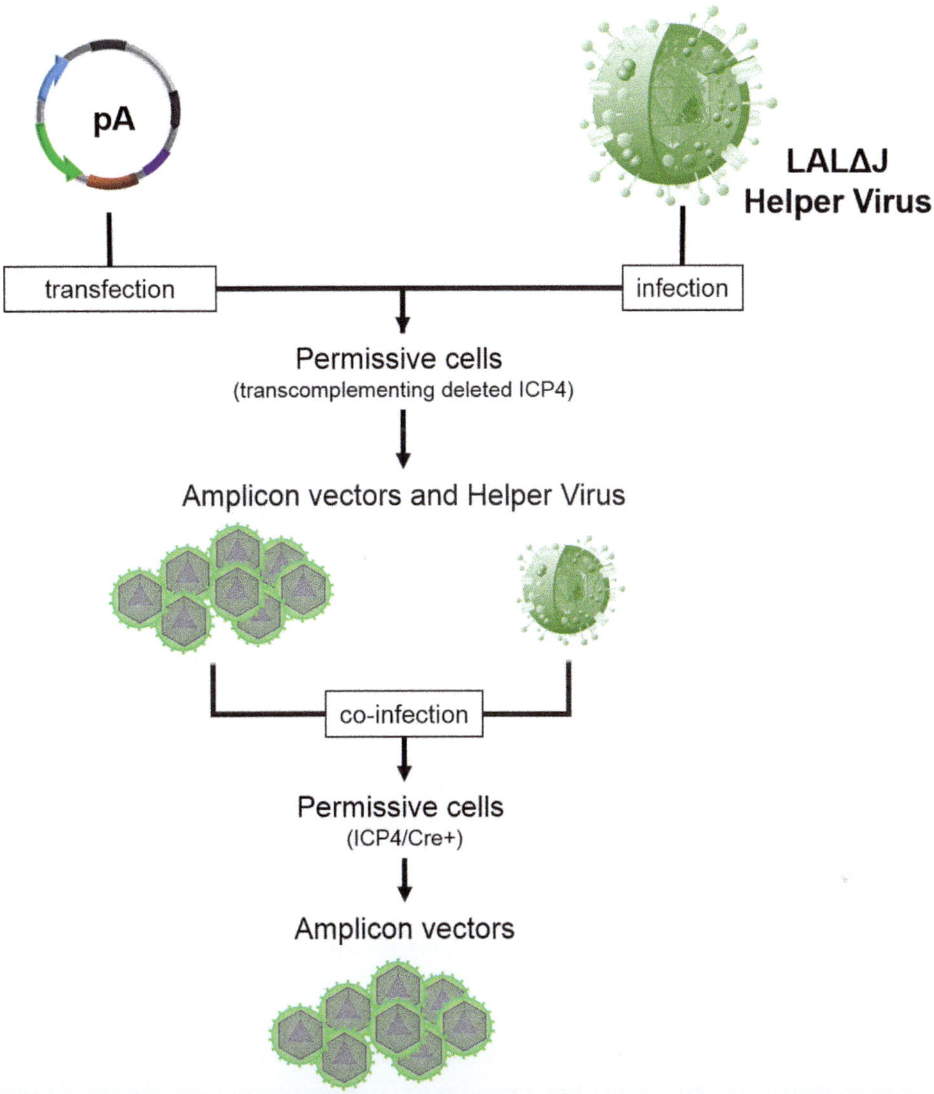


**Figure 19.5** ***Production of amplicon vectors with the Cre/loxP1 site-specific recombination system.*** ICP4-expressing Vero cells are transfected with an amplicon plasmid and superinfected with the herpes simplex virus type 1 LaLΔJ helper virus, which is a Cre-sensitive virus that lacks the essential ICP4 function. Two days postinfection, the mixed population of viral particles (amplicons and helper particles) is used to infect ICP4/Cre recombinase-expressing cells. After two more days, amplicon vectors, only barely contaminated with (replication defective) helper particles, are harvested from the infected cells culture.

"floxed" packaging signal. Nevertheless, some helper genomes can escape to Cre-recombinase action, allowing the production of some contaminating helper virus particles. For this reason, the two genes surrounding the cleavage/packaging signal, which respectively encode a virulence factor known as ICP34.5, and the essential protein ICP4, were deleted from the HSV-1-LaLΔJ helper genome.[27] Although the amplicon stocks prepared with this helper virus (in a complementing cell line expressing both Cre and ICP4 proteins) may still contain a small amount of contaminating helper particles, these particles are replication incompetent and cannot spread upon infection of target cells or tissues. The amplicon packaging process using HSV-1-LaLΔJ as helper virus includes two steps: a first one, in ICP4-complementing cells, allows the generation of large amounts of helper-contaminated amplicon vectors, while the second one, in cells expressing both ICP4 and Cre-recombinase, leads to eliminate most of the contaminating helper viruses through Cre-mediated packaging signal deletion (Fig. 19.5). Using the HSV-1-LaLΔJ helper virus system generally results in large stocks of amplicon vectors (more than $10^8$ transducing particles/ml) only barely contaminated (0.05%—0.5%) with defective, nonpathogenic helper virus particles.

## AMPLICON VECTORS IN NEUROSCIENCE (FIG. 19.6)

Different challenges to find causal relationships between neuronal molecular mechanisms and learning and memory processing have been solved by the use of various types of transgenic animals. However, CNS region, cell type, and time restrictions as specific targets have been very difficult to achieve with those approaches. (Besides these difficulties, they are very time consuming and expensive methods). Virus are extremely well-suited tools to deliver transgenes into eukaryotic cells, and HSV-1 is particularly able and efficient to infect neural cells in vitro and in vivo. HSV-1-derived amplicon vectors are nontoxic, nonmutagenic, and are able to express large amounts of information, including regulatory sequences or RNAs that could modify endogenous expression. They are now currently used to deepen the molecular basis of neuronal function and the outcoming behavior, including combined pharmacologic and transgenic approaches, and they have been delivered into distinct brain regions to investigate complex aspects of normal CNS functioning. Amplicon vectors designed to modify protein expression may carry sequences that allow overexpression or knockdown of an endogenous protein, or expression of an exogenous protein relevant for CNS functions.

When compared to other chemical, physical, or even viral vectors, in addition to their neurotropism, HSV-1 amplicons have some important advantages, mainly arising from their very large cargo capacity of almost 150 kbp. Thus, amplicons can be used to transduce a complete gene locus into neurons, containing large genomic upstream regulatory sequences and including all or most of the introns, resulting both in neuron-specific transgene transcription and in expression of all or most alternatively spliced mRNA.

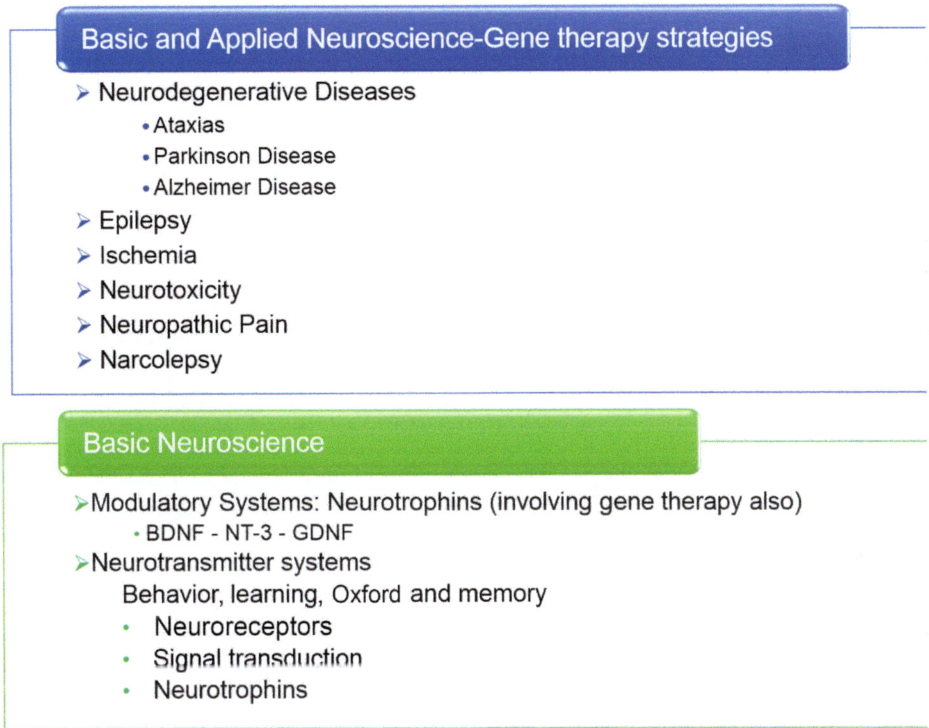


**Figure 19.6 Amplicon applications in neuroscience.**

Amplicons are unique vectors regarding these outstanding features. On the other hand, when delivering short transgenic sequences, the repetitive nature of the amplicon genome results in expression of several copies of the transduced gene, generally resulting in high-level expression of the transgenic protein. The main disadvantage of amplicons is the fact that safe vector stocks are difficult to be prepared in very large amounts. This practical shortcoming makes amplicons hardly useful for human gene therapy trials. However, as it will be developed in the following sections, our current ability to generate amplicon stocks allows these vectors to be very successfully used both for fundamental studies in the brain and for experimental gene therapy of neurologic disorders in small laboratory animals.

## AMPLICON VECTORS IN EXPERIMENTAL ANALYSIS OF BEHAVIOR, LEARNING, AND MEMORY

### Neuroreceptors

Due to their remarkable advantages, mainly their large capacity, natural neurotropism, and the facts that they are nontoxic and do not integrate into the cellular genome, amplicon vectors are chosen to modify protein expression in neural cells both in vitro and in vivo (by codifying sequences that allow overexpression or knockdown of an endogenous protein) or to express an exogenous protein (e.g., dominant-negative mutants) relevant for CNS functions, like proteins directly involved in neurotransmission and in neuron signaling. (A comprehensive review on amplicons vectors uses to study behavior can be found in Ref. Jerusalinsky et al.[28]) These vectors were used, for example, to investigate NMDA glutamate receptor (NMDAR) involvement in learning and memory. Amplicons carrying GluN1 subunit gene sequences in either sense or AS orientations (in addition to EGFP as reporter gene) were used to investigate the participation of hippocampal NMDAR by modifying the expression of that essential GluN1 subunit in the rat CNS. The ability to modify endogenous levels of GluN1 was first tested in vitro in primary neuron cultures of rat embryo cerebral cortex[29,30] and then assessed in vivo. Adult rats inoculated into the dorsal hippocampus with vectors expressing AS GluN1 performed significantly worse than control rats in an inhibitory avoidance of a foot shock task, and they did not show habituation (decreased exploratory behavior) by repeated exposure to an open field. Immunohistochemistry in serial brain slices from these animals showed that the transduced cells represented approximately 6%–7% of hippocampal pyramidal neurons in CA1 region and just about less than 1% of granule cells in the dentate gyrus, indicating that GluN1 knockdown in a small number of those neurons significantly impaired memory.[31]

We have recently developed amplicon vectors able to express AS RNA against GluN2A subunit messenger (Fig. 19.7) and other vectors to express AS RNA against GluN2B (not shown), and we have tested them in vitro in neuron culture. Preliminary results showed that AS GluN2A vectors reduced GluN2A levels by approximately 50% (measured by immunofluorescence in two independent assays, Fig. 19.7), in primary cultures of rat hippocampal neurons, without significant changes in GluN1 (not shown), making these amplicons suitable tools to selectively manipulate subunit expression in vitro, and likely in vivo.

Also, amplicons have been used to further clarify the role of alpha-amino-3-hydroxy-5-methyl-4-isoxazole-propionic acid receptor (AMPAR) in learning and memory. Most AMPARs in the nervous system contain GluR2 subunit, which allows both inward and outward currents to pass through equally well. At variance, AMPARs lacking GluR2 show a profound inward rectification; hence, when the cell membrane is depolarized, a minimal outward current can pass through these channels. Due to this feature, the incorporation of recombinant AMPARs into synapses can be functionally followed.[32] To study the AMPARs trafficking during associative fear-conditioning learning, Rumpel et al.[33] injected amplicon vectors expressing recombinant AMPARs subunits into the rat basolateral amygdala. An amplicon encoding GluR1 subunit fused with green fluorescent protein (GFP) was used to express homomeric AMPARs, which display greater inward rectification than endogenous AMPARs with GluR2. Electrophysiological detection of

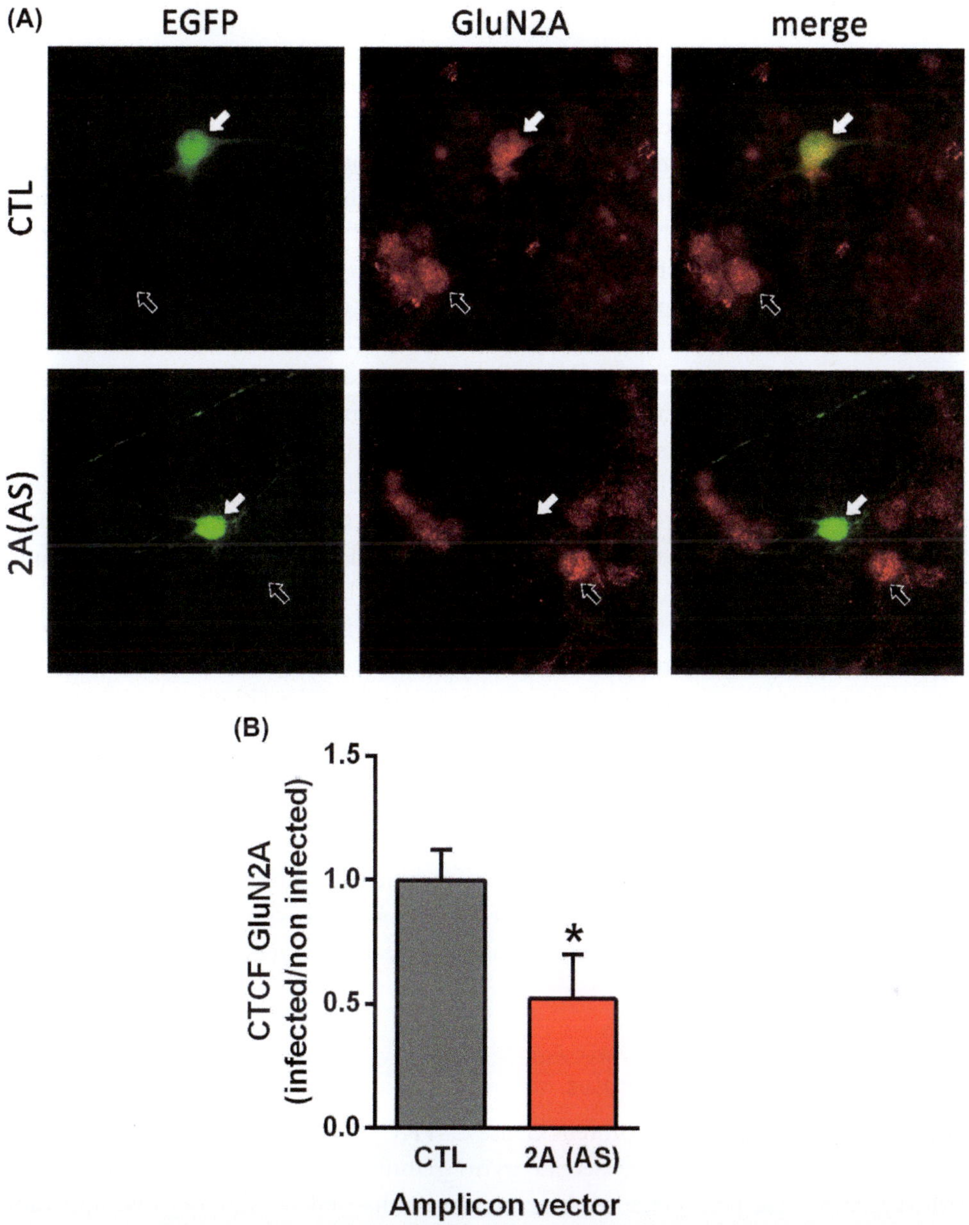


**Figure 19.7** ***Amplicon vectors carrying AS RNA sequence against GluN2A subunit of the NMDA receptor (2A(AS)) in primary cultures of rat hippocampal neurons.*** Primary hippocampal neuron mature cultures (14–17 DIV) were infected with control (CTL) or 2A(AS) amplicons (MOI = 3). At 48 h postinfection, cultures were fixed, and immunofluorescence was obtained with a rabbit anti-GluN2A primary antibody and an Alexa Fluor 568 goat antirabbit secondary antibody. Infected cells were identifiable by EGFP expression, encoded by amplicon vectors. (A) Representative immunofluorescence images 48 h after infection with CTL amplicon vectors (top row) and 2A(AS) amplicon vectors (bottom row). EGFP fluorescence (green), GluN2A immunofluorescence (red). *White arrows*: infected neurons. *Empty arrows*: noninfected neurons. Magnification: 600×. Cells infected with CTL amplicon vector and noninfected cells exhibited similar GluN2A levels, while cells infected with 2A(AS) amplicon vectors evidenced significant lower GluN2A levels than noninfected cells. (B) Bar graphs showing mean ± SEM of infected/noninfected cells ratio per field average, of GluN2A corrected total cell immunofluorescence (CTCF), in two independent assays. GluN2A level is significantly lower in cells infected with 2A(AS) amplicon, demonstrating the efficiency of this vector to knock down GluN2A in vitro. $^{*}P < .05$, One sample T-test. $N = 15$ image fields/experiment per group.

synapses undergoing plasticity was then possible by incorporation of recombinant GluR1-AMPARs into these synaptic membranes. Another amplicon vector encoding the carboxyl cytoplasmic tail of GluR1 subunit fused with GFP, which acts as a dominant negative construct to prevent synaptic incorporation of endogenous GluR1-receptors, was used to block, both in vitro and in vivo, several forms of synaptic plasticity dependent on GluR1-receptors incorporation. A third vector with only GFP sequence was used as control. Those amplicons allowed to demonstrate that fear conditioning drives AMPARs into the synapse of a large fraction of postsynaptic neurons in the basolateral amygdala. Blocking GluR1 receptor trafficking in just a few neurons (10%–20%) undergoing plasticity was sufficient to impair memories depending on this structure.[33]

Liu and coworkers reported that an amplicon expressing siRNA for the γ-Aminobutyric acid A (GABA-A) receptor α2 subunit, infused into the central nucleus of the amygdala of alcohol-preferring rats, caused profound and selective reduction of binge drinking. This behavior was associated to a decrease in both α2 subunit expression and GABA-A receptor density. The authors also found an inhibition in Toll-like receptor 4 (TLR4) expression.[34] Furthermore, an amplicon expressing TLR4 siRNA into the central amygdala, though not into the ventral pallidum, inhibited binge drinking. Nevertheless, binge drinking was effectively inhibited by an amplicon able to express siRNA against GABA-A receptor α1 subunit, infused into the ventral pallidum. Therefore, the contribution of TLR4 to binge drinking took place downstream to α2 subunit signaling in the central amygdala, though not in the ventral pallidum, strengthening the idea that this signaling would depend on specific neuroanatomic sites. Those data indicate that GABA-A α2-regulated TLR4 expression in the central amygdala contributes to binge drinking and might be crucial for early neuroadaptation in excessive drinking.[34]

## Signal Transduction Molecules

Amplicon vectors expressing a protein kinase C (PKC) β II constitutively active catalytic domain were used to transduce dentate gyrus granule cells in rats. Activation of PKC pathways in a small percentage of these neurons enhanced rat auditory discrimination reversal learning, suggesting that some auditory learning would be mediated by the hippocampus in the rat.[35]

For decades, it has been known that the cyclic adenosine monophosphate (cAMP) response element-binding protein (CREB) plays an important role in learning and memory.[36] Changes in CREB function could influence the probability for neurons in the lateral amygdala, to be recruited into a fear memory trace, suggesting a competitive model underlying memory formation. In such model, eligible neurons would be selected to participate in a memory trace as a function of their CREB activity during learning.[37] Several teams of investigators have manipulated CREB with amplicon vectors.

A wt-CREB-amplicon allowed to demonstrate that increasing CREB expression in the auditory thalamus enhances memory and leads to generalization of auditory conditioned fear, indicating that CREB-mediated plasticity in the thalamus plays a role in this cognitive process.[38] Using an amplicon vector encoding a dominant-negative mutant form of CREB (mCREB), Brightwell et al.[39] demonstrated that hippocampal overexpression of mCREB blocks long-term—though not short-term—memory for a socially transmitted food preference, thereby demonstrating that hippocampal CREB was involved in this type of memory. Using same vectors, they also showed that rats trained to make a consistent turning response in a water version of the plus maze required CREB function in the dorsolateral striatum to form a long-term memory of this strategy, and they showed that, at least in the dorsal hippocampus, it is independent of CREB function.[40]

Barrot et al.[41] using a model of protracted social isolation in adult rats, observed an increase in anxiety-like behavior and in both the latency of the onset of sexual behavior and the latency to ejaculate. In a transgenic CRE-LacZ mouse, which expresses β-gal under control of CRE, protracted social isolation reduced CRE-dependent transcription within the nucleus accumbens. The decrease in CRE-dependent transcription was mimicked in nonisolated animals by amplicon-based gene transfer of a dominant-negative mutant of CREB into the nucleus accumbens. In these mice, local inhibition of CREB activity increased anxiety-like behavior and delayed initiation of sexual behavior, without effects on the ejaculation parameters. Restoring CREB activity in the nucleus accumbens of isolated mice with amplicon vectors to overexpress wt-CREB rescued them from the anxiety phenotype as well as from the deficit in latency to initiate sexual behavior. These studies suggest a role for the nucleus accumbens in anxiety and in specific aspects of sexual behavior, providing new insights into the molecular mechanisms by which social interactions could affect brain plasticity and behavior.[41]

## NEURODEGENERATIVE DISEASES

### Ataxias

Ataxias are a group of nervous system degenerative diseases characterized by a progressive loss of movement coordination. Hereditary ataxias could be recessive or dominant. Nowadays, no treatment can effectively stop or slow down ataxia's progression.

The most prevalent form of hereditary ataxia is Frederich ataxia (FA),[42] where neurological symptoms result from neurodegeneration in the dorsal root ganglia, with loss of a large amount of sensory neurons and posterior columns, followed by degeneration in the spinocerebellar and corticospinal tracts of the spinal cord. This disease is caused by a mutation in the first intron of the frataxin gene (*frda wt*), which causes a large GAA repeat expansions (*frda mut*) and, in consequence, leads to a decrease in the level of the mitochondrial protein frataxin.

Lim et al.[43] have generated a conditional frataxin transgenic mouse with the *frda wt* gene floxed by *loxP* (*loxP-frda*). Then, the authors injected an amplicon vector carrying the CRE-recombinase transgene into the brainstem of that transgenic *loxP-frda* mouse. CRE expression causes the homologous recombination of *loxP* sites and the loss of *frda*, leading to a decrease in frataxin expression. These mice developed behavioral deficits after 4 weeks, resembling FA. In an attempt to generate an experimental treatment for FA, they injected a second amplicon vector bearing the wt *frda* cDNA. These mice showed a significant behavioral improvement 4 weeks after this second vector injection.[43]

Gomez-Sebastian et al.[44] taking advantage of the large transgene capacity of amplicons, used one of these vectors to deliver a 135-kb insert containing the full locus of the human *wt frda*, including upstream and downstream long regulatory sequences (~80 kb), to fibroblasts obtained from FA patients, expressing very low levels of frataxin. Synthesis of frataxin was confirmed by immunofluorescence in these *wt frda*-transduced frataxin-deficient cells.[44] It has also been shown that fibroblasts obtained from FA patients exhibit striking biochemical deficits, leading to increased sensitivity to oxidative stress. Functional complementation assays demonstrated restoration of the *wt* cellular phenotype in the *wt frda*-transduced cells, in response to a classical stressor like $H_2O_2$.[44]

To investigate the persistence of transgene expression, the same group injected an iBAC-HSV-1-based vector, carrying the 135-kb *wt frda* genomic DNA locus into the adult mouse cerebellum.[45] They also built up another vector, but with the *E. coli lacZ* gene inserted at the ATG start codon (iBAC-*frda*-lacZ), which was used as reporter. Direct intracranial injection of this vector into the adult mouse cerebellum resulted in a large number of cells expressing lacZ. This expression was driven by the *wt frda* locus and persisted for at least 75 days. In contrast, there was a very strong expression of the GFP from a similar vector, driven by the HSV-1 IE4/5 promoter, but with rather shorter-term persistence. Therefore, a sustained transgene expression in vivo by amplicon delivery of a very long genomic DNA locus was demonstrated for the first time. These results revealed the high potential of HSV-1 *FRDA* vectors for gene therapy of FA.

Ataxia-telangiectasia is an autosomal recessive neurodegenerative disorder due to mutations in the A-T gene (*AT*). This gene is, in the AT wt form, a kinase responsible for recognizing and correcting errors during DNA duplication before cell division. Several mutations led to an AT-mutated (ATM) protein with different degrees of activity, generating a pleiotropic phenotype characterized by cerebellar degeneration, immunodeficiency, cancer predisposition, increased radiation sensitivity, and premature aging, in correlation with the residual kinase activity of the ATM protein.[46] The *AT* cDNA is too large (~9 kb) for most of the available vectors, so it was not feasible to rescue the phenotype of cells expressing ATM. Cortes and colleagues produced an amplicon vector codifying the *AT* cDNA, to express a wt form of AT in human fibroblasts that express ATM.[47] The expression of AT wt protein was confirmed by western blot and immunofluorescence. AT kinase function was tested by in vitro phosphorylation of p53; the in vivo functionality of AT was assessed by counting the accumulation of G2/M cells after ionizing radiation.[47]

In a further study, this team constructed an amplicon encoding both the EGFP and a human FLAG-tagged-AT protein; this vector was inoculated in the cerebellum of $At^{-/-}$ mice. The number and phenotype of infected cells were assessed by EGFP fluorescence. Also, infection of thousands of cells, including Purkinje cells, in the inoculated cerebellum, was confirmed by EGFP expression. FLAG-tagged-AT expression was demonstrated at transcriptional (qRT-PCR, in situ hybridization) as well as at translational (immunoprecipitation of the full-length human protein) levels, 3 days postinoculation.[48]

The same group has developed an HSV/adeno-associated virus (AAV)-hybrid amplicon, carrying an expression cassette for the AT and EGFP cDNAs, flanked by AAV inverted terminal repeats, in an attempt to fully replace the AT gene. In the presence of the AAV Rep proteins (proteins codified in the AAV genome and required for AAV chromosome integration), this hybrid vector mediated a site-specific integration of the transgenic sequences into the AAV1 site on chromosome 19. To test functional activity of this AT vector, Cortes et al. exposed AT infected cells to ionizing radiation. Then, AT activity was assessed by specific immunofluorescence using antibodies against the phosphorylated form of AT (pAT, activated form). An increase in pAT was observed only in AT HSV/AAV infected cells and not in cells infected with a control vector.[49] The results showed that this AT HSV/AAV hybrid amplicon was able to integrate into the AAVS1 site and to achieve functional expression of human AT cDNA in vivo, in a mouse model of ATM.

## Parkinson Disease

Parkinson disease (PD) is the second most common neurodegenerative disorder; it is characterized by progressive loss of dopaminergic neurons in the substantia nigra and a decrease in the level of dopaminergic inputs to the striatum.[50] From the clinical point of view, PD is characterized by akinesia or bradykinesia, rigidity, and tremor that are directly related to dopaminergic striatal neuron loss.[51] Several studies have used amplicons in experimental settings of PD. During and collaborators were the first to report the use of amplicon vectors to deliver human tyrosine hydroxylase (TH) gene into the partially denervated striatum of 6-hydrodopamine-lesioned rats, used as a model of PD.[52] Efficient behavioral and biochemical recovery has persisted up to 1 year after gene transfer.

In an attempt to improve that vector, Sun and collaborators developed another vector able to co-express two enzymes involved in the synthesis of L-DOPA—TH and the aromatic amino acid decarboxylase (AADC)—under a modified neurofilament gene promoter that supports long-term expression in forebrain neurons.[53] This improved vector was injected in the right striatum of adult rats where PD-like symptoms had been previously induced by injection of 6-OHDA in the substantia nigra. Histologic analyses demonstrated neuronal-specific co-expression of TH and AADC from 4 days to 7 months after gene transfer. In these rats, vector injection was able to correct in about 80% the apomorphine-induced rotational behavior present in this 6-OHDA PD rat model.[53] Later on, in a series of elegant studies, Sun and coworkers further compared the activities of tissue-specific promoters to drive gene expression, particularly the promoters of TH, the neurofilament, and the vesicular glutamate transport 1 (VGLUT1).[54–59] Taking advantage of the large transgene capacity of HSV-1 vectors, they developed an amplicon that co-express the three dopamine biosynthetic enzymes (TH, GTP CH I, and AADC, a 3-genes-vector), and another one carrying all the three dopamine biosynthetic enzymes and the vesicular monoamine transporter (TH, GTP CH I, AADC, and VMAT-2, a 4-genes-vector). The effects of both vectors were compared: the 4-genes-vector supported higher levels of correction of apomorphine-induced rotational behavior than did the 3-genes-vector, and this correction was maintained for 6 months. The 4-genes-vector, though not the 3-genes-vector, supported extracellular levels of dopamine and dihydroxyphenylacetic acid, similar to those observed in normal rats, as assessed close to the injection site, and only the 4-genes-vector supported significant $K^+$-dependent release of dopamine.[55]

The effect of amplicon-mediated transduction of the dominant-negative fibroblast growth factor (FGF) receptor-1 mutant protein (FGFR1(TK-)) into the rat substantia nigra was evaluated in vivo as a possible strategy to mimic the reduced FGF signaling already documented to occur in PD. Following intranigral delivery of the FGFR1(TK-) amplicon, the number of substantia nigra neurons expressing TH was significantly reduced, leading to the conclusion that reduced FGF signaling in the substantia nigra of parkinsonian patients could play a role in the impaired dopaminergic transmission in PD.[60] In a further study, the same group analyzed the effect of ex vivo transduction of mesencephalic reaggregates with the antiapoptotic protein bcl-2, on grafted dopamine neuron survival. Using an amplicon expressing bcl-2 under the control of the TH promoter (HSV-TH9bcl-2) to transduce mesencephalic reaggregates, it was shown that, in spite of the high infection efficiency (since many cells were effectively transduced), amplicon-mediated overexpression of bcl-2 did not lead to an increase in grafted TH-immune-reactive neuron number.[61]

Mitochondrial alterations are detected in most neurodegenerative disorders and may contribute to neuronal dysfunction and demise. Rotenone or 1-methyl-4-phenyl-1,2,3,6-tetrahydropyridina (MPTP) inhibit mitochondrial complex I, causing death of dopaminergic neurons in the substantia nigra, thus providing an acute model of PD. It has been recently demonstrated that mitochondrial hexokinase II promotes neuronal survival in rotenone treated cells, and that this enzyme acts downstream of glycogen synthase kinase-3 (GSK-3), which is considered to be a critical factor in regulating neuronal cell survival and death.[62] More recently, the same group generated amplicons expressing hexokinase II. Overexpression of this protein in the substantia nigra of mice subsequently administered with rotenone or MPTP prevented neuronal cell death induced by both drugs and reduced the associated motor deficits. These results provide the first proof that hexokinase II could protect against dopaminergic neurodegeneration in vivo, and they suggest that increase of hexokinase II expression could represent a promising approach to PD treatment.[63]

## Alzheimer Disease

One of the most prevalent neurodegenerative disease in the elder human population and one of the most investigated at present is Alzheimer disease (AD). It was proposed that the peptide known as amyloid beta (Aβ) acts as a main neurotoxin leading to neurodegeneration. More precisely, a recently hypothesis states that soluble oligomers of Aβ peptide (named ADDLs, Aβ-derived diffusible ligands) bind to neuron membranes at the postsynaptic side, triggering toxic effects that finally lead to neuronal death.[64,65]

Aβ peptide is generated by degradation of the amyloid precursor protein (APP). Under physiological conditions, the main degradation pathway is the following: APP is first cleaved by α secretase and then by γ secretase, resulting in a nonamyloidogenic soluble peptide. A small part of APP is degraded by another pathway known as the amyloidogenic path: where APP is first cleaved by the β-secretase BACE-1 (instead of α-secretase) and then by the γ-secretase, generating amyloidogenic peptides (Aβ) 1–40 and 1–42 (of 40 and 42 amino acids, respectively), Aβ1-42 being the more amyloidogenic one.[66] Under certain conditions (some of them unknown) or by blocking the main degradation pathway, APP is cleaved by this amyloidogenic path. Then, Aβ peptides aggregates in soluble oligomers of 2–14 monomers (ADDLs), which can bind to the postsynaptic densities. As the concentration increases, they further aggregate into fibrils in the extracellular space and then form the typical amyloid plaques called senile plaques.[65,67–71]

At least two different groups have used amplicons in mice for "Aβ vaccination," as a possible therapeutic strategy for AD, aimed to prevent Aβ fibrils formation and/or to produce amyloid deposits removal. In a first study, transgenic Tg2576 mice overexpressing APP with the Swedish mutation (which leads to enhanced generation and extracellular deposition of the Aβ1–42 peptide) were injected with amplicons expressing either Aβ1-42 (HSVAβ) or Aβ1-42 fused to the molecular adjuvant tetanus toxin fragment C (HSVAβ/TtxFC). Although peripheral administration of both "vaccines" augmented humoral responses against Aβ and reduced Aβ deposition in the CNS, HSVAβ vaccination resulted toxic, inducing expression of proinflammatory transcripts in the mouse hippocampus.[72] Another amplicon vector, the HSV (IE)Aβ(CMV)IL-4, was built to induce co-expression of Aβ1-42 and interleukin 4 (IL-4), a cytokine that promotes the generation of Th2-like T-cell responses. A triple transgenic AD (3XTg-AD) mice, which progressively develop both amyloid and neurofibrillary tangle pathology, were vaccinated with that amplicon or with a set of control amplicons (one encoding Aβ1−42 but not IL-4 and an "empty" vector without transgenes). In 3XTg-AD mice injected with HSV (IE)Aβ(CMV)IL-4, AD-related amyloid and tau pathologies progression were significantly reduced, as compared with control vectors injected mice. Also, the expression of Th2-related Aβ-specific antibodies appeared to improve learning and memory in the Barnes maze spatial memory paradigm.[73] Therefore, these results support the use of amplicons for experimental Aβ immune therapy of AD.[74]

Neurodegenerative pathologies like AD or PD, as well as some forms of depression, have been associated to receptor-neurotransmitter systems dysfunction. L-glutamate is the major excitatory neurotransmitter in the CNS. Therefore, glutamate receptors represent an attractive molecular target in the treatment of neurodegenerative diseases, as well as in epilepsy and ischemia. Recent evidence showed that APP interacts with NMDARs.[75,76] These ionotropic glutamate receptors are tetramers made of two GluN1 obligatory subunits that are essential for receptor assembly and two different GluN2 (A-D) and/or GluN3 (A-B) regulatory subunits. The NMDAR have been involved in or associated with several neuropathologies; hence, generation of novel tools able to modify expression and composition of NMDARs should help in understanding their normal functioning (as reported in Amplicon Vectors in Experimental Analysis of Behavior, Learning, and Memory section) as well as the physiopathology. It was proposed that ADDLs bind to NMDAR or to postsynaptic complexes containing these receptors, acting as gain of function ligands.[64,65,77] By targeting such postsynaptic complexes, ADDLs activate a cascade of signals that lead to an increase in intracellular reactive oxygen species (ROS).[77] Decker and colleagues have shown that blockade of GluN1 expression by infecting primary cultured neurons with amplicons encoding an anti-GluN1 AS RNA inhibited ADDLs binding to synapses.[77] They have also shown that in those GluN1 knockdown neurons there was a great reduction in ROS formation instigated by ADDL.[77] It should be taken into account that the decrease in GluN1, which is the essential subunit for receptor assembly and for its membrane allocation, would lead to a general decrease of all NMDAR at the synaptic surface.[78] It was reported that different regulatory GluN2 subunits would also be involved in the binding of ADDLs to synaptic membranes.[79,80] Liu and colleagues suggested that increasing the activity of GluN2A and/or decreasing that of GluN2B, could reduce ADDLs-mediated cytotoxic effects in primary neuron culture.[79] Interestingly, Balducci and colleagues showed that there was an alteration in GluN2A and GluN2B subunits trafficking in mutant mice expressing an amyloidogenic human form of APP.[80] More studies on the possible specific interaction between different NMDAR subunits and Aβ peptides, both in normal and pathologic conditions, would help to further clarify the specific ADDLs-binding site.

We have recently developed amplicon vectors bearing AS sequences against GluN2A (see section Amplicon Vectors in Experimental Analysis of Behavior, Learning, and Memory section) or GluN2B. They were effective in reducing the corresponding subunit expression in vitro, in cell lines and in primary cultures of rat hippocampal neurons (Fig. 19.7, see section Amplicon Vectors in Experimental Analysis of Behavior, Learning, and Memory). We plan to use these vectors to investigate the putative involvement of those NMDAR subunits in ADDL binding and the subsequent ROS induction.

Microtubule-associated protein tau (*MAPT*) and α-synuclein (*SNCA*) play central roles in neurodegenerative disorders. Peruzzi and colleagues have recently generated amplicon-based iBAC-vectors carrying either the 143-kb *MAPT* or the 135-kb *SNCA loci*.[81] They have used these vectors to study regulation of gene expression of both *MAPT* and *SNCA*, showing functional complementation between them in cultured neurons and in organotypic brain slices. Infected neurons were able to express functional genes under physiological regulation, including the generation of multiple splice variants. In particular, multiple *MAPT* transcripts were expressed under strict developmental and cell type-specific control. Primary cultures from *Mapt(−/−)* embryos had been shown to be resistant to Aβ peptide-induced toxicity, suggesting that the tau protein might be mediating the neurotoxicity of Aβ. To test the functionality of the *MAPT* transgene, the authors examined whether the responsiveness to Aβ peptide could be restored in the *Mapt(−/−)* neurons and in organotypic brain slices. In both preparations from *Mapt(−/−)* mice infected with an amplicon vector bearing the *MAPT* transgene, the tau protein was expressed as detected by ELISA and immunocytochemistry, and the sensitivity of *Mapt(−/−)* neurons to Aβ peptide was restored.[81] As stated by the authors, the faithful retention of gene expression and phenotypic complementation by this system provides a novel and powerful approach to analyze neurological gene diseases.

## Ischemia and Hypoxia

Apoptosis plays a critical role in many neurological diseases, including stroke. To study the protective role of the antiapoptotic factor *bcl-2* in ischemia, an amplicon vector encoding bcl-2 (HSVbcl-2) was infused in rat cerebral cortex 24 h prior to ischemia induction. Expression of bcl-2 was confirmed by immunohistochemistry in animals injected with the HSVbcl2 expression vector. Tissue viability significantly increased at the injection site in HSVbcl2, but not in animals injected with a similar vector expressing *E. coli* LacZ (HSVLacZ) used as control.[82] Later, HSVbcl-2 was used to inject gerbils unilaterally into the CA1 region of the hippocampus, 24 h prior to induce transient global ischemia.[83] Results showed that the local increase in bcl-2 expression using the HSVbcl-2 vector may protect CA1 pyramidal cells from the delayed neuronal death caused by transient global ischemia, and that this happened only in the HSVbcl-2-injected hemisphere. In the same line, other group generated another bcl-2-expressing amplicon that was used to infect hippocampal cell cultures. Bcl-2-expressing vectors enhanced neuron survival after exposure to Adriamycin, glutamate, and hypoglycemia. Furthermore, dichlorofluorescein measurements indicated that there was a significant reduction in the accumulation of oxygen radicals associated with these insults.[84] Improved neuron survival could be attributed to the fact that the bcl-2 blocks nuclear apoptosis-inducing factor translocation.[85] In an experimental therapeutic approach, Lawrence and collaborators injected rats with this Bcl-2-expressing vector after ischemia induction, leading to a reduction in neuronal loss.[86] Furthermore, they showed that there was a time window (30 min−4 h after reperfusion) where the injection was effective.[86] In accordance with these results, amplicon vectors expressing the inducible heat shock protein (hsp) 72 can also attenuate cerebral ischemic injury when introduced into the rat striatum, even during postischemia.[87] Furthermore, amplicons expressing hsp72 also protected CA1 hippocampal neurons from ischemia, and this protection would be mediated, at least in part, by increased expression of bcl-2.[88]

Hypoxia caused by sleep apnea leads to long-lasting changes in neuronal circuitry.[89] Evidence from both in vitro and in vivo experimental models of intermittent hypoxia has shown neuronal death, reactive gliosis, and learning and memory impairments[90−93] that may account for cognitive and structural alterations in human patients, but the mechanism of these deleterious effects is unknown. The receptor for advanced glycation end products (RAGE) and its downstream effector nuclear factor kappa B (NF-κB) have been related to neuronal death and the conversion of astroglia to the proinflammatory neurodegenerative phenotype.[94,95] RAGE expression and its ligand S100B are increased in experimental models of sleep apnea. Angelo et al.[96] observed that NF-κB is activated in glial cells and neurons after intermittent hypoxia both in vitro, in dissociated mixed hippocampal cell cultures, and in vivo, in male Wistar rats. They performed sequential studies by "loss of function," using a combination of RAGE or S100B neutralizing antibodies, an amplicon vector able to induce the expression of RAGEΔcyto (dominant negative RAGE), and a chemical blocker of NF-κB, to investigate the contribution of the S100B/RAGE/NF-κB pathway to neuronal damage and reactive gliosis after hypoxia. The blockade of either RAGE (using the amplicon) or NF-κB (by a chemical blocker) during a critical 3-day period, when NF-κB reached maximum levels, significantly improved neuronal survival and reduced reactive gliosis. Therefore, both RAGE and NF-κB signaling are involved in neuronal alterations in sleep apnea models, and their blockage could be a strategy for preventing neuronal degeneration and reactive gliosis.[96]

## Neurotoxicity

Increases in cytoplasmic $Ca^{2+}$ concentration can lead to neurotoxicity and neuronal death. The increase of $Ca^{2+}$ could be induced by neurologic trauma associated with aging and some neurologic diseases. To prevent neurotoxic effects of cytoplasmic $Ca^{2+}$ increases, an amplicon vector that expresses the calcium-binding protein calbindin D28K (calbindin D28KHSV) was used to infect neurons, both in vitro and in vivo.[97,98] Cultured neurons infected with this vector responded to hypoglycemia and glutamatergic insults with decreased cytoplasmic $[Ca^{2+}]$ measured by microfluorometry and increased neuron survival relative to controls.[97] Furthermore, in vivo injection of calbindin D28KHSV vector in the hippocampal dentate gyrus increases neuronal survival after application of the antimetabolite 3-acetylpyridine, and increases transsynaptic neuronal survival in area CA3 following kainic acid neurotoxicity.[98]

ROS and oxidative stress damage plays an important role in neuronal death. Amplicon vectors expressing different antioxidant enzymes were used to counteract oxidative damages. Catalase and glutathione peroxidase, two enzymes involved in hydrogen peroxide degradation, were subcloned into amplicon vectors. These vectors were shown to decrease neurotoxicity induced by different agents in primary cultures of hippocampus or cerebral cortex cells.[99] A further study using amplicons to express the antioxidant enzyme Cu—Zn—SOD showed that these vectors were able to protect hippocampal neurons through the induction of glutathione peroxidase expression, though only in the case of neurons treated with sodium cyanide. The authors pointed out that the effect of the amplicons actually worsens the toxic effects of kainic acid, another classical ROS inducer, raising a cautionary note concerning gene therapy against oxidative damage.[100] Amplicons expressing glutamic acid decarboxylase (GAD67) were shown to protect non-differentiated cortical neurons from glutamate toxicity mediated by oxidative stress.[101]

It was reported that the 120 kDa HIV glycoprotein (HIV-gp120) could be neurotoxic at certain doses and that the hsp70, hsp25, or hsp90 overexpression would protect neurons from HIV-gp120 effect. Lim and collaborators overexpressed hsp70 with an ad hoc HSV-amplicon vector in cultured hippocampal neurons; in this way, they demonstrated that hsp70 overexpression protected cultured hippocampal neurons from HIV-gp120 neurotoxicity.[102]

Hypoglycemia could result in an insult for neurons. Using amplicon vectors expressing the rat brain glucose transporter, it was shown that these vectors (1) can maintain neuronal metabolism and reduce the extent of neurons loss in cultures after a period of hypoglycemia[103]; (2) protected cultured hippocampal, spinal cord, and septal neurons against various necrotic insults, including hypoglycemia, glutamate, and 3-nitropropionic acid[104]; and (3) can enhance glucose uptake in adult rat hippocampus and in hippocampal cultures.[105]

## Amplicon Vectors and Neurotrophins

Neurotrophins are a growth factors family that play important roles in the development and maintenance of the nervous system. The human brain-derived neurotrophic factor

(BDNF) is one of the most important neurotrophins. The knowledge about the complex function of BDNF in mammalian nervous system is continuously expanding; it plays a key role as mediator of activity-induced long-term potentiation in the hippocampus, as well as being involved in behavior and memory.[106] BDNF has been also implicated in neurodegenerative diseases,[107] motor diseases,[108] and fragile X syndrome.[109]

## Brain-Derived Neurotrophic Factor and Epilepsy

BDNF is widely expressed in the brain, where it exerts a key role in neuronal survival, differentiation, and plasticity.[110,111] BDNF is therefore viewed as a putative therapeutic target for several CNS disorders, particularly in epilepsy.[112,113] An important limitation for development of therapies targeting the BDNF system was the difficulty to restrict BDNF level modulation at a specific brain region. Therefore, Falcicchia et al.[114] have developed amplicon vectors capable of downregulating BDNF expression both in vitro and in vivo. They have developed two amplicons to knock down BDNF expression. One was able to express an AS RNA to target and degrade the cytoplasmic mRNA pool of BDNF, whereas the other, based on the convergent transcription technology, has been chosen to repress *BDNF* gene transcription. Both amplicon vectors proved to be effective in downregulating BDNF expression in vitro, in BDNF-expressing mesoangioblast cells. However, only the AS strategy was effective in vivo (in agreement with our experience to knocking down NMDAR subunits[29,77]) after inoculation into the hippocampus in a model of status epilepticus, where BDNF mRNA levels were strongly increased. Knocking down BDNF overexpression only in a small part of a single hippocampus cannot be expected to produce robust behavioral effects in a model in which a proconvulsant agent was systemically administered. Nevertheless, the animals treated with BDNF-AS-vectors entered convulsive status epilepticus later than those treated with a control GFP-amplicon vector. Therefore, these results demonstrate a reliable effect of amplicon vectors in knocking down gene expression in vitro and in vivo.[29,31,77]

## Neurotrophins and the Auditory System

BDNF participates in maturation and function of mammalian auditory neurons. Amplicon vectors expressing BDNF were used to assess the feasibility of gene therapy for deafness. Geschwind and collaborators used an amplicon bearing a BDNF cDNA (BDNF-amplicon) and *E. coli* β-galactosidase (β-gal) to evaluate the expression and biologic activity in established cell lines and explant cultures, prepared from spiral ganglia of the murine ear.[115] Using those BDNF-amplicons in two BDNF-responsive cell lines, PC12trkB and MG87trkB, they demonstrated efficient secretion of biologically active BDNF.[115] Also, the transduction of explanted spiral ganglia with BDNF-amplicon elicited a robust neurite outgrowth, comparable to the effect of exogenously added

BDNF.[115] Then, this BDNF-amplicon was used in mice to infect damaged spiral ganglion. Four weeks postinfection, there was a stable production of BDNF, and it supported the survival of auditory neurons, preventing their loss due to trophic factor deprivation-induced apoptosis.[116] In addition, BDNF-amplicons promoted neuronal survival at the same maximal level seen by adding exogenous BDNF in avian cochlear neuron cultures.[117]

**Other neurotrophins** were implicated in the proper function of the auditory system. Amplicons expressing neurotrophin-3 (NT-3-amplicons) were also used in murine cochlear explants model. After infection, the cochlear explants were exposed to cisplatin to induce destruction of hair cells and neurons in the auditory system. This toxicity, defined as ototoxicity, is a major dose-limiting side effect of cisplatin chemotherapy for cancer patients. Amplicon-mediated NT-3 transduction was shown to attenuate the ototoxic action of cisplatin, demonstrating the potency of NT-3 in protecting spiral ganglion neurons from degeneration.[118] Moreover, aged mice injected in the peripheral auditory system with NT-3-amplicons showed significantly more spiral ganglion surviving neurons and lower incidence of cisplatin-induced apoptosis and necrosis than those injected with a control vector.[119] Therefore, this approach seemed to be a promising treatment for prevention of chemical-induced hearing disorders and potentially for hearing degeneration due to normal aging.

## Neurotrophins in Other Neurological Disorders and Injuries

Amplicon vectors were used to compare BDNF with the glial-derived neurotrophic factor (GDNF), in protecting nigrostriatal neurons in a rat model of PD. In this study, GDNF was significantly more effective than BDNF, for both correcting behavioral deficits and protecting nigrostriatal dopaminergic neurons. The expression of both neurotrophic factors from the same vector did not show higher protection than expressing GDNF alone.[120] Later on, it was shown that intracerebral administration of this GDNF-amplicon, following occlusion of the middle cerebral artery, displayed neuroprotection in ischemic injury. Treated animals showed reduced motor deficits, and after 1 month, there was a reduction in tissue loss, in glial fibrillary acidic protein, and in caspase-3 immunostaining.[121]

In neonatal rats, the combined delivery of NT-3 and the GluN2D subunit of the NMDA receptor, which was possible thanks to the large capacity of amplicon vectors, was used to strengthen monosynaptic connections in contused cords and to induce the appearance of weak but functional multisynaptic connections in double hemi-sected cords. On the other hand, the expression of either NT-3 or GluN2D alone failed to induce appearance of synaptic responses through the hemi-sected region of newborn rats.[122] The same group has treated adult rats with the following agents: (1) anti-Nogo-A antibodies to neutralize the growth-inhibitor Nogo-A; (2) NT-3 via engineered fibroblasts to promote neuron survival and plasticity; and (3) GluN2D subunit via an amplicon vector, to elevate NMDA receptor function by reversing its $Mg^{2+}$ blockade, thereby enhancing synaptic plasticity and promoting NT-3 effects.[123] The combined treatment resulted in slightly better motor function without adverse effects (i.e., pain), suggesting that this novel combined treatment may help to improve functioning of the damaged spinal cord.[123]

### Neuropathic Pain

It is a chronic form of pain that results from an injury to the nervous system. It could be due to multiple causes, like alcoholism, chemotherapy, diabetes, facial nerve problems, among others. Amplicons expressing proenkephalin (pHSV-hPPE-lacZ, SHPZ) were used to investigate the antinociceptive effect of leuenkephalin (a product of proenkephalin processing). The amplicons were injected into the ventral periaqueductal grey[124] or into the cortex of rats.[125] Both studies showed that the injection of SHPZ, but not of a control vector only expressing the reporter gene lacZ, attenuated neuropathic pain and reduced induced hypersensitivity in rats.[124,125]

### Narcolepsy

It is a neurodegenerative sleep disorder that involves a significant loss of neurons containing the neuropeptide orexin, which is also known as hypocretin. To investigate the therapeutic potential of exogenous administration of this neuropeptide, Liu and collaborators inoculated an amplicon vector expressing pre-pro-orexin into the lateral hypothalamus of orexin knockout mice. The results showed that expression of orexin from these vectors significantly improved sleeping in these animals.[126]

## CONCLUDING REMARKS

HSV-1 amplicon vectors, as other viral vectors, are more suitable than chemical or physical methods to transduce brain neurons. This stems from the fact that neurons are very difficult to be efficiently transduced, even in in vitro cultures, and chemical transduction is most often nonselective, resulting in transduction of many different types of cells when introducing into the brain. These neurotropic viral vectors, in contrast, offer the possibility to efficiently and selectively infect neurons, including brain neurons in vivo. Moreover, since viral vectors are much more efficient in transducing neurons than chemical transfection, lower gene dosage can be used to reach significant levels of protein expression, implying a higher level of safety. Amplicon vectors are now very properly conceived, they are nontoxic and nonmutagenic, and an overwhelming amount of data has clearly shown that well-designed viral vectors are absolutely safe and preserve neuron physiology. The most relevant advantage of these amplicon vectors arises from their very large cargo capacity of almost 150 kbp. Therefore, amplicon vectors are suitable tools to study complex functions, such as learning and memory, and behavior, and they have been used both in fundamental and applied research. These vectors are also being used for experimental gene therapy in small animal models, for both inherited and acquired disorders affecting the nervous system, including neurodegenerative diseases.

## ACKNOWLEDGMENTS

ALE and DAJ would like to thank the French *CNRS*, the *Centre de Génétique et Physiologie Moléculaires et Cellulaires* (*CGPhiMC*), Universite Lyon-1, and the National Council of Science and Technology (*CONICET*) and University of Buenos Aires (*UBA*) for supporting the *International Associated Laboratory* (*LIA*) *DEVENIR* (2010–16). DAJ and MVB acknowledge the support in Argentina of the *CONICET* (*PIP* 00975), *UBA* (*UBACyT 20020130100575BA*), and *ANPCyT* (*PICT 2012–3043; PICT 2013–2221*).

## REFERENCES

1. Kwong AD, Frenkel N. Herpes simplex virus amplicon: effect of size on replication of constructed defective genomes containing eucaryotic DNA sequences. *J Virol*. 1984;51:595–603.
2. Wang S, Fraefel C, Breakefield X. HSV-1 amplicon vectors. *Meth Enzymol*. 2002;346:593–603.
3. Oehmig A, Fraefel C, Breakefield XO. Update on herpesvirus amplicon vectors. *Mol Ther*. 2004;10:630–643.
4. Roizman B, Knipe DM, Whitley R. Herpes simplex viruses. In: Knipe DM, Howley PM, et al., eds. *Fields Virology*. 5th ed. Philadelphia, PA: Lippincott, Williams & Wilkins; 2007:2501–2601.
5. Grunewald K, Cyrklaff M. Structure of complex viruses and virus-infected cells by electron cryo tomography. *Curr Opin Microbiol*. 2006;9(4):437–442.
6. Zhou ZH, Chen DH, Jakana, et al. Visualization of tegument-capsid interactions and DNA in intact herpes simplex virus type 1 virions. *J Virol*. 1999;73(4):3210–3218.
7. Oh J, Fraser NW. Temporal association of the herpes simplex virus genome with histone proteins during a lytic infection. *J Virol*. 2008;82(7):3530–3537.
8. Bataille D, Epstein AL. Herpes simplex virus type 1 replication and recombination. *Biochimie*. 1995;77(10):787–795.
9. McGeoch DJ, Dalrymple MA, Davison AJ, et al. The complete DNA sequence of the long unique region in the genome of herpes simplex virus type 1. *J Gen Virol*. 1988;69(Pt 7):1531–1574.
10. McGeoch DJ, Dolan A, Donald S, Brauer DH. Complete DNA sequence of the short repeat region in the genome of herpes simplex virus type 1. *Nucleic Acids Res*. 1986;14(4):1727–1745.
11. McGeoch DJ, Dolan A, Donald S, Rixon FJ. Sequence determination and genetic content of the short unique region in the genome of herpes simplex virus type 1. *J Mol Biol*. 1985;181(1):1–13.
12. de Silva S, Bowers WJ. Herpes virus amplicon vectors. *Viruses*. 2009;1(3):594–629.
13. Vlazny DA, Frenkel N. Replication of herpes simplex virus DNA: localization of replication recognition signals within defective virus genomes. *Proc Natl Acad Sci USA*. 1981;78(2):742–746.
14. Hardwicke MA, Schaffer PA. Cloning and characterization of herpes simplex virus type 1 oriL: comparison of replication and protein-DNA complex formation by oriL and oriS. *J Virol*. 1995;69(3):1377–1388.
15. Davison AJ, Wilkie NM. Nucleotide sequences of the joint between the L and S segments of herpes simplex virus types 1 and 2. *J Gen Virol*. 1981;55(Pt 2):315–331.
16. Deiss LP, Chou J, Frenkel N. Functional domains within the a sequence involved in the cleavage-packaging of herpes simplex virus DNA. *J Virol*. 1986;59(3):605–618.

17. Nasseri M, Mocarski ES. The cleavage recognition signal is contained within sequences surrounding an a–a junction in herpes simplex virus DNA. *Virology*. 1988;167(1):25–30.
18. McVoy MA, Nixon DE, Adler SP, Mocarsky ES. Sequences within the herpesvirus-conserved pac1 and pac2 motifs are required for cleavage and packaging of the murine cytomegalovirus genome. *J Virol*. 1998;72(1):48–56.
19. Ingvarsdottir K, Blaho JA. Role of viral chromatin structure in the regulation of herpes simplex virus 1 gene expression and replication. *Future Microbiol*. 2009;4(6):703–712.
20. DeLuca NA, McCarthy AM, Schaffer PA. Isolation and characterization of deletion mutants of herpes simplex virus type 1 in the gene encoding immediate-early regulatory protein ICP4. *J Virol*. 1985; 56(2):558–570.
21. Ozuer A, Wechuck JB, Russell B, et al. Evaluation of infection parameters in the production of replication-defective HSV-1 viral vectors. *Biotechnol Prog*. 2002;18(3):476–482.
22. Spaete RR, Frenkel N. The herpes simplex virus amplicon: a new eucaryotic defective-virus cloning-amplifying vector. *Cell*. 1982;3(1):295–304.
23. Boehmer PE, Lehman IR. Herpes simplex virus DNA replication. *Annu Rev Biochem*. 1997;66: 347–384.
24. Fraefel C, Marconi P, Epstein AL. Herpes simplex virus type 1 (HSV-1)-derived amplicon vectors for gene transfer and gene therapy. In: Lossi L, Merighi A, eds. *Neuronal Cell Death. Methods in Molecular Biology*. vol. 1254. 2015:295–316.
25. Fraefel C, Song S, Lim F, et al. Helper virus-free transfer of herpes simplex virus type 1 plasmid vectors into neural cells. *J Virol*. 1996;70:7190–7197.
26. Saeki Y, Fraefel C, Ichikawa T, Breakefield XO, Chiocca EA. Improved helper virus-free packaging system for HSV amplicon vectors using an ICP27-deleted, oversized HSV-1 DNA in a bacterial artificial chromosome. *Mol Ther*. 2001;3:591–601.
27. Zaupa C, Revol-Guyot V, Epstein AL. Improved packaging system for generation of high levels non-cytotoxic HSV-1 amplicon vectors using Cre-loxP1 site-specific recombination to delete the packaging signals of defective helper genomes. *Hum Gene Ther*. 2003;14:1049–1063.
28. Jerusalinsky D, Baez MV, Epstein AL. Herpes simplex virus type 1-based amplicon vectors for fundamental research in neurosciences and gene therapy of neurological diseases. *J Physiol Paris*. 2012; 106(1–2):2–11.
29. Adrover MF, Guyot-Revol V, Cheli VT, et al. Hippocampal infection with HSV-1-derived vectors expressing an NMDAR1 antisense modifies behavior. *Gene Brain Behav*. 2003;2(2):103–113.
30. Cheli VT, Adrover MF, Blanco C, et al. Gene transfer of NMDAR1 subunit sequences to the rat CNS using herpes simplex virus vectors interfered with habituation. *Cell Mol Neurobiol*. 2002;22(3):303–314.
31. Cheli V, Adrover M, Blanco C, et al. Knocking-down the NMDAR1 subunit in a limited amount of neurons in the rat hippocampus impairs learning. *J Neurochem*. 2006;97(suppl 1):68–73.
32. Malinow R, Malenka RC. AMPA receptor trafficking and synaptic plasticity. *Annu Rev Neurosci*. 2002;25:103–126.
33. Rumpel S, LeDoux J, Zador A, Malinow R. Postsynaptic receptor trafficking underlying a form of associative learning. *Science*. 2005;308(5718):83–88.
34. Liu J, Yang AR, Kelly T, et al. Binge alcohol drinking is associated with GABAA alpha2-regulated Toll-like receptor 4 (TLR4) expression in the central amygdala. *Proc Natl Acad Sci USA*. 2011; 108(11):4465–4470.
35. Neill JC, Sarkisian MR, Wang Y, et al. Enhanced auditory reversal learning by genetic activation of protein kinase C in small groups of rat hippocampal neurons. *Brain Res Mol Brain Res*. 2001;93(2): 127–136.
36. Kandel ER. The molecular biology of memory storage: a dialogue between genes and synapses. *Science*. 2001;294(5544):1030–1038.
37. Han JH, Kushner SA, Yiu AP, et al. Neuronal competition and selection during memory formation. *Science*. 2007;316(5823):457–460.
38. Han JH, Yiu AP, Cole CJ, Hsiang HL, Neve RL, Josselyn SA. Increasing CREB in the auditory thalamus enhances memory and generalization of auditory conditioned fear. *Learn Mem*. 2008;15(6): 443–453.

39. Brightwell JJ, Smith CA, Countryman RA, Neve RL, Colombo PJ. Hippocampal overexpression of mutant CREB blocks long-term, but not short-term memory for a socially transmitted food preference. *Learn Mem*. 2005;12(1):12–17.
40. Brightwell JJ, Smith CA, Neve RL, Colombo PJ. Transfection of mutant CREB in the striatum, but not the hippocampus, impairs long-term memory for response learning. *Neurobiol Learn Mem*. 2008; 89(1):27–35.
41. Barrot M, Wallace DL, Bolaños CA, et al. Regulation of anxiety and initiation of sexual behavior by CREB in the nucleus accumbens. *Proc Natl Acad Sci USA*. 2005;102(23):8357–8362.
42. Cossee M, Schmitt M, Campuzano V, et al. Evolution of the Friedreich's ataxia trinucleotide repeat expansion: founder effect and premutations. *Proc Natl Acad Sci USA*. 1997;94(14):7452–7457.
43. Lim F, Palomo GM, Mauritz C, et al. Functional recovery in a Friedreich's ataxia mouse model by frataxin gene transfer using an HSV-1 amplicon vector. *Mol Ther*. 2007;15(6):1072–1078.
44. Gomez-Sebastian S, Gimenez-Cassina A, Diaz-Nido J, Lim F, Wade-Martins R. Infectious delivery and expression of a 135 kb human FRDA genomic DNA locus complements Friedreich's ataxia deficiency in human cells. *Mol Ther*. 2007;15(2):248–254.
45. Gimenez-Cassina A, Wade-Martins R, Gomez-Sebastian S, Corona JC, Lim F, Diaz-Nido J. Infectious delivery and long-term persistence of transgene expression in the brain by a 135-kb iBAC-FXN genomic DNA expression vector. *Gene Ther*. 2011;18(10):1015–1019.
46. Verhagen MM, Last JI, Hogervorst FB, et al. Presence of ATM protein and residual kinase activity correlates with the phenotype in ataxia-telangiectasia: a genotype-phenotype study. *Hum Mutat*. 2012;33(3):561–571.
47. Cortes ML, Bakkenist CJ, Di Maria MV, Kastan MB, Breakefield XO. HSV-1 amplicon vector-mediated expression of ATM cDNA and correction of the ataxia-telangiectasia cellular phenotype. *Gene Ther*. 2003;10(16):1321–1327.
48. Cortes ML, Oehmig A, Perry KF, Sanford JD, Breakefield XO. Expression of human ATM cDNA in Atm-deficient mouse brain mediated by HSV-1 amplicon vector. *Neuroscience*. 2006;141(3):1247–1256.
49. Cortes ML, Oehmig A, Saydam O, et al. Targeted integration of functional human ATM cDNA into genome mediated by HSV/AAV hybrid amplicon vector. *Mol Ther*. 2008;16(1):81–88.
50. Hornykiewicz O. Basic research on dopamine in Parkinson's disease and the discovery of the nigrostriatal dopamine pathway: the view of an eyewitness. *Neurodegener Dis*. 2008;5(3–4):114–117.
51. Rodriguez-Oroz MC, Jahanshahi M, Krack P, et al. Initial clinical manifestations of Parkinson's disease: features and pathophysiological mechanisms. *Lancet Neurol*. 2009;8(12):1128–1139.
52. During MJ, Naegele JR, O'Malley KL, Geller AI. Long-term behavioral recovery in Parkinsonian rats by an HSV vector expressing tyrosine hydroxylase. *Science*. 1994;266(5189):1399–1403.
53. Sun M, Zhang GR, Kong L, et al. Correction of a rat model of Parkinson's disease by coexpression of tyrosine hydroxylase and aromatic amino acid decarboxylase from a helper virus-free herpes simplex virus type 1 vector. *Hum Gene Ther*. 2003;14(5):415–424.
54. Zhang GR, Geller AI. A helper virus-free HSV-1 vector containing the vesicular glutamate transporter-1 promoter supports expression preferentially in VGLUT1-containing glutamatergic neurons. *Brain Res*. 2010;1331:12–19.
55. Sun M, Kong L, Wang X, et al. Coexpression of tyrosine hydroxylase, GTP cyclo-hydrolase I, aromatic amino acid decarboxylase, and vesicular monoamine transporter 2 from a helper virus-free herpes simplex virus type 1 vector supports high-level, long-term biochemical and behavioral correction of a rat model of Parkinson's disease. *Hum Gene Ther*. 2004;15(12):1177–1196.
56. Zhang GR, Li X, Cao H, Zhao H, Geller AI. The vesicular glutamate transporter-1 upstream promoter and first intron each support glutamatergic-specific expression in rat postrhinal cortex. *Brain Res*. 2011;1377:1–12.
57. Zhang GR, Wang X, Yang T, et al. A tyrosine hydroxylase-neurofilament chimeric promoter enhances long-term expression in rat forebrain neurons from helper virus-free HSV-1 vectors. *Brain Res Mol Brain Res*. 2000;84(1–2):17–31.
58. Cao H, Zhang GR, Wang X, Kong L, Geller AI. Enhanced nigrostriatal neuron-specific, long-term expression by using neural-specific promoters in combination with targeted gene transfer by modified helper virus-free HSV-1 vector particles. *BMC Neurosci*. 2008;9:37.

59. Gao Q, Sun M, Wang X, Geller AI. Isolation of an enhancer from the rat tyrosine hydroxylase promoter that supports long-term, neuronal-specific expression from a neurofilament promoter, in a helper virus-free HSV-1 vector system. *Brain Res.* 2007;1130(1):1–16.
60. Corso TD, Torres G, Goulah C, et al. Transfection of tyrosine kinase deleted FGF receptor-1 into rat brain substantia nigra reduces the number of tyrosine hydroxylase expressing neurons and decreases concentration levels of striatal dopamine. *Brain Res Mol Brain Res.* 2005;139(2):361–366.
61. Sortwell CE, Bowers WE, Counts SE, et al. Effects of ex vivo transduction of mesencephalic reaggregates with bcl-2 on grafted dopamine neuron survival. *Brain Res.* 2007;1134(1):33–44.
62. Gimenez-Cassina A, Lim F, Cerrato T, Palomo GM, Diaz-Nido J. Mitochondrial hexokinase II promotes neuronal survival and acts downstream of glycogen synthase kinase-3. *J Biol Chem.* 2009;284(5): 3001–3011.
63. Corona JC, Gimenez-Cassina A, Lim F, Diaz-Nido J. Hexokinase II gene transfer protects against neurodegeneration in the rotenone and MPTP mouse models of Parkinson's disease. *J Neurosci Res.* 2010;88(9):1943–1950.
64. De Felice FG, Velasco PT, Lambert MP, et al. Abeta oligomers induce neuronal oxidative stress through an N-methyl-D-aspartate receptor-dependent mechanism that is blocked by the Alzheimer drug memantine. *J Biol Chem.* 2007;282(15):11590–11601.
65. Shankar GM, Bloodgood BL, Townsend M, Walsh DM, Selko DJ, Sabatini BL. Natural oligomers of the Alzheimer amyloid-beta protein induce reversible synapse loss by modulating an NMDA-type glutamate receptor-dependent signaling pathway. *J Neurosci.* 2007;27(11):2866–2875.
66. Thinakaran G, Koo EH. Amyloid precursor protein trafficking, processing, and function. *J Biol Chem.* 2008;283(44):29615–29619.
67. Shankar GM, Li S, Mehta TH, et al. Amyloid-beta protein dimers isolated directly from Alzheimer's brains impair synaptic plasticity and memory. *Nat Med.* 2008;14(8):837–842.
68. Haass C, Selkoe DJ. Soluble protein oligomers in neurode-generation: lessons from the Alzheimer's amyloid beta-peptide. *Nat Rev Mol Cell Biol.* 2007;8(2):101–112.
69. Klein WL. Synaptic targeting by A beta oligomers (ADDLS) as a basis for memory loss in early Alzheimer's disease. *Alzheimers Dement.* 2006;2(1):43–55.
70. Lesne S, Koh MT, Kotilinek L, et al. A specific amyloid-beta protein assembly in the brain impairs memory. *Nature.* 2006;440(7082):352–357.
71. Roselli F, Tirard M, Lu J, et al. Soluble beta-amyloid 1-40 induces NMDA-dependent degradation of postsynaptic density-95 at glutamatergic synapses. *J Neurosci.* 2005;25(48):11061–11070.
72. Bowers WJ, Mastrangelo MA, Stanley HA, Casey AE, Milo Jr LJ, Federoff HJ. HSV amplicon-mediated Abeta vaccination in Tg2576 mice: differential antigen-specific immune responses. *Neurobiol Aging.* 2005;26(4):393–407.
73. Barnes CA. Memory deficits associated with senescence: a neurophysiological and behavioral study in the rat. *J Comp Physiol Psychol.* 1979;93(1):74–104.
74. Frazer ME, Hughes JE, Mastrangelo MA, Tibbens JL, Federoff HJ, Bowers WJ. Reduced pathology and improved behavioral performance in Alzheimer's disease mice vaccinated with HSV amplicons expressing amyloid-beta and interleukin-4. *Mol Ther.* 2008;16(5):845–853.
75. Cousins SL, Hoey SE, Anne Stephenson F, Perkinton MS. Amyloid precursor protein 695 associates with assembled NR2A- and NR2B-containing NMDA receptors to result in the enhancement of their cell surface delivery. *J Neurochem.* 2009;111(6):1501–1513.
76. Hoey SE, Williams RJ, Perkinton MS. Synaptic NMDA receptor activation stimulates alpha-secretase amyloid precursor protein processing and inhibits amyloid-beta production. *J Neurosci.* 2009;29(14): 4442–4460.
77. Decker H, Jürgensen S, Adrover MF, et al. N-methyl-D-aspartate receptors are required for synaptic targeting of Alzheimer's toxic amyloid-beta peptide oligomers. *J Neurochem.* 2010;115(6):1520–1529.
78. Paoletti P. Molecular basis of NMDA receptor functional diversity. *Eur J Neurosci.* 2011;33(8): 1351–1365.
79. Liu J, Chang L, Roselli F, et al. Amyloid-beta induces caspase-dependent loss of PSD-95 and synaptophysin through NMDA receptors. *J Alzheimers Dis.* 2010;22(2):541–556.

80. Balducci C, Tonini R, Zianni E, et al. Cognitive deficits associated with alteration of synaptic metaplasticity precede plaque deposition in AbetaPP23 transgenic mice. *J Alzheimers Dis.* 2010;21(4): 1367–1381.
81. Peruzzi PP, Lawler SE, Senior SL, et al. Physiological transgene regulation and functional complementation of a neurological disease gene deficiency in neurons. *Mol Ther.* 2009;17(9):1517–1526.
82. Linnik MD, Zahos P, Geschwind MD, Federoff HJ. Expression of bcl-2 from a defective herpes simplex virus-1 vector limits neuronal death in focal cerebral ischemia. *Stroke.* 1995;26(9):1670–1674. discussion 1675.
83. Antonawich FJ, Federoff HJ, Davis JN. BCL-2 transduction, using a herpes simplex virus amplicon, protects hippocampal neurons from transient global ischemia. *Exp Neurol.* 1999;156(1):130–137.
84. Lawrence MS, Ho DY, Sun GH, Steinberg GK, Sapolsky RM. Overexpression of Bcl-2 with herpes simplex virus vectors protects CNS neurons against neurological insults *in vitro* and *in vivo*. *J Neurosci.* 1996;16(2):486–496.
85. Zhao H, Yenari MA, Cheng D, Barreto-Chang OL, Sapolsky RM, Steinberg GK. Bcl-2 transfection via herpes simplex virus blocks apoptosis-inducing factor translocation after focal ischemia in the rat. *J Cerebr Blood Flow Metabol.* 2004;24(6):681–692.
86. Lawrence MS, McLaughlin JR, Sun GH, et al. Herpes simplex viral vectors expressing Bcl-2 are neuroprotective when delivered after a stroke. *J Cerebr Blood Flow Metabol.* 1997;17(7):740–744.
87. Hoehn B, Ringer TM, Xu L, et al. Overexpression of HSP72 after induction of experimental stroke protects neurons from ischemic damage. *J Cerebr Blood Flow Metabol.* 2001;21(11):1303–1309.
88. Kelly S, Zhang ZJ, Zhao H, et al. Gene transfer of HSP72 protects cornu ammonis 1 region of the hippocampus neurons from global ischemia: influence of Bcl-2. *Ann Neurol.* 2002;52(2):160–167.
89. Feuerstein C, Naegele B, Pepin JL, Levy P. Frontal lobe-related cognitive functions in patients with sleep apnea syndrome before and after treatment. *Acta Neurol Belg.* 1997;97:96–107.
90. Decary A, Rouleau I, Montplaisir J. Cognitive deficits associated with sleep apnea syndrome: a proposed neuropsychological test battery. *Sleep.* 2000;23:369–381.
91. Engleman HM, Kingshott RN, Martin SE, Douglas NJ. Cognitive function in the sleep apnea/hypopnea syndrome (SAHS). *Sleep.* 2000;23(suppl 4):S102–S108.
92. Gozal D, Kheirandish-Gozal L. Neurocognitive and behavioral morbidity in children with sleep disorders. *Curr Opin Pulm Med.* 2007;13:505–509.
93. Gozal D, Kheirandish-Gozal L. Cardiovascular morbidity in obstructive sleep apnea: oxidative stress, inflammation, and much more. *Am J Respir Crit Care Med.* 2008;177:369–375.
94. Villarreal A, Aviles Reyes RX, Angelo MF, Reines AG, Ramos AJ. S100B alters neuronal survival and dendrite extension via RAGE-mediated NF-kappaB signaling. *J Neurochem.* 2011;117:321–332.
95. Ponath G, Schettler C, Kaestner F, et al. Autocrine S100B effects on astrocytes are mediated via RAGE. *J Neuroimmunol.* 2007;184:214–222.
96. Angelo MF, Aguirre A, Avilés Reyes RX, et al. The proinflammatory RAGE/NF-κB pathway is involved in neuronal damage and reactive gliosis in a model of sleep apnea by intermittent hypoxia. *PLoS One.* 2014;9(9):e107901.
97. Meier TJ, Ho DY, Park TS, Sapolsky RM. Gene transfer of calbindin D28k cDNA via herpes simplex virus amplicon vector decreases cytoplasmic calcium ion response and enhances neuronal survival following glutamatergic challenge but not following cyanide. *J Neurochem.* 1998;71(3):1013–1023.
98. Phillips RG, Meier TJ, Giuli LC, McLaughlin JR, Ho DY, Sapolsky RM. Calbindin D28K gene transfer via herpes simplex virus amplicon vector decreases hippocampal damage *in vivo* following neurotoxic insults. *J Neurochem.* 1999;73(3):1200–1205.
99. Wang H, Cheng E, Brooke S, Chang P, Sapolsky R. Over-expression of antioxidant enzymes protects cultured hippocampal and cortical neurons from necrotic insults. *J Neurochem.* 2003;87(6):1527–1534.
100. Zemlyak I, Nimon V, Brooke S, Moore T, McLaughlin J, Sapolsky R. Gene therapy in the nervous system with superoxide dismutase. *Brain Res.* 2006;1088(1):12–18.
101. Lamigeon C, Prod'Hon C, De Frias V, Michoudet C, Jacquemont B. Enhancement of neuronal protection from oxidative stress by glutamic acid decarboxylase delivery with a defective herpes simplex virus vector. *Exp Neurol.* 2003;184(1):381–392.

102. Lim MC, Brooke SM, Sapolsky RM. gp120 neurotoxicity fails to induce heat shock defenses, while the over expression of hsp70 protects against gp120. *Brain Res Bull.* 2003;61(2):183–188.
103. Lawrence MS, Ho DY, Dash R, Sapolsky RM. Herpes simplex virus vectors overexpressing the glucose transporter gene protect against seizure-induced neuron loss. *Proc Natl Acad Sci USA.* 1995; 92(16):7247–7251.
104. Ho DY, Saydam TC, Fink SL, Lawrence MS, Sapolsky RM. Defective herpes simplex virus vectors expressing the rat brain glucose transporter protect cultured neurons from necrotic insults. *J Neurochem.* 1995;65(2):842–850.
105. Ho DY, Mocarski ES, Sapolsky RM. Altering central nervous system physiology with a defective herpes simplex virus vector expressing the glucose transporter gene. *Proc Natl Acad Sci USA.* 1993;90(8): 3655–3659.
106. Cowansage KK, LeDoux JE, Monfils MH. Brain-derived neurotrophic factor: a dynamic gatekeeper of neural plasticity. *Curr Mol Pharmacol.* 2010;3(1):12–29.
107. Gupta VK, You Y, Gupta VB, Klistorner A, Graham SL. TrkB receptor signaling: implications in neurodegenerative, psychiatric and proliferative disorders. *Int J Mol Sci.* 2013;14(5):10122–10142.
108. He YY, Zhang XY, Yung WH, Zhu JN, Wang JJ. Role of BDNF in central motor structures and motor diseases. *Mol Neurobiol.* 2013;48(3):783–793.
109. Castren M, Castren E. BDNF in fragile X syndrome. *Neuropharmacology.* 2014;76(Pt C):729–736.
110. Cohen-Cory S, Kidane AH, Shirkey NJ, Marshak S. Brain-derived neurotrophic factor and the development of structural neuronal connectivity. *Dev Neurobiol.* 2010;70:271–288.
111. Yoshii A, Constantine-Paton M. Postsynaptic BDNF-TrkB signaling in synapse maturation, plasticity, and disease. *Dev Neurobiol.* 2010;70:304–322.
112. Simonato M, Tongiorgi E, Kokaia M. Angels and demons: neurotrophic factors and epilepsy. *Trends Pharmacol Sci.* 2006;27:631–638.
113. Simonato M, Zucchini S. Are the neurotrophic factors a suitable therapeutic target for the prevention of epileptogenesis? *Epilepsia.* 2010;51(suppl 3):48–51.
114. Falcicchia C, Trempat P, Binaschi A, et al. Silencing status epilepticus-induced BDNF expression with herpes simplex virus Type-1 based amplicon vectors. *PLoS One.* 2016;11(3):e0150995.
115. Geschwind MD, Hartnick CJ, Liu W, Amat J, Van De Water TR, Federoff HJ. Defective HSV-1 vector expressing BDNF in auditory ganglia elicits neurite outgrowth: model for treatment of neuron loss following cochlear degeneration. *Hum Gene Ther.* 1996;7(2):173–182.
116. Staecker H, Gabaizadeh R, Federoff H, Van De Water TR. Brain-derived neurotrophic factor gene therapy prevents spiral ganglion degeneration after hair cell loss. *Otolaryngol Head Neck Surg.* 1998; 119(1):7–13.
117. Garrido JJ, Alonso MT, Lim F, Carnicero E, Giraldez F, Schimmang T. Defining responsiveness of avian cochlear neurons to brain-derived neurotrophic factor and nerve growth factor by HSV-1-mediated gene transfer. *J Neurochem.* 1998;70(6):2336–2346.
118. Chen X, Frisina RD, Bowers WJ, Frisina DR, Federoff HJ. HSV amplicon-mediated neurotrophin-3 expression protects murine spiral ganglion neurons from cisplatin-induced damage. *Mol Ther.* 2001; 3(6):958–963.
119. Bowers WJ, Chen X, Guo H, Frisina DR, Federoff HJ, Frisina RD. Neurotrophin-3 transduction attenuates cisplatin spiral ganglion neuron ototoxicity in the cochlea. *Mol Ther.* 2002;6(1):12–18.
120. Sun M, Kong L, Wang X, Lu XG, Gao Q, Geller AI. Comparison of the capability of GDNF, BDNF, or both, to protect nigrostriatal neurons in a rat model of Parkinson's disease. *Brain Res.* 2005;1052(2): 119–129.
121. Harvey BK, Chang CF, Chiang YH, et al. HSV amplicon delivery of glial cell line-derived neurotrophic factor is neuroprotective against ischemic injury. *Exp Neurol.* 2003;183(1):47–55.
122. Arvanian VL, Bowers WJ, Anderson A, Horner PJ, Federoff HJ, Mendell LM. Combined delivery of neurotrophin-3 and NMDA receptors 2D subunit strengthens synaptic transmission in contused and staggered double hemisected spinal cord of neonatal rat. *Exp Neurol.* 2006;197(2):347–352.
123. Schnell L, Hunanyan AS, Bowers WJ, et al. Combined delivery of Nogo-A antibody, neurotrophin-3 and the NMDA-NR2d subunit establishes a functional 'detour' in the hemisected spinal cord. *Eur J Neurosci.* 2011;34(8):1256–1267.

124. Zou W, Huang C, Yang Y, Pan Y, Yan J, Guo Q. Microinjection of HSV-1 amplicon vector-mediated human proenkephalin into the periaqueductal grey attenuates neuropathic pain in rats. *Int J Neurosci*. 2012;122(4):189–194.
125. Zou W, Guo Q, Chen C, Yang Y, Wang E. Intrathecal herpes simplex virus type 1 amplicon vector-mediated human proenkephalin reduces chronic constriction injury-induced neuropathic pain in rats. *Mol Med Rep*. 2011;4(3):529–533.
126. Liu M, Thankachan S, Kaur S, et al. Orexin (hypocretin) gene transfer diminishes narcoleptic sleep behavior in mice. *Eur J Neurosci*. 2008;28(7):1382–1393.

# CHAPTER 20

# Cre-lox Neurogenetics: History, Present, and Future

**Joe Z. Tsien**
Augusta University, Augusta, GA, United States; Banna Biomedical Research Institute, Yunnan Academy of Science and Technology, Yunnan, China

## INTRODUCTION

Santiago Ramón y Cajal is among the first to observe and marvel at the elegance and beauty of various neurons in the brain, and wonder how they contribute to perception, emotion, memory, and behavior.[1] Structure-functional analysis of neurons began in earnest with the development of single-cell, juxtacellular-labeling, and patch-clamp techniques.[2–5] Various other technologies have further transformed the neuroscience landscape at multiple levels, ranging from genes and proteins,[6–14] to synaptic plasticity and cognitive enhancement[15–17] (Wang et al., 2009).[60]

In the later 1980s and early 1990s, following the pioneering efforts by Mario Capecchi, Oliver Smithies, and Martin Evans on gene targeting and embryonic stem (ES) cell technologies (see review by Capecchi[18]), a small number of laboratories began to generate mutant mice to define genes' function in development, cancer, or immunology.[19–22] Alcino Silva, Seth Grant, and Thomas O'Dell are among the first to apply the gene knockout approach to study the specific roles of CaMKII or Fyn kinase in plasticity and memory.[23–25] These studies set a new stage to investigate genes and brain functions. However, its limitations—such as lack of phenotypes due to genetic compensation or the developmental effects—were noticed. For example, increased spontaneous epilepsy was observed in the CaMKII knockout mice, whereas the dentate gyrus was deformed in the Fyn knockout mice. Both cases led to heated debate in the field about how the reported memory and plasticity deficits should be interpreted. With the NMDA receptor occupying center stage in plasticity and memory field, Yuqing Li and his colleagues set out to demonstrate its role in learning behavior, but they found that NMDAR1 knockout pups all died neonatally due to failed brain development, including the defective suckling reflex. For developmental neuroscientists, it was very exciting because the stereotypic whisker barrel in the thalamus failed to form.[26] Yet for cognitive and behavioral neuroscientists, it was disappointing. It further solidified the argument that deleting the gene in every cell and organ throughout development was not ideal for defining cognitive mechanisms.

*Molecular-Genetic and Statistical Techniques for Behavioral and Neural Research*
ISBN 978-0-12-804078-2, https://doi.org/10.1016/B978-0-12-804078-2.00020-9

## YOUNG AND FOOLISH

My interest in developing the conditional gene knockout technique stemmed from the specific problem at hand as I was finishing up my HHMI postdoctoral fellowship with Eric Kandel at Columbia University in the fall of 1993. From 1990 to 1993, I was testing the long-held idea, first postulated by Agranoff,[27] that long-term memory requires new protein synthesis and gene expression. Researchers were racing to identify novel genes whose expressions were regulated by brain activity. We were the first to pull out a set of genes that included a tissue-plasminogen activator, a mitogen-activated protein (MAP) kinase phosphatase, and a brain-specific immediate early gene BAD1 (published under my former name, Qian et al.[61,62] BAD1 was later also isolated by Paul Worley and named Arc.[28] While isolating these novel genes was exciting, the next logical step for me was to examine their functions in memory. Antisense, oligonucleotide-based "knock-down" methods were imprecise and not reliable, whereas gene knockout was a good choice, but its limitations and caveats were obvious.

I stumbled upon a paper by Brian Sauer on his successful demonstration of the Cre recombinase-based excision of a floxed marker gene from the circular plasmid transfected in mammalian culture cells.[29] The last paragraph of this paper put forth a big question: "Can Cre also cause recombination at lox sites located within the genome of a mammalian cell?" This led me to wonder if I could use it to develop brain subregion- and cell type-specific gene knockout and/or transgenic overexpression methods. But I knew that all the textbooks and literature described that DNA replication and recombination go hand in hand during cell division (such as meiosis or mitosis) (Fig. 20.1). This basic doctrine was imprinted into everyone's mind and was also evident from the opening

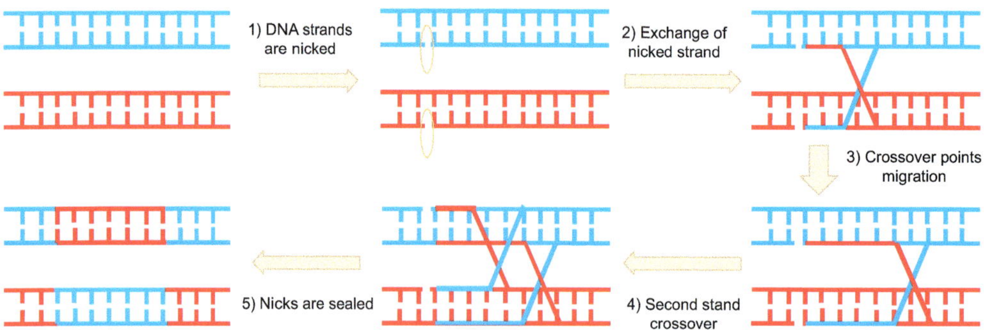


**Figure 20.1** ***DNA recombination involves complex topologic and biochemical during DNA replication.*** DNA recombination is linked with DNA replication during cell divisions to exchange genetic materials for the selection of favorable mutations or the elimination of unfavorable or deleterious mutations. Recombination also comes with DNA injury and is therefore tightly coupled with the DNA repair mechanism. Nearly all neurons in the adult brain are postmitotic to ensure their stability. This explains why brain tumors are in glial cells but not in neurons.

sentences of Sauer's paper: "The processes governing DNA recombination in mitotic mammalian cells have been the subject of intense investigation in recent years … Mitotic recombination plays a central role in the development and function of the immune system." Knowing that all neurons in the adult brain are known to be postmitotic right after birth (except a few in the dentate gyrus and olfactory bulb), any fool who set out to work on DNA recombination in the adult brain would be committing career suicide. Nature has shown that brain tumors are all in glial cells, which divide, but not in neurons. Therefore, DNA recombination as the way to create region- and cell type-specific knockout in the brain was considered to be plainly impossible. I was told that the next show was not about how and why I would fail, but where.

But I was obsessed with the thought of wanting to know the function of genes in a clear way, and only a handful of laboratories had a gene-targeting facility and ES cells for making conventional knockout mice. I asked Mario Capecchi of Utah and Susumu Tonegawa of MIT for a second postdoc position, and both said yes. When I sought Eric's advice on which one I should take, to my great surprise, Eric suggested that Susumu's lab should be a better place for trying my idea. I also got permission to use the CaMKII promoter, cloned by Mark Mayford, which becomes active after the second or third postnatal week and only expressed the forebrain principal cells such as pyramidal cells. The stage was all set, with only one minor bug: if my idea did not work out at this second postdoc stage, I would have no job.

When I arrived at MIT in the fall of 1993, I was pleased to find that Susumu did not discourage me from exploring Cre-loxP neurogenetics during our 15-s introduction. I was surprised by the zoo-like atmosphere in this laboratory of 40 postdoctorals/students. Miraculously, this survival-of-the-fittest model worked well for its productivity and, overall, MIT was exciting and refreshing to me.

To develop Cre/loxP neurogenetics, three things must happen: (1) the textbook on DNA recombination invariantly linked with DNA replication and cell division needs to be wrong; my argument was that the herpes virus (producing cold sores) infected the peripheral nerve ganglia and somehow replicated themselves (presumably without ganglia neurons division). (2) I must make various constructs and generate at least three different mouse lines, then embark on a multiyear crossing to breed them together (Fig. 20.2A–C). This means that the complex procedures and long cycles will have no quick feedback and no room for error. Back then, making even one mutant mouse was already a form of art and luck. In fact, only a few brave souls in well-equipped laboratories could do so because of its complex procedures, lengthy project cycle, expensive cost, and potential for no phenotypes nor desirable outcome at the end. A few of the lucky ones got good jobs, and the rest simply disappeared. (3) The intentional risk I brought upon myself by choosing to work on the NMDAR1 as the gene for conditional knockout instead of the BAD1/Arc will produce a bigger reward. As everyone knew, if the sites to which I inserted the loxP somehow disrupted its expression, I would have a dead pup that was to

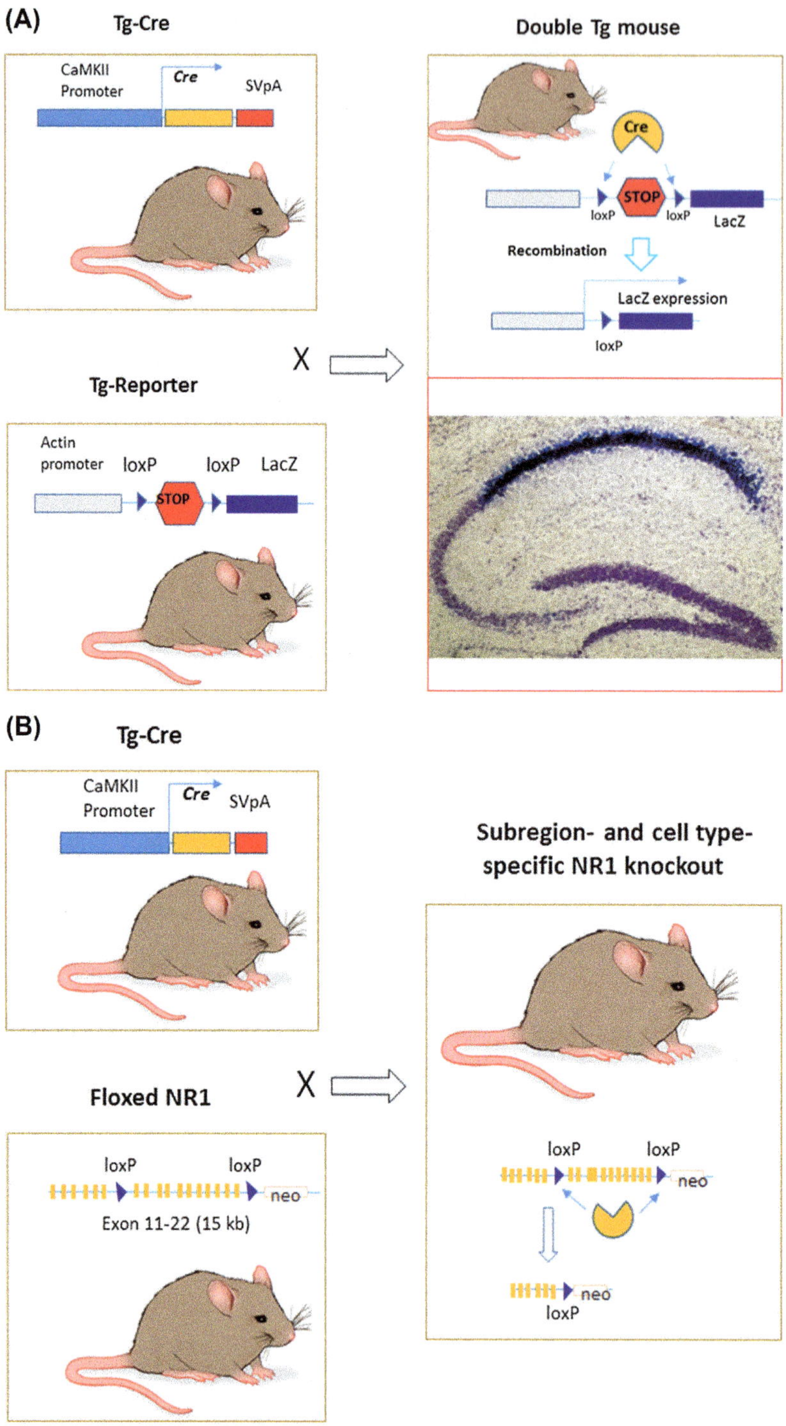
(A)
Tg-Cre
CaMKII Promoter
Cre
SVpA
Double Tg mouse
Cre
STOP
loxP
loxP
LacZ
Recombination
LacZ expression
loxP
X
Tg-Reporter
Actin promoter
loxP
loxP
LacZ
STOP
(B)
Tg-Cre
CaMKII Promoter
Cre
SVpA
Subregion- and cell type-specific NR1 knockout
Floxed NR1
X
loxP
loxP
neo
Exon 11-22 (15 kb)
loxP
loxP
neo
neo
loxP

be published in a few months by Li.[26] In contrast, unintentional disruption of the Arc from the loxP insertion would still give me a conventional knockout paper. But I convinced myself the idea was promising, and my gamble would pay off; I would at least get a badge of honor for being the *first* fool to throw himself under the bus of the textbook dogma.

The execution required complex choreography. Despite all of the chaos in the new laboratory, I found friendly colleagues willing to help me. There were David Gerber and Toshikuni Sasaoka, who showed me how to perform microinjections in fertilized eggs and blastocysts, respectively; Dongfeng Chen who helped with immune-antibody with Cre antibody; Yuqing Li with whom I discussed the NMDAR1 constructs; and Min Xu who provided me ES cells and guided me through delicate ES cell-culture procedures. I was also grateful to David Anderson at Cal Tech for the LacZ reporter and Brian Saucer at Du Pont for Cre-loxP plasmids. I completed all my constructs by January of 1994 and started production of transgenic mice and ES cell targeting of the floxed NR1 construct. I also had a hardworking undergraduate student named Cindy Tom and a technician, Jason Derwon, who assisted with the genotyping and brain sectioning.

One ominous cloud hung over my head when Klaus Rajewsky reported in July 1994 that they achieved 50% knockdown of the RNA polymerase in T cells.[30] This meant that even in dividing T cells, Cre-recombination was not efficient; either 50% of the T cells had 100% knockout of RNA polymerase or 100% of the cells had only one copy of the gene deleted, or worse, a mixed mosaic situation. Despite this, I held out the hope (after additional digging into the literature) that the transiently active promoter they used might be the culprit for the poor efficiency during T-cell development. Over the ensuing months and years, I stuck my head in the sand and labored over my Cre-loxP experiments. In the late fall of 1994, I finally got the first feedback from the experiments.

On a sunny but cold morning, I recall the great surprise when I saw the intense LacZ staining, specifically in the CA1 pyramidal cells of the hippocampus in the first Cre

**Figure 20.2** ***Cre/loxP neurogenetics for achieving region- and cell-type-specific analysis of relationship of genes, circuits, and functions in the brain.*** (A) Strategy to detect the Cre-loxP recombination in the brain. It required production of two different transgenic mice: Tg-Cre and Tg-Reporter lines, respectively. These two lines were then crossed to generate the double transgenic mice. CA1-specific recombination was detected after P19 by LacZ method. It is known that CA1 pyramidal cells undergo neurogenesis between E10 and E18 and enter the postmitotic state by P0. They are well differentiated by P7, with fully established synaptic connections. We have found that Cre/loxP recombination occurs during the middle or end of the third postnatal week in the CA1 pyramidal cells. LacZ in the CA1 pyramidal cells was detected as deep blue color on Nissl stain (in *purple-blue*) background. (B) Conditional knockout of NR1 gene in a specific cell type and region. The exon 11−21 encoding the entire transmembrane domain and C-terminus were flanked by the loxP (termed floxed) in the introns. The second loxP was followed with the neo cassette gene, which allowed for targeted ES cells selection. Luckily, insertion of loxP sites and the neo gene did not alter NR1 gene expression in the floxed homozygous mice.

transgenic line (Fig. 20.2). I could not believe my incredible luck, because the CA1 hippocampal region was the center of the universe in the eyes of many plasticity and memory researchers. Additional Cre lines confirmed similar CA1-specific recombination; then, other Cre lines showed forebrain-specific patterns. When Susumu returned from a trip to Japan, I told him about what I had found. Once he recovered from his confusion, which seemed to result from jetlag and attempting to grasp all that I had been working on, he overcame his disbelief and immediately shared his jubilation with Alcino over the phone. I asked myself why the CaMKII promoter, which was supposed to express in the forebrain, not specific to the CA1, had such a specific effect. Dongfeng Chen helped provide the clue by revealing that Cre expression was higher in the CA1 pyramidal cells. Over the following 6–8 months, I also obtained the floxed NR1 homozygous mice crossed with the Cre Tg 29.1 line. I confirmed the CA1 pyramidal cell-specific NMDA receptor knockout. Pato Huerta and I proceeded with brain slice recordings or histology, respectively, as soon as we could obtain young adult mice (~3–5 weeks). We also set aside a set of mice for collaborating with Tom McHugh, Kenny Blum, and Matthew Wilson for place-cell recordings in the NR1 mutants. Now, we know that the Cre-lox recombination in T29-1 retained the CA1 specificity only in an age-dependent manner; as Cre expression accumulated over time (starting in the 6–8th postnatal weeks), it would gradually achieve the forebrain-specific recombination (as expected from the CaMKII promoter).

By the fall of 1996, we were preparing three back-to-back manuscripts to be submitted to *Cell.* We had decided that reporting the feasibility of the Cre-lox neurogenetic would be the first, the CA1-specific NMDA receptor knockout study the second, and the place-cell characterization paper would be published as the third. This initially doomed or ignored project now, unexpectedly, resulted in multiple results, which caused back-forth arguments between Eric and Susumu about one another's intellectual contribution and authorship. The papers were eventually published back-to-back in the December 1996 issue,[31–33] with no corresponding author assigned.

## CRE-DRIVER RESOURCES FOR NEUROSCIENCES

The successful demonstration of the feasibility of Cre-lox neurogenetics has generated a firestorm in the field. Recognizing its unique usefulness for neuroscience, the NIH launched *the NIH Blueprint for Neuroscience Research,* **known as the Cre-driver project,** to create a collection of mouse strains for better defining the functions of specific cell types and neural circuits in cognitive behaviors (http://www.neuroscienceblueprint.nih.gov/factSheet/CreDriver.html). I was glad to provide my inputs during the process. The NIH chose three centers in the United States for the generation of genetically modified mice expressing Cre recombinases in the nervous system. The teams were led by Dr. Ronald Davis at Baylor College of Medicine (now at Scripps Research Institute, Florida),

Dr. Josh Huang at Cold Spring Harbor Laboratory (co-led with Sacha Nelson of Brandeis University), and Dr. Ulrich Mueller at Scripps Research Institute. This program generated more than hundreds of Cre-driver lines. The mice are available through the Mutant Mouse Regional Resource Center or the Jackson Laboratory's Cre Repository (see Table 20.1).

The NIH *Neuroscience Blueprint program* also funded the GENSAT project led by Nathaniel Heintz of The Rockefeller University to generate BAC-Cre recombinase driver mouse lines (Gong et al., 2007).[63] A total of 288 Cre lines have been generated to date (http://www.gensat.org/cre.jsp). Following these efforts, many individual laboratories and institutions also produced a variety of Cre drivers (e.g., Refs. 34,35). Altogether, over 630 Cre lines have been deposited at the Jackson Laboratory to date.

The European Union also followed the NIH initiative and launched its own Cre-driver project under the name CREATE (coordination of resources for conditional expression of mutated mouse alleles) (Table 20.1). The CREATE consortium represents a core of major European and international mouse database holders and research groups to develop a strategy for the integration and dissemination of Cre-driver strains for modeling aspects of complex human diseases in the mouse. These Cre lines can be found by contacting European Mouse Mutant Archive, Italy, or International Mouse Strain Resource (Table 20.1). In addition, the United Kingdom, Canada, and Japan have also funded several Cre-driver projects (Table 20.1).

**Table 20.1** Cre-driver mouse resources

| | | |
|---|---|---|
| Cre-driver mice Network | NIH Blueprint for Neuroscience | http://www.credrivermice.org |
| Mutant Mouse Regional Resource Center | | https://www.mmrrc.org/catalog/StrainCatalogSearchForm.jsp |
| The Jackson Laboratory's Cre Repository | | https://www.jax.org/jax-mice-and-services |
| The GENSAT project | NIH Blueprint/Rockefeller Univ. | http://www.gensat.org/cre.jsp |
| European Mouse Mutant Archive | European Union, Italy | www.emmanet.org |
| International Mouse Strain Resource | | www.findmice.org |
| Harwell (Mammalian Genetics Unit) | United Kingdom | www.har.mrc.ac.uk |
| Cre-X-mice Project | Canada | http://nagy.mshri.on.ca/cre_new/ |
| RBRC (RIKEN Biology Resource Center) | Japan | www.brc.riken.jp/lab/animal/en |

## CRE-DRIVING INTO THE FUTURE

In the past decade, one of the most exciting developments is optogenetics, which uses light stimulation to manipulate neurons[64] (Zemelman et al., 2002).[36] It has been great fun to witness optogenetics take full advantage of a decade's investment in generating a rich collection of Cre mouse lines (e.g., CaMKII::Cre, PV::Cre, SOM::Cre, D1: Cre, D2::Cre, ChAT:Cre, or DAT::Cre, etc.), enabling its widespread use in the neuroscience community, ranging from memory and decision to motor control.[37–40] However, optogenetic stimulation or the inhibition of a given type of cell in a defined region can produce unexpected disturbances across many regions in the brain.[41]

However, optogenetics was not the only game in town that benefited from Cre-loxP neurogenetics. Chemical genetics has also emerged as a powerful approach in allowing scientists to activate or suppress specific neurons. For example, one can overexpress chemically genetically modified proteins, such as kinases,[42,43] or DREADD,[44] in specific neurons or brain regions. Researchers can also use the floxed diphtheria toxin fragment A to delete specific cells using Cre lines,[45–47] and then observe the resulting phenotypes.

In addition, Cre-lox neurogenetics are used to track neural projections in the brain with retrograde virus[48,49] or CLARITY methods.[50] Jeff Lichtman and Joshua Sanes have cleverly explored the unique properties of the Cre-loxP system to randomly express different ratios of red, green, and blue derivatives of green fluorescent protein in individual neurons, a technique termed Brainbow.[51] This allowed them to flag each neuron with a distinctive color. This process has been a major contribution to the field of connectomics.

Cre-lox neurogenetics has been also combined with voltage-sensitive proteins to monitor changes in neuronal activity. Genetically encoded calcium indicators, such as the GCaMPs, have allowed researchers to infer changes in neuron activations via calcium transients.[52] Researchers are also developing other genetically encoded voltage-sensitive fluorescent proteins that can report cortical electrical responses in vitro and/or in vivo.[53,54]

Most recently, Stuart Ibsen and Sreekanth Chalasani reported an exciting way to use low-pressure ultrasound stimulation to produce microbubbles that amplify the mechanical deformations that can be detected by TRP-4, the pore-forming subunit of a mechanotransduction channel.[55] By overexpressing TRP-4 in specific neurons, the authors showed it can sensitize neurons to an ultrasound stimulus. Equally exciting, Shengjia Zhang and his colleagues recently reported another noninvasive method, namely magnetogenetics, to activate neurons in vivo using magnetic stimulation. The activation of neurons was achieved by neuronal expression of an exogenous magnetoreceptor, an iron-sulfur cluster assembly protein 1 (Isca1).[56,57] It is anticipated that these two noninvasive approaches offer additional tools to manipulate specific neurons in behaving animals. Moreover, with the recent discovery of the basic wiring and computational logic of cell assemblies,[58,59] Cre-lox neurogenetics will continue to aid neuroscientists in performing many critical experiments across different brain regions and animal species for decades to come.

## ACKNOWLEDGMENTS

This work was supported by the Georgia Research Alliance and National Institutes of Health (R01NS079774). I also would like to express my gratitude for the unique opportunity of working at the Brain Decoding Project Consortium organized by Banna Biomedical Research Institute and supported by Yunnan Science Commission (2014DG002).

## REFERENCES

1. Cajal SR. *Histology of the Nervous System.* 1909, 1910 (Trans. N. Swanson and L.W. Swanson, New York: Oxford University Press, 1995).
2. Edwards FA, Konnerth A, Sakmann B, Takahashi T. A thin slice preparation for patch clamp recordings from neurones of the mammalian central nervous system. *Pflügers Archiv.* 1989;414(5):600–612.
3. Hamill OP, Marty A, Neher E, Sakmann B, Sigworth FJ. Improved patch-clamp techniques for high-resolution current recording from cells and cell-free membrane patches. *Pflügers Archiv.* 1981;391: 85–100.
4. Klausberger T, Somogyi P. Neuronal diversity and temporal dynamics: the unity of hippocampal circuit operations. *Science.* 2008;321(5885):53–57.
5. Pinault D. A novel single-cell staining procedure performed in vivo under electrophysiological control: morpho-functional features of juxtacellularly labeled thalamic cells and other central neurons with biocytin or neurobiotin. *J Neurosci Meth.* 1996;65(2):113–136.
6. Buck L, Axel R. A novel multigene family may encode odorant receptors: a molecular basis for odor recognition. *Cell.* April 5, 1991;65(1):175–187.
7. Carraway R, Leeman SE. The isolation of a new hypotensive peptide, neurotensin, from bovine hypothalami. *J Biol Chem.* 1973;248(19):6854–6861.
8. Carlsson A. Thirty years of dopamine research. *Adv Neurol.* 1993;60:1–10.
9. Hökfelt T, et al. Occurrence of somatostatin-like immunoreactivity in some peripheral sympathetic noradrenergic neurons. *Proc Natl Acad Sci USA.* 1977;74(8):3587–3591.
10. Monyer H, et al. Heteromeric NMDA receptors: molecular and functional distinction of subtypes. *Science.* 1992;256(5060):1217–1221.
11. Moriyoshi K, Masu M, Ishii T, Shigemoto R, Mizuno N, Nakanishi S. Molecular cloning and characterization of the rat NMDA receptor. *Nature.* 1991;354(6348):31–37.
12. Noda M, et al. Primary structure of electrophorus electricus sodium channel deduced from cDNA sequence. *Nature.* 1984;312:121–127.
13. Seeburg PH, et al. The GABAA receptor family: molecular and functional diversity. *Cold Spring Harbor Symp Quant Biol.* 1990;55:29–40.
14. Südhof TC. Calcium control of neurotransmitter release. *Cold Spring Harb Perspect Biol.* 2012;4(1): a011353. https://doi.org/10.1101/cshperspect.a011353.
15. Frey S, Frey JU. 'Synaptic tagging' and 'cross-tagging' and related associative reinforcement processes of functional plasticity as the cellular basis for memory formation. *Prog Brain Res.* 2008;169:117–143.
16. Lømo T. The discovery of long-term potentiation. *Philos Trans R Soc Lond B Biol Sci.* 2003;358(1432): 617–620.
17. Tang YP, et al. Genetic enhancement of learning and memory in mice. *Nature.* 1999;401:63–69.
18. Capecchi MR. Gene targeting in mice: functional analysis of the mammalian genome for the twenty-first century. *Nat Rev Genet.* 2005;6:507–512.

19. DeChiara TM, Efstratiadis A, Robertson EJ. A growth-deficiency phenotype in heterozygous mice carrying an insulin-like growth factor II gene disrupted by targeting. *Nature*. 1990;345:78–80.
20. Koller BH, Marrack P, Kappler JW, Smithies O. Normal development of mice deficient in β2M, MHC class I. proteins, and CD8+ T cells. *Science*. 1990;248:1227–1230.
21. Thomas KR, Capecchi MR. Targeted disruption of the murine int-1 proto-oncogene resulting in severe abnormalities in midbrain and cerebellar development. *Nature*. 1990;346:847–850.
22. Zijlstra M, Li E, Sajjadi F, Subramani S, Jaenisch R. Germ-line transmission of a disrupted β2-microglobulin gene produced by homologous recombination in embryonic stem cells. *Nature*. 1989; 342:435–438.
23. Grant SG, O'Dell TJ, Karl KA, Stein PL, Soriano P, Kandel ER. Impaired long-term potentiation, spatial learning, and hippocampal development in fyn mutant mice. *Science*. 1992;258(5090): 1903–1910.
24. Silva AJ, Stevens CF, Tonegawa S, Wang Y. Deficient hippocampal long-term potentiation in α-calcium-calmodulin kinase II mutant mice. *Science*. 1992;257:201–206.
25. Silva AJ, Paylor R, Wehner JM, Tonegawa S. Impaired spatial learning in α-calcium-calmodulin kinase II mutant mice. *Science*. 1992;257:206–211.
26. Li Y, Erzurumlu RS, Chen C, Jhaveri S, Tonegawa S. Whisker-related neuronal patterns fail to develop in the trigeminal brainstem nuclei of NMDAR1 knockout mice. *Cell*. 1994;76(3):427–437.
27. Agranoff BW, Klinger PD. Puromycin effect on memory fixation in the goldfish. *Science*. 1964;146: 952–953.
28. Lyford GL, et al. Arc, a growth factor and activity-regulated gene, encodes a novel cytoskeleton-associated protein that is enriched in neuronal dendrites. *Neuron*. 1995;14(2):433–445.
29. Sauer B, Henderson N. Site-specific DNA recombination in mammalian cells by the Cre recombinase of bacteriophage P1. *Proc Natl Acad Sci USA*. 1988;85:5166–5170.
30. Gu H, Marth JD, Orban PC, Mossmann H, Rajewsky K. Deletion of a DNA polymerase β gene segment in T cells using cell type-specific gene targeting. *Science*. 1994;265:103–106.
31. McHugh TJ, Blum KI, Tsien JZ, Tonegawa S, Wilson MA. Impaired hippocampal representation of space in CA1-specific NMDAR1 knockout mice. *Cell*. 1996;87(7):1339–1349.
32. Tsien JZ, et al. Subregion- and cell type-restricted gene knockout in mouse brain. *Cell*. 1996a;87(7): 1317–1326.
33. Tsien JZ, et al. The essential role of hippocampal CA1 NMDA receptor–dependent synaptic plasticity in spatial memory. *Cell*. 1996b;87(7):1327–1338.
34. Leão RN, Mikulovic S, Leão KE, et al. OLM interneurons differentially modulate CA3 and entorhinal inputs to hippocampal CA1 neurons. *Nat Neurosci*. November 2012;15(11):1524–1530. https://doi.org/10.1038/nn. 3235. Epub October 7, 2012.
35. Madison, et al. A robust and high-throughput Cre reporting and characterization system for the whole mouse brain. *Nat Neurosci*. 2010;13:133–140.
36. Boyden ES, Zhang F, Bamberg E, Nagel G, Deisseroth K. Millisecond-timescale, genetically targeted optical control of neural activity. *Nat Neurosci*. 2005;8(9):1263–1268. https://doi.org/10.1038/nn1525.
37. Bouvier J, Caggiano V, Leiras R, et al. Descending command neurons in the brainstem that halt locomotion. *Cell*. 2015;163(5):1191–1203. https://doi.org/10.1016/j.cell.2015.10.074.
38. Kvitsiani D, Ranade S, Hangya B, Taniguchi H, Huang JZ, Kepecs A. Distinct behavioural and network correlates of two interneuron types in prefrontal cortex. *Nature*. 2013;498(7454):363–366. https://doi.org/10.1038/nature12176.
39. Madisen, et al. A toolbox of Cre-dependent optogenetic transgenic mice for light-induced activation and silencing. *Nat Neurosci*. 2012;15:793–802.
40. Senn V, Wolff SB, Herry C, et al. Long-range connectivity defines behavioral specificity of amygdala neurons. *Neuron*. 2014;81(2):428–437. https://doi.org/10.1016/j.neuron.2013.11.006.
41. Xie K, Fox GE, Liu J, Tsien JZ. 512-channel and 13-region simultaneous recordings coupled with optogenetic manipulation in freely behaving mice. *Front Syst Neurosci*. 2016a;10:48. https://doi.org/10.3389/fnsys.2016.00048. eCollection 2016.

42. Bishop AC, et al. A chemical switch for inhibitor-sensitive alleles of any protein kinase. *Nature*. September 21, 2000;407(6802):395–401.
43. Wang H, et al. Inducible protein knockout reveals temporal requirement of CaMKII reactivation for memory consolidation in the brain. *Proc Natl Acad Sci USA*. 2003;100(7):4287–4292.
44. Rogan SC, Roth BL. Remote control of neuronal signaling. *Pharmacol Rev*. 2011;63(2):291–315. https://doi.org/10.1124/pr.110.003020.
45. Brockschnieder D, Lappe-Siefke C, Goebbels S, Boesl MR, Nave KA, Riethmacher D. Cell depletion due to diphtheria toxin fragment A after Cre-mediated recombination. *Mol Cell Biol*. September 2004; 24(17):7636–7642.
46. Buch T, Heppner FL, Tertilt C, et al. Cre-inducible diphtheria toxin receptor mediates cell lineage ablation after toxin administration. *Nat Methods*. June 2005;2(6):419–426.
47. Matsumura H, Hasuwa H, Inoue N, Ikawa M, Okabe M. Lineage-specific cell disruption in living mice by Cre-mediated expression of diphtheria toxin A chain. *Biochem Biophys Res Commun*. 2004; 321(2):275–279.
48. Beier KT, et al. Circuit architecture of VTA dopamine neurons revealed by systematic input-output mapping. *Cell*. 2015;162(3):622–634.
49. Sun Y, et al. Cell-type-specific circuit connectivity of hippocampal CA1 revealed through Cre-dependent rabies tracing. *Cell Rep*. 2014;7(1):269–280.
50. Lerner TN, Shilyansky C, Davidson TJ, et al. Intact-brain analyses reveal distinct information carried by SNc dopamine subcircuits. *Cell*. 2015;162(3):635–647.
51. Livet J, Weissman TA, Kang H, et al. Transgenic strategies for combinatorial expression of fluorescent proteins in the nervous system. *Nature*. 2007;450(7166):56–62.
52. Tian L, et al. Imaging neural activity in worms, flies and mice with improved GCaMP calcium indicators. *Nat Methods*. 2009;6:875–881.
53. Akemann W, Mutoh H, Perron K, Rossier J, Knöpfel T. Imaging brain electric signals with genetically targeted voltage-sensitive fluorescent proteins. *Br J Pharmacol*. 2010;7:643–649. https://doi.org/10.1038/nmeth.1479.
54. Ghitani N, Bayguinov PO, Ma Y, Jackson MB. Single-trial imaging of spikes and synaptic potentials in single neurons in brain slices with genetically encoded hybrid voltage sensor. *J Neurophysiol*. February 15, 2015;113(4):1249–1259.
55. Ibsen S, Tong A, Schutt C, Esener S, Chalasani SH. Sonogenetics is a non-invasive approach to activating neurons in *Caenorhabditis elegans*. *Nat Commun*. 2015;6:8264.
56. Long X, Ye J, Zhao D, Zhang S-J. Magnetogenetics: remote non-invasive magnetic activation of neuronal activity with a magneto receptor. *Sci Bull*. 2015;60:2107–2119. https://doi.org/10.1007/s11434-015-0902-0. Epub September 14.
57. Wheeler MA, et al. Genetically targeted magnetic control of the nervous system. *Nat Neurosci*. 2016;19: 756–761. https://doi.org/10.1038/nn.4265.
58. Tsien JZ. A postulate on the Brain's basic wiring logic. *Trends Neurosci*. November 2015;38(11): 669–671. https://doi.org/10.1016/j.tins.2015.09.002. Epub October 16, 2015.
59. Xie K, Fox GE, Liu J, et al. Brain computation is organized via power-of-two-based permutation logic. *Front Syst Neurosci*. November 15, 2016;10:95. eCollection 2016.
60. Wang D, Cui Z, Zeng Q, Kuang H, Wang LP, Tsien JZ, Cao X. Genetic enhancement of memory and long-term potentiation but not CA1 long-term depression in NR2B transgenic rats.PLoS One. October 19, 2009;4(10):e7486. https://doi.org/10.1371/journal.pone.0007486.
61. Qian Z, Gilbert ME, Colicos MA, Kandel ER, Kuhl D. Tissue-plasminogen activator is induced as an immediate-early gene during seizure, kindling and long-term potentiation. *Nature*. 1993;361(6411): 453–457.
62. Qian Z, Gilbert M, Kandel ER. Temporal and spatial regulation of the expression of BAD2, a MAP kinase phosphatase, during seizure, kindling, and long-term potentiation. *Learn. Mem.* 1994;1: 180–188. https://doi.org/10.1101/lm.1.3.180.
63. Gong S, Doughty M, Harbaugh CR, Cummins A, Hatten ME, Heintz N, et al. Targeting Cre recombinase to specific neuron populations with bacterial artificial chromosome constructs. *J. Neurosci*. 2007; 27:9817–9823. https://doi.org/10.1523/JNEUROSCI.2707-07.2007.

64. Zemelman BV, Lee GA, Ng M, Miesenböck G. Selective photostimulation of genetically chARGed neurons. *Neuron*. 2002;33:15–22. https://doi.org/10.1016/S0896-6273(01)00574-8.

## FURTHER READING

1. Ivanova A, Signore M, Caro N, Greene NDE, Copp AJ, Martinez-Barbera JP. In vivo genetic ablation by cre-mediated expression of diphtheria toxin fragment A. *Genesis*. November 2005;43(3):129–135.
2. Hubel DH, Wiesel TN. Receptive fields, binocular interaction and functional architecture in the cat's visual cortex. *J Physiol*. 1962;160:106–154.
3. Jankowska E, Rastad J, Westman J. Intracellular application of horseradish peroxidase and its light and electron microscopical appearance in spinocervical tract cells. *Brain Res*. 1976;105:557–562.
4. Ling G, Gerard RW. The normal membrane potential of frog sartorius fibers. *J Cell Physiol*. 1949;34: 383–396.
5. Lorente de Nó R. Cerebral cortex: architecture, intracortical connections, motor projections. In: Fulton JF, ed. *Physiology of the Nervous System*. London: Oxford University Press; 1949:288–313.
6. McCrea RA, Bishop GA, Kitai ST. Intracellular staining of Purkinje cells and their axons with horseradish peroxidase. *Brain Res*. 1976;118:132–136.
7. Mountcastle VB. Modality and topographic properties of single neurons of cat's somatic sensory cortex. *J Neurophysiol*. 1957;20:408–434.
8. Scheibel ME, Scheibel AB. The organization of the nucleus reticularis thalami: a Golgi study. *Brain Res*. 1966;1:43–62.

CHAPTER 21

# Functional Analysis of Proteins Involved in Neurodegeneration Using the Model Organism *Dictyostelium*: Alzheimer's, Huntington's, and Batten Disease

**Michael A. Myre[1,a], Robert J. Huber[2], Danton H. O'Day[3,4]**
[1]University of Massachusetts Lowell, Lowell, MA, United States; [2]Trent University, Peterborough, ON, Canada; [3]University of Toronto, Toronto, ON, Canada; [4]University of Toronto Mississauga, Mississauga, ON, Canada

## INTRODUCTION

### *Dictyostelium discoideum*

Here we will review the expanding use of the model organism *Dictyostelium discoideum* as a relevant basic and biomedical system to delineate the normal function and effects of mutation to genes that cause disease in humans. Many neurodegenerative disorders, although related by their destruction of brain function, begin at the cellular level, and the pathogenic mechanisms are likely initiated through some deregulated function, expression, or mis-localization of the mutant protein. However, the many neurodegenerative disease genes are not simply expressed in neurons but are ubiquitously expressed throughout the various peripheral tissues and cell types. When considering neurodegeneration, a term used to describe the progressive loss of function or death of neurons, it is critical to understand the earliest events that lead to neuronal dysfunction to ultimately develop long-term effective therapies. Unfortunately, the normal function of many disease-linked genes remains unknown. This often affects the progress of research into how the mutation initiates disease. As a result, a variety of model organisms have been established ranging from yeast to mice that are all designed with the goal of understanding the underlying events of neurodegeneration to positively impact human health.

As a haploid organism, *Dictyostelium* is intriguing from a biologic point of view since it offers a variety of opportunities to test disease-related questions. This is largely because research can be conducted on clonal populations of cells but also as a bona fide multicellular organism (Fig. 21.1). *D. discoideum* is a soil-living eukaryotic microbe commonly referred to as the social amoeba. Upon starvation, *Dictyostelium* transitions from a population of individual cells into a multicellular organism. During this period, the cells

[a]Authors contributed equally to this article.

*Molecular-Genetic and Statistical Techniques for Behavioral and Neural Research*
ISBN 978-0-12-804078-2, https://doi.org/10.1016/B978-0-12-804078-2.00021-0

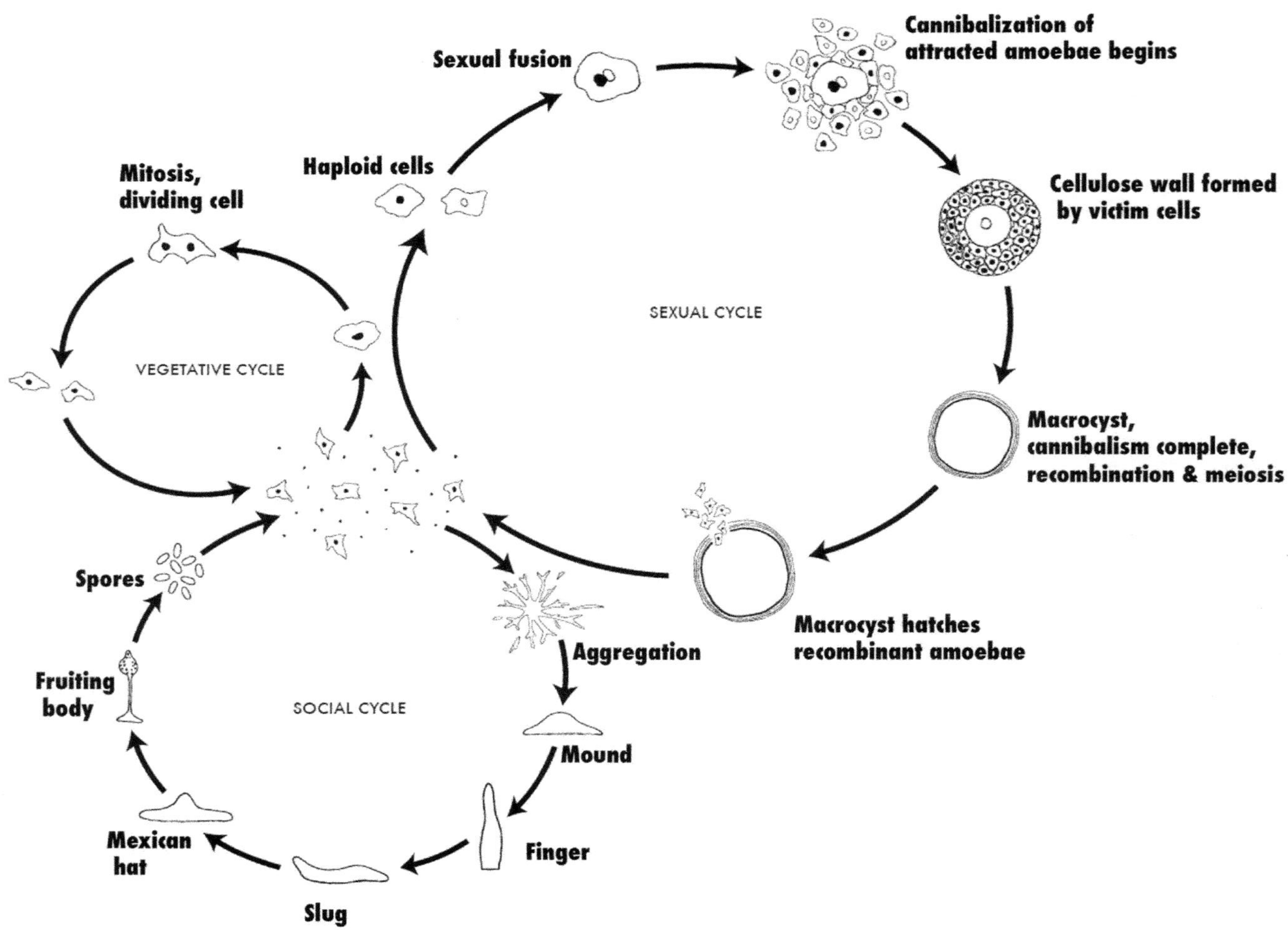


**Figure 21.1** ***The life cycle of*** **Dictyostelium discoideum.** Most of its life exists in its growth phase, as a haploid social amoeba preying upon bacteria in the soil and dividing by mitosis. Once the food source is depleted, cells enter into a 24-h multicellular developmental program. During this transition, amoebae aggregate toward secreted cAMP by chemotaxis in the thousands to form a tight mound and then enter a stage where cells begin to differentiate. Cells within the mound remain motile and are directed to differentiate, by secreted morphogens, into either prestalk or prespore cells, culminating to form a fruiting body comprised of a multicellular stalk that supports a ball of encapsulated dormant spores. *(Permissions: CC Creative Commons Attribution—Share Alike 3.0, David Brown and Joan E. Strassmann.)*

execute a series of developmental changes that proceed in defined stages over a 24-h period. Within the first 6 h of development, cells secrete and undergo chemotaxis toward cyclic adenosine $3'-5'$ monophosphate (cAMP) to form aggregation territories. The secretion of cAMP promotes a G protein-—coupled receptor signaling cascade that results in the formation of discrete mounds that contain as many as 100,000000 cells.[1,2] As development continues, cells within the mound remain motile and are directed to differentiate into either prestalk or prespore cells, leading to morphogenetic changes yielding a multicellular stalk, supporting a ball of encapsulated dormant spores.[3]

Importantly, unlike Metazoans, growth precedes development in *Dictyostelium*. As such, targeted deletions of homologous disease genes that do not affect growth can be studied in detail when cells are induced to develop synchronously with a remarkable consistency in the expression of genes across evolutionarily divergent species.[4] Upon the completion of the *Dictyostelium* genome project a number of reviews on the use of this model system to elucidate mechanisms of disease have been written, including articles on actin pathologies, mitochondrial disease, human lysosomal and trafficking disorders, host-pathogen interactions, and anticancer- drug action.[5–9]

## *Dictyostelium*: Value and Limitations of *Dictyostelium* as a Biomedical Model

For any model system to be of practical use, it should not only allow for the study of numerous basic and specialized physiological processes, but it must also do so in a cost-effective fashion. Comparison of the publication trends over the past 10 years reveals that the mammalian NIH-designated model organisms (rat and mouse) have dominated biologic research publications (Table 21.1). Although the total articles published on *Dictyostelium* is lower than the other systems, it should be noted that the *Dictyostelium* genome resource is relatively recent but has opened up many new opportunities to study genes that cause human disease. *Dictyostelium* offers great economic advantages over other eukaryotic systems as cells can grow in simple medium at room temperature, effectively limiting the need for costly growth incubators. Furthermore, *Dictyostelium* cells can grow to very high cell density (1011 1011 clonal cells) in a few days without the need for highly sophisticated equipment. Multicellular development of *Dictyostelium* is much shorter compared to mammalian systems and can be used for the rapid detection of developmental phenotypes. The haploid genome allows one to easily create gene disruptions, including the development of strains carrying multiple gene deletions. Furthermore, the cell culture conditions are simple, inexpensive, less time consuming, and not as labor intensive as they are for mammalian cells, thereby making it a highly attractive, complementary biologicbiological system. As such, *Dictyostelium*, as a model system, has been chosen by the National Institutes of Health as part of its Model Organism Initiative.

*Dictyostelium* is a powerful molecular system, as it offers most of the numerous molecular genetics techniques available for mammalian cells, including gene knockouts, gene

**Table 21.1** Total number of articles published from 2007 to 2017 using model organisms

| Model organism | PubMed keyword | Total number of articles |
|---|---|---|
| Mouse | Mouse | 623,289 |
| | *Mus musculus* | 586,420 |
| African clawed frog | African clawed frog | 7735 |
| | *Xenopus laevis* | 7697 |
| Zebrafish | Zebrafish | 20,957 |
| | *Danio rerio* | 21,160 |
| Fruit fly | Fruit fly | 36,789 |
| | *Drosophila melanogaster* | 19,433 |
| Nematode | Nematode | 26,756 |
| | Roundworm | 23,688 |
| | *Caenorhabditis elegans* | 13,303 |
| Slime mold | Slime mold | 310 |
| | Social amoeba | 251 |
| | *Dictyostelium discoideum* | 2042 |
| Yeast | Yeast | 92,073 |
| | *Saccharomyces cerevisiae* | 38,426 |
| | *Schizosaccharomyces pombe* | 4001 |

PubMed was used as the search engine, and the filter was set to search articles published within the last 10 years. Total numbers are accurate as of July 5, 2017.

knock-in, restriction enzyme-mediated mutagenesis (REMI), RNAi, and inducible gene expression, to allow a wide range of biologicbiological questions to be tested. The haploid nature of the genome allows for the unbiased discovery of direct and indirect genetic modifiers of disease through comprehensive genetic screens. The successful use of cDNA complementation/multicopy-copy suppression strategies that correct phenotypic defects is well described and can be used in combination with REMI to perform saturation screens aimed at recovering genetic modifiers.[10] REMI involves the transformation of cells with a mixture of plasmid DNA linearized with a restriction enzyme, along with a restriction enzyme capable of creating compatible cohesive ends in the genome.[11] This powerful screening technique is not available in higher organisms and, in addition, makes it possible to carry out specific genetic selection schemes, screens in which greater than a billion progeny can be generated and tested. Genetic schemes of this kind are effective at isolating potential second-site intragenic suppressor loci as well as saturating for second-site mutations that modify the phenotype of a given mutant. Clearly, this model system has no equal for establishing the networks of gene action involved in basic cell biologic processes. However, one limitation of using unicellular organisms as models for assessing the functions of genes involved in disease is that pathologies that affect specific organs cannot be as readily assessed as they can in mammalian systems. This limitation may also apply to genes that are present in unicellular organisms but are required in a more stringent manner in specialized tissues, or expressed as a variety of different isoforms in specific cell types. Although *Dictyostelium* forms a multicellular structure, it does present certain limitations with respect to the number of cell types that can be analyzed. As such, it is critical that discoveries about gene function in *Dictyostelium* must be properly assessed in the relevant mammalian cell type. But the time and money saved, coupled with the ability to generate detailed functional information not only about specific proteins but also about their functional domains and their binding partners, sets *Dictyostelium* apart as a system for the analysis of proteins linked to neurodegenerative diseases. Here, we provide some progress that has been made in the understanding of Alzheimer's, Huntington's, and Batten diseases.

## ALZHEIMER'S DISEASE

### The Involvement of Calmodulin in Alzheimer's Disease

Alzheimer's disease, the most prevalent cause of dementia, is characterized by beta amyloid plaques, neurofibrillary tangles, and neuronal dysfunction, which are widely believed to underlie the progressive neurodegenerative events associated with the disease. In spite of a strong understanding of factors and genes linked to the disease, especially for early onset (familial) Alzheimer's disease, there is no available treatment that can significantly slow or stop its unrelenting progress. While traditional approaches to finding a "cure" have focused and continue to focus on preventing the accumulation of amyloid beta and neurofibrillary tangles, more recently, attention has turned to events that precede them such as neuroinflammation and the accumulation of reactive oxygen species. With the multiple points of view that exist and continue grow, understanding the underlying causes and the search for common elements seems to be getting more difficult.

Over a dozen years ago, a central role for the small calcium-binding protein calmodulin in Alzheimer's disease was proposed.[12] This "Calmodulin Hypothesis" was a direct extension of the widely supported "Calcium Hypothesis" that presented strong evidence that dysregulation of intraneuronal calcium levels was a critical and early event in the development of the disease.[13] Since calcium primarily works by regulating calcium-binding proteins and since calmodulin is the major calcium-binding protein of all eukaryotic cells, it was logical to suggest it would be a primary target of dysregulated calcium. In support of this, the original theoretical data for the "Calmodulin Hypothesis" implied that a large number of the critical proteins in the amyloid processing pathway were calmodulin-binding proteins[12] (Fig. 21.2). Subsequent research revealed that, as predicted, β-secretase (BACE1; β-site AβPP cleaving enzyme 1), the first enzyme in the amyloid pathway, and its substrate amyloid-β precursor protein (AβPP) both bind to and are regulated by calmodulin.[14,15] Calmodulin binding to AβPP was subsequently verified by others.[16]

After BACE1, γ-secretase performs the final cleavage to produce amyloid β-peptide (Aβ). γ-Secretase consists of four main subunits: nicastrin (NCT), anterior pharynx-

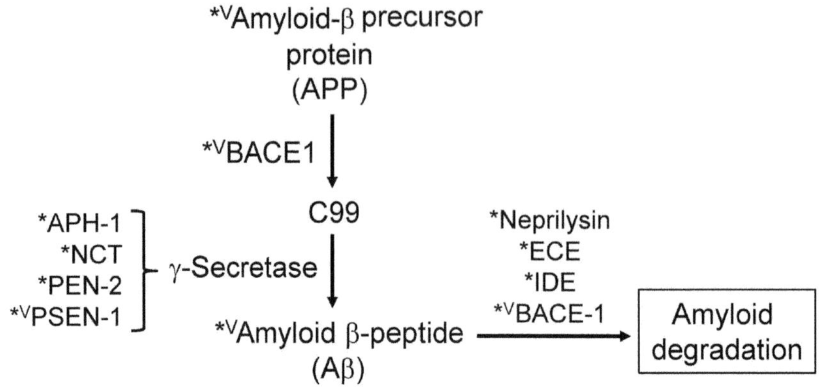


**Figure 21.2** ***Calmodulin binding proteins involved in amyloid peptide formation and turnover.*** See text for acronyms; *, calmodulin-binding domains identified; *V*, calmodulin-binding proteins experimentally verified; *Aβ*, amyloid β-peptide; *APH-1*, anterior pharynx-defective 1; BACE1, β-site AβPP cleaving enzyme 1; *ECE*, endothelin-converting enzyme; *IDE*, insulin degrading enzyme; *NCT*, nicastrin; *PEN-2*, presenilin enhancer 2; *PSEN-1*, presenilin 1.

defective 1 (APH-1), presenilin 1 (PSEN-1), and presenilin enhancer 2 (PEN-2), which assemble in the order listed.[17] PEN-2, the final subunit to be attached, facilitates auto-cleavage of PSEN-1, thereby activating the enzyme complex.[17] Each of the subunits contains putative calmodulin-binding domains, but only PSEN-1 has been experimentally validated as a true calmodulin-binding protein.[12,18]

Several enzymes that can degrade Aβ have also been proposed as calmodulin-binding proteins: neprilysin, endothelin-converting enzymes, insulin degrading enzymes, and BACE1.[19,20] Each of these Aβ degrading enzymes contain putative calmodulin-binding domains, but unlike BACE1, they have not been experimentally proven to bind to calmodulin.[21] Interestingly, calmodulin does bind to Aβ42 through a 1−8−14 motif, and when bound, Aβ cannot inhibit the brain plasma membrane $Ca^{2+}$-ATPase (PMCA) to prevent calcium entry.[22] This adds additional insight into the complex interplay between calcium dysregulation and calmodulin function.

Calmodulin-binding proteins have been linked to a diversity of other events associated with Alzheimer's disease.[23] Two of the receptors associated with Alzheimer's disease that are involved in calcium metabolism (mAchR; NMDAR) are calmodulin-binding proteins. Many genes recently identified by Genome Wide Association Studies encode putative calmodulin-binding proteins. Of these, APOE and BIN1 have been experimentally verified as calmodulin-binding proteins.[24] The apolipoprotein E ε4 allele (apoE4) is the most prevalent genetic risk factor for early onset Alzheimer's disease.

Many of the proteins that bind to calmodulin perform essential duties for the maintenance and function of synapses, which when altered in Alzheimer's disease, precede the processes of neuron death and the concomitant development of mental impairments associated with the disease. Calmodulin's involvement in the regulation of calmodulin kinase II and its potential effects on memory impairment and neurodegeneration in Alzheimer's disease has been further detailed.[25] Support for the calmodulin-regulated phosphatase calcineurin in the neuroinflammatory function of astrocytes associated with the disease has widened our understanding of calmodulin's importance.[26] Similarly, links of calmodulin to chaperone function raise the question of whether this small calcium-binding protein functions in the prion-like transformation of β-amyloid and tau that precede their fibrillar conversion.[27]

Over the years, many publications reveal that calmodulin's role in Alzheimer's disease extends much further than just the amyloid pathway. For example, calcium/calmodulin-dependent kinase II and calcineurin are known to be central to the respective phosphorylation and dephosphorylation of tau, a subject that is dealt with elsewhere.[23,25] In total, these results emphasize the multiple functions for calmodulin in Alzheimer's disease and the potential significance of this small calcium-binding protein as a target for therapeutic research, an approach already underway for Huntington's disease (HD). As new insight is gained into the mechanism of binding and binding interfaces, using approaches like molecular dynamic simulations and Markov State models, calmodulin is becoming a much more viable target for future pharmacological targeting for Alzheimer's disease and other neurodegenerative diseases.[28]

## γ-Secretase: Evolution of Structure and Function

Clearly the importance of AβPP as a substrate and understanding the mechanism that regulates Aβ production is an important focus of research; yet, there are many other known substrates of γ-secretase. Notch is a protein with a critical role in cell fate decisions that occur during development, so this line of research provides new directions of study into various cellular physiological processes, including the biology of the presenilins.[29] Furthermore, the central role of presenilin-dependent γ-secretase activity in notch signaling has been eloquently shown in many genetic experiments using various model organisms including but not limited to flies and worms.[30] In mammalian species, the situation has been more complicated largely due to gene duplications (*PSEN* and *APH-1*) and embryonic lethality resulting from knocking out either *PSEN* or *Aph-1* genes.[31] Whereas, in adult animals that have been genetically engineered to contain partial or tissue-specific knockouts of PSEN-1, they experience gastrointestinal bleeding, alterations to hair color, and dysfunctional immune properties.[32] There are now more than 90 proteins that are proposed or identified as substrates for the γ-secretases.[33] Almost all candidates are type I integral membrane proteins, and they follow the same consecutive-cleavage process outlined earlier for AβPP. This has led to the suggestion that in many cases the fundamental function of the PSEN-1 process is aimed at clearance of transmembrane domain stubs from the cell membrane. γ-Secretase and its inherent complexity of regulation, interaction with critical subunits, and other associated proteins theoretically presents a host of different mechanistic targets for therapeutic development. Therefore, most of the attention in this field has been focused on developing modulators that reduce pathogenic forms of Aβ while limiting their effect on the enzyme with respect to the processing of notch and other γ-secretase substrates.

Of particular importance regarding this chapter, the role of PSEN-dependent γ-secretase in more distantly related, and unicellular species have not been explored to any comparable extent. As briefly mentioned earlier, homologs of PSEN1/2, APH-1, PEN-2, and NCT have been studied in other non-Metazoan model organisms. The presenilins and orthologs to the other γ-secretase subunits and associated proteins were identified in *D. discoideum*, and ablation of these genes results in mutants that are viable but show a variety of phenotypic defects during development.[34,35] The *Dictyostelium* proteins are highly diverged (sequence identities ~25%–35%) in comparison with their mammalian counterparts, but have been found to be functionally equivalent.[34,35] It is crucial to note that *Dictyostelium* presenilins have very similar structural motifs also found in human presenilins, including the three amino acids (PAL) carboxy-terminal to transmembrane 8, conserved catalytic aspartic acid residues, and localization to the endoplasmic reticulum much like the mammalian presenilins. Before and elsewhere, it is revealed that mammalian PSEN1/2, APH-1, PEN-2, and NCT are proven or potential calmodulin-binding proteins.[23] In keeping with this, an analysis of the γ-secretase orthologs in *Dictyostelium* has revealed that PsenA and PEN-2 both possess traditional calcium-dependent calmodulin-binding domains, each with a diversity of binding options, while the homologs APH-1 and NCT each contain a basic, calcium-dependent calmodulin-binding domain (O'Day, unpublished). The ancestral noncatalytic- roles for presenilins across species has recently been reviewed[36] and will not be discussed in detail here. We aim to show how *Dictyostelium* represents an important research model to aid in further understanding the normal physiological role of the presenilins and γ-secretase from an evolutionary point of view in a system that has no notch or AβPP equivalent or any other known γ-secretase substrate identified thus far.[37] A number of candidate substrates have been tested as fusion-proteins without success (Myre, unpublished observations). Moreover, *Dictyostelium* also lacks the equivalent of a β-secretase or α-disintegrin and metalloproteinase domain-containing protein 10 (ADAM10), also referred to as the α-secretase.[37]

Utilizing the novel genetic approaches available for use in *Dictyostelium*, we suggest this model system might provide critical biochemical clues to help elucidate the structure-function evolution of γ-secretase components and provide a broader understanding of the regulated intramembrane cleavage events. Gaining a detailed mechanistic understanding of how γ-secretase binds and cleaves AβPP is critically important for biology and medicine. Key details on how γ-secretase regulates substrate binding have remained elusive because of a lack of high-resolution structural data, yet the most recent highly detailed atomic-resolution structure of nicastrin was solved using nicastrin from the model organism *Dictyostelium purpureum*.[38] The improved resolution obtained in this study helps us understand in greater detail the precise placement of nicastrin within the complex and further fuels speculation that nicastrin may indeed serve as a receptor of γ-secretase substrates.[39] Although, in contrast, knockout cells and siRNA of γ-secretase components suggest that presenilins and PEN-2 are sufficient for enzymatic activity.[40] Of interest and in support of these findings, ablation of the nicastrin gene appeared to have no effect on fruiting-body development in *Dictyostelium*.[35] These results suggest that nicastrin, presumably as part of the γ-secretase complex, is not required for *Dictyostelium* development, and to support this, it was also shown that *Dictyostelium* or human double-mutants for the critical aspartic acid residues in presenilin also rescued *Dictyostelium* morphological development.[35] Much more work is needed to decipher the role of each component comprising γ-secretase, but it is clear the complex exists in a variety of conformations and exhibits subtle allosteric recognition of substrate that modulates access to the catalytic site.[41] So what are the original, ancient biologic functions of γ-secretase and its individual components? *Dictyostelium* represents a novel system to help answer these and other questions surrounding the basic biomedical biology of not only the γ-secretase complex, but the nonenzymatic- roles for each of the individual components.

## Conserved Biologic Roles

Neurodegenerative diseases are often characterized by a set of cognitive, behavioral, and motor alterations that develop due to pathogenic mechanism leading to death of specific neurons. As discussed earlier, Alzheimer's disease brains contain two suspected pathological aggregates: (1) the extracellular plaques (containing Aβ) and (2) intracellular neurofibrillary tangles (containing tau protein isoforms).[42] One conceptual stance assumes that γ-secretase or PSEN itself might originally be involved in the degradation and recycling of membrane proteins.[43] From this viewpoint, the enzyme would have to act as a so-called membrane proteasome,[29] acquiring signaling functions related to γ-secretase cleavage events later in evolution.[34] This would be consistent with the observation that presenilins may regulate signaling pathways by degrading membrane or recycling endosomes. Evidence has accumulated indicating the possible role of γ-secretase in phagocytosis in *Dictyostelium*.[34] Phagocytosis, the engulfment of material by cells, is a highly conserved process that arose before the development of multicellularity. In keeping with this, γ-secretase subunits have been shown to be indispensable for proper phagocytosis. Purified phagosomal membranes from murine macrophage and *Drosophila* S2 phagocytes both show significant γ-secretase activity.[44] Microglia deficient for γ-secretase activity also exhibit aberrant phagocytosis.[45] In line with a role in phagocytosis, endocytotic nutrient uptake was also found to be altered in *Dictyostelium* mutants for γ-secretase components.[34] Taken together, all of the data across diverse species suggest that presenilins alone or in the γ-secretase complex might play a crucial role in endocytic membrane and trafficking events. What is becoming increasingly clear is that the ancient, biologic role of γ-secretase and the individual components are still beyond what we currently know from Metazoan studies. The literature suggests the possibility that presenilin-independent and γ-secretase activity may have arose before Metazoan expansion and may have independently acquired both AβPP and notch signaling. Future studies using *Dictyostelium* will undoubtedly help to understand the normal role of the γ-secretase complex and at the same time offer insight into the noncatalytic- roles of the components during phagocytosis and development.

## *Dictyostelium*: A Translational Research System

From the standpoint of developing a translational model system, certain proof-of-principle experiments must be conducted. The most critical of these revealed that *Dictyostelium* efficiently and accurately processes ectopically expressed human AβPP into the α- and β-C-terminal fragments (αCTFs, βCTFs) and Aβ40/Aβ42, whereas AβPP processing in strains that are deficient for γ-secretase subunits is blocked.[34] Furthermore, a second study showed that presenilin double-knockout mouse-blastocyst cells transfected to express either of the *Dictyostelium* presenilin genes effectively restores cleavage of mammalian notch.[35] All together the evidence thus suggests that *Dictyostelium* presenilins are functionally active components of the γ-secretase in mammalian cells and that human presenilin functionally rescues *Dictyostelium* γ-secretase mutant phenotypes.[35] The distribution of known γ-secretase substrates across species is very intriguing: all of them entirely exist in Metazoa, even though γ-secretase activity toward human AβPP and Notch have been reported in *Dictyostelium*.[34,35,46] Similarly, although genes for γ-secretase subunits have been described in plant genomes, no endogenous γ-secretase substrates have been reported.[47] With respect to *Dictyostelium*, current studies using wild-type cells that constitutively express human AβPP are being used to conduct REMI-mediated genetic suppressor screens to identify novel "sheddases" or candidate genes that alter the processing of AβPP into CTFs and Aβ. In conjunction with this, various mutant forms of AβPP such as the Swedish APP670/671 mutation, which leads to increased formation of Aβ, could also be expressed in *Dictyostelium* and subjected to genetic suppressor screens. Studies on γ-secretase and the enzyme-independent function of its components in *Dictyostelium* will impact our understanding of a critical human health-related enzyme in a model system and will influence future work toward testing the hypothesis that, in mammalian systems, blocking the function of one or more proteins identified in a *Dictyostelium* suppressor screen could be useful as a therapeutic approach in the study of Alzheimer's disease.

## HUNTINGTON'S DISEASE

### Huntingtin: Size Matters

HD is a fatal neurodegenerative disease[48] that affects muscle coordination, eventually leading to cognitive decline and death. The pathogenesis of the disease has been widely explored, yet the normal function of the protein responsible for the disease remains largely unknown. HD is a monogenic neurologic disease conferred entirely by a cytosine-adenine-guanine (CAG) repeat expansion mutation in the protein huntingtin (HTT). The CAG trinucleotide repeat within the coding region of the *htt* gene is a homopolymer stretch of glutamine (Q) residues near the amino-terminus of the protein whose length is the primary determinant of the rate of pathogenesis leading to disease onset.[48] However, research is actively attempting to identify modifier loci harboring genetic variations that alter the age at neurological onset of HD.[49] Mutant HTT confers a gain-of-function property to some aspect of the protein, but recent evidence also suggests that mutant HTT may also have loss-of-function properties that are fundamental for the survival and functioning of striatal neurons.[50] HTT is a large (350 kDa) alpha-solenoid HEAT (*h*untingtin, *e*longation factor 3, the *A* subunit of protein phosphatase 2A, and *T*OR1) repeat protein that bears little resemblance to any other known protein and whose primary function still remains unclear.[51]

The large size of HTT makes it challenging to perform biochemical experiments and develop robust structure/function assays. Consequently, few studies have been able to address the cellular functions of full-length HTT and any dysfunctions associated with the disease compared to the vast amount of literature on the amino-terminal fragment, which, in essence, removes the mutation from the context of the full-length protein. One might be inclined to argue that studying N-terminal polyQ fragments is tantamount to research on polyQ disease rather than HD. In fact, studies have revealed in the R6/2 mouse, a widely used HD model, that integration of a rearranged transgene coincides with a deletion of 5444 base pairs of host genome within a gene of unknown function, Gm12695.[52] Gm12695 is normally expressed at negligible levels in mouse brain but found to have dramatically increased levels, greater than 30-fold, in the corresponding brain tissues of R6/2 mice.[52] However, the point of this chapter is not to create controversy because much has been learned about HTT fragment toxicity from studies utilizing this N-terminal fragment model. The few genetically precise, full-length HTT knock-in mouse models for HD have aided our understanding of HTT normal function, but these mice are expensive, laborious to create, and difficult to validate. Interaction studies, again in the case of HTT and its large size, impede uncovering its normal function since many small and large-scale screening experiments use various HTT fragments as bait. There are now at least 234 proteins that have been discovered to interact with mutant huntingtin.[53] How these interacting partners ultimately contribute to understanding HTT normal function remains to be seen, and again, the interpretation of the results must be closely scrutinized as the screen was not performed with full-length HTT, which when mutated, is the actual causal agent of HD.

## *Dictyostelium* Huntingtin: Conservation of Normal Function

The *Dictyostelium* genome was screened and found to contain a single gene with sequence homology to human huntingtin. The *htt* locus is comprised of four exons, located on chromosome 2. The primary amino acid sequence firmly placed the protein within the huntingtin family, with a length of 3095 amino acids comparable to the 3144 amino acids of human huntingtin, shares 23% identity and 45% homology and the presence of numerous HEAT repeats.[54] To study the function of huntingtin in *Dictyostelium*, knockout mutants were created using homologous recombination and subsequently screened for phenotypic effects.[54] *Dictyostelium* cells deficient for HTT are viable, unlike the embryonic lethality seen in higher eukaryotes, and the protein is involved in a number of cellular processes including growth, osmoregulation, cation homeostasis, cell motility, cell shape, chemotaxis, cell-cell adhesion, and cell fate determination.[54,55] Importantly, these phenotypic deficits strongly suggest that HTT may not have a single critical function; rather, it serves as a multifunctional protein that aids in translating extracellular signals into cellular processes.

Chemotaxis, which is the directed movement of cells toward chemical attractants or away from chemical repellents, is central to the life cycle of *Dictyostelium*.[56] This form of directed migration is highly conserved and plays an important role in many biologic processes, such as embryogenesis and wound healing.[57,58] *Dictyostelium* is an ideal model system for understanding the mechanisms that control chemotaxis, so its use provides a novel approach toward understanding huntingtin function. Myre et al.[54] and subsequently Wang et al.[55] both showed that *htt*$^-$ cells displayed aberrant chemotaxis toward gradients of cAMP. Additional studies further showed that in *Dictyostelium*, huntingtin is essential for $K^+$-facilitated cAMP chemotaxis and $Ca^{2+}$ chemotaxis.[59] Curiously, this phenotypic profile closely resembled the null mutant of Nhel, a monovalent cation/$H(^+)$ exchanger, which when expressed in *htt*$^-$ cells shows aberrant intracellular localization.[59] NHE1 is necessary at the front of migrating cells for polarity and directional motility.[60] Activation of the innate immune system has been reported in HD patients, yet it is unclear whether this is due to the effect of mutant HTT or represents a secondary event in response to the degeneration of neurons. Primary HD microglial cells lines with overexpression of mutant HTT isolated from human HD myeloid cells exhibit defects in migration to chemotactic stimuli.[61] Neural stem (NS) cells isolated from embryonic wild-type (HTT(7Q/7Q)) and "knock-in" HD (HTT(140Q/140Q)) mice expressing full-length endogenous normal or mutant HTT showed impaired motility.[62] Of particular importance, Ritch et al.[62] showed that heterozygous mutant HTT(F140Q/7Q) NS cells have decreased motility compared to normal HTT(F7Q/7Q) and that this phenotype replicates those seen in HD patients, primary cell culture, or in vivo models of HD. Therefore, it appears that mutant HTT or the absence of HTT interferes with cell motility in mammalian cells. The molecular mechanisms responsible for the chemotaxis defects remain to be fully understood, yet these results suggest that the defects are in part dependent on mutant HTT expression. Interestingly, when taken together the data suggests that impaired chemotaxis in mutant HTT immune cells could be a result of loss of normal function. More studies are needed to decipher these results as immune cells are not neurons.

## Huntingtin: Regulator of Cytoskeletal Remodeling?

When *htt*$^{-}$ cells are induced to develop under nonnutrient, low ionic strength phosphate buffer (KK2; 40 mOsmol/L), the cells immediately elicit a round cell phenotype suggesting altered cytoskeletal dynamics.[54] Indeed, phalloidin staining of *Dictyostelium htt*$^{-}$ cells under these conditions reveal a disorganized cytoskeleton with a strong reduction in the levels of cortical actin. Evidence suggests that F-actin interacts aberrantly with the amino-terminus of mutant HTT.[63] Defective actin remodeling is seen in patient-derived mutant HTT cells and has also been observed in striatal cells where huntingtin levels have been reduced by siRNA, suggesting that cytoskeletal remodeling might be an ancient function of HTT.[64] Microglial studies have also shown that mutant HTT perturbs the proper regulation of actin remodeling.[61] The authors attribute this to decreased expression of cofilins and the formation of actin-cofilin rods in the nuclei of cells under stress. It is of interest to note that a type of actin rod forms in 90% of mature dormant spores during *Dictyostelium* development, and this complex organization of the actin cytoskeleton may be required for spores to achieve complete dormancy.[65] The viability of spores lacking actin rods is very low, and the shape of the spore becomes round rather than elliptical. Cells deficient for *htt*$^{-}$ in *Dictyostelium* produce round spores that cannot achieve complete dormancy.[54] The role of actin rods in this phenotype is currently under investigation. Moreover, membrane ruffling and actin polymerization are also reduced in the dendritic spines and synapses of striatal neurons from HD mouse models.[66,67] Cytoskeletal organization is needed for normal neuronal cell development and functionality, and perhaps, neurons are more sensitive to mutant HTT effects. Moreover, the effects of mutant HTT and the normal functions of HTT on actin remodeling may vary between neurons and immune cells.

## Huntingtin: Modulator of Cell Adhesion?

As we have already mentioned, *Dictyostelium htt*$^{-}$ cells display a variety of developmental abnormalities. Reis et al.[67] showed that STHdh(Q111) striatal cells have extreme deficits in calcium-dependent, N-cadherin-mediated cell adhesion and cell-substratum adhesion. *Dictyostelium htt*$^{-}$ cells completely fail to acquire EDTA-resistant homotypic cell adhesion during starvation in suspension culture.[68] In *Dictyostelium*, the gene encoding the glycoprotein csaA (contact site A/csA/gp80) is required for EDTA-resistant adhesion, a phenotype that can be rescued by constitutive, ectopic expression of csaA.[69] Expression of csaA and EDTA-resistant adhesion in *htt*$^-$ cells is grossly dysregulated but, interestingly, can be rescued by the addition of exogenous calcium to the developmental buffer but not by ectopic expression of csaA.[68] The data implies a novel role for HTT in conferring csaA-dependent, EDTA-resistant cell adhesion that warrants further investigation. Many dominant changes in cell adhesion were observed using immortalized and primary Hdh CAG knock-in mouse striatal cells.[67] Furthermore, in a cell-cluster formation assay, in which cells are dissociated, re-plated, and allowed to form aggregates over time, all the HD neural progenitor cells showed significantly decreased cell-cell adhesion properties.[70] A number of dominant, CAG length—dependent biochemical and cell biologic phenotypes such as cell adhesion have been observed in murine *Hdh* CAG knock-in ESCs and derived immortalized lines.[71]

Neural progenitors derived from HD60i and HD180i NSC lines also show less aggregation upon plating, providing further evidence for disruptions in cell adhesion that may affect cell survival and/or differentiate appropriately.[70] Similarly, *Dictyostelium htt*$^-$ cells unable to acquire EDTA-resistant adhesion during hours 3—6 of starvation in suspension culture is paralleled by a progressive impact on the developmental capacity of *htt*$^-$ cells. After 4 h of starvation in suspension culture and in stark contrast to wild-type cells, the ability of *htt*$^-$ mutants to form mature fruiting bodies is abolished.[68] Therefore, *htt*$^-$ cells otherwise capable of completing development, albeit delayed, gradually lose their ability to differentiate and develop into fruiting bodies due to the inability to make stable EDTA-resistant cell-cell contacts.

*Dictyostelium* huntingtin does not contain N-terminal glutamines, a feature that appears only in deuterostome animals first seen with two glutamines in sea urchins and expanded further in other lineages.[72] The N-terminus of huntingtin from a diverse array of lower and higher model organisms, including *Dictyostelium*, were shown to have antiapoptotic- activity when expressed in HTT-null cells.[72] However, only the N-termini from deuterostomes were able to restore N-cadherin-mediated neural cell-cell adhesion *htt*-null cells.[72] The molecular mechanism for this antiapoptotic- action remains a mystery since *Dictyostelium* does not undergo apoptosis but rather a caspase-independent form of cell death. However, it was shown that *htt*$^-$ cells, when developed as chimeric populations with wild-type cells, fail to become spores and, in contrast, were more sensitive to the signals that trigger vacuole-mediated cell death and thus invariably become dead stalk cells.[54] Therefore, as a model, *Dictyostelium* has advantages for studies on nonapoptotic- cell death in the absence of apoptotic machinery, which may illuminate mechanistic regulation of apoptosis versus other forms of cell death. The *Dictyostelium* model has helped to identify conserved functions for huntingtin. The system will continue to aid in distinguishing between which HTT functions are dependent upon polyQ-length and should assist in the future in defining the molecular mechanics regulating these cellular processes more clearly.

## NEURONAL CEROID LIPOFUSCINOSIS (BATTEN DISEASE)

### Batten Disease: The Most Common Form of Neurodegeneration in Children

NCL, also known as Batten disease, refers to a group of neurologic disorders that affect both children and adults.[73] The disease, which is the most common form of neurodegeneration in children, has a worldwide distribution and affects all ethnicities. People from Northern European populations and Newfoundland, Canada, have been reported to have the highest rates of incidence.[74–77] Clinical symptoms of NCL include intellectual impairment, epileptic seizures, the progressive loss of vision and motor function, and premature death.[78] Although classified as a rare disease, recent genetic advances now show a strong overlap of NCL phenotypes with later onset neurodegenerative diseases, including Parkinson' disease and frontotemporal dementia.[79,80] Over 430 mutations have been identified in 13 genetically distinct NCL genes.[73] Mutations in these genes cause aberrant lysosomal function and an accumulation of ceroid lipofuscin in neurons as well as other cell types outside of the central nervous system.[81]

The NCL family of proteins is comprised of lysosomal enzymes (PPT1/CLN1, TPP1/CLN2, CTSD/CLN10, CTSF/CLN13), membrane-associated proteins (DNAJC5/CLN4, KCTD7/CLN14), a soluble lysosomal protein (CLN5), a protein that is present in the secretory pathway (PGRN/CLN11), and several transmembrane proteins (CLN3, CLN6, MFSD8/CLN7, CLN8, ATP13A2/CLN12).[73,82] Understanding a protein's cellular function is central to being able to devise ways to target it, and unfortunately, the precise functions of many of the NCL proteins are still not known. Since mutations in NCL proteins cause the accumulation of ceroid lipofuscin within cells, there is a critical need to understand how these proteins function to facilitate the endocytosis and processing of material required for normal cellular function. This involves using new innovative approaches to identify the primary functions of these proteins, which will hopefully lead to targeted therapy development for this devastating and currently untreatable disease.

### Using *Dictyostelium* to Study NCL

NCL homologs have been identified and studied in several model organisms, and this work has significantly advanced our understanding of the localization and functions of NCL proteins in humans.[83–85] One such organism is the social amoeba *D. discoideum*, which contains homologs of 11 of the 13 known NCL genes.[5] In fact, *Dictyostelium* contains more homologs of NCL proteins than other commonly used model organisms including *Saccharomyces cerevisiae*, *Schizosaccharomyces pombe*, *Caenorhabditis elegans*, and *Drosophila melanogaster*.[5] Recent research on the *Dictyostelium* homologs of human TPP1/CLN2 and CLN3 indicates that this model organism provides unprecedented opportunities to resolve the normal functions of these and other NCL proteins.[86–89]

Importantly, many of the phenotypes observed in *Dictyostelium* knockout mutants are conserved in mouse and human cell models of NCL. Publically available mRNA expression data suggests different functions for the NCL homologs during the *Dictyostelium* life cycle, with some being required for growth and others for development.[5,90] Since mutations in NCL proteins cause the accumulation of ceroid lipofuscin within cells and result in nearly identical clinical manifestations, they are thought to participate in the same biologic pathway.[91] This is supported by work that has reported interactions between several NCL proteins including PPT1/CLN1, TPP1/CLN2, CLN3, CLN5, CLN6, and CLN8.[92–94] In *Dictyostelium*, many of the NCL homologs were identified in a proteomic-based analysis of the macropinocytic pathway, suggesting that they may also participate in a common pathway in this model organism.[95] In the sections to follow, we will describe how homologous recombination, REMI, expression constructs, chimera analyses, and LC-MS/MS have been used to reveal the localization and function of Tpp1/Cln2 and Cln3 in *Dictyostelium*. The significance of these results is discussed.

## The Localization and Function of Cln3 in *Dictyostelium*

The most common subtype of NCL is a juvenile form that is caused by loss-of-function mutations in ceroid lipofuscinosis neuronal 3 (CLN3).[78,82] CLN3 is a 438–amino acid multipass transmembrane protein that is conserved from yeast to human.[83–85,96] The CLN3 protein, which has been studied using a number of systems and cell models, localizes to late endosomes and the lysosomal membrane.[82] While the function of the protein has been linked to autophagy, apoptosis, intracellular trafficking, lysosomal pH homeostasis, cell cycle control, osmoregulation, and calcium homeostasis, the mechanism(s) by which CLN3 affects these processes remains unknown.[82]

The *Dictyostelium* homolog of human CLN3, Cln3, is a 421– amino acid protein predicted to contain 10–11 transmembrane domains and shares 27% identity and 45% homology to the human protein. Importantly, amino acid residues that are mutated in patients with the disease are conserved in the *Dictyostelium* homolog, as are residues that are posttranslationally- modified in the human protein (e.g., glycosylated, prenylated, myristoylated).[86] To study the function of Cln3 in *Dictyostelium*, Huber et al.[86] used the well-established method of homologous recombination to knockout the *cln3* gene.[97] The resulting cell line was extensively characterized at both the single-cell and multicellular level to uncover phenotypes during the *Dictyostelium* life cycle. During growth, loss of Cln3 alters the secretion and cleavage of the quorum-sensing proteins autocrine proliferation repressor A (AprA) and counting-factor associated protein A (CfaD).[86] Interestingly, AprA has predicted structural similarity to the human protein dipeptidyl peptidase IV, which is a chemorepellent for human neutrophils.[98] As a result of the aberrant secretion of these proteins, Cln3-deficient cells show an increased rate of

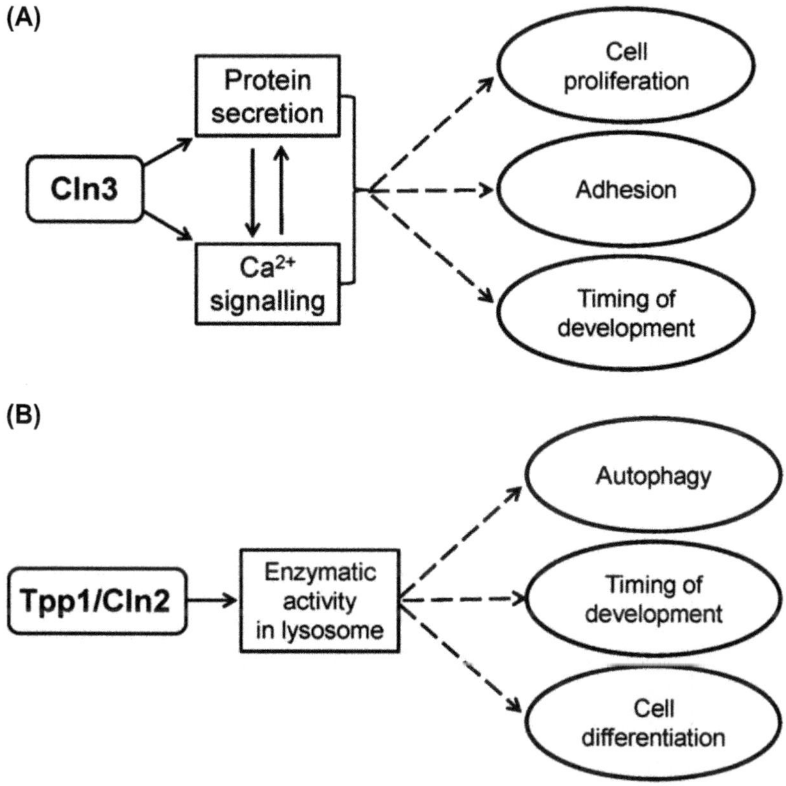


**Figure 21.3** ***Functions of Cln3 and Tpp1/Cln2 in*** **Dictyostelium.** (A) Cln3 modulates protein secretion and calcium homeostasis, which regulates cell proliferation, adhesion, and the timing of multicellular development. (B) Tpp1/Cln2 functions as a protease in the lysosome and regulates autophagy, the timing of multicellular development, and cell differentiation during fruiting body formation.

proliferation compared to wild-type cells[86] (Fig. 21.3A). Upon entering multicellular development, Cln3-deficient cells show reduced adhesion to the surface and to each other, which ultimately delays aggregation[88] (Fig. 21.3A). The decreased adhesion of *cln3*$^-$ cells correlates with alterations in the levels of two well-established cell adhesion proteins in *Dictyostelium*, contact site A (CsaA) and calcium-dependent cell adhesion molecule (CadA).[88] Importantly, this was the first evidence in any system linking CLN3 function to cell adhesion, and the phenotype has since been validated in mammalian cells lacking functional CLN3 (Huber, Atwal, Chandrachud, Cotman, unpublished observations). Finally, during the mid-to-late stages of development, *cln3*$^-$ cells develop precociously, and *cln3*$^-$ slugs display enhanced migration compared to wild-type slugs[86] (Fig. 21.3A). Importantly, these phenotypes can be rescued by re-introducing human CLN3 into *cln3*$^-$ cells, thus confirming that the function of CLN3 is evolutionarily conserved.[86]

Owing to its study in a diversity of systems, CLN3 has been reported to localize to a variety of intracellular compartments.[82] In *Dictyostelium*, proteomics-based approaches localized CLN3 to the macropinocytic pathway, along with other NCL homologs.[95] A major benefit of using *Dictyostelium* as a research organism is that the research community has access to a diversity of expression constructs that can incorporate any one of a number of tags to a protein of interest (e.g., GFP, FLAG, Myc, His, mRFPmars, etc.).[99,100] Huber et al.[86] and Huber[89] used one of these constructs to fuse Cln3 to GFP and observed that the fusion protein localizes primarily to the CV system, and to a lesser extent, compartments of the endocytic pathway and the Golgi complex. Huber and colleagues[86] also took advantage of the fully sequenced and annotated *Dictyostelium* genome to identify the minimal promoter element required to drive *cln3* expression. GFP-Cln3 expressed under control of this promoter element also localized the protein to the aforementioned subcellular compartments, thus demonstrating that the observed localization of Cln3 was not an artifact due to overexpression.

As in other protists, the *Dictyostelium* CV system functions primarily in osmoregulation.[101] Intriguingly, osmoregulatory defects have been reported in mammalian cell models lacking functional CLN3.[102–104] However, the function of this diverse organelle has also been linked to ion homeostasis, vesicular trafficking, and unconventional protein secretion (i.e., secretion of proteins that lack a signal peptide).[101,105,106] Cln3-deficiency phenotypes during development can be fully rescued by treating cells with the calcium chelator EGTA.[86,88] This observation is noteworthy considering that the *Dictyostelium* CV system is a major store of intracellular calcium.[101,107,108] These results suggest that *cln3*$^-$ cells are deficient in their ability to buffer calcium from the extracellular environment and, importantly, are consistent with studies in mammalian systems that have reported altered calcium homeostasis in the absence of functional CLN3[109–113] (Fig. 21.3A). These findings also support other reports that have linked neurodegeneration to altered calcium levels.[114–116]

In humans, neuronal signaling relies on the tight regulation of mechanisms that secrete molecules and proteins into the synaptic cleft, and there is evidence to suggest that aberrant secretion is linked to neurodegeneration.[117–119] During early *Dictyostelium* development, cells migrate in streams that relay a signal of cAMP from cell-to-cell. This signaling mechanism can be thought of as a primitive chemical synaptic system where cAMP serves as the neurotransmitter. Phosphodiesterases, which are also secreted by *Dictyostelium* cells during early development, break down cAMP just like enzymes degrade neurotransmitters in neuronal signaling. Based on its localization to the CV system and the Golgi complex, which are linked to unconventional and conventional protein secretion, respectively,[89] Huber investigated the loss of Cln3 on protein secretion during the early stages of *Dictyostelium* development. Chimeric populations of wild-type and *cln3*$^-$ cells display delayed streaming and aggregation, suggesting that signaling between *cln3*$^-$ cells is compromised.[89] In support of this observation, *cln3*$^-$ cells starved in conditioned media harvested from starving wild-type cells show near-normal timing of streaming and aggregation.[89] Based on these results, LC-MS/MS was used to reveal the protein content of conditioned media from starved cells.[89] A total of 450 proteins were detected in wild-type and *cln3*⁻ conditioned media. Of these proteins, three3 were absent in *cln3*⁻ conditioned media, and 12 were present exclusively in *cln3*⁻ conditioned media.[89] Label-free quantification revealed that 42 of the proteins were present in significantly higher amounts in *cln3*⁻ conditioned media compared to wild-type, and three3 proteins were present in significantly reduced amounts.[89] A GO (Gene Ontology) term enrichment analysis showed that a majority of the affected proteins are linked to endocytosis, vesicle-mediated transport, proteolysis, and metabolism: information that provides novel insight into how the aberrant secretion could impact early developmental events.[89] For example, proteases secreted during the early stages of *Dictyostelium* development act on the cell membrane to release components that coat the substratum and facilitate migration.[120,121] Label-free quantification revealed a number of proteases that are present in significantly altered amounts in *cln3*⁻ conditioned media compared to wild-type.[89] Therefore, it is reasonable to suggest that aberrant secretion of these proteases may disturb cell-substrate adhesion, ultimately causing a delay in the streaming and aggregation of *cln3*⁻ cells.[89] These results, combined with the localization of Cln3 to the CV system and the Golgi complex, and the effect of Cln3 loss on the extracellular levels of AprA, CfaD, and CadA, suggest that Cln3 functions in both conventional and unconventional protein secretion (Fig. 21.3A). Moreover, given that a large number of secreted proteins are present in increased amounts in *cln3*⁻ conditioned media compared to wild-type, these results indicate that loss of Cln3 results in deregulated secretion during early development. Importantly, this work provided the first evidence in any system linking CLN3 function to protein secretion. In total, research on the localization and function of Cln3 in *Dictyostelium* has provided fresh new insight into the primary function of CLN3 that is lost in affected individuals.

## The Localization and Function of Tpp1/Cln2 in *Dictyostelium*

Loss-of-function mutations in tripeptidyl peptidase 1 (TPP1/CLN2) cause a late infantile form of NCL.[78,82] In mammalian cells, TPP1/CLN2 localizes to the ER and lysosomal matrix, functions as a serine protease, and is linked to endocytosis, TNF-α-induced apoptosis, and macroautophagy.[82] While TPP1/CLN2 is highly conserved in vertebrates, homologs are not present in classical model organisms including *S. cerevisiae*, *S. pombe*, *C. elegans*, and *D. melanogaster*, which has unfortunately, limited the study of this protein.[87,122] Intriguingly, unlike other model organisms, *Dictyostelium* contains a homolog of TPP1/CLN2 that was recently characterized.[87] *Dictyostelium* Tpp1 is a 600- amino acid, 67-kDa protein that shares 36% identity, 52% homology, and many structural attributes with human TPP1/CLN2. Phillips and Gomer[87] used the well-established method of homologous recombination to knockout the *tpp1* gene in

*Dictyostelium.*[97] As in mammalian cells, Tpp1 localizes to the lysosome[82,87] (Fig. 21.3B). Autofluorescent bodies, which are the hallmark of NCL in human cells, are observed in *Dictyostelium tpp1$^-$* cells during mid-stage development.[87] Supporting its proposed enzymatic activity, *tpp1$^-$* cells show a reduced ability to cleave target substrates, and similar to Cln3, loss of Tpp1 causes cells to develop precociously[86,87] (Fig. 21.3B) Tpp1-deficient cells also display defects in autophagy and a reduced ability to form spores, the former of which is consistent with observations in mammalian systems[87] (Fig. 21.3B). To show that the localization and function of TPP1/CLN2 is evolutionarily conserved, Phillips and Gomer[87] reintroduced human TPP1/CLN2 into *tpp1$^-$* cells and showed that TPP1/CLN2 localizes to the *Dictyostelium* lysosome and restores proteolytic activity in *tpp1$^-$* cells.

In support of its localization to the lysosome, the development of *tpp1$^-$*-cells is strongly impaired when cells are developed in the presence of the lysosome-perturbing drug chloroquine.[87] Phillips and Gomer exploited our knowledge of *Dictyostelium* genetics to identify second-site mutations that suppress the aberrant development of *tpp1$^-$* cells in the presence of chloroquine. More specifically, they used REMI to generate 30,000 individual mutant clones and then monitored the development of those mutants in the presence of chloroquine.[11,87] By using this method, Phillips and Gomer observed one clone that formed fruiting bodies when developed in the presence of chloroquine. An inverse Polymerase Chain Reaction (PCR) analysis revealed that the genetic lesion was contained within the coding region of a previously uncharacterized gene, which they named *stpA* (suppressor of *tpp1*-A, DDB_G0282973).[87,123] An inverse PCR analysis uses knowledge of the sequence of the inserted plasmid, which contains a blasticidin resistance cassette (*bsr*), to isolate part of the genomic DNA at the REMI insertion site.[123] The restriction enzyme *Alu*I is used to cut the REMI-mutant DNA since its recognition site appears once within the *bsr* gene. Since *Alu*I recognition sites are not common in the *Dictyostelium* genome, this cleavage generates a fragment that encompasses a known sequence derived from the 5′ region of the *bsr* gene, as well as a fragment of DNA from the disrupted gene. The size of the fragment will depend on where the next *Alu*I recognition site appears in the *Dictyostelium* genome (current estimates are every 1050 bp). This fragment is then circularized by ligation with T4 DNA ligase, amplified by PCR using primers specific to the *bsr* gene, and sequenced to reveal the identity of the mutated gene.[123]

StpA shows limited sequence similarity to members of the oxysterol-binding family of proteins, which function in the transcriptional regulation of lipid biosynthetic genes and the trafficking of lipids (e.g., sphingolipids and sterols).[124] To validate that the mutation in *stpA* suppressed the *tpp1$^-$* phenotype, Phillips and Gomer[87] used homologous recombination to generate a double mutant cell line (*tpp1$^-$/stpA$^-$*) and then observed that the mutant cells formed morphologically normal fruiting bodies. They also observed that deletion of *stpA* restored the normal timing of aggregation when cells were starved in the presence of ammonium chloride, and development of these cells proceeded normally in the presence of chloroquine.[87] Interestingly, deletion of *stpA* was not able to restore all Tpp1-deficient phenotypes, including the reduced size and viability of *tpp1*$^-$ cells during autophagic conditions, the precocious development of *tpp1*$^-$ cells during mid-stage development, and the accumulation of autofluorescent material in *tpp1*$^-$ cells.[87] In total, these results show that loss of *stpA* can suppress the sensitivity of *tpp1*$^-$ cells to chloroquine and ammonium chloride during development, but it cannot restore all Tpp1-deficient phenotypes. Most importantly, this work demonstrates how *Dictyostelium* can be used to identify novel genetic suppressors of NCL phenotypes, which may ultimately guide future efforts to design effective therapies.

### Future Use of *Dictyostelium* as a Model System for Studying NCL

In addition to TPP1/CLN2 and CLN3, accumulated evidence indicates that *Dictyostelium* provides unprecedented opportunities to reveal the normal functions of other proteins linked to NCL. This is especially true for those proteins that have no clear homologs in other model organisms (e.g., TPP1/CLN2, CLN5). While the functions of some of the NCL homologs in *Dictyostelium* have been previously studied, many have not been investigated in relation to human disease. One clear advantage of using *Dictyostelium* to study NCL protein function is the ability to generate cell lines lacking multiple NCL homologs, which will allow researchers to assess potential functional redundancies between the NCL homologs as well as epistatic relationships between the genes.[97] This may be critically important for the study of NCL since some NCL proteins share similar expression patterns, while others have been reported to interact with each other.[92–94] The ability of human NCL proteins to rescue gene-deficiency phenotypes in *Dictyostelium* suggests that the biologic pathways regulating the functions of NCL proteins are likely conserved from *Dictyostelium* to human. In total, *Dictyostelium* provides an inexpensive and complimentary model system to investigate the localization and functions of NCL proteins, which may aid in the development of therapies to rescue the molecular defects in NCL patients.

## THE VALUE AND LIMITATIONS OF *DICTYOSTELIUM* AS A BIOMEDICAL MODEL

All eukaryotic organisms share a cellular organization made up of functionally distinct, membrane-enclosed compartments such as the nucleus, endosome, mitochondria, and endoplasmic reticulum/Golgi. In addition, similar, if not identical, mechanisms across diverse species control the cell cycle and division, the establishment of cell polarity (e.g., polarity and motility of chemotaxing *Dictyostelium*) in unicellular as well as multicellular eukaryotes. It is therefore important to not overlook the many basic molecular biologic processes that are shared by all eukaryotes, of varying complexities, and in doing so exploit the strengths of each model system. This review exemplifies how the lower organism *Dictyostelium* fulfills the necessary scientific requirements that qualify its continued and future use as an invaluable tool to understand the conserved cellular roles of genes implicated in neurodegenerative disorders. It is of fundamental importance to gain as much understanding about the earliest underlying cytopathology of any human disease to develop appropriate therapeutic targeting, increase drug efficacy, and ultimately identify a cure. This section will outline the cost effectiveness and genetic tractability of *Dictyostelium*, which has the capacity in certain circumstances to illuminate protein function and mechanisms that are conserved in higher organisms by careful, cross-species phenotypic analysis and biochemical structure/function studies.

## CONCLUSION

Several *Dictyostelium* genes share various degrees of homology and sequence identity to human genes that highlight the importance of this biomedical model system. The genome has been sequenced and made publicly available, which allows molecular and cellular biologists to examine with additional clarity the complex multifunctional aspects of gene function. Coupled with translational experiments and validation in higher eukaryotes, research that utilizes *Dictyostelium* will contribute to our understanding of neurodegeneration and also illuminate the various pathogenic mechanisms that initiate disease. This has been elegantly portrayed in *Dictyostelium* with respect to the precise production of Aβ from human AβPP. Chemotaxis is a critical cellular function with implications in immune health and embryogenesis, so the discovery that *Dictyostelium* cells deficient for HTT show highly defective chemotactic behavior and dysfunctional adhesion along with actin cytoskeleton regulation may reveal conserved roles for this protein in axon guidance or the process of neuritic extensions. In this chapter, we have discussed studies where the use of *Dictyostelium* has provided information that either supports, complements, or enhances our understanding of the defects that contribute to the pathogenesis of Alzheimer's disease, HD, and Batten disease. Indeed, bioinformatic analysis of the *Dictyostelium* genome reveals the presence of many homologous genes associated with a variety of diseases. As such, the *Dictyostelium* model system can and should be used in the cross-species analysis of human neurodegenerative disease genes to address multiple, conserved eukaryotic cellular functions.

## ABBREVIATIONS

**AD** Alzheimer's disease
**AprA** Autocrine proliferation repressor A
**AβPP** Amyloid-β precursor protein
**CadA** Calcium-dependent cell adhesion molecule
**cAMP** Cyclic $3'-5'$ monophosphate

**CfaD** Counting-factor associated protein A
**CLN3** Ceroid lipofuscinosis neuronal 3
**CsaA** Contact site A
**CV** Contractile vacuole
**HD** Huntington's disease
**HTT** Huntingtin
**NCL** Neuronal ceroid lipofuscinosis
**REMI** Restriction enzyme-mediated integration
***stpA*** Suppressor of *tpp1*-A
**TPP1/CLN2** Tripeptidyl peptidase 1

## ACKNOWLEDGMENTS

RJH is supported by a Trent University Research Grant (24613). Both RJH and MAM were supported by the Natural Sciences and Engineering Research Council of Canada to DHO'D. MAM is supported by a University of Massachusetts Lowell Research Grant (51425).

## REFERENCES

1. Parent CA, Devreotes P. Molecular genetics of signal transduction in *Dictyostelium*. *Annu Rev Biochem.* 1996;65:411–440.
2. Soderbom F, Loomis B. Cell-cell signaling during *Dictyostelium* development. *Trends Microbiol.* 1998; 6(10):402–406.
3. Williams JG, Duffy KT, Lane DP, et al. Origins of the prestalk-prespore pattern in *Dictyostelium* development. *Cell.* 1989;59(6):1157–1163.
4. Parikh A, Miranda ER, Katoh-Kurasawa M, et al. Conserved developmental transcriptomes in evolutionarily divergent species. *Genome Biol.* 2010;11(3):35.
5. Huber RJ. Using the social amoeba *Dictyostelium* to study the functions of proteins linked to neuronal ceroid lipofuscinosis. *J Biomed Sci.* 2016;23:83.
6. Clarke M. Recent insights into host-pathogen interactions from *Dictyostelium*. *Cell Microbiol.* 2010; 12(3):283–291.
7. Francione LM, Annesley SJ, Carilla-Latorre S, Escalante R, Fisher PR. The *Dictyostelium* model for mitochondrial disease. *Semin Cell Dev Biol.* 2011;22(1):120–130.
8. Maniak M. *Dictyostelium* as a model for human lysosomal and trafficking diseases. *Semin Cell Dev Biol.* 2011;22(1):114–119.
9. Myre MA. Clues to γ-secretase, huntingtin and Hirano body normal function using the model organism *Dictyostelium discoideum*. *J Biomed Sci.* 2012;10:19–41.
10. Kuspa A, Loomis WF. REMI-RFLP mapping in the *Dictyostelium* genome. *Genetics.* 1994;138(3): 665–674.
11. Kuspa A. Restriction enzyme-mediated integration (REMI) mutagenesis. *Methods Mol Biol.* 2006;346: 201–209.
12. O'Day DH, Myre MA. Calmodulin-binding domains in Alzheimer's disease proteins: extending the calcium hypothesis. *Biochem Biophys Res Commun.* 2004;230:1051–1054.
13. Khachaturian ZS. Calcium hypothesis of Alzheimer's disease and brain aging. *Ann NY Acad Sci.* 1994; 747:1–11.
14. Chavez SE, O'Day DH. Calmodulin binds to and regulates the activity of beta-secretase (BACE1). *Alzheimer's Dis Res J.* 2007;1:37–47.
15. Myre MA, Tesco G, Tanzi RE, Wasco W. Calmodulin binding to APP and the APLPs. In: *Molecular Mechanisms of Neurodegeneration: A Joint Biochemical Society/Neuroscience Ireland Focused Meeting; March 13–16.* Republic of Ireland: University College Dublin; 2005.

16. Canobbio I, Catricalà S, Balduini C, Torti M. Calmodulin regulates the non-amyloidogenic metabolism of amyloid precursor protein in platelets. *Biochim Biophys Acta.* 2011;1813:500–506.
17. Li H, Wolfe MS, Selkoe DJ. Toward structural elucidation of the γ-secretase complex. *Structure.* 2009; 17:326–334.
18. Michno K, Knight D, Campusano JM, van de Hoef D, Boulianne GL. Intracellular calcium deficits in *Drosophila* cholinergic neurons expressing wild type or FAD-mutant presenilin. *PLoS One.* 2009;4: e6904.
19. Chow VW, Mattson MP, Wong PC, Gleichmann M. An overview of APP processing enzymes and products. *Neuromol Med.* 2010;12:1–12.
20. Pacheco-Quinto J, Eckman EA. Endothelin-converting enzymes degrade intracellular β-amyloid produced within endosomal/lysosomal pathway and autophagosomes. *J Biol Chem.* 2013;288: 5606–5615.
21. O'Day DH, Myre MA. Alzheimer's disease: the calmodulin connection and β-amyloid. *Alzheimer's Dis Res Trends.* 2008;1:1–10.
22. Berrocal M, Sepulveda MR, Vazquez-Hernandez M, Mata AM. Calmodulin antagonizes amyloid-β peptides-mediated inhibition of brain plasma membrane $Ca^{2+}$-ATPase. *Biochim Biophys Acta.* 2012; 1822:961–969.
23. O'Day DH, Eshak K, Myre MA. Calmodulin binding proteins and Alzheimer's disease. *J Alzheimer's Dis.* 2015;46:553–569.
24. Berggard T, Arrigoni G, Olsson O, Fex M, Linse S, James P. 140 mouse brain proteins identified by $Ca^{2+}$-calmodulin affinity chromatography and tandem mass spectrometry. *J Proteome Res.* 2006;5: 669–687.
25. Ghosh A, Giese KP. Calcium/calmodulin-dependent kinase II and Alzheimer's disease. *Mol Brain.* 2016;8:78.
26. Fernandez A, Hervas R, Dominquez-Fraile M, et al. Blockade of the interaction of calcineurin with FOXO in astrocytes protects against amyloid-β-induced neuronal death. *J Alzheimer's Dis.* 2016;52: 1471–1478.
27. Cloutier P, Lavallee-Adam M, Faubert M, Blancehtte M, Coulombe B. A newly uncovered group of distantly related lysine methytransferases preferentially interact with molecular chaperones to regulate their activity. *PLoS Genet.* 2013;9:e1003210.
28. Shukla D, Peck A, Pande VS. Conformational heterogeneity of calmodulin binding interface. *Nat Commun.* 2016;7:10910–10919.
29. Kopan R, Ilagan MX. Gamma-secretase: proteasome of the membrane? *Nat Rev Mol Cell Biol.* June 2004;5(6):499–504. https://doi.org/10.1038/nrm1406.
30. Levitan D, Greenwald I. Facilitation of lin-12-mediated signalling by sel-12, a *Caenorhabditis elegans* S182 Alzheimer's disease gene. *Nature.* 1995;377(6547):351–354.
31. Ma G, Li T, Price DL, Wong PC. APH-1a is the principal mammalian APH-1 isoform present in gamma-secretase complexes during embryonic development. *J Neurosci.* 2005;25(1):192–198.
32. Tournoy J, Bossuyt X, Snellinx A, et al. Partial loss of presenilins causes seborrheic keratosis and autoimmune disease in mice. *Hum Mol Genet.* 2004;13(13):1321–1331.
33. Haapasalo A, Kovacs DM. The many substrates of presenilin/γ-secretase. *J Alzheimer's Dis.* 2011;25: 3–28.
34. McMains VC, Myre MA, Kreppel L, Kimmel AR. *Dictyostelium* possesses highly diverged presenilin/ gamma-secretase that regulates growth and cell-fate specification and can accurately process human APP: a system for functional studies of the presenilin/gamma-secretase complex. *Dis Mol Med.* 2010;3(9–10):581–594.
35. Ludtmann MH, Otto GP, Schilde C, et al. An ancestral non-proteolytic role for presenilin proteins in multicellular development of the social amoeba *Dictyostelium discoideum. J Cell Sci.* 2014;127(Pt 7): 1576–1584. https://doi.org/10.1242/jcs.140939. Epub January 24, 2014.
36. Otto GP, Sharma D, Williams RS. Non-catalytic roles of presenilin throughout evolution. *J Alzheimer's Dis.* 2016;52(4):1177–1187. https://doi.org/10.3233/JAD-150940.
37. Eichinger L, Pachebat JA, Glöckner G, et al. The genome of the social amoeba *Dictyostelium discoideum. Nature.* 2005;435:43–57.

38. Xie T, Yan C, Zhou R, et al. Crystal structure of the γ-secretase component nicastrin. *Proc Natl Acad Sci USA*. 2014;111(37):13349–13354. https://doi.org/10.1073/pnas.1414837111. Epub September 2, 2014.
39. Bolduc DM, Montagna DR, Gu Y, Selkoe DJ, Wolfe MS. Nicastrin functions to sterically hinder γ-secretase-substrate interactions driven by substrate transmembrane domain. *Proc Natl Acad Sci USA*. 2016;113(5):E509–E518. https://doi.org/10.1073/pnas.1512952113. Epub December 22, 2015.
40. Hu C, Xu J, Zeng L, Li T, Cui MZ, Xu X. Pen-2 and presenilin are sufficient to catalyze notch processing. *J Alzheimer's Dis*. 2017;56(4):1263–1269. https://doi.org/10.3233/JAD-161094.
41. Johnson DS, Li YM, Pettersson M, St George-Hyslop PH. Structural and chemical biology of presenilin complexes. *Cold Spring Harb Perspect Med*. 2017. https://doi.org/10.1101/cshperspect.a024067. pii: a024067.
42. Rhein V, Song X, Weisner A, et al. Amyloid-β and tau synergistically impair the oxidative phosphorylation system in triple transgenic Alzheimer's disease mice. *Proc Natl Acad Sci USA*. 2009;1: 20057–20062.
43. Lichtenthaler SF, Haass C, Steiner H. Regulated intramembrane proteolysis—lessons from amyloid precursor protein processing. *J Neurochem*. 2011;117(5):779–796. https://doi.org/10.1111/j.1471-4159.2011.07248.x. Epub April 14, 2011.
44. Jutras I, Laplante A, Boulais J, Brunet S, Thinakaran G, Desjardins M. Gamma-secretase is a functional component of phagosomes. *J Biol Chem*. 2005;280(43):36310–36317. Epub August 15, 2005.
45. Farfara D, Trudler D, Segev-Amzaleg N, Galron R, Stein R, Frenkel D. γ-Secretase component presenilin is important for microglia β-amyloid clearance. *Ann Neurol*. 2011;69(1):170–180. https://doi.org/10.1002/ana.22191. Epub November 22, 2010.
46. Smolarkiewicz M, Skrzypczak T, Michalak M, et al. Gamma-secretase subunits associate in intracellular membrane compartments in *Arabidopsis thaliana*. *J Exp Bot*. 2014;65(12):3015–3027. https://doi.org/10.1093/jxb/eru147. Epub April 10, 2014.
47. Khandelwal A, Chandu D, Roe CM, Kopan R, Quatrano RS. Moonlighting activity of presenilin in plants is independent of gamma-secretase and evolutionarily conserved. *Proc Natl Acad Sci USA*. 2007; 104:13337–13342.
48. The Huntington's Disease Collaborative Research Group. A novel gene containing a trinucleotide repeat that is expanded and unstable on Huntington's disease chromosomes. *Cell*. 1993;72:971–983.
49. Genetic Modifiers of Huntington's Disease (GeM-HD) Consortium. Identification of genetic factors that modify clinical onset of Huntington's disease. *Cell*. 2015;162(3):516–526. https://doi.org/10.1016/j.cell.2015.07.003.
50. Cattaneo E, Zuccato C, Tartari M. Normal huntingtin function: an alternative approach to Huntington's disease. *Nat Rev Neurosci*. 2005;6(12):919–930.
51. Zuccato C, Valenza M, Cattaneo E. Molecular mechanisms and potential therapeutical targets in Huntington's disease. *Physiol Rev*. 2010;90(3):905–981. https://doi.org/10.1152/physrev.00041.2009.
52. Jacobsen JC, Erdin S, Chiang C, et al. Potential molecular consequences of transgene integration: the R6/2 mouse example. *Sci Rep*. 2017;7:41120. https://doi.org/10.1038/srep41120.
53. Kaltenbach LS, Romero E, Becklin RR, et al. Huntingtin interacting proteins are genetic modifiers of neurodegeneration. *PLoS Genet*. 2007;3(5):e82.
54. Myre MA, Lumsden AL, Thompson MN, Wasco W, MacDonald ME, Gusella JF. Deficiency of huntingtin has pleiotropic effects in the social amoeba *Dictyostelium discoideum*. *PLoS Genet*. 2011;7: e1002052.
55. Wang Y, Steimle PA, Ren Y, et al. *Dictyostelium* huntingtin controls chemotaxis and cytokinesis through the regulation of myosin II phosphorylation. *Mol Biol Cell*. 2011;22(13):2270–2281. https://doi.org/10.1091/mbc.E10-11-0926. Epub May 11, 2011.
56. Devreotes P, Janetopoulos C. Eukaryotic chemotaxis: distinctions between directional sensing and polarization. *J Biol Chem*. 2003;278:20445–20448.
57. Garcia GL, Parent CA. Signal relay during chemotaxis. *J Microsc*. 2008;231:529–534.
58. Ridley AJ, Schwartz MA, Burridge K, et al. Cell migration: integrating signals from front to back. *Science*. 2003;302:1704–1709.

59. Wessels D, Lusche DF, Scherer A, Kuhl S, Myre MA, Soll DR. Huntingtin regulates Ca(2+) chemotaxis and K(+)-facilitated cAMP chemotaxis, in conjunction with the monovalent cation/H(+) exchanger Nhe1, in a model developmental system: insights into its possible role in Huntington's disease. *Dev Biol.* 2014;394:24–38.
60. Frantz C, Karydis A, Nalbant P, Hahn KM, Barber DL. Positive feedback between Cdc42 activity and $H^+$ efflux by the Na-H exchanger NHE1 for polarity of migrating cells. *J Cell Biol.* 2007;179(3):403–410.
61. Kwan W, Träger U, Davalos D, et al. Mutant Huntington impairs immune cell migration in Huntington disease. *J Clin Invest.* 2012;122(12):4737–4747. https://doi.org/10.1172/JCI64484. Epub November 19, 2012.
62. Ritch JJ, Valencia A, Alexander J, et al. Multiple phenotypes in Huntington disease mouse neural stem cells. *Mol Cell Neurosci.* 2012;50(1):70–81. https://doi.org/10.1016/j.mcn.2012.03.011. Epub April 6, 2012.
63. Angeli S, Shao J, Diamond MI. F-actin binding regions on the androgen receptor and huntingtin increase aggregation and alter aggregate characteristics. *PLoS One.* 2010;5(2):e9053.
64. Munsie L, Caron N, Atwal RS, et al. Mutant huntingtin causes defective actin remodeling during stress: defining a new role for transglutaminase 2 in neurodegenerative disease. *Hum Mol Genet.* 2011;20(10):1937–1951.
65. Sameshima M, Kishi Y, Osumi M, Mahadeo D, Cotter DA. Novel actin cytoskeleton: actin tubules. *Cell Struct Funct.* 2000;25(5):291–295.
66. Lynch G, Kramar EA, Rex CS, et al. Brain-derived neurotrophic factor restores synaptic plasticity in a knock-in mouse model of Huntington's disease. *J Neurosci.* 2007;27(16):4424–4434. https://doi.org/10.1523/JNEUROSCI.5113-06.
67. Reis SA, Thompson MN, Lee JM, et al. Striatal neurons expressing full-length mutant huntingtin exhibit decreased N-cadherin and altered neuritogenesis. *Hum Mol Genet.* 2011;20(12):2344–2355. https://doi.org/10.1093/hmg/ddr127.
68. Thompson MN, MacDonald ME, Gusella JF, Myre MA. Huntingtin supplies a csaA-independent function essential for EDTA-resistant homotypic cell adhesion in *Dictyostelium discoideum*. *J Hunting Dis.* 2014;3(3):261–271. https://doi.org/10.3233/JHD-140112.
69. Wong LM, Siu CH. Cloning of cDNA for the contact site A glycoprotein of *Dictyostelium discoideum*. *Proc Natl Acad Sci USA.* 1986;83(12):4248–4252.
70. HD iPSC Consortium. Induced pluripotent stem cells from patients with Huntington's disease show CAG-repeat-expansion-associated phenotypes. *Cell Stem Cell.* 2012;11(2):264–278. https://doi.org/10.1016/j.stem.2012.04.027. Epub June 28, 2012.
71. Jacobsen JC, Gregory GC, Woda JM, et al. HD CAG-correlated gene expression changes support a simple dominant gain of function. *Hum Mol Genet.* 2011;20(14):2846–2860. https://doi.org/10.1093/hmg/ddr195. Epub May 2, 2011.
72. Sardo VL, Zuucato C, Gaudenzi G, et al. An evolutionary recent neuroepithelial cell adhesion function of huntingtin implicates ADAM10-Ncadherin. *Nat Neurosci.* 2012;15(5):713–721. https://doi.org/10.1038/nn.3080.
73. Mole SE, Cotman SL. Genetics of the neuronal ceroid lipofuscinoses (Batten disease). *Biochim Biophys Acta.* 2015;1852:2237–2241.
74. Andermann E, Jacob JC, Andermann F, Carpenter S, Wolfe L, Berkovic S. The Newfoundland aggregate of neuronal ceroid-lipofuscinosis. *Am J Med Genet.* 1988;5:111–116.
75. Ju W, Zhong R, Moore S, et al. Identification of novel CLN2 mutations shows Canadian specific NCL2 alleles. *J Med Genet.* 2002;39:822–825.
76. Moore SJ, Buckley DJ, MacMillan A, et al. The clinical and genetic epidemiology of neuronal ceroid lipofuscinosis in Newfoundland. *Clin Genet.* 2008;74:213–222.
77. Williams RE. Appendix 1: NCL incidence and prevalence data. In: Mole SE, Williams RE, Goebel HH, eds. *The Neuronal Ceroid Lipofuscinoses (Batten Disease).* 2nd ed. Oxford: Oxford University Press; 2011:361–365.
78. Schulz A, Kohlschütter A, Mink J, Simonati A, Williams R. NCL diseases - clinical perspectives. *Biochim Biophys Acta.* 2013;1832:1801–1806.

79. Dearborn JT, Harmon SK, Fowler SC, et al. Comprehensive functional characterization of murine infantile Batten disease including Parkinson-like behavior and dopaminergic markers. *Sci Rep.* 2015;5:12752.
80. Deng H, Xiu X, Jankovic J. Genetic convergence of Parkinson's disease and lysosomal storage disorders. *Mol Neurobiol.* 2015;51:1554–1568.
81. Radke J, Stenzel W, Goebel HH. Human NCL neuropathology. *Biochim Biophys Acta.* 2015;1852: 2262–2266.
82. Cárcel-Trullols J, Kovács AD, Pearce DA. Cell biology of the NCL proteins: what they do and don't do. *Biochim Biophys Acta.* 2015;1852:2242–2255.
83. Phillips SN, Muzaffar N, Codlin S, et al. Characterizing pathogenic processes in Batten disease: use of small eukaryotic model systems. *Biochim Biophys Acta.* 2006;1762:906–919.
84. Bond M, Holthaus SM, Tammen I, Tear G, Russell C. Use of model organisms for the study of neuronal ceroid lipofuscinosis. *Biochim Biophys Acta.* 2013;1832:1842–1865.
85. Faller KM, Gutierrez-Quintana R, Mohammed A, et al. The neuronal ceroid lipofuscinoses: opportunities from model systems. *Biochim Biophys Acta.* 2015;1852:2267–2278.
86. Huber RJ, Myre MA, Cotman SL. Loss of Cln3 function in the social amoeba *Dictyostelium discoideum* causes pleiotropic effects that are rescued by human CLN3. *PLoS One.* 2014;9:e110544.
87. Phillips JE, Gomer RH. Partial genetic suppression of a loss-of-function mutant of the neuronal ceroid lipofuscinosis-associated protease TPP1 in *Dictyostelium discoideum. Dis Mol Med.* 2015;8(2):147–156. https://doi.org/10.1242/dmm.018820. Epub December 24, 2014.
88. Huber RJ, Myre MA, Cotman SL. Aberrant adhesion impacts early development in a *Dictyostelium* model for juvenile neuronal ceroid lipofuscinosis. *Cell Adhes Migrat.* 2017;11:399–418.
89. Huber RJ. Loss of Cln3 impacts protein secretion in the social amoeba *Dictyostelium. Cell Signal.* 2017; 35:61–72.
90. Rot G, Parikh A, Curk T, Kuspa A, Shaulsky G, Zupan B. DictyExpress: a *Dictyostelium discoideum* gene expression database with an explorative data analysis web-based interface. *BMC Bioinf.* 2009; 10:265.
91. Persaud-Sawin DA, Mousallem T, Wang C, Zucker A, Kominami E, Boustany RM. Neuronal ceroid lipofuscinosis: a common pathway? *Pediatr Res.* 2007;61:146–152.
92. Lyly A, von Schantz C, Heine C, et al. Novel interactions of CLN5 support molecular networking between neuronal ceroid lipofuscinosis proteins. *BMC Cell Biol.* 2009;10:83.
93. Fabritius AL, Vesa J, Minye HM, Nakano I, Kornblum H, Peltonen L. Neuronal ceroid lipofuscinosis genes, CLN2, CLN3 and CLN5 are spatially and temporally co-expressed in a developing mouse brain. *Exp Mol Pathol.* 2014;97:484–491.
94. Scifo E, Szwajda A, Debski J, et al. Drafting the CLN3 protein interactome in SH-SY5Y human neuroblastoma cells: a label-free quantitative proteomics approach. *J Proteome Res.* 2013;12:2101–2115.
95. Journet A, Klein G, Brugière S, et al. Investigating the macropinocytic proteome of *Dictyostelium* amoebae by high-resolution mass spectrometry. *Proteomics.* 2012;12:241–245.
96. The International Batten Disease Consortium. Isolation of a novel gene underlying Batten disease, CLN3. *Cell.* 1995;82:949–957.
97. Faix J, Linkner J, Nordholz B, Platt JL, Liao XH, Kimmel AR. The application of the Cre-loxP system for generating multiple knock-out and knock-in targeted loci. *Methods Mol Biol.* 2013;983:249–267.
98. Herlihy SE, Tang Y, Phillips JE, Gomer RH. Functional similarities between the *Dictyostelium* protein AprA and the human protein dipeptidyl-peptidase IV. *Protein Sci.* 2017;26:578–585.
99. Levi S, Polyakov M, Egelhoff TT. Green fluorescent protein and epitope tag fusion vectors for *Dictyostelium discoideum. Plasmid.* 2000;44:231–238.
100. Veltman DM, Akar G, Bosgraaf L, Van Haastert PJ. A new set of small, extrachromosomal expression vectors for *Dictyostelium discoideum. Plasmid.* 2009;61:110–118.
101. Plattner H. Contractile vacuole complex—its expanding protein inventory. *Int Rev Cell Mol Biol.* 2013;306:371–416.
102. Stein CS, Yancey PH, Martins I, Sigmund RD, Stokes JB, Davidson BL. Osmoregulation of ceroid neuronal lipofuscinosis type 3 in the renal medulla. *Am J Physiol Cell Physiol.* 2010;298: C1388–C1400.

103. Getty A, Kovács AD, Lengyel-Nelson T, et al. Osmotic stress changes the expression and subcellular localization of the Batten disease protein CLN3. *PLoS One*. 2013;8:e66203.
104. Tecedor L, Stein CS, Schultz ML, Farwanah H, Sandhoff K, Davidson BL. CLN3 loss disturbs membrane microdomain properties and protein transport in brain endothelial cells. *J Neurosci*. 2013;33: 18065–18079.
105. Kinseth MA, Anjard C, Fuller D, Guizzunti G, Loomis WF, Malhotra V. The Golgi-associated protein GRASP is required for unconventional protein secretion during development. *Cell*. 2007; 130:524–534.
106. Sriskanthadevan S, Brar SK, Manoharan K, Siu CH. Ca(2+)-calmodulin interacts with DdCAD-1 and promotes DdCAD-1 transport by contractile vacuoles in *Dictyostelium* cells. *FEBS J*. 2013;280: 1795–1806.
107. Moniakis J, Coukell MB, Janiec A. Involvement of the $Ca^{2+}$-ATPase PAT1 and the contractile vacuole in calcium regulation in *Dictyostelium discoideum*. *J Cell Sci*. 1999;112:405–414.
108. Malchow D, Lusche DF, Schlatterer C, De Lozanne A, Müller-Taubenberger A. The contractile vacuole in $Ca^{2+}$-regulation in *Dictyostelium*: its essential function for cAMP-induced $Ca^{2+}$-influx. *BMC Dev Biol*. 2006;6:31.
109. Boriack RL, Bennett MJ. CLN-3 protein is expressed in the pancreatic somatostatin-secreting delta cells. *Eur J Paediatr Neurol*. 2001;5(suppl A):99–102.
110. An Haack K, Narayan SB, Li H, Warnock A, Tan L, Bennett MJ. Screening for calcium channel modulators in CLN3 siRNA knock down SH-SY5Y neuroblastoma cells reveals a significant decrease of intracellular calcium levels by selected L-type calcium channel blockers. *Biochim Biophys Acta*. 2011; 1810:186–191.
111. Warnock A, Tan L, Li C, An Haack K, Narayan SB, Bennett MJ. Amlodipine prevents apoptotic cell death by correction of elevated intracellular calcium in a primary neuronal model of Batten disease (CLN3 disease). *Biochem Biophys Res Commun*. 2013;436:645–649.
112. Chandrachud U, Walker MW, Simas AM, et al. Unbiased cell-based screening in a neuronal cell model of Batten disease highlights an interaction between $Ca^{2+}$ homeostasis, autophagy, and CLN3 protein function. *J Biol Chem*. 2015;290:14361–14380.
113. Chang JW, Choi H, Kim HJ, et al. Neuronal vulnerability of CLN3 deletion to calcium-induced cytotoxicity is mediated by calsenilin. *Hum Mol Genet*. 2007;16:317–326.
114. Vardjan N, Verkhratsky A, Zorec R. Astrocytic pathological calcium homeostasis and impaired vesicle trafficking in neurodegeneration. *Int J Mol Sci*. 2017;18:E358.
115. Popugaeva E, Pchitskaya E, Bezprozvanny I. Dysregulation of neuronal calcium homeostasis in Alzheimer's disease - a therapeutic opportunity? *Biochem Biophys Res Commun*. 2017;483:998–1004.
116. Raymond LA. Striatal synaptic dysfunction and altered calcium regulation in Huntington disease. *Biochem Biophys Res Commun*. 2017;483:1051–1062.
117. Chung WC, Linscott ML, Rodriguez KM, Stewart CE. The regulation and function of fibroblast growth factor 8 and its function during gonadotropin-releasing hormone neuron development. *Front Endocrinol*. 2016;7:114.
118. Gonçalves SA, Macedo D, Raquel H, et al. shRNA-based screen identifies endocytic recycling pathway components that act as genetic modifiers of alpha-synuclein aggregation, secretion and toxicity. *PLoS Genet*. 2016;12:e1005995.
119. Pérez M, Cuadros R, Hernández F, Avila J. Secretion of full-length tau or tau fragments in a cell culture model. *Neurosci Lett*. 2016;634:63–69.
120. Rossomando EF, Maldonado B, Crean EV, Kollar EJ. Protease secretion during onset of development in *Dictyostelium discoideum*. *J Cell Sci*. 1978;30:305–318.
121. Dimond RL, Burns RA, Jordan KB. Secretion of Lysosomal enzymes in the cellular slime mold, *Dictyostelium discoideum*. *J Biol Chem*. 1981;256:6565–6572.
122. Wlodawer A, Li M, Gustchina A, Oyama H, Dunn BM, Oda K. Structural and enzymatic properties of the sedolisin family of serine-carboxyl peptidases. *Acta Biochim Pol*. 2003;50:81–102.
123. Keim M, Williams RS, Harwood AJ. An inverse PCR technique to rapidly isolate the flanking DNA of *Dictyostelium* insertion mutants. *Mol Biotechnol*. 2004;26:221–224.
124. Olkkonen VM, Li S. Oxysterol-binding proteins: sterol and phosphoinositide sensors coordinating transport, signaling and metabolism. *Prog Lipid Res*. 2013;52:529–538.

# CHAPTER 22

# The Role of Human Endogenous Retroviruses (HERVs) in the Pathologies of the Nervous System

**Sulev Kõks**[1,2,3], **Gea Kõks**[1,3]
[1]University of Tartu, Tartu, Estonia; [2]Estonian University of Life Sciences, Tartu, Estonia; [3]Prion OÜ, Tartu, Estonia

## INTRODUCTION

The human genome contains almost half of its size, 45%, repetitive retroelements or transposable elements (TEs). TEs are derived from ancient viruses and the result of viral genome integration that occurred a long time ago. TEs can be divided into two major classes (Table 22.1): transposons (3% of the genome) and retrotransposons or retroelements (42% of the genome).[1] Retroelements are divided into long terminal repeat (LTR) and non-LTR elements. Non-LTR elements are the most abundant and are divided into short interspersed elements (SINE) and long interspersed elements (LINE). SINEs are Alu and MIR repeats without independent amplification and protein coding capacity, and they are also considered nonautonomous TEs.[2] Alu repeats are the most common elements in the human genome, and they make up around 15% of the human genome.[3] LINE elements contain autonomous L1 and L2 elements (LINE repeats). The LTR class consists of elements of the retroviral origin, and this class is more heterogenous (Table 22.2). One group of LTR elements are human endogenous retroviruses (HERVs), and they occupy around 8% of our genome.[4] These viruses are derived from ancient infections of retroviruses that were reverse transcribed

**Table 22.1** Overview of the transposable elements (TEs) in the human genome[2]

| TEs (45% of genome) | | |
|---|---|---|
| ***DNA-transposons 2.8%*** | | |
| Retrotransposons or retroelements 42.2% | Non-LTR, 33.9% | SINE (Alu, MIR)<br>LINE (L1, L2)<br>Pseudogenes |
| | LTR, 8.3% | Class I ERV + MER4<br>Class II ERV<br>Class II ERV<br>Others (MST, MLT) |

Percentage of the elements in the genome is given in brackets.

*Molecular-Genetic and Statistical Techniques for Behavioral and Neural Research*
ISBN 978-0-12-804078-2, https://doi.org/10.1016/B978-0-12-804078-2.00022-2

**Table 22.2** Characteristics of subgroups of LTR retroelements[2]

| Element | Characteristics |
|---|---|
| Class I ERV | Similar to type C or $\gamma$ retroviruses |
| Class II ERV | Similar to type B or $\beta$ retroviruses |
| Class III ERV | Also called ERV-L, limited similarity to spumaviruses |
| MER4 | Nonautonomous class I ERV |
| MST | Named after restriction enzyme site MstII, MaLR |
| MLT | Mammalian LTR transposon, MaLR |

and integrated into the genome of ancestral host animals. These viral genomes were fixed and inherited tens of millions of years ago.[1,5] Almost all of the HERVs have lost their infectivity. However, presence of the *env* gene in HERVs confers their potential to spread between cells. LTR class also contains HERV-like repeat elements that are derived from HERVs (SINE-R) and contain one-copy of LTR element. Finally, there are sequences with LTR-like features but without any homologous proviral structure. Repbase is a central curated database for repetitive elements in which more than 200 families of LTR-containing retroelements are defined and grouped into six superfamilies.[6] These families are class I ERVs (similar to type C retroviruses), class II ERV (similar to type B retroviruses), class III ERV (distantly similar to spuma retroviruses), MER4 (nonautonomous class I-related ERVs), MST (common MstII restriction site), and MLT (mammalian LTR transposon). MST and MLT are both part of the larger nonautonomous mammalian apparent LTR retrotransposon (MaLR) superfamily.[2]

TEs form the basis of the "junk DNA" that has been without any understandable function in the recent past. Generally, it has been suggested that about half of our genome is derived from the TEs, and they do not perform any specific function. New analyses have estimated that more than two-thirds of our genome has resulted from incorporation of viral DNA.[7] However, during evolution, our genome has "domesticated" several TEs to fulfill different physiological functions, like V(D)J recombination in the immune system or cellular fusion during placental development and even cellular protection against retroviral infections.[8–10] Taken together, it is obvious that many TEs have had substantial impact on shaping our genomes and changing our phenotypes. Analysis of human genes has found that TEs are overrepresented in the mRNAs of rapidly evolving mammalian genes and potentially have gene regulation function.[11] On the other hand, HERVs and other LTRs are normally suppressed, but they can potentially act as transcriptional regulators and this way activate malignant disorders or autoimmune responses.[5,12] The importance of HERVs and LTRs in gene regulatory networks has been verified in several studies.[13] The role of HERVs has been shown in embryonic development, in immune regulation, and in the pathogenesis of complex disorders.[14–16] Most remarkably, HERVs are involved in different diseases including melanoma, amyotrophic lateral sclerosis (ALS), multiple sclerosis (MS), schizophrenia (SCH), bipolar disorder, autism spectrum disorder (ASD), and attention deficit hyperactivity disorder (ADHD).[17–19] The present chapter gives an overview of the role of HERVs in the neurological and psychiatric disorders.

## HUMAN ENDOGENOUS RETROVIRUSES IN NEUROLOGICAL DISEASES

To appreciate the potential impact of repetitive elements on the genome function, it is important to mention that roughly one-quarter of human promoter regions contain TEs.[20] TEs produce negative impact by causing mutations in genomes, but it is also recognized that TEs are involved in the regulation of gene expression.[21,22] In the latter case, TEs should have significant impact in the pathogenesis of diseases. The first study where TEs were described as causal agents for human diseases was published in 1988. In this paper, two separate insertions of L1s were described to cause hemophilia A by disrupting the factor VII gene.[23] Thereafter, TEs were found to cause muscular dystrophy, neurofibromatosis, and colon cancer.[24–26] LINE-1-induced insertional mutagenesis in the myc locus has been shown to cause breast cancer.[25] In a more recent study, aberrantly activated LTR of the MaLR family (THE1B) was found to induce the expression of *CSF1R* in Hodgkin lymphoma.[27] Therefore, TEs can contribute to the pathogenesis of diseases directly via mutagenesis or by the regulation of the gene expression networks. From this perspective, two of the most well-studied neurological disorders are MS and ALS. Summarized information is given in Table 22.3.

### Multiple Sclerosis

MS is associated with demyelination in the CNS leading to disturbed sensory, motor, autonomic, and cognitive functions and fatigue. The etiology of MS remains debated, but pathogenesis involves chronic inflammation and autoimmune mechanisms.[28] The role of viruses on the pathogenesis of this disease has been suspected for a long time.[28,29] The first discovery of reverse transcriptase activity of the cell line isolated

**Table 22.3** Neurological disorders with the involvement of transposable elements (TEs) and type of evidence

| Disorder | TEs | Evidence |
|---|---|---|
| Multiple sclerosis | HERV-W (MSRV) | Induction of immune response, expression, experimental |
| | HERV-H (Fc1) | Association study |
| | HERV-K (K13, K18, K113) | Association study, expression |
| Amyotrophic lateral sclerosis | HERV-K | Expression, experimental |
| | HERV-W | Controversial evidence |

from cerebrospinal fluid (CSF) of a patient with MS indicated that this activity was not related to the common "suspects" HTLV-I or HIV-1 and HIV-2.[30] Later, the viral particles were isolated and designated as "MS-associated" retrovirus (MSRV).[31] Additional molecular characterization of the MSRV identified its similarity to the endogenous retroviral sequence ERV9.[32] MSRV was found to be expressed not in all, but in many MS patients, and it was found in noncellular RNA from the plasma and CSF of the patients.[32] The MSRV-related transcript contains *gag*, *pol*, and *env* genes, and it is expressed in the placenta. This new HERV family was named HERV-W.[33] Subsequently, the role of HERV-W in MS was repeatedly confirmed in different independent studies.[34–37] More importantly, HERV-W (MSRV) is not only associated with the presence of MS, but in the CSF, it correlates with the clinical progression of the disease.[35,36,38] Presence of the HERV-W sequences in the CSF is a marker for rapid disease progression.[36–38]

The molecular mechanism of the infectivity of HERV-W has been studied for more than a decade. HERV-W has been shown to activate innate immunity with subsequent release of proinflammatory cytokines by using a Toll-like receptor 4 (TLR4) agonist effect.[39] TLR4-activated macrophages introduce T-lymphocytes that are required to turn HERV-W envelope (ENV) protein into superantigen, and to ignite the downstream immune cascade leading to production of the IL-1β, IL-6, or TNF-α.[39] In addition, dendritic cells are activated and Th1-like responses are developed. Both TLR4 and CD14 receptors are involved in the proinflammatory effects of ENV protein.[39] The pathogenicity of HERV-W was further analyzed in relation to the T-lymphocyte-mediated immunopathology.[40] HERV-W particles induced polyclonal T-lymphocyte response with a bias in the Vβ16 chain usage as a surface receptor. This means that HERV-W triggers an abnormal immune response that is characteristic of superantigens.[40] In addition to the peripheral immune system, further studies indicated the role of HERV-W in the differentiation of the oligodendroglial precursor cells. More precisely, HERV-W particles induced the proinflammatory cytokines and nitric oxide synthase in oligodendroglial precursors cells expressing TLR4.[41] In addition to the in vitro studies, animal experiments also confirmed the impact of HERV-W in the induction of immune response.[40] More precisely, mice immunized with the HERV-W showed clear experimental allergic encephalomyelitis.[42] In vivo experiments confirmed that HERV-W is a highly potent agonist for the TLR4, and this interaction leads to the proinflammatory activity and impairs differentiation of the oligodendroglial precursor cells to mature myelinating cells.[43] Taken together, the role of HERV-W in pathophysiology of MS has been demonstrated in clinical observational studies, as well as in vitro and in vivo laboratory experimental studies. Based on these findings, a humanized antibody, GNbAC1, against HERV-W ENV extracellular domain was developed to treat patients with MS.[43–45] This antibody has been successful in preclinical and clinical studies, showing low toxicity and good tolerability.[45] Preclinical studies indicated no safety risks, so Phase I, the first-in-human randomized, double-blind, placebo-controlled, dose-escalation study was performed.[44] In 33 male subjects, only minor and nonspecific adverse effects were recorded, and this allowed launching of the Phase II program.[44] A Phase 2a study was performed as a single-blind, placebo-controlled, randomized study in 10 MS patients.[45–47] All patients tolerated the GNbAC1, and treatment significantly reduced transcription of the MSRV. Moreover, a recent study indicated that HERV-W is also involved in chronic inflammatory demyelinating polyradiculoneuropathy and that it induces inflammatory damage of Schwann cells.[48] The pathogenic effect of HERV-W was inhibited by the GNbAC1, suggesting a broad utility for the antibody.[48] The story of HERV-W is an elegant example of how discovery may progress from the identification of a potentially infectious agent to the development of a promising treatment option for a disease that has been without any possible cure. This is probably the first endogenous retrovirus that has been used as a specific therapeutic target for the development of antibodies.

In addition to the HERV-W, genetic association studies have found increased odds ratio in the locus for the HERV-Fc1 gene.[49] HERV-Fc1 is closely related to the HERV-H, and more distantly to the HERV-W and HERV-K.[49] The association of HERV-Fc1 with MS was repeated in another independent study.[50] Moreover, highly significant genetic interaction between HERV-Fc1 and HERV-K13 loci was described.[50] In one study using postmortem tissues, increased mRNA level of HERV-K and HERV-W was found.[51] Taken together, the case of MS excellently exemplifies the pathogenetic potential and impact of HERVs in our genomes. While the involvement of HERV-W in MS is clearly demonstrated, the role of HERVs in other diseases needs further analysis.

## Amyotrophic Lateral Sclerosis

ALS is a progressive and fatal neurodegenerative disease that is characterized by the loss of upper (in the cerebral cortex or brain stem) and lower (in the spinal cord or cranial nerves) motor neurons, leading to increasing weakness.[52] ALS is sporadic disease with the prevalence of five in 100,000, and viral involvement in its aetiology has been suspected for many years, mainly on the basis of similarities with poliomyelitis.[53] The cause of ALS is not known, but the role of HERVs has been suggested recently. The first pioneering studies analyzed peripheral blood sera from patients with motor neuron diseases. In one study, sera from the 56 patients were analyzed for reverse transcriptase activity by using product-enhanced reverse transcriptase assay (PERT).[53] Indeed, in 33 patients (59%) cell-free reverse transcriptase activity was detected compared to only three of the controls.[53] This activity was not caused by the HIV-1, HIV-2, HTLV-I, HTLV-II, HRV-5, or human foamy virus. In another study, 30 patients with sporadic ALS were analyzed for reverse transcriptase activity in blood sera.[54] The authors applied a similar serum purification protocol combined with a sensitive PERT assay. Fourteen patients were PERT positive (47%), a statistically significant difference compared to the genetically related and unrelated controls.[54] Increased RT activity on sera of HIV-negative ALS patients was further confirmed by an additional study.[55] In this study, RT activity was measured also in CSF, but the difference (39% vs. 19%) was not statistically significant.[55] All these early studies used biochemical detection of RT activity, so the identity of retroviruses was not discoverable. Application of RNA analysis with RT-PCR subsequently answered this question and identified increased expression of the HERV-K in the autopsy brains.[56] Authors also found that HERV-K is expressed in cortical neurons of the ALS patients. All these studies were descriptive and were designed to show association between RT activity and ALS. A recent study identified causative connection between HERV-K and ALS.[18] First, authors found increased expression of the HERV-K in cortical and spinal neurons of ALS patients. Afterward, they showed causal relationship between the expression of HERV-K and neuronal pathologies in cellular models as well as in mice using transgenic technology.[18] Transgenic mice expressing the HERV-K developed progressive motor dysfunction accompanied by selective loss of volume of the motor cortex and decreased synaptic activity. Transgenic mice had injuries in the anterior horn of the spinal cord, and this was associated with muscle atrophy.[18] Moreover, authors identified that expression of HERV-K is regulated by TDP-43, a protein that is overexpressed in ALS and is known to form cytoplasmic aggregates. HERV-K has five binding sites for the TDP-43, and as TDP-43 seems to be involved in other neurodegenerative disorders (e.g., frontotemporal dementia), the potential pathophysiological impact of HERV-K is likely not limited to ALS.[18]

## HUMAN ENDOGENOUS RETROVIRUSES IN PSYCHIATRIC DISEASES

Out of all psychiatric diseases, SCH and bipolar disease are the most frequently studied disorders with regard to endogenous retroviruses. But some information is also available on autism. A general overview is given in Table 22.4.

### Schizophrenia

In the past, it has been suggested that exogenous viruses and retroviral infections are responsible for SCH, but there were no studies proposing endogenous retroviruses to play any role. Some early studies suggested that one of the pathogenic factors for SCH could be retroviral infection, or that the disease is induced by quantal changes in "virogene" or "schizovirus" within our genome.[57–59] These early studies proposed that endogenous retroviral integration to the human genome near the gene critical for SCH leads to the disruption of this gene and development of disease.[57] These hypotheses were too complex to challenge during their time, but recent developments in molecular

**Table 22.4** Psychiatric disorders with the involvement of transposable elements (TEs) and type of evidence

| Disorder | TEs | Evidence |
|---|---|---|
| Schizophrenia | SZRV-1 (schizovirus, HERV-W, MSRV) | DNA cloning, expression |
| | SZRV-2 | Differential cloning |
| | HERV-K (K115) | Variations, expression |
| | L1 elements | More copies |
| Bipolar disorder | HERV-I (IP-T47D) | Expression |
| | HERV-9 (Seq63) | Expression |
| | HERV-L | Expression |
| | NMWV7 | Expression |
| | HERV-K (K10, KC4) | Expression, treatment |
| | HERV-W | Expression, treatment |
| | HERV-Fb | Expression, treatment |
| Autism | HERV-H | Expression |
| | HERV-K | Expression |
| | HERV-W | Expression |
| | L1 | Expression |
| ADHD | HERV-H | Expression, treatment |

technologies allowed performing better experimental testing. One of the first studies used representational difference analysis (RDA) of blood DNA from three pairs (two male and one female) of monozygotic twins discordant for SCH.[60] Each affected member met the DSM-III-R criteria for SCH. The unaffected twins had no psychotic or affective illness. For the independent testing and analysis of the isolated DNA fragments, DNA from five additional pairs of affected individuals and their unrelated matched controls was extracted. This study gave very complex results. Authors identified 128 sequences in the genomes of the affected persons that were grouped in four contigs.[60] The first contig was designated SZRV-1, and this sequence was similar to the sequence found to be implicated in MS (MSRV, ERV-9, or HERV-W).[32] The identified sequence (HERV-W) was later found to be differentially expressed in cell-free CSF of individuals with SCH.[61] Analysis of CSFs and postmortem brain tissue from the SCH patients revealed significant increase of transcription of HERV-W family of endogenous retroviruses.[61,62] The results from the second RDA of DNA revealed a sequence named SZRV-2.[63] SZRV-2 is confined to 12q13, and this site was also shared with the SZRV-1. Southern blot analysis indicated that this region contains tandem repeats and is known to harbor other viral elements.[63] Interestingly, this locus contains several genes potentially important for the pathogenesis of SCH, thus supporting the original hypothesis of retroviral integration near the "SCH gene."[63] Further supporting the initial findings, HERV-W was also found in the plasma of the individuals with recent-onset SCH.[64]

Another study analyzed sera from 58 SCH patients and identified the ERV9 family retrovirus in 20 patient samples.[65] No ERV9 was detected in control persons. This result was further confirmed with Western blot analysis and also by positive immunoreactivity in the sera of patients.[65]

In addition to the HERV-W, the expression of HERV-K has been identified in the cDNA libraries of the postmortem brains of the SCH patients.[66] HERV-K belongs to an interesting group of HERVs, because it is a full-length provirus, and is probably capable for reinfections.[67] Transcriptional activity of HERV-K has been found in the brains of SCH and bipolar patients.[68] HERV-K115 is mapped to the 8p23.1 region, which is related to SCH susceptibility.[69] A case-control study did not find any significant associations between SCH patients and variations in the HERK-K115 gene.[70] However, it was concluded that variations in the HERV-K115 could modify the onset of the disease.[70] Variations in the HERV-K18 were found to be associated with type 2 diabetes (T2D) in SCH patients.[71] More precisely, the haplotype defined by two polymorphisms (7086 and 8146 C/T) in the envelope region was highly associated with T2D in SCH patients. The established odds ratio was 9.0 (95% CI 2.3−34.7, $P < .001$) after adjustment for age, gender, race, and current therapy with clozapine.[71] This finding was partially confirmed in a Danish population with 750 cases.[72] They did not find any association between HERV-K18 and SCH or T2D in patients with SCH. However, in younger SCH patients, there was a trend for association with T2D, a finding that was in agreement with previous results.[71,72] Genetic interaction between SCH and T2D based on HERVs is plausible, as several studies have found comorbidity between SCH and T2D.[73,74]

Peripheral activation of endogenous viruses has also been studied in several investigations. For instance, HERV-W antigenemia was described in SCH patients.[75] In addition, activation of HERV-W sequences has been analyzed by several additional research papers.[76,77] In the plasma of 42 out of 118 individuals suffering from recent-onset SCH, activation of expression of the sequences homologous to HERV-W *env* gene was found. Overexpression of the HERV-W in glioma cell line was found to induce upregulation of brain-derived neurotrophic factor, neurotrophic tyrosine kinase receptors type 2 (*NTRK2* or *TrkB*), and dopamine receptor D3 genes.[77] Authors concluded that transcriptional activation of HERV-W causes upregulation of the genes implicated in SCH. Similarly, HERV-W LTR has been described in the regulatory region of the GABA receptor B1 gene (*GABBR1*).[76] Similar insertion of the human ERV in the enhancer has been shown for the *PRODH* gene.[78] The *PRODH* gene encodes proline dehydrogenase, and it is involved in the synthesis of neurotransmitters. Hypomethylation of this enhancer induced higher expression of the *PRODH* gene, and this could have an impact in brain development.[78]

While all these previous studies were more focused on specific HERVs or a specific insertion locus, recent technologies enable a much wider approach. Comprehensive analysis of L1 retrotransposition was performed in neuronal genome in postmortem samples from different mental diseases.[79] Clear increase in the L1 retrotransposition was found in SCH. In major depression and bipolar disorder samples, the increased L1 content was borderline in terms of statistical significance.[79] SCH brains showed robust differences, and these differences were confirmed in an independent sample of additional 34 brains. Authors used whole-genome sequencing to identify genomic regions for the L1 insertions, and they found that it localized preferentially to synapse- and SCH-related genes. This study also showed that hyperactive L1 retrotransposition can be triggered after immune activation, so environmental factors can contribute to the susceptibility of SCH.[79] In one metagenomics study, 23 drug-naïve, first-onset SCH patients were analyzed. The authors did not find any evidence for the activation of endogenous retroviruses in the sera of these patients.[80]

Taken together, HERV-W and HERV-K seem to be the most common HERVs found to be associated with SCH.

## Bipolar Disorder

Bipolar disorder is studied less intensively compared to the previously described diseases. Samples from bipolar disorder patients are usually studied along with SCH samples. In one large-scale study, 50 members of 20 HERV families were analyzed in 215 brain samples.[68] The authors of this study found HERV-IP-T47D (HERV-I family) to be reduced in bipolar samples compared to those of SCH. Seq63 from the ERV9 family was reduced in bipolar samples compared to controls. HERV-K10 from the HML-2 family was upregulated in both SCH and bipolar samples compared to the control. And finally, NMWV7 from the HML-7 subclass and HERV-L were downregulated compared to controls and SCH.[68] Interestingly, some HERVs had similar expression profiles between diseases, but other HERVs exhibited disease-specific idiosyncratic profiles. In another study using immunohistochemistry, significant downregulation of the HERV-W in different brain areas of bipolar patients was described.[81] This reduced expression in immunohistochemical staining could indicate neuronal loss in the brains of bipolar patients, potentially resulting from altered expression of HERV during the illness.[81] However, this hypothesis needs further clarification.

Some studies have analyzed the effect of treatment on the expression of HERVs. Valproic acid (VLP) is one of the most commonly used drugs for bipolar disorder. The effect of VLP on the expression of different HERV families in different human cell lines (glioblastoma, neuroblastoma, neural stem cell line) has been studied.[82] VLP dose-dependently upregulated transcription of several HERVs from different subclasses. VLP upregulated HERVs from class I (HERV-E, HERV-H, HERV-W, ERV9, and HERV-F) and from class II (HML-3, HML-4, HML-6, HML-9, HML-10). The stronger influence of VLP was established for HERV-Fb, ERV9, Seq26 (HML-3), and HERV-KC4/HML-10.[82] In postmortem brain samples, HML-2 and ERV9 were upregulated in patients with bipolar disease without treatment compared to the control samples, and VLP treatment did not affect this result.[82]

In summary, existing evidence suggests that HERVs do play roles in bipolar disorder, and traditional pharmacotherapies may affect expression of HERVs.

## Other Psychiatric Diseases

Autism or ASD is a complex disorder associated with altered brain development. ASD is clinically variable. However, an often observed feature is impaired social interaction with peers, impaired communication, and stereotypical and rigid patterns of behavior.[83] The prevalence of ASD is increasing, and the increasing societal burden necessitates more research.[84] Despite decades of studies, the etiology of ASD is still unknown. Accumulating evidence suggests that ASD is a multifactorial disease with a strong genetic basis.[84] Twin studies have found high heritability (more than 90%). Nevertheless, the exact genetic loci and mechanisms underlying the disease have still not been identified. As a new development, an increasing number of studies have described positive association between copy number variants (CNVs) and ASD.[85] In the light of the emerging role of CNVs, analysis of TEs in ASD becomes feasible. RNA from peripheral blood mononuclear cells (PBMCs) of 28 Caucasian children was analyzed for the presence of retroviral RNA from four HERV families (HERV-E, HERV-H, HERV-K, HERV-W).[83] The expression of HERV-H was higher in ASD patients, and expression of HERV-W was lower in ASD patients than in controls. Expression on HERV-K was similar, and the expression of HERV-E was absent.[83] In vitro stimulation of the PBMCs from the ASD patients resulted in higher upregulation of the HERV-H expression compared to controls.[83] Moreover, authors described negative correlation between the HERV-H expression and the Psychoeducational Profile-3 score.[83]

In an additional analysis performed by the same group, an independent sample of Albanian ASD patients was studied.[86] HERV-W was found upregulated, while HERV-K and HERV-W were downregulated in ASD patients compared to controls.[86] Elevated HERV-W expression in autistic children suggests that infection or inflammation during pregnancy could be an etiological risk factor for the development of ASD.[86]

Another recent study was performed on frozen postmortem cerebellar cortex samples from 13 autistic patients and 13 control individuals.[87] Expression analysis of L1 ORF1 and ORF2 was performed with QRT-PCR. L1 expression was found to be significantly increased in the brains of autistic patients, which was inversely correlated with GSH/GSSG ratio.[87] Positive correlation between L1 expression and FOXO3 was also established.[87] However, the impact of L1 expression to neuronal

CNVs and genomic instability requires additional research with innovative genomic technologies.

Similarly to ASD, HERVs may impact the development of ADHD. The expression of retroviral RNAs of three HERV families (H, K, and W) were analyzed in 30 patients with ADHD and 30 healthy controls using PBMCs.[88] The only significantly increased expression found in ADHD patients was for HERV-H. Expression of HERV-K and HERV-W remained unchanged. ADHD has a complex etiology, and the authors concluded that HERVs may present a possible link between the environmental, biologic, and genetic factors contributing to the development of this disease. In another study with a single ADHD patient, HERV-H expression in PBMCs was evaluated before and after treatment with methylphenidate (MPH).[89] Compared to healthy controls, HERV-H was highly and significantly overexpressed in the patients before treatment, but after 6 months of treatment with MPH the expression of HERV-H was normalized to the level of that of controls.[89] This study gives support to the idea that expression changes in HERVs can be pathogenic, and treatments can modify HERV expression and relieve the symptoms.

## CONCLUSIONS

Almost half of the human genome contains TEs, and the impact of these elements on physiology and pathophysiology is not understood yet. However, growing data suggest that TEs play significant roles in a number of CNS disorders. The best studied example is MS, where a clear role of HERV-W in disease development has been demonstrated. Moreover, treatment with antibodies against HERV-W aborts the destruction of neurons and stops progression of the disease.[45] Another elegant example is the involvement of HERV-K in ALS.[18] This chapter summarized existing evidence on how HERVs and other TEs could impact our health, and how understanding this impact may help us develop treatment either by development of new drugs or by repurposing already existing drugs to target the activity of endogenous retroviruses.

Taking together accumulated existing evidence, I suggest that HERV elements represent a missing link between environmental and genetic factors underlying numerous pathogenic processes of complex human diseases. Notably, for the majority of complex diseases, we do not have an obvious and validated etiological explanation. Activation of dormant retroviruses in our genome is "the enemy within."[90] Endogenous retroviruses constitute a new class of pathogens different from classical environmental and infectious microbes, arising from the host's genome itself.[90] Understanding their function and effects thus may open entire new research avenues and therapeutic applications.

## REFERENCES


1. Bannert N, Kurth R. Retroelements and the human genome: new perspectives on an old relation. *Proc Natl Acad Sci USA*. 2004;101(suppl 2):14572–14579.
2. Medstrand P, van de Lagemaat LN, Mager DL. Retroelement distributions in the human genome: variations associated with age and proximity to genes. *Genome Res*. 2002;12(10):1483–1495.
3. Kazazian Jr HH, Moran JV. The impact of L1 retrotransposons on the human genome. *Nat Genet*. 1998;19(1):19–24.
4. Hughes JF, Coffin JM. Evidence for genomic rearrangements mediated by human endogenous retroviruses during primate evolution. *Nat Genet*. 2001;29(4):487–489.
5. Katoh I, Kurata S. Association of endogenous retroviruses and long terminal repeats with human disorders. *Front Oncol*. 2013;3:234.
6. Jurka J. Repbase update: a database and an electronic journal of repetitive elements. *Trends Genet*. 2000; 16(9):418–420.
7. Solyom S, Kazazian Jr HH. Mobile elements in the human genome: implications for disease. *Genome Med*. 2012;4(2):12.
8. Zhou L, Mitra R, Atkinson PW, Hickman AB, Dyda F, Craig NL. Transposition of hAT elements links transposable elements and V(D)J recombination. *Nature*. 2004;432(7020):995–1001.
9. Ponferrada VG, Mauck BS, Wooley DP. The envelope glycoprotein of human endogenous retrovirus HERV-W induces cellular resistance to spleen necrosis virus. *Arch Virol*. 2003;148(4): 659–675.
10. Blond JL, Lavillette D, Cheynet V, et al. An envelope glycoprotein of the human endogenous retrovirus HERV-W is expressed in the human placenta and fuses cells expressing the type D mammalian retrovirus receptor. *J Virol*. 2000;74(7):3321–3329.
11. van de Lagemaat LN, Landry JR, Mager DL, Medstrand P. Transposable elements in mammals promote regulatory variation and diversification of genes with specialized functions. *Trends Genet*. 2003; 19(10):530–536.
12. Dolei A. Endogenous retroviruses and human disease. *Expet Rev Clin Immunol*. 2006;2(1):149–167.
13. Chuong EB, Elde NC, Feschotte C. Regulatory evolution of innate immunity through co-option of endogenous retroviruses. *Science*. 2016;351(6277):1083–1087.
14. Goke J, Lu X, Chan YS, et al. Dynamic transcription of distinct classes of endogenous retroviral elements marks specific populations of early human embryonic cells. *Cell Stem Cell*. 2015;16(2): 135–141.
15. Suntsova M, Garazha A, Ivanova A, Kaminsky D, Zhavoronkov A, Buzdin A. Molecular functions of human endogenous retroviruses in health and disease. *Cell Mol Life Sci*. 2015;72(19):3653–3675.
16. Wang J, Xie G, Singh M, et al. Primate-specific endogenous retrovirus-driven transcription defines naive-like stem cells. *Nature*. 2014;516(7531):405–409.
17. Singh S, Kaye S, Francis N, et al. Human endogenous retrovirus K (HERV-K) rec mRNA is expressed in primary melanoma but not in benign naevi or normal skin. *Pigm Cell Melanoma Res*. 2013;26(3): 426–428.
18. Li W, Lee MH, Henderson L, et al. Human endogenous retrovirus-K contributes to motor neuron disease. *Sci Transl Med*. 2015;7(307), 307ra153.
19. Morandi E, Tarlinton RE, Gran B. Multiple sclerosis between genetics and infections: human endogenous retroviruses in monocytes and macrophages. *Front Immunol*. 2015;6:647.
20. Jordan IK, Rogozin IB, Glazko GV, Koonin EV. Origin of a substantial fraction of human regulatory sequences from transposable elements. *Trends Genet*. 2003;19(2):68–72.
21. Kashkush K, Feldman M, Levy AA. Transcriptional activation of retrotransposons alters the expression of adjacent genes in wheat. *Nat Genet*. 2003;33(1):102–106.
22. Hamdi HK, Nishio H, Tavis J, Zielinski R, Dugaiczyk A. Alu-mediated phylogenetic novelties in gene regulation and development. *J Mol Biol*. 2000;299(4):931–939.
23. Kazazian Jr HH, Wong C, Youssoufian H, Scott AF, Phillips DG, Antonarakis SE. Haemophilia A resulting from de novo insertion of L1 sequences represents a novel mechanism for mutation in man. *Nature*. 1988;332(6160):164–166.

24. Holmes SE, Dombroski BA, Krebs CM, Boehm CD, Kazazian Jr HH. A new retrotransposable human L1 element from the LRE2 locus on chromosome 1q produces a chimaeric insertion. *Nat Genet.* 1994; 7(2):143–148.
25. Morse B, Rotherg PG, South VJ, Spandorfer JM, Astrin SM. Insertional mutagenesis of the myc locus by a LINE-1 sequence in a human breast carcinoma. *Nature.* 1988;333(6168):87–90.
26. Miki Y, Nishisho I, Horii A, et al. Disruption of the APC gene by a retrotransposal insertion of L1 sequence in a colon cancer. *Cancer Res.* 1992;52(3):643–645.
27. Lamprecht B, Walter K, Kreher S, et al. Derepression of an endogenous long terminal repeat activates the CSF1R proto-oncogene in human lymphoma. *Nat Med.* 2010;16(5):571–579, 571 pp. following 579.
28. Christensen T. HERVs in neuropathogenesis. *J Neuroimmune Pharmacol.* 2010;5(3):326–335.
29. Koprowski H, DeFreitas EC, Harper ME, et al. Multiple sclerosis and human T-cell lymphotropic retroviruses. *Nature.* 1985;318(6042):154–160.
30. Perron H, Geny C, Laurent A, et al. Leptomeningeal cell line from multiple sclerosis with reverse transcriptase activity and viral particles. *Res Virol.* 1989;140(6):551–561.
31. Perron H, Lalande B, Gratacap B, et al. Isolation of retrovirus from patients with multiple sclerosis. *Lancet.* 1991;337(8745):862–863.
32. Perron H, Garson JA, Bedin F, et al. Molecular identification of a novel retrovirus repeatedly isolated from patients with multiple sclerosis. The Collaborative Research Group on multiple sclerosis. *Proc Natl Acad Sci USA.* 1997;94(14):7583–7588.
33. Blond JL, Beseme F, Duret L, et al. Molecular characterization and placental expression of HERV-W, a new human endogenous retrovirus family. *J Virol.* 1999;73(2):1175–1185.
34. Mameli G, Astone V, Arru G, et al. Brains and peripheral blood mononuclear cells of multiple sclerosis (MS) patients hyperexpress MS-associated retrovirus/HERV-W endogenous retrovirus, but not Human herpesvirus 6. *J Gen Virol.* 2007;88(Pt 1):264–274.
35. Arru G, Mameli G, Astone V, et al. Multiple sclerosis and HERV-W/MSRV: a multicentric study. *Int J Biomed Sci.* 2007;3(4):292–297.
36. Sotgiu S, Arru G, Mameli G, et al. Multiple sclerosis-associated retrovirus in early multiple sclerosis: a six-year follow-up of a Sardinian cohort. *Mult Scler.* 2006;12(6):698–703.
37. Serra C, Sotgiu S, Mameli G, Pugliatti M, Rosati G, Dolei A. Multiple sclerosis and multiple sclerosis-associated retrovirus in Sardinia. *Neurol Sci.* 2001;22(2):171–173.
38. Sotgiu S, Serra C, Mameli G, et al. Multiple sclerosis-associated retrovirus and MS prognosis: an observational study. *Neurology.* 2002;59(7):1071–1073.
39. Rolland A, Jouvin-Marche E, Viret C, Faure M, Perron H, Marche PN. The envelope protein of a human endogenous retrovirus-W family activates innate immunity through CD14/TLR4 and promotes Th1-like responses. *J Immunol.* 2006;176(12):7636–7644.
40. Perron H, Jouvin-Marche E, Michel M, et al. Multiple sclerosis retrovirus particles and recombinant envelope trigger an abnormal immune response in vitro, by inducing polyclonal Vbeta16 T-lymphocyte activation. *Virology.* 2001;287(2):321–332.
41. Kremer D, Schichel T, Forster M, et al. Human endogenous retrovirus type W envelope protein inhibits oligodendroglial precursor cell differentiation. *Ann Neurol.* 2013;74(5):721–732.
42. Perron H, Dougier-Reynaud HL, Lomparski C, et al. Human endogenous retrovirus protein activates innate immunity and promotes experimental allergic encephalomyelitis in mice. *PLoS One.* 2013;8(12): e80128.
43. Madeira A, Burgelin I, Perron H, Curtin F, Lang AB, Faucard R. MSRV envelope protein is a potent, endogenous and pathogenic agonist of human toll-like receptor 4: relevance of GNbAC1 in multiple sclerosis treatment. *J Neuroimmunol.* 2016;291:29–38.
44. Curtin F, Lang AB, Perron H, et al. GNbAC1, a humanized monoclonal antibody against the envelope protein of multiple sclerosis-associated endogenous retrovirus: a first-in-humans randomized clinical study. *Clin Ther.* 2012;34(12):2268–2278.
45. Curtin F, Perron H, Kromminga A, Porchet H, Lang AB. Preclinical and early clinical development of GNbAC1, a humanized IgG4 monoclonal antibody targeting endogenous retroviral MSRV-Env protein. *MAbs.* 2015;7(1):265–275.

46. Derfuss T, Curtin F, Guebelin C, et al. A phase IIa randomized clinical study testing GNbAC1, a humanized monoclonal antibody against the envelope protein of multiple sclerosis associated endogenous retrovirus in multiple sclerosis patients - a twelve month follow-up. *J Neuroimmunol.* 2015;285:68–70.
47. Derfuss T, Curtin F, Guebelin C, et al. A phase IIa randomised clinical study of GNbAC1, a humanised monoclonal antibody against the envelope protein of multiple sclerosis-associated endogenous retrovirus in multiple sclerosis patients. *Mult Scler.* 2015;21(7):885–893.
48. Faucard R, Madeira A, Gehin N, et al. Human endogenous retrovirus and neuroinflammation in chronic inflammatory demyelinating polyradiculoneuropathy. *EBioMedicine.* 2016;6:190–198.
49. Nexo BA, Christensen T, Frederiksen J, et al. The etiology of multiple sclerosis: genetic evidence for the involvement of the human endogenous retrovirus HERV-Fc1. *PLoS One.* 2011;6(2):e16652.
50. Nexo BA, Villesen P, Nissen KK, et al. Are human endogenous retroviruses triggers of autoimmune diseases? Unveiling associations of three diseases and viral loci. *Immunol Res.* 2016;64(1):55–63.
51. Johnston JB, Silva C, Holden J, Warren KG, Clark AW, Power C. Monocyte activation and differentiation augment human endogenous retrovirus expression: implications for inflammatory brain diseases. *Ann Neurol.* 2001;50(4):434–442.
52. Christensen T. Human endogenous retroviruses in neurologic disease. *Acta Pathol Microbiol Immunol Scand.* 2016;124(1–2):116–126.
53. Andrews WD, Tuke PW, Al-Chalabi A, et al. Detection of reverse transcriptase activity in the serum of patients with motor neurone disease. *J Med Virol.* 2000;61(4):527–532.
54. Steele AJ, Al-Chalabi A, Ferrante K, Cudkowicz ME, Brown Jr RH, Garson JA. Detection of serum reverse transcriptase activity in patients with ALS and unaffected blood relatives. *Neurology.* 2005;64(3):454–458.
55. MacGowan DJ, Scelsa SN, Imperato TE, Liu KN, Baron P, Polsky B. A controlled study of reverse transcriptase in serum and CSF of HIV-negative patients with ALS. *Neurology.* 2007;68(22):1944–1946.
56. Douville R, Liu J, Rothstein J, Nath A. Identification of active loci of a human endogenous retrovirus in neurons of patients with amyotrophic lateral sclerosis. *Ann Neurol.* 2011;69(1):141–151.
57. Crow TJ. The continuum of psychosis and its implication for the structure of the gene. *Br J Psychiatry.* 1986;149:419–429.
58. O'Reilly RL, Singh SM. Retroviruses and schizophrenia revisited. *Am J Med Genet.* 1996;67(1):19–24.
59. Fuller Torrey E, Yolken RH. Familial and genetic mechanisms in schizophrenia. *Brain Res Brain Res Rev.* 2000;31(2–3):113–117.
60. Deb-Rinker P, Klempan TA, O'Reilly RL, Torrey EF, Singh SM. Molecular characterization of a MSRV-like sequence identified by RDA from monozygotic twin pairs discordant for schizophrenia. *Genomics.* 1999;61(2):133–144.
61. Karlsson H, Bachmann S, Schroder J, McArthur J, Torrey EF, Yolken RH. Retroviral RNA identified in the cerebrospinal fluids and brains of individuals with schizophrenia. *Proc Natl Acad Sci USA.* 2001;98(8):4634–4639.
62. Yolken RH, Karlsson H, Yee F, Johnston-Wilson NL, Torrey EF. Endogenous retroviruses and schizophrenia. *Brain Res Brain Res Rev.* 2000;31(2–3):193–199.
63. Deb-Rinker P, O'Reilly RL, Torrey EF, Singh SM. Molecular characterization of a 2.7-kb, 12q13-specific, retroviral-related sequence isolated by RDA from monozygotic twin pairs discordant for schizophrenia. *Genome.* 2002;45(2):381–390.
64. Karlsson H, Schroder J, Bachmann S, Bottmer C, Yolken RH. HERV-W-related RNA detected in plasma from individuals with recent-onset schizophrenia or schizoaffective disorder. *Mol Psychiatry.* 2004;9(1):12–13.
65. Huang W-J, Liu Z-C, Wei W, Wang G-H, Wu J-G, Zhu F. Human endogenous retroviral pol RNA and protein detected and identified in the blood of individuals with schizophrenia. *Schizophr Res.* 2006;83(2–3):193–199.
66. Kim HS, Wadekar RV, Takenaka O, et al. SINE-R.C2 (a *Homo sapiens* specific retroposon) is homologous to CDNA from postmortem brain in schizophrenia and to two loci in the Xq21.3/Yp block linked to handedness and psychosis. *Am J Med Genet.* 1999;88(5):560–566.
67. Belshaw R, Dawson AL, Woolven-Allen J, Redding J, Burt A, Tristem M. Genomewide screening reveals high levels of insertional polymorphism in the human endogenous retrovirus family HERV-K(HML2): implications for present-day activity. *J Virol.* 2005;79(19):12507–12514.

68. Frank O, Giehl M, Zheng C, Hehlmann R, Leib-Mosch C, Seifarth W. Human endogenous retrovirus expression profiles in samples from brains of patients with schizophrenia and bipolar disorders. *J Virol.* 2005;79(17):10890–10901.
69. Suarez BK, Duan J, Sanders AR, et al. Genomewide linkage scan of 409 European-ancestry and African American families with schizophrenia: suggestive evidence of linkage at 8p23.3-p21.2 and 11p13.1-q14.1 in the combined sample. *Am J Hum Genet.* 2006;78(2):315–333.
70. Otowa T, Tochigi M, Rogers M, Umekage T, Kato N, Sasaki T. Insertional polymorphism of endogenous retrovirus HERV-K115 in schizophrenia. *Neurosci Lett.* 2006;408(3):226–229.
71. Dickerson F, Rubalcaba E, Viscidi R, et al. Polymorphisms in human endogenous retrovirus K-18 and risk of type 2 diabetes in individuals with schizophrenia. *Schizophr Res.* 2008;104(1–3):121–126.
72. Nyegaard M, Demontis D, Thestrup BB, et al. No association of polymorphisms in human endogenous retrovirus K18 and CD48 with schizophrenia. *Psychiatr Genet.* 2012;22(3):146–148.
73. Gragnoli C, Reeves GM, Reazer J, Postolache TT. Dopamine-prolactin pathway potentially contributes to the schizophrenia and type 2 diabetes comorbidity. *Transl Psychiatry.* 2016;6:e785.
74. Foley DL, Mackinnon A, Morgan VA, et al. Common familial risk factors for schizophrenia and diabetes mellitus. *Aust NZ J Psychiatry.* 2016;50(5):488–494.
75. Perron H, Mekaoui L, Bernard C, Veas F, Stefas I, Leboyer M. Endogenous retrovirus type W GAG and envelope protein antigenemia in serum of schizophrenic patients. *Biol Psychiatry.* 2008;64(12):1019–1023.
76. Hegyi H. GABBR1 has a HERV-W LTR in its regulatory region—a possible implication for schizophrenia. *Biol Direct.* 2013;8:5.
77. Huang W, Li S, Hu Y, et al. Implication of the env gene of the human endogenous retrovirus W family in the expression of BDNF and DRD3 and development of recent-onset schizophrenia. *Schizophr Bull.* 2011;37(5):988–1000.
78. Suntsova M, Gogvadze EV, Salozhin S, et al. Human-specific endogenous retroviral insert serves as an enhancer for the schizophrenia-linked gene PRODH. *Proc Natl Acad Sci USA.* 2013;110(48):19472–19477.
79. Bundo M, Toyoshima M, Okada Y, et al. Increased l1 retrotransposition in the neuronal genome in schizophrenia. *Neuron.* 2014;81(2):306–313.
80. Canuti M, van Beveren NJM, Jazaeri Farsani SM, et al. Viral metagenomics in drug-naive, first-onset schizophrenia patients with prominent negative symptoms. *Psychiatry Res.* 2015;229(3):678–684.
81. Weis S, Llenos IC, Sabunciyan S, et al. Reduced expression of human endogenous retrovirus (HERV)-W GAG protein in the cingulate gyrus and hippocampus in schizophrenia, bipolar disorder, and depression. *J Neural Transm.* 2007;114(5):645–655.
82. Diem O, Schaffner M, Seifarth W, Leib-Mosch C. Influence of antipsychotic drugs on human endogenous retrovirus (HERV) transcription in brain cells. *PLoS One.* 2012;7(1):e30054.
83. Balestrieri E, Arpino C, Matteucci C, et al. HERVs expression in autism spectrum disorders. *PLoS One.* 2012;7(11):e48831.
84. Aldinger KA, Plummer JT, Qiu S, Levitt P. SnapShot: genetics of autism. *Neuron.* 2011;72(2), 418–418.e411.
85. Salyakina D, Cukier HN, Lee JM, et al. Copy number variants in extended autism spectrum disorder families reveal candidates potentially involved in autism risk. *PLoS One.* 2011;6(10):e26049.
86. Balestrieri E, Cipriani C, Matteucci C, et al. Transcriptional activity of human endogenous retrovirus in Albanian children with autism spectrum disorders. *N Microbiol.* 2016;39(3):228–231.
87. Shpyleva S, Melnyk S, Pavliv O, Pogribny I, Jill James S. Overexpression of LINE-1 retrotransposons in autism brain. *Mol Neurobiol.* 2017.
88. Balestrieri E, Pitzianti M, Matteucci C, et al. Human endogenous retroviruses and ADHD. *World J Biol Psychiatry.* 2014;15(6):499–504.
89. D'Agati E, Pitzianti M, Balestrieri E, Matteucci C, Sinibaldi Vallebona P, Pasini A. First evidence of HERV-H transcriptional activity reduction after methylphenidate treatment in a young boy with ADHD. *N Microbiol.* 2016;39(3):237–239.
90. Engel ME, Hiebert SW. The enemy within: dormant retroviruses awaken. *Nat Med.* 2010;16(5):517–518.

CHAPTER 23

# Optogenetics Dissection of Sleep Circuits and Functions

**Cornelia Schöne, Antoine Adamantidis**
Department for BioMedical Research, University of Bern, Bern, Switzerland

## INTRODUCTION

To understand the sufficiency of a neuronal circuit to trigger a particular behavioral response and to show its necessity for natural intrinsic behavior is a central challenge in modern neuroscience. This will provide us with a causal link to understand how activity and wiring patterns of an identified neuron population relate to behavior and cognitive processes. Gain- and loss-of-function assays are commonly used to establish such causality. In the field of neuroscience, lesion, electrical, genetic, and pharmacological manipulations have substantially contributed to our understanding of the brain's neuronal function and anatomic structures relevant for specific behaviors. A major difficulty, however, has been to combine high spatial and genetic selectivity with high temporal precision for measuring neural activity to build a bridge between neuronal identities, neuronal firing, and behavior. Addressing this difficulty has only become feasible since the rise of optogenetic tools, which allow the modulation of cellular physiology with millisecond precision through light pulses for acute and chronic interventions (Fig. 23.1A). At the heart of optogenetics are newly developed gene-targeting strategies and light-targeting devices that allow for robust and selective targeting, imaging, and controlling modalities of individual neurons, as well as large brain-wide ensembles. The optical and molecular strategies that comprise optogenetic tools have greatly expanded in the last decade and now include a wide range of genetically encoded, light-sensitive proteins for modulating cellular physiology. These methods include altering biochemical pathways, enzymatic activity and gene expression, excitability, and firing patterns of neurons. In addition to these perturbational tools, a battery of genetically encoded indicators of cellular activity allows the investigator to record intrinsic activity patterns from identified neurons. Correlation of such patterns with behavior helps with initial identification of behaviorally relevant motifs. Optogenetic stimuli can then be fine-tuned in a more naturalistic way by playing back previously recorded patterns and analyzing their effects on behavior. Here, we will summarize these constantly improving optogenetic strategies and further describe how this technique has been used to unravel sleep circuits and sleep function.

*Molecular-Genetic and Statistical Techniques for Behavioral and Neural Research*
ISBN 978-0-12-804078-2, https://doi.org/10.1016/B978-0-12-804078-2.00023-4

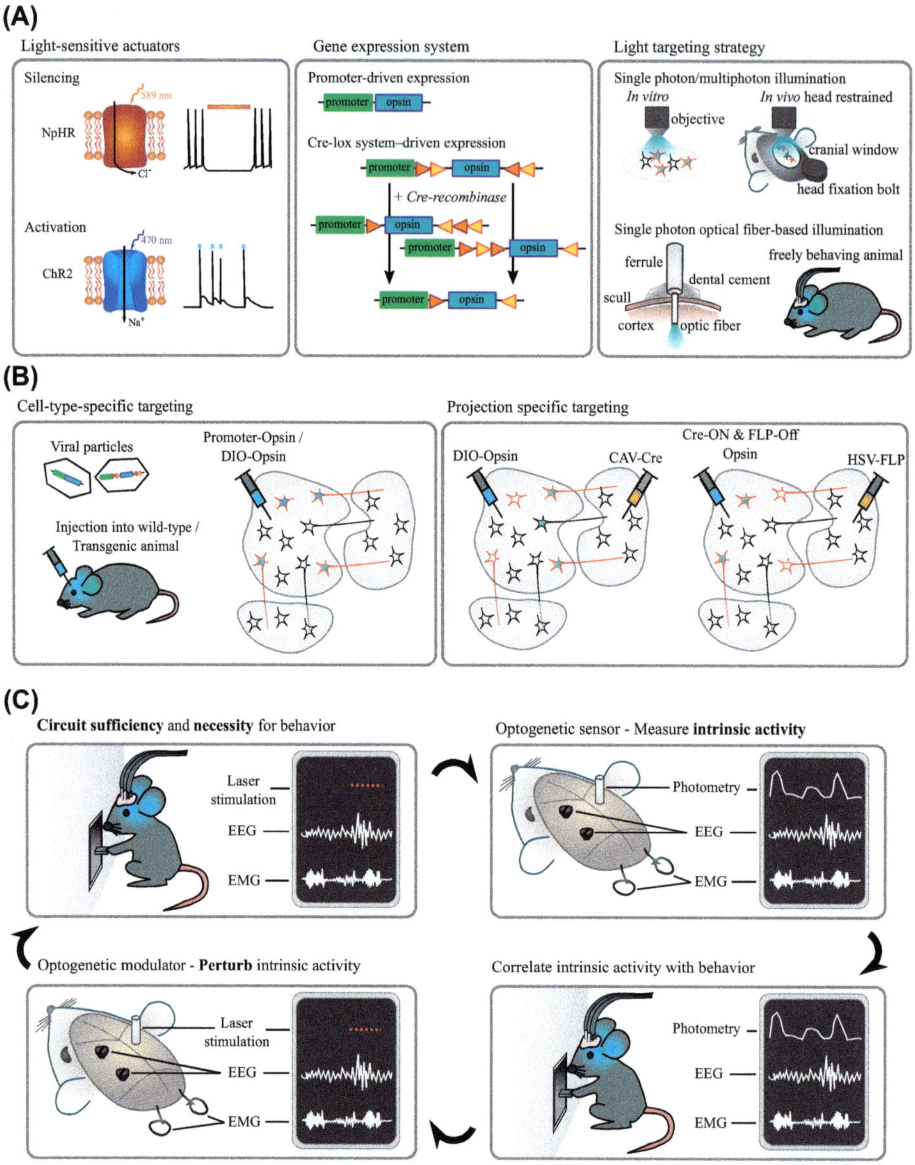


**Figure 23.1** (A) Main components of optogenetic experiments. **Left box:** Light-activated membrane currents allow silencing (Top) and activation (Bottom) of neuronal activity with ms precision. Silencing mediated by optical activation of NpHR is mediated through chloride influx, while excitation through cation channel channelrhodopsin-2 (ChR2) is mostly attributed to sodium influx. Respective effects on neuronal firing are shown on the right. **Middle box:** Two commonly used expression systems. Opsins can either be directly driven by the promoter of a marker protein of the cell population of interested (Top), or the opsin can be expressed using the Cre-lox system (Bottom). The Cre-lox system requires expression of Cre recombinase confined to the cells of interest. In a first step of recombination of the double floxed (*triangles*) inverted open reading frame of the opsin becomes reverted, and

## PRINCIPLES FOR OPTOGENETIC DISSECTION OF NEURAL ACTIVITY

The first proof of principle studies to show that light could be used to control transmembrane voltages and record activity in living cells by using genetically targeted expression of optogenetic probes were conducted in cultured cells such as *Xenopus laevis* oocytes and primary neurons in vitro.[1–6] Since these pioneering studies the optogenetic approach was soon applied in living organisms such as *Caenorhabditis elegans*, *Drosophila*, and rodents to control behavior (see Fig. 23.1).[7–10] The discovery of novel proteins derived from bacterial and algae light-sensitive pumps and ion channels and their further development has given rise to probes with improved membrane targeting, toxicity, light sensitivity, photocurrents (activators and inhibitors) or quantum yield (fluorescent sensors), kinetics, and compatibility with gene expression systems. These modern optogenetic activators, inhibitors, and indicators now allow the design of loss-of-function, gain-of-function, and observational approaches for neural circuit interrogation in a wide variety of experimental models ranging from cell culture to freely moving rodents and nonhuman primates.[11,12]

### Genetically Encoded Indicators, Inhibitors, and Activators of Neuronal Activity

#### *Genetically Encoded Sensors of Neural Activity*

Calcium is a ubiquitous second messenger that plays an essential role in excitable cells, from signal transduction to plasticity. During action potential firing, voltage-gated

subsequently the double-flox sites are inactivated to allow stable expression. **Right box:** Illustration of different light targeting strategies. For in vitro and in vivo head restrained applications, single and multiphoton microscopes can be used to shape light (Top). Free-running mice are most commonly tethered to an optic fiber patch cord that can be mated to an optic fiber cemented on the scull and targeted to the brain region of interest. (B) Combination of different viral targeting strategies allows for cell-type and projection-specific labelling with different sites of injection. Left: Promotor-driven and Cre-lox-driven expression can be used for cell-type targeting (*red cells*). Right: Projection-specific labelling uses the property of some viruses to infect axons and abuse retrograde transport machinery to reach the nucleus for expression. Right, Top: CAV-Cre projection-specific labelling is combined with a Cre-dependent opsin injected in the somatic location. Right, Bottom: Cre and FLP-dependent opsins allow combination of cell-type- and projection-site-specific expression. Here HSV-FLP is injected into projection site. Interestingly, this method can be used to exclude or include specific projection targets (e.g., FLP-Off vs. FLP-On). These methods can further be refined by implanting optical stimulation fibers onto the somata versus projection targets of interest. (C) For dissection of circuit correlates of behavior, the circuit is first screened for activation patterns, e.g., using photometry to measure population calcium signals and then correlated with a behavior. This allows calibrating optogenetic perturbations, so activation patterns mimic naturally occurring patterns, while inhibition will be most effective during natural high activity. Optogenetically evoked activity can be compared to intrinsic activity, e.g., by combining a blue light–sensitive calcium sensor with a red-shifted opsin to record activity while stimulating.

calcium sensors open to enable calcium signalling cascades secondary to electrical signalling, while glutamate excitation of NMDA receptors allows synaptic calcium entry, crucial for synaptic plasticity processes such as long-term potentiation.[13] Action potential firing and synaptic inputs can therefore be assessed by measuring relative levels of intracellular calcium.[14,15] Fluorescent calcium sensors have been a successful tool for measuring intracellular calcium levels of subcellular compartments, individual neurons, and neuronal ensembles in vitro and in vivo (Reviewed in Ref. 16).

Initially, acetoxymethyl ester forms of small molecule organic calcium indicators, such as Fura-2, were used to systematically label neurons.[17] Such dyes responded to changes in calcium levels with changes in peak emission wavelength or intensity.[18,19] While powerful in sensitivity and emission, these were lacking genetic specificity. The first genetically encoded calcium indicator (GECI) was described as early as 1997 by Tsien et al.[2] The first "breakthrough" GECI, GCamP3, did not arrive until 2009 but showed substantial improvements in dynamic range, baseline fluorescence, and calcium affinity.[20] Similar to previous GCamp versions, GCamp3 was a fusion protein between a circularly permuted, green fluorescent protein, calmodulin (CaM), and a $Ca^{2+}$/CaM-binding peptide, of which now multiple improved versions exist.[20–25] Calcium binding affects the conformational structure of GCamp3 and induces a change in fluorescence intensity that allows detection of even single action potentials. GCamp6 variants provide the best signal-to-noise ratio so far, reaching capabilities of small molecule calcium sensor OGB-1.[25]

To understand how distinct neuron populations act together the activity of these needs to be recorded simultaneously. Development of calcium sensors with red-shifted excitation and emission spectra promised simultaneous imaging of green and red GECIs that can be targeted to separate neuron populations.[26,27] The photons emitted from red sensors would then be separable from the more traditional green GECIs through specifically tailored light filters, and their distinct excitation spectra allow to independently activate each sensor by shining light of only a specific narrow band wavelength at a time. In addition, red shifting would also improve limitations of short-wavelength light-scattering properties and toxicity to cells, and it avoids the overlap in excitation spectra of blue light–sensitive excitatory opsin variants. First red variants consisted similarly to green versions of a circularly permuted red fluorescent protein (cpRFP), calmodulin (CaM), and the $Ca^{2+}$/CaM-binding peptide. Multiple stepwise improved versions based on mRuby or mApple red fluorescent proteins have been described.[27–31] Dana et al. described novel further improved versions, the mRuby-based jRCamp1a and b and the mApple-based jRGECO1a that showed comparable sensitivity to calcium as GCamp6 variants.[32]

Alongside the development of calcium indicators with improved quantum yields and stability or altered excitation/emission spectra, a novel class of photoconvertible calcium indicators allows the investigator to "highlight" spatially restricted cells for calcium imaging.[33–35] Fosque et al. designed a novel indicator named CaMPARI, which undergoes efficient and irreversible light-induced green-to-red photoconversion only when intracellular calcium levels are also elevated, thus allowing activity-dependent tagging of neurons for calcium imaging.[36] This has enabled all-optical connectivity mapping in acute brain slices.[37] In vivo, cells that are activated during a specific behavioral task could then be followed in other conditions, which might be a key tool, for example, to understand how daytime activity is reflected or replayed during sleep.

Neuronal activity consists of suprathreshold action potentials and subthreshold variation of transmembrane voltage. While calcium signals can only reflect suprathreshold activity in neurons (depolarizations that are strong enough to open voltage-gated calcium channels), information on subthreshold activity including excitatory versus inhibitory synaptic inputs leading to excitatory postsynaptic potential or inhibitory postsynaptic potential is lost. Genetically encoded voltage indicators (GEVIs) were engineered to overcome this limitation of GECIs. However, one major challenge has been that action potentials only last a few milliseconds, which would require fast acquisition rates to be captured reliably. Such short acquisition times require particularly bright and photostable indicators. Since GEVIs report on transmembrane voltage, these need to be confined to membranes, in contrast to GECIs that fill the cytosol and sense intracellular calcium that remains elevated for dozens to hundreds of milliseconds after an action potential, thus enabling easier detection at a higher signal-to-noise ratio. The first GEVI was described in 1997 by Siegel and Isacoff[38] and many followed. The currently brightest and fastest GEVIs are the Ace-mNeon series described by Gong et al.[39] Hochbaum et al.[40] produced an expression vector that combines QuasAr voltage indicators with excitatory opsin expression (Optopatch), which enabled these investigators to optically control and read neuronal activity simultaneously. A Cre-dependent "Floxopatch" driver line enables brain-wide expression of Optopatch in cells by crossing with a specific Cre-driver line.[41] However, temporally precise monitoring of transmembrane voltages across large fields of view still remains challenging.

The recent development of fluorescence recording techniques with single-cell (wearable miniature microscope) and population (fiber photometry using one-pixel optical fiber) resolution in free-running animals has allowed to link activity of genetically defined neuron populations with natural behaviors.[42–45] The use of graded index (GRIN) lenses and optical fibers allows calcium imaging from previously inaccessible deep brain tissues such as the hypothalamus or brainstem, which can be combined with sleep-wake recordings. For example, imaging through GRIN lenses allowed investigators to identify neurons in the laterodorsal tegmental nucleus that were maximally active during wakefulness (GABAergic) or REM sleep (glutamatergic).[46] Fiber photometry also enabled investigators to measure population activity of brainstem tyrosine hydroxylase—positive dopamine cells across sleep-wake state transitions.[47] With these novel possibilities, we can expect the field of in vivo optical imaging (not only of sleep circuits) to substantially grow in coming years, in particular with the ability to perform experiments in freely moving animals, which allows for dissection of single-cell and population dynamics across natural sleep-wake behavior. A recent review on optical brain imaging techniques for measuring fluorescence signals can be found in Yang and Yuste, while a detailed review on fluorescent indicator proteins can be found in Rodriguez et al.[48,49]

### *Channelrhodopsin-2 Variants for Fast Excitation*

Optogenetic neuronal membrane excitation/depolarization represents a powerful tool for probing neuronal circuit control of a specific behavior. A major milestone was reached in 2002, when the genetic code for two single-component light-sensitive ion channels termed channelrhodopsins were isolated: the green light (510 nm)—sensitive proton permeable channelrhodopsin-1 (ChR1) and the blue light (470 nm)—activated cation channel channelrhodopsin-2 (ChR2).[4,5] ChR1 suffered from weak temporal activation and deactivation kinetics, while ChR2 significantly depolarized cells with millisecond precision. This provided a first promising tool for the control of transmembrane potentials of single cells. The feasibility of this approach for evoking neuronal action potentials was first demonstrated by Boyden et al. who used lentiviral vectors to transduce cultured hippocampal neurons with a ChR2-GFP fusion protein. GFP-labelled cells could be reversibly and repeatedly depolarized by short blue light pulses at 473 nm with no additional cofactors necessary.[6] Importantly, neuronal intrinsic electrical properties were not significantly affected by expression of ChR2, suggesting that the approach is compatible with physiological studies of neuronal activity in healthy neurons.[6]

One drawback was the slow decay kinetics (>10 ms) of ChR2, which led to slow repolarization with generation of double spikes or poor optical control with depolarization plateaus at higher stimulation frequencies above 30 Hz. The first ultrafast variant with E123T mutation overcame this limitation and was called ChETA.[50] Compared to wild-type ChR2, ChETA expressing fast-spiking cortical interneurons showed markedly improved fidelity to high-frequency stimulation (40—200 Hz), while still evoking only single spikes with brief 1—2 millisecond light pulses. A more recent slightly red-shifted variant called Chronos shows equally fast kinetics, but superior light sensitivity and high frequency fidelity up to at least 60 Hz (peak excitation near ~500 nm).[51] Thus, the improved light-controlled activation and deactivation properties of novel ChR2 variants allow reliable control of neuronal firing up to at least 200 Hz that exceed the gamma range (30—150 Hz), while avoiding duplicate spikes and plateau potentials of the original ChR2 variant.

A number of mutant channels have been engineered to alter and improve different deficiencies of ChR2, including adaptation of the codon sequence to fit mammalian expression systems (humanized ChR2). Table 23.1 summarizes these novel variants. For example, the original ChR2 and first fast variants such as ChETA suffered from their small single-channel conductance that resulted only in subthreshold depolarization and

**Table 23.1** Opsin variants for optogenetic activity control

| Opsin | Type | Peak wavelength | Off-kinetic | References |
|---|---|---|---|---|
| ***Excitatory opsins*** | | | | |
| LiGluR | Cation channel | 380 nm | 92 ms | 150–152 |
| ChIEF | Cation channel | 450 nm | 9.8 ms | 60,153 |
| ChR2 | Cation channel | 460–470 nm | 9.2 ms | 5,6,50 |
| ChR2 (H134R) | Cation channel | 470 nm | 18 ms | 10 |
| ChEF | Cation channel | 470 nm | 26 ms | 60,153 |
| ChR2 (E123A) | Cation channel | 470 nm | 4 ms | 50,54 |
| ChR2 (E123T/T159C) | Cation channel | 470 nm | 10 ms | 54 |
| CatCh | Cation channel | 474 nm | 16 ms | 154 |
| ChD | Cation channel | 480 nm | 7.8 ms | 60,153,155 |
| ChR2 (E123T) | Cation channel | 490 nm | 4.2 ms | 50,54 |
| ChR1 | Cation channel | 500 nm | ~50 ms | 4 |
| Chronos | Cation channel | 500 nm | 3.6 ms | 51 |
| ChRGR | Cation channel | 505 nm | ~5 ms | 5,155–157 |
| VCOMET | Cation channel | 530 nm | 100 ms | 153 |
| $C1V1_{TT}$ | Cation channel | 535 nm | 34 ms | 55,124 |
| C1V1 | Cation channel | 540 nm | 85–156 ms | 55,158,159 |
| VChR1 | Cation channel | 545 nm | 132–150 ms | 55,59,155–157 |
| ReaChR | Cation channel | 590 nm | 137 ms | 153 |
| Chrimson | Cation channel | 590 nm | 21 ms | 51 |
| ChrimsonR | Cation channel | 590 nm | 16 ms | 51 |
| ***Bistable opsins*** | | | | |
| **Opsin** | **Type** | **On/off wavelength** | **Off-kinetic** | **References** |
| SFO | Cation channel | 470 nm/590 nm | 42s–6.9 min | 54,124 |
| SSFO | Cation channel | 470 nm/590 nm | 29 min | 54,124 |
| VChR1-SFO/SSFO | Cation channel | 560 nm/390 nm | 32 s–5 min | 54,55,63,124 |
| SwiChR | $Cl^-$ pump | 470 nm/632 nm | 7 s | 65 |
| ***Inhibitory opsins*** | | | | |
| **Opsin** | **Type** | **Peak wavelength** | **Off-kinetic** | **References** |
| BLINK1 | $K^+$ channel | (455 nm) | 51 s | 71 |
| iC1C2 | $Cl^-$ pump | 470 nm | 24 ms | 65 |

*Continued*

**Table 23.1** Opsin variants for optogenetic activity control—cont'd

| | | | | |
|---|---|---|---|---|
| GtACR1 | $Cl^-$ channel | 515 nm | (?) | 67,68 |
| I1K6NaR | $Na^+$ outward pump | 528 nm | (?) | 69 |
| MAC | $Cl^-$ pump | 550 nm | (?) | 61 |
| Arch/T | Proton pump | 566 nm | 19/9 ms | 11,61,63 |
| eBR | Proton pump | 568 nm | 19 ms | 160 |
| eNpHR3.0 | $Cl^-$ pump | 590 nm | 4.2 ms | 62,63 |
| Jaws | $Cl^-$ pump | 604 nm | ~6 ms | 66,161 |
| KR2 | $Na^+$ outward pump | 526 nm | 7.1 ms | 69,70 |
| $KR2_{K+}$ | $K^+/Na^+$ outward pump | 526 nm | (?) | 69 |

Adapted from Yizhar et al. 2011[124], Lin et al. (2009)[153], and Fenno et al. (2011)[162].

spike failures at lower expression levels and light intensities or in larger (low input resistance) neurons. However, high expression levels or long-term high-intensity light stimulation can alter cell physiology.[52] Additionally, long-term activation is limited by sustained illumination, since this leads to depolarization-dependent deactivation of ChR2 channels. Berndt et al. developed a channelrhodopsin variant ChR2(T159C) that responded even to low light levels with very large currents.[53] ChR2(T159C)-mediated blue light stimulation could evoke spike trains with light intensities of 1.9 mW/mm$^2$, a low light intensity that did not evoke spikes using wild-type ChR2 or ChETA opsins. In addition to the T159C mutation, strong photocurrents were also achieved with channel mutants H134R and L132C. Unfortunately, in all three variants the mutation resulted in slow kinetics and temporally imprecise firing, when stimulation was performed at high frequencies. The highly efficient variant ChR2(ET/TC) combined the T159C and ChETA E123T mutations to achieve both large photocurrents and fast kinetics. Strikingly, it could evoke spiking up to at least 60 Hz in labelled hippocampal neurons.[53] The enhancement of photoactivated current and light sensitivity further improved methodological restrictions of expression, intrinsic neuronal excitability, and light delivery, while allowing for precisely timed and light-driven spiking at high frequencies that approach physiological relevance.

By contrast, slow decay kinetics were brought to an extreme in so-called step-function opsins, which were engineered to produce even longer photocycles to alter excitability on longer time scales. Mutation of ChR2 at C128 resulted in decay kinetics of up to 100 s (Ref. 54), while Yizhar et al. later generated a stable step-function opsin (SSFO) variant that combined ChR2 C128T/D156A mutations to achieve decay constants of around 30 min.[54,55] Thus, cells expressing these slow mutants can be depolarized to threshold and near-threshold potentials for seconds to tens of minutes using blue light. This increase in neuronal excitability with nonsynchronized action potential firing may reflect more naturalistic firing patterns as compared to both electrical or optogenetic activation with fast ChR2 variants that can promote artificial synchronization of stimulated cells. SSFOs are opened quickly within a few milliseconds of 473-nm blue light illumination, while green to red light reverses the channel state. This enables temporally precise switching between states with high and low excitability through activation and deactivation of genetically defined target neurons. Moreover, chronic constant excitation would not be achievable with fast variants, as long-term illumination of brain structures may produce heating of brain tissue, thereby altering intrinsic cellular properties and neuronal physiology, which could lead to significant damage.[56–58]

Similar to using red GECIs instead of green GECIs, using red light for activating optogenetic probes would allow the investigator to use less light power and penetrate deeper into brain tissue compared to blue. The first red-shifted excitatory channelrhodopsin, VChR1, with a peak activation at 545-nm red light, was cloned from the colonial alga *Volvox carteri*.[59] However, its limiting properties were poor membrane targeting, slow channel decay kinetics, and small photocurrents. A variant of VChR1 termed C1V1 improved membrane expression, but it suffered from low peak and steady-state photocurrents with similar levels of light intensity. The red light–activated channelrhodopsin ReaChR with a peak excitation at 610 nm is significantly red-shifted by 140 nm compared to ChR2, and it also demonstrates large photocurrents compared to previous red-shifted opsins.[60] Strikingly, ReaChR-mediated red light stimulation of motor cortex as well as deep brain structures allowed to drive vibrissae movement. To avoid invasive optic fiber implants the authors also used a skull-mounted LED-based system, thus minimizing damage to brain tissues. However, similarly to blue variants, increase in photocurrents was associated with long decay constants and temporal imprecision of action potential firing. Chrimson and ChrimsonR show significantly improved time-off kinetics compared to previous red-shifted variants, which is associated with increased fidelity at low frequencies (<10 Hz for Chrimson, and up to 40 Hz for ChrimsonR).[51] The availability of blue and red light–sensitive probes has opened endless combinatorial approaches for separate excitation in two distinct neuron populations simultaneously.

### *Halorhodopsins and Bacteriorhodopsins for Fast Neural Silencing*

In contrast to traditional pharmacological or lesion-based interventions, fast optogenetic silencing can provide reversible, specific, and fast blockade of intrinsic spiking activity of genetically targeted cell types, allowing for accurate dissection of neural circuits underlying behavior. Multiple versions of halorhodopsin-type chloride pumps and bacteriorhodopsin-type proton pumps have been used widely to produce light-activated hyperpolarization through inward flux of chloride or outward flux of protons, respectively.[61–63]

A first mammalian codon-optimized variant of *Natronomas pharaonis* halorhodopsin (NpHR) rendered cultured hippocampal neurons light sensitive to 590 nm, which induced hyperpolarization during the illumination period. eNpHR2.0 improved expression levels and surface membrane trafficking of the protein. A neurite trafficking sequence was added in version eNpHR3.0, resulting in the widespread membrane expression throughout the soma and dendrites of transfected neurons. Hyperpolarization could be induced with red-shifted light up to 680 nm and could average 40–50 mV with photocurrents of up to 5 nA[62,63] However, prolonged activation of eNpHR 3.0 may alter chloride equilibrium potentials, resulting in a polarity shift of GABAergic synaptic potentials, as well as increased rebound excitation following cessation of light activation.[64] Therefore, experiments must follow appropriate control conditions to test for possible shift in ions reversal potential and modification of synaptic transmission.

The bacteriorhodopsin archaerhodospin-3 (Arch) from *Halorubrum sodomense* (see Table 23.1) avoids disturbances of chloride equilibrium, since Arch-mediated hyperpolarization acts through expulsion of protons from the cell.[61] While Arch is activated by low-intensity yellow-green light, it supports long-term and reversible inhibition through photocurrents of up to 900 pA in vitro. At lights off, photocurrents terminate rapidly, although slightly more slowly compared to eNpHR3.0 ($\tau_{off}$, eNpHR3.0: 4.2 ms, Arch and ArchT: ~9 ms).[61,62] ArchT, a related homolog of Arch from *Halorubrum* strain TP009, showed a threefold increase in sensitivity to yellow-green light, with far larger photocurrents induced between 1 and 10 $mW/mm^2$ and enhanced axonal targeting.[11] Importantly, in vivo ArchT-mediated silencing resulted in a near-complete silencing of recorded cortical neurons in awake macaque monkeys, supporting its suitability for loss-of-function studies.[11] Although proton pumps can result in increases in pH levels, changes in somatic pH have been reported as small and self-limiting, and therefore well tolerated in neurons.[61] Further, ArchT-mediated silencing does not increase rebound excitation as compared to NphR3.0.[64] In summary, ArchT has been suggested as an effective, reliable, and safe alternative opsin to use of eNpHR3.0 for neural silencing. However, since these effects largely depend on cell types and circuits, a systematic assessment of control conditions is warranted.

Opsins sensitive to different light bands allow perturbation and/or observation of distinct circuits simultaneously. In this respect, the blue light–sensitive proton pump Mac derived from *Leptosphaeria maculans* allows the investigator to hyperpolarize neurons with a peak activation wavelength of 470 nm.[61] The blue light–sensitive opsin called iC1C2 was engineered by converting an existing blue light–sensitive excitatory cation channel into an anion channel permeable to chloride.[65] Blue light activation of iC1C2 induces robust outward currents with the reversal potential of chloride (−64 mV).[65] These two blue light–sensitive inhibitory opsins thus might nicely complement yellow-green light–sensitive inhibitory ArchT and eNpHR variants and other red-shifted opsins mentioned earlier. Moreover, due to the similarity of iC1C2 with excitatory ChRs and SSFOs, this channel can be converted into a step-waveform inhibitory ChR (SwiChR) by mutation of cysteine-167.[65] This produces outward currents for up to 7 s upon brief blue light activation due to its slow channel inactivation kinetics, and thus allows sustained photoinhibition without constant blue light illumination.

In contrast to blue light–shifted opsins, red shifting allows deeper penetration of light into the tissue and therefore allows less-proximal positioning of light sources. Further development of the red light–sensitive chloride-conducting hyperpolarizing opsin Halo57, derived from *Haloarcula salinarum* (strain Shark), led to a novel inhibitory opsin called Jaws.[66] Jaws produces approximately three times bigger outward photocurrents than NpHR3.0 and ArchT, while maintaining similar fast kinetics as previous inhibitory opsins.[66] The striking light sensitivity of Jaws allowed blocking nearly all spiking activity in brain regions up to 3 mm away from the cranial surface by using transcranial red light stimulation. Thus, by combining ultra-light-sensitive opsins and transcranial light stimulation, proximal damage of brain tissue through light fibers can be avoided, while greatly expanding the brain volume that can be targeted.

Some optogenetic circuit mapping experiments may require optogenetic silencing of axon terminals. In a recent study, Mahn et al. investigated the effects of presynaptic silencing of excitatory layer 4 thalamocortical synapses through eArch3.0, eNpHR3.0, and opsins.[67] They found that synaptic optical silencing reduced, but did not completely block evoked transmission when light was applied briefly during electrical stimulation in the thalamocortical fiber tract, suggesting that more powerful inhibitory opsins may be required for complete presynaptic silencing. Moreover, long-term silencing over minutes with eArch3.0 increased the rate of spontaneously occurring excitatory and inhibitory postsynaptic currents that could be explained by intracellular alkalization as it is reduced by co-application of L-Lactate to stabilize pH. The authors further tested a member of a new family of light-gated anion channel rhodopsins, GtACR1, for their efficiency in blocking synaptic release, which has been shown to robustly suppress somatic action potentials.[67,68] However, brief activation of synapto-axonic GtACR1 produced short-latency EPSCs, similar to those evoked by electrical stimulation. eNpHR3.0 might thus represent the currently most suitable candidate for presynaptic suppression of neurotransmitter release. Opsins that do not affect intracellular pH or chloride, for example, outward sodium pump and light-gated potassium channels, might be other candidate opsins to improve efficacy of fast optogenetic inhibition of transmitter release with reduced side effects.[69–71]

## Genetic Targeting Strategies for Opsin Expression

### *Promoter-Based Specificity in Transgenic Expression of Optogenetic Tools*

In general, the amount of mRNA transcribed for a specific gene is tightly regulated by genetic regulatory elements. Noncoding DNA sequences build transcription factor recognition sites, chromatin remodeling proteins regulate the 3D folding structure of the chromatin, and together with other regulatory elements, control accessibility of specific genes to the transcription machinery.[72] Similarly, the efficacy of opsin expression depends on the genetic targeting strategy used. Opsins can be linked to a specific gene promoter sequence that enables gene expression in the presence of matching regulatory proteins, thereby restricting labelling to cell types with known genetic identities. To deliver a promoter-opsin vector to postmitotic cells such as neurons, recombinant adeno-associated-virus (AAV), lentivirus, retrovirus, and herpes simplex virus (HSV) have been used successfully for gene transfer to brain tissues (reviewed in Ref. 73). Recombinant viral vectors are, in essence, genetically engineered viruses in which a minimal set of viral genes has been retained to allow host cell entry, transport to the nucleus, and expression of the transgene, while eliminating virulence functions such as replication and cytotoxicity.[74] Since the genetic sequence that is possible to package into a viral vector backbone is restricted by size (~4.7 kb AAV, ~9 kb lentivirus), while genomic promoter sequences range from hundreds of bases to several thousands of bases, full-length promoters can usually not be inserted into a viral vector backbone.[75,76] Minimal promoter sequences of 0.2–1.5 kb can be used for optogenetic targeting, but only a few such promoters that retained sufficient cell-type specificity could be identified to date (see Refs. 12,77–79). Lentiviral vectors allow for larger backbone vectors, and they are therefore effective where a minimal promoter sequence lies below 3.5 kb.[12,80] Weak transgene expression levels may limit the utility of minimal promoter sequences in viral vector-based optogenetics when high expression levels for effective light-based control over targeted neurons is required. In such cases, alternative methods exist that include classical transgenic mouse engineering.

Transgenic (multiple random insertion) or knock-in (locus-specific single insertion) recombinant DNA technologies are less limited by the size of the promoter than viral targeting strategies. The knock-in approach is generally considered more specific and stable, due to the locus-specific single insertion of the transgene. However, transgenic mice usually receive multiple opsin gene copies and produce higher expression levels with sufficient photocurrents to facilitate optogenetic manipulation. Since the first transgenic mouse model expressing ChR2 under the Thy1 promoter was reported in 2007, additional mouse lines that express ChR2 in GABAergic neurons (VGAT-ChR2, Pvalb-ChR2), single olfactory glomeruli (M72-ChR2), cholinergic neurons (ChAT-ChR2), and others have been generated (see also www.jax.org).[81–83] However, transgenic mouse models may yield insufficient opsin levels for reliable optogenetic perturbation. For example, optical manipulations of distal axon boutons were reported to require higher light intensities.[84] Importantly, knock-in strategies may perturb the expression of the endogenous gene and cause haplo-insufficiency (a reduction in overall expression of the gene into which the transgene has been targeted), which in consequence may significantly change synaptic and physiological properties and animal behavior.[85] Some genetic mouse models may also target multiple neural circuits simultaneously, and thus strongly limit the selectivity of the optogenetic approach. For example, transgenic VGAT-ChR2 animals show restricted ChR2 expression in all neurons with active

VGAT promoter; however, this includes most inhibitory neurons in the brain. Locally restricting optical stimulation to specific brain areas or nuclei is possible. However, in addition to local cell bodies, this will usually also stimulate axon terminals and fibers *en passant* from unrelated neurons, and compromise the specificity of the optical perturbation. Recombinase-based methods or multivirus circuit/connectivity-based targeting allow for combinatorial refinement of transgene expression and dramatically improve targeting of subclasses of neurons that share anatomic (projection sites), activity, or genetic features (see later in text).

### Recombinase-Mediated Expression of Optogenetic Tools

DNA recombinase enzymes such as Cre or FLP recombinases detect specific flanking sequences that, depending on their orientation, allow for excision or rotation of the flanked DNA segment.[86] This mechanism can be used to activate/deactivate a gene depending on the presence of the matching recombining enzyme, which allows for recombinase-dependent viral vector delivery of opsins. The genetic specificity arises from linking recombinase expression to a cell-type-specific endogenous promoter, for example, in a viral vector or transgenic animal (see earlier in text). Importantly, the genomic copies necessary for recombinase function are very low, while sufficient genomic copies of recombinase-dependent opsin require a higher level of expression usually reached by viral transduction or Cre-reporter mice.[87] For example, ChR2 can be targeted to somatostatin-containing interneurons (SST) by crossing a Cre-dependent reporter mouse expressing ChR2 with SST-Cre mice to label mostly Martinotti cells.[87] This approach can be extended to mouse lines carrying the FLP recombinase, which recognizes sequence elements that are incompatible with the Cre recombinase. More complex "boolean logic" gates have been described, utilizing both Cre- and FLP-dependent expression cassettes that allow Cre-on, FLP-on, Cre-off, FLP-off, and all combinations of dual-recombinase logic.[88] Due to the long use of Cre-transgenic mouse lines, many of them have been well characterized with often specific expression patterns observed in particular brain areas. This continuously expanding repertoire of transgenic animals therefore provides a resource applicable to many scientific questions (see also www.jax.org).

### Anterograde Circuit Mapping

Neural circuit dissection is one of the most widely used applications of optogenetic techniques. Recombinase-enhanced expression of membrane-bound and fluorescently labelled ChR2 allows the investigator to visualize and optogenetically stimulate neuronal long-range projections by restricting the optical stimulus to the cell terminals, thus allowing one to target a subpopulation of neurons that is defined by their projection target (see later in text and Fig. 23.1B). Optogenetic activation combined with electrophysiological recording at neuronal projections allows functional anterograde circuit mapping.[89,90] This can highlight postsynaptic circuit components not only in acute brain slice preparations, but also in vivo, and it allows the probing of transmitter identities and their release dynamics.[91–95]

Neural circuit dissection can be refined by limiting opsin labelling to neurons with specific projection targets, thus eliminating possible activation of neurons projecting elsewhere. This can be achieved by using viral vectors such as HSV, type 2 canine adenovirus (CAV2), or the recently developed AAV2-retro virus, which produce robust transduction of neurons through their axon and presynaptic terminals.[96–100] When injected at anatomic projection sites, this enables projection-specific optogenetic modulation of neurons or even "tagging" of neurons through combined photostimulation and electrophysiological recording.[97,98,101,102] Reports exist that HSV and CAV2 may target distinct neuronal populations, and CAV2 may also transduce axons *en passant*.[103,104] Thus, complementary anatomic controls are necessary to rule out false-positive and -negative results in circuit-based expression experiments.

### Retrograde Circuit Mapping

Contrary to projection-based tagging of neurons, which is focused on studying the outputs of a circuit, retrograde monosynaptic circuit-based targeting allows to map brain-wide synaptic inputs to a defined neuron population or even a single neuron.[105–107] The approach uses the ability of rabies particles to retrogradely cross chemical synapses. A recombinant rabies virus backbone based on rabies SAD B19 strain was engineered to express fluorescent proteins (or opsins) and is devoid of its natural glycoprotein (SADΔG) required to build functional virus particles.[108] Thus, for successful transsynaptic spread, SADΔG rabies requires separate delivery of the G gene to host cells. Since nonhost cells are devoid of the G gene, ΔG rabies transsynaptic tracing is limited to a single synapse. To genetically define host-cell identity, rabies infection can be restricted to neurons expressing avian-specific TVA receptor by pseudotyping the SADΔG rabies with the envelope protein from avian sarcoma and leukosis virus. Rabies glycoprotein and TVA receptor genes can then be expressed using Cre-dependent AAVs for targeting a specific population as host cells.[92,109–111] Rabies-based circuit tracing thus provides a tool for efficient monosynaptic retrograde tracing from genetically identified neurons such as REM sleep-promoting melanin-concentrating-hormone (MCH) neurons or wake-promoting hypocretins/orexins neurons.[111] While onset of expression with this technique is fast, rabies transduction only leaves a 2-week window for experimentation, as it is toxic to neurons, making this system incompatible with long-term studies.[99]

### Activity-Based Tagging of Neuronal Ensembles

Advances in neuroimaging techniques have led to definitions of "neuronal ensembles" that are based on their spatiotemporal patterns of activity, which arise from synaptic and circuit connectivity rather than by genetic or anatomic properties.[16] Conventional optogenetic strategies use single-photon activation of opsins, which enables activation of large neuron populations but lacks ensemble-specific single-cell resolution. Novel two-photon activation strategies have overcome this limitation, but they are still technically challenging and incompatible with unrestrained behavior.[16] Genetic strategies have been developed to induce opsin expression only in neurons active during a given time span. Thus, perturbation of a specific neuronal ensemble can be achieved through genetically restricted opsin expression and activation as opposed to optically restricted activation through patterning of excitation light. Immediate early genes, such as c-fos or Arc/Arg3.1, are expressed in neurons that have recently been activated (minutes to hours) and have been exploited as cellular markers of behavior or stimulus-induced activity. Since the characterization of the genetic elements responsible for the activity-dependent transcription of immediate early genes, the same mechanisms were used to label neurons based on cellular activity.[112,113] Fast and inducible expression of optogenetic actuators is achieved via TRE-ttA system that can be linked to immediate early gene expression.[114] This multicomponent expression system consists of an AAV delivered expression cassette (containing the ChR2 gene) controlled by the tet-response element (TRE). TRE requires binding of the tetracycline transactivator (tTA) to induce transcription. C-fos-tTA transgenic mice allow for restricted expression of tTA in c-fos positive neurons.[115] Two versions of tTA exist. In the tet-OFF system, tTA binds to TRE only in the absence of doxycycline. This requires chronic doxycycline administration that is interrupted for the labelling experiment. The advantage of the tet-ON system is that doxycycline administration is only required during the labelling experiment. However, in this case, even in the absence of doxycycline, weak labelling/unspecific optogenetic targeting may occur. Such systems have been used successfully to trigger opsin expression in activated hippocampal neurons during aversive learning and during appetitive experience, which thus allows selective activation of memory engrams.[115,116] Similarly, a drug-inducible form of Cre recombinase ($Cre^{ERT2}$) can be expressed under the promoters of c-fos or Arc, to replace the functionality of TRE-ttA system.[117,118] Such Fos-TRAP and Arc-TRAP mice allow brain-wide "TRAPing" of active neurons with any Cre-dependent transgene.[118] Based on the Arc promoter, E-SARE viral vectors provide a similar tool for activity-dependent optogenetic targeting in nontransgenic animals, when only local labelling is required.[119] Thus, a range of tools has successfully allowed to tag activated neurons based on immediate early gene transcription. However, it remains crucial to characterize the time window of temporal integration for activity-dependent expression, transcription efficacy, and baseline expression levels for every target, expression system, and behavioral paradigm of interest.

## Technical Considerations

Optogenetic tools have found many applications from stimulating/silencing and imaging at subcellular resolution in vitro and in vivo levels. Each of these experiments requires light delivery at appropriate optical resolution and intensities to activate opsins, with minimal side effects on the preparation. In general, two strategies for illumination can be followed: (1) spatially restricted illumination, which can be achieved with single-cell and subcellular resolution in certain conditions, for example, by using two-photon illumination and holographic light patterning approaches,[48,120] (2) or broad illumination, which requires restriction of opsin expression to genetically identified cells or subcellular compartments. Many optogenetic stimulation/silencing experiments in vivo use implantable optic fibers that can be plugged to high-power light sources such as lasers and LEDs, and thus require tethering of animals to light fiber cables; however, wireless approaches have also been developed.[121,122] Optical light fiber implants result in physical damage along the fiber tract, which can alter neuronal morphology, and cause transient inflammation.[123] They further provide broad, but spatially limited illumination, with the power of light decaying fast with increasing distance from the tip of the fiber due to light scattering and adsorption properties of brain tissue.[124]

Therefore, careful assessment of light power needed for reliable opsin activation in a specific brain structure and volume is recommended. The combination of highly light-sensitive opsins, which produce large photocurrents (e.g., Jaws), can allow for noninvasive photostimulation while concurrently modulating the activity of a large population of cells.[66] Spatially and genetically restricted targeting strategies are then necessary to compensate for the lack of optical resolution in this approach (see Genetic Targeting Strategies for Opsin Expression section).

Importantly, after careful selection of light and opsin targeting strategies, proof of principle experiments need to be performed to test the viability, reliability, and selectivity of the opsin targeting strategy (e.g., double labelling experiments of fixed brain tissue) and to test the functionality of opsin expression for the expected effect on intrinsic activity, best through the chosen method for light delivery. Opsin functionality can be tested with complementary methods such as electrophysiological measurement of evoked or suppressed spiking activity in vitro or in vivo (see Fig. 23.2E–G). It is further recommended to pattern optogenetic stimulations to match intrinsic activity patterns to produce naturalistic activity. See Kim et al. for methods that complement optogenetic approaches and Allen et al. for a detailed assessment of possible side effects in optogenetic experiments.[52,125]

## OPTOGENETIC DISSECTION OF SLEEP CIRCUITS

Rhythms in the brain range from fast gamma oscillations to circadian and ultradian rhythms. Circadian rhythms shape sleep-wake rhythmicity to match 24-h days, while sleep behavior itself consists of cycles between different sleep phases. Circadian and sleep cycles together produce diurnal or nocturnal rhythms in a living organism. Sleep-wake cycles are produced by coordinated activation and inhibition of widely distributed neural

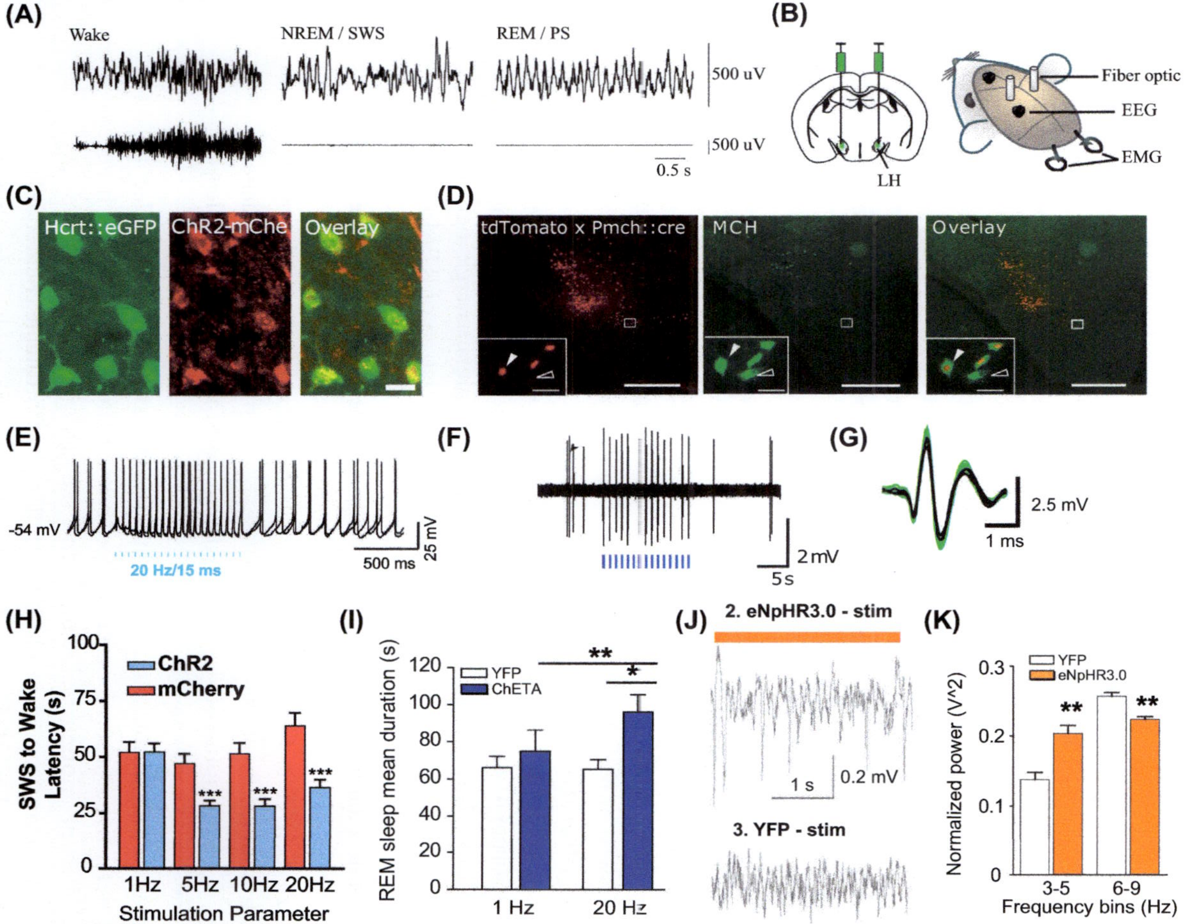


**Figure 23.2** Optogenetic manipulation of sleep states. (A) Representative examples of mouse EEG (top) and EMG (bottom) recordings of wake, NREM, and REM states. (B) Example configuration of EEG/EMG and optic fiber implants for bilateral stimulation. (C) Example micrographs of lateral hypothalamic Hcrt:eGFP cells (*green*) co-labeled with lentivirally delivered ChR2-mCherry (*red*) visualized through double immunohistochemistry. (D) Example micrographs of lateral hypothalamic MCH neurons (*green*), labelled with tdTomato through cross of Pmch:cre with respective reporter line. (E) Acute brain slice whole-cell patch-clamp recording of hcrt cell expressing ChR2. The cell synchronizes spikes to the light stimulation at 20 Hz. Overlay of two traces are shown. (F) In vivo recording of multiunit activity shows time-locked spikes to blue light stimulation. Time-locked events are overlaid in (G). (H) Latencies of SWS to wake transitions upon blue light stimulation of ChR2-mCherry or mCherry-expressing hcrt cells. ChR2-mCherry activation reduced the latency to wakefulness, when stimulating above 5 Hz. Control mCherry remained unchanged. (I) REM sleep duration of YFP- or ChETA-expressing MCH neurons. 20-Hz stimulation of ChR2 mice increased the REM sleep duration, while control animals remained unchanged. (J) Example EEG traces during optogenetic inhibition of MCH neurons (top) and control (bottom), applied during REM sleep. Inhibition of MCH cells increased slow frequency oscillations (3–5 Hz) and reduced theta range (6–9 Hz) oscillations, as quantified in (K).

networks in the brain, which to date have not been fully understood. These networks are strongly modulated by internal and external factors to ensure that behavioral state is appropriate to current hormonal, homeostatic, metabolic, circadian, and environmental needs.

Mammalian sleep-wake cycles consist of extended periods of wakefulness, followed by NREM (slow-wave) sleep, which can be followed by either wakefulness or less frequently by REM (paradoxical) sleep. REM sleep is usually followed by wakefulness. These states are classically discriminated by electroencephalogram (EEG) and electromyogram (EMG) recordings (Fig. 23.2A), by their characteristic oscillatory components. These global network oscillations arise from the temporal and spatial summation of electric fields generated by synchronized neuronal activity. Initiation, maintenance, and termination of sleep-wake states may require highly specialized neural circuits to produce the network oscillations characteristic for each state. However, these states are not discrete, as arousal and consciousness levels as well as oscillations are variable within individual states or sleep epochs.[126]

Since the epidemic of encephalitis lethargica, in which affected patients slept most of the day, was associated with a lesion in the midbrain and posterior hypothalamus, it was postulated that these regions contain key neurons for the production of arousal and wakefulness.[127] Since then, much work has been done to disentangle the different circuits involved (as reviewed in Ref. 126). Some of the most intensively studied neurons are brainstem and hypothalamic monoaminergic neurons, basal forebrain cholinergic neurons, and orexin/hypocretin neurons, which mainly promote wakefulness, while neurons in the ventrolateral preoptic area (VLPO) and median preoptic nucleus (MnPO) mainly promote NREM sleep.[126] On the other hand, glutamatergic neurons in the sublaterodorsal nucleus of the pons are essential for the generation of REM sleep.

Here, we will focus on a few studies to highlight how optogenetics has been (and can be) used to understand sleep circuits and function. Optogenetic sleep studies have helped to build causal relationships between neuronal identities, their activity, and their effects on sleep-wake behavior, for example, through optogenetic manipulation of sleep-wake state transitions,[12] sleep-state maintenance,[128] sleep oscillations,[129] sleep-dependent learning,[129] sleep homeostasis,[130] or synaptic transmission between sleep-wake neurons,[94] as reviewed subsequently.

The hypothalamus contains many distinct cell populations that are involved in sleep-wake regulation. Lateral hypothalamic orexins/hypocretins neurons ($LH^{Hcrt/Ox}$) have long been implicated in promoting wakefulness and stabilizing sleep-wake states, as the loss of a functional orexin system produced narcolepsy with cataplexy in rodents, dogs, and humans (reviewed in Ref. 131). In the first study to apply optogenetics in vivo, Adamantidis et al. targeted ChR2-mCherry to $LH^{Hcrt/Ox}$ neurons using a lentivirus (Fig. 23.2B).[12] Importantly, the authors confirmed that ChR2-mCherry expression was restricted to $LH^{Hcrt/Ox}$ neurons by using Hcrt:eGFP mice that express eGFP under the prepro-orexin promoter (Fig. 23.2C). The functionality of opsin stimulation was tested using whole-cell patch-clamp recordings in vitro, which allows the measurement of jitter, delay, and probability of time-locked action potentials (Fig. 23.2E). By optical stimulation of $LH^{Hcrt/Ox}$ neurons at frequencies between 5 and 30 Hz during NREM and REM sleep the authors induced a reduction of the latency of transition to wakefulness in ChR2-mCherry animals as compared to control animals, expressing only the fluorophore mCherry (Fig. 23.2H).[12] This optically induced reduction in sleep-to-wake latency was further blocked by a single dose of an Hcrt receptor 1 antagonist, suggesting that indeed transmitters released by $LH^{Hcrt/Ox}$ cells and its respective receptors are responsible for the wake-promoting effect. In contrast, 1-Hz light pulses did not change sleep-wake latencies significantly (Fig. 23.2H). This study established, for the first time, a causal link between frequency-dependent electrical activity of $LH^{Hcrt/Ox}$ neurons and induction of wakefulness from sleep. The $LH^{Hcrt/Ox}$ wake-promoting effect was independent of circadian time, but it was suppressed by high sleep pressure induced through several hours of sleep-deprivation.[130] This highlights that sleep homeostasis critically influences sleep-wake actuators and that both are ultimately intermingled, while circadian rhythms possibly perform a suppressive/permissive role for certain sleep-wake events. Poststimulation immunohistochemistry for c-fos allowed investigators to assess the identities of activated neurons downstream of $LH^{Hcrt/Ox}$ cells, and it suggested that tuberomammillary histamine neurons and locus coeruleus norepinephrine neurons (LC-NE) could mediate the arousal promoting effect of $LH^{Hcrt/Ox}$ neurons.[163] Since optogenetic inhibition of norepinephrine cells during the activation of LH-OH neurons reduced the arousal promoting effects of $LH^{Hcrt/Ox}$ stimulation, it was confirmed that LC-NE mediates part of this arousal effect.[163] These early findings demonstrated the reliability of optogenetics for functional dissection of sleep-wake circuits to understand the dynamics of sleep-to-wake transitions and arousal systems in general, while opening a new era of optogenetic sleep research for systematic manipulation of specific features of sleep in the mammalian brain.

Intermingled with $LH^{Hcrt/Ox}$ neurons, neurons containing MCH have been implicated in REM sleep behavior.[132,133] For example, correlative evidence stems from functional neuroanatomic experiments that showed increased fos labelling in MCH neurons during REM sleep rebound, in line with in vivo juxtacellular recordings indicating that MCH neurons maximally discharge during REM sleep and icv injections of MCH peptides, which induced REM and to a lesser extent NREM sleep.[132,133] To establish a causal relationship between temporally precise MCH activity and REM sleep, Jego et al. used acute optogenetic activation of MCH neurons (ChETA, SSFO) at the onset of REM sleep.[128] This extended the duration of REM, but not non-REM sleep episodes, while acute stimulation of MCH neurons during NREM sleep facilitated the transitions toward REM sleep.[128,134] These findings indicate a role for MCH neurons in initiation and maintenance of REM sleep. In line with this, in vivo activation of

MCH neuron terminals in TMN or medial septum also extended the duration of REM sleep episodes and chronic optogenetic activation over hours of MCH cells increase sleep.[128,135] In contrast, their acute silencing (eNpHR3.0, ArchT) reduced the frequency and amplitude of hippocampal theta rhythm, without affecting REM sleep duration, while their ablation did not affect REM sleep, but increased wakefulness with reduced NREM.[128,134] Thus, MCH neuron activation is sufficient for REM sleep control; however, it is not necessary.

Similar to optogenetic inhibition of MCH cells, optogenetic inhibition of downstream medial septal GABAergic (MS-GABA) neurons also suppressed hippocampal theta rhythm (Fig. 23.3B,C).[129] For optical targeting of MS-GABA neurons, VGAT Cre mice were injected with AAV containing floxed ArchT-eYFP gene into the medial septum and implanted with an optical fiber for light delivery at 594 nm (Fig. 23.3A). To assess sleep-wake states and hippocampal theta, mice were also instrumented with EEG/EMG electrodes and LFP electrodes placed in CA1. This allowed to identify REM states for suppression (and measurement) of theta rhythm time-locked to the occurrence of REM (Fig. 23.3B and C). The authors then used this as a tool to probe behavior in animals that had normal amount of REM sleep but without theta rhythm, and assessed the impact on learning and memory consolidation (Fig. 23.3E). Specifically, the animals were trained in contextual memory tasks (novel object recognition and fear conditioning), after which the animals were allowed to sleep while theta was suppressed during REM during the 4 h following the training session. During the memory retrieval session 24 h later, MS-GABA ArchT-eYFP animals were less able to retrieve contextual memory specifically (Fig. 23.3F—H), while other memories remained intact. Moreover, suppression of MS-GABA activity outside of REM sleep or light stimulation of control animals had no effect on memory consolidation, highlighting the specific role of REM sleep theta rhythms for contextual memory consolidation. These results highlight how optogenetics can be used to modulate brain rhythms directly through control of key neurons and to help us understand their role in cognitive functions such as learning.

Many of the neurons that have been implicated in the control of sleep-wake behavior are monoaminergic or peptidergic neurons. These markers allow the genetic identification of such neurons; however, this does not usually represent the only transmitter expressed in these neurons.[136] For example, in vitro activation of MCH neuron terminals induced $GABA_A$-mediated inhibitory postsynaptic currents in wake-promoting histaminergic neurons of the tuberomammillary nucleus (TMN-HA), while the role of endogenous MCH peptides for synaptic transmission remains unclear.[128] Similarly, it was unclear what the synaptic influence of co-expressed hypocretin/orexin peptides and glutamate is in LH-OH neurons. To address this, Schöne et al. used optogenetics to study the release of hcrt peptide and glutamate from hcrt neurons by recording monosynaptically connected histamine cells in an in vitro preparation.[94] The authors identified that LH-OH neuron optical stimulation evoked hypocretin receptor 2—dependent slow

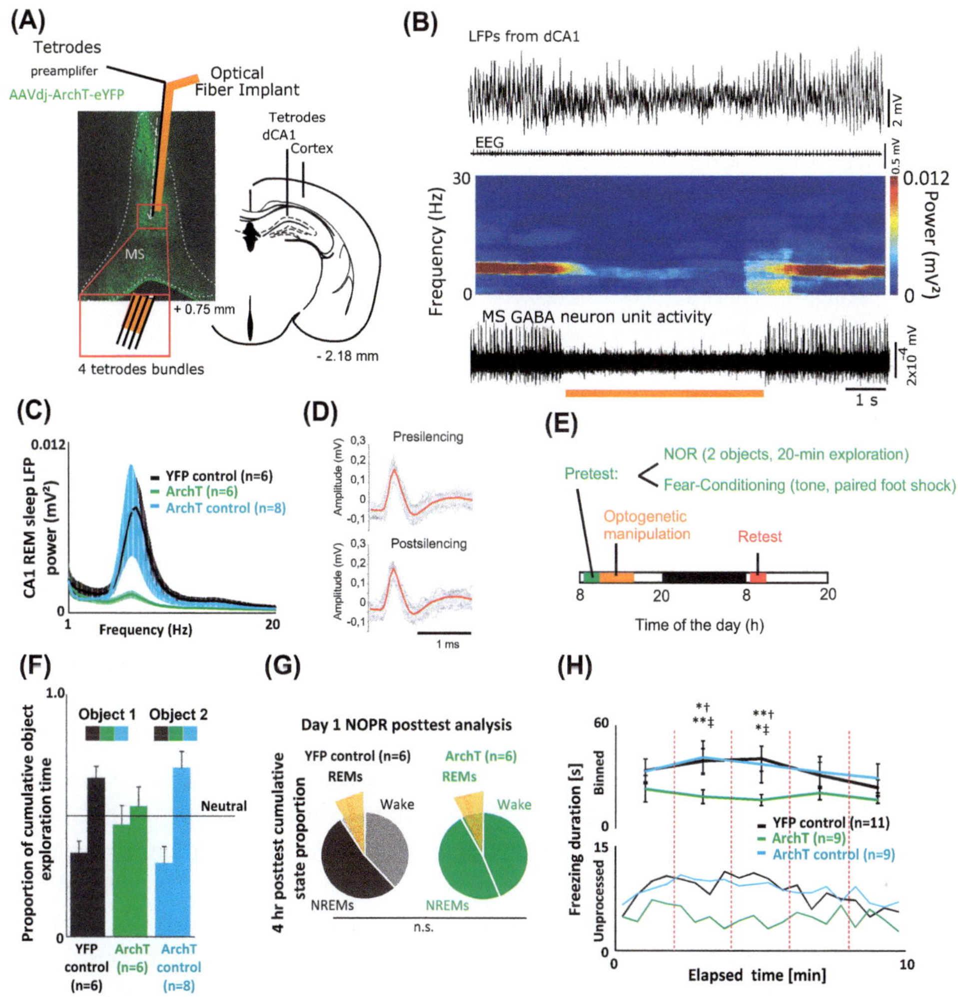


**Figure 23.3** Hippocampal theta rhythm is required for contextual memory consolidation. (A): Illustration AAV delivered Cre-dependent ArchT-eYFP expression in medial septum of VGAT:cre mice, with respective location of fiberoptic and tetrode implants. Tetrode implantation in dCA1 allowed for direct measurement of hippocampal theta oscillations. (B) Optogenetic silencing of MSGABA neurons (bottom trace) suppresses REM sleep–typical theta oscillations in dCA1, as shown in example LFP trace (top) and frequency decomposition (middle). (C) Averaged power spectra of ArchT and YFP control animals stimulated during REM, as well as ArchT controls, which were left unstimulated. (D) Action potential spike waveforms remain unchanged by ArchT-mediated optical silencing. (E) Optogenetic perturbation of contextual memory formation. Mice were trained in novel object recognition or fear-conditioning tasks, followed by 4 h of optogenetic sleep manipulation. Mice were then retested on the memory tasks 1 day after. (F) Control mice spend a substantially bigger proportion of time exploring the novel object 2, compared to the previously presented object 1 24 h after unperturbed sleep. In contrast, inhibition of MSGABA neurons (ArchT group), caused animals to spend equal amounts on object 1 and 2, indicating both appear equally novel. (G) Importantly, overall sleep time spent in each sleep state over the 4-h period following the training session was not different between ArchT and control groups. (H) Contextual fear memory caused increased freezing behavior in control animals after unperturbed sleep. However, animals undergoing MSGABA inhibition did not show such increase, indicating that the fear context was not learned.

postsynaptic currents and glutamatergic AMPA receptor–mediated fast postsynaptic currents. Blocking glutamate-driven spiking did not affect LH-OH-driven spiking and vice versa, suggesting isolation of outputs. Importantly, glutamate-driven spiking accounted for fast changes in histamine firing, while peptide-mediated transmission produced slow-onset, long-lasting increase in postsynaptic firing. Therefore, co-transmission enabled fast and long-lasting transmission in LH-OH to TMN-HA circuits, which may be translated to other sleep-wake circuits. In general, this may represent parallel, nonredundant control signals through downstream effectors.[94]

## PERSPECTIVES

In summary, a diversity of optogenetic tools, opsin, and light targeting strategies have enabled detailed dissection of the circuits involved in sleep-wake state transitions, sleep-state maintenance, sleep oscillations, and sleep-dependent learning, sleep homeostasis, or synaptic transmission between sleep-wake neurons; some of these studies are reviewed earlier in the text.[12,94,128–130,137–142] A detailed review of optogenetic sleep studies can be found in Shiromani et al.[143] Closed-loop optogenetic experiments have been used for few neuroscience research questions so far,[144–146] but they would provide powerful tools for dissecting the role and function of specific sleep events such as spindles and slow waves. This enables the manipulation of specific sleep events or behaviors as they occur, if fast enough algorithms for online detection are developed. In humans, online detection of sleep-slow waves has been used to trigger auditory stimulation, transcranial direct-current stimulation (tDCS), or transcranial magnetic stimulation at different phases of the oscillation, which enabled targeted enhancement or reduction of sleep-slow wave rhythms and phase-coupled spindle activity.[147,148] Further, enhancement of sleep-slow waves improved consolidation of declarative memory.[147] In the future, similar approaches would enable circuit-specific closed-loop optogenetic manipulations of sleep events in animal models.[149]

## REFERENCES

1. Nagel G, Möckel B, Büldt G, Bamberg E. Functional expression of bacteriorhodopsin in oocytes allows direct measurement of voltage dependence of light induced $H^+$ pumping. *FEBS Lett.* 1995; 377(2):263–266. http://www.ncbi.nlm.nih.gov/pubmed/8543064.
2. Tsien RY, Miyawaki A, Llopis J, et al. Fluorescent indicators for $Ca^{2+}$ based on green fluorescent proteins and calmodulin. *Nature.* 1997;388(6645):882–887. https://doi.org/10.1038/42264.
3. Zemelman BV, Lee GA, Ng M, Miesenböck G. Selective photostimulation of genetically chARGed neurons. *Neuron.* 2002;33(1):15–22. http://www.ncbi.nlm.nih.gov/pubmed/11779476.
4. Nagel G, Ollig D, Fuhrmann M, et al. Channelrhodopsin-1: a light-gated proton channel in green algae. *Science.* 2002;296(5577):2395–2398. https://doi.org/10.1126/science.1072068.
5. Nagel G, Szellas T, Huhn W, et al. Channelrhodopsin-2, a directly light-gated cation-selective membrane channel. *Proc Natl Acad Sci USA.* 2003;100(24):13940–13945. https://doi.org/10.1073/pnas.1936192100.

6. Boyden ES, Zhang F, Bamberg E, Nagel G, Deisseroth K. Millisecond-timescale, genetically targeted optical control of neural activity. *Nat Neurosci*. 2005;8(9):1263–1268. https://doi.org/10.1038/nn1525.
7. Bellmann D, Richardt A, Freyberger R, et al. Optogenetically induced olfactory stimulation in Drosophila larvae reveals the neuronal basis of odor-aversion behavior. *Front Behav Neurosci*. 2010;4:27. https://doi.org/10.3389/fnbeh.2010.00027.
8. Gradinaru V, Mogri M, Thompson KR, Henderson JM, Deisseroth K. Optical deconstruction of parkinsonian neural circuitry. *Science*. 2009;324(5925):354–359. https://doi.org/10.1126/science.1167093.
9. Zhang F, Wang L-P, Brauner M, et al. Multimodal fast optical interrogation of neural circuitry. *Nature*. 2007;446(7136):633–639. https://doi.org/10.1038/nature05744.
10. Nagel G, Brauner M, Liewald JF, Adeishvili N, Bamberg E, Gottschalk A. Light activation of channelrhodopsin-2 in excitable cells of *Caenorhabditis elegans* triggers rapid behavioral responses. *Curr Biol*. 2005;15(24):2279–2284. https://doi.org/10.1016/j.cub.2005.11.032.
11. Han X, Chow BY, Zhou H, et al. A high-light sensitivity optical neural silencer: development and application to optogenetic control of non-human primate cortex. *Front Syst Neurosci*. 2011;5:18. https://doi.org/10.3389/fnsys.2011.00018.
12. Adamantidis AR, Zhang F, Aravanis AM, Deisseroth K, de Lecea L. Neural substrates of awakening probed with optogenetic control of hypocretin neurons. *Nature*. 2007;450(7168):420–424. https://doi.org/10.1038/nature06310.
13. Catterall WA. Voltage-gated calcium channels. *Cold Spring Harb Perspect Biol*. 2011;3(8):a003947. https://doi.org/10.1101/cshperspect.a003947.
14. Smetters D, Majewska A, Yuste R. Detecting action potentials in neuronal populations with calcium imaging. *Methods*. 1999;18(2):215–221. https://doi.org/10.1006/meth.1999.0774.
15. Yasuda R, Nimchinsky EA, Scheuss V, et al. Imaging calcium concentration dynamics in small neuronal compartments. *Sci Signal*. 2004;2004(219):pl5. https://doi.org/10.1126/stke.2192004pl5.
16. Carrillo-Reid L, Yang W, Miller JK, et al. Imaging and optically manipulating neuronal ensembles. *Annu Rev Biophys*. 2017;46(March). https://doi.org/10.1146/annurev-biophys-070816-033647.
17. Yuste R, Katz LC. Control of postsynaptic $Ca^{2+}$ influx in developing neocortex by excitatory and inhibitory neurotransmitters. *Neuron*. 1991;6(3):333–344. http://www.ncbi.nlm.nih.gov/pubmed/1672071.
18. Grynkiewicz G, Poenie M, Tsien RY. A new generation of $Ca^{2+}$ indicators with greatly improved fluorescence properties. *J Biol Chem*. 1985;260(6):3440–3450. http://www.ncbi.nlm.nih.gov/pubmed/3838314.
19. Cobbold PH, Rink TJ. Fluorescence and bioluminescence measurement of cytoplasmic free calcium. *Biochem J*. 1987;248(2):313–328. http://www.ncbi.nlm.nih.gov/pubmed/3325037.
20. Tian L, Hires SA, Mao T, et al. Imaging neural activity in worms, flies and mice with improved GCaMP calcium indicators. *Nat Methods*. 2009;6(12):875–881. https://doi.org/10.1038/nmeth.1398.
21. Ohkura M, Matsuzaki M, Kasai H, Imoto K, Nakai J. Genetically encoded bright $Ca^{2+}$ probe applicable for dynamic $Ca^{2+}$ imaging of dendritic spines. *Anal Chem*. 2005;77(18):5861–5869. https://doi.org/10.1021/ac0506837.
22. Tallini YN, Ohkura M, Choi B-R, et al. Imaging cellular signals in the heart in vivo: cardiac expression of the high-signal $Ca^{2+}$ indicator GCaMP2. *Proc Natl Acad Sci USA*. 2006;103(12):4753–4758. https://doi.org/10.1073/pnas.0509378103.
23. Akerboom J, Rivera JDV, Guilbe MMR, et al. Crystal structures of the GCaMP calcium sensor reveal the mechanism of fluorescence signal change and aid rational design. *J Biol Chem*. 2009;284(10):6455–6464. https://doi.org/10.1074/jbc.M807657200.
24. Akerboom J, Chen T-W, Wardill TJ, et al. Optimization of a GCaMP calcium indicator for neural activity imaging. *J Neurosci*. 2012;32(40):13819–13840. https://doi.org/10.1523/JNEUROSCI.2601-12.2012.
25. Chen T-W, Wardill TJ, Sun Y, et al. Ultrasensitive fluorescent proteins for imaging neuronal activity. *Nature*. 2013;499(7458):295–300. https://doi.org/10.1038/nature12354.

26. Oheim M, van't Hoff M, Feltz A, Zamaleeva A, Mallet J-M, Collot M. New red-fluorescent calcium indicators for optogenetics, photoactivation and multi-color imaging. *Biochim Biophys Acta*. March 2014. https://doi.org/10.1016/j.bbamcr.2014.03.010.
27. Akerboom J, Carreras Calderón N, Tian L, et al. Genetically encoded calcium indicators for multi-color neural activity imaging and combination with optogenetics. *Front Mol Neurosci*. 2013; 6(March):2. https://doi.org/10.3389/fnmol.2013.00002.
28. Zhao Y, Araki S, Wu J, et al. An expanded palette of genetically encoded $Ca^{2+}$ indicators. *Science*. 2011;333(6051):1888−1891. https://doi.org/10.1126/science.1208592.
29. Wu J, Abdelfattah AS, Miraucourt LS, et al. A long Stokes shift red fluorescent $Ca^{2+}$ indicator protein for two-photon and ratiometric imaging. *Nat Commun*. 2014;5:5262. https://doi.org/10.1038/ncomms6262.
30. Inoue M, Takeuchi A, Horigane S, et al. Rational design of a high-affinity, fast, red calcium indicator R-CaMP2. *Nat Methods*. 2015;12(1):64−70. https://doi.org/10.1038/nmeth.3185.
31. Shaner NC, Campbell RE, Steinbach PA, Giepmans BNG, Palmer AE, Tsien RY. Improved monomeric red, orange and yellow fluorescent proteins derived from Discosoma sp. red fluorescent protein. *Nat Biotechnol*. 2004;22(12):1567−1572. https://doi.org/10.1038/nbt1037.
32. Dana H, Mohar B, Sun Y, et al. Sensitive red protein calcium indicators for imaging neural activity. *Elife*. 2016;5:e12727. https://doi.org/10.7554/eLife.12727.
33. Hoi H, Matsuda T, Nagai T, Campbell RE. Highlightable $Ca^{2+}$ indicators for live cell imaging. *J Am Chem Soc*. 2013;135(1):46−49. https://doi.org/10.1021/ja310184a.
34. Berlin S, Carroll EC, Newman ZL, et al. Photoactivatable genetically encoded calcium indicators for targeted neuronal imaging. *Nat Methods*. 2015;12(9):852−858. https://doi.org/10.1038/nmeth.3480.
35. Ai M, Mills H, Kanai M, et al. Green-to-red photoconversion of GCaMPHoffman RM, ed. *PLoS One*. 2015;10:1−11(9). https://doi.org/10.1371/journal.pone.0138127.
36. Fosque BF, Sun Y, Dana H, et al. Neural circuits. Labeling of active neural circuits in vivo with designed calcium integrators. *Science*. 2015;347(6223):755−760. https://doi.org/10.1126/science.1260922.
37. Zolnik TA, Sha F, Johenning FW, et al. All-optical functional synaptic connectivity mapping in acute brain slices using the calcium integrator CaMPARI. *J Physiol*. 2017;595(5):1465−1477. https://doi.org/10.1113/JP273116.
38. Siegel MS, Isacoff EY. A genetically encoded optical probe of membrane voltage. *Neuron*. 1997;19(4): 735−741. https://doi.org/10.1016/S0896-6273(00)80955-1.
39. Gong Y, Huang C, Li JZ, et al. High-speed recording of neural spikes in awake mice and flies with a fluorescent voltage sensor. *Science*. 2015;350(6266):1361−1366. https://doi.org/10.1126/science.aab0810.
40. Hochbaum DR, Zhao Y, Farhi SL, et al. All-optical electrophysiology in mammalian neurons using engineered microbial rhodopsins. *Nat Methods*. 2014;11(8):825−833. https://doi.org/10.1038/nmeth.3000.
41. Lou S, Adam Y, Weinstein EN, et al. Genetically targeted all-optical electrophysiology with a transgenic cre-dependent optopatch mouse. *J Neurosci*. 2016;36(43):11059−11073. https://doi.org/10.1523/JNEUROSCI.1582-16.2016.
42. Ghosh KK, Burns LD, Cocker ED, et al. Miniaturized integration of a fluorescence microscope. *Nat Methods*. 2011;8(10):871−878. https://doi.org/10.1038/nmeth.1694.
43. Ziv Y, Ghosh KK. Miniature microscopes for large-scale imaging of neuronal activity in freely behaving rodents. *Curr Opin Neurobiol*. 2015;32:141−147. https://doi.org/10.1016/j.conb.2015.04.001.
44. Cai DJ, Aharoni D, Shuman T, et al. A shared neural ensemble links distinct contextual memories encoded close in time. *Nature*. 2016;534(7605):115−118. https://doi.org/10.1038/nature17955.
45. Resendez SL, Jennings JH, Ung RL, et al. Visualization of cortical, subcortical and deep brain neural circuit dynamics during naturalistic mammalian behavior with head-mounted microscopes and chronically implanted lenses. *Nat Protoc*. 2016;11(3):566−597. https://doi.org/10.1038/nprot.2016.021.
46. Cox J, Pinto L, Dan Y, Reid RC, McCarley RW. Calcium imaging of sleep−wake related neuronal activity in the dorsal pons. *Nat Commun*. 2016;7:10763. https://doi.org/10.1038/ncomms10763.

47. Eban-Rothschild A, Rothschild G, Giardino WJ, Jones JR, de Lecea L. VTA dopaminergic neurons regulate ethologically relevant sleep-wake behaviors. *Nat Neurosci*. 2016;19(10):1356–1366. https://doi.org/10.1038/nn.4377.
48. Yang W, Yuste R. In vivo imaging of neural activity. *Nat Methods*. 2017;14(4):349–359. https://doi.org/10.1038/nmeth.4230.
49. Rodriguez EA, Campbell RE, Lin JY, et al. The growing and glowing toolbox of fluorescent and photoactive proteins. *Trends Biochem Sci*. 2017;42(2):111–129. https://doi.org/10.1016/j.tibs.2016.09.010.
50. Gunaydin LA, Yizhar O, Berndt A, Sohal VS, Deisseroth K, Hegemann P. Ultrafast optogenetic control. *Nat Neurosci*. 2010;13(3):387–392. https://doi.org/10.1038/nn.2495.
51. Klapoetke NC, Murata Y, Kim SS, et al. Independent optical excitation of distinct neural populations. *Nat Methods*. 2014;11(3):338–346. https://doi.org/10.1038/nmeth.2836.
52. Allen BD, Singer AC, Boyden ES. Principles of designing interpretable optogenetic behavior experiments. *Learn Mem*. 2015;22(4):232–238. https://doi.org/10.1101/lm.038026.114.
53. Berndt A, Schoenenberger P, Mattis J, et al. High-efficiency channelrhodopsins for fast neuronal stimulation at low light levels. *Proc Natl Acad Sci USA*. 2011;108(18):7595–7600. https://doi.org/10.1073/pnas.1017210108.
54. Berndt A, Yizhar O, Gunaydin LA, Hegemann P, Deisseroth K. Bi-stable neural state switches. *Nat Neurosci*. 2009;12(2):229–234. https://doi.org/10.1038/nn.2247.
55. Yizhar O, Fenno LE, Prigge M, et al. Neocortical excitation/inhibition balance in information processing and social dysfunction. *Nature*. 2011;477(7363):171–178. https://doi.org/10.1038/nature10360.
56. Christie IN, Wells JA, Southern P, et al. fMRI response to blue light delivery in the naïve brain: implications for combined optogenetic fMRI studies. *Neuroimage*. 2013;66. https://doi.org/10.1016/j.neuroimage.2012.10.074.
57. Hodgkin AL, Katz B. The effect of temperature on the electrical activity of the giant axon of the squid. *J Physiol*. 1949;109(1–2):240–249. http://www.ncbi.nlm.nih.gov/pubmed/15394322.
58. Moser E, Mathiesen I, Andersen P. Association between brain temperature and dentate field potentials in exploring and swimming rats. *Science*. 1993;259(5099):1324–1326. http://www.ncbi.nlm.nih.gov/pubmed/8446900.
59. Zhang F, Prigge M, Beyrière F, et al. Red-shifted optogenetic excitation: a tool for fast neural control derived from *Volvox carteri*. *Nat Neurosci*. 2008;11(6):631–633. https://doi.org/10.1038/nn.2120.
60. Lin JY, Knutsen PM, Muller A, Kleinfeld D, Tsien RY. ReaChR: a red-shifted variant of channelrhodopsin enables deep transcranial optogenetic excitation. *Nat Neurosci*. 2013;16(10). https://doi.org/10.1038/nn.3502.
61. Chow BY, Han X, Dobry AS, et al. High-performance genetically targetable optical neural silencing by light-driven proton pumps. *Nature*. 2010;463(7277):98–102. https://doi.org/10.1038/nature08652.
62. Gradinaru V, Thompson KR, Deisseroth K. eNpHR: a *Natronomonas* halorhodopsin enhanced for optogenetic applications. *Brain Cell Biol*. 2008;36(1–4):129–139. https://doi.org/10.1007/s11068-008-9027-6.
63. Mattis J, Tye KM, Ferenczi EA, et al. Principles for applying optogenetic tools derived from direct comparative analysis of microbial opsins. *Nat Methods*. 2012;9(2):159–172. https://doi.org/10.1038/nmeth.1808.
64. Raimondo JV, Kay L, Ellender TJ, Akerman CJ. Optogenetic silencing strategies differ in their effects on inhibitory synaptic transmission. *Nat Neurosci*. 2012;15(8):1102–1104. https://doi.org/10.1038/nn.3143.
65. Berndt A, Lee SY, Ramakrishnan C, Deisseroth K. Structure-guided transformation of channelrhodopsin into a light-activated chloride channel. *Science*. 2014;344(6182):420–424. https://doi.org/10.1126/science.1252367.
66. Chuong AS, Miri ML, Busskamp V, et al. Noninvasive optical inhibition with a red-shifted microbial rhodopsin. *Nat Neurosci*. 2014;17(8):1123–1129. https://doi.org/10.1038/nn.3752.

67. Mahn M, Prigge M, Ron S, Levy R, Yizhar O. Biophysical constraints of optogenetic inhibition at presynaptic terminals. *Nat Neurosci.* 2016;19(4):554–556. https://doi.org/10.1038/nn.4266.
68. Govorunova EG, Sineshchekov OA, Janz R, Liu X, Spudich JL. Natural light-gated anion channels: a family of microbial rhodopsins for advanced optogenetics. *Science.* 2015;349(6248):647–650. https://doi.org/10.1126/science.aaa7484.
69. Hoque MR, Ishizuka T, Inoue K, et al. A Chimera $Na^+$-Pump rhodopsin as an effective optogenetic silencerBarnes S, ed. *PLoS One.* 2016;11(11):e0166820. https://doi.org/10.1371/journal.pone.0166820.
70. Kato HE, Inoue K, Abe-Yoshizumi R, et al. Structural basis for $Na^+$ transport mechanism by a light-driven $Na^+$ pump. *Nature.* 2015;521(7550):48–53. https://doi.org/10.1038/nature14322.
71. Cosentino C, Alberio L, Gazzarrini S, et al. Engineering of a light-gated potassium channel. *Science.* 2015;348(6235):707–710. https://doi.org/10.1126/science.aaa2787.
72. Lemon B, Tjian R. Orchestrated response: a symphony of transcription factors for gene control. *Gene Dev.* 2000;14(20):2551–2569. https://doi.org/10.1101/GAD.831000.
73. Kantor B, Bailey RM, Wimberly K, Kalburgi SN, Gray SJ. Methods for gene transfer to the central nervous system. *Adv Genet.* 2014;87:125–197. https://doi.org/10.1016/B978-0-12-800149-3.00003-2.
74. Verma IM, Weitzman MD. Gene therapy: twenty-first century medicine. *Annu Rev Biochem.* 2005; 74(1):711–738. https://doi.org/10.1146/annurev.biochem.74.050304.091637.
75. Naldini L, Blömer U, Gallay P, et al. In vivo gene delivery and stable transduction of nondividing cells by a lentiviral vector. *Science.* 1996;272(5259):263–267. http://www.ncbi.nlm.nih.gov/pubmed/8602510.
76. Wu Z, Yang H, Colosi P. Effect of genome size on AAV vector packaging. *Mol Ther.* 2010;18(1): 80–86. https://doi.org/10.1038/mt.2009.255.
77. Dittgen T, Nimmerjahn A, Komai S, et al. Lentivirus-based genetic manipulations of cortical neurons and their optical and electrophysiological monitoring in vivo. *Proc Natl Acad Sci USA.* 2004;101(52): 18206–18211. https://doi.org/10.1073/pnas.0407976101.
78. Nathanson JL, Yanagawa Y, Obata K, Callaway EM. Preferential labeling of inhibitory and excitatory cortical neurons by endogenous tropism of adeno-associated virus and lentivirus vectors. *Neuroscience.* 2009;161(2):441–450. https://doi.org/10.1016/j.neuroscience.2009.03.032.
79. Rahim AA, Wong AMS, Howe SJ, et al. Efficient gene delivery to the adult and fetal CNS using pseudotyped non-integrating lentiviral vectors. *Gene Ther.* 2009;16(4):509–520. https://doi.org/10.1038/gt.2008.186.
80. Jasnow AM, Rainnie DG, Maguschak KA, Chhatwal JP, Ressler KJ. Construction of cell-type specific promoter lentiviruses for optically guiding electrophysiological recordings and for targeted gene delivery. *Methods Mol Biol (Clifton, NJ).* 2009;515:199–213. https://doi.org/10.1007/978-1-59745-559-6_13.
81. Arenkiel BR, Peca J, Davison IG, et al. In vivo light-induced activation of neural circuitry in transgenic mice expressing Channelrhodopsin-2. *Neuron.* 2007;54(2):205–218. https://doi.org/10.1016/j.neuron.2007.03.005.
82. Zhao S, Ting JT, Atallah HE, et al. Cell type–specific channelrhodopsin-2 transgenic mice for optogenetic dissection of neural circuitry function. *Nat Methods.* 2011;8(9):745–752. http://www.ncbi.nlm.nih.gov/pubmed/21985008.
83. Smear M, Resulaj A, Zhang J, Bozza T, Rinberg D. Multiple perceptible signals from a single olfactory glomerulus. *Nat Neurosci.* 2013;16(11):1687–1691. https://doi.org/10.1038/nn.3519.
84. Jackman SL, Beneduce BM, Drew IR, Regehr WG. Achieving high-frequency optical control of synaptic transmission. *J Neurosci.* 2014;34(22):7704–7714. https://doi.org/10.1523/JNEUROSCI.4694-13.2014.
85. Kolisnyk B, Guzman MS, Raulic S, et al. ChAT-ChR2-EYFP mice have enhanced motor endurance but show deficits in attention and several additional cognitive domains. *J Neurosci.* 2013;33(25): 10427–10438. https://doi.org/10.1523/JNEUROSCI.0395-13.2013.
86. Taniguchi H, He M, Wu P, et al. A resource of cre driver lines for genetic targeting of GABAergic neurons in cerebral cortex. *Neuron.* 2011;71(6):995–1013. https://doi.org/10.1016/j.neuron.2011.07.026.

87. Madisen L, Mao T, Koch H, et al. A toolbox of Cre-dependent optogenetic transgenic mice for light-induced activation and silencing. *Nat Neurosci*. 2012;15(5):793–802. https://doi.org/10.1038/nn.3078.
88. Fenno LE, Mattis J, Ramakrishnan C, et al. Targeting cells with single vectors using multiple-feature Boolean logic. *Nat Methods*. 2014;11(7):763–772. https://doi.org/10.1038/nmeth.2996.
89. Petreanu L, Huber D, Sobczyk A, Svoboda K. Channelrhodopsin-2-assisted circuit mapping of long-range callosal projections. *Nat Neurosci*. 2007;10(5):663–668. https://doi.org/10.1038/nn1891.
90. Arenkiel BR, Klein ME, Davison IG, Katz LC, Ehlers MD. Genetic control of neuronal activity in mice conditionally expressing TRPV1. *Nat Methods*. 2008;5(4):299–302. https://doi.org/10.1038/nmeth.1190.
91. Tye KM, Prakash R, Kim S-Y, et al. Amygdala circuitry mediating reversible and bidirectional control of anxiety. *Nature*. 2011;471(7338):358–362. https://doi.org/10.1038/nature09820.
92. Lammel S, Lim BK, Ran C, et al. Input-specific control of reward and aversion in the ventral tegmental area. *Nature*. 2012;491(7423). https://doi.org/10.1038/nature11527.
93. Schone C, Cao ZFH, Apergis-Schoute J, Adamantidis A, Sakurai T, Burdakov D. Optogenetic probing of fast glutamatergic transmission from hypocretin/orexin to histamine neurons *In Situ*. *J Neurosci*. 2012;32(36):12437–12443. https://doi.org/10.1523/JNEUROSCI.0706-12.2012.
94. Schöne C, Apergis-Schoute J, Sakurai T, Adamantidis A, Burdakov D. Coreleased orexin and glutamate evoke nonredundant spike outputs and computations in histamine neurons. *Cell Rep*. 2014;7(3): 697–704. https://doi.org/10.1016/j.celrep.2014.03.055.
95. Williams RH, Chee MJS, Kroeger D, et al. Optogenetic-mediated release of histamine reveals distal and autoregulatory mechanisms for controlling arousal. *J Neurosci*. 2014;34(17):6023–6029. https://doi.org/10.1523/JNEUROSCI.4838-13.2014.
96. Fink DJ, Glorioso JC. Engineering herpes simplex virus vectors for gene transfer to neurons. *Nat Med*. 1997;3(3):357–359. https://doi.org/10.1038/nm0397-357.
97. Lima SQ, Hromádka T, Znamenskiy P, Zador AM. PINP: a new method of tagging neuronal populations for identification during in vivo electrophysiological recordingNitabach MN, ed. *PLoS One*. 2009;4(7):e6099. https://doi.org/10.1371/journal.pone.0006099.
98. Soudais C, Laplace-Builhe C, Kissa K, Kremer EJ. Preferential transduction of neurons by canine adenovirus vectors and their efficient retrograde transport in vivo. *FASEB J*. 2001;15(12): 2283–2285. https://doi.org/10.1096/fj.01-0321fje.
99. Wickersham IR, Lyon DC, Barnard RJO, et al. Monosynaptic restriction of transsynaptic tracing from single, genetically targeted neurons. *Neuron*. 2007;53(5):639–647. https://doi.org/10.1016/j.neuron.2007.01.033.
100. Tervo DG, Hwang B-Y, Viswanathan S, et al. A designer AAV variant permits efficient retrograde access to projection neurons. *Neuron*. 2016;92(2):372–382. https://doi.org/10.1016/j.neuron.2016.09.021.
101. Xiong Q, Znamenskiy P, Zador AM. Selective corticostriatal plasticity during acquisition of an auditory discrimination task. *Nature*. 2015;521(7552):348–351. https://doi.org/10.1038/nature14225.
102. Znamenskiy P, Zador AM. Corticostriatal neurons in auditory cortex drive decisions during auditory discrimination. *Nature*. 2013;497(7450):482–485. https://doi.org/10.1038/nature12077.
103. Senn V, Wolff SBE, Herry C, et al. Long-range connectivity defines behavioral specificity of amygdala neurons. *Neuron*. 2014;81(2):428–437. https://doi.org/10.1016/j.neuron.2013.11.006.
104. Schwarz LA, Miyamichi K, Gao XJ, et al. Viral-genetic tracing of the input-output organization of a central noradrenaline circuit. *Nature*. 2015;524(7563):88–92. https://doi.org/10.1038/nature14600.
105. Marshel JH, Mori T, Nielsen KJ, Callaway EM. Targeting single neuronal networks for gene expression and cell labeling in vivo. *Neuron*. 2010;67(4):562–574. https://doi.org/10.1016/j.neuron.2010.08.001.
106. Vélez-Fort M, Rousseau CV, Niedworok CJ, et al. The stimulus selectivity and connectivity of layer six principal cells reveals cortical microcircuits underlying visual processing. *Neuron*. 2014;83(6): 1431–1443. https://doi.org/10.1016/j.neuron.2014.08.001.

107. Wertz A, Trenholm S, Yonehara K, et al. Single-cell-initiated monosynaptic tracing reveals layer-specific cortical network modules. *Science*. 2015;349(6243):70–74. https://doi.org/10.1126/science.aab1687.
108. Fadai-Ghotbi B, Natelson B, Tsiang H, Etessami R, Ceccaldi P-E, Conzelmann K-K. Spread and pathogenic characteristics of a G-deficient rabies virus recombinant: an in vitro and in vivo study. *J Gen Virol*. 2000;81(9):2147–2153. https://doi.org/10.1099/0022-1317-81-9-2147.
109. Pollak Dorocic I, Fürth D, Xuan Y, et al. A whole-brain atlas of inputs to serotonergic neurons of the dorsal and median raphe nuclei. *Neuron*. 2014;83(3):663–678. https://doi.org/10.1016/j.neuron.2014.07.002.
110. Weissbourd B, Ren J, DeLoach KE, Guenthner CJ, Miyamichi K, Luo L. Presynaptic partners of dorsal raphe serotonergic and GABAergic neurons. *Neuron*. 2014;83(3):645–662. https://doi.org/10.1016/j.neuron.2014.06.024.
111. González JA, Iordanidou P, Strom M, Adamantidis A, Burdakov D, Figure S. Awake dynamics and brain-wide direct inputs of hypothalamic MCH and orexin networks. *Nat Commun*. 2016;7(c):11395. https://doi.org/10.1038/ncomms11395.
112. Reijmers LG, Perkins BL, Matsuo N, Mayford M. Localization of a stable neural correlate of associative memory. *Science*. 2007;317(5842):1230–1233. https://doi.org/10.1126/science.1143839.
113. Kawashima T, Okuno H, Bito H. A new era for functional labeling of neurons: activity-dependent promoters have come of age. *Front Neural Circuits*. 2014;8:37. https://doi.org/10.3389/fncir.2014.00037.
114. Gossen M, Freundlieb S, Bender G, Muller G, Hillen W, Bujard H. Transcriptional activation by tetracyclines in mammalian cells. *Science*. 1995;268(5218). http://science.sciencemag.org/content/268/5218/1766/tab-pdf.
115. Liu X, Ramirez S, Pang PT, et al. Optogenetic stimulation of a hippocampal engram activates fear memory recall. *Nature*. 2012;484(7394):381–385. https://doi.org/10.1038/nature11028.
116. Ramirez S, Liu X, Lin P-A, et al. Creating a false memory in the hippocampus. *Science*. 2013;341(6144):387–391. https://doi.org/10.1126/science.1239073.
117. Root CM, Denny CA, Hen R, Axel R. The participation of cortical amygdala in innate, odour-driven behaviour. *Nature*. 2014;515(7526):269–273. https://doi.org/10.1038/nature13897.
118. Guenthner CJ, Miyamichi K, Yang HH, Heller HC, Luo L. Permanent genetic access to transiently active neurons via TRAP: targeted recombination in active populations. *Neuron*. 2013;78(5):773–784. https://doi.org/10.1016/j.neuron.2013.03.025.
119. Kawashima T, Kitamura K, Suzuki K, et al. Functional labeling of neurons and their projections using the synthetic activity–dependent promoter E-SARE. *Nat Methods*. 2013;10(9):889–895. https://doi.org/10.1038/nmeth.2559.
120. Packer AM, Russell LE, Dalgleish HWP, Häusser M. Simultaneous all-optical manipulation and recording of neural circuit activity with cellular resolution in vivo. *Nat Methods*. 2015;12(2):140–146. https://doi.org/10.1038/nmeth.3217.
121. Park SI, Brenner DS, Shin G, et al. Soft, stretchable, fully implantable miniaturized optoelectronic systems for wireless optogenetics. *Nat Biotechnol*. 2015;33(12):1280–1286. https://doi.org/10.1038/nbt.3415.
122. Montgomery KL, Yeh AJ, Ho JS, et al. Wirelessly powered, fully internal optogenetics for brain, spinal and peripheral circuits in mice. *Nat Methods*. 2015;12(10):969–974. https://doi.org/10.1038/nmeth.3536.
123. Polikov VS, Tresco PA, Reichert WM. Response of brain tissue to chronically implanted neural electrodes. *J Neurosci Methods*. 2005;148(1):1–18. https://doi.org/10.1016/j.jneumeth.2005.08.015.
124. Yizhar O, Fenno LE, Davidson TJ, Mogri M, Deisseroth K. Optogenetics in neural systems. *Neuron*. 2011;71(1):9–34. https://doi.org/10.1016/j.neuron.2011.06.004.
125. Kim CK, Adhikari A, Deisseroth K. Integration of optogenetics with complementary methodologies in systems neuroscience. *Nat Rev Neurosci*. 2017;18(4):222–235. https://doi.org/10.1038/nrn.2017.15.
126. Scammell TE, Arrigoni E, Lipton JO. Neural circuitry of wakefulness and sleep. *Neuron*. 2017;93(4):747–765. https://doi.org/10.1016/j.neuron.2017.01.014.

127. von Economo C. Sleep as a problem of localization. *J Nerv Ment Dis*. 1930;71(March):1–5.
128. Jego S, Glasgow SD, Herrera CG, et al. Optogenetic identification of a rapid eye movement sleep modulatory circuit in the hypothalamus. *Nat Neurosci*. 2013;16(September):1–10. https://doi.org/10.1038/nn.3522.
129. Boyce R, Glasgow SD, Williams S, Adamantidis A. Causal evidence for the role of REM sleep theta rhythm in contextual memory consolidation. *Science*. 2016;352(6287):812.
130. Carter ME, Adamantidis A, Ohtsu H, Deisseroth K, de Lecea L. Sleep homeostasis modulates hypocretin-mediated sleep-to-wake transitions. *J Neurosci*. 2009;29(35):10939–10949. https://doi.org/10.1523/JNEUROSCI.1205-09.2009.
131. Ohno K, Sakurai T. Orexin neuronal circuitry: role in the regulation of sleep and wakefulness. *Front Neuroendocrinol*. 2008;29(1):70–87. https://doi.org/10.1016/j.yfrne.2007.08.001.
132. Verret L, Goutagny R, Fort P, et al. A role of melanin-concentrating hormone producing neurons in the central regulation of paradoxical sleep. *BMC Neurosci*. 2003;4(1):19. https://doi.org/10.1186/1471-2202-4-19.
133. Hassani OK, Lee MG, Jones BE. Melanin-concentrating hormone neurons discharge in a reciprocal manner to orexin neurons across the sleep-wake cycle. *Proc Natl Acad Sci USA*. 2009;106(7):2418–2422. https://doi.org/10.1073/pnas.0811400106.
134. Tsunematsu T, Ueno T, Tabuchi S, et al. Optogenetic manipulation of activity and temporally controlled cell-specific ablation reveal a role for MCH neurons in sleep/wake regulation. *J Neurosci*. 2014;34(20):6896–6909. https://doi.org/10.1523/JNEUROSCI.5344-13.2014.
135. Konadhode RR, Pelluru D, Blanco-Centurion C, et al. Optogenetic stimulation of MCH neurons increases sleep. *J Neurosci*. 2013;33(25). https://doi.org/10.1523/JNEUROSCI.1225-13.2013.
136. Schöne C, Burdakov D. Glutamate and GABA as rapid effectors of hypothalamic "peptidergic" neurons. *Front Behav Neurosci*. 2012;6:81. https://doi.org/10.3389/fnbeh.2012.00081.
137. Carter ME, Yizhar O, Chikahisa S, et al. Tuning arousal with optogenetic modulation of locus coeruleus neurons. *Nat Neurosci*. 2010;13(12):1526–1533. https://doi.org/10.1038/nn.2682.
138. Ozen Irmak S, de Lecea L. Basal forebrain cholinergic modulation of sleep transitions. *Sleep*. 2014;37(12):1941–1951. https://doi.org/10.5665/sleep.4246.
139. Weber F, Chung S, Beier KT, Xu M, Luo L, Dan Y. Control of REM sleep by ventral medulla GABAergic neurons. *Nature*. 2015;526(7573). https://doi.org/10.1038/nature14979.
140. Van Dort CJ, Zachs DP, Kenny JD, et al. Optogenetic activation of cholinergic neurons in the PPT or LDT induces REM sleep. *Proc Natl Acad Sci USA*. 2015;112(2):584–589. https://doi.org/10.1073/pnas.1423136112.
141. Xu M, Chung S, Zhang S, et al. Basal forebrain circuit for sleep-wake control. *Nat Neurosci*. 2015;18(11). https://doi.org/10.1038/nn.4143.
142. Tsunematsu T, Kilduff TS, Boyden ES, Takahashi S, Tominaga M, Yamanaka A. Acute optogenetic silencing of orexin/hypocretin neurons induces slow-wave sleep in mice. *J Neurosci*. 2011;31(29):10529–10539. https://doi.org/10.1523/JNEUROSCI.0784-11.2011.
143. Shiromani PJ, Peever JH. New neuroscience tools that are identifying the sleep-wake circuit. *Sleep*. 2017;40(4). https://doi.org/10.1093/sleep/zsx032.
144. Paz JT, Davidson TJ, Frechette ES, et al. Closed-loop optogenetic control of thalamus as a tool for interrupting seizures after cortical injury. *Nat Neurosci*. 2012;16(1):64–70. https://doi.org/10.1038/nn.3269.
145. Krook-Magnuson E, Szabo GG, Armstrong C, Oijala M, Soltesz I. Cerebellar directed optogenetic intervention inhibits spontaneous hippocampal seizures in a mouse model of temporal lobe epilepsy. *eNeuro*. 2014;1(1). https://doi.org/10.1523/ENEURO.0005-14.2014.
146. Siegle JH, Wilson MA. Enhancement of encoding and retrieval functions through theta phase-specific manipulation of hippocampus. *Elife*. 2014;3. https://doi.org/10.7554/eLife.03061.
147. Ngo H-VV, Martinetz T, Born J, Mölle M. Auditory closed-loop stimulation of the sleep slow oscillation enhances memory. *Neuron*. 2013;78(3):545–553. https://doi.org/10.1016/j.neuron.2013.03.006.
148. Bellesi M, Riedner BA, Garcia-Molina GN, Cirelli C, Tononi G. Enhancement of sleep slow waves: underlying mechanisms and practical consequences. *Front Syst Neurosci*. 2014;8:208. https://doi.org/10.3389/fnsys.2014.00208.

149. Bartic C, Battaglia FP, Wang L, Nguyen TT, Cabral H, Navratilova Z. A multichannel recording system with optical stimulation for closed-loop optogenetic experiments. *Methods Mol Biol (Clifton, NJ)*. 2016;1408:333–344. https://doi.org/10.1007/978-1-4939-3512-3_23.
150. Volgraf M, Gorostiza P, Numano R, Kramer RH, Isacoff EY, Trauner D. Allosteric control of an ionotropic glutamate receptor with an optical switch. *Nat Chem Biol*. 2006;2(1):47–52. https://doi.org/10.1038/nchembio756.
151. Szobota S, Gorostiza P, Del Bene F, et al. Remote control of neuronal activity with a light-gated glutamate receptor. *Neuron*. 2007;54(4):535–545. https://doi.org/10.1016/j.neuron.2007.05.010.
152. Levitz J, Popescu AT, Reiner A, Isacoff EY. A toolkit for orthogonal and in vivo optical manipulation of ionotropic glutamate receptors. *Front Mol Neurosci*. 2016;9:2. https://doi.org/10.3389/fnmol.2016.00002.
153. Lin JY, Lin MZ, Steinbach P, Tsien RY. Characterization of engineered channelrhodopsin variants with improved properties and kinetics. *Biophys J*. 2009;96(5):1803–1814. https://doi.org/10.1016/j.bpj.2008.11.034.
154. Kleinlogel S, Feldbauer K, Dempski RE, et al. Ultra light-sensitive and fast neuronal activation with the $Ca^{2+}$-permeable channelrhodopsin CatCh. *Nat Neurosci*. 2011;14(4):513–518. https://doi.org/10.1038/nn.2776.
155. Wang H, Sugiyama Y, Hikima T, et al. Molecular determinants differentiating photocurrent properties of two channelrhodopsins from chlamydomonas. *J Biol Chem*. 2009;284(9). https://doi.org/10.1074/jbc.M807632200.
156. Wen L, Wang H, Tanimoto S, et al. Opto-current-clamp actuation of cortical neurons using a strategically designed channelrhodopsin. *PLoS One*. 2010;5(9):e12893. https://doi.org/10.1371/journal.pone.0012893.
157. Kawazoe Y, Yawo H, Kimura KD. A simple optogenetic system for behavioral analysis of freely moving small animals. *Neurosci Res*. 2013;75(1):65–68. https://doi.org/10.1016/j.neures.2012.04.011.
158. Erbguth K, Prigge M, Schneider F, Hegemann P, Gottschalk A. Bimodal activation of different neuron classes with the spectrally red-shifted channelrhodopsin chimera C1V1 in *Caenorhabditis elegans*. Nitabach MN, ed. *PLoS One*. 2012;7(10):e46827. https://doi.org/10.1371/journal.pone.0046827.
159. Prakash R, Yizhar O, Grewe B, et al. Two-photon optogenetic toolbox for fast inhibition, excitation and bistable modulation. *Nat Methods*. 2012;9(12):1171–1179. https://doi.org/10.1038/nmeth.2215.
160. Gradinaru V, Zhang F, Ramakrishnan C, et al. Molecular and cellular approaches for diversifying and extending optogenetics. *Cell*. 2010;141(1):154–165. https://doi.org/10.1016/j.cell.2010.02.037.
161. Madisen L, Garner AR, Shimaoka D, et al. Transgenic mice for intersectional targeting of neural sensors and effectors with high specificity and performance. *Neuron*. 2015;85(5):942–958. https://doi.org/10.1016/j.neuron.2015.02.022.
162. Fenno L, Yizhar O, Deisseroth K. The development and application of optogenetics. *Annu. Rev. Neurosci*. 2011;34:389–412.
163. Carter ME, Brill J, Bonnavion P, Huguenard JR, Huerta R, de Lecea L. Mechanism for Hypocretin-mediated sleep-to-wake transitions. *Proceedings of the National Academy of Sciences of the United States of America*. 2012;109(39):E2635–E2644. https://doi.org/10.1073/pnas.1202526109.

CHAPTER 24

# The Use of DREADDs for Dissecting the Contribution of Cellular and Neural Circuit Mechanisms in Models of Neurodegenerative Disease

**Puneet Sharma[1], Ilse S. Pienaar[1,2]**
[1]Imperial College London, London, United Kingdom; [2]University of Sussex, Falmer, United Kingdom

## DESIGNER RECEPTORS EXCLUSIVELY ACTIVATED BY DESIGNER DRUGS

A technology abbreviated as DREADDs (designer receptors exclusively activated by designer drugs) uses synthetically derived receptors and selective, otherwise inert, exogenous ligands to transiently activate or inactivate targeted neuronal types within specific brain regions. A range of transfer strategies (but principally using viral vector infection methods) are used for delivering DREADD receptors into neural tissue, with stereotaxic microinjection of the virus into a particular location that refines spatial specificity of DREADD expression.

Designer Receptors Exclusively Activated by Designer Drugs section of the current chapter will serve readers with a basic overview of the principles underlying the workings of DREADDs (Fig. 24.1), but it does not intend to be exhaustive as to the cellular and molecular mechanisms underlying application of DREADDs. Interested readers seeking greater details pertaining to DREADDs' molecular mechanisms of action are referred to several excellent review articles, e.g., by Armbruster et al.[1]; Ferguson and Neumaier[2]; Rogan and Roth[3]; Sternson and Roth;[4] and Roth.[5] Designer Receptors Exclusively Activated by Designer Drugs section also reviews recent advances to the DREADD toolbox. Neurodegenerative Disease section seeks to highlight the tremendous potential held by the DREADD approach in many research areas relating to neurodegenerative disease, by highlighting recently developed approaches and applications. In particular, after an overview of the causes and features of two prominent neurodegenerative diseases, namely Alzheimer disease (AD) and Parkinson disease (PD), we review highlights from studies that have applied DREADD technologies to answer questions relating to both diseases, by making use of preclinical models, presented as a series of case studies. The findings from such studies pave the way toward use of DREADDs for relatively noninvasive, selective, reversible responses by neuronal populations and circuits, by means of a dose-responsive orally administered agonist that targets the human brain. Hence, the

*Molecular-Genetic and Statistical Techniques for Behavioral and Neural Research*
ISBN 978-0-12-804078-2, https://doi.org/10.1016/B978-0-12-804078-2.00024-6

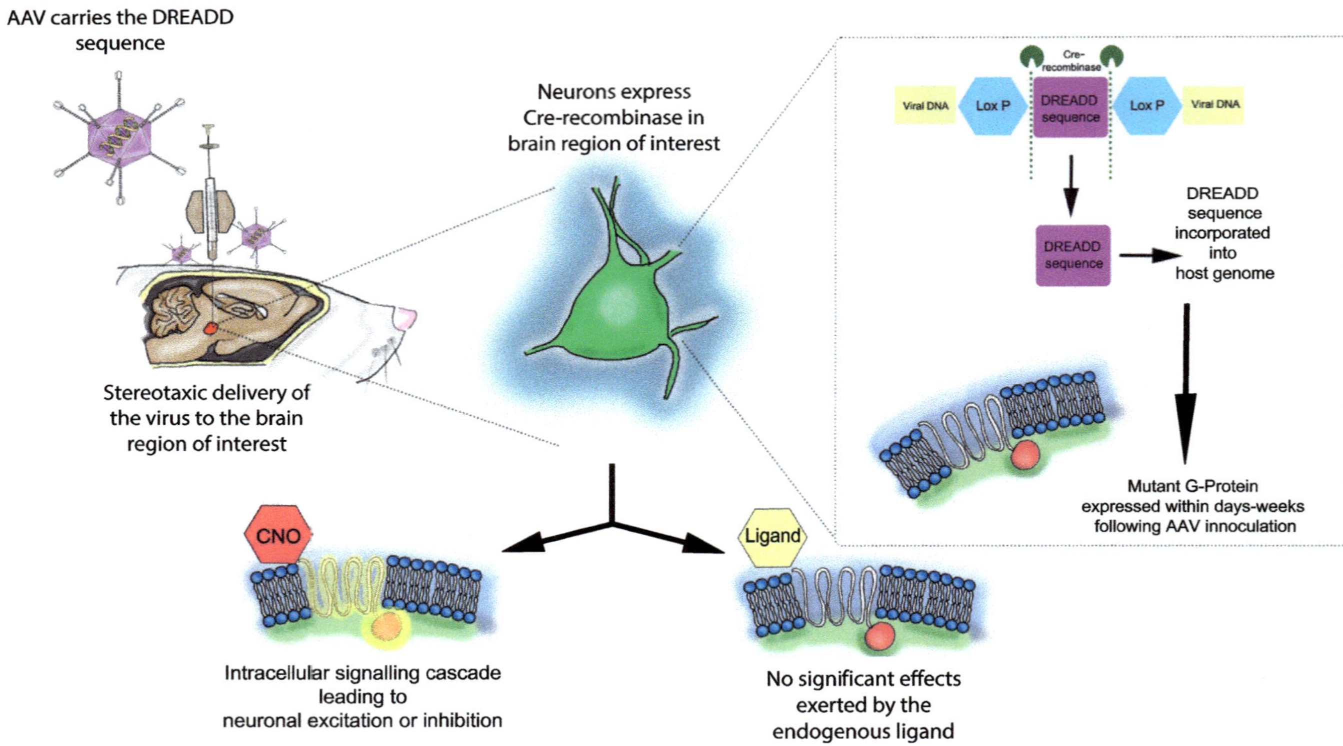


**Figure 24.1** Overview of the principles underlying the chemogenetic method. Although many variations to this technique now exist (see main body of text), a popular and validated approach is to combine stereotaxy with Cre-lox systems to achieve highly selective neuronal expression of the DREADD receptor. The figure illustrates the stereotaxic delivery of an adeno-associated virus vector to a target population of cells. Only cells in close proximity to the injection site are exposed to viral particles. Of these cells, only those expressing Cre recombinase will be able to excise and use the DREADD sequence, flanked as it is by LoxP sites and otherwise contained within the viral DNA. Transgenic animals express Cre recombinase under a given promotor (e.g., ChAT); hence, only cells (i.e., cholinergic cells, in this example) can free the sequence for action by the cell's translational machinery. Within weeks of viral exposure, select Cre recombinase–expressing cells will express a mutated G protein–coupled receptor. The artificially expressed receptor can then be modulated by a designer ligand (commonly clozapine-N-oxide) but is otherwise inert to endogenous neurotransmitters.

chapter ends (Translational Potential of Chemogenetics for Treating Neurodegenerative section) by reflecting on the potential that DREADDs offer to aid in the development of novel therapeutic treatments against neurodegenerative disease.

## DREADDs: An Experimentally Useful GPCR Signaling Platform

DREADDs entail use of receptor proteins. These derive from targeted single nucleotide mutagenesis of the DNA encoding for G protein–coupled receptors (GPCRs), resulting in synthetic receptors. Upon being expressed within neuronal cell membranes, they react with high potency to otherwise inert ligands such as clozapine-N-oxide (CNO), but they remain unresponsive to endogenous ligands. DREADDs offer distinct advantages for achieving either gain-of-function or loss-of-function effects (depending on the DREADD construct's design) over alternative technologies for manipulating neurons' functions, including that of optogenetics. One advantage is that, similar to standard pharmacotherapies, CNO can be administered systemically and crosses the blood-brain barrier (BBB) to activate DREADDs expressed in the brain.

Commonly used DREADD variants include human (h)M4Di and hM3Dq. Both these versions of DREADDs contain the same point mutations (Y149 $C^{3.33}$/A239 $G^{5.46}$) that abolish receptor affinity for the native ligand, acetylcholine (ACh), but they allow receptor binding and subsequent activation by the small pharmacologically inert molecule, CNO.[1] hM4Di is an engineered version of the human M4 muscarinic ACh receptor that in humans is encoded for by the *CHRM4* gene. Wild-type M4 (and M2) muscarinic cholinergic receptors signal primarily by activating Gi/o-signaling proteins,[6] to function as inhibitory autoreceptors for ACh. Therefore, when hM4Di is activated by CNO, membrane hyperpolarization occurs by means of decreased cyclic adenosine monophosphate (cAMP) signaling, as well as increased activation of inwardly rectifying potassium channels.[1,3] Physiologically, this results in the temporary suppression of neuronal activity. In addition, activation of hM4Di inhibits vesicle release at synaptic terminals (similar to the function of endogenous Gi-coupled receptors in the brain), to henceforth silence synapses.[7] In this regard, Stachniak and others[8] illustrated the usefulness of a pharmacologically selective approach that utilizes hM4Di-DREADD for suppressing presynaptic neurotransmitter release for cell type-specific and axon projection-selective functional analysis of diverse neural circuits. In this case, the investigators made use of hM4Di-DREADD to localize and functionally manipulate circuit connections from the hypothalamus to the midbrain, responsible for feeding behavior.

In contrast, native M3 receptors couple to $G\alpha q$ protein to signal,[9] and since its introduction as an experimental tool, it has provided insight into the functions of a variety of cell types, including neurons, glia, pancreatic b cells, and hepatocytes (reviewed by Urban & Roth[10]). Since Gq signaling triggers the release of calcium from intracellular stores, activation of hM3Dq receptors by CNO induces depolarization of the membrane potential of neurons that express hM3Dq, resulting in increased neuronal firing rate.[1]

## Recent Advances in the DREADD Toolkit

The following section describes exciting progress recently made, which allows DREADDs to address longstanding unanswered questions relating to intact brain functions and the cellular mechanisms underlying perturbed brain functions that manifest as recognizable neurological disease.

### *DREADD Alleles That Permit for Increased and More Effective Expression of DREADD in Diverse Cell Types*

A major technological advance was recently made by Sciolino and colleagues,[11] who developed Cre- and Flp/Cre responsive knock-in allele mouse lines, which permit for noninvasive targeting of hM3Dq to the cell body and dendrites of genetically defined cells. Although application of this method is limited, since expression of the knock-in construct is restricted to neurons and brain regions for which promotors are available, such somatodendritic localization of hM3Dq is deemed an improvement over traditional viral vector-mediated methods for delivering the construct to brain tissue. This is since traditional methods require stereotaxic (and hence, invasive) delivery of the DREADD, which may result in significant damage to the overlying parenchyma and even risk death for experimental animals undergoing the procedure. The method put forward by Sciolino and others[11] utilizes recombinase-responsive hM3Dq alleles to offer a noninvasive alternative for activating specific, genetically defined, neurons. In a proof-of-principle study, these investigators revealed that CNO-induced anxiety-like behavior and suppressed locomotion in a dose-dependent manner to reproduce behavioral phenotypes that previous studies[12] had shown can be evoked by optogenetic stimulation.

### *DREADDs Activated By Non-CNO Ligands*

Use of CNO in DREADD applications is slightly disadvantaged in that in both nonhuman primates and humans, at least a minimal amount, it is reverse metabolized into the psychoactive molecule clozapine,[13] the commonly reported side effects of this drug's use being sedation, hypotension, and anticholinergic syndrome, while myocarditis is an infrequently seen fatal side effect. Hence, the identification and development of different ligands by which to activate DREADDs is an important issue, for DREADDs to not only increase their facilitative roles in the multiplexed dissection of neural circuitry and behavior, but also for such tools to realize their translational potential.

Recently, a study by Gomez and colleagues[14] claimed that the in vivo DREADD actuator consists of clozapine. Here, it was shown that CNO binds to DREADD, but at low affinity, whereas clozapine-DREADD binding occurred with high affinity. Based on their findings, the authors recommended the use of subthreshold levels of clozapine, instead of high doses of CNO, to minimize the chance of producing non-DREADD-mediated effects. CNO's potential for back-metabolism to clozapine, especially in non-rodent species, has long been recognized, but this was deemed to be at low levels, even in humans (10% or less).[15] Previous findings also question the results by Gomez and colleagues[14] that suggest that CNO functions as a very poor DREADD chemical actuator, with Armbruster and others[1] showing that all known Gq-DREADDs are activated by even low nM levels of CNO.

Furthermore, the study by Gomez and others[14] reported that upon systemic CNO injections, converted clozapine readily enters the brain and occupies expressed DREADDs, whereas CNO itself was impeded from penetrating the CNS. Although this position on the ability of CNO to effective cross the BBB is similar to previous work when assessed acutely after systemic administration, Bender and others[16] reported that cerebral uptake of both compounds is almost identical at 60 min after systemic administration. Taken together, the paper by Gomez and colleagues[14] has been met with polarized views by researchers working in the DREADD field, and at the time of writing this chapter, a definitive stance as to CNO-DREADD usage by the research leaders in the field was awaited. However, the need to include relevant controls, e.g., by including CNO-receiving animals that express only an irrelevant protein (i.e., green fluorescent protein (GFP)) and for a CNO-only DREADD-free control group when designing DREADD studies, remains.[5]

Efforts are ongoing at developing new non-CNO chemical actuators to avoid any potential risk for the in vivo back-metabolism of CNO. This includes Compound 21 that is unlikely to back-metabolize to clozapine, and which the developers claim possesses limited off-target activity and very high selectivity for activating hM3Dq compared to other GPCRs.[13] A water-soluble version of Compound 21 has recently been made commercially available by the life science company Hello Bio.

Recently, Vardy and colleagues[17] used the human kappa-opioid receptor's (hKOR) structure to develop a mutated version of a Gi-coupled DREADD, for decreasing neuronal activity via exclusive activation by the otherwise inert ligand salvinorin B (SalB), an inactive metabolite of the KOR-selective agonist salvinorin A (SalA). SalB displayed good CNS penetrability[17] and selectivity for the expressed hKOR DREADD, which remained unresponsive to dynorphin, the endogenous ligand to KOR. SalB administration to mice did not induce any of the behavioral effects typically seen in response to KOR activation, e.g., analgesia and anhedonia.

Based on previous reports that showed that the D138A mutation in hKOR abolishes the binding of all known KOR agonists without affecting either the affinity or potency of the KOR agonist SalA,[18,19] the group created a hKOR D138N mutant. This was expressed using a Cre recombinase—dependent adeno-associated virus (AAV), which targeted the mutated hKOR DREADD to specific neuronal populations in different *cyclic* recombinase (Cre)-driver mouse lines. The study's investigators performed several behavioral assays to determine whether peripheral administration of SalB produces a behavioral phenotype, by targeting three distinct neuronal groups. Results include a SalB dose-dependent increase in locomotor activity, with vehicle that produced no effect, when hKOR DREADD was expressed in GABA (gamma-aminobutyric acid)ergic neurons projecting from the ventral tegmental area (VTA), similarly to what has been reported before.[20] The functional use of a hKORD DREADD-SalB combination was also validated in rats, with Marchant and others[21] showing behavioral effects when manipulating brain circuits involved in motivated behavior. Vardy and others[17] performed whole-cell patch-clamp recordings on acutely prepared brain slices taken from the mice to verify the ability of the novel hKOR DREADD to generate hyperpolarization of neurons, upon binding to SalB.

The investigators further demonstrated how KOR can be used in conjunction with either hM3Dq or hM4Di to achieve bidirectional modulation of neuronal activity in different brain areas in the same animal, allowing for investigations of diverse physiological systems through the use of multiplexed chemogenetic actuators. This so-called "multiplexing" of DREADDs where the activity of genetically defined neuronal groups is either increased or decreased simultaneously or sequentially in the same animal was achieved by expressing both excitatory hM3Dq and inhibitory KOR DREADD in the same animal. The development of this combined approach represents an important advance in the DREADD field, as it permits for bidirectional control over neuronal activity. However, understanding as to the timeline and kinetics at play in persistent CNO-hM3Dq signaling compared to more transient SALB-KORD effects is needed before these DREADD-ligand pairings can be effectively combined to manipulate neuronal activity in a physiologically relevant manner. This is since SalB mediates rapid, short-lasting effects lasting ~5 min, whereas CNO evokes delayed, long-lasting effects that last ~60 min instead. This large difference that the drugs exert on the temporal dynamics of neuronal activity may deem SalB more informative as to the behavioral effects following *transient* inactivation of neurons, while CNO-based activation of DREADD could inform more on the behavioral effects associated with chronic activation of neurons,[22] with acute versus chronic activation of neurons that can result in strikingly different behavioral consequences.[23] Additionally, because of the limited solubility of SalB, analogs of SalB should be developed that show improved water solubility.

### Caged DREADD Ligands Offer Spatiotemporal Precise Control

The temporal dynamics of DREADD activation depend upon respective agonists' pharmacokinetic properties. For instance, plasma levels of CNO (given at recommended dosages of 0.1–0.3 mg/kg) peak within 30 min following peripheral administration, but then decline sharply over the next 2 h[24]; however, CNO continues to exert behavior effects for a further 6 h.[9] Thus, DREADD technology is the preferred choice in experimental conditions that require neuronal activity to be manipulated over relatively long periods of time (hours to days), whereas optogenetics is preferred when neuronal activity needs to be manipulated over shorter time intervals (seconds to milliseconds). Temporal resolution of DREADD can be increased by using a photoactivatable or "caged" ligand such as CNO (which has been produced by Prof. Bryan Roth's lab and is available on a collaborative basis) to modulate the activity of microbial opsins, which are endogenous ion channels in neurons. The "caging" technique involves rendering a molecule biologically inert ("caged") through chemical modification using a photoremovable protecting group. Illumination induces a concentration increase of the caged molecule, which binds to a cellular receptor to then either switch the downstream cellular process on or off.[25]

In animals that had been surgically implanted with an optic cable by which to deliver light to a brain region of interest, the photocaged ligand serves as an optogenetic actuator, providing higher temporal precision[26] than can be offered by standard DREADD-CNO neuronal activation. However, since optogenetic techniques are already so well established, and invasive surgery is unavoidable in either technique, optogenetics may be considered the strategy of choice for inducing a reliable phenotype with precise spatiotemporal control.

### *Genetically Modified Rodents Aid Cell-Specific Expression of DREADD*

Delivery of an AAV carrying a promoter for transgene expression within defined neuronal subpopulations is often limited, as promoter sequences are often too large to package into a vector, particularly into an AAV, which holds a packaging capacity of only ~4.7–5.4 kilo bases.[27] In addition, cell type-specific promoters often drive expression to only suboptimal levels.[28] This is inadequate for experimental purposes, which entails driving neuronal activity to assess for functional sufficiency, or silence neuronal activity, to evaluate their necessity. To overcome these limitations, genomic insertion of transgenes to create cell-specific transgenic mice and rats[29] have made possible the restricted expression of a DREADD to specific neuronal groups into a restricted brain region of interest to manipulate and functionally assess diverse cell types. For instance, both transgenic mice and more recently transgenic rats also have been generated that express Cre recombinase that are driven by promoters of CNS-specific genes by using a bacterial artificial chromosome (BAC) DNA construct containing large fragments of genomic DNA flanking the desired gene.

For instance, Gong and others[30] created 10 mouse transgenic lines, including BAC-Cre constructs that included two genes encoding the neurotransmitter-synthesizing enzymes, tyrosine hydroxylase (*Th*) and choline acetyl transferase (*ChAT*), as well as two genes encoding the neurotransmitter transporters, namely for the serotonin (*Slc6a4*) and norepinephrine (*Slc6a2*). In addition, a BAC-Cre driver line was generated that expressed Cre specifically in Purkinje neurons of the cerebellum, using a BAC-Cre construct that included the gene for Purkinje cell protein 2 (*Pcp2*). A CKLF-like

MARVEL transmembrane dopamine containing 5 (cmtm5) BAC-Cre line was also generated, with cmtm5 being a myelinating oligodendrocyte cell-specific marker.

For creating these transgenic lines, previously established BAC engineering is a rapid and effective method for creating these transgenic mice by pronuclear injection.[31] Briefly, this entailed inserting an intron containing either Cre recombinase or a ligand-dependent recombinase consisting of a fusion protein between the mutated ligand binding domain of the human estrogen receptor and Cre recombinase, which is activated to 4-hydroxy-tamoxifen, but is unresponsive to the endogenous estrogen ligand, 17 beta-oestradiol.[32] This was followed by addition of a long polyadenylate tail into the BAC vector. For this, a vector carrying the selective markers as well as a homology arm taken from the locus being targeted, for guiding recombination, was used. The founders' offspring were then crossed to reporter mice targeted to the ROSA26 locus that express enhanced GFP following Cre-mediated excision of a stop sequence.[33] Adult offspring mice of all lines were characterized for functional recombination of the reporter locus with eGFP immunohistochemistry.

In general, opinion exists among neuroscientists that experimental rats could offer a greater repertoire of behavior than mice, with a number of behavioral tasks that have been optimized for rats, including those assessing olfactory discrimination,[34] certain auditory tasks,[35] as well as visual attention, and also working memory.[36] Witten and others[37] used the promoters *Th* and *ChAT* to generate genetically restricted recombinase-driver rat lines that are suitable for driving gene expression in dopaminergic, noradrenergic, and cholinergic neurons. This work represents a significant step forward, as the rat lines offer an experimental means for selective targeting of genetically defined cell types. To demonstrate the utility of these experimental animal models for clarifying the relationship between neuron activation and behavioral functions, the group applied optogenetic targeting to neuromodulate Cre-expressing cells in vivo. *Th::Cre* rats were stereotaxically injected with a Cre-dependent virus in dopaminergic and, in a separate experiment, in noradrenergic neural structures. Optogenetics-based illumination of ChR2-expressing catecholaminergic neurons in freely behaving transgenic rats (compared to wild-type littermates) resulted in a positive causal relationship between the firing of dopaminergic neurons located in a midbrain structure, the VTA and positive reinforcement, whereby the likelihood of a behavioral response increases due to the outcome of that response.[38] This outcome suggests that VTA dopaminergic neurons can be regarded as sufficient for driving intracranial self-stimulation (ICSS), an operant paradigm in which subjects' operant responding is maintained by means of exogenous pulses of electrical stimulation applied to a specific brain area.

The *Th::Cre* model was validated with both in vitro and in vivo electrophysiological recordings applied to ChR2-expressing TH neurons, and acute slice patch-clamp recordings of neurons in the nucleus basalis expressing ChR2-YFP, presumed to be cholinergic neurons, with light-evoked activation that evoked reliable neural activity in both models.

Alternatively, transgenic rodents have been created that express excitatory or inhibitory DREADDs in different neuronal subtypes or glia.[39] Synthetic ligands then activate the expressed DREADD to facilitate neuronal control. This approach bypasses the need to perform invasive surgery on the rats for intracranial delivery of viral constructs carrying the DREADD transgene, promoter, and a fluorescent reporter.[40] This method holds several advantages, especially relating to behavioral neuroscience experiments. In this regard, Frumberg and others[41] showed that brain surgery could result in persistent, injury-related deficit in the performance of a cognitive task in rats. Although chemogenetic applications lag behind optogenetic tools for translation to human neuromodulation, the minimal invasiveness associated with the use of chemogenetics (optogenetic applications conventionally require invasive optical fiber implants) may make DREADDs a more suitable candidate for ultimate clinical translation.

## NEURODEGENERATIVE DISEASE

Neurodegenerative diseases are currently incurable and debilitating conditions that result in progressive degeneration and the eventual death of neuronal cells. This heterogeneous group of disorders includes frontotemporal dementia, amyotrophic lateral sclerosis, multiple sclerosis, Machado-Joseph disease, Huntington disease, and amyloid polyneuropathy, in addition to AD and PD, the latter two being the most common types of neurodegenerative diseases. The last few decades have seen intense research on multiple fronts, advancing our understanding of the pathologies affecting cellular processes in neurodegenerative disease. This knowledge continues to lay the foundation for identifying multiple therapeutic targets to subsequently develop promising multitarget agents that either exhibit protective effects or slow down the morbidity of these diseases. After providing a brief overview as to the risk factors and neuropathologic characteristics of AD and PD, the current section will focus on studies where DREADDs have been applied to animal and cellular models of these diseases.

### Alzheimer Disease

AD is a progressive dementia caused by loss of neurons, concomitant with the presence of two main microscopic neuropathologic hallmarks, namely extracellular amyloid plaques and intracellular neurofibrillary tangles. The main proteinaceous constituent of amyloid plaques is the neurotoxic peptide amyloid beta (Aβ) that has been cleaved sequentially from a larger precursor protein (APP) by two enzymes, namely β-secretase (also called BACE1) and γ-secretase. Neurofibrillary tangles comprise mainly of the protein tau that binds microtubules. Uncoupling of tau from microtubules results in aggregation into tangles, which in turn inhibits transport and results in microtubule disassembly.

Considerable evidence has been gathered from autopsied AD brains that basal forebrain cholinergic neurons are lost.[42,43] The loss of this cell population decreases the concentration of ACh, which plays a critical role in modulating information processing. Previous work also showed decreased levels of the ACh-synthesizing enzyme, choline acetyltransferase (ChAT), in AD patients' brains (measured by means of a cortical brain biopsy), with the loss that correlated with decreased cognitive function, as measured by a neuropsychological test battery.[142] On the other hand, activity of acetylcholinesterase (AChE), an enzyme responsible for hydrolyzing ACh, is decreased in the brains of AD patients.[44] Intriguingly, AChE has been deemed as causative in AD, by directly interacting with Aβ, resulting in Aβ fibrillogenesis.[45] AD also associates with progressive degeneration of monoamine neurotransmitter systems, which include dopamine, noradrenaline, adrenaline, and serotonin. Specifically, evidence was provided that noradrenergic neurons located in the locus coeruleus undergo pathologic changes in the brains of AD patients, including accumulation of tau protein and progressing through to a profound loss of such neurons.[46] In addition, in AD patients, the loss of serotonergic neurons in the dorsal raphe nucleus, a bilateral, heterogeneous brainstem structure, has been described, as well as dysfunction of such neurons' axonal terminals, locating to the neocortex.[47] Lastly, numerous studies reported that neurons classed as "glutamatergic," which utilize the amino acid glutamate as their principal neurotransmitter, essential to memory formation through processes such as long-term potentiation, are also adversely affected during the course of AD.[48] For example, in AD specimens, glutamate-immunoreactive neocortical pyramidal neurons of layers V and III contain neurofibrillary tangles.[49] Reduced expression of hippocampal glutamate transporters in AD has also been described, to reinforce the notion that glutamate clearance from the synaptic cleft is impaired.[50] Hence, current pharmacologic approaches to treatment of AD are based on targeting the selective vulnerability of the cholinergic, noradrenergic, serotonergic, and glutamatergic neuronal systems.

Although age remains the best-known risk factor for developing AD, a recent meta-analysis of 323 studies confirmed the importance of additional modulators, including hypertension, smoking, and hypocholesterolemia. Other modifiable risk factors are depression and low educational attainment. In contrast, dietary exposure to folate, vitamin E/C, and coffee were deemed to be protective against AD.[51] Genetic susceptibility factors have also been demonstrated. In this regard, linkage analysis has identified mutations in genes encoding for *APP*/presenilin 1 (*PS1*) and presenilin 1 and 2 (*PSEN1, PSEN2*), which were all deemed to cause early-onset AD. Thus far, apolipoprotein E-ε4 allele is the only confirmed genetic risk factor for late-onset AD, although genome-wide association studies have identified more than 20 genetic loci that associate with late-onset AD. Mutations in the genes *PLD3* (phospholipase D3 gene), *TREM2* (Triggering Receptor Expressed on Myeloid Cells 2), *UNC5C* (Unc-5 Netrin Receptor C), *AKAP9* (A-kinase anchor protein 9), and *ADAM10* (a disintegrin and metalloproteinase) may carry an intermediate risk for developing AD. For a detailed overview of the genetic basis of AD, we recommend that the interested reader consults the review article by Giri and colleagues.[52]

## Application of DREADDs to Preclinical Models of AD

A number of studies thus far have utilized targeted expression of DREADDs as a tool by which to enhance or suppress neuronal activity of discrete neuronal populations in models of neurodegenerative disease (Fig. 24.2). However, beyond their ability to simply *interrogate* neural network function, DREADDs also offer the realistic prospect of *therapeutic* modulation. Here, we summarize studies that have applied DREADDs for gaining control of circuits of which the constituent neurons "encode" for learning and memory functions. When applied to preclinical models of AD, this approach has provided valuable insights into disease mechanisms and presented opportunities to test therapeutic approaches.

**Case study 1: Reduced formation and enhanced clearance of fibrillar amyloid plaques in DREADD-recipient neurites.**

The secretion of Aβ peptides from neurons has been shown to occur in an activity-dependent manner,[53] while human and animal studies that made use of amyloid positron emission tomography imaging showed that cortical regions most susceptible to Aβ deposition in either aged or AD subjects display constantly high basal brain activities at resting state.[54,55] The first experimental evidence that chronic synaptic hyperactivity causally relates to Aβ deposition and that this correlation may be linked to the pathogenesis of AD was provided in a study that utilized optogenetics to chronically stimulate APP transgenic mice, revealing that upregulation of neural activity exacerbates Aβ deposition.[56]

Yuan and Grutzendler[57] recently implemented a chemogenetic approach to test the hypothesis that chronic modulation of neural activity could prevent formation of fibrillar amyloid plaques and also aid clearance. Excitatory (hM3D) or inhibitory (hM4D) DREADD was injected into the subarachnoid space or CA1 hippocampal subfield, using two AD transgenic mouse models, namely the PS/APP mouse model that carries the KM670/671/NL and DeltaE9 mutations of the human PS1 gene under the prion protein complex promoter, and the 5XFAD mouse model, which overexpresses both the human APP gene harboring KM670/671/NL, V717I, and I716V mutations and human PS1 harboring M146L and L286V mutations under the Thy1 promoter.

For analysis, a distinction was drawn between smaller (with this population that was also the most abundant) plaques, which measured 1−6 μm in diameter, diffuse amyloid deposits, histochemically identified as being anti-Aβ positive, but thioflavin S negative, and larger plaques (15 μm in diameter), with small and diffuse plaques that were assumed to be the most recently formed. Inhibitory DREADD activation markedly reduced total diffuse amyloid deposit, but it exerted no change on the level of larger plaques. From this result, it would appear that neural activity influences the formation of new plaques but

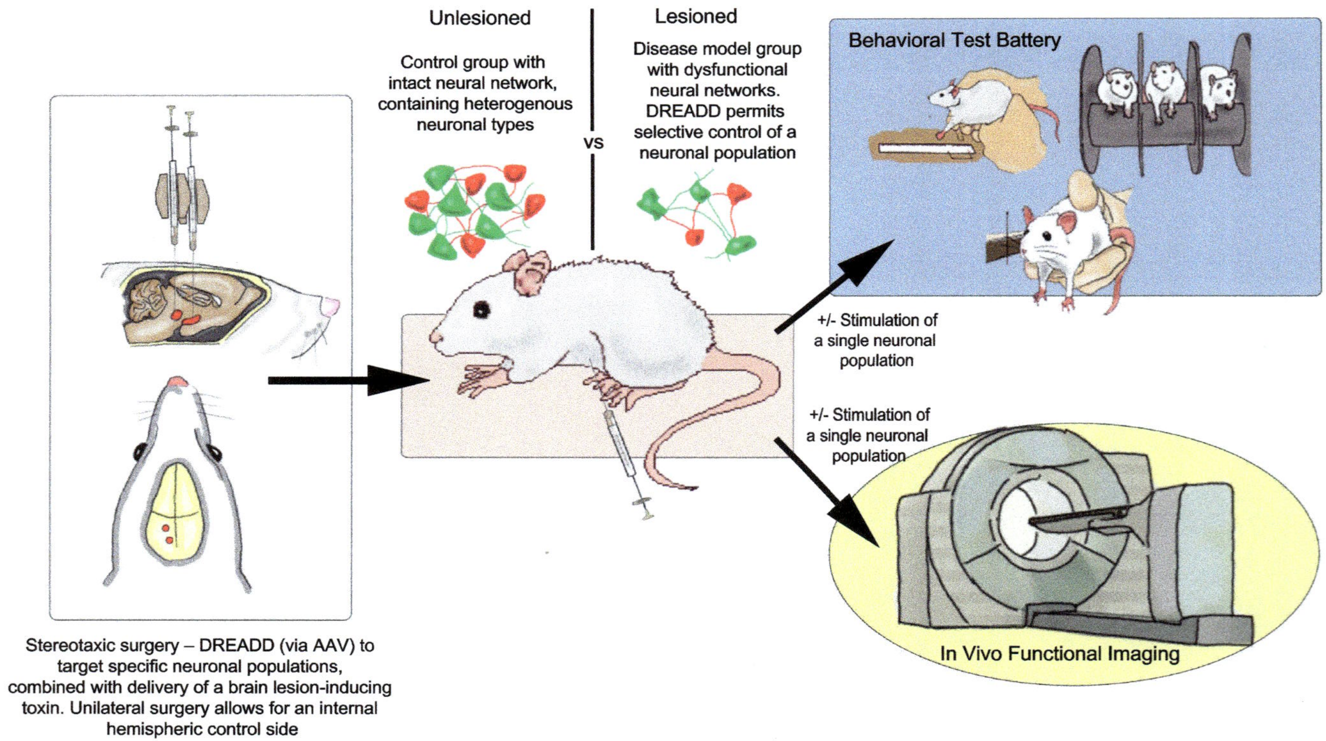


**Figure 24.2** Use of chemogenetics for investigating the causes and putative treatments of neurodegenerative diseases. A workflow is given to illustrate investigations relating to Parkinson disease (PD); however, a similar experimental approach can be applied for studying other neurodegenerative diseases. During initial stereotaxic surgery a second injection of toxin may be delivered to a separate brain region to create an animal model of disease. For example, lactacystin (a proteasome inhibitor) can be injected into the substantia nigra to create a unilateral rodent model of PD, leaving one hemisphere intact. Concomitant injection of the adeno-associated virus containing the DREADD construct (Fig. 24.1) creates a neuronal pool under selective chemogenetic control. Subsequently, the effects of stimulation (or inhibition) of these neurons in both control (nondisease) and lesioned (diseased) states can be determined, both behaviorally and through in vivo imaging. This technique is particularly advantageous in behavioral studies, as animal test subjects are stimulated through a peripheral injection of the designer drug, leaving an animal unencumbered for behavioral test batteries (whereas a mounted headstage for delivery of optogenetic pulse stimulation might have impeded behavioral performance). The effects of such precise neuronal control may be subtle and can be overwhelmed by intersubject variation; internal controls serve a useful role in keeping the number of animal test subjects to an ethical minimum.

does not deplete existing ones. Next, the group determined for effects by neural activity change on the size of the so-called "Aβ halo," which describes the dense central core of fibrillar amyloid plaque, surrounded by a less compact peripheral halo of oligomeric and protofibrillar Aβ.[58] Inhibition of neural activity reduced the diameter of the halo, for all plaque size classifications, emphasizing the study's principal finding that a reduction of neural activity prevents formation of plaques. Condello and colleagues[59] showed that an Aβ peptide halo's size correlates with the level of axonal dystrophy and synapse loss observed in the vicinity of the plaques; hence the investigators asked if inhibition of neural activity influences the density of neuronal synapses in the vicinity of amyloid plaques and/or if such modulation of neuronal activity could alter the magnitude of dystrophic axonal abnormalities. In the inhibitory DREADD-expressing regions, the level of axonal dystrophy surrounding plaques was markedly reduced, in contrast with hM3D-CNO-induced activation, which rather exacerbated the degree of axonal dystrophy seen around plaques. Finally, to further explore the postulation that neuronal activity can alter Aβ peptide release at presynaptic terminals and also within somato-dendritic fields,[60] the authors explored the relative contribution made by presynaptic versus postsynaptic sites to accumulation of amyloid plaques. In this regard, previous work suggested that this may occur either through synaptic vesicular mechanisms[61] or nonsynaptic neuropeptide-like secretion.[62] After 30 days of continuous CNO treatment, the authors observed decreased burden of amyloid plaque in the thalamus, compared to mice that expressed hM3D, with such animals that instead showed a significant increase in plaques; the thalamus comprised the projection target of DREADD-expressing neurons, with the cell bodies that were located within cortical layer V. Taken together, the study elegantly illustrates the long-term consequences of modulating neuronal activity on both amyloid deposition and also axonal and synaptic abnormalities surrounding the plaques. The findings also point the way toward potential therapeutic application of inhibitory DREADDs for treating amyloid burden and associated synaptic pathology in AD by reducing chronically elevated neuronal activity.

**Case study 2: DREADD-mediated locus coeruleus activation restores memory function in mice modelling Down syndrome and AD.**

The locus coeruleus (LC), a nucleus located in the brain stem, is primarily responsible for the body's responses to panic and stress. As the main source of noradrenergic input to memory-relevant regions such as the hippocampus,[63] it also serves to modulate hippocampal synaptic plasticity.[64] During the early stages of AD, LC neurons undergo degeneration.[65,66] A large body of evidence points out the importance of the LC in the development of AD. For instance, it has been shown that administration of DSP4 (N-(2-chloroethyl)-N-ethyl-2-bromobenzylamine hydrochloride), a selective neurotoxin for LC-noradrenergic neurons in rodent and bird brains,[67] can aggravate amyloid pathology in AD mouse models.[68] Others showed that the LC can be regarded as a therapeutic target for AD, with studies showing that LC-enhancing drugs can slow the progression of

AD pathology, including amyloid aggregation, oxidative stress, inflammation, and spatial memory function.[69,70] In addition, recent work revealed that noradrenaline released from the LC, when a subject is engaged in an arousing, mentally challenging task, could serve to protect neurons against cellular damage.[71] This finding helps to explain why education and engaging careers seemingly prevent cognitive decline in the elderly.[72,73]

Individuals with Down syndrome (trisomy 21) exhibit high comorbidity with AD[74,75] and similar AD-related neuropathology.[74] In a series of studies, Fortress and others[76] provided evidence that targeting the LC-NE pathway offers a plausible target for new therapeutic agents in treating DS-AD and idiopathic AD. The group showed that hM3Dq-CNO-mediated LC activation restores memory function in middle-aged Ts65Dn mice, with such animals that contain a triplication of the centromere and proximal end of chromosome 17[77] and undergo progressive loss of LC-noradrenergic neurons, commencing at 6 months of age,[78,79] similar to what is seen in the brains of Down syndrome and AD patients.[65,66] Spontaneous hyperactivity had previously been reported in Ts65Dn mice.[80]

The hM3Dq designer receptor, with the transgene's expression that is driven by the PRSx8 promoter[81] and with a C-terminal HA1.1 tag inserted to identify the transfected LC-noradrenaline neurons, was delivered bilaterally to the LC in middle-aged (12-month-old) Ts65Dn mice. Animals were assessed for spontaneous motor activity (which included horizontal and vertical activity) before and after CNO injections. Mice were also subjected to the novel object recognition task, which entails presenting the animal with two similar objects during an initial session, to then replace one of the two objects by a new object during a second session. The amount of time the mouse takes to explore the new object provides an index of recognition memory.[82] Lastly, the mice were assessed in the spontaneous alternation plus maze, an index of hippocampus-dependent spatial working memory.[83]

The study revealed that acute administration of CNO to activate expressed hM3D reduced hyperactivity in Ts65Dn mice, while also increasing the time the mice spent in the center in the spontaneous locomotion box. It would thus appear that hyperactivity and anxiety deficits may have been reduced or even recovered following LC-enhancing DREADD activation. In addition, LC-noradrenergic neuronal activation via DREADD also improved consolidation as well as retrieval of hippocampal-dependent memory, in agreement with previous studies that also demonstrated that the noradrenergic system fulfills a role for maintaining intact cognition.[78,84]

In the second part of the study, Fortress and others[76] made use of the noradrenergic prodrug L-DOPS, which is often prescribed for treatment of neurogenic orthostatic hypotension,[85] to validate the significance the LC-noradrenergic system plays in hippocampal functions in this mouse model. The effects of long-term L-DOPS administration were compared to that of acute activation of the LC-noradrenergic neurons via hM3D-CNO. Effects were compared in terms of hippocampus-dependent behavioral tasks, with results showing that long-term systemic dosing with L-DOPS further reduces baseline deficits in this rodent model. As β1-adrenergic receptors that express within the hippocampus are required for memory retrieval,[86] and since Ts65Dn mice exhibit reduced expression of β1-adrenergic receptors in the hippocampus due to the progressive loss of LC-NE innervation,[78] the authors postulated that this could be the signaling mechanism by which the excitatory DREADD receptors achieved an improved memory performance in the Ts65Dn mice. Thus, examination of the density of expressed β1-adrenergic receptors within the hippocampi found a reduction in mice subjected to the L-DOPS treatment regime. Efforts placed for understanding how cellular mechanisms that lie downstream of β1-adrenergic receptor signaling facilitates hippocampus-dependent memory in the aging Ts65Dn mouse model promise to assist discovery of novel drug targets for treating both AD and DS-AD.

## Parkinson Disease

PD is the second most common neurodegenerative condition after AD, with an estimated prevalence of 0.3% in the general population, but 1%–2% in those older than 60 years of age.[87,88] PD is characterized by four cardinal clinical signs, namely tremor at rest, muscle rigidity, slowness of movement (bradykinesia), and postural and gait impairment. The underlying neuropathology is attributable, in large part, to the extrapyramidal motor system, particularly the basal ganglia (subcortical nuclei controlling voluntary action) and associated deep brain nuclei.

In recent times the nonmotor symptoms of PD have justifiably received increasing attention in both research and therapeutic circles. Although common, PD's nonmotor features have historically been under appreciated and include olfactory dysfunction, rapid eye movement sleep behavior disorder (RBD), constipation, depression, and pain, which precede the onset of motor symptoms by several years.[89] To date, the neuroanatomic and neurochemical substrates for these symptoms remain largely unknown.

While no method prevents PD in humans (clinically proven neuroprotective agents for PD are not yet available), well-tolerated pharmacotherapies are available to improve motor and nonmotor PD-related deficits to maintain the best possible quality of life for patients. Antiparkinsonian medications include levodopa, an amino acid precursor of dopamine, which acts by replenishing depleted striatal dopamine. Levodopa is typically given in combination with a peripheral (and therefore incapable of penetrating the BBB) dopamine-decarboxylase inhibitor (e.g., co-carbidopa) to limit peripheral distribution and side effects of dopamine. In cases where greater symptomatic relief is required, monoamine oxidase B (MAO-B) inhibitors (e.g., selegiline) are usually added to the therapeutic regime, since inhibiting the activity of MAO-B (an enzyme responsible for metabolizing dopamine) results in increased levels of dopamine. Should abnormal involuntary movements (dyskinesia) develop, a common complication of long-term levodopa use, a catechol-O-methyltransferase (COMT) is often used to extend the duration of levodopa activity and thereby manage levodopa-induced dyskinesias. However, deep brain stimulation (DBS) therapy (discussed subsequently) is regarded as the most efficient treatment for drug-induced dyskinesias, while offering the further advantage of reducing the dose of levodopa required for achieving therapeutic effect, thereby reducing drug-related side effects. Treatment of nonmotor symptoms of PD remains largely inadequate. A systemic review deemed only four treatments as being effective: pramipexole for treating depressive symptoms, clozapine for alleviating psychosis, rivastigmine against dementia, and botulinum toxin as an antisialorrhea medication.[90] For an overview of pharmacological treatment options available to PD patients for providing symptomatic relief of motor and nonmotor symptoms, the interested reader is referred to review articles, e.g., Refs. 91,92.

DBS is an effective therapeutic option for a select group of patients, particularly those suffering from levodopa-related dyskinesias and tremor. Conventional DBS targets include the internal globus pallidus (GPi) and the subthalamic nucleus (STN). The latter perhaps offers a greater reduction in medication doses, but the former seems better tolerated in those with preexisting psychiatric history. Overall, meta-analysis demonstrates similar therapeutic effects between the two targets.[93] However, although the efficacy of STN-DBS to improve segmental motor symptoms has been robustly demonstrated in several clinical trials and meta-analyses,[93–96] other motor features such as akinesia and freezing of gait remain relatively resistant to STN-DBS or may even worsen.[97,98] Alternative DBS targets are under investigation, of which the pedunculopontine nucleus (PPN) is a leading candidate for reversing severe, medically intractable gait and postural impairments.[99] Studies suggested that surgical intervention to augment PPN activity could restore normal function.[100]

The majority of PD cases are not heritable, i.e., have environmental causes, and are termed sporadic. A number of risk factors have been associated with sporadic PD, with age being the most important, but exposure to pesticides and other toxins, as well as undergoing an oophorectomy (surgical removal of one or both ovaries) have also been documented as potential causative factors.[101] Early-onset PD, i.e., PD developing before the age of 40 years, is often heritable. Among heritable cases, 10%–15% of patients show mendelian autosomal inheritance, a monogenetic form of PD (reviewed in Ref. 102). The inheritance pattern is either autosomal dominant, with disease-causing variants found within the genes *SNCA* (α-synuclein), *LRRK2* (leucine-rich repeat kinase 2), and VPS35 (vacuolar sorting protein 35), or it is classified as autosomal recessive, affecting the genes encoding for *PARK2* (parkin), *PINK1* (PTEN-induced putative kinase 1), *ATP13A2* (DJ1, also known as PARK7), *ATP13A2* (PARK9), *FBX07* (PARK15), and *PLA2G6* (group VI phospholipases A2). A familial factor transmitted through *non-mendelian* modes of inheritance[103] has also been linked to PD etiology. However, this remains relatively rare, with only a single study that found evidence for nonmendelian transmission in PD patients, with the authors identifying linkage at five distinct chromosomal regions.[104] Other candidate genes that have been associated with parkinsonism include *Nurr1*,[105] *synphylin-1* isoforms,[106] and *POLG* (polymerase gamma)[107]; however, it remains unclear whether these are biologically significant.

A 2006 report by the US National Institute for Neurological Disorders and Stroke (NINDS)[108] estimated that approximately 50,000 new PD cases are diagnosed in the United States each year. The economic costs of these brain disorders are correspondingly large, encompassing not only the cost of treatment, but also of lost productivity of patients and their caregivers, for whom looking after chronically disabled family members often represent a source of practical, emotional, and financial burden.

In an aging population, the prevalence of age-related diseases such as AD and PD is expected to increase substantially in coming decades, as exemplified by predictions from Dorsey and others,[109] who estimated that in the United States alone, the prevalence of PD will at least double by 2030. Consequently, the need for greater understanding as to the disease causes, to subsequently develop new experimental therapeutics, has never been greater.

## Application of DREADDs to Preclinical Models of PD

**Case study 1: Recombinase-driver rats restrict DREADD expression to PPN cholinergic neurons and reveal such neurons responsible for PD-related postural instability and gait impairment.**

In PD, some promise has been shown through precise neuromodulation of circuits involving both degenerated dopaminergic and nondopaminergic basal ganglia neurons (reviewed by Ref. 110). Moreover, in the domain of PD research, DREADD technology is being increasingly applied for dissecting the neural circuitry of various PD-related symptoms, in the realistic hope that such findings will translate into a viable therapy.

In the postmortem brains of PD patients, PPN cholinergic neurons appear disproportionately susceptible to cell death, compared to other PPN neuronal subtypes.[111] Since destruction of PPN cholinergic neurons in PD associates with the onset and progression of axial symptoms, such as postural instability and gait freezing, the PPN represents a DBS target[112]; however, in the PPN's heterogeneous cell populations, the neuronal substrate responsible for both the axial-related symptoms of PD and for correcting these through PPN-DBS has not been established. In response, our group[113] utilized *ChAT::Cre* rats[37] to permit cholinergic-specific neuromodulation by CNO activation of Cre-dependent excitatory DREADD, spatially restricted to the PPN through stereotaxic infusion of the viral construct. Rats were rendered parkinsonian by means of a unilateral, intranigral injection of lactacystin, a proteasome inhibitor, with this rat model of PD that had previously been characterized by our group.[114] The toxin-induced parkinsonian rats were compared to sham-lesioned control animals. Lactacysin induced substantial degeneration of both the SNpc dopaminergic and PPN cholinergic neurons, to a similar extent that is seen in clinical PD. CNO administration for activating remaining PPN cholinergic neurons in this rat model of PD resulted in a dramatic reversal of several motor symptoms associated with PD, with behavioral tests that assessed postural instability, gait, sensorimotor integration, forelimb akinesia, and general motor activity. Taken together, the study revealed that select activation of cholinergic PPN neurons in parkinsonian brains can remedy several motor symptoms associated with PD. In addition, the study showed that a DREADD approach applied to a transgenic Cre recombinase rat system offers a minimally invasive means for controlling mammalian brain function, to provide a template for other experimental work aimed at generating insight as to the relative contribution made by specific neuronal subtypes to gain-of-function or loss-of-function during neuropathologies.

**Case study 2: Basal ganglia pathway-specific Gq-DREADD reveals new insights into the mechanisms underlying the therapeutic and dyskinetic effects of levodopa treatment.**

Dopamine precursor drugs, dopamine agonists, and inhibitors of dopamine catabolism are the main therapeutic options, as previously mentioned. Levodopa, a dopamine precursor, which serves as an effective and well-tolerated dopamine replacement agent, remains the most potent drug for controlling PD symptoms, particularly those relating to bradykinesia. Levodopa was introduced as an antiparkinsonian medication in the early 1970s; however, the limitations relating to its use soon became apparent, with its use associating with significant complications, including a "wearing off" effect, with the most serious being levodopa-induced motor fluctuations, including dyskinesias, following approximately 5 years of continuous treatment.[115] However, despite concerns relating to long-term use of levodopa, these are outweighed by concerns for complications resulting from surgical delivery of the device for delivering DBS.

For the past 20 years, the prevailing model of basal ganglia function has deemed that two neural circuits that originate from populations of striatal medium spiny neurons (MSNs) exert opposing effects on different output structures, with such projection pathways that have opposite effects on movement.[116] MSNs projecting from the striatum to the GPi and pars reticulata of the substantia nigra (SN) form the "striatonigral" neural circuit, also known as the "direct" pathway, its activation which facilitates voluntary movement. Such neurons contain dopamine receptor subtype 1 (D1) receptors. On the other hand, the "striatopallidal" neural circuit, alternatively known as the "indirect" pathway, consists of striatal MSNs that express dopamine receptor subtype 2 (D2) receptors that these neurons project indirectly to innervate the external globus pallidus (GPe) and STN. D1 receptors are coupled to a G-protein subunit and one of the three main families of G-proteins, namely $G_s$ alpha ($G\alpha s$), which activates a cAMP-dependent pathway by activating intracellular signaling of the enzyme adenylyl cyclase. In contrast, D2 receptors couple to $G\alpha_i$, another major family of G-proteins, with signal transduction occurring through inhibition of adenylyl cyclase, thus reducing cAMP levels. Striatal MSNs are inhibitory in nature and use GABA, the main inhibitory neurotransmitter within the CNS, as the principal neurotransmitter. Direct and indirect pathway GABAergic MSNs exert fundamentally different cellular responses to extracellular dopamine, through opposing effects on basal ganglia output nuclei resulting in opposite behavioral effects. In this regard, activation of the direct pathway facilitates movement, while activation of the indirect pathway suppresses movement. In PD, dopaminergic neurons that innervate the basal ganglia degenerate. The use of neurotoxin-based animal models of PD in experimental studies showed that excitability of the direct and indirect pathways is disrupted, to favor suppression of movement and therefore manifest as poverty of movement (hypokinesia) and slowness of movement (bradykinesia), which characterizes the movement of PD patients.[116] For a more detailed overview of the workings of the classical model of basal ganglia, comprising the "direct" and "indirect" pathways and how dysfunction of these could manifest as PD-related symptoms, see Gerfen and Surmeier.[117]

A study published by Angela Cenci's group (2017) utilized hM3Dq stimulation in 6-hydroxydopamine 6-OHDA-lesioned mice, a well-established model to mimic aspects of clinical PD in rodents (for a review on the 6-OHDA rodent model of PD, see Ref. 118). Specifically, in this study, mice were unilaterally injected with 6-OHDA, a neurotoxin that selectively degenerates catecholaminergic neurons (which includes those synthesizing dopamine, hence dopaminergic neurons), with the toxin delivered via stereotaxic means to the medial forebrain bundle, resulting in complete denervation of the dopaminergic nigrostriatal pathway, and hence striatal dopamine depletion. Unilateral lesioning of the nigrostriatal pathway produces side-biased motor impairments that reflect the motor deficits seen in PD. To restrict DREADD expression to either the direct or indirect pathway in vivo, two genotypes of heterozygous BAC transgenic mice were utilized, expressing Cre under control of either the D1 receptor or D2 receptor promoter.

The study generated significant insights with regard to the role of the direct and indirect pathways in directing normal basal ganglia function, but also how striatal dopamine depletion causes hypokinesia and bradykinesia in PD patients. The study went further by elucidating on the neural mechanisms underlying levodopa-induced dyskinesia, with several studies that have hypothesized that the pathologic pathway of this disabling side effect of prolonged levodopa therapy involves degeneration of indirect MSNs (reviewed by Ref. 119).

Using pathway-specific Gq-DREADD stimulation, the study found that specific activation of the direct and indirect spinal projection neurons exerted an opposite modulatory effect on motor activity displayed by both nonlesioned and dopamine-denervated mice. Moreover, the head, body, and tail of the mice, placed in an open field arena, were tracked by means of a video-tracking system. Using this measure of horizontal and vertical activity in nonlesioned mice, stimulation of the direct pathway exerted opposite effects to indirect pathway stimulation: in direct stimulation, horizontal and vertical activity increased, while reducing ipsilateral turns and increasing the number of contralateral rotations. Interestingly, the authors noted that chemogenetic stimulation (of the direct pathway) in parkinsonian mice yielded greater effects in these motor parameters compared to intact animals. Indirect pathway stimulation in parkinsonian animals produced barely noticeable motor changes. This result suggests that dopaminergic nigrostriatal lesions (with resultant striatal dopaminergic denervation, a pathological hallmark of parkinsonian brains and perhaps underlying enhanced activity of the striatopallidal neurons[120]) do not obstruct the effects upon motor behavior caused by chemogenetic stimulation of the striatonigral neurons in parkinsonian mice.

To determine whether pathway-specific DREADD stimulation could simulate the motor-recovery effects of levodopa in the hemiparkinsonian dopamine-denervated mice, both genotypes were administered CNO for activating either the nigrostriatal or striatopallidal neurons, followed by therapeutically relevant doses of levodopa. The mice were subjected to the "cylinder" test, which assesses the mouse's ability to spontaneously make use of both forelimbs for vertically exploring the inner wall of a glass cylinder.[121] As the mice had sustained a unilateral lesion of the nigrostriatal pathway, a forelimb use asymmetry score was obtained for each individual mouse, by expressing the results from the limb contralateral to the lesion. Mice were also rated periodically over a period of 180 min, following injection of CNO, on three subtypes of abnormal involuntary movements (axial, limb, and orofacial),[122] using a severity scale that is based on the duration and persistence of each dyskinetic behavior.

Inhibitory pathway stimulation inhibited both the hypokinesia and the prodyskinetic actions of levodopa, but prevented levodopa-induced motor (contralateral forelimb use) benefits from manifesting in the "cylinder" test. On the other hand, CNO-induced activation of the striato-nigral neurons fully restored forelimb use symmetry in the mice, mimicking the effects of levodopa. Taken together, this set of results mirrors results achieved following optogenetic stimulation of these neuronal systems in mice.[123]

Hence, these findings essentially validate previous reports that hyperactivity of the direct pathway underlie both the therapeutic and pathologic (mainly manifesting as dyskinesia) effects of dopamine restoration therapy,[123,124] with levodopa forming the mainstay of such treatment in PD. A further interesting finding from this work was that combination of DREADD-based stimulation of the direct pathway with pharmacologic stimulation of D2 receptors, through use of the D2 agonist quinpirole, thereby mimicking the effects of levodopa on both the direct and indirect pathway, reproduced all the features of levodopa-induced dyskinesia. Separately given, both quinpirole and CNO induced dyskinesia, albeit only moderately, but peaked at different time points following drug administration. Their combined treatment elicited a synergistic prodyskinetic effect, with similar axial, limb, and orofacial test scores to those achieved when the highest levodopa dose was given.

This work contributes toward better understanding the relative roles played by the direct and indirect neuronal circuits of the basal ganglia for facilitating well-coordinated movements, but also when such movements are disrupted, as seen in parkinsonism. In particular, the study is timely, due to increased questioning of the original proposal that stated that dopamine exerts opposite effects on neurons constituting the direct and indirect pathway. This includes a critical reappraisal by Calabresi and colleagues[125] of the classical model, pointing out the potential significance of the discovery that the direct and indirect pathways interact at the levels of the striatum and globus pallidus (GP) by means of collaterals and interneuronal synaptic circuits. This suggests that basal ganglia circuits operate vastly more complex than what the original simplistic circuit model proposed.[126] Of particular value is the study's translational message that treatment of levodopa-related hyperkinesia might be improved by targeting not only neurons of the direct, but also the indirect pathway. In contrast to previous work, the author discovered through the use of pathway- and cell type-specific DREADD stimulation that MSNs that constitute the indirect pathway can also fulfill a fundamental role in the onset and maintenance of levodopa-induced dyskinesia. In this regard, the current findings suggest that treatments that target both the direct and indirect pathway offer an advantage for alleviating PD-related motor symptoms and levodopa-induced dyskinesia, over ones that target a single pathway only.

**Case study 3: Dopaminergic neuron progenitors expressing DREADD are functionally integrated into existing neural circuits and induce motor recovery in parkinsonian mice.**

Neurodegenerative disease is characterized pathologically by the progressive loss of neurons, thus the ultimate therapeutic aims are to replace lost neuronal populations, for such newly formed/transplanted neurons to form synapses that functionally connect the neurons and to ultimately form neural networks that are similar to the ones lost as a result of disease. Interest in cell replacement therapies for human CNS disorders dates back to the late 1970s, when experimental results provided proof-of-principle that intrastriatal grafts of fetal mesencephalic tissue, rich in dopaminergic neurons, induced functional recovery in rats with neurotoxin-induced lesions of the nigrostriatal dopaminergic system.[127] Extensive animal experimental studies performed since this time have provided additional evidence that neuronal replacement and partial reconstruction of neuronal circuitry is feasible (reviewed by Ref. 128).

Recently, Chen and colleagues[129] showed that human pluripotent stem cell (hPSC)—derived neurons, differentiated and engrafted into functional human midbrain dopaminergic neurons, can be simultaneously chemogenetically controlled. hPSC cells were engineered (using the genome editing tool CRISPR (clustered regularly interspaced short palindromic repeats)) to express either excitatory hM3Dq or inhibitory hM4Di receptors. CNO was administered for 3 consecutive days to mice that had sustained an intranigral, unilateral lesion with 6-OHDA. Operated-on animals were then tested using several motor behavioral assays, including apomorphine (a dopamine receptor agonist)-induced rotation, which induces lesioned animals to rotate spontaneously in an asymmetric, contralateral direction.[130] The battery of tests also included the "cylinder" test, as well as the accelerating version of the rotarod test, which assesses neuromuscular coordination.[131] The testing paradigm consisted of 2 training days, during which time the speed of the mechanically driven rotarod was incrementally adjusted. Testing took place on the third day. The latency of the mouse to fall from the balance beam into a holding chamber was used as an index of performance.

The results of the study provide evidence that the use of DREADDs for controlling the physiological activity of engrafted human cells could increase the long-term therapeutic success of human stem cell transplantation for treating neurodegenerative disease. Specifically, graft cells expressing hM4Di can be hyperpolarized through the use of CNO, to inhibit excessive overcompensation by the engrafted cells. On the other hand, should it be required to increase the excitatory activity of the grafted cells' function, a hM3Dq-CNO approach can be applied.[129] Essentially, as DREADD-mediated modulation is reversible, this study represents a step closer toward precision medicine in that the feasibility to alter activity of the transplanted human cells can be utilized to improve the therapeutic outcome or mitigate unwanted side effects associated with the transplant therapy.

In PD, loss of dopaminergic innervation leads to hyperactivity in the GPi, which serves as the main output nucleus of the basal ganglia. This results in profound disturbance in the function of motor circuits to provide the therapeutic rationale for application of DBS to the GPi, and it is believed to act by reducing this pathologic activity.[132] Hence, destabilization of neuronal network activity and the resulting destabilization that results from aberrant alterations in neuronal network activity and associated compensatory responses is a recognized mode by which motor dysfunctions seen in PD patients and cognitive impairments associated with AD, at least in part, are instigated and worsen over time.

## TRANSLATIONAL POTENTIAL OF CHEMOGENETICS FOR TREATING NEURODEGENERATIVE DISEASE

In summary, the discovery that inert small molecules can be used for controlling neuronal activity via highly selective receptor-ligand pairing, termed DREADDs, provides exciting opportunities for deconstructing complex behavior (especially higher order ones, such as learning, memory, and feeding), which arises from equally complex and heterogeneous neural networks. DREADDs also offer new avenues for developing therapies and possibly neuroprotective agents, for instance against AD and PD patients. However, before the use of DREADDs in clinical practice can become a reality, considerable refinement of the technology is required, not least by overcoming the challenge of stably expressing DREADD within human neurons by means of viral-mediation.

Recent efforts to apply optogenetics as a therapy for vision restoration in patients diagnosed with retinal degenerative disease emphasize the translational potential of neuromodulatory techniques. In this regard, the privately held biopharmaceutical company RetroSense Therapeutics (recently acquired by the global pharmaceutical company Allergan) announced success in an initial clinical trial. Here the aim was to monitor blind patients for adverse effects after receiving an AAV serotype 2-based vector to deliver cDNA encoding a truncated version of the opsin channelrhodopsin-2 (ChR2) to retinal ganglion cells (RGC), responsible for passing visual signals from photoreceptors to the brain, via intravitreal injection. This approach allowed for targeted optogenetic stimulation of the RGCs, as ChR2, an algal protein derived from *Chlamydomonas reinhardtii*, is a light-activated cation channel capable of inducing depolarization and action potentials in neurons. The opsin-expressing cells fire when stimulated with blue light, resulting in the passage of visual information to the brain. Proof-of-principle studies showed that transducing inner retinal neurons restored vision in homozygous rd1 (rd1/rd1) mice.[133] In this animal model of photoreceptor degeneration, mice harbor a null mutation in the beta-subunit of rod phosphodiesterase,[134] similar to certain types of retinitis pigmentosa seen in humans.[135] Subsequent work revealed restoration of visuobehavioral responses by light-activation of ChR2-expressing retinal bipolar cells in the rd1 mouse model.[136]

Although it is not expected that patients receiving such optogenetic gene therapy will fully recover their vision, the clinical trials will be considered a success if participants gain the ability to navigate independently or even recognize faces.[137]

DREADDs certainly have the potential to be clinically translated, perhaps more so than optogenetics, with the latter methods that require implantation of rigid light-emitting devices. A critical advantage that DREADDs hold over alternative neural modifiers such as optogenetics are their ability to provide noninvasive temporal control of neuronal signaling, hence wisely eliminating the need for the implantation of light-delivering prosthetics. In experimental animals, this holds clear advantages, as it allows for assessing the effects of neural modulation on overt animal behavior while being minimally invasive. This has been particularly useful in our experience when evaluating the significance of altered behavior in highly heterogeneous neural structures (such as those found in the brainstem), behaviors which if restricted by prostheses may have been difficult to adequately observe, measure, or localize in the presence of restriction from cranial prostheses.

The widespread use of chemogenetic approaches to animal models of disease highlights that DREADDs could offer relatively noninvasive, selective, reversible inhibition or excitation of neuronal populations and circuits in neurodegenerative disease patients. Recent efforts have investigated the role of DBS in modulating nonmotor neuronal circuits, including disturbances impacting mood and cognition, for example, in AD (recently reviewed by Ref. 138). However, the electrostatic effects of DBS are only crudely localized in comparison to the size of target regions, which creates difficulty in identifying precise mechanisms of action at the network level. This in turn leads to challenges in refining DBS or creating alternatives. Chemogenetics, through *exclusive* targeting of a specific neuronal population (and its downstream pathway), offers the promise of unparalleled precision and possibly fewer side effects compared to the more diffuse effects of DBS. Clearly the landscape is set for further DREADD-based evaluation of ultraprecise therapeutic modulation. Although relative advantages of a chemogenetic approach over more traditional interventions, such as DBS or neuropharmaceuticals, may need to be ascertained in the human clinic, data from basic science investigations indicate that a DREADD approach for targeting a select neuronal population might offer increased efficacy and decrease adverse effects. However, the main obstacle to translation of DREADD techniques to human therapy remains the ability to safely and consistently express an exogenous protein in human tissue. In this regard, promising experiments revealed that DREADDs are well tolerated in nonhuman primates.[139] However, such gene therapy should first become routine before it can realize its clinical potential and transcend its current use, namely that of being an excellent tool for improving our understanding of the biologic basis of behavior.

Small molecule therapeutics offer first line treatments for debilitating neurodegenerative disorders, including AD and PD. The striking similarity that DREADDs hold with conventional therapeutics allow for crossover of insights deriving from experiments that utilize DREADDs to the neurophysiological phenomena responsible for therapeutic efficacy. This is since, as GPCRs, DREADDs represent a class of drug targets that are modulated either directly or indirectly by 36% of currently approved drugs.[140] In addition, small molecular ligands for activating expressed DREADDs exhibit drug-like pharmacokinetics, while such chemical actuators are inexpensive and safe. The development of KOR sets the stage for chemogenetics to reach the next stage of the technology's development by instigating the discovery and development of ligands, other than CNO, to activate expressed DREADD receptors. In particular, mutated KOR is activated by the otherwise inert ligand SalB to result in decreased neuronal activity, but with KOR-expressing neurons remaining unresponsive to the endogenous ligand dynorphin.[17]

Despite SalB still being limited as a DREADD ligand, being a poorly water-soluble active pharmaceutical (solubility is only achieved in 100% DMSO (dimethyl sulfoxide)),[40] which might associate with low or erratic bioavailability, such efforts are critical steps toward developing DREADD tools for clinical application. This is since CNO is disadvantaged, with the drug undergoing a degree of back-transformation to generate the psychoactive molecule clozapine in nonhuman and human primates.[15,141]

DREADDs could emerge as a way to potentially treat a variety of diseases. In this chapter, we demonstrated how two neurodegenerative disorders are exceptional candidates for DREADD-based intervention: AD and PD. This final section overviewed some of the technical issues faced for successful translation of DREADD technology to humans, including gene therapy and drug delivery difficulties. DREADDs present powerful tools for application into biotechnological and therapeutic arenas, if current efforts to overcome its inherent technical limitations prove successful.

## ABBREVIATIONS

**ACh** Acetylcholine
**AChE** Acetylcholinesterase
**AAV** Adeno-associated virus
**Aβ** Amyloid beta
**APP** Amyloid precursor protein
**BAC** Bacterial artificial chromosome
**BBB** Blood-brain barrier
**COMT** Catechol-*O*-methyltransferase
**ChR2** Channelrhodopsin-2
**ChAT** Choline acetyltransferase
**cmtm5** CKLF-like MARVEL transmembrane dopamine containing 5
**CNO** Clozapine-N-oxide
**CRISPR** Clustered regularly interspaced short palindromic repeats
**cAMP** Cyclic adenosine monophosphate
**Cre** Cyclic recombinase
**DMSO** Dimethyl sulfoxide
**DREADDs** Designer Receptors Exclusively Activated by Designer Drugs
**GPe** External GP
**GABA** Gamma-aminobutyric acid
**GP** Globus pallidus
**GPCRs** G protein–coupled receptors
**GFP** Green fluorescent protein
**hKOR** Human kappa-opiod receptor
**hPSC** Human pluripotent stem cell
**GPi** Internal GP
**ICSS** Intracranial self-stimulation
**KORD** Kappa-opioid receptor as template
**LC** Locus coeruleus
**MSNs** Medium spiny neurons
**MAO-B** Monoamine oxidase B
**PLD3** Phospholipase D3 gene
**PSEN1/2** Presenilin 1/2
**Pcp2** Purkinje cell protein 2
**RBD** Rapid eye movement sleep behavior disorder
**RGC** Retinal ganglion cells
**SalB** Salvinorin B
**6-OHDA** 6-hydroxydopamine
**SN** Substantia nigra
**TREM2** Triggering receptor expressed on myeloid cells 2
**Th** Tyrosine hydroxylase
**UNC5C** Unc-5 netrin receptor C
**VTA** Ventral tegmental area

## REFERENCES

1. Armbruster BN, Li X, Pausch MH, Herlitze S, Roth BL. Evolving the lock to fit the key to create a family of G protein-coupled receptors potently activated by an inert ligand. *Proc Natl Acad Sci USA*. 2007;104:5163–5168.
2. Ferguson SM, Neumaier JF. Grateful DREADDs: engineered receptors reveal how neural circuits regulate behavior. *Neuropsychopharmacology*. 2012;37:296–297.
3. Rogan SC, Roth BL. Remote control of neuronal signaling. *Pharmacol Rev*. 2011;63:291–315.
4. Sternson SM, Roth BL. Chemogenetic tools to interrogate brain functions. *Annu Rev Neurosci*. 2014; 37:387–407.
5. Roth BL. DREADDs for neuroscientists. *Neuron*. 2016;89:683–694.
6. Hulme EC, Birdsall NJ, Buckley NJ. Muscarinic receptor subtypes. *Annu Rev Pharmacol Toxicol*. 1990; 30:633–673.
7. Zhu H, Roth BL. Silencing synapses with DREADDs. *Neuron*. 2014;82:723–725.
8. Stachniak TJ, Ghosh A, Sternson SM. Chemogenetic synaptic silencing of neural circuits localizes a hypothalamus→midbrain pathway for feeding behavior. *Neuron*. 2014;82:797–808.
9. Alexander GM, Rogan SC, Abbas AI, et al. Remote control of neuronal activity in transgenic mice expressing evolved G protein-coupled receptors. *Neuron*. 2009;63:27–39.
10. Urban DJ, Roth BL. DREADDs (designer receptors exclusively activated by designer drugs): chemogenetic tools with therapeutic utility. *Annu Rev Pharmacol Toxicol*. 2015;55:399–417.
11. Sciolino NR, Plummer NW, Chen YW, et al. Recombinase-dependent mouse lines for chemogenetic activation of genetically defined cell types. *Cell Rep*. 2016;15:2563–2573.
12. McCall JG, Al-Hasani R, Siuda ER, et al. CRH engagement of the locus coeruleus noradrenergic system mediates stress-induced anxiety. *Neuron*. 2015;87:605–620.
13. Chen X, Choo H, Huang XP, et al. The first structure-activity relationship studies for designer receptors exclusively activated by designer drugs. *ACS Chem Neurosci*. 2015;6:476–484.
14. Gomez JL, Bonaventura J, Lesniak W, et al. Chemogenetics revealed: DREADD occupancy and activation via converted clozapine. *Science*. 2017;357:503–507.
15. Jann MW, Lam YW, Chang WH. Rapid formation of clozapine in Guinea-pigs and man following clozapine-N-oxide administration. *Arch Int Pharmacodyn Ther*. 1994;328:243–250.
16. Bender D, Holschbach M, Stocklin G. Synthesis of n.c.a. carbon-11 labelled clozapine and its major metabolite clozapine-N-oxide and comparison of their biodistribution in mice. *Nucl Med Biol*. 1994; 21:921–925.
17. Vardy E, Robinson JE, Li C, et al. A new DREADD facilitates the multiplexed chemogenetic interrogation of behavior. *Neuron*. 2015;86:936–946.
18. Kane BE, Nieto MJ, McCurdy CR, Ferguson DM. A unique binding epitope for salvinorin A, a non-nitrogenous kappa opioid receptor agonist. *FEBS J*. 2006;273:1966–1974.
19. Vardy E, Mosier PD, Frankowski KJ, et al. Chemotype-selective modes of action of κ-opioid receptor agonists. *J Biol Chem*. 2013;288:34470–34483.
20. Van Zessen R, Phillips JL, Budygin EA, Stuber GD. Activation of VTA GABA neurons disrupts reward consumption. *Neuron*. 2012;73:1184–1194.
21. Marchant NJ, Whitaker LR, Bossert JM, et al. Behavioral and physiological effects of a novel kappa-opioid receptor-based DREADD in rats. *Neuropsychopharmacology*. 2016;41:4012–4409.
22. Whissell PD, Tohyama S, Martin LJ. The use of DREADDs to deconstruct behavior. *Front Genet*. 2016;7:70.
23. Urban DJ, Zhu H, Marcinkiewcz CA, et al. Elucidation of the behavioral program and neuronal network encoded by dorsal raphe serotonergic neurons. *Neuropsychopharmacology*. 2016;41: 1404–1415.
24. Guettier JM, Gautam D, Scarselli M, et al. A chemical-genetic approach to study G protein regulation of metal cell function in vivo. *Proc Natl Acad Sci USA*. 2009;106:19197–19202.
25. Ellis-Davies GCR. Caged compounds: photorelease technology for control of cellular chemistry and physiology. *Nat Methods*. 2007;4:619–628.

26. Mourot A, Tochitsky I, Kramer RH. Light at the end of the channel: optical manipulation of intrinsic neuronal excitability with chemical photoswitches. *Front Mol Neurosci.* 2013;6:5.
27. Grieger JC, Samulski RJ. Packaging capacity of adeno-associated virus serotypes: impact of larger genomes on infectivity and postentry steps. *J Virol.* 2005;79:9933–9944.
28. Gompf HS, Budygin EA, Fuller PM, Bass CE. Targeted genetic manipulations of neuronal subtypes using promoter-specific combinatorial AAVs in wild-type animals. *Front Behav Neurosci.* 2015;9:152.
29. Madisen L, Mao T, Koch H, et al. A toolbox of Cre-dependent optogenetic transgenic mice for light-induced activation and silencing. *Nat Neurosci.* 2012;15:793–802.
30. Gong S, Doughty M, Harbaugh CR, et al. Targeting cre recombinase to specific neuron populations with bacterial artificial chromosome constructs. *J Neurosci.* 2007;27:9817–9823.
31. Gong S, Yang XW, Li C, Heintz N. Highly efficient modification of bacterial artificial chromosome (BAC) using novel shuttle vectors containing the R6Kgamma origin of replication. *Genome Res.* 2002; 12:1992–1998.
32. Feil R, Wagner J, Metzler D, Chambon P. Regulation of Cre recombinase activity by mutated estrogen receptor ligand-binding domains. *Biochem Biophys Res Commun.* 1997;237:752–757.
33. Soriano P. Generalized lacZ expression with the ROSA26 Cre reporter strain. *Nat Genet.* 1999;21: 70–71.
34. Uchida N, Mainen ZF. Speed and accuracy of olfactory discrimination in the rat. *Nat Neurosci.* 2003;6: 1224–1229.
35. Otazu GH, Tai LH, Yang Y, Zador AM. Engaging in an auditory task suppresses responses in auditory cortex. *Nat Neurosci.* 2009;12:646–654.
36. Chudasama Y, Robbins TW. Dopaminergic modulation of visual attention and working memory in the rodent prefrontal cortex. *Neuropsychopharmacology.* 2004;29:1628–1636.
37. Witten IB, Steinberg EE, Lee SY, et al. Recombinase-driver rat lines: tools, techniques, and optogenetic application to dopamine-mediated reinforcement. *Neuron.* 2011;72:721–733.
38. Steinberg EE, Boivin JR, Saunders BT, Witten IB, Deisseroth K, Janak PH. Positive reinforcement mediated by midbrain dopamine neurons requires D1 and D2 receptor activation in the nucleus accumbens. *PLoS One.* 2014;9:e94771.
39. Farrell MS, Pei Y, Wan Y, et al. A Gαs DREADD mouse for selective modulation of cAMP production in striatopallidal neurons. *Neuropsychopharmacology.* 2013;38:854–862.
40. Smith KS, Bucci DJ, Luikart BW, Mahler SV. DREADDs: use and application in behavioral neuroscience. *Behav Neurosci.* 2016;130:137–155.
41. Frumberg DB, Fernando MS, Lee DE, Biegon A, Schiffer WK. Metabolic and behavioural deficits following a routine surgical procedure in rats. *Brain Res.* 2007;1144:209–218.
42. Whitehouse PJ, Price DL, Storable RG, Clark AW, Coyle JT, Delon MR. Alzheimer's disease and senile dementia: loss of neurons in the basal forebrain. *Science.* 1982;215:1237–1239.
43. Pepeu G, Grazia Giovannini M. The fate of the brain cholinergic neurons in neurodegenerative diseases. *Brain Res.* 2017;1670:173–184.
44. Perry AK, Perry RH, Blessed G, Tomlinson BE. Necropsy evidence of central cholinergic deficits in senile dementia. *Lancet.* 1977;1:189.
45. Inestrosa NC, Alvarez A, Pérez CA, et al. Acetylcholinesterase accelerates assembly of amyloid-beta-peptides into Alzheimer's fibrils: possible role of the peripheral site of the enzyme. *Neuron.* 1996;16: 881–891.
46. Gannon M, Che P, Chen Y, Joao K, Robertson ED, Wang Q. Noradrenergic dysfunction in Alzheimer's disease. *Front Neurosci.* 2015;9:220.
47. Palmer AM, DeKosky ST. Monoamine neurons in aging and Alzheimer's disease. *J Neural Transm Gen Sect.* 1993;91:135–159.
48. Revert TJ, Baker GB, Jhamandas J, Kar S. Glutamate system, amyloid β peptides and tau protein: functional interrelationships and relevance to Alzheimer disease pathology. *J Psychiatr Neurosci.* 2013;38:6–23.
49. Kowall NW, Beal MF. Glutamate-, glutaminase-, and taurine-immunoreactive neurons develop neurofibrillary tangles in Alzheimer's disease. *Ann Neurol.* 1991;29:162–167.

50. Jacob CP, Koutsilieri E, Bartl J, et al. Alterations in expression of glutamatergic transporters and receptors in sporadic Alzheimer's disease. *J Alzheim Dis*. 2007;11:97–116.
51. Xu W, Tan L, Wang HF, et al. Meta-analysis of modifiable risk factors for Alzheimer's disease. *J Neurol Neurosurg Psychiatry*. 2015;86:1299–1306.
52. Giri M, Zhang M, Lü Y. Genes associated with Alzheimer's disease: an overview and current status. *Clin Interv Aging*. 2016;11:665–681.
53. Cirrito JR, Yamada KA, Finn MB, et al. Synaptic activity regulates interstitial fluid amyloid-β levels in vivo. *Neuron*. 2005;48:913–922.
54. Buckner RL, Snyder AZ, Shannon BJ, et al. Molecular, structural, and functional characterization of Alzheimer's disease: evidence for a relationship between default activity, amyloid, and memory. *J Neurosci*. 2005;25:7709–7717.
55. Bero AW, Bauer AQ, Stewart FR, et al. Bidirectional relationship between functional connectivity and amyloid-β deposition in mouse brain. *J Neurosci*. 2012;32:4334–4340.
56. Yamamoto K, Tanei Z, Hashimoto T, et al. Chronic optogenetic activation augments aβ pathology in a mouse model of Alzheimer disease. *Cell Rep*. 2015;11:859–865.
57. Yuan P, Grutzendler J. Attenuation of β-amyloid deposition and neurotoxicity by chemogenetic modulation of neural activity. *J Neurosci*. 2016;36:632–641.
58. Lowe J, Mirra SS, Hyman BT, Dickson DW. Ageing and dementia. In: Love Seth, Louis David, Ellison David W, eds. *Greenfield's Neuropathology*. 8th ed. vol. 2. CRC Press; 2008.
59. Condello C, Yuan P, Schain A, Grutzendler J. Microglia constitute a barrier that prevents neurotoxic protofibrillar Aβ42 hotspots around plaques. *Nat Commun*. 2015;6:6176.
60. Wei W, Nguyen LN, Kessels HW, Hagiwara H, Sisodia S, Malinow R. Amyloid beta from axons and dendrites reduces local spine number and plasticity. *Nat Neurosci*. 2010;13:190–196.
61. Cirrito JR, Kang JE, Lee J, et al. Endocytosis is required for synaptic activity-dependent release of amyloid-beta in vivo. *Neuron*. 2008;58:42–51.
62. Ludwig M, Leng G. Dendritic peptide release and peptide-dependent behaviours. *Nat Rev Neurosci*. 2006;7:126–136.
63. Loy R, Koziell DA, Lindsey JD, Moore RY. Noradrenergic innervation of the adult rat hippocampal formation. *J Comp Neurol*. 1980;189:699–710.
64. Stanton PK, Sarvey JM. Depletion of norepinephrine, but not serotonin, reduces long-term potentiation in the dentate gyrus of rat hippocampal slices. *J Neurosci*. 1985;5:2169–2176.
65. Mann DM, Yates PO, Marcyniuk B. A comparison of nerve cell loss in cortical and subcortical structures in Alzheimer's disease. *J Neurol Neurosurg Psychiatry*. 1986;49:310–312.
66. German DC, Manaye KF, White 3rd CL, et al. Disease-specific patterns of locus coeruleus cell loss. *Ann Neurol*. 1992;32:667–676.
67. Ross SB, Stenfors C. DSP4, a selective neurotoxin for the locus coeruleus noradrenergic system. A review of its mode of action. *Neurotox Res*. 2015;27:5–30.
68. Kalinin S, Gavrilyuk V, Polak PE, et al. Noradrenergic deficiency in brain increases beta-amyloid plaque burden in an animal model of Alzheimer's disease. *Neurobiol Aging*. 2007;28:1206–1214.
69. Kalinin S, Polak PE, Lin SX, Sakharkar AJ, Pandey SC, Feinstein DL. The noradrenaline precursor L-DOPS reduces pathology in a mouse model of Alzheimer's disease. *Neurobiol Aging*. 2012;33:1651–1663.
70. Hammerschmidt T, Kummer MP, Terwel D, et al. Selective loss of noradrenaline exacerbates early cognitive dysfunction and synaptic deficits in APP/PS1 mice. *Biol Psychiatry*. 2013;73:454–463.
71. Mather M, Harley CW. The locus coeruleus: essential for maintaining cognitive function and the aging brain. *Trends Cognit Sci*. 2016;20:214–226.
72. Carlson MC, Saczynski JS, Rebok GW, et al. Exploring the effects of an "everyday" activity program on executive function and memory in older adults: experience corps. *Gerontol*. 2008;48:793–801.
73. Park DC, Lodi-Smith J, Drew L, et al. The impact of sustained engagement on cognitive function in older adults: the synapse project. *Psychol Sci*. 2014;25:103–112.
74. Krinsky-McHale SJ, Devenny DA, Kittler P, Silverman W. Selective attention deficits associated with mild cognitive impairment and early stage Alzheimer's disease in adults with Down syndrome. *Am J Ment Retard*. 2008;113:369–386.

75. Glasson EJ, Dye DE, Bittles AH. The triple challenges associated with age-related comorbidities in Down syndrome. *J Intellect Disabil Res*. 2014;58:393–398.
76. Fortress AM, Hamlett ED, Vazey EM, et al. Designer receptors enhance memory in a mouse model of Down syndrome. *J Neurosci*. 2015;35:1343–1353.
77. Davisson MT, Schmidt C, Akeson EC. Segmental trisomy of murine chromosome 16: a new model system for studying Down syndrome. *Prog Clin Biol Res*. 1990;360:263–280.
78. Salehi A, Faizi M, Colas D, et al. Restoration of norepinephrine-modulated contextual memory in a mouse model of Down syndrome. *Sci Transl Med*. 2009;1:7ra17.
79. Lockrow J, Boger H, Bimonte-Nelson H, Granholm AC. Effects of long-term memantine on memory and neuropathology in Ts65Dn mice, a model for Down syndrome. *Behav Brain Res*. 2011;221: 610–622.
80. Escorihuela RM, Fernandez-Teruel A, Vallina IF, et al. A behavioral assessment of Ts65Dn mice: a putative Down syndrome model. *Neurosci Lett*. 1995;199:143–146.
81. Hwang DY, Carlezon Jr WA, Isacson O, Kim KS. A high-efficiency synthetic promoter that drives transgene expression selectively in noradrenergic neurons. *Hum Gene Ther*. 2001;12:1731–1740.
82. Leger M, Quiedeville A, Bouet V, et al. Object recognition test in mice. *Nat Protoc*. 2013;8: 2531–2537.
83. Savage LM, Chang Q, Gold PE. Diencephalic damage decreases hippocampal acetylcholine release during spontaneous alternation testing. *Learn Mem*. 2003;10:242–246.
84. Dang V, Medina B, Das D, et al. Formoterol, a long-acting β2 adrenergic agonist, improves cognitive function and promotes dendritic complexity in a mouse model of Down syndrome. *Biol Psychiatry*. 2014;75:179–188.
85. Isaacson SH, Skettini J. Neurogenic orthostatic hypotension in Parkinson's disease: evaluation, management, and emerging role of droxi-dopa. *Vasc Health Risk Manag*. 2014;10:169–176.
86. Murchison CF, Zhang XY, Zhang WP, Ouyang M, Lee A, Thomas SA. A distinct role for norepinephrine in memory retrieval. *Cell*. 2004;117:131–143.
87. Hirtz D, Thurman DJ, Gwinn-Hardy K, Mohamed M, Chaudhuri AR, Zalutsky R. How common are the "common" neurologic disorders? *Neurology*. 2007;68:326–337.
88. Alves G, Forsaa EB, Pedersen KF, Gjerstad MD, Larsen JP. Epidemiology of Parkinson's disease. *J Neurol*. 2008;255:S18–S32.
89. Tolosa E, Gaig C, Santamaría J, Compta Y. Diagnosis and the premotor phase of Parkinson disease. *Neurology*. 2009;72:S12–S20.
90. Seppi K, Weintraub D, Coelho M, et al. The Movement Disorder Society evidence-based medicine review update: treatments for the non-motor symptoms of Parkinson's disease. *Mov Disord*. 2011;26: S42–S80.
91. Kakkar AK, Dahiya N. Management of Parkinson's disease: current and future pharmacotherapy. *Eur J Pharmacol*. 2015;750:74–81.
92. Lotia M, Jankovic J. New and emerging medical therapies in Parkinson's disease. *Expet Opin Pharmacother*. 2016;17:895–909.
93. Liu Y, Li W, Tan C, et al. Meta-analysis comparing deep brain stimulation of the globus pallidus and subthalamic nucleus to treat advanced Parkinson disease. *J Neurosurg*. 2014;121(3):709–718.
94. Williams A, Gill S, Varma T, et al. Deep brain stimulation plus best medical therapy versus best medical therapy alone for advanced Parkinson's disease (PDSURG Trial): a randomised, open-label trial. *Lancet Neurol*. 2010;9:581–591.
95. Weaver FM, Follett K, Stern M, et al. Bilateral deep brain stimulation vs best medical therapy for patients with advanced Parkinson disease: a randomized controlled trial. *J Am Med Assoc*. January 7, 2009; 301:63–73.
96. Deuschl G, Schade-Brittinger C, Krack P, et al. German Parkinson Study Group, Neurostimulation Section. A randomized trial of deep-brain stimulation for Parkinson's disease. *N Engl J Med*. 2006; 355(9):896–908.
97. Limousin P, Krack P, Pollak P, et al. Electrical stimulation of the subthalamic nucleus in advanced Parkinson's disease. *N Engl J Med*. 1998;339:1105–1111.

98. Kleiner-Fisman G, Fisman DN, Sime E, Saint-Cyr JA, Lozano AM, Lang AE. Long-term follow up of bilateral deep brain stimulation of the subthalamic nucleus in patients with advanced Parkinson disease. *J Neurosurg*. 2003;99:489–495.
99. Wilcox RA, Cole MH, Wong D, Ciyne T, Silburn P, Kerr G. Pedunculopontine nucleus deep brain stimulation produces sustained improvement in primary progressive freezing of gait. *J Neurol Neurosurg Psychiatry*. 2011;82:1256–1259.
100. Ngoga D, Mitchell R, Kausar J, Hodson J, Harries A, Pall H. Deep brain stimulation improves survival in severe Parkinson's disease. *J Neurol Neurosurg Psychiatry*. 2014;85:17–22.
101. Elbaz A, Moisan F. Update in the epidemiology of Parkinson's disease. *Curr Opin Neurol*. 2008;21: 454–460.
102. Hernandez DG, Reed X, Singleton AB. Genetics in Parkinson disease: mendelian versus non-Mendelian inheritance. *J Neurochem*. 2016;139(Suppl 1):59–74.
103. Zareparsi S, Taylor TD, Harris EL, Payami H. Segregation analysis of Parkinson disease. *Am J Med Genet*. 1998;80:410–417.
104. Scott WK, Nance MA, Watts RL, et al. Complete genomic screen in Parkinson disease: evidence for multiple genes. *J Am Med Assoc*. 2001;286:2239–2244.
105. Le WD, Xu P, Jankovic J, et al. Mutations in NR4A2 associated with familial Parkinson disease. *Nat Genet*. 2003;33:85–89.
106. Marx FP, Holzmann C, Strauss KM, et al. Identification and functional characterization of a novel R621C mutation in the synphilin-1 gene in Parkinson's disease. *Hum Mol Genet*. 2003;12: 1223–1231.
107. Davidzon G, Greene P, Mancuso M, et al. Early-onset familial parkinsonism due to POLG mutations. *Ann Neurol*. 2006;59:859–862.
108. National Institute of Neurological Disorders, Stroke (NINDS). *Parkinson's Disease: Challenges, Progress, and Promise*; 2006. Retrieved from https://www.ninds.nih.gov/Disorders/All-Disorders/NINDS-Parkinsons-Disease-Information-Page/Parkinsons-Disease-Challenges.
109. Dorsey ER, Constantinescu R, Thompson JP, et al. Projected number of people with Parkinson disease in the most populous nations, 2005 through 2030. *Neurology*. 2007;68:384–386.
110. Sharma P, Pienaar IS. Pharmacogenetic and optical dissection for mechanistic understanding of Parkinson's disease: potential utilities revealed through behavioural assessment. *Neurosci Biobehav Rev*. 2014;47:87–100.
111. Pienaar IS, Elson JL, Racca C, Nelson G, Turnbull DM, Morris CM. Mitochondrial abnormality associates with type-specific neuronal loss and cell morphology changes in the pedunculopontine nucleus in Parkinson disease. *Am J Pathol*. 2013a;183:1826–1840.
112. Morita H, Hass CJ, Moro E, Sudhyadhom A, Kumar R, Okun MS. Parkinson study group functional neurosurgical working group. Pedunculopontine nucleus stimulation: where are we now and what needs to be done to move the field forward? *Front Neurol*. 2014;5:243.
113. Pienaar IS, Gartside SE, Sharma P, et al. Pharmacogenetic stimulation of cholinergic pedunculopontine neurons reverses motor deficits in a rat model of Parkinson's disease. *Mol Neurodegen*. 2015;10:47.
114. Pienaar IS, Harrison IF, Elson JL, et al. An animal model mimicking pedunculopontine nucleus cholinergic degeneration in Parkinson's disease. *Brain Struct Funct*. 2013b;220:479–500.
115. Jankovic J. Motor fluctuations and dyskinesias in Parkinson's disease: clinical manifestations. *Mov Disord*. 2005;20:S11–S16.
116. Albin RL, Young AB, Penney JB. The functional anatomy of basal ganglia disorders. *Trends Neurosci*. 1989;12:366–375.
117. Gerfen CR, Surmeier DJ. Modulation of striatal projection systems by dopamine. *Annu Rev Neurosci*. 2011;34:441–466.
118. Simola N, Morelli M, Carta AR. The 6-hydroxydopamine model of Parkinson's disease. *Neurotic Res*. 2007;11:151–167.
119. Feyder M, Bonito-Oliva A, Fisone G. L-DOPA-induced dyskinesia and abnormal signaling in striatal medium spiny neurons: focus on dopamine D1 receptor-mediated transmission. *Front Behav Neurosci*. 2011;5:71.

120. Mallet N, Ballion B, Le Moine C, Gonon F. Cortical inputs and GABA interneurons imbalance projection neurons in the striatum of parkinsonian rats. *J Neurosci.* 2006;26:3875–3884.
121. Schallert T, Fleming SM, Leasure JL, Tillerson JL, Bland ST. CNS plasticity and assessment of forelimb sensorimotor outcome in unilateral rat models of stroke, cortical ablation, parkinsonism and spinal cord injury. *Neuropharmacology.* 2000;39:777–787.
122. Lundblad M, Picconi B, Lindgren H, Cenci MA. A model of L-DOPA-induced dyskinesia in 6-hydroxydopamine lesioned mice: relation to motor and cellular parameters of nigrostriatal function. *Neurobiol Dis.* 2004;16:110–123.
123. Kravitz AV, Freeze BS, Parker PR, et al. Regulation of parkinsonian motor behaviours by optogenetic control of basal ganglia circuitry. *Nature.* 2010;466:622–626.
124. Bateup HS, Santini E, Shen W, et al. Distinct subclasses of medium spiny neurons differentially regulate striatal motor behaviors. *Proc Natl Acad Sci USA.* 2010;107:14845–14850.
125. Calabresi P, Picconi B, Tozzi A, Ghiglieri V, Di Filippo M. Direct and indirect pathways of basal ganglia: a critical reappraisal. *Nat Neurosci.* 2014;17:1022–1030.
126. Cazorla M, de Carvalho FD, Chohan MO, et al. Dopamine D2 receptors regulate the anatomical and functional balance of basal ganglia circuitry. *Neuron.* 2014;81:153–164.
127. Stenevi U, Björklund A, Svendgaard NA. Transplantation of central and peripheral monoamine neurons to the adult rat brain: techniques and conditions for survival. *Brain Res.* 1976;114:1–20.
128. Björklund A, Landfall O. Replacing dopamine neurons in Parkinson's disease: how did it happen? *J Parkinson's Dis.* 2017;7(Suppl 1):S23–S33.
129. Chen Y, Xiong M, Dong Y, et al. Chemical control of grafted human PSC-derived neurons in a mouse model of Parkinson's disease. *Cell Stem Cell.* 2016;18:817–826.
130. Ungerstedt U, Arbuthnott GW. Quantitative recording of rotational behavior in rats after 6-hydroxy-dopamine lesions of the nigrostriatal dopamine system. *Brain Res.* 1970;24:485–493.
131. Cappuccino BR, Harris LW, Anderson DR, Lennox WJ, Gales V, Dawson JS. Use of the accelerating rotarod for assessment of motor performance decrement induced by potential anticonvulsant compounds in nerve agent poisoning. *Drug Chem Toxicol.* 1992;15:177–201.
132. Dostrovsky JO, Hutchison WD, Lozano AM. The globus pallidus, deep brain stimulation, and Parkinson's disease. *Neuroscientist.* 2002;8:284–290.
133. Bi A, Cui J, Ma YP, et al. Ectopic expression of a microbial-type rhodopsin restores visual responses in mice with photoreceptor degeneration. *Neuron.* 2006;50:23–33.
134. Bowes C, Li T, Danciger M, Baxter LC, Applebury ML, Farber DB. Retinal degeneration in the rd mouse is caused by a defect in the beta subunit of rod cGMP-phosphodiesterase. *Nature.* 1990;347:677–680.
135. McLaughlin ME, Sandberg MA, Berson EL, Dryja TP. Recessive mutations in the gene encoding the beta-subunit of rod phosphodiesterase in patients with retinitis pigmentosa. *Nat Genet.* 1993;4:130–134.
136. Lagali PS, Balya D, Awatramani GB, et al. Light-activated channels targeted to ON bipolar cells restore visual function in retinal degeneration. *Nat Neurosci.* 2008;11:667–675.
137. Francis PJ, Mansfield B, Rose S. Proceedings of the first international optogenetic therapies for vision symposium. *Transl Vis Technol.* 2013;2:4.
138. Xu DS, Ponce FA. Deep brain stimulation for Alzheimer's disease. *Curr Alzheimer Res.* 2017;14:356–361.
139. Eldridge MA, Lerchner W, Saunders RC, et al. Chemogenetic disconnection of monkey orbitofrontal and rhinal cortex reversibly disrupts reward value. *Nat Neurosci.* 2016;19:37–39.
140. Klabunde T, Hessler G. Drug design strategies for targeting G-protein-coupled receptors. *Chembiochem.* 2002;3:928–944.
141. Chang WH, Lin SK, Lane HY, et al. Reversible metabolism of clozapine and clozapine N-oxide in schizophrenic patients. *Prog Neuropsychopharmacol Biol Psychiatry.* 1998;22:723–739.
142. Baskin YK, Dietrich WD, Green EJ. Two effective behavioral tasks for evaluating sensorimotor dysfunction following traumatic brain injury in mice. *J Neurosci Methods.* 2003;129:87–93.

## FURTHER READING


1. Alpaca C, Andreoli L, Irene S, Jakobsson J, Fieblinger T, Cenci MA. Chemogenetic stimulation of striatal projection neurons modulates responses to Parkinson's disease therapy. *J Clin Invest.* 2017;127:720–734.
2. Davis KL, Mohs RC, Marin D, et al. Brain choline acetyltransferase and mental function in Alzheimer disease. *J Am Med Assoc.* 1999;281:1401–1406.
3. Katzman R. The prevalence and malignancy of Alzheimer disease: a major killer. *Arch Neurol.* 1976;33:217–218.
4. Nagy A. Cre recombinase: the universal reagent for genome tailoring. *Genesis.* 2000;26:99–109.
5. Schrag A, Schott JM. Epidemiological, clinical, and genetic characteristics of early-onset parkinsonism. *Lancet Neurol.* 2006;5:355–363.

SECTION VI

# Ethical Considerations

CHAPTER 25

# Genes and Human Behavior: Ethical Implications

**Jon Beckwith[1], Robin Pierce[2]**
[1]Harvard Medical School, Boston, MA, United States; [2]Tilburg University, Tilburg, The Netherlands

## INTRODUCTION

The last 50 years have seen a significant change among biologic, biomedical, and behavioral scientists in their acceptance of ethical and social constraints on their research. The events responsible for these changes include the period of social and political activism of scientific and medical researchers beginning in the 1960s; the establishment in the 1970s of the National Commission for the Protection of Human Subjects in Biomedical and Behavioral Research, which issued its important "Belmont Report"[1]; the year-long moratorium on recombinant DNA research to consider its potential dangers in 1975; and the formation of the Working Group on Ethical, Legal, and Social Implications of the Human Genome Project established in 1989 by the project head, James Watson. While several of these changes were initiated in the United States, in one form or another, they became widely accepted in many countries. These road marks were not just temporary guidelines for biologic research. It is almost de rigueur today for professional scientific societies that cover new research fields such as systems biology and synthetic biology to have readily established committees on the ethics and social impact of their science. In some cases, discussions of certain specific social/ethical issues such as dual-use research is even required by funding agencies. The recent U.S. National Academy of Sciences Conference on the social/ethical issues in the pursuit of the new genetic technique of gene drive was exhaustingly examined, involving researchers from around the world and academics from many fields, in addition to the biologic sciences.

Deliberations at this level, while important, are limited in the scope of what they can affect and direct. No one would argue that all potential damaging consequences of research have been successfully confronted, as there are still areas where the impact of the publication of a single scientific paper may have destructive effects on individuals or groups. Here, we would like to point out such examples in the field of human behavioral genetics. In particular, we review a series of scientific reports that have taken hold and lasted for many decades in the public's mind well after the scientific community has dismissed the reports as invalid. In many cases, these faulty ideas have influenced social norms. These examples raise questions about whether there is a responsibility of the

*Molecular-Genetic and Statistical Techniques for Behavioral and Neural Research*
ISBN 978-0-12-804078-2, https://doi.org/10.1016/B978-0-12-804078-2.00025-8

scientific community and/or individual scientists to intervene in such situations. We argue that it is difficult to expect scientists to respond or anticipate potential dangers of this sort without a sufficient exposure to the history of science in their education. Becoming sensitized to such potential problems by understanding past examples of faulty science's impact better positions scientists to minimize and confront these situations that they see in their own fields.

## HENRY HERBERT GODDARD AND THE KALLIKAK FAMILY

Our first example of a scientific claim that persisted well beyond its ill-deserved shelf-life goes back to the beginning of the 20th century. In 1912, the eminent psychologist Henry Herbert Goddard published "The Kallikak Family: A Study in the Heredity of Feeble-Mindedness."[2] He described the siring by a man named Martin Kallikak of two families of children as a result of his relationships with two women. Goddard himself invented the pseudonymous surname Kallikak, which included the two Greek roots Kallos (beautiful) and Kakos (bad). This implicit labeling of the two sets of children was done to highlight the characterization of one of the women as having "outstanding character" (Kallos) and the other woman as supposedly "feeble-minded" (Kakos). Goddard reported that the children born from the "high quality" mother showed the same superior characteristics as that mother, while the latter union produced what he called "a race of defective degenerates."

With the exciting rediscovery of Gregor Mendel's theory of genetics in 1900, it became common to label people as genetically defective if they showed presumed undesirable qualities such as low intelligence or criminality. The eugenics movement in the United States looked to genetics as providing support for their proposals to sterilize or segregate socially the "unfit" and to ban those of different, supposedly defective, ethnic groups immigrating into the United States.

Goddard became prominent in the eugenics movement in the United States in the first few decades of the 20th century, and his dramatic description of the two sets of Kallikak children became one of the strongest supports for the eugenicists.[3,4] But, reexaminations of Goddard's research revealed that he and his coworkers used arbitrary techniques in developing their judgment of a person's qualities.[4] Often, Goddard's assistants were asked to estimate a person's IQ score simply by examining the looks of the subjects being studied. He and his colleagues used other extraordinarily subjective evaluations of the mental capacities and character of its live subjects. Furthermore, the evaluations of the character of deceased members of families being studied were based on the retrospective reports by surviving family members of the behavior of those individuals.

Although some scientists were critical of Goddard's scientific approaches, the myth of the Kallikaks retained its hold on the public imagination. As decades went by after the publication of "The Kallikak Family," and despite the scientific exposés that pointed out the low quality of the research, these rejections in academic journals had little effect and never reached the public. Thus, the Kallikaks remained the popularly accepted symbol of the importance of genetics in human social behavior and aptitudes. As a child growing up in the 1940s, one of us (J.B.) heard his mother and others talk of the Kallikak family as a valid indicator of the inheritance of traits like intelligence or bad behavior. The work even remained reputable among some scientists as late as the 1961 appearance of a new edition of a psychology textbook by Garrett and Bonner.[5] This widely used text continued to report Goddard's Kallikak story, presenting it in discussions of genetics and behavior. Furthermore, in 1977, the NBC television network premiered a situation comedy entitled "The Kallikaks," who had been described as a "lying and cheating" "white trash" family.[6] Thus, the myth of the Kallikaks still continued to be familiar at least to some in the United States until 65 years after the publication of Goddard's book.

## THE RISE AND FALL OF THE CRIMINAL CHROMOSOME

The development of techniques for visualizing human chromosomes in the 1950s opened an important era in biomedical sciences. Very quickly, discovery of extra chromosomes, lost chromosomes, and chromosomal deletions and insertions were observed in some individuals. In some cases, these observable chromosomal variants were correlated with new syndromes. This progress led the research group of Edinburgh geneticist Patricia Jacobs to ask whether criminal behavior might be associated with variations in chromosome numbers. To pursue this question the group examined the karyotypes of the inmates at a State Hospital for the Criminally Insane. In 1965, the authors presented their findings in Nature Magazine that 3.5% of the inmates of this Edinburgh institution carried an extra Y chromosome.[7] The article, entitled "Aggressive Behavior, Mental Subnormality and the XYY Males," suggested that the 3.5% finding likely represented an unusually high proportion of males with an extra Y chromosome in this hospital for the criminally insane. Without any further detail and no direct knowledge of the frequency of XYY males in the general population, this study attracted considerable attention by the media upon its publication. The idea that a double dose of the male Y chromosome in these individuals could seem naively to fit with highly aggressive behavior must have strengthened the idea in many people.

However, a mass murder that took place in Chicago in 1966 vastly amplified the media and public belief that XYY men were violent criminals.[8] The murderer, Richard Speck, was quickly caught. However, because of communication difficulties between journalists and a scientist and a geneticist's speculation to a reporter, the news spread that Speck was an XYY male. Headlines such as "Chromosomes and Crime" and "Born Bad" appeared in newspapers and magazines. But, we now know Speck was an XY not XYY male. This was demonstrated when, out of curiosity, a Vanderbilt University geneticist Eric Engel requested a sample of Speck's blood and examined it for his chromosome makeup. Although Engel found Speck to be an XY male, newspapers continued to refer to Speck as an XYY male. Engel asked to be able to identify Speck's chromosomes again with a new sample. He still only observed the XY chromosomal makeup of Speck. Finally, Engel was successful in getting the Associated Press to distribute a report of Engel's finding. However, it was never covered by major newspapers or news magazines. In 1972, Engel published an article in the *American Journal of Mental Deficiency Research* where, yet again, he reported the same finding, and expressed being "amazed" that the misinformation about Speck had persisted, despite the chromosome analysis.[9]

Within several years after the publication of the paper by Jacobs et al., other researchers summarized information that had accumulated in the interim indicating that the links of the extra Y chromosome to aggression and criminality were weak.[10] XYY males in institutions (including those of the original Jacobs study) were not aggressive and had committed crimes against property, not people. The frequency of XYY males institutionalized in subsequent studies never came close to the high frequency seen in the original Jacobs study. The review concluded that behavior of the XYY male was probably not very different from that of XY males. In 1976, a study of a large number of XYY males living in the general population supported the conclusions of the 1974 review.[11] In 1982, 17 years after the publication of the original paper, Dr. Jacobs, dismayed by the media coverage of her study, stated, "In retrospect, I should not have used the words 'aggressive behavior' in the title of my paper and should not have described the institution as a place for 'the treatment of individuals with dangerous, violent or criminal propensities.'"[12]

Despite the criticisms, failed replications, and eventual dismissal of the idea of the criminal chromosome by scientists, the myth of the XYY male persisted, spreading through the culture as an apparent leading example of genetic roots for criminality, much like the Kallikaks had become an exemplar of intelligence and criminality's roots in heredity. In high school biology to medical school psychiatry textbooks, sometimes illustrated with Richard Speck's picture, students continued to learn well into the 1970s that XYY males are super aggressive and prone to criminality. A series of British novels about "The XYY man," a superspy for MI5, was turned into a Granada TV series in 1976.[13] Two movies "featured" XYY males, Italian horror movie director Dario Argento's "The Cat O'Nine Tails" in 1971 and "Alien"[3] in 1993. In the latter, Sigourney Weaver lands her damaged space ship on a planet inhabited by XYY males who have been exiled from Earth. Their leader tells Weaver that they are a bunch of "thieves, murderers, rapists and child molesters … all scum." In the same year that "Alien"[3] appeared, an episode of the TV show "Law and Order" featured a story of a 14-year-old boy with an extra Y chromosome who beat his friend to death. In 1985, Harvard professors James

Q. Wilson and Richard J. Herrnstein used the supposed link of XYY to criminality to support arguments in their book "Crime and Human Nature."[14] Some high school biology classes in the United States have been teaching the validity of the XYY-criminality connection well into the 21st century.[15]

In addition to the entertainment aspects of the prolongation of the XYY story, the films, TV, and books, and its continuation into social policy proposals, the wide acceptance of claims about the extra Y chromosome has also had its direct social impact. In the 1970s, in at least two states, juveniles from poor areas were screened for the XYY genetic makeup.[16] About 5000 African-American children in Baltimore were screened for the so-called criminal chromosome. To this day, this screening project is well-known, and it has contributed to the general mistrust by the black community of medical and police representatives.[17,18] Studies in the United States and other countries screened newborn baby boys for the extra Y chromosome in research that posed significant ethical problems, including stigmatization. Even broader claims that criminality was genetically determined, not the result of poor environments, were bolstered by the media reports on XYY males.[19]

The research done since the 1970s indicates that the only characteristics that may be applied to XYY males generally is that they are taller and tend to exhibit slower learning curves in certain subjects. One of the most recent reports studied the development of XYY males from birth along with XY and XXY males as controls.[20] The authors present "evidence for a slightly increased liability to antisocial behavior in XYY men." However, controlling for the various factors that might have influenced behavior of the XYY males, the authors conclude that "IQ contributes significantly to the likelihood of having a conviction, but that social class and karyotype [the extra Y chromosome] does not."

Now, many years of identifying XYY males from screening has yielded broad information about their development and lives. None of this new information indicates significant criminal behavior among XYY males. Rather the parents of XYY children have been incorporated into organizations for parents of children who exhibit broad-spectrum autism. XYY males are considered by some as exhibiting high-functioning autism.[21] Ironically, given the XYY history, among the materials circulated by parent organizations for children with the extra Y chromosome are booklets with titles such as "Jack and His Extra Y" picturing a proudly smiling young XYY boy on its cover. The word "criminality" never arises in all of this literature except to state that this characterization of the XYY male is invalid.[22] As a further awareness of the long-lived misrepresentation of XYY males, the National Organization for Rare Disorders states that XYY males generally go on to lead full, healthy, and normal lives.[23] Patricia Jacobs and her coworkers have continued research on individuals with sex chromosome trisomy. In a review of a number of studies, they note that XYY males had mean IQs that were within broadly normal limits, had marked difficulties in speech and language, motor skills, and educational achievement, but nevertheless lived independently.[24]

Like the Kallikaks, the XYY story has had an extraordinarily long shelf-life, having been perpetuated for at least 40 years after the original report by Jacob and her colleagues. Both Eric Engel the cytologist and Patricia Jacobs the geneticist expressed disappointment and regret that the story had been blown up to such proportions. But their subsequent statements in scientific journals never received any attention, while the original "criminal chromosome" version of the story persisted. One possible sign that, after all these years, the tide of misinformation in the public arena may have begun to turn is a 2007 episode of the television series CSI Miami in which the crime investigators come to *learn* that the original claim for a determinative role of the extra Y chromosome in criminal behavior has been disproven.

## MONOAMINE OXIDASE A: THE WARRIOR GENE?

The story of the "warrior gene" is a quintessential example of the persistence of questionable science capturing the public imagination and surfacing in ways that are not justified given the problems with the underlying research. The story has gone through different stages with expanding versions of the meaning of the monoamine oxidase A (MAOA) gene.

In 1993, reports published in the *American Journal of Human Genetics*[25] and *Science Magazine*[26] told of studies on a family in The Netherlands showing that an inactive version of the gene for the enzyme MAOA was present in several of the male members of a family. These men exhibited extremely aggressive behavior and some degree of mental retardation. In both articles, authors Hans Brunner and his colleagues hinted at a possible generalization of this finding from this one family to the larger problem of aggression in society, concluding that "... should a mutation in the MAOA structural gene be identified in our family, this will have implications for the study of the biological mechanisms underlying disturbed aggression regulation in general ..."[25] The mutation was identified in the *Science* paper.

Based on these results, a *Science Magazine* reporter suggested that "it might be possible to identify people who are prone to violent acts by screening for MAOA gene mutations ..."[26] Popular science magazines and the general media picked up the hint.[27] Prominently featured articles appeared in the popular press, including an article in *Newsweek* entitled "The genetics of bad behavior."[28] This general news magazine with a circulation of over 3 million published this article accompanied by a graphic showing a violent confrontation between Palestinians and Israelis, implying a genetic basis for world strife, in general. This notion of genes for violence later surfaced in a legal case in Georgia in which Xandra Breakefield, one of the authors of the *Science* paper, was called to testify regarding the genetic basis of aggressive and violent behavior.[29] Although the legal appeal to genetics was unsuccessful in this case, the rapid spread of supposed genetic knowledge to the public and the ways that it can be used and misused became apparent. To their credit, the scientists involved in the study had come forward to express concern about media representations of their reports. Dr. Breakefield was sufficiently dismayed by the publicity to announce in 1994 that she would no longer work on links between violence and genes (personal communication). Dr Brunner later flatly stated that "... the notion of an 'aggression gene' does not make sense."[30] The mutation in the MAOA gene in this family was shown to completely eliminate enzyme activity by causing termination of the translation of the protein before it was completed. No instances of such a deleterious mutation in the human MAOA gene have been found since.

Nevertheless, the progress toward the idea of MAOA's connection with aggression did not end here. Rather, it gained new impetus 10 years later in 2002 with the publication by Caspi and coworkers of a study in *Science Magazine* on individuals who were found to have variants of the MAOA gene that resulted in either increased levels or decreased levels of the MAOA enzyme.[31] When the researchers examined adult males with a variant of the MAOA gene termed MAOA-L (L for low activity), they found that those males who had been maltreated as children were more likely to develop antisocial or violent behavior in comparison to children with the high-activity MAOA variants. The conclusions of the authors were necessarily very different from the previous MAOA study in that the outcome for these males appeared to be due to a combination of genetic and environmental (G × E) effects.

The media coverage of the Caspi et al. study was mixed, with some reporting on the "gene/abuse" link and others deterministically referring only to the genetic component of the studies, with headlines such as "Study Finds Genetic Link to Violence" and "Criminals Share a Common Genetic Flaw."[32] Subsequently, a number of researchers published their efforts to replicate the study of Caspi and coworkers. These subsequent replication attempts yielded mixed results, as follows.[32] Some studies reached similar conclusions, another reported a "non-significant trend" toward the Caspi conclusion, and yet others failed to replicate the findings, including the gene—environment interaction. At least two studies found a contradictory inverse relationship, that is, an association between the *high-activity genotype* and aggression in males. These fascinating results raised the possibility that other factors, whether genetic, cultural, or environmental that varied between societies, influenced the outcome of replication efforts.

Within a few years, a group at Vanderbilt University's medical school and its law school suggested that the information available on MAOA, the effects of variants of MAOA, and child abuse on human behavior might be relevant to some legal cases. The authors discussed their experience in genotyping criminal defendants, including for the MAOA variants, and presenting genetic information at criminal trials.[33]

The notion of a "warrior gene" entered the spotlight in 2006 when researchers Lea and Chambers in New Zealand conducted a study involving the Māori, an indigenous people.[34] The researchers sought to understand gene variants of monoamine oxidase as a way of identifying elevated risk for alcohol and tobacco dependence. Based on a finding that a variant of the MAOA gene, MAOA-L, was present in 56% of the 17 Māori males in the sample, the researchers drew the conclusion that an evolutionary mechanism was at work, asserting that positive selection for MAOA-L accounted for the high prevalence of this gene variant in the Māori male population. The authors suggested that the selection may have resulted when the "Polynesian males who embark on long, dangerous canoe voyages and engaged in (and survived) war with other islander tribes carried the AGCCG haplotype, coupled with the 3-repeat allele of MAO-A30bp-rpt [an MAOA variant], to Aotearoa (New Zealand) where they both increased in frequency due to rapid population growth."[34]

When the researchers presented their study at the 11th International Congress of Human Genetics in Brisbane, Australia, in 2006, the media pounced on it, reporting that Māori men were genetically predisposed to "violence, criminal acts, and risky behavior."[34a] In an interview with the Australian Associated Press, Rod Lea, lead author of the study, apparently extended this unsubstantiated evolutionary hypothesis regarding a genetic basis for the Māori as "fearless warriors" to one supporting a predisposition to increased criminality. With the coverage of this study, the notion of a "warrior gene" entered the public lexicon, and the unfortunate and troubling moniker of the "warrior gene" became an accessible explanation and justification for attributing certain negative or socially undesirable behavioral patterns to a specific group of people.[35]

Several in the scientific community were quick to point out that the science simply did not support the claims that the authors had made.[36] Under even modest scrutiny, critical problems became evident. With a total sample of 46 Māori males, each having at least one Māori parent, the researchers reported a frequency of the low variant allele of 56% in the Māori males tested as compared to 34% in Caucasian males, as reported in an earlier study by Lea and Chambers. The study went on to report that in a subsample of 17 Māori males (selected because they had eight Māori great grandparents and thus reduced European admixture), the AGCCG haplotype frequency was increased in carriers of the "functional" 3-repeat (3R) allele compared to non-Māori carriers. The authors then stated that "this finding in itself is evidence of positive (natural) selection acting at the MAO-A gene." On the basis of 17 (out of 46) Māori males in the sample having the 3R allele, the researchers concluded the existence of a genetic basis for a complex sociocultural behavior intersecting with complex historical circumstances. With such astounding simplicity, coupled with a remarkable paucity of supporting data, it is little wonder that this study inspired so much controversy.[36a]

The generalizability of conclusions about an entire population from such a small sample is problematic, at best. In fact, the only study that had identified an association (and had done so within the context of a gene—environment interaction) was conducted with Caucasians.[37] Making the assumption that an allele functions in precisely the same manner in different populations is unjustified. Third, a core assumption by Lea and colleagues was that MAOA-L is associated with aggression in Māori males when, in fact, there are no reports of genetic epidemiological experiments that have tested for an association between MAOA and aggression, and moreover, there is no direct evidence that supports a claim that the MAOA 3R variant actually confers "warrior-like" qualities on Māori males.[37]

An additional source of criticism focused on the social implications of the research and the handling by the media. A study reporting a genetic predisposition to aggressive behavior and criminality among the Māori in an existing social context of pervasive stereotyping of that ethnic group regarding violent behavior based on such thin evidence seems, at best, irresponsible. Among the numerous public statements decrying the declaration of a Māori "warrior gene," an article appeared in *MAI Review* (*Māori and Indigenous Review*), "'Warrior genes' *and the disease of being Māori*," in which the author, anticipating possible implications of wide acceptance of the study's findings, points out that "ascribing Māori domestic violence to genomic expression of a MAO gene raises the possibility that being Māori could then be placed in the abnormal category of syndrome and disease and subject to therapeutic interventions."[37] Asserting a genetic basis for negative behavior in a particular ethnic group had long been viewed with suspicion, in part due to the eugenic reverberations that it almost invariably invokes in many places in the world. Furthermore, ethical issues begin to emerge with the making of such claims based on such weak science. Yet, as the "warrior gene" story went on to show, simplistic and largely unsubstantiated inferences made from very limited scientific findings can quickly become mythical in their reach and uptake.[37]

The insidiousness of the scientific reports of a "warrior gene" was exacerbated by the media. Not only was there no epidemiological evidence to justify this genetic link between aggression and MAOA-3R, further concerns arise in that this genetic basis for violence was ascribed to the Māori population when other ethnicities referenced in the study demonstrated an even higher frequency of the MAOA-3R. Notably, little was made of the fact that an even higher percentage of the 3R variant (77%) was found among Chinese males, and yet, no claims were made about a Chinese predisposition to criminality stemming from a "warrior gene."[37]

Indeed, in 2009, *New Scientist* reported on one such study with the headline, "Gangsta gene identified in US teens."[38] This study, published in *Comprehensive Psychiatry*, reported that based on data drawn from the National Longitudinal Study of Adolescent Health for 2000 teens, those with the MAOA-L gene were not only twice as likely to join a gang, but they were also more likely to use weapons than those without this variant.[39] The problems with this study abound. Not only were both gang membership and use of weapon based on self-report, there was no epidemiological genetic basis for an association between MAOA-L and these phenomena, should they actually have occurred. Essentially, the study asserted a genetic basis for two events, the circumstances and factual occurrence of which were neither corroborated nor scientifically established. Moreover, the sample involved exclusively African-Americans, giving rise to two concerns: (1) the only association between MAOA-L and aggression had been found in a Caucasian population in New Zealand, and (2) the generalizability of a study based on an exclusively African-American sample is limited, even if the foundational association between MAOA-L and aggression had been established. Moreover, the study also showed that the majority of MAOA-L carriers were not gang members.

The announcement of a "warrior gene" in 2006 by Lea and colleagues also generated a line of questionable progeny that somehow led to the attempted introduction of evidence of MAOA-L on behalf of the defendant. In 2006, Davis Bradley Waldroup was accused of violently murdering his wife's friend and injuring his wife in a bloody rage. Initially charged with first degree murder, Waldroup's defense offered "evidence" of diminished responsibility in the form of the MAOA-L gene along with an assertion of childhood abuse. Quite possibly the first US case in which genetics was used to influence a finding of legal responsibility, the court then reduced the charge of first degree murder to one of voluntary manslaughter. Even the most respected of news outlets covered the story and discussed the possibility of a "warrior gene." A 2010 NPR broadcast, entitled "Can your genes make you murder," examined the Waldroup case and featured interviews with some of the jurors as well as a defense expert, William Bernet of Vanderbilt University. Echoing the as yet unreplicated Caspi study, Bernet stated, "His [Waldroup] genetic makeup, combined with his history of child abuse, together created a vulnerability that he would be a violent adult."[40] Bernet and colleagues had been called in on a number of occasions to genotype defendants and testify about genetic information in criminal trials.[34] The Waldroup case was their first successful attempt to employ the MAOA-L variant to influence guilt or sentencing. NPR had also conducted interviews with jurors who had served on the Waldroup case. Their statements about their deliberations left little doubt that genetic evidence had played a role in their findings. In fact, one juror commented, "Some people without this [gene] would react totally different than he would … A diagnosis is a diagnosis, it's there … A bad gene is a bad gene."[40]

In Italy in 2010, in the sentencing phase of a trial of a man convicted of murder, the court reduced his sentence from 9 to 8 years because "the accused had tested positive for genetic variants that made him particularly prone to be aggressive under stressful circumstances …"[41] Notably, this decision provoked an editorial from the European Journal of Human Genetics presented on behalf of the Public and Professional Policy Committee of the European Society of Human Genetics in which they concluded that "there is no scientific support to declare that gene variants, claimed to predispose to aggression, would make the carriers incapable of repressing an aggressive behavior and thus unable to choose appropriate socially acceptable behaviours."[41]

In 2011, a documentary was shown on the National Geographic Channel entitled "Born to Rage: Inside the Warrior Gene." This popular science outlet explored the notion that some people may be "born to rage" based on an increasing number of studies purportedly supporting this hypothesis. Now available on DVD, the persistence of the notion of a warrior gene is well poised for a long shelf-life despite the holes in the science. The popular media, perhaps less concerned with science credibility, presented the phenomenon of a warrior gene with zest and flair. Reaching an even larger audience, the warrior gene was the subject of the U.S. popular daytime talk show "Dr Phil" in April 2011.[42] This show featured several individuals who viewed themselves as "rageaholics." As part of the program, these guests were tested for MAOA-L, and all had a positive result, leading to the show's conclusion that "inside a rageaholic's DNA, a warrior gene may be pulling the strings." This popular talk show with an audience of roughly 4.8 million also invited viewers to get themselves tested for the warrior gene and referred them to the website of a genetic testing lab where they could order the test. The perpetuation of bad science found an accomplice in the world of entertainment, the full consequences of which may only be beginning to surface in these exceptional court cases.

It is encouraging to note that the most recent case in which genetic evidence was proffered by the defense was rejected by the court. In 2015, Anthony Yepez was charged with murder and sought to raise a genetic defense, but the judge refused to permit testimony about a warrior gene and its possible impact. Apparently, the judge felt "iffy" about the reliability of science to prove what it proposed to prove and supported the state's request to disallow the testimony.[43,44]

The life of small studies can rarely be predicted, but the saga of the warrior gene highlights several key points relating to the responsibility of the scientific community and others to remain vigilant regarding unjustified uses of questionable science, to provide clear explanations of the limitations, and to correct misleading portrayals of the science. Although it is true that a bell cannot be unrung, stepping up to provide a reasoned critique of science headed down a spurious path, as some scientists have done regarding the warrior gene, is an important function of the scientist as citizen and professional, especially when the stakes are so high. The story of the warrior gene is, in part, remarkable for its longevity, continued vibrancy, and social impact. Indeed, the studies persist.

Early in the 21st century, BBC-TV introduced its UK version of a Law and Order series of shows. One of the episodes in the year 2009, "Unloved," describes a young boy who kills another boy. This episode was based directly on the version of the earlier

U.S. Law and Order that had featured an XYY boy (see earlier). However, ironically, this remake could no longer use the U.S. entire script for the murderer since XYY was no longer considered the "criminal chromosome." Instead, the script writers gave the young killer an MAOA-L variant, the new "poster boy" for criminality.

## BOYS, GIRLS, AND MATHEMATICAL ABILITY

In 1980, researchers at the Studies of Mathematically Precocious Youth of Johns Hopkins University reported in *Science Magazine* on mathematical test performance differences between boys and girls. The impact of this study was to reverberate through the subsequent decades. In their article entitled "Sex Differences in Math Ability: Fact or Artifact?,"[45] Julian Stanley and Camilla Benbow reported that they had given the SAT math test to seventh and eighth grade students who were mathematically precocious and analyzed differences between performances of the boys and girls. The most striking finding was that, among the highest level scorers on the test (700–800), boys outnumbered girls 13 to 1. The authors sought to measure environment/social factors that were responsible for this dramatic difference between the boys and girls scores. To do this, they considered two possible confounding factors that might have influenced the students differentially. They (1) quantified the number of mathematically based courses that the students had taken before taking the test and (2) questioned the students about their interest in math of all of those students. Finding no differences in relative interest in math and no different mathematic background between boys and girls, along with the dramatic difference in SAT scores, the authors concluded their *Science* article: "... it seems likely that putting one's faith in boy-versus-girl socialization processes as the only permissible explanation of the sex differences in mathematics is premature."

*Science Magazine* highlighted this study in their News and Views section with an article entitled "Math and Sex: Are Girls Born with Less Ability?" and with a subheadline "A Johns Hopkins Group Says 'Probably': Others are not so Sure."[46] Strengthening the impression that the authors favored social explanations rather than biologic ones for differences in achievements of boys and girls, Dr. Benbow responded to *Science* in an interview saying "Women would be better off accepting their differences and working to encourage girls to achieve as much as they can than to constantly blame their lesser achievements in mathematics solely on social factors."[46]

A media blitz followed the article and authors' comments. *Time Magazine*'s article, "The Gender Factor in Math: A New Study Says Males May Be Naturally Abler than Females" was accompanied by a cartoon in which a schoolgirl is unable to multiply two single-digit numbers while a boy completes the multiplication of a four- and three-digit numbers.[47] *Newsweek's* article entitled its report "Do Males Have a Math Gene?"[48] Articles in both the weekly and Sunday *New York Times* also reported on the study immediately upon its publication. This story, in its most determinist form, spread very rapidly to magazines such as *Playboy*, *Reader's Digest*, and the popular science magazines that were being published in that era.[49]

However, not surprisingly, there had been extensive research on many other socialization factors, none of which Benbow and Stanley had considered possible confounding factors in their search for explanation of the boy-girl difference.[50] Numerous articles were being and had been published in educational, psychology, sociological, and other academic journals that pointed to the kinds of toys boys and girls played with, social attitudes that lead to boys' disdain of girls who were "science whizzes," the influence of parental and school teachers' preconceptions of differences in girls' and boys' aptitudes in math, as well as prejudices that girls interested in mathematics faced in college and beyond. Benbow and Stanley had hardly exhausted possible explanatory social factors by comparing only the number of courses taken by boys and girls and assessing expressed interest in math of boys and girls. As usual, critical responses to the conclusions of the Hopkins group were restricted largely to academic journals, not receiving any coverage by sources available to the public. It was the "male math gene" idea that persisted in the public's mind. Studies done to estimate the impact of the extensive publicity for this study reported a negative influence on parental attitudes and on the attitudes of school students themselves toward girls' math abilities.[50]

Twenty-five years later, an indication of the continuing impact of the Benbow-Stanley study surfaced in 2005. President of Harvard University, Larry Summers, in a widely reported speech, offered possible explanations for the low number of women in mathematics and related fields. For one of these explanations, he alluded to scientific evidence supporting a biologic explanation for this difference, stating that "... in the special case of science and engineering, there are issues of intrinsic aptitude ..."[51] Elsewhere, Summers cited as backing for this statement psychology Professor Steven Pinker's book "The Blank Slate" in which the Benbow-Stanley study is referred to as evidence for the "intrinsic aptitude" explanation.[52] Interestingly, Summers comments were deplored by many of the Harvard faculty and may have contributed, along with other issues, to his resignation in 2006.

However, shortly after Summers' controversial comments, the Johns Hopkins University study of mathematically precocious youth released data it had been collecting since 1980. In the years between 1980 and 2005, the great difference between boys' and girls' performance on the math tests had progressively largely disappeared. The 13:1 (male to female) ratio of 1980 had dropped to 4:1 in 1997 and to 2.8:1 in 2005.[53,54] Clearly, this improvement in girls' performance was not due to changes in girls' genes in the intervening years. The altered social environment began in the 1970s in which the women's movement in the United States was successful in pushing for improved educational environments for girls studying math and science. Furthermore, during that period, there was an increased awareness and understanding of the social factors that prevented women's advancement in math and science. These ideas did spread culturally and resulted in an increase in the number of women in these fields. Consistent with this social explanation, recent cross-cultural studies have presented evidence that women's achievements in math and science are correlated with overall improved attitudes toward gender of the society in which they live.[55–57]

Despite this significant change in the achievements of women in mathematics, there was an absence of any of the standard media coverage of these findings and of other studies that could have publicly undercut the basis for the conclusions of Benbow and Stanley of 1980. There were no News and Views stories in *Science Magazine*, no newspaper reports, no headlines in *Time, Newsweek*, or *Playboy*. It seems that refutations of what were once highly newsworthy stories and, in cases such as this, ones that influenced public attitudes are not newsworthy themselves. The authors are aware of one exception, which was an article that appeared, importantly, in the *Chronicle of Higher Education*.[54] Thus, the overblown response to the 1980 paper by Benbow and Stanley lasted in impact in the public arena without any effective response until 25 years later, ironically from the same scientific project, but without the publicity that was needed. It may be that social changes have ultimately begun to moderate the negative impact to some degree.

## SMART GENES, STUPID GENES

In 2005, *Science Magazine* published an article reporting the discovery of a human gene variant that was proposed to be important for the ongoing evolution of the human brain.[58] The researchers, led by Bruce Lahn of the University of Chicago, found that most humans have one of two versions (polymorphisms) of a gene called microcephalin. The microcephalin gene had previously attracted attention when researchers found that complete deletion of this gene led to a rare condition in which babies are born with very small brains. However, the specific function of the protein encoded by the microcephalin gene is unknown.

Lahn and his coworkers asked what the frequency was of these two polymorphisms in peoples from different parts of the world. They found that one of these variants was quite common in peoples of sub-Saharan Africa, and the other was found more often in populations in the rest of the world. They then asked whether one or the other polymorphism had been evolutionarily selected for in the populations where it was common, thus conferring some advantage on that population. To answer this question, they used a computer program to determine whether the prevalence of one of the polymorphisms in particular populations was due to genetic selection in the evolution of modern humans. They reported that results from the computer analysis indicated that the polymorphism found more commonly outside of Africa had been *selected for* sometime between 14,000 and 60,000 years ago. They then went on to point out that this period was one in which "modern human behavior, such as art and the use of symbolism" arose. As a result, in their conclusion, they speculated that the polymorphism found more commonly in Europeans and other populations, and not in sub-Saharan Africans, may have enhanced one or more of several traits including "brain size" and "cognition." These speculations on their findings were readily interpreted by others to suggest the possibility that sub-Saharan Africans were not as culturally advanced as other peoples for genetic reasons.

In the same issue of *Science*, Lahn's paper was highlighted with a "News of the Week" article entitled "Are Human Brains Still Evolving? Brain Genes Show Signs of Selection."[59] Despite the cautious wording of the "News" article, Lahn's study was picked up by many newspapers with reports that included Nicholas Wade's *New York Times* article, "Brains May Still be Evolving, Studies Hint,"[60] and the *Baltimore Sun*'s, "Two Evolving Genes May Allow Humans to Become Smarter."[61] *Discover Magazine* deemed the report one of the "top discoveries" of 2005. And without waiting for confirmation of the speculations by Lahn and coworkers, those with an eye on potential social implications of the study joined in with comments that were widely reported. John Derbyshire in *National Review Online* stated, "Our cherished national dream of a well-mixed and harmonious meritocracy ... may be unattainable."[62] To the *New Scientist*, University of Wisconsin anthropologist John Hawkes claimed that "Whatever advantage these genes give, some groups have it and some don't. This has to be the worst nightmare of people who believe strongly there are no differences in brain function between groups."[63]

This rapid extension from the scientists' preliminary results and expansive speculation to commentators' claims of relevance to issues of equality is disturbing, but not unusual given the other examples to which we have referred. Nevertheless, it is hard to see how one is justified in making such leaps of logic in the absence of any knowledge of what the microcephalin gene does other than the extreme effects of its complete deletion. Nothing is known about the protein product of the gene, about its function in the brain or in other parts of the body, and about whether there are any effects of the two polymorphisms on the expression of the gene. At the time of publication, the authors had not sought evidence for differences in brain size or brain function in individuals carrying one or the other polymorphism. Moreover, during the vast time range of 14,000–60,000 years ago, which the authors concluded was when one of the polymorphisms had been selected for, many other changes in human living conditions and culture were occurring other than the rise of art and symbolism. A focus on those specific features of the evolution of human culture seemed a particular stretch. Finally, other human population geneticists questioned the validity of the specific computer program used by Lahn's group as a means of concluding anything about the evolutionary selection of one of the polymorphisms.[64] Using what they claimed to be a more sophisticated computational approach, these latter scientists concluded that this polymorphism had not been selected for.

However, within a year, all of these criticisms became irrelevant when other scientists, trying to test the speculative suggestions of Lahn and coworkers on brain size and cognition, showed that they could not be correct. Specifically, these groups reported that there was no difference in brain size nor in IQ test performance between people with one or the other polymorphism.[65,66] Moreover, Dr. Lahn stated to a reporter that he tried determining the version of his own microcephalin gene and reported that the experiment's outcome was "blurry," and that "it wasn't looking good."[67] Much of this information was reported in the second news article in *Science*.[68]

In this case, it seemed as though the corrective features of the scientific process had worked. Scientists had reported on new findings, and others tested the theories of the original paper and showed that the original speculation on issues of intelligence and brain size were incorrect. Yet, missing again is the communication to the public of studies that falsified aspects of the original *Science* paper. Other than the *Science Magazine* article, the only media reports we could find on these refutations were an article in the *Wall St. Journal*[67] and one from the Australian Broadcasting Company that proudly pointed out that some of the researchers who were coauthors of a refutational article were Australian.

## BIOLOGIC RACE REVISITED OR NOT

This last scientific study to be discussed here is one that has fueled a controversy over the question of race and genetics and differs significantly from the other cases we have described in this chapter. Specifically, the authors of the research have repeatedly emphasized their concerns that their work has been distorted to justify the argument for existence of biologic races. In 2002, Noah Rosenberg and his coworkers at Stanford University published an article in *Science Magazine* in which they sought to determine whether one could group people by variations (markers) in their DNA in a way that correlated with those people's geographic regions of origin.[69] There was no simple way to group people since there were no "private genetic markers" that were specific to different regions of the world. Instead, the researchers found that the sophisticated computer program STRUCTURE could sift through the genomic markers (microsatellites) that *varied in their frequency* in peoples from different geographic regions. They found that probabilistically they could indicate the likelihood that a subject came from a particular region. The researchers programmed the computer to go through the data several times to see if they could neatly assort the individuals into two, three, four, five, or six geographic groups. When the data sorted people into five groups, it was noticeable that those groups fell roughly into the same geographic regions that were proposed by one of the earliest scientists (Ludwig von Blumenthal) to claim the existence of five races. However, the Stanford team themselves never used the word race in their paper. The publication of the paper was accompanied in *Science* by a "Perspectives" article by two geneticists, who explained the research in simpler terms and also did not refer to any implications relevant to questions of race.[70] However, a report on the *Science* paper by *New York Times* science writer Nicholas Wade stated that the researchers "had concluded that people belong to five principal groups corresponding to the major geographical regions of the world: African, European, Asian, Melanesian and the Americans."[71]

Wade misrepresented the paper in that he referred to the sorting of peoples into five groups as the only grouping in the *Science* paper, ignoring the others that were formed when the computer was asked to sort people into two, three, four, and six groups. In subsequent papers published in academic journals and books by members of the Stanford group, the authors presented in a clearer way how the analysis worked, thereby refuting Wade's interpretation. Later, as far more extensive human data from around the world became available, the Stanford group and others used the same STRUCTURE program. With data that was closer to saturating more and more regions from around the world, other researchers found that people could be assorted rather neatly into larger numbers of groups, in one case as many as 18.[72]

Still, it was not long before this striking work on populations of the world was distorted by those with white supremacist views to support their claims that the world was divided into races and that African-Americans were an inferior race. The authors of the Stanford study had not drawn any such conclusions. Nevertheless, whether it was from word of mouth or the initial reports of Wade and his later book, at least the *White Nationalist American Renaissance Magazine* and Stormfront began to follow publications in human population genetics that suggested that contemporary genetics would support their racist views.[73,74]

These views were strengthened even more with the publication in 2014 of Nicholas Wade's book, *A Troublesome Inheritance: Genes, Race and Human History*.[75] Wade argued that those peoples who emigrated "out of Africa" some 50,000–100,000 years ago may have undergone positive genetic selection to evolve new social behaviors. These abilities, he suggested, would have allowed them to build social institutions that were necessary for survival in the change from hunter-gatherer lifestyles to more settled communities. Those peoples who remained in Africa would not have undergone this genetic selection, Wade speculates, leaving them without the genes for advanced culture. Among the reviewers of Wade's book, scientists were overwhelmingly critical of Wade's arguments. In particular, one scientific reviewer was critical of Wade's claim that he was only speculating, although he would repeatedly make declarative statements of fact for genetics of behavior of different peoples.[76]

In response to a review (critical) of the Wade book that appeared in the *New York Times Book Review*, Rosenberg, a coauthor of the *Science* paper and a few other population geneticists gathered the signatures of ~140 researchers in the field who protested what they considered the hijacking of their work by Wade. They stated that Wade's book was a "misappropriation of research from our field to support arguments about differences among human societies ... [and state that Wade] juxtaposes an incomplete and inaccurate account of our research on human genetic differences with speculation that recent natural selection has led to worldwide differences in IQ test results, political institutions and economic development. We reject Wade's implication that our findings substantiate his guesswork. They do not. We are in full agreement that there is no support from the field of population genetics for Wade's conjectures."[77]

We believe that the unusually strong response from a scientific community to counter Wade's misrepresentation of their research has been an important step. Nevertheless, there were aspects of the original Rosenberg et al. paper in 2002 that may have triggered faulty media coverage to begin with. While the accompanying Perspectives article in *Science Magazine* was helpful, the paper itself was difficult to follow and easily subject to misunderstanding, whether for journalists or even for many scientists. Further, the graphics, which showed blocks of different color representing the different geographic regions, could appear at a glance to show relatively solid blocks of genetic identity among peoples from different geographic regions. This was certainly not the intention of the authors to give this impression, as that interpretation was known to be invalid.

Many quite effective articles have presented this population genetics research and its limitations in clearer terms, in particular in making the point that the study was directed toward assessing ancestry not race. However, these responses written to clear up misunderstandings or misrepresentations were all published in scientific journals or in academic books that did not reach the broad public. Articles that best explain the complexities, although published several years after the original 2002 paper, included two by authors of the 2002 paper.[78,79] There have been two ventures in which groups have made an effort to reach the public with interpretations of the significance of the Rosenberg et al. paper in simpler language and which counter the misrepresentations of this genetic research. In the first case, a group that includes the two authors of this chapter has developed and published a set of dialogs around the subjects of race, genetics, and ancestry[80] for high school or college classrooms after presenting them twice at the Cambridge Science Festival in Cambridge, Massachusetts. In the second case, an educator with graduate degrees in genetics and education has prepared classroom approaches to changing the standard way of presenting genetics so that the significance of the population genetic studies are much clearer. This presentation shows how current knowledge from this population genetics research does not support the concept of biologic races.[81]

## CONCLUSION

When the conclusions of a scientific study are found to be incorrect or even fraudulently generated, scientists will often point out that science is inherently a self-correcting process. If the problematic conclusions of a study are of any interest, other scientists will either try to repeat the experiments and/or test the predictions of those conclusions. If a sufficient number of these tests fail, the original scientific claims are rejected and are no longer part of accepted scientific information. However, the examples described in this chapter show that this verification process is not necessarily perfect in an important way. That is, while scientists may have come to reject a specific scientific theory, the public, more often than not, does not learn of that rejection. When this sequence of events occurs in the case of sciences that explore human social behavior, the conclusions of the faulty science may remain in the public realm for decades. Studies of this sort have influenced social norms and governmental policies that harm the poor, minorities, and women, groups often already marginalized in society.

What changes in the scientific community could lead to stemming the negative social impact of studies of the kind described here? First, we suggest a change in the education of students of science that would make it more likely for scientists to be prepared to handle such situations. Specifically, we propose that a required component of this education is a history of science that goes beyond simply presenting a sequence of how science progresses. Rather, students could learn of those instances in the past when scientific conclusions or speculations were either purposefully or unexpectedly misrepresented, leading to harmful consequences. Such a history would also include those instances where studies clearly born of innocent motives are misused by others. These lessons in critical thinking may also help scientists to examine their own work in new ways, taking into consideration its social implications, including how the work could be misinterpreted or misused. This education can also be effective when it is part of student mentoring.

A scientist educated to be more sensitive to these issues could give attention to aspects of their work that could be distorted. Adding sections to papers that directly address why the science is inappropriate for certain applications could restrain exaggerations of the implications by others. This kind of approach to preparing papers on potentially sensitive subjects has been discussed in several journal articles where combinations of scientists and bioethicists have suggested guidelines for how to write papers that deal with matters of race.[82–84] Such guidelines for other such subjects could be institutionalized by journals, institutions, and professional societies.

However, even if a research paper has been written to make its implications crystal clear, some readers may still be willful enough to reject these "clarifications" of the paper, for instance, to promote their own social agenda. Of course, the simplest response to such a situation is for the authors of the original study to write an article in academic journals and hope the response ultimately reaches the public. But nearly all of the examples brought up in this chapter indicate that rebuttal in academic journals to misrepresentations of the science alone is not likely to find its way to the public. Academic publications rarely achieve the attention of the more dramatic public claims that may be made by those who chose to misrepresent the science. But there are ways, depending on the case, of reaching the media, e.g., via op-ed pieces, letters to the newspapers, or effective letters or short pieces in journals. In some cases, it may be possible for those trying to correct misinformation to connect with a scientific society to which a scientist belongs as a path to publicly confront the public misuse of their work. Scientific societies are increasingly becoming involved in ethical issues in science.

Scientists like people of any other group have very different and strong convictions that may unconsciously influence their work. They may overinterpret their own results and, thus, promote their conclusions in ways that contribute to public discussions on social policies they favor. While we have made suggestions for how scientists with concerns about the misuse of their own work could insure its proper use, in a case where a scientist is pushing the social implications of their work on the basis of their overinterpretation, who is it that could point out the problems and, at least, catalyze a debate over the work?

Other sources for the correcting or countering of misrepresentation of science are the scientists, individuals, or groups who have their own research careers and have committed themselves to explore work of others in their field that they deem to be problematic. One of the most remarkable of these scientists is the late Stephen Jay Gould, who devoted much of his life to exposing the misuse and/or misrepresentation of science that supported discriminatory and harmful effects on individuals and peoples. As Gould stated in his book *The Mismeasure of Man*, in his section "Learning by debunking," "It [debunking] must use more adequate biology to drive out fallacious ideas."[4]

The reader may have noticed that the set of problematic examples of scientific research on behavior and human aptitudes described in this chapter nearly all include prominent articles in the journals *Science* or *Nature*. This observation raises the question of whether scientific journals ought to play a role in insuring from the beginning that the transmission to the public of an unwarranted conclusion or speculation with harmful social consequences is prevented. This requires, of course, that the editors and reviewers themselves be more sensitized to the potential fallout of the publicity for relevant scientific articles. Editors of journals could peruse articles for aspects that could be easily misrepresented and thus have harmful effects. To this end, they could suggest alterations to authors that admit the limitations of the findings in the papers and, in some instances, include statements of what the data do not show. Journals could solicit reviewers of papers on the kinds of subjects indicated here who have expertise in social/ethical issues, in addition to the usual referees who evaluate the science presented in the paper.

Clearly, the news media, their science reporters, and their editors present the most popular and direct version of the transmission of scientific information to the public. Certainly, pressures in journalism may lead to sensationalism in media coverage. However, some research on scientific journalism suggests that the dramatic stories and headlines that appear in popular coverage of science may be less a result of the science writers' overinterpretation than the fault of the researcher whose comments to varying degrees inflate the significance, possible inferences, or implications arising from the study.[85,86] Interacting with journalists is another skill for scientists to learn to make clear the implications and nonimplications of their work.[87]

Just as a number of scientists have been active publicly in presenting and defending such subjects as evolutionary theory, climate change, and vaccination as being well-based on strong science, so can scientists themselves intervene when science is misrepresented to the public to support harmful social policies.

Our current systems (e.g., reward and evaluation), practices, and norms are built upon an enduring belief that good science rises to the top. However, for a variety of reasons, this outcome has been shown frequently not to be the case. Sometimes the system works in precisely the opposite fashion as a result of conflicting incentives (e.g., fame, competition, or confirmation of ideological biases). Furthermore, the pervasiveness of a networked society allows for the rapid dissemination of reports, regardless of scrutiny applied, thus making all the more urgent the shifting of norms that serve to minimize publication of inaccurate, poorly conceived, or substantiated science. Norms and practices that minimize this community-damaging consequence need to be developed and integrated into standard practice. Whether the failing is at the level of the researcher, media reporting, or existing social biases, the scientific community can take steps toward socially responsible practice of science such as those suggested here.

## REFERENCES

1. The National Commission for the Protection of Human Subjects of Biomedical and Behavioral Research, Office of the Secretary Department of Health, Education, and Welfare, United States Government. *The Belmont Report: Ethical Principles and Guidelines for the Protection of Human Subjects of Research*; 1979. https://www.hhs.gov/ohrp/regulations-and-policy/belmont-report/.
2. Goddard HH. *The Kallikak Family: A Study in the Heredity of Feeble-mindedness*. New York: MacMillan; 1912.
3. Zenderland L. *Measuring Minds: Henry Herbert Goddard and the Origins of American Intelligence Testing*. Cambridge: Cambridge University Press; 1998.
4. Gould SJ. *The Mismeasure of Man*. 2nd ed. New York: Norton; 1996:351–353.
5. Garrett H, Bonner J. *General Psychology*. New York: MacMillan; 1961.
6. IMDB. http://www.imdb.com/title/tt0075521/reviews?ref_=tt_ql_3.
7. Jacobs PA, Brunton M, Melville MM, Brittain RP, McClermont WF. Aggressive behavior, mental sub-normality and the XYY male. *Nature*. 1965;208:1351–1352.
8. Beckwith J. *Making Genes, Making Waves: A Social Activist in Science*. Cambridge: Harvard University Press; 2002:119–120.
9. Engel E. The making of an XYY. *Am J Mental Defic Res*. 1972;77:123–127.

10. Borgaonkar S, Shah SA. The XYY chromosome male—or syndrome. *Prog Med Genet.* 1974;10: 135–222.
11. Witkin HA, Mednick SA, Schulsinger F, et al. Criminality in XYY and XXY men. *Science.* 1976;193: 547–555.
12. Jacobs PA. The William Allan memorial award address: human population cytogenetics; the first twenty-five years. *Am J Hum Genet.* 1982;34:689–698.
13. Royce K. *The XYY Man.* New York: Avon; 1973.
14. Wilson J, Herrnstein RJ. *Crime and Human Nature.* New York: Simon and Schuster; 1985.
15. Based on survey of graduate students in a class at Harvard Medical School given by the co-author of this chapter (JB) up until 2014.
16. Pyeritz R, Schreier RH, Madansky C, Miller L, Beckwith J. The XYY male: the making of a myth. In: *Ann Arbor Science for the People.* Minneapolis: Burgess; 1977:86–100Biology as a Social Weapon.
17. Skloot R. *The Immortal Life of Henrietta Lacks.* New York: Broadway Paperbacks; 2010:167.
18. Washington HA. *Medical Apartheid.* New York: Broadway Books; 2006:319.
19. Hastings Center. *The XYY Controversy: Researching Violence and Genetics.* Hastings Center Report Special Supplement; 1980. http://www.jstor.org/stable/3560454.
20. Gotz MJ, Johnstone EC, Ratcliffe G. Criminality and antisocial behavior in unselected men with sex chromosome abnormalities. *Psych Med.* 1999;29:953–962.
21. Navon D, Eyal G. Looping genomes: diagnostic change and the genetic makeup of the autism population. *Am Soc Social.* 2016;121:1416–1471.
22. AXYS Association for X and Y Chromosome Variations (AXYS). http://www.genetic.org/Knowledge/BooksandResources.aspx.
23. XYY Syndrome - NORD (National Organization for Rare Disorders). http://rarediseases.org/rare-diseases/xyy-syndrome/.
24. Leggett V, Jacobs P, Nation K, Scerif G, Bishop DV. Neurocognitive outcomes of individuals with a sex chromosome trisomy: XXX, XYY, or XXY: a systematic review. *Dec Med Child Neurol.* 2010;52: 119–219.
25. Brunner HG, Nelen M, Breakefield XO, Ropers HH, van Oost BA. Abnormal behaviour associated with a point mutation in the structural gene for monoamine oxidase A. *Science.* 1993;262:578–583.
26. Brunner H, Nelen M, van Zandvoort P, et al. X-linked borderline mental retardation with prominent behavioral disturbance: phenotype, genetic localization, and evidence for disturbed monoamine metabolism. *Am J Hum Genet.* 1993;52:1032–1039.
27. Angier N. Gene tied to male violence is studied. *New York Times.* October 22, 1993.
28. Morrell V. Evidence found for a possible 'aggression gene'. *Science.* 1993;260:1722.
29. Cowley G, Hall C. The genetics of bad behaviour. *Newsweek.* November 1993;1:57.
30. Felsenthal E. Man's genes made him kill, his lawyers claim. *Wall St J.* November 1, 1994:B1–B5.
31. Brunner HG. MAOA deficiency and abnormal behavior: perspectives on an association. In: *Ciba Foundation Symposium Genetics of Criminal and Antisocial Behavior Wiley, Chichester.* 1994:155–164.
32. Caspi A, McClay J, Moffitt TE, et al. Role of genotype in the cycle of violence in maltreated children. *Science.* 2002;297:851–854.
33. Morris C, Shen A, Pierce K, Beckwith J. Deconstructing violence. In: Krimsky S, Gruber J, eds. *Biotechnology in Our Lives.* New York: Skyhorse Publishing; 2013:233–244.
34. Bernet W, Vnencak-Jones L, Farahany F, Montgomery SA. Bad nature, bad nurture and testimony regarding MAOA and SLC6A4 genotyping at murder trials. *J Forensic Sci.* 2007;52:1–10;
    a. Christchurch Press, August 9, 2006.
35. Lea R, Chambers G. Monoamine oxidase, addiction, and the "warrior" gene hypothesis. *NZ Med J.* 2007;120:U2441.
36. Perbal L. The 'warrior gene' and the Maori people: the responsibility of the geneticists. *Bioethics.* 2013; 27:382–387;
    a. Merriman T, Cameron V. Risk-taking: behind the warrior gene story. *NZ Mus J.* 2007;120:U2440.
37. Hook GR. "Warrior genes" and the disease of being Māori. *MAI Rev.* 2009:2.

38. Callaway E. *The New Scientist. Daily News*. Gangsta gene identified in US teens; June 19, 2009. https://www.newscientist.com/article/dn17337-gangsta-gene-identified-in-us-teens?DCMP=OTC-rss&nsref=online-news.
39. Beaver K, DeLisi M, Vaughn MG, Barnes JC. Monoamine oxidase A genotype is associated with gang membership and weapon use. *Comp Psych*. 2010;51(2):130–134.
40. Hagerty B. *Can Your Genes Make You Murder?*. National Public Radio; July 1, 2010. http://www.npr.org/templates/story/story.php?storyId=128043329.
41. Forzano FP, Borry A, Cambon-Thomsen SV, et al. Italian appeal court: a genetic predisposition to commit murder. *Eur J Hum Genet*. 2010;8:519–521.
42. Born to Rage. Dr.Phil. http://www.drphil.com/shows/1626/. Originally aired July 26, 2011.
43. Stiny A. Judge throws out 'warrior gene' defense. *Albuquerque J*. January 30, 2015.
44. Haywood P. Judge rejects warrior gene testimony in murder trial. *The New Mexican*; January 19, 2015. http://santafenewmexican.com/news/local_news/judge-rejects-warrior-gene-testimony-in-murder-trial/article_67b49eb2-23ce-59a4-9de8-0cf0c2ab444a.html.
45. Benbow C, Stanley J. Sex differences in mathematical ability: fact or artifact? *Science*. 1980;210:1262–1264.
46. Kolata GB. Math and sex: are girls born with less ability? *Science*. 1980;210:1234–1235.
47. The gender factor in math. A new study says males may be naturally abler than females. *Time*. December 15, 1980:57.
48. Williams DA, King P. Do males have a math gene? *Newsweek*. December 15, 1980:73.
49. Beckwith B. He man and she woman: cosmo and Playboy groove on genes. *Columbia J Rev*. 1982;12:48–51, 50.
50. Beckwith J. Gender and math performance: does biology have implications for educational policy? *J Edu*. 1983;165:158–174.
51. Summers LH. *Remarks at NBER Cference on Diversifying the Science & Engineering Workforce, Office of the President*. Harvard University; January 14, 2005. http://www.president.harvard.edu/speeches/summers_2005/nber.php.
52. Hemel J. Sociologist cited by summers calls his talk uninformed. *Harvard Crimson*; January 14, 2005. http://www.thecrimson.com/article.aspx?ref=505363.
53. Hyde JS, Mertz JE. Gender, culture, and mathematics performance. *Proc Natl Acad Sci USA*. 2009;106:8801–8807.
54. Monastersky R. Primed for numbers. *Chron High Educ*. March 4, 2006:A1.
55. Brody L, Millis C. Talent search res.: what have we learned? *High Abil Stud*. 2005;16:97–111.
56. Guiso L, Monte FP, Sapienza P, Zingales L. Diversity culture gender and math. *Science*. 2008;320:1164–1165.
57. Nosek BA, Smyth FL, Sriram N, et al. National differences in gender-science stereotypes predict national sex differences in science and math achievement. *Proc Natl Acad Sci USA*. 2009;106:10593–10597.
58. Evans PD, Gilbert SL, Mekel-Bobrov N. Microcephalin, a gene regulating brain size, continues to evolve adaptively in humans. *Science*. 2005;309:1717–1720.
59. Balter M. Evolution: are human brains still evolving? brain genes show signs of selection. *Science*. 2005;309:1662–1663.
60. Wade N. Brain may still be evolving, studies hint. *New York Times*; September 9, 2000. http://query.nytimes.com/gst/fullpage.html?res=9B01E1DE1331F93AA3575AC0A9639C8B63&sec=health.
61. Kotulak R. Two evolving genes may allow humans to become smarter. *Baltimore Sun*. September 9, 2005. Telegraph Section:3A.
62. Derbyshire J. The spectre of difference. *Nat Rev Online*; November 7, 2005. http://www.johnderbyshire.com/Opinions/HumanSciences/specterofdifference.htm.
63. Inman M. Human brains enjoy ongoing evolution. *New Scientist*. September 9, 2005. http://www.newscientist.com/article/dn7974-human-brains-enjoy-ongoing-evolution.html.
64. Timpson N, Heron J, Smith D, et al. Comment on papers by Evans, et al. and Mekel-Bobrov et al. on evidence for positive selection of MCPH1 and ASPM. *Science*. 2007;317:1036.

65. Mekel-Bobrov N, Posthuma D, Gilbert S, et al. The ongoing adaptive evolution of ASPM and Microcephalin is not explained by increased intelligence. *Hum Mol Genet.* 2007;16:600–608.
66. Rushton JP, Vernon PA, Bons TA. No evidence that polymorphisms of brain regulator genes Microcephalin and ASPM are associated with general mental ability, head circumference or altruism. *Biol Lett.* 2007;3:157–160.
67. Regalado A. Head examined: scientist's study of brain genes sparks a backlash. *Wall St J*; June 16, 2006. https://online.wsj.com/article/SB11504065329081636.html.
68. Balter M. Bruce Lahn profile: links between brain genes evolution and cognition challenged. *Science.* 2006;314:1872.
69. Rosenberg NA, Pritchard JK, Weber JL, et al. Genetic structure of human populations. *Science.* 2002; 298:2381–2384.
70. King MC, Motulsky A. Mapping human history. *Science.* 2002;298:2342–2343.
71. Wade N. Gene study identifies five main populations. *New York Times.* December 20, 2002.
72. Campbell MC, Tishkoff SA. The evolution of human genetic and phenotypic variation in Africa. *Curr Biol.* 2010;20:R166–R173. https://doi.org/10.1016/j.cub.2009.11.050.
73. http://boards.straightdope.com/sdmb/archive/index.php/t-358328.html; (2006)
74. Taylor J. Nicholas Wade takes on the regime. *American Renaissance*; March 2, 2014. https://www.amren.com/features/2014/03/attack-on-the-regime/.
75. Wade N. *A Troublesome Inheritance: Genes, Race and Human History*. Penguin; 2014.
76. Orr HA. Stretch genes. *New York Review of Books.* June 5, 2014.
77. Coop G, Eisen M, Nielsen R, Prseworski M, Rosenberg NA. *Letters: a troublesome inheritance.* In: *The Sunday New York Times Book Review*. August 8, 2014.
78. Feldman MW, Lewontin RC. Race ancestry and medicine. In: Koenig BA, Lee SS-J, Richardson SS, eds. *Revisiting Race in a Genomic Age*. New Brunswick: Rutgers University Press; 2008:89–101.
79. Rosenberg NA. Population-genetic perspective on the similarities and differences among worldwide human populations. *Hum Biol.* 2011;83:659–684.
80. Beckwith J, Bergman K, Carson M, et al. Using dialogues to explore genetics ancestry and race. *Am Biol Teach.* 2017;79:525–537.
81. Donovan BM. Learned inequality: racial labels in the biology curriculum can affect the development of racial prejudice. *J Res Sci Teach.* 2017;54:379–411. https://doi.org/10.1002/tea.21370.
82. Lee SS-J, Mountain J, Koenig K, et al. The ethics of characterizing difference: guiding principles on using racial categories in human genetics. *Genome Biol.* 2008;9:404. https://doi.org/10.1186/gb-2008-9-7-404.
83. Vitti JJ, Cho MK, Tishkoff SA, Sabeti PA. Human evolutionary genomics:ethical and interpretive issues. *Trends Genet.* 2012;3:137–145. https://doi.org/10.1016/j.tig.2011.12.001.
84. Race, Ethnicity and Genetics Working Group. The use of racial ethnic and ancestral categories in human genetics research. *Am J Hum Genet.* 2005;77:519–532.
85. Bubela TM, Caulfield TA. Do the print media "hype" genetic research? A comparison of newspaper stories and peer-reviewed research papers. *CMAJ.* 2004;170:1399–1407.
86. Condit C. Commentary on Bubela and Caulfield article. *CMAJ.* 2004;170:1415–1416.
87. Condit C. How geneticists can help reporters get their story right. *Nat Rev Genet.* 2007;8:815–820.

# CHAPTER 26

# Ethical Considerations for Animal Use in Behavioral and Neural Research

**David Hanwell**
University of Toronto, Toronto, ON, Canada

## INTRODUCTION

The main aim of this chapter is to provide researchers with an overview of ethical considerations to be made prior to using animals in science (such as neuroscience and behavioral research). These are discussed in the context of the animal ethics review and oversight infrastructure that is required at most institutions to ensure that animals are used in an ethical manner and in compliance with applicable laws, regulations, and guidelines. Another aim is to have the chapter serve as a practical guide as to what details are to be determined and provided to ethics committees for review and approval. By furthering the knowledge and appreciation of the ethics review process, the time to protocol approval may be shortened. To begin, there is discussion of how the researcher should first consider if/what animal use is appropriate to address the scientific questions, considering the negative impact it may have on animals (i.e., perform a "harm-benefit analysis"). The 3R tenets of replacement, reduction, and refinement are then introduced as principles that are to be applied to reduce the "harm" as much as possible. Next, an overview of the regulatory framework commonly in place in some form at most institutions is described to impart better understanding of where institutional policies and procedures originate. Following is discussion of the basics of the protocol review process: the committees that perform the review, the considerations made, and the information that is needed prior to approval being given. The chapter concludes with a description of the oversight that is in place for after a protocol has been approved; this includes discussion of common mechanisms to ensure research is proceeding in practice as it was approved and to improve the quality of the research and animal care, wherever possible, based on experience gained as research progresses.

## ETHICAL CONSIDERATIONS FOR ANIMAL USE IN SCIENCE

Ideally, animals would not be used in research at all given that they often experience some degree of pain and/or distress. Accordingly, it is ethically unacceptable to not consider the use of nonanimal models. As will be discussed later, evidence of this consideration being made is often a requirement prior to approval being granted to use animals

*Molecular-Genetic and Statistical Techniques for Behavioral and Neural Research*
ISBN 978-0-12-804078-2, https://doi.org/10.1016/B978-0-12-804078-2.00026-X

in research. It is accepted that in many areas of research, however, no alternatives exist that can suitably replace the complexity of a whole organism, so complete avoidance of animals is not (currently) feasible in these cases. After the existence of appropriate nonanimal models has been considered and implemented where possible (e.g., initial in vitro toxicity testing of novel drugs[1]), consideration is to be given to which animal species is the most appropriate to use. In neural and behavioral research, the frequent aim is to gain further insights into normal human functioning and the pathophysiology in various disorders and what treatments are effective, so consideration needs to be given as to which species is most likely to result in findings that are translatable to humans. Generally speaking, "the use of animals in research, teaching, and testing is acceptable only if it promises to contribute to understanding of fundamental biologic principles, or to the development of knowledge that can reasonably be expected to benefit humans or animals."[2] Once these benefits are identified, a "harm-benefit analysis" should be undertaken in which anticipated harm is morally weighed against the benefits. Note that "harm" does not only refer to experimental interventions but many other potential sources of pain/distress. Factors to be considered include the following:[3]

- the species and numbers of animals involved
- the type, frequency, and duration of adverse effects
- the methods used to control adverse effects
- the way in which the experiment will be terminated, i.e., the "humane endpoint" (terminating an experiment as early as possible to avoid unnecessary suffering while still reaching the scientific objectives)
- any expected mortality
- the type of housing (single or social) and environmental enrichment
- the whole life experience of the animal, including any significant transportation before, during, or after procedures and any prior use or planned reuse

A dilemma is that the more representative a model is for studying human biology, the closer it tends to be evolution-wise and thus more ethically challenging to use that species.[4] For example, chimpanzees have long been considered the "gold standard" for studying human hepatitis viruses due to the translatability of study findings to humans, but because they are so closely related to humans, efforts are underway to develop "lower" species that can replace them.[5–7] The public entrusts the scientific community to determine what level of harm to animals is justified in the pursuit of scientific discovery that aims to benefit people.[8–10] Relating to common experimental paradigms and models used in neurobehavioral research, it has been questioned "…whether indeed the combination of these test systems and strains is genuinely believed to deliver the best possible results in anxiety research or is simply based on tradition, conservatism and lack of exploration of alternative models—or acceptability to journals and their equally conservative referees."[4] This suggests that in some cases, the harm may not be adequately justified if the model used is not likely to yield relevant findings. It is therefore important to carefully consider animal model choice and to educate the public why a chosen animal model(s) is appropriate for the research.

In summary, performing a harm-benefit analysis is warranted to ensure that the benefit of using animals in research reasonably outweighs the burden it will impose on them. In doing this, it also ensures accountability to the public that provides most funding for animal research. To improve the harm-benefit ratio, consideration is given as to how the harm can be maximally reduced. It is recognized, however, that calculating a ratio based on determined levels of benefit and harm is difficult as these are not entirely objective/measurable parameters. And what ratio is considered acceptable? As a guide, several ways in which a harm-benefit analysis may be performed have been summarized by a joint working group of the American Association of Laboratory Animal Science and the Federation of European Laboratory Animal Science Associations, along with a detailed example of how an analysis can be performed on a protocol.[11,12] It is important that there be continued discussion among scientists, policy makers, and the public and that the outcomes of these discussions be taken into consideration when determining whether or not the benefits sufficiently outweigh the harm to justify the use of animals.

## THE 3RS

The 3Rs refer to replacement, reduction and refinement and represent the foundation of ethical use of animals in research. The concept of the "3Rs" was proposed by W.M.S. Russell and R.L. Burch in their 1959 book *The Principles of Humane Experimental Technique.*[13] The 3Rs serve as guiding principles that should be considered and implemented whenever applicable throughout all stages of experimental planning as well as after work begins. Most, if not all, ethics review committees ask scientists how each of the 3Rs were considered in their proposal, as well as on a regular basis (e.g., at time of amendment and/ or renewal). Next, more details are provided for each of the 3 Rs.

### Replacement

As discussed, nonanimal models should be used for all or part of the research whenever possible. Examples of replacements include the use of computer simulations or in vitro systems such as cell lines. These are forms of "absolute replacement," whereas the use of a less sentient (possibly lower-order) animal species comprise "relative replacement" strategies when less pain and distress is expected relative to use of higher-order species. The harm-benefit analysis described earlier should be factored in when deciding on the most appropriate model. In neuroscience, many different models have been used that range from higher-order vertebrates such as nonhuman primates, cats, rodents, and zebrafish to (relatively) lower-order and possibly less sentient invertebrates such as pond snails, roundworms, and fruit flies.[14–17]

## Reduction

Reduction refers to any strategy that will result in fewer animals being used to obtain sufficient data to answer the research question, or in maximizing the information obtained per animal and thus potentially limiting or avoiding the subsequent use of additional animals, without compromising animal welfare.[18] Note that the use of too few animals is also not desirable because potential differences in groups may go undetected, thus also wasting animals. Some reduction strategies include having a biostatistician review the proposed study design and plans for data analysis to ensure they are appropriate; this avoids a requirement for repeat studies requiring more animals.[19] A power analysis can be performed using existing data (e.g., in the literature) and/or data from a small pilot study to determine minimum group sizes based on the degree of variability.[20] Consider using both sexes in studies; this not only reduces animal use but also better enables potential sex differences in neurobiology to be identified. For example, per a review on NIH development of a policy on this topic "a study demonstrated that mice with XY chromosomes in the central nervous system had greater neurodegeneration than did those with XX chromosomes. The findings have important implications for other sex-skewed neurological conditions, including Parkinson's disease, schizophrenia and stroke" and also "the most widely accepted Multiple Sclerosis animal model — rodent experimental autoimmune encephalomyelitis (EAE) — has revealed that sex differences in MS are related to both reproductive and nonreproductive factors."[21] Assuming animal welfare is not overly compromised, another strategy includes serially measuring (e.g., by imaging) individual animals rather than using (or euthanizing) separate cohorts at each time point (note: the use of serial measurements is appropriate if the data points are not treated as independent and are analyzed using repeated measures statistical tests such as repeated measures ANOVA, or if they are averaged to reduce variation possibly associated with measurement error). Others include surplus animals of one researcher that would otherwise be euthanized (e.g., due to wrong genotype or sex) that may be suitable for another to use, and multiple tissues obtained from each animal and/or are shared among researchers. Regarding the latter, several initiatives have been developed to increase animal/tissue sharing to achieve organized and larger-scale reduction in animal use.[22,23]

Another example of a strategy that (indirectly) reduces animal use is providing detailed descriptions in publications on topics not typically included such as housing and husbandry, adverse events, how animals were allocated to different groups, and the time of day procedures were carried out. By doing so, reproducibility of studies increases, thereby maximizing the output of published research and minimizing animal use in (unnecessary) repeat studies. Lastly, it is important to also disclose negative results, i.e., lack of a significant effect, when such finding is obtained, just as much as to publish on positive (significant) effects. Describing such negative findings will reduce unnecessary replication of research by scientists hoping to find positive results where others have already shown they likely do not exist. To provide formal guidance on what is useful to record and later publish, the ARRIVE (Animal Research: Reporting of In Vivo Experiments) and PREPARE (Planning Research and Experimental Procedures on Animals: Recommendations for Excellence) guidelines should be consulted.[24,25]

### Refinement

In general, refinement refers to any modifications in procedures or practices that minimize pain and/or distress. From a researcher's perspective, there may be a tendency to focus on how experimental procedures can be refined to further reduce pain and/or distress, but the principle is to be applied to all aspects and stages of the animal's life, including before and after use in studies. Examples of refinement include seeking less stressful transport methods to/within the institution and habituating animals to handling and/or procedures to reduce the level of restraint required during studies. Creating a more "natural" or enriched environment should also be sought, as this results in demonstration of more species-appropriate behaviors (e.g., freely move about without restriction of body position, digging, climbing, grooming, social interactions with conspecifics). Other refinement examples include reducing pain by altering the type, frequency, dosage, and/or combination of drugs used for anesthesia and analgesia, as well as adding or modifying nonpharmacological practices to reduce animal discomfort (e.g., improved maintenance of body temperature and hydration during the perioperative period). Importantly, as the final disposition of most animals used in science is euthanasia, methods should be continually sought that produce a more humane death. That is, techniques that "induce the most rapid and painless and distress-free death possible."[26]

It is generally recognized that unalleviated animal pain and distress can skew data and that refinements can also improve the quality and reproducibility of scientific studies.[27–29] However, variations in types of refinements, such as the drugs that are used to manage pain or the type of enriched housing provided, can all affect research outcomes, so it is important to include refinement details in publications.[30]

## HUMANE ENDPOINTS

A fundamental principle that must be considered is that no animal is to experience any *unnecessary* pain and/or distress. Although successful refinement of procedures results in less pain and/or distress, complete elimination may not be possible. In these cases, an acceptable upper level of pain/distress that is inherently part of the research needs to be established. For example, at least a minor degree of discomfort is expected after an invasive surgery despite the continued refinement of anesthetic and analgesic use. There may be cases, however, where the degree of pain/distress demonstrated by animals exceeds what is expected. When this point is reached, humane intervention must occur to reduce the pain/distress back to an acceptable level. If this cannot be accomplished (e.g., by discontinuing exposure of the animal to an experimental stressor or by administering additional or a different analgesic), then animals must be humanely euthanized as soon as possible to minimize the duration of this excessive pain/distress.

To determine what appropriate humane endpoints are—that is, the point at which intervention occurs to eliminate pain/distress or minimize it to an acceptable level—the type of procedures and their degree of invasiveness need to be considered. Any expected compromise to animal welfare should be described in detail with plans for how humane endpoints will be detected for each and what action will be taken if reached. The appearance and behavior of animals is typically monitored during the critical period(s) following procedures to evaluate their status. Descriptions of species-specific behaviors that are indicative of pain/distress can be included in an ethogram scoring system that can be used to more objectively and consistently evaluate the condition of animals on study. A threshold score may be determined, so actions are taken, depending on the types of behaviors displayed and other evaluative criteria (e.g., body condition, temperature, hydration status, percent weight loss). The timing and frequency of monitoring must be established, so humane endpoints are detected as early as possible.

The goal is to refine the humane endpoint(s), so the earliest indicator(s) that an animal's welfare is becoming compromised is determined, so humane intervention occurs as soon as possible. This minimizes the degree of pain/distress experienced by the animals. An important consideration is what the scientific aim of the study is and whether an earlier intervention still permits the aim to be achieved. Ideally the aim is achieved prior to the onset of pain or distress. The earliest endpoint indicators should be validated as truly predictive of the expected decline in animal welfare; if not, interventions that are performed too early or inappropriately may result in animals being prematurely taken off study, given treatment, or euthanized.

There are many resources available that can be used as aids when establishing humane endpoints.[31–33] In addition, institutional veterinary staff should be consulted, so strategies can be developed collaboratively and tailored to the study. This will help ensure that the proposed humane endpoint plan is sufficiently detailed and is appropriate considering the study proposed. After some experience is gained with the model, it may be possible to further refine the humane endpoints, so pain/distress is reduced even more.

## REGULATORY OVERSIGHT FRAMEWORK

Animal use in science involves an oversight framework at multiple levels: national and/or regional, and institutional. Although the structure and functioning of these differ to some degree, the approach is largely harmonized globally and encompasses the 3Rs principles described earlier.[34] This section describes the "typical" approach taken by many countries. At the national level, many countries have legislation relating to animal use in research, and this is usually accompanied by regulations and/or guidelines. In the European Union, for example, all member countries must follow Directive 2010/63/EU on the protection of animals used for scientific purposes, which took full effect in 2013. In the United States, federal legislation known as the Animal Welfare Act regulates the treatment of animals in research. For countries without national/regional legislation a national body may issue guidelines (e.g., the Canadian Council on Animal Care in Canada), and internationally, several organizations establish standards and guidelines that countries may use when developing a framework (e.g., International Council for Laboratory Animal Science and the World Organisation for Animal Health (OIE)).

At the institutional level, the responsibility of local oversight of animal use in research is commonly undertaken by an ethics committee (known as Animal Ethics Committee, Institutional Animal Care and Use Committee, Animal Care Committee). Their existence is usually mandated by external regulators and accrediting bodies to act as a local "quality control" mechanism, ensuring applicable legislation and/or guidelines are appropriately implemented at the institution for all aspects of an animal use program.[35–37] This results in external regulators serving in a "quality assurance" role, holding the ethics committees accountable. In addition, an institutional official or senior administrator is typically responsible for the overall program, and for ensuring that the ethics committees are properly supported to continually achieve their mandate.

## IACUCS

The ethics committee, hereon referred to as the Institutional Care and Use Committee (IACUC), is tasked with overseeing the entire animal care and use program, which includes ensuring proper functioning of the IACUC itself; animal facilities; and programs for training, postapproval monitoring, occupational health and safety, and veterinary care. Day-to-day responsibilities in each of these animal care and use program components are delegated to competent staff, many of whom are ex officio IACUC members. The diverse membership of IACUCs enables multiperspective review of the ethical merit of the project and most often includes veterinarians, committee coordinators, facility managers, scientists, institutional nonanimal users, and members of the public. The latter should not have a history of animal use nor have any affiliation with the institution to avoid a potential conflict of interest. The majority of the committee's work involves animal protocol review (discussed later), but they also remain up to date on and reform, when necessary, various aspects of the program by developing or revising local guidelines and standard operating procedures; visiting animal facilities; addressing compliance issues; and preparing for inspections.

IACUCs interpret the laws and guidelines that set the required standards for animal care and use and determine how best to implement them at their institution: approaches that are appropriate for one institution may not be for another. In other words, locally established IACUC standards influence how animal research is conducted at the institution, but all institutions within a given region must comply with the same laws and/or external regulatory guidelines. Where there is flexibility in how a standard is met, a number of different approaches may be possible. Standards that can be determined in this manner are known as "performance standards." For example, it is standard that aseptic technique be used during surgery to reduce the risk of postoperative infection. How asepsis is maintained throughout a given surgical procedure is up to the institution (i.e., the IACUC and researcher). Use of a sterile surgical pack for each animal would certainly be ideal for reducing risk of contamination but is costly; instead, a bead sterilizer may be acceptable to allow multiple use of instruments using "sterile tips" technique to maintain asepsis between certain numbers of animals.[38]

In contrast, a "standard or guideline that specifies in detail a method, technology, or technique for achieving a desired outcome" is known as an "engineering standard." A common engineering standard is cage size; a specified minimum amount of space is often prescribed for each species in regulations. For example, in the United States, the National Research Council's "Guide for the Care and Use of Laboratory Animals" recommends that all mice be housed in cages that are at least 5 in (12.7 cm) high, and for mice 10–15 g in weight, provide at least 8 $in^2$ (51 $cm^2$) of floor space per animal.[36] Any exception to this would require IACUC approval (e.g., based on validated performance indices of well-being).

Both types of standards are important, as engineering standards set a benchmark that holds institutions accountable for achieving a desired outcome related to animal welfare, while performance-based standards enable context to be considered to what method or technique is most appropriate to achieve the desired outcome. It is important that IACUCs are flexible and do not over interpret legislation and guidelines nor impose engineering standards when there is evidence that a standard may be met in a more efficient and economical manner without compromise to animal welfare. Institutions should establish a culture of collaboration between IACUCs and researchers in which solutions to meet standards are sought that maximize research productivity in a manner that avoids unnecessary regulatory burden. Indeed, self-imposed regulatory burden at institutions has been recognized as a common issue that does not only apply to animal-based research.[39–42] IACUCs must remain up to date with new scientific knowledge, and with it, changing standards; this necessitates regular review of institutional policies and practices, so they remain appropriate.

## ETHICS REVIEW OF PROPOSED ANIMAL USE

The normal scope of animal species requiring ethics review is all vertebrates and higher-order invertebrates, such as cephalopods (e.g., octopus, squid, and cuttlefish).[43] The IACUC review enables the IACUC to perform a harm-benefit analysis, consider how the researcher has implemented the 3Rs to minimize any expected harm, and ensure that all applicable laws and regulations will be met. Laws and/or guidelines usually require prior IACUC review and approval of proposed animal use in most countries/regions, but also do most research funding sources and academic journals. Protocol approval ranges from 1−5 year periods, depending on the country/region, and it may be subject to interim rereview by the IACUC. There is usually a requirement for the IACUC to perform ongoing monitoring of animal use activities postapproval (described later in Postapproval Monitoring section).

As the focus of the IACUC review is on the ethics of the proposed animal use, the scientific merit is evaluated at arm's length from the IACUC (usually prior to IACUC review). The use of grant funding that was awarded based on scientific merit is often accepted as sufficient evidence of scientific justification of using animals. Where evidence of scientific merit review is lacking, an independent peer review process (again, independent of the IACUC) can be both established and used by the institution. Note that the IACUC may address scientific points during the review, however, but these comments are usually made through the lens of animal welfare, as poor scientific study design or inappropriate scientific protocols leads to low-integrity data and thus wasted animals (which goes against the 3R principle of reduction).

## PLANNING ANIMAL USE PROTOCOLS

Regarding IACUC assessment of anticipated harm and implementation of the 3Rs, a primary focus is on refinement strategies. But for the IACUC to identify potential harms and decide on the appropriateness of proposed pain/distress mitigation strategies, a clear understanding of the plans for care, husbandry, housing, and experiments is needed. The following are topics for which the IACUC typically requests information in protocol applications:[44,35,45]

- a rationale and purpose of the proposed use of animals
- a clear and concise sequential description of the procedures involving the use of animals
- availability or appropriateness of the use of less invasive procedures, other species, isolated organ preparation, cell or tissue culture, or computer simulation
- justification of the species and number of animals proposed
- prevention of unnecessary duplication of experiments

- nonstandard housing and husbandry requirements
- impact of the proposed procedures on the animals' well-being
- appropriate sedation, analgesia, and anesthesia
- conduct of surgical procedures
- postprocedural care and observation
- description and rationale for anticipated or selected endpoints
- criteria and process for timely intervention or euthanasia if painful or stressful outcomes are anticipated
- method of euthanasia or disposition of animals
- adequacy of training and experience of personnel and roles and responsibilities of the personnel involved
- use of hazardous materials and provision of a safe working environment

For a protocol to be approved by an IACUC, sufficient detail in each of the aforementioned topics (and throughout) must be provided in a clear manner, so the IACUC understands how and why animals are needed in the research. It is equally important to determine how information is to be provided to the IACUC, as it is to determine what details are relevant. Overall, the use of lay language and avoidance of technical terms helps the IACUC to understand the proposal. When there are multiple stages or interventions involved with a project, a visual aid, such as a flow chart, often helps the committee better understand what animals will be subjected to and thus evaluate the harm potential. In addition, to the subsequently suggested considerations, other literature specific to planning neurobehavioral studies is available for consultation.[46,47]

Although it may be common practice to use certain species of animal as a model for the type of research being proposed, a statement should be made explaining why the species chosen is more suitable than using (possibly less sentient) lower-order species and nonanimal models. Including these statements indicates that the 3R principle of replacement has been considered and implemented whenever possible. As discussed earlier in the chapter, the use of a lower-order species is not necessarily appropriate; it is acceptable to use a higher-order species as a model if it is predicted to provide more relevant data. Recall, however, that the more relevant a species is to study humans, the higher the sentience and thus potential for harm, particularly when used in invasive projects. In these cases, the IACUC must be sufficiently satisfied that the perceived benefit(s) described outweighs the unavoidable harm that the research will likely cause to the more sentient animals.

## DEGREE OF INVASIVENESS

IACUCs may be required to gauge the degree of harm expected by grading according to their "ethical invasiveness," with procedures that cause more pain and distress graded higher (e.g., none to minimal, mild, moderate, severe). It is important to note that the level of ethical invasiveness is not necessarily related to the degree of physical invasiveness that the research entails, but rather the amount of pain and distress to which the animal is potentially subjected. For example, an abdominal surgery (laparotomy) is physically invasive, but if it is performed on an animal that is at a surgical plane of anesthesia and which is eventually euthanized rather than recovered from anesthesia, then the only distress the animal feels is during the anesthetic induction, so the ethical invasiveness would be low (i.e., graded as "none to minimal"). However, if the animal recovers from the surgery, then the pain and distress after recovery would be high, so the ethical invasiveness would also be high (i.e., graded as "severe"). Committees review protocol submissions proportionally, with more attention paid to protocols that are more invasive. Thus, sufficient detail needs to be provided as to why any expected pain/distress is justified and how it will be minimized (i.e., through consideration of the 3R of refinement). The risk of compromised animal welfare is higher if painful/distressful procedures are IACUC approved without sufficient consideration of plans for the implementation of appropriate mitigation strategies.

## DETERMINING ANIMAL NUMBERS REQUIRED

When requesting animal numbers in the protocol the number to be requested in part depends on what types of animal use the IACUC is required to report to external regulators.[48] Generally, justification for the total animal number request should be supported by a breakdown by experiment, groups, group sizes, sex, strain, age, and time points. If breeding animals, indicate the number required to be generated for studies and those to be maintained as ongoing breeders. If only fractions of the offspring have the correct genotype, phenotype, and/or sex, this should be explained and estimates factored into the total number request. As with all 3Rs, consideration and implementation of reduction strategies need to be demonstrated; these include direct measures such as keeping good breeding practices and records to minimize production of surplus animals and not unnecessarily duplicating studies. Indirect reduction can be achieved through consistent use of animal care and use practices that follow established standards. By doing so, for example, will further minimize the avoidable removal of animals from studies due to health issues (e.g., postoperative infection due to poor aseptic technique), as well as reduce data variability, which in turn will reduce required group sizes to detect group differences, and thus total animal use. Also, it is worth repeating that unless the researcher has expertise in designing animal studies, consideration should be given to consulting with a biostatistician early in the experimental design phase. Not only will this help avoid unnecessary animal use but also ensure that bias is avoided, appropriate controls are included, and proper statistical tests are used for data analysis to ensure quality and integrity.

## DESCRIPTIONS OF EXPERIMENTAL AND ANIMAL CARE DETAILS

When describing experimental procedures, provision of specific details is important, so the IACUC can determine whether the current plan is ethically acceptable or if revisions are required prior to approval. The IACUC needs to understand the harm that each animal may experience from start to finish (such as from birth or import of the animal to euthanasia) by considering the invasiveness of each intervention, how many each may undergo, and how often, through to how euthanasia will be performed. For complicated protocols, providing a schematic of how the various experiments, groups of animals, interventions of each group (e.g., injections, sampling, surgeries, water/food deprivation, behavior test, imaging, and euthanasia), in what order, time between each, is often helpful in this regard. If performing surgeries, note that usually only one "major" surgical procedure is permitted in which an animal is recovered unless the IACUC approves more based on scientific justification. This prevents any one animal from experiencing too much cumulative pain/distress. Generally, major survival surgery is any "surgical intervention that penetrates and exposes a body cavity... has the potential for inducing permanent physical or physiologic impairment, and/or any procedure associated with orthopaedics or extensive tissue dissection or transection."[49] Similarly, the IACUC usually sets limits on the number of multiple "minor" survival procedures per the cumulative harm expected. At some point, it is decided that it is more ethically acceptable to use more animals with each undergoing fewer procedures (and thus less cumulative harm).

If experiments will be performed within lab space outside of vivaria, prior IACUC approval based on justification is often required. Justifiable exceptions may include, for example, the inability to readily and repeatedly move specialized equipment (e.g., for behavior assessment or imaging) into the vivarium if there is no dedicated space available for it.

## STANDARD OPERATING PROCEDURES

When describing procedures in the protocol, many IACUCs promote that researchers seek and reference any applicable institutional standard operating procedures (SOPs) that have been previously IACUC approved. The referencing and use of SOPs ensures the IACUC that the procedure will be consistently performed in a manner that has been deemed by the IACUC to meet acceptable standards. Other benefits of consistent application of SOPs are less variable data and use as a tool for training new research and animal care staff. If the researcher plans to deviate from the SOP(s), specific details of deviations should be provided with justification. If researchers set out to develop an SOP (or are asked to by the IACUC), ideally it is done in collaboration with the IACUC (usually delegated to veterinary staff) to help ensure that the IACUC will approve it. Conversely, researchers should be involved when IACUCs are developing SOPs, so they both meet IACUC standards and still enable the researcher to adequately perform the experiments (i.e., remain practical). Overall, referencing institutional SOPs when describing experimental and animal care plans facilitates more efficient protocol writing and can shorten time to protocol approval since these have already been approved by the IACUC.

Although sufficiently detailed procedures are required for IACUCs to envision what will happen, flexibility should be included in the protocol, whenever possible, to reduce the need for additional amendments when minor changes to experimental design are desired post approval. IACUCs should promote this approach, as it lessens the administrative burden on both parties. For example, rather than specifying only one set of parameters for volume, route, and time points for initial administration of a candidate substance and/or the collection of blood or other bodily fluids, all potential routes, volumes, and frequencies to be used can be listed. As described earlier, an institutional SOP for administration of substances, for example, might be available that can be referenced, so the IACUC is ensured that the range used for each parameter is acceptable for the species, age, and size. References for the administration of substances and sampling of blood are available for consultation during the planning stages.[50–52]

## ADVERSE EFFECTS AND MONITORING

The safety profile of any administered substances such as chemicals, biohazardous agents, and radioactivity should be provided, if known, in relation to the planned dosages to be given to animals. This information enables the IACUC to see if animals may experience pain/distress due to administration of a substance above the maximum tolerated dose (MTD). If some degree of toxicity is expected, it needs to be described with scientific justification as to why lower (safer) doses cannot be used. In these cases, or if the MTD is unknown, a strategy for minimizing the duration and severity of any adverse effects is required. Some additional resources for MTD study design and implementation of the 3Rs are referenced here.[1,53]

In general, it should be stated, whenever possible, for each type of experimental intervention whether or not there is potential for any adverse effects. If available, supporting references should also be provided to substantiate these statements. Where adverse effects are probable or expected, specific descriptions should be provided as to how and when they will manifest (e.g., changes in appearance, behavior, activity level, and body weight), along with measures to be taken to reduce the duration and magnitude as much as possible. That is, state the humane endpoint(s) for each, when they may occur, and corresponding actions to be taken if they occur. If it is not yet known when adverse effects will occur, animal monitoring should begin at relatively high frequency until experience is gained as to what is appropriate. Frequency can then be reduced outside of critical periods. Seeking to establish more timely intervention results in less pain and distress and thus represents humane endpoint refinement. Institutional guidelines relating to humane endpoint determination and/or veterinary staff should be consulted for this.

In neurobehavioral research, many experimental paradigms exist that necessitate clear descriptions of humane endpoints due to the pain and/or distress they may cause. Explanations should be provided that describe how the detrimental effects will be limited to the minimal level required to obtain the experimental objectives. Examples include these:

- electric shocks
- surgery (e.g., craniotomies, peripheral nerve ligations)
- food or water restriction
- induction of pain and/or withholding of analgesics (in studies of pain)
- forced running or swimming
- isolation or crowding, removal of environmental enrichment
- disease models (e.g., multiple sclerosis, Parkinson disease, amyotrophic lateral sclerosis)
- sleep deprivation
- drug dependency and withdrawal symptoms
- noise (e.g., startle reflex)
- thermal gradients (for pain assessment).

Many types of experimental interventions require changes to the normal husbandry, housing, or care practices; justification is to be provided to the IACUC. Example exceptions in neurobehavioral studies include single housing of social species (e.g., due to surgery or for measuring activity/behaviors of individual animals) or altering length of light cycle (e.g., sleep deprivation).

## ANIMAL MODELS: WELFARE, ETHICS, AND SAFETY

Animal models used in research are selected based on their possession of characteristics that are relevant to the field of study. Characteristics relate to those inherent to a given species and those acquired by strains or stocks within a species through selective breeding or molecular genetic engineering (e.g., using Crispr/Cas9 technology). Models can be further manipulated by altering animals physically (e.g., by surgically creating neuroanatomic lesions) and/or chemically (e.g., MPTP treatment to induce parkinsonism) and/or genetically (e.g., using viral vectors). In cases where the phenotype of the animal has been altered by genetic, physical, or chemical means, the impact on the animal's welfare must be considered, and when adversely affected, addressed. Namely, if any condition develops that involves pain and/or distress, clear plans for special care and humane endpoints are needed. As mentioned before, details as to what and when signs are expected to develop and plans for early detection and intervention need to be predetermined as much as possible. If the effects are not known, experience with the model should be used to develop earlier humane endpoints and refine care practices whenever possible. The protocol is to be updated to reflect any improvements made.

In addition to potential adverse effects on animal welfare, genetically modified animals and/or their gametes may pose a risk of disrupting ecosystems if they are released into the wild and propagate, so native species are negatively impacted or eliminated. For example, the use of Crispr/Cas9 technology has been used to demonstrate that a construct could be inserted into a mosquito malaria vector that results in female sterility. If released into the wild, it could "modify insect populations over a timeframe of a few years."[54] In this case, the release would be intentional to eliminate the mosquito vector from areas where malaria is endemic; but if reduction or elimination of this species in the wild was not desired, unintentional release could result in this all the same. Thus, it is important to know how to properly contain and dispose of animals and germline materials to prevent their release. It is an ethical responsibility of scientists to consider the risks and consequences and act accordingly. Where possible as a precaution, scientists should genetically modify animals in such a way that should animals escape, the risk of mutations propagating within the wild is low.

When viral vectors such as lentivirus or adeno-associated virus are used to deliver foreign genes into animals (e.g., via stereotaxic injection into regions of the brain) to study their neurobiologic effects, it is similarly important to follow proper containment, decontamination, and disposal protocols established at the institution to prevent release of these biohazardous agents into the environment. Most viral vectors in use have a high safety profile since they are often replication deficient and nonpathogenic. However, as the viral vectors can integrate into the host genome and the frequency and location at which the recombinant viral vectors insert into the host genome is unknown, there is always a nonzero risk that exposure to the virus will result in insertional mutagenesis and cancer.[55] As to how long animals pose a risk due to virus shedding post infection, it is usually not known; however, studies are being done to better characterize viral vectors, so more evidence-based guidelines for safe use can be created.[56,57]

## BREEDING ANIMALS

If animal colonies will be maintained by breeding, practices must ensure that the number of animals generated for experiments and continued colony maintenance are minimized to as few as necessary. This requires good record-keeping practices, particularly if numerous lines are being maintained and crossed. As the generation of new genetically engineered lines often results in the production of many surplus animals (of incorrect genotype), it should be demonstrated to the IACUC that such strains are unavailable elsewhere.[58] Once created, it is recommended that consideration be given to cryopreserving gametes or embryos of each line to safeguard it in the event of colony loss (e.g., due to mis-crossing, breeding cessation, loss of phenotype, or disease outbreak); reduce animal numbers generated when not in use; and limit genetic drift.[58,59] If animals are to be genotyped, "whenever there is the need to both identify an individual and collect a DNA sample for genotyping, a method should be chosen that meets both goals."[60] For genetically engineered mice, examples include tissue obtained by ear biopsy and distal phalanx biopsy (at postnatal day 7).[60] Each line created should be named using standard nomenclature; resources are available to assist with this.[61,62] This is important, as it provides such information as animal "strain(s) involved, the gene(s) that have been targeted and/or inserted, the type of genetic modification(s) that are associated to the strain, and the laboratory where the strain was produced. A correct annotation enables all these relevant details to be passed to future users, while ensuring that they can correctly interpret the origin of the line, its history and its current status."[63]

## EUTHANASIA

The method(s) to be used to euthanize animals should be chosen from those recognized as acceptable for the species, age, and weight of the animal. The IACUC likely has institutional euthanasia guidelines that can be consulted. These are often based on such documents as the "AVMA Guidelines for the Euthanasia of Animals" (US) or the "CCAC guidelines on: euthanasia of animals used in science" (Canada).[26,64] If a method not recognized as acceptable due to increased likelihood of pain and distress if not done properly (e.g., a physical method such as cervical dislocation without prior anesthetic), the IACUC will require justification. Regardless of the method used, only competent personnel are to perform the procedures. As for many other procedures performed as part of the research, institutional SOPs may be available that describe how to properly perform methods. Veterinary staff should be consulted in advance of protocol submission if/when there is uncertainty about the most appropriate method to use and/or when available options may interfere with the research.

## POSTAPPROVAL MONITORING

Once a protocol has been approved by the IACUC, and the animal work begins, there are usually several mechanisms in place to monitor how the animal-based research is going and to ensure it is proceeding as approved on the protocol. Collectively, the various mechanisms are commonly referred to as a postapproval monitoring (PAM) program. Continued oversight of animal use activities is often mandated by regulations and guidelines; however, flexibility is often granted to institutions as to how the PAM program functions.[44,35] The PAM program involves ongoing communication between researchers and IACUCs, veterinary and animal care on all aspects of animal care and use. Through constant monitoring, communication, and collaboration among these stakeholders, deficiencies can be identified and improvements made as part of a continuous feedback loop. In doing so, the animal care and use program successfully operates in which research is well-facilitated, good animal health and welfare is achieved, and (as a result) all regulatory requirements are met.

Common components of a PAM program include these:

- regular assessment of animals by trained veterinary technical staff
- veterinarian(s) trained in laboratory animal medicine available for consultation by both research and animal care staff to address any unexpected animal health and welfare issues that develop, and to help with model development and procedural issues
- visits by the IACUC to all animal facilities and spaces where animal use occurs outside of these
- systematic and formal review of researcher's animal use[65]

As mentioned at the beginning of the chapter, the IACUC can be seen as the "quality control" mechanism by overseeing the animal care and use program and ensuring that regulatory requirements are met on a day-to-day basis. Regulators of animal research (e.g., USDA, CCAC, AAALAC) serve in a "quality assurance" role to maintain public trust in animal-based research by holding institutions accountable when laws and/or guidelines are not met. Accordingly, for animal research to continue to be publicly supported and funded, researchers must work together with IACUCs, and in turn, institutions must work together with regulators to meet or exceed regulatory requirements.

## CONCLUSION

Animals play an integral role in many areas of neuroscience and behavioral research. Society requires that they only be used minimally when needed and that they be treated ethically and humanely throughout. To facilitate this, many regulators require that IACUCs review proposals prior to animal use to ensure, among other things, that the 3Rs have been applied. Postapproval monitoring of ongoing animal use serves to find opportunities to further implement the 3Rs, so both the quality of animal care and welfare and the integrity of the research they are used for are further optimized.

## REFERENCES


1. Parasuraman S. Toxicological screening. *J Pharmacol Pharmacother.* April 2011;2(2):74–79.
2. "CCAC Policy Statement On: Ethics of Animal Investigation." [Online]. Available: http://www.ccac.ca/en_/standards/policies/policy-ethics_animal_investigation. [Accessed: 21-Jul-2017].
3. Griffin G, Clark JM, Zurlo J, Ritskes-Hoitinga M. Scientific uses of animals: harm-benefit analysis and complementary approaches to implementing the three Rs. *Rev Sci Tech.* April 2014;33(1):265–272.
4. Lee G, Illes J, Ohl F, eds. *Ethical Issues in Behavioral Neuroscience.* vol. 19. Berlin, Heidelberg: Springer Berlin Heidelberg; 2015.
5. Dorner M, Horwitz JA, Donovan BM, et al. Completion of the entire hepatitis C virus life cycle in genetically humanized mice. *Nature.* September 2013;501(7466):237–241.
6. Bukh J. Animal models for the study of hepatitis C virus infection and related liver disease. *Gastroenterology.* May 2012;142(6), 1279–1287.e3.

7. Tesfaye A, Stift J, Maric D, Cui Q, Dienes H-P, Feinstone SM. Chimeric mouse model for the infection of hepatitis B and C viruses. *PLoS One*. October 2013;8(10):e77298.
8. van der Worp HB, Howells DW, Sena ES, et al. Can animal models of disease reliably inform human studies? *PLoS Med*. March 2010;7(3):e1000245.
9. Jucker M. The benefits and limitations of animal models for translational research in neurodegenerative diseases. *Nat Med*. November 2010;16(11):1210–1214.
10. Moran CJ, Ramesh A, Brama PAJ, O'Byrne JM, O'Brien FJ, Levingstone TJ. The benefits and limitations of animal models for translational research in cartilage repair. *J Exp Orthop*. December 2016;3(1):1.
11. Brønstad A, Newcomer CE, Decelle T, Everitt JI, Guillen J, Laber K. Current concepts of harm-benefit analysis of animal experiments - report from the AALAS-FELASA Working Group on Harm-Benefit Analysis – Part 1. *Lab Anim*. June 2016;50(1 Suppl):1–20.
12. Laber K, Newcomer CE, Decelle T, Everitt JI, Guillen J, Brønstad A. Recommendations for addressing harm-benefit analysis and implementation in ethical evaluation – report from the AALAS-FELASA Working Group on Harm-Benefit Analysis – Part 2. *Lab Anim*. June 2016;50(1 Suppl):21–42.
13. Russell W, Burch R. *The Principles of Humane Experimental Technique*. London: Methuen; 1959.
14. Gerlai R, Poshusta TL, Rampersad M, et al. Forward genetic screening using behavioral tests in zebrafish: a proof of concept analysis of mutants. *Behav Genet*. January 2017;47(1):125–139.
15. Lu TZ, Kostelecki W, Sun CLF, Dong N, Pérez Velázquez JL, Feng Z-P. High sensitivity of spontaneous spike frequency to sodium leak current in a Lymnaea pacemaker neuron. *Eur J Neurosci*. December 2016;44(12):3011–3022.
16. Wang L, Sato H, Satoh Y, Tomioka M, Kunitomo H, Iino Y. A gustatory neural circuit of *Caenorhabditis elegans* generates memory-dependent behaviors in Na(+) chemotaxis. *J Neurosci*. February 2017; 37(8):2097–2111.
17. Feuillette S, Delarue M, Riou G, et al. Neuron-to-neuron transfer of FUS in Drosophila primary neuronal culture is enhanced by ALS-associated mutations. *J Mol Neurosci*. May 2017;62(1):114–122.
18. "CCAC 3Rs." [Online]. Available: http://3rs.ccac.ca/en/about/three-rs.html. [Accessed: 22-Jul-2017].
19. Festing MFW, Altman DG. Guidelines for the design and statistical analysis of experiments using laboratory animals. *ILAR J*. 2002;43(4):244–258.
20. Charan J, Kantharia ND. How to calculate sample size in animal studies? *J Pharmacol Pharmacother*. October 2013;4(4):303–306.
21. Clayton JA, Collins FS. Policy: NIH to balance sex in cell and animal studies. *Nature*. May 2014; 509(7500):282–283.
22. Speirs V. Animal research: share surplus animal tissue. *Nature*. June 2015;522(7555):156.
23. Morrissey B, Blyth K, Carter P, et al. The sharing experimental animal resources, coordinating holdings (SEARCH) framework: encouraging reduction, replacement, and refinement in animal research. *PLoS Biol*. January 2017;15(1):e2000719.
24. Kilkenny C, Browne WJ, Cuthill IC, Emerson M, Altman DG. Improving bioscience research reporting: the ARRIVE guidelines for reporting animal research. *PLoS Biol*. June 2010;8(6):e1000412.
25. Smith AJ, Clutton RE, Lilley E, Hansen KEA, Brattelid T. PREPARE: guidelines for planning animal research and testing. *Lab Anim*. 2017. https://doi.org/10.1177/0023677217724823, 2367721772482.
26. Leary S, Anthony R, Cartner S, et al. *AVMA Guidelines for the Euthanasia of Animals: 2013 Edition Members of the Panel on Euthanasia*. 2013.
27. Bayne K, Würbel H. The impact of environmental enrichment on the outcome variability and scientific validity of laboratory animal studies. *Rev Sci Tech*. April 2014;33(1):273–280.
28. Hood RD. *Developmental and Reproductive Toxicology: A Practical Approach*. Informa Healthcare; 2012.
29. Stuart SA, Robinson ESJ. Reducing the stress of drug administration: implications for the 3Rs. *Sci Rep*. November 2015;5(1):14288.
30. Carbone L, Austin J. Pain and laboratory animals: publication practices for better data reproducibility and better animal welfare. *PLoS One*. May 2016;11(5):e0155001.
31. Bhasin J, Latt R, Macallum E, et al. *CCAC Guidelines On: Choosing an Appropriate Endpoint in Experiments Using Animals for Research, Teaching and Testing*. 1998.

32. How to determine humane endpoints for research animals. *Lab Anim (NY)*. December 2015;45(1):19.
33. "Humane Endpoints | NC3Rs." [Online]. Available: http://www.nc3rs.org.uk/humane-endpoints. [Accessed: 22-Jul-2017].
34. Guillen J. *Laboratory Animals: Regulations and Recommendations for Global Collaborative Research*. Amsterdam, Boston: Academic Press; 2014.
35. "CCAC Terms of Reference for Animal Care Committees."
36. National Research Council (U.S.), Committee for the Update of the Guide for the Care and Use of Laboratory Animals, Institute for Laboratory Animal Research (U.S.). *Guide for the Care and Use of Laboratory Animals*. National Academies Press; 2011.
37. "Institutional Animal Care and Use Committees | Animal Welfare Information Center | NAL | USDA." [Online]. Available: https://www.nal.usda.gov/awic/institutional-animal-care-and-use-committees. [Accessed: 22-Jul-2017].
38. ACLAM position statement on rodent surgery. *J Am Assoc Lab Anim Sci*. November 2016;55(6): 822–823.
39. Thulin JD, Bradfield JF, Bergdall VK, et al. The cost of self-imposed regulatory burden in animal research. *FASEB J*. August 2014;28(8):3297–3300.
40. Haywood JR, Greene M. Avoiding an overzealous approach: a perspective on regulatory burden. *ILAR J*. 2008;49(4):426–434.
41. *Findings of the FASEB Survey on Administrative Burden*. 2013.
42. Arvizu DE, Droegemeier KK, Córdova FA, et al. *Reducing Investigators' Administrative Workload for Federally Funded Research Task Force on Administrative Burdens*. 2014.
43. Fiorito G, Affuso A, Anderson DB, et al. Cephalopods in neuroscience: regulations, research and the 3Rs. *Invert Neurosci*. 2014;14(1):13–36.
44. N. R. C. (US) C. for the U. of the G. for the Animals and Care and Use of Laboratory. *Guide for the Care and Use of Laboratory Animals*. US: National Academies Press; 2011.
45. Silk SB, Hampton LL, Brown PA. What investigators need to know about the use of animals. *ILAR J*. January 2014;54(3):324–328.
46. B. Research. *Guidelines for the Care and Use of Mammals in Neuroscience*. The National Academies Press; 2003.
47. *Guiding Principles for Behavioural Laboratory Animal Science*. 2013.
48. "Canadian Council on Animal Care Instructions for Completion of the CCAC Animal Use Data Form (AUDF)."
49. Brown MJ, Pearson PT, Tomson FN. Guidelines for animal surgery in research and teaching. AVMA panel on animal surgery in research and teaching, and the ASLAP (American Society of Laboratory Animal Practitioners). *Am J Vet Res*. September 1993;54(9):1544–1559.
50. V Turner P, Brabb T, Pekow C, Vasbinder MA. Administration of substances to laboratory animals: routes of administration and factors to consider. *J Am Assoc Lab Anim Sci*. September 2011;50(5): 600–613.
51. Turner PV, Pekow C, Vasbinder MA, Brabb T. Administration of substances to laboratory animals: equipment considerations, vehicle selection, and solute preparation. *J Am Assoc Lab Anim Sci*. September 2011;50(5):614–627.
52. "Blood Sampling | NC3Rs." [Online]. Available: https://www.nc3rs.org.uk/3rs-resources/blood-sampling. [Accessed: 22-Jul-2017].
53. Chapman K, Sewell F, Allais L, et al. A global pharmaceutical company initiative: an evidence-based approach to define the upper limit of body weight loss in short term toxicity studies. *Regul Toxicol Pharmacol*. October 2013;67(1):27–38.
54. Hammond A, Galizi R, Kyrou K, et al. A CRISPR-Cas9 gene drive system targeting female reproduction in the malaria mosquito vector *Anopheles gambiae*. *Nat Biotechnol*. 2015;34(1):78–83.
55. Chandler RJ, Sands MS, Venditti CP. Recombinant adeno-associated viral integration and genotoxicity: insights from animal models. *Hum Gene Ther*. 2017;28(4):314–322.
56. Reuter JD, Fang X, Ly CS, Suter KK, Gibbs D. Assessment of hazard risk associated with the intravenous use of viral vectors in rodents. *Comp Med*. 2012;62(5):361–370.
57. NIH. *NIH Guidelines for Research Involving Recombinant or Synthetic Nucleic Acid Molecules*. 2016.

58. Agca Y. Genome resource banking of biomedically important laboratory animals. *Theriogenology*. November 2012;78(8):1653–1665.
59. Mazur P, Leibo SP, Seidel GE. Cryopreservation of the germplasm of animals used in biological and medical research: importance, impact, status, and future directions. *Biol Reprod*. January 2008;78(1): 2–12.
60. Bonaparte D, Cinelli P, Douni E, et al. FELASA guidelines for the refinement of methods for genotyping genetically-modified rodents: a report of the Federation of European Laboratory Animal Science Associations Working Group. *Lab Anim*. July 2013;47(3):134–145.
61. "ZFIN Zebrafish Nomenclature Guidelines – General Information – ZFIN Community Wiki." [Online]. Available: https://wiki.zfin.org/display/general/ZFIN+Zebrafish+Nomenclature+Guidelines. [Accessed: 22-Jul-2017].
62. "MGI-Mouse Nomenclature Home Page." [Online]. Available: http://www.informatics.jax.org/mgihome/nomen/index.shtml. [Accessed: 22-Jul-2017].
63. Montoliu L, Whitelaw CBA. Using standard nomenclature to adequately name transgenes, knockout gene alleles and any mutation associated to a genetically modified mouse strain. *Transgenic Res*. April 2011;20(2):435–440.
64. Charbonneau R, Niel L, Olfert E, et al. *CCAC Guidelines on: Euthanasia of Animals Used in Science*. 2010.
65. Vanderford DA, Doss S, Banks RE. A retrospective review of postapproval monitoring at a large academic institution. *Lab Anim (NY)*. October 2015;44(10):395–401.

# INDEX

'*Note*: Page numbers followed by "f" indicate figures, "t" indicate tables and "b" indicate boxes.'

## C

## F

## G

## J



## K



## L

## M

## O

## S

## T